Operations Research Proceedings

GOR (Gesellschaft für Operations Research e.V.)

More information about this series at http://www.springer.com/series/722

Janis S. Neufeld • Udo Buscher • Rainer Lasch •
Dominik Möst • Jörn Schönberger
Editors

Operations Research Proceedings 2019

Selected Papers of the Annual International Conference of the German Operations Research Society (GOR), Dresden, Germany, September 4-6, 2019

Springer

Editors
Janis S. Neufeld
Faculty of Business and Economics
TU Dresden
Dresden, Germany

Udo Buscher
Faculty of Business and Economics
TU Dresden
Dresden, Germany

Rainer Lasch
Faculty of Business and Economics
TU Dresden
Dresden, Germany

Dominik Möst
Faculty of Business and Economics
TU Dresden
Dresden, Germany

Jörn Schönberger
Faculty of Transportation and Traffic
TU Dresden
Dresden, Germany

ISSN 0721-5924 ISSN 2197-9294 (electronic)
Operations Research Proceedings
ISBN 978-3-030-48438-5 ISBN 978-3-030-48439-2 (eBook)
https://doi.org/10.1007/978-3-030-48439-2

This Springer imprint is published by the registered company Springer Nature Switzerland AG.
The registered company address is: Gewerbestrasse 11, 6330 Cham, Switzerland

Preface

OR2019, the joint annual scientific conference of the national Operations Research Societies of Germany (GOR), Austria (ÖGOR), and Switzerland (SVOR), was held at the Technische Universität Dresden on September 3–6, 2019. The School of Civil and Environmental Engineering of Technische Universität Dresden supported it, and both the Faculty of Transport and Traffic Science and the Faculty of Business and Economics acted as the hosts of OR2019.

After more than 1 year of preparation, OR2019 provided a platform for more than 600 experts in operations research. OR2019 was the host for guests from more than 30 countries. The scientific program comprised 3 invited plenary talks (including the presentation of the winner of the GOR science award), 7 invited semi-plenary talks, and more than 400 contributed presentations.

The Operations Research 2019 proceedings present a carefully reviewed and selected collection of full papers submitted by OR2019 participants. This selection of 99 manuscripts reflects the large variety of themes and the interdisciplinary position of operations research. It demonstrates that operations research is able to contribute to the solution of the large problems of our time. In addition, it shows that senior researchers, postdocs, and PhD students as well as graduate students cooperate to find answers needed to cope with recent as well as future challenges. Both theory building and its application fruitfully interact.

We say thank you to all the people who contributed to the successful OR2019 event: the international program committee, the invited speakers, the contributors of the scientific presentations, our sponsors, the GOR, the ÖGOR, the SVOR, more of 50 stream chairs, and all our session chairs. In addition, we express our sincere gratitude to the staff members from TU Dresden who joined the organizing committee and spent their time in the preparation and the execution of OR2019.

Dresden, Germany — Janis S. Neufeld
Dresden, Germany — Udo Buscher
Dresden, Germany — Rainer Lasch
Dresden, Germany — Dominik Möst
Dresden, Germany — Jörn Schönberger
January 2020

Contents

Part I GOR Awards

Analysis and Optimization of Urban Energy Systems 3
Kai Mainzer

Optimization in Outbound Logistics—An Overview 11
Stefan Schwerdfeger

Incorporating Differential Equations into Mixed-Integer Programming for Gas Transport Optimization 19
Mathias Sirvent

Scheduling a Proportionate Flow Shop of Batching Machines 27
Christoph Hertrich

Vehicle Scheduling and Location Planning of the Charging Infrastructure for Electric Buses Under the Consideration of Partial Charging of Vehicle Batteries 35
Luisa Karzel

Data-Driven Integrated Production and Maintenance Optimization 43
Anita Regler

Part II Business Analytics, Artificial Intelligence and Forecasting

Multivariate Extrapolation: A Tensor-Based Approach 53
Josef Schosser

Part III Business Track

Heuristic Search for a Real-World 3D Stock Cutting Problem 63
Katerina Klimova and Una Benlic

Part IV Control Theory and Continuous Optimization

Model-Based Optimal Feedback Control for Microgrids with Multi-Level Iterations 73
Robert Scholz, Armin Nurkanovic, Amer Mesanovic, Jürgen Gutekunst, Andreas Potschka, Hans Georg Bock, and Ekaterina Kostina

Mixed-Integer Nonlinear PDE-Constrained Optimization for Multi-Modal Chromatography 81
Dominik H. Cebulla, Christian Kirches, and Andreas Potschka

Sparse Switching Times Optimization and a Sweeping Hessian Proximal Method 89
Alberto De Marchi and Matthias Gerdts

Toward Global Search for Local Optima 97
Jens Deussen, Jonathan Hüser, and Uwe Naumann

First Experiments with Structure-Aware Presolving for a Parallel Interior-Point Method 105
Ambros Gleixner, Nils-Christian Kempke, Thorsten Koch, Daniel Rehfeldt, and Svenja Uslu

A Steepest Feasible Direction Extension of the Simplex Method 113
Biressaw C. Wolde and Torbjörn Larsson

Convex Quadratic Mixed-Integer Problems with Quadratic Constraints 123
Simone Göttlich, Kathinka Hameister, and Michael Herty

Part V Decision Theory and Multiple Criteria Decision Making

The Bicriterion Maximum Flow Network Interdiction Problem in s-t-Planar Graphs 133
Luca E. Schäfer, Tobias Dietz, Marco V. Natale, Stefan Ruzika, Sven O. Krumke, and Carlos M. Fonseca

Assessment of Energy and Emission Reduction Measures in Container Terminals using PROMETHEE for Portfolio Selection 141
Erik Pohl, Christina Scharpenberg, and Jutta Geldermann

Decision-Making for Projects Realization/Support: Approach Based on Stochastic Dominance Rules Versus Multi-Actor Multi-Criteria Analysis 149
Dorota Górecka

Part VI Discrete and Integer Optimization

A Stochastic Bin Packing Approach for Server Consolidation with Conflicts 159
John Martinovic, Markus Hähnel, Waltenegus Dargie, and Guntram Scheithauer

Optimal Student Sectioning at Niederrhein University of Applied Sciences 167
Steffen Goebbels and Timo Pfeiffer

A Dissection of the Duality Gap of Set Covering Problems 175
Uledi Ngulo, Torbjörn Larsson, and Nils-Hassan Quttineh

Layout Problems with Reachability Constraint 183
Michael Stiglmayr

Modeling of a Rich Bin Packing Problem from Industry 191
Nils-Hassan Quttineh

Optimized Resource Allocation and Task Offload Orchestration for Service-Oriented Networks 199
Betül Ahat, Necati Aras, Kuban Altınel, Ahmet Cihat Baktır, and Cem Ersoy

Job Shop Scheduling with Flexible Energy Prices and Time Windows 207
Andreas Bley and Andreas Linß

Solving the Multiple Traveling Salesperson Problem on Regular Grids in Linear Time 215
Philipp Hungerländer, Anna Jellen, Stefan Jessenitschnig, Lisa Knoblinger, Manuel Lackenbucher, and Kerstin Maier

The Weighted Linear Ordering Problem 223
Jessica Hautz, Philipp Hungerländer, Tobias Lechner, Kerstin Maier, and Peter Rescher

Adaptation of a Branching Algorithm to Solve the Multi-Objective Hamiltonian Cycle Problem 231
Maialen Murua, Diego Galar, and Roberto Santana

Part VII Energy and Environment

Combinatorial Reverse Auction to Coordinate Transmission and Generation Assets in Brazil: Conceptual Proposal Based on Integer Programming 241
Laura S. Granada, Fernanda N. Kazama, and Paulo B. Correia

A Lagrangian Decomposition Approach to Solve Large Scale Multi-Sector Energy System Optimization Problems 249
Andreas Bley, Angela Pape, and Frank Fischer

Operational Plan for the Energy Plants Considering the Fluctuations in the Spot Price of Electricity 257
Masato Dei, Tomoki Fukuba, Takayuki Shiina, and K. Tokoro

Design of an Electric Bus Fleet and Determination of Economic Break-Even 265
Marius Madsen and Marc Gennat

Tradeoffs Between Battery Degradation and Profit from Market Participation of Solar-Storage Plants 273
Leopold Kuttner

On the Observability of Smart Grids and Related Optimization Methods 281
Claudia D'Ambrosio, Leo Liberti, Pierre-Louis Poirion, and Sonia Toubaline

Part VIII Finance

A Real Options Approach to Determine the Optimal Choice Between Lifetime Extension and Repowering of Wind Turbines 291
Chris Stetter, Maximilian Heumann, Martin Westbomke, Malte Stonis, and Michael H. Breitner

Measuring Changes in Russian Monetary Policy: An Indexed-Based Approach 299
Nikolay Nenovsky and Cornelia Sahling

Part IX Graphs and Networks

Usage of Uniform Deployment for Heuristic Design of Emergency System 309
Marek Kvet and Jaroslav Janáček

Uniform Deployment of the p-Location Problem Solutions 315
Jaroslav Janáček and Marek Kvet

Algorithms and Complexity for the Almost Equal Maximum Flow Problem 323
R. Haese, T. Heller, and S. O. Krumke

Exact Solutions for the Steiner Path Cover Problem on Special Graph Classes 331
Frank Gurski, Stefan Hoffmann, Dominique Komander, Carolin Rehs, Jochen Rethmann, and Egon Wanke

Subset Sum Problems with Special Digraph Constraints 339
Frank Gurski, Dominique Komander, and Carolin Rehs

Part X Health Care Management

A Capacitated EMS Location Model with Site Interdependencies 349
Matthias Grot, Tristan Becker, Pia Mareike Steenweg, and Brigitte Werners

Online Optimization in Health Care Delivery: Overview and Possible Applications 357
Roberto Aringhieri

Part XI Logistics and Freight Transportation

On a Supply-Driven Location Planning Problem 367
Hannes Hahne and Thorsten Schmidt

Dispatching of Multiple Load Automated Guided Vehicles Based on Adaptive Large Neighborhood Search 375
Patrick Boden, Hannes Hahne, Sebastian Rank, and Thorsten Schmidt

Freight Pickup and Delivery with Time Windows, Heterogeneous Fleet and Alternative Delivery Points 381
Jérémy Decerle and Francesco Corman

Can Autonomous Ships Help Short-Sea Shipping Become More Cost-Efficient? 389
Mohamed Kais Msakni, Abeera Akbar, Anna K. A. Aasen, Kjetil Fagerholt, Frank Meisel, and Elizabeth Lindstad

Identification of Defective Railway Wheels from Highly Imbalanced Wheel Impact Load Detector Sensor Data 397
Sanjeev Sabnis, Shravana Kumar Yadav, and Shripad Salsingikar

Exact Approach for Last Mile Delivery with Autonomous Robots 405
Stefan Schaudt and Uwe Clausen

A Solution Approach to the Vehicle Routing Problem with Perishable Goods 413
Boris Grimm, Ralf Borndörfer, and Mats Olthoff

Part XII Optimization Under Uncertainty

Solving Robust Two-Stage Combinatorial Optimization Problems Under Convex Uncertainty 423
Marc Goerigk, Adam Kasperski, and Paweł Zieliński

Production Planning Under Demand Uncertainty: A Budgeted Uncertainty Approach 431
Romain Guillaume, Adam Kasperski, and Paweł Zieliński

Robust Multistage Optimization with Decision-Dependent Uncertainty 439
Michael Hartisch and Ulf Lorenz

Examination and Application of Aircraft Reliability in Flight Scheduling and Tail Assignment 447
Martin Lindner and Hartmut Fricke

Part XIII OR in Engineering

Comparison of Piecewise Linearization Techniques to Model Electric Motor Efficiency Maps: A Computational Study 457
Philipp Leise, Nicolai Simon, and Lena C. Altherr

Support-Free Lattice Structures for Extrusion-Based Additive Manufacturing Processes via Mixed-Integer Programming 465
Christian Reintjes, Michael Hartisch, and Ulf Lorenz

Optimized Design of Thermofluid Systems Using the Example of Mold Cooling in Injection Molding 473
Jonas B. Weber, Michael Hartisch, and Ulf Lorenz

Optimization of Pumping Systems for Buildings: Experimental Validation of Different Degrees of Model Detail on a Modular Test Rig ... 481
Tim M. Müller, Lena C. Altherr, Philipp Leise, and Peter F. Pelz

Optimal Product Portfolio Design by Means of Semi-infinite Programming 489
Helene Krieg, Jan Schwientek, Dimitri Nowak, and Karl-Heinz Küfer

Exploiting Partial Convexity of Pump Characteristics in Water Network Design 497
Marc E. Pfetsch and Andreas Schmitt

Improving an Industrial Cooling System Using MINLP, Considering Capital and Operating Costs 505
Marvin M. Meck, Tim M. Müller, Lena C. Altherr, and Peter F. Pelz

A Two-Phase Approach for Model-Based Design of Experiments Applied in Chemical Engineering 513
Jan Schwientek, Charlie Vanaret, Johannes Höller, Patrick Schwartz, Philipp Seufert, Norbert Asprion, Roger Böttcher, and Michael Bortz

Assessing and Optimizing the Resilience of Water Distribution Systems Using Graph-Theoretical Metrics 521
Imke-Sophie Lorenz, Lena C. Altherr, and Peter F. Pelz

Part XIV Production and Operations Management

A Flexible Shift System for a Fully-Continuous Production Division 531
Elisabeth Finhold, Tobias Fischer, Sandy Heydrich, and Karl-Heinz Küfer

Capacitated Lot Sizing for Plastic Blanks in Automotive Manufacturing Integrating Real-World Requirements 539
Janis S. Neufeld, Felix J. Schmidt, Tommy Schultz, and Udo Buscher

Facility Location with Modular Capacities for Distributed Scheduling Problems 545
Eduardo Alarcon-Gerbier

Part XV Project Management and Scheduling

Diversity of Processing Times in Permutation Flow Shop Scheduling Problems 555
Kathrin Maassen and Paz Perez-Gonzalez

Proactive Strategies for Soccer League Timetabling 563
Xiajie Yi and Dries Goossens

Constructive Heuristics in Hybrid Flow Shop Scheduling with Unrelated Machines and Setup Times 569
Andreas Hipp and Jutta Geldermann

A Heuristic Approach for the Multi-Project Scheduling Problem with Resource Transition Constraints 575
Markus Berg, Tobias Fischer, and Sebastian Velten

Time-Dependent Emission Minimization in Sustainable Flow Shop Scheduling 583
Sven Schulz and Florian Linß

Analyzing and Optimizing the Throughput of a Pharmaceutical Production Process 591
Heiner Ackermann, Sandy Heydrich, and Christian Weiß

A Problem Specific Genetic Algorithm for Disassembly Planning and Scheduling Considering Process Plan Flexibility and Parallel Operations 599
Franz Ehm

Project Management with Scarce Resources in Disaster Response 607
Niels-Fabian Baur and Julia Rieck

Part XVI Revenue Management and Pricing

Capacitated Price Bundling for Markets with Discrete Customer Segments and Stochastic Willingness to Pay: A Basic Decision Model 617
Ralf Gössinger and Jacqueline Wand

Insourcing the Passenger Demand Forecasting System for Revenue Management at DB Fernverkehr: Lessons Learned from the First Year 625
Valentin Wagner, Stephan Dlugosz, Sang-Hyeun Park, and Philipp Bartke

Tax Avoidance and Social Control 633
Markus Diller, Johannes Lorenz, and David Meier

Part XVII Simulation and Statistical Modelling

How to Improve Measuring Techniques for the Cumulative Elevation Gain upon Road Cycling 643
Maren Martens

A Domain-Specific Language to Process Causal Loop Diagrams with R 651
Adrian Stämpfli

Deterministic and Stochastic Simulation: A Combined Approach to Passenger Routing in Railway Systems 659
Gonzalo Barbeito, Maximilian Moll, Wolfgang Bein, and Stefan Pickl

Predictive Analytics in Aviation Management: Passenger Arrival Prediction 667
Maximilian Moll, Thomas Berg, Simon Ewers, and Michael Schmidt

Part XVIII Software Applications and Modelling Systems

Xpress Mosel: Modeling and Programming Features for Optimization Projects 677
Susanne Heipcke and Yves Colombani

Part XIX Supply Chain Management

The Optimal Reorder Policy in an Inventory System with Spares and Periodic Review 687
Michael Dreyfuss and Yahel Giat

Decision Support for Material Procurement 693
Heiner Ackermann, Erik Diessel, Michael Helmling, Christoph Hertrich, Neil Jami, and Johanna Schneider

Design of Distribution Systems in Grocery Retailing 701
Andreas Holzapfel, Heinrich Kuhn, and Tobias Potoczki

A Comparison of Forward and Closed-Loop Supply Chains 707
Mehmet Alegoz, Onur Kaya, and Z. Pelin Bayindir

Part XX Traffic, Mobility and Passenger Transportation

Black-Box Optimization in Railway Simulations 717
Julian Reisch and Natalia Kliewer

The Effective Residual Capacity in Railway Networks with Predefined Train Services 725
Norman Weik, Emma Hemminki, and Nils Nießen

A Heuristic Solution Approach for the Optimization of Dynamic Ridesharing Systems 733
Nicolas Rückert, Daniel Sturm, and Kathrin Fischer

Data Analytics in Railway Operations: Using Machine Learning to Predict Train Delays 741
Florian Hauck and Natalia Kliewer

Optimization of Rolling Stock Rostering Under Mutual Direct Operation 749
Sota Nakano, Jun Imaizumi, and Takayuki Shiina

The Restricted Modulo Network Simplex Method for Integrated Periodic Timetabling and Passenger Routing 757
Fabian Löbel, Niels Lindner, and Ralf Borndörfer

Optimizing Winter Maintenance Service at Airports 765
Henning Preis and Hartmut Fricke

Coping with Uncertainties in Predicting the Aircraft Turnaround Time at Airports 773
Ehsan Asadi, Jan Evler, Henning Preis, and Hartmut Fricke

Strategic Planning of Depots for a Railway Crew Scheduling Problem 781
Martin Scheffler

Periodic Timetabling with Flexibility Based on a Mesoscopic Topology 789
Stephan Bütikofer, Albert Steiner, and Raimond Wüst

Capacity Planning for Airport Runway Systems 797
Stefan Frank and Karl Nachtigall

Data Reduction Algorithm for the Electric Bus Scheduling Problem 805
Maros Janovec and Michal Kohani

Crew Planning for Commuter Rail Operations, a Case Study on Mumbai, India 813
Naman Kasliwal, Sudarshan Pulapadi, Madhu N. Belur, Narayan Rangaraj, Suhani Mishra, Shamit Monga, Abhishek Singh, S. G. Sagar, P. K. Majumdar, and M. K. Jagesh

Part I
GOR Awards

Analysis and Optimization of Urban Energy Systems

Kai Mainzer

Abstract Cities and municipalities are critical for the success of the energy transition and hence often pursue their own sustainability goals. However, there is a lack of the required know-how to identify suitable combinations of measures to achieve these goals.

The RE^{3}ASON model allows automated analyses, e.g. to determine the energy demands as well as the renewable energy potentials in an arbitrary region. In the subsequent optimization of the respective energy system, various objectives can be pursued—e.g. minimization of discounted system expenditures and emission reduction targets. The implementation of the model employs various methods from the fields of geoinformatics, economics, machine learning and mixed-integer linear optimization.

The model is applied to derive energy concepts within a small municipality. By using stakeholder preferences and multi-criteria decision analysis, it is shown that the transformation of the urban energy system to use local and sustainable energy can be the preferred alternative from the point of view of community representatives.

Keywords Urban energy systems · Renewable energy potentials · Mixed-integer linear optimization

1 Introduction

Many cities and municipalities are aware of their importance for the success of the energy system transformation and pursue their own sustainability goals. However, smaller municipalities in particular often lack the necessary expertise to quantify local emission reduction potentials and identify suitable combinations of technological measures to achieve these goals.

K. Mainzer (✉)
Karlsruhe Institute of Technology (KIT), Karlsruhe, Germany
e-mail: kai.mainzer@kit.edu

J. S. Neufeld et al. (eds.), *Operations Research Proceedings 2019*, Operations Research Proceedings, https://doi.org/10.1007/978-3-030-48439-2_1

There exist a number of other models that are intended for supporting communities with finding optimal energy system designs. However, many of these models fail to consider local renewable energies and demand side measures such as building insulation. Additionally, almost all of these models require exogenous input data such as energy demand and local potentials for renewable energy, which have to be calculated beforehand. This leads to these models being applied usually only within a single case study and not being easily transferable to other regions.

For an energy system model to be useful to such communities, it should:

- Determine the transformation path for the optimal design of the urban energy system, taking into account the specific objectives of the municipality
- Consider sector coupling, in particular synergy effects between the electricity and heat sectors
- Consider interactions between technologies and optimal technology combinations
- provide intuitive operation via a graphical user interface, ensuring ease-of-use
- Implement automated methods for model endogenous determination of the required input data, especially energy demand and local energy resources
- Use freely available, "open" data
- Provide strategies for coping with computational complexity at a high level of detail

2 The RE^{3}ASON Model

In order to adequately meet these requirements, the RE^{3}ASON (Renewable Energy and Energy Efficiency Analysis and System OptimizatioN) model was developed. The model consists of two parts: the first part provides transferable methods for the analysis of urban energy systems, which are described in Sect. 2.1. The second part of the model uses these methods and the data obtained for the techno-economic optimization of the urban energy system and is described in Sect. 2.2. For a more detailed model description, the reader is referred to [1].

2.1 Modelling the Energy Demand and Renewable Energy Potentials

The RE^{3}ASON model provides transferable methods that can be used to determine the energy infrastructure and the local building stock, the structure of electricity and heat demand and the potentials and costs of climate-neutral energy generation from photovoltaics, wind power and biomass within a community (c.f. Fig. 1).

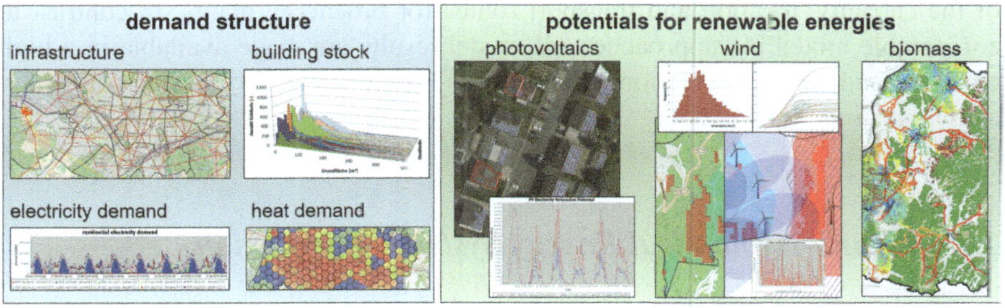

Fig. 1 Exemplary illustration of the methods for modelling the energy demand structure (left) and the potentials for renewable energies (right)

Fig. 2 Possible module placement for new PV systems with lowest possible electricity production costs (blue/white) and roofs that are already covered with PV modules (red border). Own depiction with map data from Bing Maps [2]

The unique feature of these methods lies in the use and combination of public data, which are available nationwide and freely. For this reason, the developed model is transferable in contrast to previous models, so that arbitrary German, with some of the methods also international municipalities, can be analyzed with comparatively small effort. For the implementation of these features, different methods were used, e.g. from the fields of geoinformatics, radiation simulation, business administration and machine learning.

This approach enables, for example, a very detailed mapping of the local potentials for renewable energies, which in several respects goes beyond the capabilities of previous modelling approaches. For example, the model allows, for the first time, the consideration of potentially new as well as existing PV plants solely on the basis of aerial photographs (c.f. Fig. 2). Other notable new methods are the consideration of the effects of surface roughness and terrain topography on the power generation from wind turbines as well as the automated determination of the capacity, location and transport routes for bioenergy plants. In contrast to comparable modelling approaches, all model results are made available in a high temporal and spatial resolution, which opens up the possibility of diverse further analyses based on these data.

2.2 *Techno-economical Energy System Optimization*

Furthermore, a mixed integer linear optimization model was developed, which uses the data determined with the previous methods as input and determines the optimal design of the urban energy system from an overall system perspective. Various objectives and targets can be selected (c.f. Chap. 3), whereupon the model determines the required unit dispatch as well as investment decisions in energy conversion technologies over a long-term time horizon. The structure of the model is shown in Fig. 3.

The optimization model is implemented using GAMS and generates about 6 million equations and 1.5 million variables (13,729 of which are binaries). On a 3.2 GHz, 12 core machine with 160 GB RAM, depending on the chosen objective, it can take between 7 and up to 26 h to solve within an optimality gap of 2.5%, using CPLEX. The processing time can be reduced significantly (by up to 95%) however, by providing valid starting solutions from previous runs for subsequent alternatives.

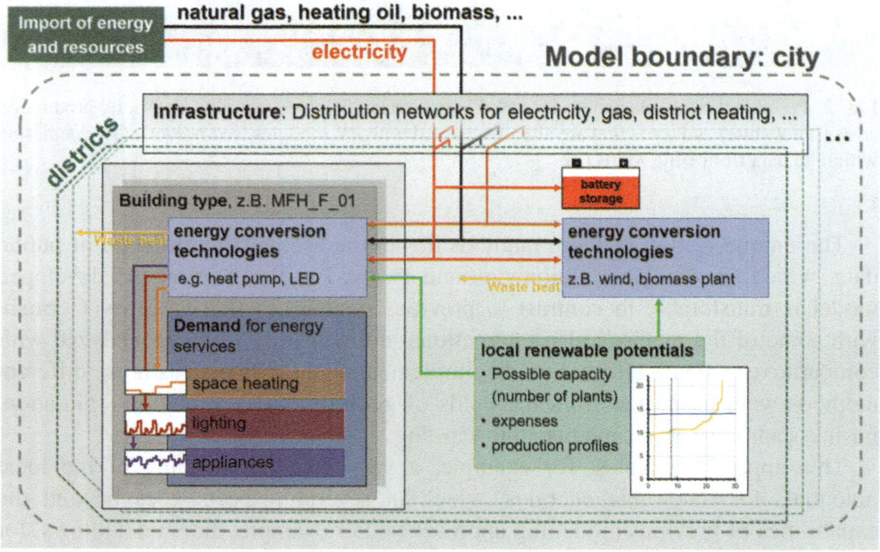

Fig. 3 Structure of the developed optimization model

3 Model Application and Results

In order to demonstrate the transferability of the model, it has been applied to a heterogeneous selection of municipalities in Germany. For details about this application, the reader is referred to [1], Sect. 6.1. The model has further been applied in a more detailed case study to the municipality of Ebhausen, Germany. This chapter represents an excerpt from this analysis, more details can be found in [3].

Ebhausen had a population of about 4700 in 2013, consists of four distinct districts and has a rather low population density of 188 people per km^2 (compared to the German average of 227). It is dominated by domestic buildings and a few small commercial premises, but no industry.

In this case study, the community stakeholders have been involved by means of three workshops that have been held within the community during the project. The workshop discussions revealed three values: economic sustainability, environmental sustainability, and local energy autonomy. A total of eight alternatives for the 2030 energy system have been identified to achieve these values: three scenarios which minimize only one single criteria, namely costs (A-1), emissions (A-2) and energy imports (A-3). Based on A-2, three additional emission-minimizing scenarios have been created, which additionally restrict costs to a maximum surplus of 10% (A-2a), 20% (A-2b) and 50% (A-2c) as compared to the minimum costs in A-1. In A-2c, (net) energy imports have additionally been forbidden. Based on A-3, two additional energy import minimizing scenarios have been created which also restrict costs to a 10% (A-3a) and 20% (A-3b) surplus. Table 1 shows results for the portfolio of technological measures as derived from the optimization model, Fig. 4 shows a comparison of the target criteria for the eight examined alternatives.

It is clear that these results differ substantially from one another. Alternative A-1 implies moderate PV and wind (three turbines) capacity additions, insulation

Table 1 Portfolios of technological measures associated with the eight alternatives

#	PV capacity (MW)	Wind capacity (MW)	Dominant heating system	Insulation[a]	Appliances[b]
A-1	2.0	6.0	Gas boiler	2	50
A-2	1.7	2.0	Pellet heating	3	100
A-2a	1.5	8.0	Heat pump	2/3	90
A-2b	0.6	8.0	Heat pump	2/3	90
A-2c	24.9	0	Heat pump	3	90
A-3	23.9	0	Heat pump	3	100
A-3a	18.3	6.0	Heat pump	2	30
A-3b	24.8	0	Heat pump	2	40

[a]Dominant level of building insulation employed, i.e. from 1 (low) to 3 (high), whereby 2/3 implies a roughly 50/50 split
[b]Fraction (%) of highest standard, i.e. A+++

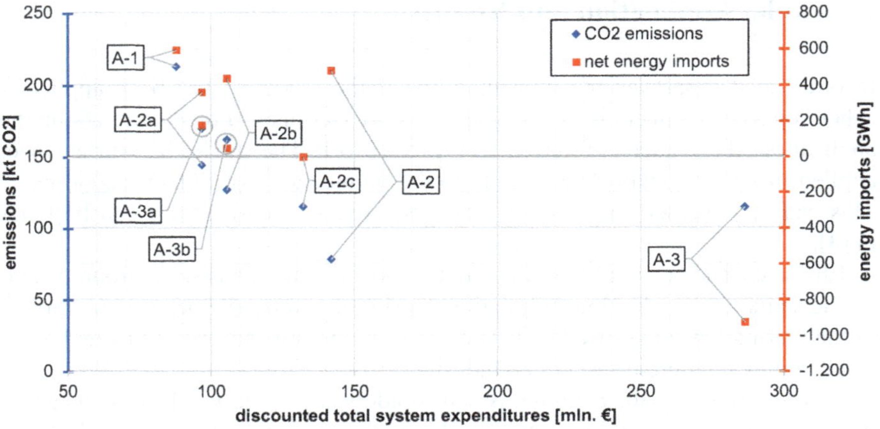

Fig. 4 Comparison of the target criteria in the eight alternatives examined

and electrical appliance improvements, with a mixture of heating systems including gas boilers and heat pumps, as well as some electric storage heaters. In contrast, the alternatives A-2 and A-3 have rather extreme results. The former translates into a moderate PV and wind capacity, maximum efficiency levels of insulation and appliances, and heating dominated by pellets boilers. The latter alternative involves a very high level of PV capacity, no wind capacity, maximum efficiency levels of insulation and electrical appliances, as well as heating systems dominated by heat pumps. The other five alternatives represent a compromise between these extremes, some of which have additional constraints. It is thus possible to quantify the additional level of energy autonomy and/or CO_2 emissions that can be achieved by increasing the total costs from the absolute minimum to 110%, 120% or even 150% of this value. It becomes apparent that significant emission reductions can be achieved with only minor additional costs (cf. Fig. 4). For example, allowing for a 10% increase in total costs leads to a 51% (of total achievable) emission reduction or a 27% net import reduction, and allowing for a 20% increase in costs leads to a 64% (of totally achievable) emission reduction or a 36% of net import reduction. In addition, these relatively small relaxations in the permissible costs result in substantially different energy systems.

In the second workshop with the community representatives, inter-criteria preferences (i.e. weights) for the three values economic sustainability, environmental sustainability and local energy autonomy were elicited. The results show that the alternatives A-1 (Costs first) and A-3 (Imports first) are outperformed by the other alternatives and that the remaining six alternatives achieve very similar performance scores, of which alternative A-2c (Emissions focus with cost and import constraints) achieves the highest overall performance score for the elicited preference parameters.

4 Conclusion

The results highlight the importance of automated methods for the analysis of the urban energy system as well as of the participatory approach of involving the key stakeholders in order to derive feasible energy concepts for small communities. While the first enables also smaller communities with no expertise in energy system modelling to gain transparency about their current and possible future energy systems, the latter makes it possible to take their specific values and objectives into account, ultimately enabling them to achieve their sustainability goals.

References

1. Mainzer, K.: Analyse und Optimierung urbaner Energiesysteme: Entwicklung und Anwendung eines übertragbaren Modellierungswerkzeugs zur nachhaltigen Systemgestaltung. Dissertation, Karlsruhe (2019)
2. Microsoft. Bing aerial images: orthographic aerial and satellite imagery. https://www.bing.com/maps/aerial
3. McKenna, R., Bertsch, V., Mainzer, K., Fichtner, W.: Combining local preferences with multi-criteria decision analysis and linear optimization to develop feasible energy concepts in small communities. Eur. J. Oper. Res. (2018)

Optimization in Outbound Logistics—An Overview

Stefan Schwerdfeger

Abstract In the era of e-commerce and just-in-time production, an efficient supply of goods is more than ever a fundamental requirement for any supply chain. In this context, this paper summarizes the author's dissertation about optimization in outbound logistics, which received the dissertation award of the German Operations Research Society (GOR) in the course of the OR conference 2019 in Dresden. The structure of the thesis is introduced, the investigated optimization problems are described, and the main findings are highlighted.

Keywords Logistics · Order picking · Transportation · Machine scheduling

1 Introduction

A few clicks, this is all it takes to get almost everything for our daily requirements. A banana from Ecuador or Australian wine, Amazon Fresh has everything available and, as members of Amazon Prime, we receive our orders within the next few hours. Naturally, the internet is not limited to exotic foodstuff, and customers are free to order everywhere, but it is a basic requirement of today's commercial life that everything is available, at any time and everywhere. To achieve this, distribution centers mushroom all over the world, and trucks sneak along the street networks everywhere. In Germany, for instance, there are about 3 million registered trucks, which transported almost 79% of all goods and contributed 479 million tkm (tonne-kilometers) in 2017 [15]. Thus, efficient logistics processes have become the backbone of the modern society.

In this context, the paper on hand takes a look at a subdiscipline of logistics, which focuses on all logistics activities after the finished product leaves production until it is finally handed over to the customer. This subdiscipline is called outbound

S. Schwerdfeger (✉)
Lehrstuhl für Management Science, Friedrich-Schiller-Universität Jena, Jena, Germany
e-mail: Stefan.Schwerdfeger@uni-jena.de; https://www.mansci.uni-jena.de/

J. S. Neufeld et al. (eds.), *Operations Research Proceedings 2019*, Operations Research Proceedings, https://doi.org/10.1007/978-3-030-48439-2_2

logistics, which is an umbrella term for a variety of processes, among which we focus on order picking (Sect. 2), transportation (Sect. 3), and the question of how to balance logistics workload in a fair manner (Sect. 4). The following sections summarize papers, which are part of the cumulative dissertation [10]. Given the limited space of this paper, however, we only roughly characterize the problems treated as well as the results gained and refer the interested reader to the respective papers for details.

2 Order Picking

The first part of the dissertation investigates order picking problems and is based on the papers [3, 11]. Order picking is the process of collecting products from an available assortment in a specified quantity defined by customer orders. In the era of e-commerce, warehouses face significant challenges, such as a huge amount of orders demanding small quantities, free shipping, and next-day (or even same-day) deliveries [2].

Thus, it is not surprising that there is a bulk of optimization problems in the area of order picking, such as layout design and warehouse location at the strategic level, as well as zoning, picker routing, order batching, and order sequencing at the operational level [5–7]. In this context, the first part of the thesis is devoted to two sequencing problems, which are highly operational planning tasks and thus demand fast solution procedures. While the papers both seek for a suited order sequence, they differ in their underlying storage system (A-Frame dispenser [3] vs. high-bay rack [11]) and thus their specific problem setting, resulting in different constraints and objectives. In general, however, they can (unanimously) be defined as follows. Given a set of orders $\mathcal{J} = \{1, \ldots, n\}$, we seek for an order sequence π, which is a permutation of set $\mathcal{J}$ and a picking plan $x(\pi) \in \Omega$ minimizing the total picking costs $F(x(\pi))$ (caused by, e.g., picking time, wages, and picking errors):

$$\text{Minimize } F(x(\pi)) \tag{1}$$

$$\text{s.t. } x(\pi) \in \Omega. \tag{2}$$

where Ω denotes the set of feasible solutions.

Within our investigations, it turned out that (1)–(2) is NP-hard for both problems [3, 11], so that determining an optimal solution for an increasing problem size may become difficult. However, the good news is that our complexity analysis further revealed that the remaining sub-problems are solvable in polynomial time when the order sequence π is given, i.e., the corresponding picking plan $x(\pi)$ can efficiently be determined. Thus, we utilized this property to develop effective solution approaches. To obtain an initial solution, a straightforward greedy heuristic turned out as a promising approach determining good solutions within negligible computational time. More precisely, the procedure starts with an empty sequence and in each iteration, it appends the order increasing the objective value least. In this way, a first solution is generated as a starting point for further improvements due to tailor-made metaheuristics or exact algorithms. In [3], we propose a multi-start local search procedure using various neighborhoods to alter the sequence and improve the solution. The algorithm shows very effective within our computational study and robust against an increasing problem size, so that even instances of real-world size can be solved. In [11], a branch-and-bound procedure is introduced where each branch of the tree corresponds to a (partial) order sequence. With the developed solution techniques on hand, we further examine the gains of optimization. We compare the picking performance due to the optimized sequences and benchmark them with traditional first-come-first-serve sequences. Our results reveal a huge potential for improvement over the status quo.

More precisely, in [3], workers continuously refill the A-Frame dispenser. However, whenever the workforce is not able to timely replenish the SKUs, additional utility workers take over to guarantee an error-free order picking process. For the solutions obtained by our optimization approaches in only about 7% of all replenishment events a utility worker have to step in. Therefore, a reduction of 75% of the required utility workers can be achieved. On contrary, in [11], we showed that our tailor-made solution procedures decreased the overall order picking time up to 20% for a high-bay rack with crane-supplied pick face where pickers move along the pick face to collect all SKUs demanded.

3 Transportation

The second part of the dissertation examines transportation problems and is based on the papers [1, 12]. Broadly speaking, the general task of transportation is to transfer goods or people from A to B. What sounds so simple, is often actually a complex planning problem, especially when facing real-world requirements, such as free shipping and tight time windows. From a company's point of view, transportation (and other logistics activities such as order picking) is a vital part of their supply chain, but not a value-adding one. Therefore, it is crucial to keep the costs low and plan wisely.

In the world of operations research, transportation of goods is most often associated with the well-known vehicle routing problem, and a vast amount of literature is available that considers a plethora of different problem settings. Given the current requirements of e-commerce, however, just-in-time (JIT) deliveries, i.e., the supply with goods at the point of time they are needed, are of particular interest. Thus, we focus on the JIT-concept in both papers [1, 12] and investigate a basic network structure of a single line where goods have to be transported from A to B.

In [12], the task is to find a feasible assignment of goods to tours and tours to trucks, such that each delivery is in time and the fleet size is minimized, i.e., we allow multiple subsequent tours of the trucks. Thus, we face with the problem to assemble truck deliveries and to schedule the trucks' departure times at the same time. To solve this NP-hard problem, we develop an effective binary search method. The binary search approach is based on tight bounds on the optimal fleet size. Therefore, we identify relaxations that can be solved in polynomial time to obtain lower bounds. The upper bound is determined by dividing the problem into two subproblems, where the first one is solved by a polynomial time approximation scheme to determine the truckloads/tours at the first step. Afterward, at the second step, the procedure applies an optimal polynomial time approach to assign the loads/tours to trucks and to schedule the respective departure times (see [12]). Finally, we used our algorithms for a sensitivity analysis. It shows that the driving distance, the level of container standardization, and the time windows of the JIT-concept have a significant impact on the potential savings promised by optimized solutions [12].

In [1], we determine the trucks' departure times for the example of truck platooning. A platoon describes a convoy of trucks that drive in close proximity after each other. The front truck controls the followers and due to the reduced gap in between them, the aerodynamic drag reduces and less fuel is consumed. Platooning is not only beneficial for economic reasons (e.g., lower fuel costs), but also for environmental ones (e.g., less pollution). In [1], the complexity status for different problem settings (cost functions, time windows, platoon lengths) is investigated. The presented polynomial time procedures provide a suited basis to tackle large-sized problems efficiently. Furthermore, the sensitivity analysis reveals that each of the investigated influencing factors (diffusion of platooning technology, willingness-to-wait, platoon length) strongly impacts the possible fuel savings and a careful planning is required to exploit the full potential of the platooning technology. In particular, our investigations showed that the possible amount of saved fuel might be considerably lower than the technical possible ones. This is indeed a critical result, as our single road problem setting supports the platooning concept, as finding (spatially and temporally) suited partners becomes increasingly unlikely for entire road networks. However, in light of technological developments such as autonomous driving, substantial savings (driver wages) can be achieved justifying the investment costs of truck platooning technology [1].

4 Workload Balancing

The last part of the dissertation examines balancing problems and is based on the papers [9, 13, 14]. Generally speaking, the purpose is to assign a set of tasks $\mathcal{J} = \{1, \dots, n\}$, each defined by its workload p_j ($j \in \mathcal{J}$), to a set of resources $\mathcal{I} = \{1, \dots, m\}$ in order to obtain an even spread of workload. Let $x_{ij} = 1$, if task $j \in \mathcal{J}$ is assigned to resource $i \in \mathcal{I}$ and $x_{ij} = 0$ otherwise, then the problem can be defined as

$$\text{Minimize } F(C_1, \ldots, C_m) \tag{3}$$

$$\text{s.t.} \sum_{j=1}^{n} p_j x_{ij} = C_i \qquad i = 1, \ldots, m \tag{4}$$

$$\sum_{i=1}^{m} x_{ij} = 1 \qquad j = 1, \ldots, n \tag{5}$$

$$x_{ij} \in \{0, 1\} \qquad i = 1, \ldots, m;\ j = 1, \ldots, n \tag{6}$$

where F represents different alternative (workload balancing) objectives, such as the makespan $C_{\max} = \max_{i \in \mathcal{I}} C_i$, its counterpart $C_{\min} = \min_{i \in \mathcal{I}} C_i$ (see [9]), the difference $C_\Delta = C_{\max} - C_{\min}$, or the sum of squared workloads $C^2 = \sum_{i \in \mathcal{I}} C_i^2$ (see [13, 14]). These problems are also known as machine scheduling problems and despite of their simple structure, they are mostly NP-hard.

Problems like (3)–(6) often occur as subproblems or relaxed problem versions. For instance, assume that there is a set of orders, which must be assembled in a warehouse. As a result of the picking problem, we obtain a set of picking tours $\mathcal{J}$ with length p_j ($j \in \mathcal{J}$), which must be assigned to the set of pickers $\mathcal{I}$. One goal might be efficiency, so that we aim to minimize the total makespan for picking the given order set, i.e. $C_{\max}$, or we seek for a fair share of work, i.e. C^2. Moreover, assume that there is a picking tour having to retrieve three washing machines, whereas during a second tour only a pair of socks has to be picked. Then, p_j might represent the ergonomic stress of a picking tour $j \in \mathcal{J}$ instead of its length and applying the objective C^2 yields an ergonomic distribution of work.

A major drawback of traditional objectives such as $C_{\max}$, $C_{\min}$, or C_Δ is that they only take the boundaries into account. Thus, the workload of the remaining machines is only implicitly addressed and may even contravene the balancing issue. Thus, all machines must be considered explicitly, e.g., by C^2. Previous approaches tackled this problem with local search algorithms where the neighborhood is defined by interchanging jobs between pairs of machines. In the case of two machines, however, all the above objectives are equal and the procedure, thus, does not adequately target the general balancing objective. Therefore, we develop an effective solution method to optimally solve the three-machine case [13] and the m-machine case in general [14]. Moreover, we propose a suited local search algorithm to solve C^2 [13].

In [14], we further demonstrate that the developed exact and heuristic solution procedures can handle a vast amount of objectives ($C_{\max}$, $C_{\min}$, C_Δ, and other related objectives). Furthermore, tests [13, 14] show a significant improvement compared to benchmark instances of [4, 8, 13] and the off-the-shelf solver Gurobi. The proposed algorithms do not only improve on the solution quality, but they were also capable to reduce the computation time by several orders of magnitude.

5 Conclusion

The thesis [10] is equally devoted to solving important practical problem settings and deriving theoretical findings for selected problems from the field of outbound logistics. We translate problems motivated from business practice into mathematical optimization problems and, based on their complexity status, develop suited solution procedures capable of handling large-sized instances within short response times. Thus, these algorithms can be applied for operational and tactical planning tasks, such as order sequencing in warehousing, the scheduling of truck deliveries, or workload balancing.

It can be concluded, that sophisticated optimization is a suitable tool to considerably improve over the simple rules-of-thumb that are usually applied in business practice. Technological developments are often merely a first step to gain competitive advantages but optimization deciding on an intelligent application of the technologies is almost always equally important. For instance, in our platooning example, test applications show that fuel savings up to 10% or even 15% are technologically possible. However, this requires that, at any time, a suited platooning partner is available. Therefore, sophisticated matching algorithms to form platoons are needed which constitutes a delightful task for optimization.

References

1. Boysen, N., Briskorn, D., Schwerdfeger, S.: The identical-path truck platooning problem. Transp. Res. B: Methodol. **109**, 26–39 (2018)
2. Boysen, N., De Koster, R., Weidinger, F.: Warehousing in the e-commerce era: a survey. Eur. J. Oper. Res. **277**, 396–411 (2019)
3. Boywitz, D., Schwerdfeger, S., Boysen, N.: Sequencing of picking orders to facilitate the replenishment of A-frame systems. IISE Trans. **51**, 368–381 (2019)
4. Cossari, A., Ho, J.C., Paletta, G., Ruiz-Torres, A.J.: A new heuristic for workload balancing on identical parallel machines and a statistical perspective on the workload balancing criteria. Comput. Oper. Res. **39**, 1382–1393 (2012)
5. De Koster, R., Le-Duc, T., Roodbergen, K.J.: Design and control of warehouse order picking: a literature review. Eur. J. Oper. Res. **182**, 481–501 (2007)
6. Gu, J., Goetschalckx, M., McGinnis, L.F.: Research on warehouse operation: a comprehensive review. Eur. J. Oper. Res. 177, 1–21 (2007)
7. Gu, J., Goetschalckx, M., McGinnis, L.F.: Research on warehouse design and performance evaluation: a comprehensive review. Eur. J. Oper. Res. **203**, 539–549 (2010)
8. Ho, J.C., Tseng, T.L.B., Ruiz-Torres, A.J., López, F.J.: Minimizing the normalized sum of square for workload deviations on m parallel processors. Comput. Ind. Eng. **56**, 186–192 (2009)
9. Lawrinenko, A., Schwerdfeger, S., Walter, R.: Reduction criteria, upper bounds, and a dynamic programming based heuristic for the k_i-partitioning problem. J. Heuristics **24**, 173–203 (2018)
10. Schwerdfeger, S.: Optimization in outbound logistics. Ph.D. thesis, FSU Jena, 2018
11. Schwerdfeger, S., Boysen, N.: Order picking along a crane-supplied pick face: the SKU switching problem. Eur. J. Oper. Res. **260**, 534–545 (2017)

12. Schwerdfeger, S., Boysen, N., Briskorn, D.: Just-in-time logistics for far-distant suppliers: scheduling truck departures from an intermediate cross docking terminal. OR Spectr. **40**, 1–21 (2018)
13. Schwerdfeger, S., Walter, R.: A fast and effective subset sum based improvement procedure for workload balancing on identical parallel machines. Comput. Oper. Res **73**, 84–91 (2016)
14. Schwerdfeger, S., Walter, R.: Improved algorithms to minimize workload balancing criteria on identical parallel machines. Comput. Oper. Res. **93**, 123–134 (2018)
15. Statistisches Bundesamt: Statistisches Jahrbuch—Deutschland und Internationales. Statistisches Bundesamt, Wiesbaden. https://www.destatis.de/DE/Themen/Querschnitt/Jahrbuch/statistisches-jahrbuch-aktuell.html (2018). Accessed 29 June 2019

Incorporating Differential Equations into Mixed-Integer Programming for Gas Transport Optimization

Mathias Sirvent

Abstract The article summarizes the findings of my Ph.D. thesis finished in 2018; see (Sirvent, Incorporating differential equations into mixed-integer programming for gas transport optimization. FAU University Press, Erlangen, 2018). For this report, we specifically focus on one of the three new global decomposition algorithms, which is used to solve stationary gas transport optimization problems with ordinary differential equations. Moreover, we refer to the promising numerical results for the Greek natural gas transport network.

Keywords (Mixed-integer) (non)linear programming · Optimization with differential equations · Simulation based optimization · Decomposition methods · Stationary gas transport optimization

1 Motivation

Natural gas is one of the most important energy sources. Figure 1 shows a graph representing the distribution of power generation in Germany depending on the commodity from 1990 to 2018. While the proportion of renewable energy has massively risen, electricity from nuclear power, black coal, and brown coal has dropped. The proportion of natural gas remains constant and is politically supported for the transition period towards an emission-free world. The reason for this is, amongst others, that the inherent fluctuation of the renewable energy production can be stabilized by gas-fired power plants that start up and shut down quickly. Moreover, energy production with natural gas emits less greenhouse gas compared to coal; see [2].

Consequently, the transportation through gas networks is an essential task and gives rise to gas transport problems. Such optimization problems involve

M. Sirvent (✉)
Friedrich-Alexander-Universität Erlangen-Nürnberg, Discrete Optimization, Erlangen, Germany
e-mail: sirvent@posteo.de

J. S. Neufeld et al. (eds.), *Operations Research Proceedings 2019*, Operations Research Proceedings, https://doi.org/10.1007/978-3-030-48439-2_3

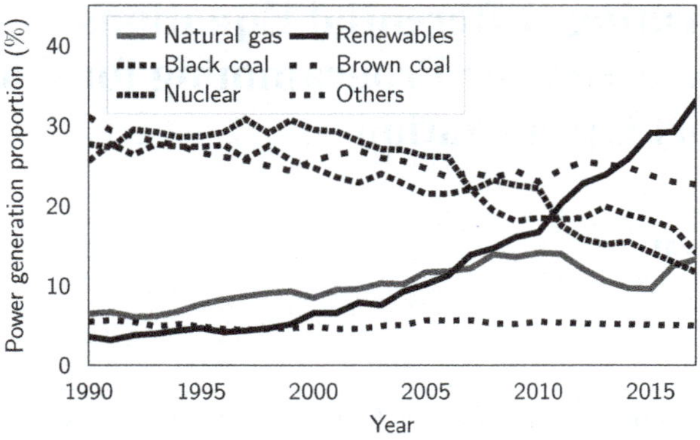

Fig. 1 Gross power generation in Germany; see [1]

discrete decisions to switch network elements. The physical behavior of natural gas is described by differential equations. Thus, when dealing with gas transport optimization, MIPs constrained by differential equations become relevant.

2 A Decomposition Method for MINLPs

Note that significant parts of this chapter are published; see [3, 4]. We define

$$\min_{x} \quad c^\top x \tag{1a}$$

$$\text{s.t.} \quad Ax \geq b, \quad x^- \leq x \leq x^+, \quad x_{\mathcal{C}} \in \mathbb{R}^{|\mathcal{C}|}, \quad x_{\mathcal{I}} \in \mathbb{Z}^{|\mathcal{I}|}, \tag{1b}$$

$$x_{i_2} = f_i(x_{i_1}) \quad \forall i \in [\sigma]. \tag{1c}$$

In more detail, $c^\top x$ is a linear objective function with $c \in \mathbb{R}^n$. Moreover, $x_i = (x_{i_1}, x_{i_2})$ denotes a pair of variables for all variable index pairs $i = (i_1, i_2) \in [\sigma]$ with $i_1, i_2 \in \mathcal{C}$ and a finite set $[\sigma] := \{0, \ldots, \sigma\}$. A function $f_i : \mathbb{R} \to \mathbb{R}$ couples the variables x_{i_1} and x_{i_2}, i.e., $x_{i_2} = f_i(x_{i_1})$ holds for all $i \in [\sigma]$. Problem (1) is a nonconvex MINLP because of the nonlinear equality constraints in (1c). A common assumption is that nonlinear functions are factorable. In this case, every MINLP can be transformed into Problem (1). Solving MINLPs is a highly active field of research and an extensive literature overview is given in [5].

Optimizing over the full constraint set of Problem (1) is expensive. Our idea is to decompose the problem into a master problem and a subproblem. The solutions of the subproblems are used to generate piecewise linear relaxations of the feasible set of (1c). The master problem uses these relaxations and iteratively gets better MIP relaxations of Problem (1) to solve. At the end, we obtain a globally optimal solution of Problem (1) or a proof of infeasibility.

We assume that the bounds x^- and x^+ of Problem (1) give rise to a-priorily known compact boxes $\Omega_i := [x_i^-, x_i^+] := [x_{i_1}^-, x_{i_1}^+] \times [x_{i_2}^-, x_{i_2}^+] \subseteq \mathbb{R}^2$ for all $i \in [\sigma]$ such that the graph of f_i is contained in Ω_i. The algorithm constructs a sequence of subsets $(\Omega_i^k)_k \subseteq \Omega_i$ for all $i \in [\sigma]$ such that Ω_i^k converges to the graph of f_i for iterations $k \to \infty$. For a better handling, we define $\mathrm{gr}(f_i)$ with $\mathrm{gr}(f_i) := \{x_i \in \mathbb{R}^2 : x_{i_2} = f_i(x_{i_1})\}$. Note that Ω_i^k is supposed to form a relaxation of $\mathrm{gr}(f_i) \cap [x_i^-, x_i^+]$ for all $i \in [\sigma]$. We assume that Ω_i^k are finite unions of polytopes

$$\Omega_i^k := \bigcup_{j \in [r_i^k - 1]} \Omega_i^k(j) \tag{2}$$

for all $i \in [\sigma]$ where $\Omega_i^k(j)$ are polytopes for all j in some finite index set $[r_i^k - 1]$.

The master problem is defined over the relaxed sets Ω_i^k and the subproblems are used to refine these relaxations. With these preparations, we are now in the position to state the kth master problem

$$\begin{aligned} \min_x \quad & c^\top x \\ \text{s.t.} \quad & Ax \geq b, \quad x^- \leq x \leq x^+, \quad x_{\mathcal{C}} \in \mathbb{R}^{|\mathcal{C}|}, \quad x_{\mathcal{I}} \in \mathbb{Z}^{|\mathcal{I}|}, \\ & x_i \in \Omega_i^k \quad \forall i \in [\sigma] \end{aligned} \tag{M(k)}$$

that we solve to global optimality providing a solution $\tilde{x}^k$.

If the master problem is infeasible, Problem (1) is infeasible because of the relaxation property. On the other hand, if the master problem's result is already an feasible solution of Problem (1), we are done. If this is not the case, we need to improve our approximation. To this end, we consider the kth subproblem providing a new linearization point on the graph of f_i. With this at hand, the subproblem of the kth iteration reads

$$\begin{aligned} \psi^2(\tilde{x}^k) := \min_x \quad & \|x - \tilde{x}^k\|_2^2 \\ \text{s.t.} \quad & x_i \in \mathrm{gr}(f_i) \qquad \forall i \in [\sigma]. \end{aligned} \tag{S(k)}$$

The solutions of (S(k)) are denoted by $\mathring{x}^k$. By construction, the subproblem (S(k)) has a nonempty feasible set. Moreover, the subproblem (S(k)) can be split up such that single subproblems

$$\begin{aligned} \psi_i^2(\tilde{x}_i^k) := \min_{x_i} \quad & \|x_i - \tilde{x}_i^k\|_2^2 \\ \text{s.t.} \quad & x_i \in \mathrm{gr}(f_i) \end{aligned} \tag{S(i, k)}$$

can be solved in parallel in every iteration k of the algorithm. We remark that the objective values $\psi_i^2(\widetilde{x}_i^k)$ of the subproblems (S(i, k)) are a natural measure of feasibility for every $i \in [\sigma]$. Thus, we define ℓ_2-ε-feasibility.

Definition 1 (ℓ_2-ε-Feasibility) Let $\varepsilon > 0$. A solution $\widetilde{x}^k$ of the master problem (M(k)) is called *ℓ_2-ε-feasible* if $\psi_i(\widetilde{x}_i^k) \leq \varepsilon$ for all $i \in [\sigma]$.

We assume that the nonlinear constraints in (1c) are not given analytically and focus on the following assumption.

Assumption 1 *We have an oracle that evaluates $f_i(x_{i_1})$ and $f_i'(x_{i_1})$ for all $i \in [\sigma]$. Furthermore, all f_i are strictly monotonic, strictly concave or convex, and differentiable with a bounded first derivative on $x_{i_1}^- \leq x_{i_1} \leq x_{i_1}^+$.*

Master Problem The master problem (M(k)) is a relaxation of Problem (1) and it is supposed to be an MIP. Instead of (1c), which is nonlinear, we take a piecewise linear relaxation of its feasible set $\mathrm{gr}(f_i) \cap [x_i^-, x_i^+]$ for all $i \in [\sigma]$ into account to specify Ω_i^k. Beforehand, we strengthen the bounds of the involved variables using Assumption 1. The original bounds $x_i^- \leq x_i \leq x_i^+$ are updated by $l_i \leq x_i \leq u_i$ with $l_{i_1} := \max\left\{x_{i_1}^-, f_i^{-1}(x_{i_2}^-)\right\}$, $u_{i_1} := \min\left\{x_{i_1}^+, f_i^{-1}(x_{i_2}^+)\right\}$, $l_{i_2} := \max\left\{x_{i_2}^-, f_i(x_{i_1}^-)\right\}$, and $u_{i_2} := \min\left\{x_{i_2}^+, f_i(x_{i_1}^+)\right\}$ for all $i \in [\sigma]$.

For the piecewise linear relaxation of $\mathrm{gr}(f_i) \cap [l_i, u_i]$, we use a combination of the incremental method and classical outer approximation; see [6–8]. To construct the set Ω_i^k within the kth master problem (M(k)), we assume that there are values $x_{i_1}^{k,j} \in \mathbb{R}$ for all $j \in [r_i^k]$ with $l_{i_1} =: x_{i_1}^{k,0} < x_{i_1}^{k,1} < \cdots < x_{i_1}^{k,r_i^k} := u_{i_1}$ for all $i \in [\sigma]$. Now, as Assumption 1 gives an oracle for $f_i(x_{i_1})$ and $f_i'(x_{i_1})$, there are values $f_i(x_{i_1}^{k,j}) =: x_{i_2}^{k,j} \in \mathbb{R}$ with $l_{i_2} =: x_{i_2}^{k,0} < x_{i_2}^{k,1} < \cdots < x_{i_2}^{k,r_i^k} := u_{i_2}$ and $f_i'(x_{i_1}^{k,j}) \in \mathbb{R}$ with $f_i'(x_{i_1}^{k,0}) > f_i'(x_{i_1}^{k,1}) > \cdots > f_i'(x_{i_1}^{k,r_i^k})$ for all $j \in [r_i^k]$ and for all $i \in [\sigma]$. The inequalities hold because of the strictly increasing and strictly concave functions. We abbreviate

$$\mathcal{L}_i^k := \left\{x_{i_1}^{k,0}, x_{i_1}^{k,1}, \ldots, x_{i_1}^{k,r_i^k-1}, x_{i_1}^{k,r_i^k}\right\},$$

$$\mathcal{C}_i^k := \left\{x_{i_2}^{k,0}, x_{i_2}^{k,1}, \ldots, x_{i_2}^{k,r_i^k-1}, x_{i_2}^{k,r_i^k}\right\},$$

$$\mathcal{G}_i^k := \left\{f_i'(x_{i_1}^{k,0}), f_i'(x_{i_1}^{k,1}), \ldots, f_i'(x_{i_1}^{k,r_i^k-1}), f_i'(x_{i_1}^{k,r_i^k})\right\}.$$

The sets Ω_i^k for the kth master problem (M(k)) are uniquely defined by the sets $\mathcal{L}_i^k$, $\mathcal{C}_i^k$, and $\mathcal{G}_i^k$ and are modeled as an MIP; see Fig. 2 on the right. Note that an MIP can be solved in finite time, i.e., a standard MIP solver can compute a global optimal solution or prove infeasibility in finite time.

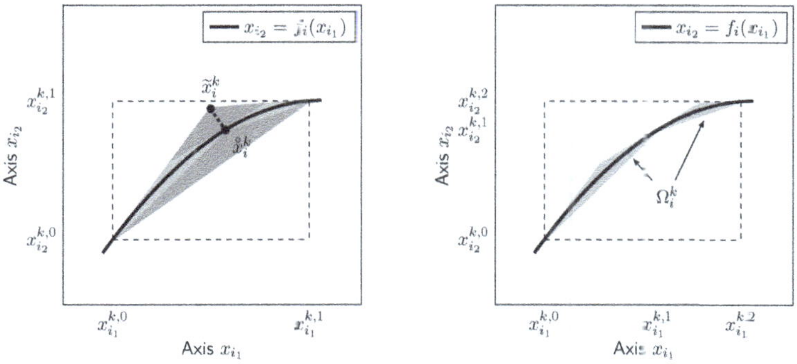

Fig. 2 Subproblem (S(i, k)) on the left and master problem (M(k)) on the right

Subproblem Given a solution $\tilde{x}_i^k$ of (M(k)), the subproblems (S(i, k)) are solved to either determine that the solution found by (M(k)) is close enough to the original feasible set, or alternatively, to provide a new point $\mathring{x}_i^k$ with its derivative to be added to $\mathcal{L}_i^{k+1}$, $\mathcal{C}_i^{k+1}$, and $\mathcal{G}_i^{k+1}$, respectively. A subproblem (S(i, k)) that determines the closest points on gr(f_i) with respect to the given master problems' solutions $\tilde{x}_i^k$ is illustrated in Fig. 2 on the left. The subproblems have a nonempty feasible set by construction. Moreover, the objective values of the subproblems are a natural measure for feasibility; see Definition 1.

Remarks A manual of the algorithm can be found in [4, Algorithm 4.1]. The algorithm is correct in the sense that it terminates with a globally optimal ℓ_2-ε-feasible point of Problem (1) or with the indication of infeasibility. The proof can be found in [3, 4], e.g., in [4, Theorem 4.4.7]. We use the algorithm to solve stationary gas transport optimization problems for the Greek natural gas transport network; see Fig. 3. Detailed results can be found in [3, 4], e.g., in [4, Chapter 4.9.3]. Note that we formulate and discuss additional assumptions and algorithms in [4, 9]. A detailed classification of the algorithm in the light of existing literature can be found in [3, 4], e.g., in [4, Chapter 2 and Chapter 4.7].

3 Scientific Contributions of the Thesis

In Chap. 4 of the thesis, three new global algorithms to solve MIPs constrained by differential equations are presented. Note that Chap. 2 of this report goes into the details of the first algorithm. A typical solution approach transforms the differential equations to linear constraints. The new global algorithms do not rely on this transformation and can work with less information about the underlying differential equation constraints. In an iterative process, MIPs and small nonlinear programs are solved alternately and the correct and finite terminations of the

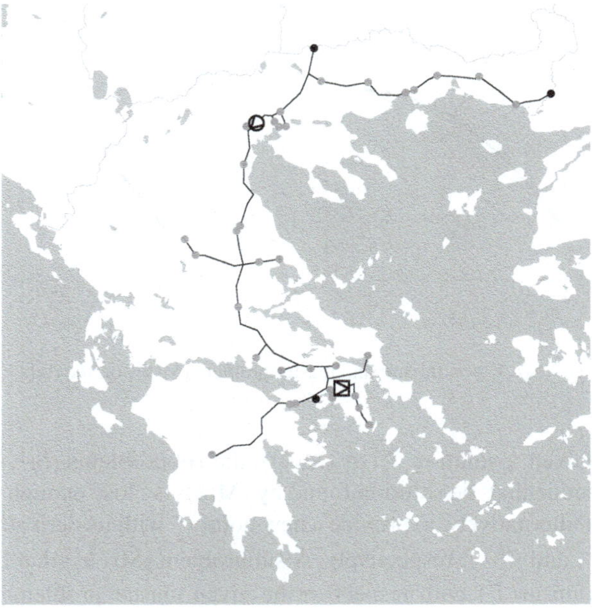

Fig. 3 Greek natural gas transport network with 3 entry nodes (black), 45 exit nodes (gray), 1 control valve (black symbol in the south), 1 compressor machine (black symbol in the north), and 86 pipes (black)

algorithms are proven. An extensive theoretical framework that distinguishes the assumptions on the constraints is set up. The developments allow to solve stationary gas transport optimization problems with ordinary differential equations. In this sense, promising numerical results for the Greek natural gas transport network are shown. Furthermore, the way for more general simulation-based algorithms is paved. Further details about the algorithm are published in international journals; see [3, 9].

In Chap. 5 of the thesis, an instantaneous control algorithm for transient gas network optimization with partial differential equations is presented. A new and specific discretization scheme that allows to use MIPs inside of the instantaneous control algorithm is developed for the example of gas. Again, promising numerical results that illustrate the applicability of the approach are shown. Detailed and dynamic results can be found on public videos.[1] These findings pave the way for more research in the field of transient gas network optimization, which, due to its hardness, is often disregarded in the literature. Note that further details about the instantaneous control algorithm are published; see [10].

[1] https://youtu.be/6F74WZ0CZ7Y and https://youtu.be/4c85DeaAhsA.

Moreover, an essential output of the thesis is the supply of the Greek gas network data for the community; see Fig. 3. It has been processed in a cooperation with the University of Thessaloniki after a common research project; see [11]. In the meantime, the data is provided as an instance of GasLib; see [12].

References

1. AG Energiebilanzen e.V.: Arbeitsgemeinschaft Energiebilanzen e.V (2019). http://www.ag-energiebilanzen.de/
2. Wagner, H.-J., Koch, M.K., Burkhardt, J., Große Böckmann, T., Feck, N., Kruse, P.: CO_2-Emissionen der Stromerzeugung—Ein ganzheitlicher Vergleich verschiedener Techniken. Fachzeitschrift BWK **59**, 44–52 (2007)
3. Gugat, M., Leugering, G., Martin, A., Schmidt, M., Sirvent, M., Wintergerst, D.: Towards simulation based mixed-integer optimization with differential equations. Networks **72**, 60–83 (2018)
4. Sirvent, M.: Incorporating Differential Equations into Mixed-Integer Programming for Gas Transport Optimization. FAU University Press, Erlangen (2018)
5. Belotti, P., Kirches, C., Leyffer, S., Linderoth, J., Luedtke, J., Mahajan, A. Mixed-integer nonlinear optimization. Acta Numer. **22**, 1–131 (2013)
6. Markowitz, H.M., Manne, A.S.: On the solution of discrete programming problems. Econometrica **25**, 84–110 (1957)
7. Duran, M.A., Grossmann, I.E.: An outer-approximation algorithm for a class of mixed-integer nonlinear programs. Math. Program. **36**, 307–339 (1986)
8. Fletcher, R., Leyffer, S.: Solving mixed integer nonlinear programs by outer approximation. Math. Program. **66**, 327–349 (1994)
9. Schmidt, M., Sirvent, M., Wollner, W.: A decomposition method for MINLPs with Lipschitz continuous nonlinearities. Math. Program. **178**, 449–483 (2019)
10. Gugat, M., Leugering, G., Martin, A., Schmidt, M., Sirvent, M., Wintergerst, D. MIP-based instantaneous control of mixed-integer PDE-constrained gas transport problems. Comput. Optim. Appl. **70**, 267–294 (2018)
11. Sirvent, M., Kanelakis, N., Geißler, B., Biskas, P.: Linearized model for optimization of coupled electricity and natural gas systems. J. Mod. Power Syst. Clean Energy **5**, 364–374 (2017)
12. Schmidt, M., Aßmann, D., Burlacu, R., Humpola, J., Joormann, I., Kanelakis, N., Koch, T., Oucherif, D., Pfetsch, M.E., Schewe, L., Schwarz, R., Sirvent, M.: GasLib—a library of gas network instances. Data **2**, 1–18 (2017)

Scheduling a Proportionate Flow Shop of Batching Machines

Christoph Hertrich

Abstract In this paper we investigate the problem to schedule a proportionate flow shop of batching machines (PFB). We consider exact and approximate algorithms for tackling different variants of the problem. Our research is motivated by planning the production process for individualized medicine. Among other results we present the first polynomial time algorithm to schedule a PFB for any fixed number of machines. We also study the online case where each job is unknown until its release date. We show that a simple scheduling rule is two-competitive. For the special case of two machines we propose an algorithm that achieves the best possible competitive ratio, namely the golden section.

Keywords Proportionate flow shop · Batching machines · Permutation schedules · Dynamic programming · Online algorithms

1 Motivation and Problem Description

A frequent feature of industrialized production processes is the usage of so-called *batching machines*, which can process a certain number of products (or jobs) simultaneously. However, once started, newly arriving jobs have to wait until the machine has finished processing the current set of jobs. If jobs arrive over time at such a machine, an interesting planning problem occurs. Every time the machine becomes idle, one has to decide how long to wait for the arrival of more jobs at the cost of delaying the already available jobs. In this paper we investigate a setup where several batching machines are arranged in a linear, flow shop like production

C. Hertrich (✉)
Technische Universität Berlin, Berlin, Germany

Technische Universität Kaiserslautern, Kaiserslautern, Germany

Fraunhofer Institute for Industrial Mathematics (ITWM), Kaiserslautern, Germany
e-mail: hertrich@math.tu-berlin.de

J. S. Neufeld et al. (eds.), *Operations Research Proceedings 2019*, Operations Research Proceedings, https://doi.org/10.1007/978-3-030-48439-2_4

process. This model is inspired by a particular application in pharmaceutical production, where individualized medicine is produced to order. The single steps are performed by high-end machines, like, for example, pipetting robots, which can handle the step for multiple end products at the same time. Sung et al. [9] describe a variety of other applications including, for instance, semiconductor manufacturing.

Formally, the manufacturing process studied in this paper is structured in a *flow shop* manner, where each step is handled by a single, dedicated machine. Each *job* J_j, $j = 1, 2, \ldots, n$, has to be processed by *machines* $M_1, M_2, \ldots, M_m$ in order of their numbering. A job is only available for processing at machine M_i, $i = 2, 3, \ldots, m$, when it has finished processing on the previous machine M_{i-1}.

We consider the problem variants with and without *release dates* $r_j \geq 0$, $j = 1, 2, \ldots, n$. The processing of job J_j on the first machine may not start before the job has been released. The case without release dates may be modeled by setting $r_j = 0$ for all $j = 1, 2, \ldots, n$.

Processing times are job-independent, meaning that each machine M_i, $i = 1, 2, \ldots, m$, has a fixed *processing time* p_i, which is the same for every job when processed on that machine. In the literature, a flow shop with machine- or job-independent processing times is sometimes called a *proportionate* flow shop [7].

Recall that, as a special feature of our application, machines in the flow shop may be able to handle multiple jobs at the same time. These kind of machines are called *(parallel) batching machines* and a set of jobs processed simultaneously on some machine is called a *batch* on that machine. All jobs in one batch on some machine M_i have to start processing on M_i at the same time. In particular, all jobs in one batch have to be available for processing on M_i before the batch can be started. Any batch on machine M_i is processed for exactly p_i time units, independent of its size. This distinguishes parallel batching machines from *serial batching machines*, where the processing time of a batch increases with the number of individual jobs it contains. Each machine M_i, $i = 1, 2, \ldots, m$ has a *maximum batch size* b_i, which is the maximum number of jobs a batch on machine M_i may contain.

Given a feasible schedule S, we denote by $c_{ij}(S)$ the *completion time* of job J_j on machine M_i. For the completion time of job J_j on the last machine we also write $C_j(S) = c_{mj}(S)$. If there is no confusion which schedule is considered, we may omit the reference to the schedule and simply write c_{ij} and C_j.

We consider the following widely used objective functions: makespan $C_{\max}$, (weighted) total completion time $\sum(w_j)C_j$, maximum lateness $L_{\max}$, (weighted) total tardiness $\sum(w_j)T_j$, and (weighted) number of late jobs $\sum(w_j)U_j$. In the context of approximations and online algorithms we are also interested in maximum flow time $F_{\max}$ and total flow time $\sum F_j$, where the flow time of a job is $F_j = C_j - r_j$. Note that all these objective functions are *regular*, that is, nondecreasing in each job completion time C_j.

Using the standard three-field notation for scheduling problems, our problem is denoted as

$$Fm \mid r_j, p_{ij} = p_i, p\text{−batch}, b_i \mid f,$$

where f is one of the objective functions mentioned above. We refer to the described scheduling model as *proportionate flow shop of batching machines* and abbreviate it by PFB.

The main part of this paper deals with the *offline* version, where all data is known in advance. However, we are also interested in the *online* version, where each job is unknown until its release date. In particular, this means that the total number n of jobs remains unknown until the end of the scheduling process.

Previous research on scheduling a PFB has been limited to heuristic methods [9] and exact methods for the special case of two machines [1, 8, 10]. Until now, there have been no positive or negative complexity results for a PFB with more than two machines.

2 Permutation Schedules

Our approach to scheduling a PFB relies on the concept of permutation schedules. In a *permutation schedule* the order of the jobs is the same on all machines of the flow shop. If there exists an optimal schedule which is a permutation schedule with a certain ordering σ of the jobs, we say that *permutation schedules are optimal* and call σ an *optimal job ordering*.

Suppose that for some PFB problem permutation schedules are optimal. Then the scheduling problem can be split into two parts: (1) find an optimal ordering σ of the jobs and (2) for each machine, partition the job set into batches in accordance with the ordering σ such that the resulting schedule is optimal.

Generalizing results from [8–10], we prove the following theorems in [5].

Theorem 1 *For a PFB to minimize $C_{\max}$ or $\sum C_j$, permutation schedules are optimal and any earliest release date order is an optimal ordering of the jobs.*

There exist examples showing that Theorem 1 is not valid for traditional objective functions involving weights or due dates. However, if all jobs are released simultaneously, it turns out that permutation schedules are always optimal.

Theorem 2 *For a PFB without release dates, permutation schedules are optimal for any objective function.*

While it might still be difficult to determine the optimal job permutation, there are several objective functions for which this can be achieved efficiently.

Theorem 3 *Consider a PFB without release dates. Any ordering by nonincreasing weights is optimal for minimizing $\sum w_j C_j$ and any earliest due date order is optimal for minimizing $L_{\max}$ and $\sum T_j$.*

3 Dynamic Program

Once we have fixed a job permutation σ, the next goal is to compute an optimal schedule with jobs ordered by σ. For simplicity, suppose jobs are already indexed by σ, in which case we have to find a permutation schedule with job order $J_1, J_2, \ldots, J_n$. In this section we sketch out how this can be achieved in polynomial time for any fixed number m of machines. For more details, see [5]. We describe a dynamic program that makes use of the important observation that, for a given machine M_i, it suffices to consider a restricted set of possible job completion times on M_i that is not too large. This is formalized by the following lemma, which generalizes an observation by Baptiste [2].

Lemma 4 *For regular objective functions, there exists an optimal schedule in which each job completes processing on machine M_i at a time $c_{ij} \in \Gamma_i$, where*

$$\Gamma_i = \left\{ r_{j'} + \sum_{i'=1}^{i} \lambda_{i'} p_{i'} \;\middle|\; j' \in [n],\ \lambda_{i'} \in [n] \text{ for } i' \in [i] \right\}.$$

Note that $|\Gamma_i| \leq n^{i+1} \leq n^{m+1}$, which is polynomial in n if m is fixed.

Proof *(Idea)* Observe that for regular objective functions, there is an optimal schedule without unnecessary idle time. This means that every batch on M_i is started either when its last job completes M_{i-1} or at the completion time of the previous batch on M_i. The claim follows inductively from this property. □

The dynamic program schedules the jobs one after the other until all jobs are scheduled. It turns out that, in order to decide how J_{j+1} can be added to a schedule that already contains J_1 to J_j, the only necessary information is, first, the completion time of J_j on every machine, and second, the size of the batch containing J_j on every machine. Therefore, we define

$$g(j, t_1, t_2, \ldots, t_m, k_1, k_2, \ldots, k_m)$$

to be the minimum objective value of a partial schedule for jobs J_1 to J_j such that, for all $i = 1, \ldots, m$, J_j is processed on M_i with completion time $c_{ij} = t_i$ in a batch of size k_i. Then one can compute the g-values for an index $j + 1$ recursively from those for index j. The optimal objective value is the minimum of all g-values for $j = n$ and the optimal schedule can be determined by backtracking. Due to Lemma 4 the range of the parameters t_i can be chosen to be Γ_i, while the range for parameters k_i is $\{1, 2, \ldots, b_i\}$ with $b_i \leq n$. Therefore, the total number of g-values to compute is bounded polynomially in n if m is fixed. More precisely, the following theorem holds.

Theorem 5 *Consider a PFB instance with a constant number of m machines and a regular sum or bottleneck objective function. Then, for a given ordering of the jobs, the best permutation schedule can be found in time* $\mathcal{O}(n^{m^2+5m+1})$.

Combining Theorem 5 with Theorems 1 and 3, we obtain

Corollary 6 *For any fixed number m of machines, the problems*

$$Fm \mid r_j, p_{ij} = p_i, p\text{−batch}, b_i \mid f$$

with $f \in \{C_{\max}, \sum C_j\}$ *and*

$$Fm \mid p_{ij} = p_i, p\text{−batch}, b_i \mid f$$

with $f \in \{\sum w_j C_j, L_{\max}, \sum T_j\}$ *can be solved in polynomial time* $\mathcal{O}(n^{m^2+5m+1})$.

Moreover, in the case without release dates, the same statement can be shown for the (weighted) number of late jobs [6]. This is done by modifying the dynamic program slightly such that it finds the optimal job ordering and a corresponding optimal schedule simultaneously.

4 Approximations and Online Algorithms

While the algorithm of the previous section runs in polynomial time for fixed m, its running time is rather impractical already for small values of m. Moreover, in practice, jobs arrive over time and are usually unknown until their release date. Therefore, simple (online) scheduling strategies with a provable performance guarantee are of interest.

In this section we focus on the objective functions $C_{\max}$, $\sum C_j$, $F_{\max}$, and $\sum F_j$. One can show that Theorem 1 is also valid for the latter two objectives. Hence, assuming that jobs are indexed in earliest release date order, we may consider only schedules with job order $J_1, J_2, \ldots, J_n$ on all machines. In order to show approximation ratios and competitiveness of our algorithms, we use a lower bound c^*_{ij} for the completion time c_{ij} of J_j on M_i in any feasible schedule. Note that due to the processing time on each machine, we obtain

$$c_{(i+1)j} \geq c_{ij} + p_{i+1} \tag{1}$$

for all $i = 1, 2, \ldots, m-1$, $j = 1, 2, \ldots, n$. Secondly, the batch capacity on each machine in combination with the fixed permutation yield

$$c_{i(j+b_i)} \geq c_{ij} + p_i \tag{2}$$

for all $i = 1, 2, \ldots, m$, $j = 1, 2, \ldots n - b_i$. Therefore, with starting values $c^*_{0j} = r_j$ for $j = 1, 2, \ldots, n$ and $c^*_{ij} = -\infty$ for $i = 1, 2, \ldots, m$, $j \leq 0$, we recursively define

$$c^*_{ij} = \max\{c^*_{(i-1)j}, c^*_{i(j-b_i)}\} + p_i.$$

Inductive application of (1) and (2) yields the following lemma.

Lemma 7 *Any feasible permutation schedule with job order $J_1, J_2, \ldots, J_n$ satisfies $c_{ij} \geq c^*_{ij}$ for all $i = 1, 2, \ldots, m$ and $j = 1, 2, \ldots, n$.*

This bound has three interesting interpretations, namely as solution of a hybrid flow shop problem, as linear programming relaxation of a mixed-integer program, and as a knapsack dynamic programming algorithm. For details, see [5].

Next, we define the *Never-Wait* algorithm. On each machine M_i, a new batch is immediately started as soon as M_i is idle and there are jobs available. The size of the batch is chosen as minimum of b_i and the number of available jobs. Note that this is in fact an online algorithm. In [5] we prove the following theorem.

Theorem 8 *The Never-Wait algorithm is a two-competitive online algorithm for minimizing $C_{\max}$, $\sum C_j$, $F_{\max}$, and $\sum F_j$ in a PFB.*

It can be shown that, with respect to $C_{\max}$, the competitive ratio of the Never-Wait algorithm is in fact not better than 2. Interestingly, the opposite strategy, namely always waiting for a full batch, yields no constant approximation guarantee at all.

For a PFB with only one machine, that is, a single batching machine with identical processing times, it is known that, with respect to $C_{\max}$, the best possible competitive ratio of a deterministic online algorithm is $\varphi = \frac{1+\sqrt{5}}{2} \approx 1.618$ [3, 4, 11]. This also implies that φ is a lower bound for the competitive ratio in a PFB with arbitrarily many machines. In [5] we provide a specific algorithm for a PFB with two machines matching this bound.

Theorem 9 *There is a φ-competitive online algorithm to minimize $C_{\max}$ and $\sum C_j$ in a PFB with $m = 2$ machines.*

5 Further Research

An open question is to establish the precise complexity status of scheduling a PFB if the number m of machines is part of the input. Another direction for further research is closing the gap between φ and 2 for the competitive ratio of scheduling a PFB online with more than two machines.

Acknowledgments I would like to thank my advisors Sven O. Krumke from TU Kaiserslautern and Heiner Ackermann, Sandy Heydrich, and Christian Weiß from Fraunhofer ITWM, Kaiserslautern, for their continuous, excellent support.

References

1. Ahmadi, J.H., Ahmadi, R.H., Dasu, S., Tang, C.S.: Batching and scheduling jobs on batch and discrete processors. Oper. Res. **40**(4), 750–763 (1992)
2. Baptiste, P.: Batching identical jobs. Math. Methods Oper. Res. **52**(3), 355–367 (2000)
3. Deng, X., Poon, C.K., Zhang, Y.: Approximation algorithms in batch processing. J. Comb. Optim. **7**(3), 247–257 (2003)
4. Fang, Y., Liu, P., Lu, X.: Optimal on-line algorithms for one batch machine with grouped processing times. J. Comb. Optim. **22**(4), 509–516 (2011)
5. Hertrich, C.: Scheduling a proportionate flow shop of batching machines. Master thesis, Technische Universität Kaiserslautern (2018). http://nbn-resolving.de/urn:nbn:de:hbz:386-kluedo-54968.
6. Hertrich, C., Weiß, C., Ackermann, H., Heydrich, S., Krumke, S.O.: Scheduling a proportionate flow shop of batching machines (2020). arXiv:2006.09872
7. Panwalkar, S.S., Smith, M.L., Koulamas, C.: Review of the ordered and proportionate flow shop scheduling research. Naval Res. Logist. **60**(1), 46–55 (2013)
8. Sung, C.S., Kim, Y.H.: Minimizing due date related performance measures on two batch processing machines. Eur. J. Oper. Res. **147**(3), 644–656 (2003)
9. Sung, C.S., Kim, Y.H., Yoon, S.H.: A problem reduction and decomposition approach for scheduling for a flowshop of batch processing machines. Eur. J. Oper. Res. **121**(1), 179–192 (2000)
10. Sung, C.S., Yoon, S.H.: Minimizing maximum completion time in a two-batch-processing-machine flowshop with dynamic arrivals allowed. Eng. Optim. **28**(3), 231–243 (1997)
11. Zhang, G., Cai, X., Wong, C.K.: On line algorithms for minimizing makespan on batch processing machines. Naval Res. Logist. **48**(3), 241–258 (2001)

Vehicle Scheduling and Location Planning of the Charging Infrastructure for Electric Buses Under the Consideration of Partial Charging of Vehicle Batteries

Luisa Karzel

Abstract To counteract the constantly increasing CO_2 emissions, especially in local public transport, more environmentally friendly electric buses are intended to gradually replace buses with combustion engines. However, their current short range makes charging infrastructure planning indispensable. For a cost-minimal allocation of electric vehicles to service trips, the consideration of vehicle scheduling is also crucial. This paper addresses the modeling and implementation of a simultaneous solution method for vehicle scheduling and charging infrastructure planning for electric buses. The Savings algorithm is used to construct an initial solution, while the Variable Neighborhood Search serves as an improvement heuristic. The focus is on a comparison between partial and complete charging processes of the vehicle battery within the solution method. An evaluation based on real test instances shows that the procedure implemented leads to large cost savings. Oftentimes, the consideration of partial charging processes is superior to the exclusive use of complete charging processes.

Keywords Electric vehicle scheduling problem · Charging infrastructure planning · Variable neighborhood search

1 Introduction

In order to counteract the constantly increasing CO_2 emissions, especially in local public transport, Germany is replacing high-pollutant diesel buses with electric buses in pilot projects, as these are locally emission-free and therefore more environmentally friendly. However, this advantage of local zero-emissions also

L. Karzel (✉)
Freie Universität Berlin, Berlin, Germany
e-mail: luisa@karzel.com

J. S. Neufeld et al. (eds.), *Operations Research Proceedings 2019*, Operations Research Proceedings, https://doi.org/10.1007/978-3-030-48439-2_5

entails a number of disadvantages. For example, the purchase costs for an electric bus today are almost twice as high as for a diesel bus, and it also has a much shorter range, which can only be increased by means of charging stations within the route network[3, 12].

In order to counteract the disadvantages of the short range, the high acquisition costs and the required charging infrastructure, efficient planning is required for the scheduling of electric vehicles and for the construction of a charging infrastructure within the route network. The two resulting problems of vehicle scheduling and charging infrastructure planning for electric buses are to be solved as cost-effectively as possible.

In the research community, the two problems are often considered separately and solved individually; the finished charging infrastructure plan serves as a fixed input for the vehicle scheduling solution for electric buses. However, in order to exploit the dependencies between the choice of charging stations in the network and the solution of vehicle scheduling planning for electric buses, a simultaneous consideration of both optimization problems is necessary.

This paper therefore presents the modelling and implementation of a simultaneous solution method for the vehicle scheduling and charging infrastructure planning for electric buses. The focus also lies on an aspect that has so far received little attention in research, namely the consideration of partial charging processes of the vehicle battery within the solution method. Although partial charging of the vehicle battery is technically possible, due to high complexity many studies only allow complete charging of the vehicle battery at a charging station (for examples see [1, 8, 9]). Partial charging processes can generate degrees of freedom, expand the solution space and thus bring savings potential, since there are more possibilities for charging strategies that would not be possible in terms of time if the battery were to be fully charged.

2 Modelling and (Meta-) Heuristic Solution Method

Since vehicle scheduling for electric buses, also called *Electric Vehicle Scheduling Problem* (E-VSP), is a special case of the *Vehicle Scheduling Problem* (VSP), the modeling of the E-VSP is based on the VSP. The VSP involves the cost-minimized assignment of service trips to the number of available vehicles [4]. The following restrictions must be complied with:

- Each service trip must be covered exactly once
- Each vehicle begins and ends a vehicle rotation in the same depot
- There must be no time overlaps between the service trips within one vehicle rotation

While the general VSP does not yet have any range restrictions to consider because the fuel buses have a sufficiently large range, this becomes a central problem with the use of electric buses. The E-VSP is defined as a VSP with route and charging time restrictions, as the batteries of the electric vehicles have only very small capacities and the charging times are limited due to the fact that all service trips within a single vehicle rotation are subject to departure and arrival times [10]. Due to the range limitation, additional restrictions must now be complied with:

- The residual energy of a vehicle battery must never drop to zero or exceed its maximum capacity.
- The battery of a vehicle can only be recharged at charging stations within the road network.

With the simultaneous solution of vehicle scheduling and charging infrastructure planning for electric buses, the total costs, consisting of the costs for the schedule and the development of the infrastructure with charging stations, are to be minimized.

Both the E-VSP and the location planning of charging stations are part of the class of NP-hard problems. The proof of the NP-hardness for the single depot E-VSP is given by Ball [2], for the location planning of charging stations one is referred to [11]. Since both sub-problems are considered NP-hard respectively, the simultaneous consideration is at least as difficult to solve as the more difficult of the two sub-problems. For this reason, the simultaneous consideration of the E-VSP with the location planning of charging stations is likely to be a problem that is difficult to solve and will therefore be solved with the aid of a metaheuristic.

For the construction of a first initial solution, the Savings algorithm is used, while the Variable Neighborhood Search (VNS) is used as an improvement heuristic. The VNS makes use of the advantages of a variable neighborhood size as a diversification strategy and combines this with a local search for intensification [5, 7], see Algorithm 1.

Algorithm 1 Variable neighborhood search source: Gendreau and Potvin [6]

```
 1: function VNS(x, k_max, t_max)
 2:     t ← 0
 3:     while t < t_max do
 4:         k ← 1
 5:         repeat
 6:             x' ← shaking(x, k)
 7:             x'' ← bestImprovement(x')
 8:             x, k ← neighborhoodChange(x, x'', k)
 9:         until k = k_max
10:         t ← CpuTime()
11:     end while
12:     return x
13: end function
```

Given is an initial solution x and an initial neighborhood size k_i (line 4) on the basis of which a local search is started using a BestImprovement method (line 7), which determines the local optimum of the neighborhood (x'') If no improvement can be achieved with the given neighborhood size k_i, the neighborhood is extended and raised to the size k_j in order to apply the BestImprovement method again within this new neighborhood. This method of neighborhood enlargement, also called NeighborhoodChange (line 8), is always used when the current neighborhood can no longer be used to achieve improvements. As soon as an improvement is found or the maximum neighborhood size is reached, the neighborhood is reset to the initial size k_i and the cycle restarts. To give the VNS a stochastic influence, a so-called Shaking function (line 6) is used within the cycle, which at the beginning and after each neighborhood magnification randomly changes the considered solution based on the current neighborhood (x'). This should make it possible to explore new areas of the solution space and corresponds to an explorative approach. The three methods of BestImprovement, NeighborhoodChange and Shaking are performed cyclically until a previously defined termination criterion (line 3) occurs and the current solution is returned as the best solution.

Both the Savings algorithm and the VNS can allow partial charging processes. Thus the effectiveness of partial charging can be investigated and a comparison between the use of partial and complete charging processes can take place.

3 Evaluation

For the evaluation of the method, ten real test instances of different size and topography are used, which are solved by the solution method. The ten test instances range in size from 867 to 3067 service trips and 67–209 stops. All test instances showed significant improvements compared to the initial solution.

Overall, the solutions are improved by an average of 8.56% with the help of VNS, the highest percentage savings being 16.76%,[1] the lowest 3.27%, which clearly highlights the benefits of VNS. The primary factor here is the saving of charging stations, while vehicles are only rarely saved. Furthermore, for the termination criterion $iter_{max}$[2] and the maximum neighborhood size k_{max}[3] parameter tests are conducted to determine the optimal setting of both parameters. It was shown that the best results with the highest parameter assignments of $iter_{max} = 100{,}000$ and $k_{max} = 50\%$ were found and that these settings can be used for future instances.

The focus, however, is on the comparison between the exclusive use of full charges during the charging processes for the electric buses and the use of partial charges. Table 1 shows this comparison; the better solution of the two variants is highlighted for each instance. The table shows that in five out of ten cases the

[1]This corresponds to a saving of 5.9 million monetary units.

[2]The maximum number of iterations is selected as the termination criterion.

[3]For the maximum neighborhood size, the number of vehicle rotations used within the initial solution is used. For example, if $k_{max} = 20\%$, the neighborhood can increase to 20% of the rotations used.

Table 1 Comparison of the final solutions after 100,000 iterations and a maximum neighborhood size of 50% with the use of full charging or partial charging

	Improvement with full charging				Improvement with partial charging			
	Total costs (million)	Vehicle #	CS #[a]	Operating costs (thousand)	Total costs (million)	Vehicle #	CS #	Operating costs (thousand)
t867	29.96	73	3	13.08	**29.71**	**73**	**2**	**13.01**
t1135	34.47	83	5	19.07	**33.67**	**81**	**5**	**19.19**
t1296	24.61	59	4	13.61	**23.56**	**57**	**3**	**13.51**
t2633	**59.95**	**146**	**6**	**46.18**	61.25	148	8	47.38
t3067	**73.35**	**177**	**10**	**47.71**	73.85	177	12	46.55
t10710A	**38.72**	**93**	**6**	**27.27**	38.88	94	5	28.45
t10710B	43.58	107	3	32.43	**41.58**	**102**	**3**	**31.66**
t10710C	40.88	99	5	26.53	**39.52**	**95**	**6**	**24.97**
t10710D	**36.07**	**87**	**5**	**22.41**	36.17	86	7	21.29
t10710E	**29.17**	**71**	**3**	**19.14**	29.97	73	3	19.34

[a]Number of charging stations

method with partial charging generated a better solution than the method with full charging. On average, the solution with partial charges is 3.07% better than the variant using only full charges. The biggest difference is at instance t10710B with 4.59%, which corresponds to a total value of 2 million monetary units. For a total of five instances, the procedure with full charging is better, the average difference between the two is 1.23%. Therefore, partial charging is preferable to the procedure with exclusive use of full charging in many cases. The instances where a worse solution was achieved are solutions with long empty runs with high consumption of energy. A sub-function within the solution method cannot optimally handle such empty runs at present, as it uses a greedy algorithm for runtime reasons. However, this can be prevented by using a complete search, which would increase the runtime but ultimately generate a higher solution quality.

In summary, it can be said that the procedure implemented in this paper for the simultaneous solution of vehicle scheduling and charging infrastructure planning for electric buses offers considerable savings potential in terms of total costs and that the procedure is also suitable for larger instances. The use of partial charges within the charging process for charging the electric buses generally allows for a larger solution space and in many cases generates better solutions than with the exclusive use of full charges. Therefore, it should also be considered in further research works in order to increase the degree of reality of the optimization problem and to generate solutions with a higher quality.

4 Conclusion

Pilot projects launched in Germany to increase the use of electric buses in local public transport show the possibilities and applicability of alternative driving technologies in order to reduce CO_2 emissions. The local emission-free operation of electric buses makes them an attractive alternative to conventional buses with combustion engines, but this also comes with a shortened range and higher acquisition costs. Added to this are the costs of setting up a charging infrastructure within the network in order to be able to increase the range of the electric buses. To overcome these disadvantages as good as possible, optimised planning of vehicle scheduling and the charging infrastructure is required. Within this Master thesis, a simultaneous solution method was developed, which generated a common vehicle deployment and charging infrastructure plan. The Savings method was used as a construction heuristic and the Neighborhood Search variable was selected as the improvement heuristic. Both heuristics allowed the use of partial charging during a charging process.

Although the solution process implemented in this master thesis achieves significant improvements and significantly reduces the overall costs compared to the initial solution, the process should be further refined and expanded in future works. The primary focus should be on the implementation of a complete search within the sub-function of the solution method in order to exploit the savings potential of partial charging processes to the maximum. In addition, the model for the solution method is based on some simplifying assumptions, which should be resolved in future works in order to guarantee a higher degree of realism.

If the extensions mentioned above are applied in future works, an excellent solution method can be developed which is very realistic and can probably be used for a large number of instances.

References

1. Adler, J.D., Mirchandani, P.B.: The vehicle scheduling problem for fleets with alternative-fuel vehicles. Transp. Sci. **51**(2), 441–456 (2016)
2. Ball, M.: A comparison of relaxations and heuristics for certain crew and vehicle scheduling problems. In: ORSA/TIMS Meeting, Washington (1980)
3. Berliner Verkehrsbetriebe: E-bus Berlin: Hab den wagen voll geladen (2015). Accessed 11 Sept 2018
4. Bunte, S., Kliewer, N.: An overview on vehicle scheduling models. Public Transp. **1**(4), 299–317 (2009)
5. Dréo, J., Pétrowski, A., Siarry, P., Taillard, E.: Metaheuristics for Hard Optimization: Methods and Case Studies. Springer, Berlin (2010)
6. Gendreau, M., Potvin, J.-Y.: Handbook of Metaheuristics, vol. 2. Springer, Berlin (2010)
7. Hansen, P., Mladenović, N.: Variable neighborhood search: principles and applications. Eur. J. Oper. Res. **130**(3), 449–467 (2001)
8. Li, J.-Q.: Transit bus scheduling with limited energy. Transp. Sci. **48**(4), 521–539 (2014)
9. Reuer, J., Kliewer, N., Wolbeck, L.: The electric vehicle scheduling problem: a study on time-space network based and heuristic solution approaches. In: Proceedings of the 13th Conference on Advanced Systems in Public Transport (CASPT), Rotterdam (2015)
10. Wang, H., Shen, J.: Heuristic approaches for solving transit vehicle scheduling problem with route and fueling time constraints. Appl. Math. Comput. **190**(2), 1237–1249 (2007)
11. Yang, J., Sun, H.: Battery swap station location-routing problem with capacitated electric vehicles. Comput. Oper. Res. **55**, 217–232 (2015)
12. ÜSTRA Hannoversche Verkehrsbetriebe Aktiengesellschaft: Stadtbus, n.d. Accessed 11 Sept 2018

Data-Driven Integrated Production and Maintenance Optimization

Anita Regler

Abstract We propose a data-driven integrated production and maintenance planning model, where machine breakdowns are subject to uncertainty and major sequence-dependent setup times occur. We address the uncertainty of breakdowns by considering various covariates and the combinatorial problem of sequence-dependent setup times with an asymmetric Traveling Salesman Problem (TSP) approach. The combination of the TSP with machine learning optimizes the production planning, minimizing the non-value creating time in production and thus, overall costs. A data-driven approach integrates prediction and optimization for the maintenance timing, which learns the influence of covariates cost-optimal via a mixed integer linear programming model. We compare this approach with a sequential approach, where an algorithm predicts the moment of machine failure. An extensive numerical study presents performance guarantees, the value of data incorporated into decision models, the differences between predictive and prescriptive approaches and validates the applicability in practice with a runtime analysis. We show the model contributes to cost savings of on average 30% compared to approaches not incorporating covariates and 18% compared to sequential approaches. Additionally, we present regularization of our prescriptive approach, which selects the important features, yielding lower cost in 30% of the instances.

Keywords Data-driven optimization · Traveling salesman problem · Prescriptive analytics · Condition-based maintenance · Machine learning

---

A. Regler (✉)
Logistics & Supply Chain Management, TUM School of Management,
Technical University Munich, Munich, Germany

J. S. Neufeld et al. (eds.), *Operations Research Proceedings 2019*, Operations Research Proceedings, https://doi.org/10.1007/978-3-030-48439-2_6

1 Introduction

We consider a manufacturing environment of an one-line, multiple-product production system that faces two challenges: (i) Due to the significant differences between the products, high sequence-dependent setup times account for non-value creating downtime and (ii) the significant amount of unplanned machine breakdowns, which leads to supply shortages, lost profits and thus, customer dissatisfaction.

For an optimized production plan, the setup time and the uncertain breakdowns need to be minimized to generate more output, better utilize the capacities of the production lines and reduce the time to delivery from customer orders, leading to an improvement of customer satisfaction. In order to cope with these challenges, an integration of production and maintenance planning is needed, that does not only minimize the setup cost, but also takes into account the trade-off between breakdown costs and the additional maintenance costs, caused by frequent scheduling.

By addressing the challenge of breakdowns, predictive maintenance can, when appropriately planned, reduce machine downtime by detecting unexpected trends in feature data (e.g., sensor data), which may contain early warnings on pattern changes. Predictive maintenance can ensure the availability, reliability, and safety of the production systems. It generates profits through an undisrupted production system, optimizing cost, quality, and throughput simultaneously. However, predictive maintenance does not account for the underlying structure of the optimization problem, which might yield suboptimal production and maintenance decisions. This asks for prescriptive analytics approaches that integrate prediction and optimization.

In the course of this research we answer the following questions:

How to integrate production and maintenance scheduling for a holistic production optimization model? How can the decision maker efficiently use data of past observations of breakdowns and covariates to solve the problem? Which performance guarantees does the decision maker have and how do these scale with various problem parameters? What is the value of capturing the structure of the optimization problem when making predictions? How is the applicability of the models in practice?

2 Mathematical Formulation

This research proposes a data-driven optimization approach for integrated production and maintenance planning, where machine breakdowns are subject to uncertainty and major sequence-dependent setup times occur.

We address the uncertainty of breakdowns by considering various covariates such as sensor signals and the combinatorial problem of sequence-dependent setup times with an asymmetric TSP approach [1]. The combination of the TSP with machine learning, to simultaneously optimize the production schedule and maintenance timing, minimizes the non-value creating time in production lines and thus, the overall costs. We apply this by defining a maintenance node in the TSP graph. Furthermore, we train data-driven thresholds based on a modified proportional hazard model from condition-based maintenance. The threshold includes covariates, such as sensor data (vibration, pressure, etc.), whose impact is learned directly from data using the empirical risk minimization principle from learning theory ([2], p. 18).

Rather than conducing prediction and optimization sequentially, our data-driven approach integrates them and learns the impact of covariates cost-optimal via a mixed integer linear programming model to account for the complex structures of optimization models. We compare this approach with a sequential approach, where an algorithm predicts the moment of machine failure. The integrated prescriptive algorithm considers the costs during training, which significantly influences the decisions as the models are trained on a loss function consisting of both maintenance and breakdown costs, whereas the predictive approach is trained on forecasting errors not incorporating any kind of costs. Our prescriptive approach is based on principles of data-driven literature, which is applied to different problems such as the Newsvendor Problem [3–5], portfolio management [6, 7], the minimization of logistics costs in retail [8] or commodity procurement [9].

To our prescriptive model the general notation (Table 1) is applied.

The parameters α and β_m are furthermore out-of-sample not decision variables, but parameters.

In order to integrate the dimension t of the covariate observations to the time used for production jobs, variables x_{ijt} and C_{ijt} have the dimension t. They are only set up in the regarding production cycle t, where maintenance is scheduled, and the job is part of the production slot. For all other t, where no maintenance is set up, the variables are set to zero. t is also used to separate and define the different production slots/cycles.

The target of the optimization models is the minimization of the costs, arising throughout the production system. Therefore, we state the following linear decision rules for x_{ijt}, y_t and z_t:

- For every $i = 1, \ldots, n$, $j = 2, \ldots, n$ and $t = 1, \ldots, k$, x_{ijt} is set up, whenever the edge (i, j) is in the graph and product j is scheduled after product i in production slot t.
- x_{i1t} equals one and maintenance is set up after job i for precisely one predecessor job, if z_t is set to one in t for every $i = 2, \ldots, n$ and $t = 1, \ldots, k$.
- z_t is set to one if the machine age in t plus the threshold function exceeds zero. Another interpretation is when the age is higher than the absolute value of the threshold function $\alpha + \sum_{m=1}^{l} \beta_m F_{mt}$ for every $t = 1, \ldots, k$ This is in line with the hazard function from proportional hazard models.
- y_t is set to one, whenever a breakdown occurs, and no maintenance is done in t, which accounts for a penalty setup for every $t = 1, \ldots, k$.

Table 1 Notation for the prescriptive production planning model

Sets	
$t = 1, \ldots, T$	Time frame/time steps for the sensor data. Each time frame accounts for one observation of every covariate at a certain point in time
$i, j = 1, \ldots, n$	n is the number of jobs to be scheduled. The combination (i, j) is defined as the edges between job i (predecessor) and job j (successor)
$m = 1, \ldots, M$	Set of covariates of type m
Parameters	
BM	Sufficient big number
b_t	= 1, if the machine breaks in t, 0 otherwise
a_t	Age of the machine in time frame t
c^b	Costs for one breakdown of the machine
c^p	Cost per unit time of production
c^m	Costs per maintenance setup
q_{ij}	Sum of setup and production time for j if scheduled after i
F_{mt}	Value of covariate m (numerical value of sensor observation like temperature, pressure or vibration) in time frame t
Decision Variables	
y_t	= 1, if a breakdown occurs and no maintenance is set up in t, 0 otherwise
z_t	= 1, if maintenance is set up in t, 0 otherwise
x_{ijt}	= 1, if product j is produced after job i in production cycle ending in time frame t, 0 otherwise
C_{ijt}	Completion time of job j following job i when set up in cycle ending in t
α	Intercept/feature independent term of the threshold function
β_m	Coefficient for covariate m of the threshold function

Prescriptive production planning model:

$$\min \sum\nolimits_{i=1}^{n} \sum_{j=2}^{n} \left(x_{ijt} \bullet q_{ij}\right) \bullet c^p + \sum\nolimits_{t=1}^{k} z_t \bullet c^m + \sum\nolimits_{t=1}^{k} y_t \bullet c^b \tag{1}$$

Subject to:

$$\sum\nolimits_{j=2}^{n} x_{1jt} = z_t \quad \forall t = 1, \ldots, T \tag{2}$$

$$\sum\nolimits_{j=2}^{n} x_{j1t} = z_t \quad \forall t = 1, \ldots, T \tag{3}$$

$$\sum\nolimits_{j=1}^{n} \sum\nolimits_{t=1}^{k} x_{ijt} = 1 \quad \forall i = 2, \ldots, n \tag{4}$$

$$\sum_{j=1}^{n}\sum_{i=1}^{k} x_{jit} = 1 \quad \forall i = 2, \ldots, n \tag{5}$$

$$c_{1j} \bullet x_{1jt} \leq C_{1jt} \quad \forall j = 2, \ldots, n; \quad t = 1, \ldots, T \tag{6}$$

$$\sum_{i=1}^{n} C_{ijt} + \sum_{k=1}^{n} \left(q_{jk} \bullet x_{jkt}\right) \leq \sum_{k=1}^{n} C_{jkt} \quad \forall j = 2, \ldots, n; \quad t = 1, \ldots, T \tag{7}$$

$$\alpha + \sum_{m=1}^{l} \beta_m F_{mt} + a_t \leq BM \bullet z_t \quad \forall t = 1, \ldots, T \tag{8}$$

$$-a_t - \left(\alpha + \sum_{m=1}^{l} \beta_m F_{mt}\right) \leq BM \bullet (1 - z_t) \quad \forall t = 1, \ldots, T \tag{9}$$

$$y_t \geq st_t - z_t \quad \forall t = 1, \ldots, T \tag{10}$$

$$C_{ijt} \geq 0 \quad \forall i, j = 1, \ldots, n; \quad t = 1, \ldots, T \tag{11}$$

$$x_{ijt} \in \{0, 1\} \quad \forall i, j = 1, \ldots, n; t = 1, \ldots, T \tag{12}$$

$$y_t, z_t \in \{0, 1\} \quad \forall t = 1, \ldots, T \tag{13}$$

The objective function (1) minimizes the overall costs. It includes the production costs c^p, the sum of the maintenance costs c^m and the sum of the breakdown costs c^b multiplied with binary setup variables. Constraints (2) and (3) set—for the t in which z_t equals one—the maintenance node (node one) to one, over the sum of all production jobs as a successor or predecessor jobs. Constraints (4) and (5) ensure, that every production job (2, …, n) is set up exactly once. The completion times C_{ijt} are calculated with the Eqs. (6) and (7). Constraints (8), (9) and (10) are the prescriptive part of the model. This part is learning in-sample the intercept and the covariate coefficients for each of the sensors and represents the decision rules out-of-sample. Constraints (8) and (9) determine the maintenance setup decision. The two constraints ensure, that maintenance is set up, whenever the threshold control constraints are reached (8). If this function is not greater than zero it is not allowed set up maintenance (9). Constraint (10) sets up the penalty/breakdown costs whenever a machine breakdown occurs, and no maintenance is done. This constraint is as well used for the learning in-sample as a penalty constraint for wrong decisions. Out-of-sample are the βs and α given as parameters and the age and the state of the machine calculated. Equation (11) sets the continuous variables C_{ijt} greater equal zero. Equations (12) and (13) are setting x_{ijt} and y_t, z_t as binary variables.

3 Results

In an extensive numerical study, we present the value of data incorporated into decision models and validate the applicability in practice with a runtime analysis. We examine the predictive and prescriptive model and compare these to a small data approach that does not incorporate covariates when optimizing a time-based threshold and the perfect foresight optimum to state cost deviations to the ex-post optimal decisions.

Not having an infinite amount of data leads in theory to a bias, as the algorithms do not have the information to determine the cost-optimal parameters. As stated by the asymptotic optimality theorem, the solution converges to the perfect foresight optimum, if given an infinite amount of data [6]. The numerical results for the finite sample bias show that our prescriptive approach tends to the perfect foresight optimum (below 1% deviation) at a considerable low amount of 1500 historical observations (predictive approach 10% deviation, small data approach 30% deviation). The challenge of the generalization error—the generalizability of the in-sample decision to out-of-sample data [3]—is most prominent with a high number of covariates and a low number of observations, causing risks for the decision maker. This is addressed with the lasso regularization extension in order to select the decision-relevant features and regulate against overfitting. This approach yields lower cost in 80% of the instances compared to the approach without regularization.

The sensitivity to the cost structure of the prescriptive model while learning is the significant difference to the predictive model. The prescriptive model adjusts the decisions according to the associated costs of breakdowns and maintenance, while the predictive model proposes the same decision regardless the costs, which leads to additional risks. This translates into cost savings of 50%, considering a ratio of 1/25 of maintenance to breakdown costs.

The overall runtimes for the training of the predictive approach (2500 observations 0.02 s) are significantly lower than of the prescriptive runtime (346 s), which shows the trade-off between runtimes and robust decisions. By considering the results of cost deviation, below 1% at a training size of 1500 with a training runtime of 18 s, the model is applicable in practice. The optimization of the sequence-dependent setup times and the scheduling of 1 month with 60 jobs on a conservative choice of machine has a runtime of less than half an hour with two maintenance setups and is therefore applicable in practice as well.

4 Conclusion and Managerial Insights

Overall, we find that the prescriptive model contributes to cost savings of on average 30% compared to approaches not incorporating covariates and 18% compared to predictive approaches. This shows the high importance of the covariates in the maintenance context, as the small data approach never captures the true nature of the machine state. Furthermore, it shows the potential in capturing the optimization problem when making predictions.

We conclude, the data-driven integrated production and maintenance optimization model is suitable to solve the challenges presented and can significantly reduce costs in the production environment.

Acknowledgement The author thanks Prof. Dr. Stefan Minner and Dr. Christian Mandl from the chair of Logistics & Supply Chain Management (Technical University Munich) for supervising the thesis and their support, as well as Richard Ranftl for sharing real-world context and technical development support.

References

1. Dantzig, G., Fulkerson, R., Johnson, S.: Solution of a large-scale travelling-salesman problem. Oper. Res. **2**(4), 363–410 (1954)
2. Vapnik, V.N.: The Nature of Statistical Learning Theory. Springer Science+Business Media, New York, NY (1995)
3. Ban, G.-Y., Rudin, C.: The big data newsvendor: practical insights from machine learning. Oper. Res. **67**(1), 90–108 (2019)
4. Beutel, A.-L., Minner, S.: Safety stock planning under causal demand forecasting. Int. J. Prod. Econ. **140**(2), 637–645 (2012)
5. Oroojlooy, A., Snyder, L., Takác, M.: Applying deep learning to the newsvendor problem. arXiv:1607.02177 (2017)
6. Ban, G.-Y., Gallien, J., Mersereau, A.: Dynamic procurement of new products with covariate information: the residual tree method. Manuf. Service Oper. Manag. (2018). Forthcoming
7. Elmachtoub, A. N., Grigas, P.: Smart "predict, then optimize". arXiv:1710.08005v2 (2017)
8. Taube, F., Minner, S.: Data-driven assignment of delivery patterns with handling effort considerations in retail. Comput. Oper. Res. **100**, 379–393 (2018)
9. Mandl, C., Minner, S.: Data-driven optimization for commodity procurement under price uncertainty. Working Paper. Technical University Munich (2019)

Part II
Business Analytics, Artificial Intelligence and Forecasting

Multivariate Extrapolation: A Tensor-Based Approach

Josef Schosser

Abstract Tensor extrapolation attempts to integrate temporal link prediction and time series analysis using multi-linear algebra. It proceeds as follows. Multi-way data are arranged in the form of tensors, i.e., multi-dimensional arrays. Tensor decompositions are then used to retrieve periodic patterns in the data. Afterwards, these patterns serve as input for time series methods. However, previous approaches to tensor extrapolation are limited to special cases and typical applications of link prediction.

The paper at hand connects state-of-the-art tensor decompositions with a general class of state-space time series models. In doing so, it offers a useful framework to summarize existing literature and provide various extensions to it. Moreover, it overcomes the boundaries of classical link prediction and examines the application requirements in traditional fields of time series analysis. A numerical experiment demonstrates the superiority of the proposed method over univariate extrapolation approaches in terms of forecast accuracy.

Keywords Forecast accuracy · Multi-linear algebra · Temporal link prediction · Tensor decomposition · Time series analysis

1 Introduction

Forecasts from univariate time series models have proven to be highly accurate in many application fields. However, univariate specifications are limited in the sense that they are unable to capture dynamic inter-relationships between variables of interest. In order to account for these associations, the paper at hand employs tensor extrapolation, a method developed for the purpose of temporal link prediction. We provide various extensions to the current state of tensor extrapolation and adapt it for

J. Schosser (✉)
University of Passau, Passau, Germany
e-mail: josefschosser01@gmail.com

J. S. Neufeld et al. (eds.), *Operations Research Proceedings 2019*, Operations Research Proceedings, https://doi.org/10.1007/978-3-030-48439-2_7

use in typical fields of time series analysis. An empirical application demonstrates that our approach is able to improve forecast accuracy. However, the model may prove useful in other contexts. Contemporary data sets are more fine-grained in resolution than traditional data, often indexing individual users (customers, etc.) or items (products, etc.) instead of aggregating at the group level. Tensor extrapolation captures this level of detail and generates forecasts simultaneously. Therefore, it promises to enhance both predictive quality and related decisions.

2 Background

This section gives the relevant concepts from linear and multi-linear algebra. In addition, it shows the current state of tensor extrapolation and explains our extensions.

2.1 *Linear and Multi-Linear Algebra*

Multi-way arrays, or *tensors*, are multi-dimensional collections of numbers. The dimensions are known as *ways*, *orders*, or *modes* of a tensor. Using this terminology, scalars, vectors, and matrices can be interpreted as zero-order, first-order, and second-order tensors, respectively. Tensors of order three and higher are called higher-order tensors (cf. [7]). In the simplest high-dimensional case, the tensor can be thought of as a "data cube" (see [8]). This case should be formalized in the following: Let $I, J, K \in \mathbb{N}$ represent index upper bounds, i.e., the number of entities in the modes of interest; a third-order tensor is denoted by $\underline{\mathbf{X}} \in \mathbb{R}^{I \times J \times K}$. The modes of $\underline{\mathbf{X}}$ are referred to as mode A, mode B, and mode C, respectively. Throughout this section, we will restrict ourselves to third-order tensors for reasons of simplicity. Nevertheless, the concepts introduced naturally extend to tensors of order four and higher.

Tensor-based methods originally appeared in the 1920s, but only relatively recently gained increased attention in computer science, statistics, and related disciplines. The *Candecomp/Parafac* (*CP*) decomposition is one of the most common tensor factorizations. In matrix formulation, it is given by

$$\mathbf{X}_A \approx \mathbf{A}\mathbf{I}_A \left(\mathbf{C} \otimes \mathbf{B}\right)^T , \tag{1}$$

where $\mathbf{X}_A$ and $\mathbf{I}_A$ represent the matricizations of $\underline{\mathbf{X}}$ and $\underline{\mathbf{I}}$, respectively, in mode A (cf. [6]). Thereby, $\underline{\mathbf{I}}$ denotes the third-order unit superdiagonal array whose elements $i_{r'r''r'''}$ equal one when $r' = r'' = r'''$ and zero otherwise. The symbol $\otimes$ denotes the so-called Kronecker product. Given a factorization with R components, the matrices $\mathbf{A}$ (for the mode A), $\mathbf{B}$ (for the mode B), and $\mathbf{C}$ (for the mode C) are of sizes $(I \times R)$, $(J \times R)$, and $(K \times R)$, respectively. A visualization can be found in Fig. 1.

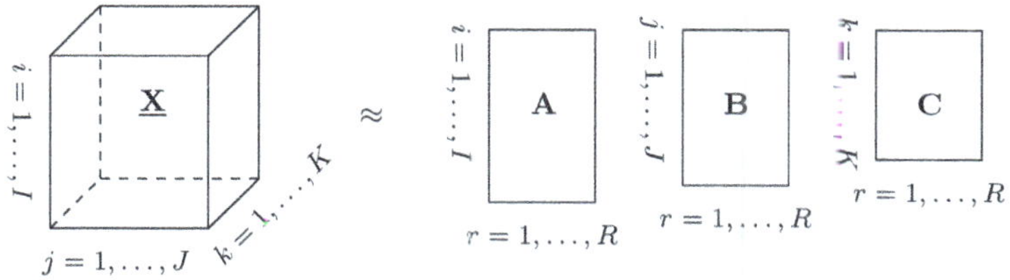

Fig. 1 CP decomposition of third-order tensor $\underline{\mathbf{X}} \in \mathbb{R}^{I \times J \times K}$ into the component matrices $\mathbf{A} \in \mathbb{R}^{I \times R}$, $\mathbf{B} \in \mathbb{R}^{J \times R}$, and $\mathbf{C} \in \mathbb{R}^{K \times R}$

CP is fitted to the data by minimizing the so-called Frobenius norm of approximation errors,

$$\|\mathbf{E}_A\|^2 = \|\mathbf{X}_A - \mathbf{A}\mathbf{I}_A (\mathbf{C} \otimes \mathbf{B})^T\|^2, \tag{2}$$

with respect to **A**, **B**, and **C**. This can be done by means of an *Alternating Least Squares* (*ALS*) algorithm, which alternatingly updates every component matrix keeping fixed the remaining parameter matrices upon convergence. Under mild conditions, the CP model gives unique solutions up to permutations and scalings (cf. [8]). There are several techniques for determining the number of components R in CP decompositions. The most prominent heuristic for model selection is the so-called *Core Consistency Diagnostic* (*CORCONDIA*), which assesses the appropriateness of the model applied in quantifying its degree of superdiagonality (cf. [1]). In practical applications, the CP model is often favored due to its ease of interpretation (cf. [8]).

2.2 *Tensor Extrapolation: Literature and Extensions*

The relevant literature introduces *tensor extrapolation* as a means to temporal link prediction. That means, link data (more precisely, a sequence of observed networks) for T time steps are assumed to be given as input. The goal is to predict the relations at future times $T + 1, T + 2, \ldots, T + L$. Without loss of generality, our exposition is limited to the case where the network snapshots can be represented in the form of a matrix. Here, tensors provide a straightforward way to integrate the temporal dimension. Consequently, a third-order data array $\underline{\mathbf{X}}$ of size $(I \times J \times T)$ is given, with time being modeled in mode C. Extrapolation entails computing the tensor $\hat{\underline{\mathbf{X}}}$ of size $(I \times J \times L)$ that includes estimates concerning future links. The approach is based on the use of tensor decomposition and exponential smoothing. It proceeds as follows (see also Algorithm 1). The multi-way data array is decomposed applying Candecomp/Parafac (CP) factorization. Each of the component matrices gives information about one mode, i.e., detects latent structure in the data. For further processing, the “time” component matrix **C** of size $(T \times R)$ is converted into a set of column vectors $\mathbf{c}_r$. These vectors capture different periodic patterns, e.g., seasons or trends. The periodic patterns discovered are used as input for exponential smoothing techniques. In doing so, forecasts are obtained and arranged as columns $\hat{\mathbf{c}}_r$ of the new "time" component matrix $\hat{\mathbf{C}}$ of size $(L \times R)$. Subsequently, the tensor $\hat{\underline{\mathbf{X}}}$ can be calculated; it contains estimates concerning future links.

There are two fundamental papers that combine tensor decomposition and temporal forecasting methods in the context of link prediction. Both articles treat binary data (indicating the presence or absence of relations of interest). Dunlavy et al. [2] use the temporal profiles computed by CP as a basis for the so-called *Holt-Winters Method*, i.e., exponential smoothing with additive trend and additive seasonality. Spiegel et al. [11] apply CP decomposition in connection with simple exponential smoothing, i.e., exponential smoothing without trend and seasonality. In each case, the model parameters are deliberately set. To date, the introduced basic types are used largely unchanged. However, empirical studies document the demand for a much broader range of extrapolation techniques (cf. [3]). In addition, it is advisable to optimize the model parameters based on suitable criteria.

Algorithm 1 Tensor extrapolation

Input $\underline{\mathbf{X}}$, R

Output $\hat{\underline{\mathbf{X}}}$

CP decomposition: $\mathbf{A}, \mathbf{B}, \mathbf{C} \leftarrow \underline{\mathbf{X}}, R$

Separate column vectors: $\mathbf{c}_1, \mathbf{c}_2, \ldots, \mathbf{c}_R \leftarrow \mathbf{C}$

for each $\mathbf{c}_r \in 1, \ldots, R$ **do**

  Exponential smoothing: $\hat{\mathbf{c}}_r \leftarrow \mathbf{c}_r$

end for

Merge column vectors: $\hat{\mathbf{C}} \leftarrow \hat{\mathbf{c}}_1, \hat{\mathbf{c}}_2, \ldots, \hat{\mathbf{c}}_R$

Calculate tensor: $\hat{\underline{\mathbf{X}}} \leftarrow \mathbf{A}, \mathbf{B}, \hat{\mathbf{C}}$

Given these shortcomings, our contribution is twofold. First, we resort to an automatic forecasting procedure based on a general class of state-space models subsuming all standard exponential smoothing methods (cf. [5]). Model selection and parameter optimization are "individual" (cf. [3, p. 648], [9, p. 153]), meaning that they are based on each single periodic pattern contained in a column of the matrix $\mathbf{C}$. Second, in contrast to the above-mentioned papers, we apply the methodology to real-valued data. We thus go beyond the boundaries of traditional link prediction and investigate the conditions for use in typical application fields of time series analysis. Rooted in the spirit of operations research, our modifications are designed to be of immediate practical relevance: Using real-world data, we demonstrate the superiority of our method over traditional extrapolation approaches in terms of forecast accuracy.

3 Application

Our empirical study resorts to data provided by the United Nations. Trade data between countries of the world are available at the UN Comtrade website http://comtrade.un.org/. We select a set of 30 countries, mostly large or developed countries with high gross domestic products and trade volumes. As a trade category, we choose "exports of goods". We analyze monthly data that includes the years 2012 through 2016. Since the exports of a country to itself are not defined, we obtain in total 870 time series with 60 observations each. The data are split into an estimation sample and a hold-out sample. The latter consists of the twelve most recent observations, which are kept for evaluating the forecast accuracy. Estimation sample and hold-out sample are arranged in tensor form. We thus obtain $\underline{\mathbf{X}}^{est}$ of size $(30 \times 30 \times 48)$ and $\underline{\mathbf{X}}^{hold}$ of size $(30 \times 30 \times 12)$, respectively. The methods under consideration are implemented, or trained, on the estimation sample. The forecasts are produced for the whole of the hold-out sample and arranged as tensor $\hat{\underline{\mathbf{X}}}$ of size $(30 \times 30 \times 12)$. Finally, forecasts are compared to the actual withheld observations.

We use two popular performance measures from the field of time series analysis, the *Mean Absolute Percentage Error* (*MAPE*), and the *Symmetric Mean Absolute Percentage Error* (*sMAPE*). MAPE is positively skewed in the case of series with small scale; sMAPE tries to fix this problem (cf. [4, 12]). In the following, we should be aware of the fact that the time series are of very different scale. This is directly visible when comparing the mean exports from China to the US $(32, 139, 813, 285$ US-\$) and those from Austria to Malaysia $(47, 115, 981$ US-\$). Since CP minimizes squared error, distortions may result. Consequently, data preprocessing may be necessary. Moreover, it is important to determine the number of components R in CP decompositions. We apply CORCONDIA to narrow down the number of designs examined (cf. [1]).

Table 1 displays our main results. We use univariate extrapolation as a baseline. Here, each time series is extrapolated separately. Consequently, any inter-link dependencies are ignored. In general, the errors are relatively high and exhibit considerable skewness. For both measures, the best-performing method is highlighted with its result in bold.

If raw data are used, the forecasting accuracy of tensor extrapolation is inferior to that of univariate extrapolation. A simple preprocessing in the form of a centering across the time mode (compare [6]) changes the situation. Now, tensor extrapolation outperforms univariate extrapolation. Moreover, we normalize the data. In the course of this, the time series are rescaled from the original range so that all values are within the range of zero and one. Normalization gives good results, but cannot beat centering. Additionally, rescaling alters dependencies in the sense that an adjustment of the number of components R is required. Our results continue to hold if the training period is reduced to 36 observations (for more details, see [10]).

Obviously, there are dynamic inter-relationships between the variables of interest. Univariate extrapolation cannot capture these dependencies. In contrast, tensor

Table 1 Forecasting accuracy in terms of MAPE and sMAPE

Method	MAPE	sMAPE
Univariate extrapolation	277.2906	25.7766
Raw data		
Multivariate extrapolation ($R = 4$)	1090.8720	51.7728
Multivariate extrapolation ($R = 5$)	1023.8131	51.6427
Multivariate extrapolation ($R = 6$)	788.3220	59.1561
Multivariate extrapolation ($R = 7$)	767.4262	51.3555
Multivariate extrapolation ($R = 8$)	805.6478	62.9822
Centered data		
Multivariate extrapolation ($R = 4$)	266.5636	28.6726
Multivariate extrapolation ($R = 5$)	266.8292	**21.5501**
Multivariate extrapolation ($R = 6$)	196.3126	22.9681
Multivariate extrapolation ($R = 7$)	**194.0047**	26.7026
Multivariate extrapolation ($R = 8$)	200.7392	35.3210
Normalized data		
Multivariate extrapolation ($R = 2$)	219.3151	26.4533
Multivariate extrapolation ($R = 3$)	216.1289	25.5509
Multivariate extrapolation ($R = 4$)	221.5137	25.4884
Multivariate extrapolation ($R = 5$)	244.6676	25.8155
Multivariate extrapolation ($R = 6$)	242.7884	25.2513

The estimation sample includes the years 2012–2015, the hold-out sample covers the year 2016

extrapolation identifies them and improves forecast accuracy. That means, accounting for the relational character of the data pays off.

References

1. Bro, R., Kiers, H.A.L.: A new efficient method for determining the number of components in PARAFAC models. J. Chemom. **17**(5), 274–286 (2003). https://doi.org/10.1002/cem.801
2. Dunlavy, D.M., Kolda, T.G., Acar, E.: Temporal link prediction using matrix and tensor factorizations. ACM Trans. Knowl. Discovery Data **5**(2), e10 (2011). https://doi.org/10.1145/1921632.1921636
3. Gardner, E.: Exponential smoothing: the state of the art—part II. Int. J. Forecasting **22**(4), 637–666 (2006). https://doi.org/10.1016/j.ijforecast.2006.03.005
4. Hyndman, R.J., Koehler, A.B.: Another look at measures of forecast accuracy. Int. J. Forecasting **22**(4), 679–688 (2006). https://doi.org/10.1016/j.ijforecast.2006.03.001
5. Hyndman, R.J., Koehler, A.B., Snyder, R.D., Grose, S.: A state space framework for automatic forecasting using exponential smoothing. Int. J. Forecasting **18**(3), 439–454 (2002). https://doi.org/10.1016/S0169-2070(01)00110-8
6. Kiers, H.A.L.: Towards a standardized notation and terminology in multiway analysis. J. Chemom. **14**(3), 105–122 (2000). https://doi.org/10.1002/1099-128X(200005/06)14:3<105::AID-CEM582>3.0.CO;2-I
7. Kolda, T.G., Bader, B.W.: Tensor decompositions and applications. SIAM Rev. **51**(3), 455–500 (2009). https://doi.org/10.1137/07070111X

8. Papalexakis, E.E., Faloutsos, C., Sidiropoulos, N.D.: Tensors for data mining and data fusion: models, applications, and scalable algorithms. ACM Trans. Intell. Syst. Technol. **8**(2), e16 (2016). https://doi.org/10.1145/2915921
9. Petropoulos, F., Makridakis, S., Assimakopoulos, V., Nikolopoulos, K.: 'Horses for Courses' in demand forecasting. Eur. J. Oper. Res. **237**(1), 152–163 (2014). https://doi.org/10.1016/j.ejor.2014.02.036
10. Schosser, J.: Multivariate extrapolation: a tensor-based approach. Working Paper, University of Passau (2019).
11. Spiegel, S., Clausen, J., Albayrak, S., Kunegis, J.: Link prediction on evolving data using tensor factorization. In: New Frontiers in Applied Data Mining: PAKDD 2011 International Workshops, pp. 100–110 (2012). https://doi.org/10.1007/978-3-642-28320-8_9
12. Tofallis, C.: A better measure of relative prediction accuracy for model selection and model estimation. J. Oper. Res. Soc. **66**(8), 1352–1362 (2015). https://doi.org/10.1057/jors.2014.103

Part III
Business Track

Heuristic Search for a Real-World 3D Stock Cutting Problem

Katerina Klimova and Una Benlic

Abstract Stock cutting is an important optimisation problem which can be found in many industries. The aim of the problem is to minimize the cutting waste, while cutting standard-sized pieces from sheets or rolls of a given material. We consider an application of this problem arising from the packing industry, where the problem is extended from the standard one or two dimensional definition into the three dimensional problem. The purpose of this work is to help businesses determine the sizes of boxes to purchase so as to minimize the volume of empty space of their packages. Given the size of a real-world problem instances, we present an effective Adaptive Large Neighbourhood Search heuristic that is able to decrease the volume of empty space by an average of 22% compared to the previous approach used by the business.

Keywords Cutting and packing · Adaptive neighborhood search · Heuristics

1 Introduction

Stock cutting is a well-known optimisation problem arising from important practical applications. It consists in cutting standard-sized pieces of stock material (e.g., paper rolls or sheet metal) so as to minimize the amount of wasted material. According to a study conducted by a leading international packing company, 50% of the packing volume is air. Considering that Amazon alone dispatched over 5 billion orders in 2017, the potential for packing improvement is massive. In the ideal case scenario, each order would be packed into a custom-made box that fits its

K. Klimova (✉)
Satalia, Camden, London, UK
e-mail: kat@satalia.com

U. Benlic
School of Electrical and Automation Engineering, East China Jiaotong University, Nanchang, China

J. S. Neufeld et al. (eds.), *Operations Research Proceedings 2019*, Operations Research Proceedings, https://doi.org/10.1007/978-3-030-48439-2_8

dimensions. However, this is generally impossible from the practical stance as the packing process would get significantly slower and costly—a suitable box would need to be produced for each order.

The purpose of this work is to determine the three-dimensions of a given finite number of box types available for packing to help businesses reduce the amount of empty packing volume. As stock cutting is known to be NP-complete [1], we propose an Adaptive Large Neighbourhood Search heuristic based on different repair and destroy move operators. The heuristic iterates between a destruction and a repairer phase. The destruction phase is the diversification mechanism which consists in removing a subset of items (elements) from a given complete solution. This is based on fast destroy operators to escape from local optima, while the repairer phase is the intensification mechanism that makes use of greedy move operators to lead the search towards new quality solutions. Experimental results on a set of real-world instances show an average decrease of around 22% in air volume compared to the solutions used by the business.

2 Literature Review

Different formulations and applications of the cutting problem have been studied in the literature since the 60s. The first problem definition was the minimization of cost for cutting a given number of lengths of material from stock material of a given cost. A linear programming approach for this problem was proposed by Gilmore and Gomory [2]. Even though the problem was first defined as one dimensional, the definition was soon extended to consider two dimensions. For instance, the work by Gilmore and Gomory [3] presents a solution to multistage cutting stock problems with two or more dimensions. More recently, Belov et al. [4] proposed a branch-and-cut-and-price algorithm for one-dimensional stock cutting and two-dimensional two-stage cutting. In [5], Hifi presented a combination of dynamic programming and hill climbing for the two-dimensional stock cutting problem. Both one and two dimensional stock cutting problems can be frequently found in practice, from cutting wires to cutting boxes and corrugated paper.

Despite its practical applications in the packing industry, only limited research has been done on the three-dimensional stock cutting problem [6], while more attention has been devoted to the closely related 2D and 3D packing problem that consists in packing items into minimal number of containers [7]. We present the first heuristic approach based on the Adaptive Large Neighborhood Search [8] framework for the 3D stock cutting problem.

3 Formal Definition

The problem considered in this work is encountered at almost every online shipping company, where a decision has to be made on the sizes of packing boxes that the business needs to order so as to minimize the volume of empty space of their packages. For practical reasons, the maximum number of different box types (sizes) must not be exceeded, which is generally from three to twenty box types for the majority of businesses.

Given a large number of orders consisting of different items, the problem is then to determine the box types (dimensions) that the business needs to purchase, along with their corresponding quantities, while ensuring that the permitted limit of box types is not exceeded. We further take into account the common practice of item consolidation (placement) into a single box with the aim to minimize empty volume. These consolidated items then form a single object of a cumulative volume. To determine the dimensions of this object, we rotate every item such that x and z are the longest and the shortest dimensions respectively. The longest x dimensions across each item to consolidate becomes the x dimension of the new consolidated object. We determine the y dimension of the new object in the same manner, while the z dimension is determined given the new x and y dimensions and the cumulative volume.

Let n be the maximum number of box types and let $I = \{1 \dots, m\}$ be a set of m historical orders with corresponding dimensions $s^I_{x,i}, s^I_{y,i}, s^I_{z,i}, i \in I$, such that $s_{x,i} \geq s_{y,i} \geq s_{z,i}$. The volume v of an item is then computed as $s_{x,i} * s_{y,i} * s_{z,i}$.

The 3D cutting problem consists in (1) determining the x, y, z dimensions of each box b, represented by the decision variables $s^B_{d,b} \geq 0, d \in \{x, y, z\}, b = 1 \dots n$; and (2) in determining the assignment of each item $i \in I$ to boxes, where $u_{b,i} \in 0, 1, b = 1 \dots n, i \in I$ is a binary variable that indicates if item i is assigned to box b. The complete mathematical model is given below.

$$\min \quad \sum_{i \in I, b \in B} u_{b,i}(v^B_b - v^I_i), s.t. \tag{1}$$

$$\sum_{b \in 1..n} u_{b,i} = 1, \forall i \in I \tag{2}$$

$$s^B_{x,b} \geq s^B_{y,b} \geq s^B_{z,b}, \forall b = 1 \dots n \tag{3}$$

$$s^B_{d,b} \geq u_{b,i} s^I_{d,i}, \forall i \in I, b = 1 \dots n, d \in \{x, y, z\} \tag{4}$$

$$u_{b,i} \in \{0, 1\}, \forall i \ I, b = 1 \dots n \tag{5}$$

$$s^B_{d,b} \in \mathbb{N}, \forall b = 1 \dots n, d \in \{x, y, z\} \tag{6}$$

Equation (1) defines the objective which is to minimize the difference between the box volume v^B and the item volume v^I if item is assigned to the given box. Equation (2) ensures that each order is assigned to exactly one box type, while Eq. (3) ensures that dimensions x and z are the largest and the shortest box dimensions respectively. Equation (4) ensures that each item fits the box assigned to it.

Although the above formulation could be linearized, the problem still remains hard to solve for the existing exact solvers. The reason for this is the definition of problem where input can be millions of items which need to be assigned to one of tens of boxes while applied constraints leave the search space too large.

4 Proposed Approach

4.1 General Framework

A solution to the 3D stock cutting problem can be represented as an array S of integers, where each element of the array corresponds to an item $i \in I$, while $S(i)$ is the box $1 \leq b \leq n$ assigned to item i. Starting from a random assignment of items to boxes, the proposed algorithm iterates between a destroy and a repair procedure, where the destroyer consists in deallocating a selection of items from the solution for the purpose of diversification, while the repairer reconstructs the partial solution by reallocating all the items removed in the destroyer phase. A distinguishing feature of the proposed Adaptive Large Neighborhood Search (ALNS) approach is the use of multiple move operators during both the destroyer and the repairer phase.

Let $\mathcal{M} = \{(m_1^d, m_1^r), \ldots, (m_k^d, m_k^r)\}$ be the set of combinations (pairs), where m^d and m^r are the move operators used in the next destroyer and repairer phase respectively. Each iteration of ALNS first consists in adaptively selecting a pair $(m^d, m^r) \in \mathcal{M}$ as described in Sect. 4.3. The algorithm then proceeds by applying α moves with operator m^d to the current solution to diversify the search, followed by α moves with operator m^r to reconstruct the solution, where α is a parameter that controls the diversification strength ($\alpha = 100$ in our experiments). Finally, the algorithm updates the best recorded solution if the solution following the repair phase constitutes an improvement. The main algorithmic framework of the proposed ALNS is given in Algorithm 1.

4.2 Move Operators

Move operators are the key element of a Large Neighborhood Search algorithm. We distinguish between two types of move operators—destroyers and repairers. Given a complete solution, each move of a destroyer deallocates an item from

Algorithm 1 ALNS framework

$S \Leftarrow buildInitialSolution$
$\mathcal{M} \leftarrow \{(m_1^r, m_1^d), \ldots, (m_k^r, m_k^d)\}$ /*set of move operator pairs*/
$S_{best} \Leftarrow S$
while Stopping condition is not met **do**
  $(m^r, m^d) \leftarrow selectMoveOperatorPair(M)$
  $S \Leftarrow destroy(S, m^d, \alpha)$
  $S \Leftarrow repair(S, m^r, \alpha)$
  if $cost(S_{best}) > cost(S)$ **then**
    $S_{best} \Leftarrow S$
  end if
end while

its allocated box type leading to a partial solution. Given a partial solution, each move of a repairer reassigns an item to a box. Since escaping from local minima is especially difficult for very large data sets with small number of box types available, the number of destroyers for our ALNS exceeds the number of repairers.

The proposed approach makes use of five types of destroy operators: (1) random operator consists in deallocating from a solution a randomly selected item; (2) best operator consists in removing from the solution an item with the largest volume of empty space; (3) smaller container operator removes the smallest item from a randomly selected box type; (4) larger container removes the largest item from a randomly selected box type; and (5) clustered operator deallocates from the solution an item from a selected cluster, where a cluster is formed of α items of similar dimensions.

Three move operators are used during the repairer phase: (1) random operator that assign a deallocated item to a randomly selected box type; (2) best operator that assigns a deallocated item to the best fitting box type so as to minimize the volume of empty space; and (2) dimension-fixed repairer that assigns a deallocated item to a box type only if the assignment does not lead to a change in the box dimensions.

4.3 Adaptive Procedure for Operator Selection

Given five destroy and three repair operators, the number of operator combinations in $\mathcal{M}$ (see Algorithm 1) is fifteen. Before the first iteration of the destroy/repair phase, each pair $p_k \in \mathcal{M}$ has an equal probability $p_k = 1/|\mathcal{M}|$ of selection. This probability is then adaptively updated based on the performance of the selected operator pair at the end of the ALNS iteration. Let $times(k)$, $k \in \mathcal{M}$ be the number of times that operator pair p was used by ALNS, and let $score(k)$ be the number of times that the solution obtained after an application of k is better than the solution from the previous ALNS iteration in terms of the objective value. The updated probability p_k of using k in the next ALNS iteration is determined as $p_k = v_k/q$, where $v_k = p_k * (1 - \epsilon) + \epsilon * (score(k)/times(k))$, and $q = \sum_{k \in \mathcal{M}} p_k$. ϵ is a parameter that takes a value in the range [0, 1].

Table 1 Results of 10 independent runs

Data set	Scenario	Orders	Templates	Best total (m^3)	Avg total (m^3)	Avg void/order (L)
1	1	277,000	4	1482	24,415	6.53
1	2	277,000	9	757	918	2.73
2	1	2,100,000	4	145,514	148,556	69.29
2	2	2,100,000	9	143,952	146,711	68.55

A move operator pair $n \in \mathcal{M}$ to be used in the next iteration of ALNS is then determined using the well-known roulette selection strategy based on its selection probability p_n. To avoid premature convergence towards a single move operator pair, a pair $p_k \in \mathcal{M}$ is selected at random with a probability γ, where γ is a parameter.

5 Computational Results

This section presents computational results on two real-world data instances and two scenarios. First scenario's maximum number of available box types is limited to four types and second scenario's limit is nine types. Unfortunately, we are unable to disclose any details on the used data instances or the actual solutions used by the business.

We perform 10 independent runs for each instance and scenario, where each run is limited to 20 min that was deemed acceptable for the client. For each case, Table 1 shows the best and the average total volume of empty space across all the runs, as well as the average void in liters per order. We include average void per order as it was one of main KPIs, the value in table represents this value averaged over the 10 runs.

In case of the first data instance, the dimension of the largest box for scenario 1 is $600 \times 590 \times 420$ mm with 44,501 orders ($\sim$16%) larger than $330 \times 280 \times 265$, and $600 \times 592 \times 590$ mm for scenario 2 with 771,040 ($\sim$37%) of items being larger than $444 \times 374 \times 195$ mm. It is important to note that the data sets are strongly heterogeneous in dimensions.

6 Conclusion

This paper presents the first application of Adaptive Large Neighborhood Search (ALNS) framework to a real-work 3D stock cutting problem that arises from online shipping industry. The key elements of ALNS is a set of destroy and repair move operators that are selected in a probabilistic and adaptive manner. The proposed approach has been adopted by our client (a leading packing company) and is able to report a reduction in the total volume of empty space of their packages by around 22% on average compared to their previous solution.

References

1. Blazewicz, M., Drozdowski, M., Boleslaw, S., Walkowiak, R.: Two dimensional cutting problem: basic complexity results and algorithms for irregular shapes. Found. Control Eng. **14**(4), (1989)
2. Gilmore, P.C., Gomory, R.E.: A linear programming approach to the cutting-stock problem. Oper. Res. **9**(6), 849–859 (1961)
3. Gilmore, P.C., Gomory, R.E.: Multistage cutting stock problems of two and more dimensions. Oper. Res. **13**(1), 94–120 (1965)
4. Belov, G., Guntram S.: A branch-and-cut-and-price algorithm for one-dimensional stock cutting and two-dimensional two-stage cutting. Eur. J. Oper. Res. **171**(1), 85–106 (2006)
5. Hifi, M.: Dynamic programming and hill-climbing techniques for constrained two-dimensional cutting stock problems. J. Comb. Optim. **8**(1), 65–84 (2004)
6. De Queiroz, T.A., et al.: Algorithms for 3D guillotine cutting problems: unbounded knapsack cutting stock and strip packing. Comput. Oper. Res. **39**(2), 200–212 (2012)
7. Martello, S., Pisinger, D., Vigo, D.: The three-dimensional bin packing problem. Oper. Res. **48**(2), 256–267 (2000)
8. Ropke, S., Pisinger, D.: An adaptive large neighborhood search heuristic for the pickup and delivery problem with time windows. Transp. Sci. **40**(4), 455–472 (2006)

Part IV
Control Theory and Continuous Optimization

Model-Based Optimal Feedback Control for Microgrids with Multi-Level Iterations

Robert Scholz, Armin Nurkanovic, Amer Mesanovic, Jürgen Gutekunst, Andreas Potschka, Hans Georg Bock, and Ekaterina Kostina

Abstract Conventional strategies for microgrid control are based on low level controllers in the individual components. They do not reflect the nonlinear behavior of a coupled system, which can lead to instabilities of the whole system. Nonlinear model predictive control (NMPC) can overcome this problem but the standard methods are too slow to guarantee sufficiently fast feedback rates. We apply Multi-Level Iterations to reduce the computational expenses to make NMPC real-time feasible for the efficient feedback control of microgrids.

Keywords Nonlinear model predictive control · Optimal control · Power engineering · Microgrid

1 Introduction

In the context of the energy transition, the use of renewable energy sources (RES) has increased significantly over the last years. Most of the RES are small and connected to medium or low voltage grids. The high number of RES is a rising challenge for the current control paradigm of the utility grid. Microgrids are small electrical networks with heterogeneous components and they are considered to become a key technology to facilitate the integration of RES, because they allow to cluster local components as a single controllable part of a larger electrical network.

R. Scholz (✉) J. Gutekunst · A. Potschka · H. G. Bock · E. Kostina
Interdisciplinary Center for Scientific Computing (IWR), Heidelberg University, Heidelberg, Germany
e-mail: robert.scholz@iwr.uni-heidelberg.de

A. Nurkanovic · A. Mesanovic
Siemens AG, Munich, Germany
e-mail: armin.nurkanovic@siemens.com

J. S. Neufeld et al. (eds.), *Operations Research Proceedings 2019*, Operations Research Proceedings, https://doi.org/10.1007/978-3-030-48439-2_9

However, the effective operation of microgrids is considered to be extremely difficult. The main challenge is to keep the frequency and voltage within the tight operational limits even for demanding load scenarios. State of the art methodology relies on a hierarchical control structure, comprising proportional-integral- and droop-based controllers on different levels. However, experience shows that high penetration of RES is pushing this control paradigm to its limits [1].

For this reason, we consider a different control approach based on nonlinear model predictive control (NMPC). A dynamic model of the whole system is used to compute the optimal feedback control. In contrast to PI-controllers, this allows to take into account the complete nonlinear dynamics of the coupled system and thus promises to have excellent stabilization properties even for demanding load scenarios. NMPC is a well-established control strategy, but the direct application to microgrids is hardly possible, because the complete solution of the underlying optimal control problems takes too much time to react to sudden disturbances.

To address this problem, we propose the use of the so-called Multi-Level Iteration scheme [2], which eliminates the need to solve the underlying optimal control problems until convergence. This reduces the feedback delay and increases feedback rates drastically and makes NMPC feasible for the control of microgrids.

2 Nonlinear Model Predictive Control

In the classical framework of NMPC, we divide the simulation horizon into a sequence of sampling points $0 = t_0 \leq \cdots \leq t_k \leq \cdots \leq t_f$. In every sampling interval the state of the system x_k is fed to the controller and a feedback signal u_{x_k} is computed. In the traditional NMPC setting, the following optimal control problem (OCP) is solved in every sampling interval:

$$\min_{x(\cdot),z(\cdot),u(\cdot)} \quad \int_{t_k}^{t_k+T} L(x(t), z(t), u(t))\mathrm{d}t \tag{1a}$$

$$\text{s.t.} \quad \dot{x}(t) = f(x(t), z(t), u(t)), 0 = h(x(t), z(t), u(t)), \tag{1b}$$

$$x(t_k) = x_k, \qquad t \in [t_k, t_k + T], \tag{1c}$$

$$x^{\mathrm{lo}} \leq x(t) \leq x^{\mathrm{up}}, z^{\mathrm{lo}} \leq z(t) \leq z^{\mathrm{up}}, u^{\mathrm{lo}} \leq u(t) \leq u^{\mathrm{up}}. \tag{1d}$$

The prediction horizon T is assumed to be constant. The differential and algebraic states $x(t) \in \mathbb{R}^{n_x}$ and $z(t) \in \mathbb{R}^{n_z}$ and the control $u(t) \in \mathbb{R}^{n_u}$ are subject to the DAE system (1b) with initial value set to the current system state x_k (1c). The objective is of Lagrange type with integrand L. The NMPC feedback signal applied in the interval $[t_k, t_{k+1})$ is the first part of the solution $u_{x_k}(t) = u_k^*(t)$.

We use the direct multiple shooting discretization, introduced by Bock [2], to transform the infinite dimensional problem (1) to a finite dimensional, structured nonlinear program (NLP) with the variable w which collects the discretized state and control variables:

$$\min_{w} l(w) \qquad \text{s.t.} \qquad b(w) + Ex_k = 0, \qquad w^{\text{lo}} \le w \le w^{\text{up}}. \tag{2}$$

Here l is the discretized objective function (1a), the function b together with the constant matrix E represent the discretized DAE system (1b) with the initial value embedding constraint [3] (1c) and w^{lo} and w^{up} are the lower and upper bounds on states and controls. A sequential quadratic programming (SQP) algorithm can be used to solve this NLP. Starting from an initial guess of the primal-dual solution, a sequence of iterates $(w_j, \lambda_j, \mu_j)_{j\in\mathbb{N}}$ is generated by solving quadratic programs (QP) of the following form:

$$\min_{\Delta w} \frac{1}{2}\Delta w^{\top} A \Delta w + a^{\top}\Delta w \quad \text{s.t.} \quad \begin{cases} b(w_j) + Ex_k + B\Delta w = 0, \\ w^{\text{lo}} \le \Delta w + w_j \le w^{\text{up}}. \end{cases} \tag{3}$$

A represents either the exact Hessian of the Lagrange function $\nabla_w^2 \mathcal{L}(w_j, \lambda_j, \mu_j)$ of the NLP or an approximation of it. The evaluation of the objective gradient at the current iterate w_j is given by $a = \nabla_w l(w_j)$, the evaluation of the constraints by $b = b(w_j)$ and its Jacobian by $B = \nabla b(w_j)^{\top}$. The iterate is updated with the primal-dual solution $(\Delta w^{QP}, \lambda^{QP}, \mu^{QP})$ of QP (3):

$$w_{j+1} = w_j + \Delta w^{QP}, \qquad \lambda_{j+1} = \lambda^{QP}, \qquad \mu_{j+1} = \mu^{QP}. \tag{4}$$

Under mild assumptions, local quadratic convergence of the SQP method is guaranteed. This can be exploited such that provided once the iterates are sufficiently close to the solution, only one iteration per sampling time is sufficient to obtain excellent approximations to the true optimal solution [3]. A further reduction of computation time is achieved with the so called Real-Time Iterations (RTI), introduced by Diehl [2, 3]. Because the initial value x_k enters only linearly in QP (3), almost all data for setting up the QP can be prepared based on a current iterate (w_j, λ_j, μ_j), even before the current system state x_k is available. As soon as x_k is available, only the QP solution step is left to generate a feedback signal.

3 Multi-Level Iteration

Although with the Real-Time Iteration scheme, the computational effort necessary in each sampling time is reduced from solving the complete NLP to setting up and solving one QP, there still remains a lot of computational effort. To set up the QP (3), the constraints, the objective gradient, the constraint Jacobian and the Lagrange Hessian (corresponding to the data b, a, B, A) have to be computed for each iterate (w_j, λ_j, μ_j). Multi-Level Iterations are a way to drastically reduce this computational effort and thus speed up the feedback generation process.

Table 1 Computations and update formulas for the QP data for the different levels

	Necessary computations				Update formula for QP data			
Level	$b(w_j)$	$a(w_j)$	$B(w_j)$	$A(w_j, \lambda_j)$	b	a	B	A
D	✓	✓	✓	✓	$b(w_j)$	$a(w_j)$	$B(w_j)$	$A(w_j, \lambda_j)$
C	✓	✓	(✓)[a]	✗	$b(w_j)$	$a(w_j) + (\bar{B}_C^\top - B(w_j)^\top)\lambda_j$	$\bar{B}_C$	$\bar{A}_C$
B	✓	✗	✗	✗	$b(w_j)$	$\bar{a}_B + \bar{A}_B(w_j - \bar{w}_B)$	$\bar{B}_B$	$\bar{A}_B$
A	✗	✗	✗	✗	$\bar{b}_A$	$\bar{a}_A$	$\bar{B}_A$	$\bar{A}_A$

[a]Only the vector-matrix product $\lambda^\top B$ needs to be computed in an adjoint fashion

The main idea behind the MLI scheme is based on the fact that Newton-type methods (such as the SQP method described in the previous section) do not require the exact computations of derivatives in A and B to remain locally convergent. This can be exploited to avoid the expensive evaluation of the Hessian and the Jacobian in every iteration.

Instead, we update the different components of QP (3) in four hierarchical levels, with descending computational complexity. Each level stores a reference point $(\bar{w}, \bar{\lambda}, \bar{\mu})$ and the corresponding QP data $\bar{b}, \bar{a}, \bar{B}$ and $\bar{A}$. Every level is working on its own set of iterates, which are independent of the other levels.

The amount of updated information decreases from a full update in Level D to no update at all in Level A. Table 1 explains which data exactly is computed in each iteration and how the QP data is updated with the new available computations. Level D corresponds to a full SQP step, Level C can be interpreted as *Optimality iterations* because updates still contain new objective gradient information. Level B can be interpreted as *Feasibility iterations* because the updates still contains a computation of the constraints $b(w_j)$ and Level A is essentially linear MPC with linearization at the reference point $(\bar{w}_A, \bar{\lambda}_A, \bar{\mu}_A)$. A detailed description of the levels and its interaction can be found in [4].

4 Scenario and Model Description

We apply MLI to a microgrid system comprising of a diesel generator (DG), a photovoltaic plant (PV) and a passive PQ-load. The DG consists of a synchronous generator (SG), actuated by a diesel engine (Fig. 1). Thereby, the speed of the DG is controlled by a proportional controller. Both the diesel engine, as well as the speed controller are modeled with the standard IEEE DEGOV1 model. For voltage control, an automatic voltage regulator (AVR) is included, which follows a proportional feedback law. It is modeled with the standard IEEE AC5A model. The algebraic states in this model originate from the algebraic power flow equations, as well as from algebraic states in the SG model. The setpoints for frequency ω_{ref} and voltage V_{ref} serve as control variables of the NMPC controller. We aim to steer the frequency $\omega(t)$ and the voltage $V(t)$ to the nominal value 1 p.u. at the load and prevent peaks that violate the operational limits of $\pm 10\%$ voltage and frequency deviation from

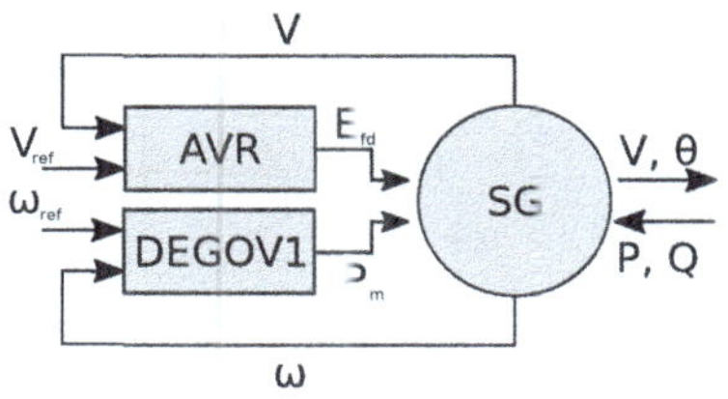

Fig. 1 Diesel generator with primary controllers

the nominal value. To achieve this, we use a Lagrange type objective function $L(x(t), z(t), u(t)) = \|\omega(t) - 1\|^2 + \|V(t) - 1\|^2$. To simulate the intermittent behavior of the PV, we consider a sudden decrease of the PV production from 100% to 5%, lasting 20 s, which corresponds to a cloud passing over the PV plant. During this period the generator needs to compensate the active power shortage. The simulation has an overall length of 30 s.

5 Numerical Results

We discretize OCP (1) with two multiple shooting intervals and the prediction horizon is fixed at $T = 2$ s. The length of the first shooting interval corresponds to the sampling time and the second to the rest of the prediction horizon. The numerical simulations are carried out with the NMPC framework MLI [4], written in MATLAB. For integration and sensitivity generation, the SolvIND integrator suite is used and the QPs are solved by qpOASES [5].

We compare our proposed MLI-controller with a typical state-of-the-art control setup for small microgrids: The DEGOV1 is equipped with an integral controller for the frequency for steady-state error elimination with a settling time of approximately 20 s. The frequency setpoint is updated every 500 ms. The setpoint V_{ref} for the AVR is kept constant for the complete simulation time.

Level D allows us to compute reference values on a sampling grid with 500 ms, which leads to a significantly lower initial peak and a lower settling time compared to the traditional control setup. However, with an accurate integration, this scheme is not real time feasible, since the maximal computation time is over 7 s. To reduce the computation time of level D below 0.5 s, a fixed step-size integrator on a coarse grid could be used. But this degrades the performance of the controller to such an extent that the advantage vanishes almost completely. To overcome this downside, we propose to use level C, B or A instead. They are real time feasible, even with sampling times below 500 ms. Figure 2 shows the performance of level C, B and A in comparison to the traditional control approach. *Level C* uses a sampling time of 200 ms and is able to steer the frequency and voltage to the nominal value without an offset. Since no updates on the sensitivities are used in *level B*, it is possible to operate the level B controller with a sampling time of 100 ms. The system settles significantly faster with a lower initial peak, but with a voltage offset to the nominal

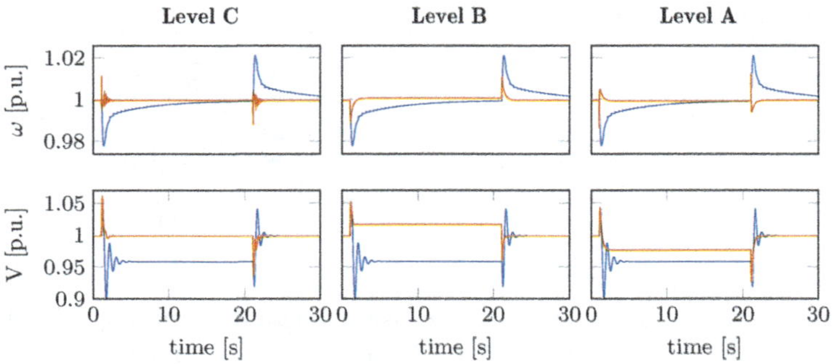

Fig. 2 Tracking of frequency and voltage for level C, B and A. The performance of the state-of-the-art integral controller is shown by the blue trajectory. All levels stabilize the system faster with a smaller initial peaks

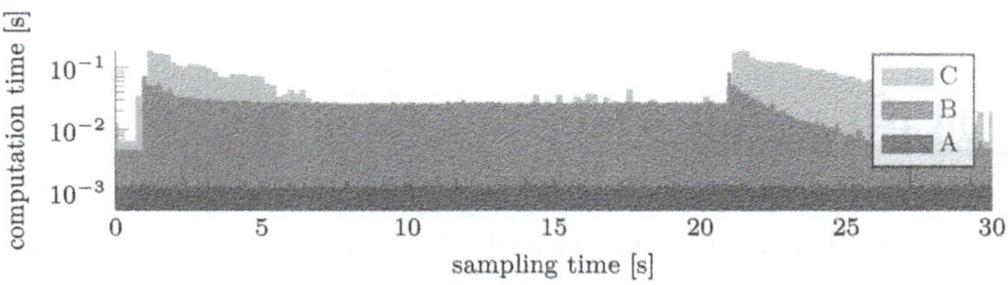

Fig. 3 Computation time for every iteration. The sampling time for level A is 5 ms, for level B 100 ms and for level C 200 ms. All levels are real time feasible

value. However, since level B is guaranteed to converge to a feasible point, the operational limits are satisfied. In *level A* no integration of the dynamical system is involved and therefore it is possible to reduce the sampling time to 5 ms. These short sampling times allow for a control feedback with the lowest initial peak and the shortest settling time, even though the system is in a slightly suboptimal state during the power shortage. From a theoretical point of view, it is not possible to ensure, that the bounds are satisfied, but in this case, the offset is significantly lower than the traditional control approach. In Fig. 3 the computation times of schemes with constant use of Phase A, B and C are depicted over the complete simulation horizon. The computation time for all three levels is lower than the sampling time and therefore the methods are real time feasible. The maximal iteration time for level A is 3.2 ms, for level B 80 ms and for level C 185 ms.

6 Conclusion

We presented a novel microgrid controller based on MLI. We used an example microgrid, modeled as a differential algebraic equation system, to perform numerical experiments and compare the performance with the traditional control approach.

All presented levels stabilized frequency and voltage fast and more reliably. In addition to that the sampling time was significantly reduced. In the future, we would like to consider bigger microgrids, MLI with parallel levels and simultaneous online state-estimation.

References

1. Ilić, M., Jaddivada, R., Miao, X.: Modeling and analysis methods for assessing stability of microgrids. In: 20th IFAC World Congress, vol. 50, pp. 5448–5455 (2017). IFAC-PapersOnLine
2. Bock, H.G., Diehl, M., Kostina, E., Schlöder, J.: Constrained optimal feedback control of systems governed by large differential algebraic equations. In: Real-Time PDE-Constrained Optimization, pp. 3–22. SIAM (2007)
3. Diehl, M.: Real-Time Optimization for Large Scale Nonlinear Processes. Dissertation, Heidelberg University (2001)
4. Wirsching, L.: Multi-level iteration schemes with adaptive level choice for nonlinear model predictive control. Dissertation, Heidelberg University (2018)
5. Ferreau, H.J., Kirches C., Potschka A., Bock H.G., Diehl, M.: qpOASES: a parametric active-set algorithm for quadratic programming. In: Mathematical Programming Computation, vol. 6, pp. 327–363 (2014)

Mixed-Integer Nonlinear PDE-Constrained Optimization for Multi-Modal Chromatography

Dominik H. Cebulla, Christian Kirches, and Andreas Potschka

Abstract Multi-modal chromatography emerged as a powerful tool for the separation of proteins in the production of biopharmaceuticals. In order to maximally benefit from this technology it is necessary to set up an optimal process control strategy. To this end, we present a mechanistic model with a recent kinetic adsorption isotherm that takes process controls such as pH and buffer salt concentration into account. Maximizing the yield of a target component subject to purity requirements leads to a mixed-integer nonlinear optimal control problem constrained by a partial differential equation. Computational experiments indicate that a good separation in a two-component system can be achieved.

Keywords Optimal control · PDE-constrained optimization · Mixed-integer programming · Chromatography

Mathematics Subject Classification (2010) 34H05, 35Q93, 90C11

1 Introduction

Still being the method of choice in the downstream processing of biopharmaceuticals, chromatography undergoes major developments [1]. A strategy to reduce the production costs of biopharmaceuticals is to replace multiple complementary chromatography steps by fewer multi-modal chromatography (MMC) steps in combination with an optimal process control strategy.

D. H. Cebulla (✉) · C. Kirches
Institute for Mathematical Optimization, TU Braunschweig, Braunschweig, Germany
e-mail: d.cebulla@tu-bs.de; c.kirches@tu-bs.de

A. Potschka
Interdisciplinary Center for Scientific Computing (IWR), Heidelberg University, Heidelberg, Germany
e-mail: potschka@iwr.uni-heidelberg.de

J. S. Neufeld et al. (eds.), *Operations Research Proceedings 2019*, Operations Research Proceedings, https://doi.org/10.1007/978-3-030-48439-2_10

We present a mechanistic partial differential equation (PDE) model for column liquid chromatography, focusing on a kinetic isotherm that describes the adsorption behavior in MMC and takes buffer salt concentration and pH into account, the latter incorporated in a novel fashion. Maximizing the yield of a target component subject to purity requirements leads to a formulation which belongs to the challenging class of (nonlinear) mixed-integer PDE-constrained optimization (MIPDECO) problems that attain much interest in current research [2–4]. Based on an experimental setup we investigate the optimized control and model response trajectories. We conclude with a short summary.

2 A Mechanistic Model for Multi-Modal Chromatography

In liquid phase chromatography, a liquid solvent (mobile phase) is pumped through a column which is packed with adsorbent particles (stationary phase). Components of a mixture are separated by exploiting their different adsorption behaviors, which usually can be altered by modification of the process conditions, e.g. by shifting the pH or changing the concentration of buffer salts.

We employ a transport-reactive-dispersive model which is stated as a PDE in one spatial dimension and describes, for every component $i \in \{1, \ldots, n_{\text{comp}}\}$, the molar concentration in the mobile phase $c_i = c_i(t, x)$, liquid particle phase $c_{p,i} = c_{p,i}(t, x)$, and adsorbed particle phase $q_i = q_i(t, x)$, respectively, in dependence of time $t \in [0, T]$ and axial position in the column $x \in [0, L]$,

$$\begin{aligned} \frac{\partial c_i}{\partial t} &= -\frac{v(t)}{\varepsilon_b}\frac{\partial c_i}{\partial x} + D_{\text{ax}}\frac{\partial^2 c_i}{\partial x^2} - \frac{1-\varepsilon_b}{\varepsilon_b}\frac{3}{r_p}k_{\text{eff},i}\left(c_i - c_{p,i}\right), \\ \frac{\partial c_{p,i}}{\partial t} &= -\frac{(1-\varepsilon_p)}{\varepsilon_p}\frac{\partial q_i}{\partial t} + \frac{3}{\varepsilon_p r_p}k_{\text{eff},i}\left(c_i - c_{p,i}\right). \end{aligned} \tag{1}$$

The axial dispersion coefficient D_{ax} depends on the flow velocity $v(t)$,

$$D_{\text{ax}} = D_{\text{ax}}(v(t)) \approx 2\, r_p\, \lambda_{D_{\text{ax}}}\, v(t)\, \varepsilon_b^{-1},$$

see [5], with $\lambda_{D_{\text{ax}}} \approx 1$ under typical conditions. For a description of model parameters, we refer to [5].

The adsorption kinetics $\partial q_i/\partial t$ are derived from an isotherm for MMC, which is a combination of isotherms for ion-exchange chromatography (IEX) and hydrophobic interaction chromatography (HIC) based on work by Mollerup [6], Nfor et al. [7], and Deitcher et al. [8]. It has the underlying (informal) chemical equilibrium (compare [7, 8]),

$$\mathrm{P}_{\text{sol},i} + \nu_i\,(\mathrm{LS}) + n_i\mathrm{L} \rightleftharpoons \mathrm{P}_{\text{ads},i} + \nu_i\mathrm{S} + n_i\xi\mathrm{W}.$$

The equilibrium describes that the ith solute protein $P_{sol,i}$ interacts simultaneously with $\nu_i = z_i/z_s$ adsorbed salt ions LS (also called counter-ions) and n_i hydrophobic ligands L. Thus, the protein gets adsorbed ($P_{ads,i}$), ν_i salt ions S are released from the stationary phase and ξ water molecules W are displaced from the protein and ligand for every protein–ligand contact.

Following the derivations in [6–8], we arrive at the highly nonlinear adsorption kinetics

$$\frac{\partial q_i}{\partial t} = k_{\text{kin},i}\Bigg[k_{\text{eq},i}\Big(\Lambda_{\text{IEX}} - \sum_{j=1}^{n_{\text{comp}}} (z_j + \varsigma_j)\, q_j\Big)^{\nu_i} \Big(\Lambda_{\text{HIC}} - \sum_{j=1}^{n_{\text{comp}}} (n_j + \delta_j)\, q_j\Big)^{n_i} \tag{2}$$

$$\exp\left(\kappa_{\text{P},i}\, c_{p,i} + (\kappa_{\text{salt},i} - \rho\, \xi n_i)\, c_{p,\text{salt}}\right) c_{p,i} - \tilde{c}^{n_i} q_i \left(z_s\, c_{p,\text{salt}}\right)^{\nu_i}\Bigg] .$$

The implicitly introduced salt component $c_{p,\text{salt}}$, which we formally give the index 0, obeys (1); due to electroneutrality constraints we have

$$\frac{\partial q_{\text{salt}}}{\partial t} = -z_s{}^{-1} \sum_{j=1}^{n_{\text{comp}}} z_j \frac{\partial q_j}{\partial t} = -\sum_{j=1}^{n_{\text{comp}}} \nu_j \frac{\partial q_j}{\partial t} .$$

To incorporate the pH, we propose to add a new component that accounts for the concentration of H^+ ions. Neglecting units, we then have

$$\text{pH} = \text{pH}(t, x) := -\log_{10} c_{p,\text{H}^+}(t, x) .$$

For simplicity, we assume that H^+ ions do not get adsorbed, hence $\partial q_{\text{H}^+}/\partial t = 0$. Furthermore, with given parameters $z_{i,1}$ and $z_{i,2}$, we incorporate the pH dependency of the binding charges z_i according to [6],

$$z_i = z_i(t, x) = z_{i,1} \times \log \text{pH}(t, x) + z_{i,2} .$$

The mass-balanced boundary conditions are due to Danckwerts [9] and read for all components (including salt and pH) as

$$\frac{\partial c_i}{\partial x}(t, 0) = \frac{v(t)}{\varepsilon_b D_{\text{ax}}}\left(c_i(t, 0) - c_{\text{in},i}(t)\right), \quad \frac{\partial c_i}{\partial x}(t, L) = 0, \quad t \in (0, T) , \tag{3}$$

with inlet concentrations $c_{\text{in},i}(t)$, whereas the initial conditions are given by

$$\begin{aligned} &c_i(0, x) = 0, && c_{p,i}(0, x) = 0, && q_i(0, x) = 0, \\ &c_{\text{salt}}(0, x) = c_{\text{salt,init}}, && c_{p,\text{salt}}(0, x) = c_{\text{salt,init}}, && q_{\text{salt}}(0, x) = \Lambda_{\text{IEX}}, \\ &c_{\text{H}^+}(0, x) = 10^{-\text{pH}_{\text{init}}}, && c_{p,\text{H}^+}(0, x) = 10^{-\text{pH}_{\text{init}}}, && q_{\text{H}^+}(0, x) = 0, \end{aligned} \tag{4}$$

for $x \in (0, L)$, where $c_{\text{salt,init}}$ is the initial salt concentration and pH_{init} is the initial pH chosen by the experimenter.

3 An Optimal Control Problem for Chromatography

Multiple performance and economic criteria exist for a chromatographic process, see [5], but we lay our focus on yield and purity. To this end, we define

$$m_{\text{coll},i}(t) = \int_0^t \dot{V}(\tau)\, c_i(\tau, L)\, \text{coll}(\tau)\, \mathrm{d}\tau \tag{5}$$

as the collected amount of component i. The volumetric flow rate is given by $\dot{V}(t)$ and can be computed from $v(t)$. The control $\text{coll}(t) \in \{0, 1\}$ indicates whether we collect the eluate at time t or not. Yield Yi_i and purity Pu_i are then defined as

$$\text{Yi}_i(t) = \frac{m_{\text{coll},i}(t)}{\int_0^t \dot{V}(\tau)\, c_{\text{in},i}(\tau)\, \mathrm{d}\tau}, \qquad \text{Pu}_i(t) = \frac{m_{\text{coll},i}(t)}{\sum_{j=1}^{n_{\text{comp}}} m_{\text{coll},j}(t)},$$

and our goal will be to maximize the yield of a target component under given lower bounds on its purity.

To set up the optimal control problem (OCP), we subsume all PDE states from the model presented in Sect. 2 in a state vector $u(t, x) \in \mathbb{R}^{n_u}$, all controls (volumetric flow rate, component inlet, salt inlet, pH inlet, and eluate collection) in a vector $\mathbf{q}(t) \in \mathbb{R}^{n_\mathbf{q}}$, and we set $p = (\text{pH}_{\text{init}}, c_{\text{salt,init}})$, see (4). We omit the model parameters as they are assumed to be known.

Denoting the target component by $*$ we arrive at the PDE-constrained nonlinear mixed-integer OCP

$$\begin{aligned} \max_{u(\cdot), \mathbf{q}(\cdot), p, T} \quad & \text{Yi}_*(T) \\ \text{s.t.} \quad & e(u; \mathbf{q}) = 0, \qquad u(0, x; p) = u_0(x, p), \\ & \text{Pu}_*(T) \geq \text{Pu}_{\min,*}, \quad \underline{\mathbf{q}} \leq \mathbf{q}(t) \leq \overline{\mathbf{q}}, \quad \text{coll}(t) \in \{0, 1\}, \end{aligned} \tag{6}$$

where the constraints are (left to right, top to bottom) the PDE model (1) with boundary conditions (3), initial values (4), the purity constraint, bounds on the controls, and the binary constraint on the eluate collection, respectively. Note that since $\text{coll}(t)$ is a binary control and enters (5) linearly, OCP (6) is already in partially outer convexified form, compare [10].

4 Computational Experiments

To solve the OCP (6), we first discretize the PDE model in space with a weighted essentially non-oscillatory (WENO) scheme, compare [11], thus replacing the PDE by a system of ordinary differential equations (ODEs). The binary control $\text{coll}(t)$ is relaxed to the interval [0, 1]. We use CasADi [12] to employ a direct single shooting approach to transfer the resulting OCP into a nonlinear programming (NLP) problem; for further details we refer to the code templates provided by CasADi. As an integrator we use IDAS [13]; the NLP problem is solved with IPOPT [14] whose desired convergence tolerance is set to $\texttt{tol} = 10^{-4}$. The prototypical implementation was done with MATLAB.

4.1 Experimental Setup

Our setup is based on two components to be separated from each other; the *first* component is the target. We fix $T = 10\,\text{min}$ and divide the time horizon into 20 control intervals on which the controls are given by a constant value. The volumetric flow rate is $\dot{V} \equiv 0.5\,\text{mL}\,\text{min}^{-1}$ and the feed inlet is given by $c_{\text{in},1/2}(t) = 0.01\,\text{mmol}\,\text{L}^{-1}$ for $0 \le t \le 2$, otherwise it is zero. The controls $c_{\text{in,salt}}(t)$, $c_{\text{in},\text{H}^+}(t)$, and $\text{coll}(t)$ are subject to optimization.

The model parameters for (1) are $L = 25\,\text{mm}$, $\lambda_{D_{\text{ax}}} = 10$, $\varepsilon_b = 0.35$, $\varepsilon_p = 0.8$, $r_p = 0.02\,\text{mm}$, $k_{\text{eff,salt}} = 0.1\,\text{mm}\,\text{min}^{-1}$, $k_{\text{eff},\text{H}^+} = 0.1\,\text{mm}\,\text{min}^{-1}$, and $k_{\text{eff},1/2} = 0.15/0.05\,\text{mm}\,\text{min}^{-1}$. Isotherm parameters (2) are $k_{\text{eq},1/2} = 20/15$, $k_{\text{kin},1/2} = 20/10\,(\text{mol}\,\text{L}^{-1})^{-(\nu_{1/2}+n_{1/2})}\,\text{min}^{-1}$, $z_s = 1.0$, $z_{1/2,1} = 0.2/1.0$, $z_{1/2,2} = -0.2/-1.0$, $\varsigma_{1/2} = \delta_{1/2} = 75/100$, $\Lambda_{\text{IEX}} = \Lambda_{\text{HIC}} = 0.2\,\text{mol}\,\text{L}^{-1}$, $n_{1/2} = 1.0/1.2$, $\xi = 16.0$, $\rho = -0.015\,\text{L}\,\text{mol}^{-1}$, $\tilde{c} = 1\,\text{mol}\,\text{L}^{-1}$, $\kappa_{\text{P},1/2} = 100/150\,\text{L}\,\text{mol}^{-1}$, and $\kappa_{\text{salt},1/2} = 2.0/0.5\,\text{L}\,\text{mol}^{-1}$.

The parameters were chosen to reflect typical values that occur in column liquid chromatography. In [5], such values are reported for parameters occurring in (1), although we use a higher axial dispersion coefficient $\lambda_{D_{\text{ax}}}$ to represent higher non-idealities in the column. The isotherm parameters reflect values that have been reported e.g. in [7, 8]. The first component has only a very small binding charge, contrary to the second component whose binding charge is also more dependent on pH. Due to higher values chosen for $k_{\text{eq},1}$ and $k_{\text{kin},1}$, the first component is more strongly affected by the adsorption process than the second.

4.2 Discussion of Numerical Results

The resulting concentration profiles at column outlet and the corresponding pH control for $\text{Pu}_{\text{min},1} = 0.99$ are depicted in Fig. 1. We limit the presentation to

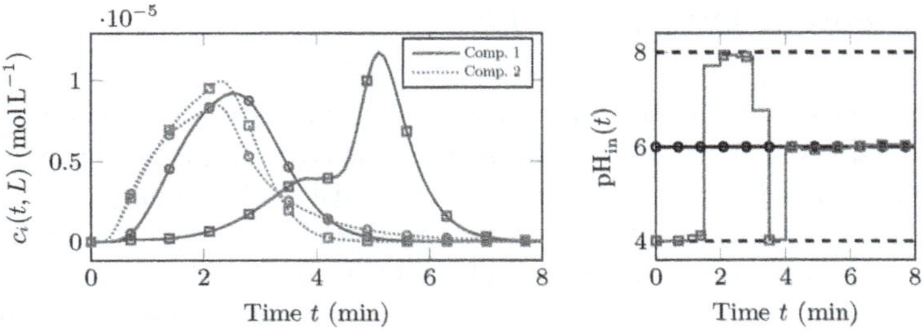

Fig. 1 Initial (◯) and optimized (□) concentration profile (left) and pH (right) for a minimum purity of 0.99. Bounds on pH are depicted by dashed lines

$t \in [0, 8]$ as both components are completely eluted then. The two components can be separated well, despite initially having a large overlap area. The optimized yield decreases with increasing purity requirements, ranging from 95.6% for a purity of 80% to 76.2% for a purity of 99%.

Initially, the salt inlet stays at $1.0\,\mathrm{mol\,L^{-1}}$, switching to $0.2\,\mathrm{mol\,L^{-1}}$ after $t = 4$ min, whereas the pH inlet shows a more varying behavior. It is indeed the interplay between salt concentration and pH which leads to a superior separation behavior, justifying the incorporation of both quantities as process controls.

The relaxed optimal eluate collection $\mathrm{coll}(t)$ is already a binary control with $\mathrm{coll}(t) = 1$ for $t \geq 4$ and zero elsewhere. This means that no further actions have to be taken regarding the mixed-integer constraint. However, in case of non-integrality we must enforce this constraint, e.g. by carefully choosing an appropriate rounding strategy as described in [10, 15].

5 Conclusions

We presented a mechanistic model for MMC with a recent kinetic adsorption isotherm, emphasizing the incorporation of salt concentration and, especially, pH. We then described a mixed-integer nonlinear PDE-constrained OCP where the yield of a target component is maximized subject to lower bounds on its purity. Our numerical results suggest that incorporation of pH is an important factor for a good separation of components, resulting in a high yield of the target component even under strong purity requirements.

Acknowledgments The authors gratefully acknowledge support by the German Federal Ministry for Education and Research, grant no. 05M17MBA-MOPhaPro.

References

1. Rathore, A.S., Kumar, D., Kateja, N.: Recent developments in chromatographic purification of biopharmaceuticals. Biotechnol. Lett. **40**(6), 895–905 (2018)
2. Geißler, B., Kolb, O., Lang, J., Leugering, G., Martin, A., Morsi, A. Mixed integer linear models for the optimization of dynamical transport networks. Math. Methods Oper. Res. **73**(3), 339–362 (2011)
3. Hante, F.M., Sager, S.: Relaxation methods for mixed-integer optimal control of partial differential equations. Comput. Optim. Appl. **55**(1), 197–225 (2013)
4. Buchheim, C., Kuhlmann, R., Meyer, C.: Combinatorial optimal control of semilinear elliptic PDEs. Comput. Optim. Appl. **70**(3). 641–675 (2018)
5. Schmidt-Traub, H., Schulte, M., Seidel-Morgenstern, A. (eds.): Preparative Chromatography, 2nd edn. Wiley-VCH, Weinheim (2012)
6. Mollerup, J.M.: A review of the thermodynamics of protein association to ligands, protein adsorption, and adsorption isotherms. Chem. Eng. Technol. **31**(6), 864–874 (2008)
7. Nfor, B.K., Noverraz, M., Chilamkurthi, S., Verhaert, P.D.E.M., van der Wielen, L.A., Ottens, M.: High-throughput isotherm determination and thermodynamic modeling of protein adsorption on mixed mode adsorbents. J. Chromatogr. A **1217**(44), 6829–6850 (2010)
8. Deitcher, R.W., Rome, J.E., Gildea, P.A., O'Connell, J.P., Fernandez, E.J.: A new thermodynamic model describes the effects of ligand density and type, salt concentration and protein species in hydrophobic interaction chromatography. J. Chromatogr. A **1217**(2), 199–208 (2010)
9. Danckwerts, P.V.: Continuous flow systems. Distribution of residence times. Chem. Eng. Sci **2**(1), 1–13 (1953)
10. Sager, S., Bock, H.G., Diehl, M.: The integer approximation error in mixed-integer optimal control. Math. Program. **133**(1), 1–23 (2012)
11. von Lieres, E., Andersson, J.: A fast and accurate solver for the general rate model of column liquid chromatography. Comput. Chem. Eng. **34**(8), 1180–1191 (2010)
12. Andersson, J.A.E., Gillis, J., Horn, G., Rawlings, J.B., Diehl, M.: CasADi: a software framework for nonlinear optimization and optimal control. Math. Program. Comput. **11**(1), 1–36 (2019)
13. Hindmarsh, A.C., Brown, P.N., Grant, K.E., Lee, S.L., Serban, R., Shumaker, D.E., Woodward, C.S.: SUNDIALS: suite of nonlinear and differential/algebraic equation solvers. ACM Trans. Math. Softw. **31**(3), 363–396 (2005)
14. Wächter, A., Biegler, L.T.: On the implementation of an interior-point filter line-search algorithm for large-scale nonlinear programming. Math. Program. **106**(1), 25–57 (2006)
15. Manns, P., Kirches, C.: Improved regularity assumptions for partial outer convexification of mixed-integer PDE-constrained optimization problems. ESAIM: Control Optim. Calculus Var. **26**, 32 (2020)

Sparse Switching Times Optimization and a Sweeping Hessian Proximal Method

Alberto De Marchi and Matthias Gerdts

Abstract The switching times optimization problem for switched dynamical systems, with fixed initial state, is considered. A nonnegative cost term for changing dynamics is introduced to induce a sparse switching structure, that is, to reduce the number of switches. To deal with such problems, an inexact Newton-type arc search proximal method, based on a parametric local quadratic model of the cost function, is proposed. Numerical investigations and comparisons on a small-scale benchmark problem are presented and discussed.

Keywords Switched dynamical systems · Switching time optimization · Sparse optimization · Cardinality · Proximal methods

MSC 2010: 90C26, 90C53, 49M27

1 Introduction

We focus on the switching times optimization (STO) problem for switched dynamical systems, which consists in computing the optimal time instants for changing the system dynamics in order to minimize a given objective function. A cost term penalizing changes of the continuous dynamics, whose sequence is given, is added to encourage a sparse switching structure. In this paper, for the sake of simplicity and without loss of generality, we consider problems with autonomous dynamical systems, cost functions in Mayer form and fixed final time. Building upon a cardinality-based formulation of the switching cost [3], in Sect. 2 an equivalent composite nonconvex, nonsmooth optimization problem is introduced, which is amenable to proximal methods [5, 6]. In Sect. 3 we propose a novel proximal arc search method, which builds upon both proximal gradient and Newton-type

A. De Marchi (✉) · M. Gerdts
Universität der Bundeswehr München, Neubiberg, Germany
e-mail: alberto.demarchi@unibw.de; matthias.gerdts@unibw.de

J. S. Neufeld et al. (eds.), *Operations Research Proceedings 2019*, Operations Research Proceedings, https://doi.org/10.1007/978-3-030-48439-2_11

methods, aiming at *fast* and *safe* iterates. Numerical tests in Sect. 4 show that it consistently performs well compared to established methods on several instances of a benchmark problem.

2 Problem

Let us consider a time interval $[0, T]$, with final time $T > 0$, and a dynamical system switching between $N > 1$ modes, with initial state $\mathbf{x}_0 \in \mathbb{R}^n$. Consider switching times $\boldsymbol{\tau} = (\tau_1, \ldots, \tau_{N+1})^\top$ and switching intervals $\boldsymbol{\delta} = (\delta_1, \ldots, \delta_N)^\top$, satisfying $0 = \tau_1 \leq \tau_2 \leq \ldots \leq \tau_{N+1} = T$ and $\delta_i = \tau_{i+1} - \tau_i$ for $i = 1, \ldots, N$. Hence, the set Δ of feasible vectors $\boldsymbol{\delta}$ is the simplex of size T in $\mathbb{R}^N$. Our goal is to find feasible switching intervals $\boldsymbol{\delta}^\star$ minimizing an objective functional in composite form, consisting of a Mayer term m and a switching cost term S, weighted by a scalar $\sigma > 0$. The STO problem reads

$$\begin{aligned} \underset{\boldsymbol{\delta} \in \Delta}{\text{minimize}} \quad & m(\mathbf{x}(T)) + \sigma S(\boldsymbol{\delta}) \\ \text{subject to} \quad & \dot{\mathbf{x}}(t) = \mathbf{f}_i(\mathbf{x}(t)), \qquad t \in [\tau_i, \tau_{i+1}), \qquad i = 1, \ldots, N \\ & \mathbf{x}(0) = \mathbf{x}_0 \end{aligned} \tag{1}$$

with each $\mathbf{f}_i : \mathbb{R}^n \to \mathbb{R}^n$ assumed differentiable [9]. The cost $S(\boldsymbol{\delta})$ can be expressed as the cardinality of the support of vector $\boldsymbol{\delta}$, for any $\boldsymbol{\delta} \in \Delta$, that is, the number of nonzero elements in $\boldsymbol{\delta}$, as proposed in [3]. The direct single shooting approach yields a reformulation of problem (1) without constraints, even though it may be at a disadvantage compared to the multiple shooting approach [7]. Due to initial conditions and dynamics in (1), a unique state trajectory $\mathbf{x}_\delta$ is obtained for any feasible $\boldsymbol{\delta} \in \Delta$, and the smooth term M can be defined as $M(\boldsymbol{\delta}) := m(\mathbf{x}_\delta(T))$. Then, problem (1) can be equivalently rewritten as a finite dimensional problem, namely

$$\underset{\boldsymbol{\delta} \in \Delta}{\text{minimize}} \quad M(\boldsymbol{\delta}) + \sigma S(\boldsymbol{\delta}) \tag{$\mathcal{P}_\sigma$}$$

which is composite nonsmooth nonconvex with a compact convex feasible set.

3 Methods

Let us consider the finite dimensional optimization problem $\mathcal{P}_\sigma$ with $\sigma > 0$. This can be handled by proximal methods [1, 5, 6], which in general require at least the gradient of the smooth term M and the proximal operator of the nonsmooth term S. Feasibility can be ensured at each iteration by considering the constraints in the proximal operator itself, so that the proximal point is always feasible [3]. Instead, for $\sigma = 0$, problem $\mathcal{P}_\sigma$ turns into a standard nonlinear program (NLP). Even in this case, standard NLP solvers may end up in local minimizers of $\mathcal{P}_\sigma$, as STO problems are often nonconvex [7].

Remark 1 (Smooth Cost and Gradient) Evaluating the gradient of the smooth term M requires computing the sensitivity of the state trajectory $\mathbf{x}_\delta(T)$ [4]. This can be achieved, e.g., by using the sensitivity equation or by linearization of the dynamics over a background time grid and direct derivation. In the numerical tests the latter approach is adopted, which can readily give second-order information too; for more details refer to [9].

Remark 2 (Proximal Operator) Given $\sigma > 0$, the proximal operator for problem $\mathcal{P}_\sigma$ is a possibly set-valued mapping [6], defined as

$$\operatorname{prox}_\gamma(\mathbf{x}) = \arg\min_{\mathbf{u}\in\Delta} \left\{ \sigma S(\mathbf{u}) + \frac{1}{2\gamma} \|\mathbf{u} - \mathbf{x}\|_2^2 \right\}, \quad \text{for any } \gamma > 0. \tag{2}$$

For $\Delta = \mathbb{R}^N$ and $\Delta = \mathbb{R}^N_{\geq 0}$, the proximal point can be expressed analytically and computed entrywise in that the optimization problem is separable. Instead, for the simplex-constrained case, entrywise optimization is not possible due to the coupling among entries. An efficient method for its evaluation is discussed and tested in [2], with accompanying code, and adopted in [3].

3.1 Sweeping Hessian Proximal Method

Let us consider a composite function $\phi := f + g$ and the problem of finding a vector $\mathbf{x}$ minimizing $\phi(\mathbf{x})$, provided an initial guess $\mathbf{x}_0$, with function f smooth, function g possibly extended real-valued, and ϕ lower bounded: further assumptions are discussed below. We propose a Sweeping HEssian ProXimal (SHEPX) method, which is an iterative proximal arc search method, inspired by the proximal arc search procedure in [5] and the averaging line search in [8]. At the k-th iteration, $k = 0, 1, 2, \ldots$, we build a local, parametric, quadratic model $\check{f}_k^t$ of the smooth term f around the current vector $\mathbf{x}_k$, namely

$$\check{f}_k^t(\mathbf{x}) := f(\mathbf{x}_k) + \nabla f(\mathbf{x}_k)^\top (\mathbf{x} - \mathbf{x}_k) + \frac{1}{2}(\mathbf{x} - \mathbf{x}_k)^\top B_k^t (\mathbf{x} - \mathbf{x}_k) \tag{3}$$

with B_k^t a symmetric matrix. Parameter t allows to generate a family of quadratic models, depending on B_k^t, which we define as a weighted combination

$$B_k^t := t B_k + \frac{1-t}{t} I, \quad t \in (0, 1], \tag{4}$$

between the identity matrix I and a symmetric matrix B_k which models the curvature of f in a neighborhood of $\mathbf{x}_k$; this can be the exact Hessian $\nabla^2 f(\mathbf{x}_k)$ or, e.g., a BFGS approximation [5]. Given (3) and (4), the method generates sequences $\{t_k\}_k$, $\{\mathbf{x}_k\}_k$ such that each update is a solution to a composite subproblem, namely

$$\mathbf{x}_{k+1} = \arg\min_{\mathbf{x}} \left\{ \check{f}_k^{t_k}(\mathbf{x}) + g(\mathbf{x}) \right\}, \tag{5}$$

which is amenable to (accelerated) proximal gradient methods. Concurrently, a backtracking arc search procedure finds $t_k = \beta^{i_k}$, $\beta \in (0, 1)$, with i_k the lowest nonnegative integer such that the sufficient descent condition

$$\phi(\mathbf{x}_{k+1}) < \phi(\mathbf{x}_k) - \frac{\eta}{2} t_k \|\mathbf{x}_{k+1} - \mathbf{x}_k\|_2^2 \tag{6}$$

is satisfied, for some $\eta \geq 0$. Warm-starting the composite subproblems (5) could greatly reduce the computational requirements; however, this issue is not further developed in the following, where the current vector $\mathbf{x}_k$ is chosen as initial guess.

Remark 3 Lee et al. [5] adopted a backtracking line search procedure to select a step length that satisfies a sufficient descent condition, given a search direction obtained with $B_k^t := B_k$. Also, they mentioned a proximal arc search procedure, which has some benefits and drawbacks over the line search, such as the fact that an arc search step is an optimal solution to a subproblem but requires more computational effort. As a model for the proximal arc search, they considered $B_k^t := B_k/t$ [5, Eq. 2.20], for decreasing values of $t \in (0, 1]$, in place of (4).

For $t \to 0^+$, the model proposed in (4) yields $B_k^t \approx I/t$, which corresponds to what is assumed by proximal gradient methods. Hence, for sufficiently small $t > 0$, solutions to subproblem (5) converge on the proximal gradient step, with stepsize controlled by t, with no need to additionally estimate the Lipschitz constant of ∇f [5, 6]. On the other hand, for $t = 1$, the second-order information is fully exploited, as $B_k^1 = B_k$, possibly accelerating convergence. Thanks to these features, SHEPX seamlessly combines proximal gradient and Newton-type methods, exploiting faster convergence rate of the latter while retaining the convergence guarantees of the former [1, 5, 6]. Adopting a quasi-Newton scheme for B_k and adaptive stopping conditions for subproblems (5), as discussed in [5], makes SHEPX an inexact Newton-type proximal arc search method.

Remark 4 A detailed analysis and further development of the algorithm are ongoing research. Currently, we are interested in the requirements for having global convergence to a (local) minimizer. To this end, the forward-backward envelope could be used as a merit function to select updates with sufficient decrease, as in [8, Eq. 9], to handle nonconvex problems.

4 Numerical Results

We consider several instances of an exemplary problem and adopt different methods and variants to solve them: FISTA, an accelerated proximal gradient method [1], PNOPT, a proximal Newton-type line search method [5], and SHEPX, the aforementioned sweeping Hessian proximal method. Both exact Hessian and BFGS approximation are tested. As initial guess for problem $\mathcal{P}_\sigma$ with $\sigma > 0$, we use the solution to $\mathcal{P}_\sigma$ with $\sigma = 0$, obtained via the `fmincon` MATLAB routine, with interior-point method and initial guess $\delta_i = T/N$, $i = 1, \ldots, N$. We stress that, in general, as both terms in the composite cost function are nonconvex, only local minimizers can be detected. The results are obtained with MATLAB 2018b, on Ubuntu 16.04, with Intel Core i7-8700 3.2 GHz and 16 GB of RAM.

The Fuller's control problem has a solution which shows chattering behaviour, making it a small-scale benchmark problem [7]. We consider $N = 40$ modes, and the i-th dynamics read $\dot{x}_1 = x_2$, $\dot{x}_2 = v_i$, with the discrete-valued control v_i taking values in the given sequence $\{v^1, v^2, v^3, v^4, v^1, v^2, \ldots\}$, with values $v^1 = 1$, $v^2 = 0.5$, $v^3 = -1$ and $v^4 = -2$. Initial state $\mathbf{x}_0 = (0.01, 0)^\top$ and final time $T = 1$ are fixed. The cost functional, $\int_0^T x_1^2(t)\mathrm{d}t + \|\mathbf{x}(T) - \mathbf{x}_0\|_2^2$, can be transformed in Mayer form by augmenting the dynamics. We choose the background time grid with 100 time points [9], a maximal number of iterations (200, or 1000 for FISTA, for a fair comparison, because it is a first-order method and does not consider second-order information), and a stepsize tolerance ($\|\boldsymbol{\delta}_{k+1} - \boldsymbol{\delta}_k\|_2 < 10^{-6}$). For SHEPX, we set $\beta = 0.1$ and $\eta = 0$.

Table 1 summarizes the solutions found for different values of the switching cost σ, in terms of cost and cardinality of $\boldsymbol{\delta}^\star$. Statistics regarding the optimization process are also reported, such as required iterations and time. In Fig. 1 the state trajectories are depicted for two cases, highlighting the sparsity-inducing effect of the switching cost. The results show that SHEPX performs similarly to FISTA and better than PNOPT in terms of solution quality. We argue the line search procedure adopted by PNOPT is detrimental for cardinality optimization problems, which benefit from updating by solving a proximal subproblem. Also, SHEPX requires much less iterations than FISTA, meaning that some second-order information is exploited. Interestingly, the quasi-Newton variant of PNOPT seems to work better than the one with exact Hessian, while it holds the opposite for SHEPX. The latter might be able to exploit the second-order information which the former cannot handle with the line search, for which the positive-definite approximation obtained via BFGS is beneficial.

5 Outlook

We proposed a proximal Newton-type arc search method for dealing with cardinality optimization problems. Numerical tests on a sparse switching times optimization problem with switching cost have demonstrated the viability of the approach. A

Table 1 Solutions and computational performances, with different methods, for switching cost $\sigma \in \{10^{(i/3)-3} \mid i = 0, 1, 2, 3\}$

σ	Method	Cost value	Cardinality	Iterations	CPU time [s]
0.001	Initial guess	0.0400	40	402	4.85
	FISTA	0.0340 {0.0340}	34 {34}	200* {1000*}	5.90 {28.16}
	PNOPT	**0.0150** (0.0340)	15 (34)	200* (6)	3.85 (**0.30**)
	SHEPX	0.0330 (0.0340)	33 (34)	17 (148)	0.42 (3.84)
0.0022	Initial guess	0.0880	40	402	4.81
	FISTA	0.0726 {0.0726}	33 {33}	200* {1000*}	5.96 {27.72}
	PNOPT	0.0220 (0.0311)	10 (14)	200* (14)	3.90 (**0.43**)
	SHEPX	**0.0176** (0.0229)	8 (10)	52 (200*)	1.56 (5.08)
0.0046	Initial guess	0.1840	40	402	4.89
	FISTA	0.0329 {**0.0236**}	7 {5}	200* {351}	5.39 {8.86}
	PNOPT	0.0330 (0.0470)	7 (10)	200* (5)	3.96 (0.37)
	SHEPX	**0.0236** (0.0333)	5 (7)	12 (200*)	**0.28** (5.15)
0.01	Initial guess	0.4000	40	402	4.82
	FISTA	0.0509 {**0.0306**}	5 {3}	200* {449}	5.24 {11.14}
	PNOPT	0.0513 (0.0712)	5 (7)	200* (4)	3.90 (0.36)
	SHEPX	**0.0306** (0.0515)	3 (5)	10 (200*)	**0.26** (4.99)

Variant with more iterations in { }, and with BFGS in ( ). Symbol * denotes that the iteration limit is reached. Boldface highlights best cost value and CPU time

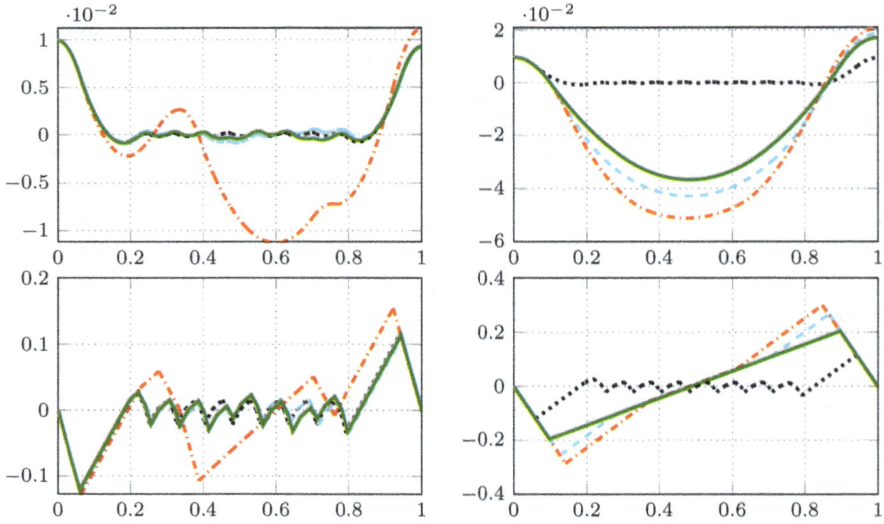

Fig. 1 Differential states x_1 (top) and x_2 (bottom) versus time t, for switching cost $\sigma = 0.001$ (left) and $\sigma = 0.01$ (right): initial guess (dotted black), FISTA (200 iterations, dashed blue), PNOPT (dash-dotted orange) and SHEPX (solid green)

comparison to other proximal methods, in terms of computation time and solution quality, has shown its effectiveness. Future research needs to further analyze the proposed method and to extend the present work to a more general class of problems. In particular, we aim at embedding proximal methods in the augmented Lagrangian framework for dealing with constraints and eventually tackling mixed-integer optimal control problems.

References

1. Beck, A., Teboulle, M.: A fast iterative shrinkage-thresholding algorithm for linear inverse problems. SIAM J. Imag. Sci. **2**(1), 183–202 (2009). https://doi.org/10.1137/080716542
2. De Marchi, A.: Cardinality, Simplex and Proximal Operator (2019). https://doi.org/10.5281/zenodo.3334538
3. De Marchi, A.: On the mixed-integer linear-quadratic optimal control with switching cost. IEEE Control Syst. Lett. **3**(4), 990–995 (2019). https://doi.org/10.1109/LCSYS.2019.2920425
4. Gerdts, M.: Optimal Control of ODEs and DAEs. De Gruyter (2011). https://doi.org/10.1515/9783110249996
5. Lee, J.D., Sun, Y., Saunders, M.A.: Proximal Newton-type methods for minimizing composite functions. SIAM J. Optim. **24**(3), 1420–1443 (2014). https://doi.org/10.1137/130921428
6. Parikh, N., Boyd, S.: Proximal algorithms. Found. Trends® Optim. **1**(3), 127–239 (2014). https://doi.org/10.1561/2400000003
7. Sager, S.: Numerical methods for mixed-integer optimal control problems, Ph.D. thesis. University of Heidelberg, Heidelberg (2005). Interdisciplinary Center for Scientific Computing
8. Stella, L., Themelis, A., Sopasakis, P., Patrinos, P.: A simple and efficient algorithm for nonlinear model predictive control. In: 56th IEEE Conference on Decision and Control (CDC), pp. 1939–1944. IEEE, New York (2017). https://doi.org/10.1109/CDC.2017.8263933
9. Stellato, B., Ober-Blöbaum, S., Goulart, P.J.: Second-order switching time optimization for switched dynamical systems. IEEE Trans. Autom. Control **62**(10), 5407–5414 (2017). https://doi.org/10.1109/TAC.2017.2697681

Toward Global Search for Local Optima

Jens Deussen, Jonathan Hüser, and Uwe Naumann

Abstract First steps toward a novel deterministic algorithm for finding a minimum among all local minima of a nonconvex objective over a given domain are discussed. Nonsmooth convex relaxations of the objective and of its gradient are optimized in the context of a global branch and bound method. While preliminary numerical results look promising further effort is required to fully integrate the method into a robust and computationally efficient software solution.

Keywords Nonconvex optimization · McCormick relaxation · Piecewise linearization

1 Motivation and Prior Work

We consider the problem of finding a minimum among all local minima of a nonconvex function $f : [x] \to \mathbb{R}$ over a given domain $[x] = ([\underline{x_i}, \overline{x_i}])_{i=0}^{n-1} \subseteq \mathbb{R}^n$ (a "box" in $\mathbb{R}^n$) for moderate problem sizes ($n \sim 10$). Formally,

$$\min_{x \in [x]} f(x) \quad \text{s.t. } \nabla f(x) = 0 \,. \tag{1}$$

A branch and bound method is employed. It decomposes the domain recursively into smaller subdomains. A subdomain is discarded as soon as a known lower bound for its function values exceeds a given upper bound on the global optimum. Such bounds can be computed, for example, using interval arithmetic. Local convergence

J. Deussen (✉) · J. Hüser · U. Naumann
Informatik 12: Software and Tools for Computational Engineering, RWTH Aachen University, Aachen, Germany
e-mail: deussen@stce.rwth-aachen.de, info@stce.rwth-aachen.de
http://stce.rwth-aachen.de

J. S. Neufeld et al. (eds.), *Operations Research Proceedings 2019*, Operations Research Proceedings, https://doi.org/10.1007/978-3-030-48439-2_12

of the method is defined as the size of the bounding boxes of potential global optima undercutting a given threshold. Global convergence requires all remaining subdomains to satisfy this local convergence property. See [7] for further details on branch and bound using interval arithmetic.

In [2] the basic interval branch and bound algorithm was extended with additional criteria for discarding subdomains. For example, interval expansions of the gradient of the objective over subdomains containing local optima need to contain the zero vector in $\mathbb{R}^n$:

$$0 \notin \nabla f([x]) \,. \tag{2}$$

A subdomain can be discarded if any component fulfills either $\overline{[\nabla f(x)]_i} < 0$ or $\underline{[\nabla f(x)]_i} > 0$. A corresponding test for concavity was proposed. The required gradients and Hessian vector products can be evaluated with machine accuracy using Algorithmic Differentiation (AD) [4, 12]. The adjoint mode of AD allows for the gradients and Hessian vector products to be evaluated at a constant multiple of the computational cost of evaluating the objective. Application to various test problems yielded reductions in the number of branching steps.

Bounds delivered by interval arithmetic can potentially be improved by using McCormick relaxations. The latter provide continuous piecewise differentiable convex underestimators and concave overestimators of the objective and of its derivatives. Subgradients of the under-[over-]estimators yield affine relaxations whose minimization [maximization] over the given subdomain results in potentially tighter bounds; see also Fig. 1. Refer to [11] for further details.

Implementation of the above requires seamless nesting of McCormick, interval, and AD data types. Corresponding support is provided, for example, by our AD software dco/c++} [9] which relies heavily on the type genericity and metaprogramming capabilities of C++.

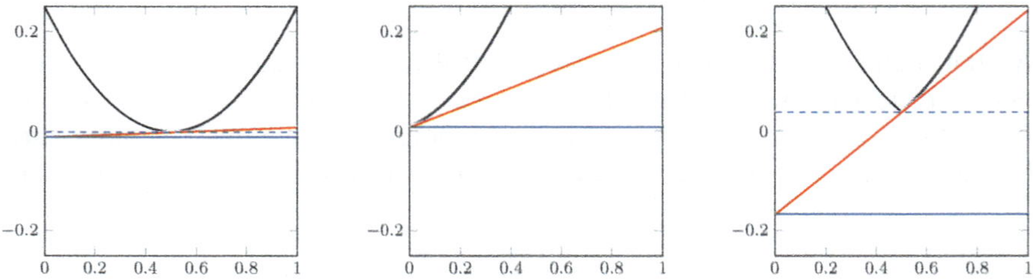

Fig. 1 Lower bound obtained by linearization (solid red line) close to the minimum of the convex McCormick relaxation (solid blue line): Smooth minimum in the interior (left), minimum at boundary (middle) and sharp minimum (right); tight lower bound dotted

2 Better Bounds

The main contribution of this work comes as improved bounds facilitating earlier discard of subdomains in the context of branch and bound. We use McCormick relaxations of the objective f and of its gradient ∇f over the individual subdomains generated by the branch and bound algorithm. Extension to the concavity test proposed in [2] is the subject of ongoing work. In the following we will use $g(x)$ as a substitute of the objective $f(x)$ or its gradient components $[\nabla f(x)]_i$. Minimization [maximization] of the continuous, convex [concave], potentially only piecewise differentiable $\check{g}(x)$ $[\hat{g}(x)]$ may require a nonsmooth optimization method, for example, based on piecewise linearization. Affine relaxations are constructed at the approximate local minima based on corresponding subgradients. Their minimization over the given subdomains yield linear programs the solutions of which are expected to improve the respective interval bounds. The computation of subgradients of McCormick relaxations in adjoint mode AD is discussed in [1].

In order to find a tight lower bound of f on $[x]$ it is desirable to minimize its convex relaxation $\min_{x\in[x]} \check{f}(x)$. The gradient test involves both, the minimization of the underestimator as well as the maximization of the overestimator.

We present two different approaches. The first method is straight forward to implement and is expected to work well in the case of smooth $\check{g}$ but potentially performs poorly for nonsmooth $\check{g}$. The second method explicitly treats nonsmoothness by a combination of successive piecewise linearization and a quasi-Newton method. While it is expected to obtain tight lower bounds on general McCormick relaxations it is less trivial to implement robustly.

2.1 *Projected Gradient Descent*

If $\check{g}$ has a Lipschitz continuous gradient then projected gradient descent is a feasible method for finding a good lower bound. We find an approximation to the global minimum of the McCormick relaxation in a finite number of steps via the iteration

$$x^{k+1} = \text{proj}(x^k - \alpha \nabla \check{g}(x^k), [x])$$

where α is an appropriately chosen step size and $\text{proj}(y, [x])$ is the Euclidean projection of y onto $[x]$. Given strong convexity the projected gradient descent iteration converges linearly to the global optimum $x^\star$.

After iterating for a finite number of K steps to obtain x^K we will generally not have $x^K = x^\star$ up to machine precision and hence possibly $\check{g}(x^K) > \check{g}(x^\star)$. As for a convex function the first-order Taylor expansion is a lower bound it suffices to minimize the linearization on the boundary to get a lower bound up to machine precision

$$x^l = \arg\min_{x \in [x]} \breve{g}(x^K) + \nabla \breve{g}(x^K)(x - x^K)\ .$$

Finding x^l is cheap since the optimum is the corner of $[x]$ pointed to by the negative gradient.

If the global optimum $x^\star$ lies in the interior of $[x]$ the gradient norm is small for approximations of $x^\star$, i.e. we will have $\|\nabla \breve{g}(x^K)\|_2 \ll 1$ and hence $\breve{g}(x^K) + \nabla \breve{g}(x^K)(x^l - x^K)$ will be smaller than but close to $\breve{g}(x^\star)$ (see the left of Fig. 1). If for some dimension the global optimum lies at the boundary then choosing the value of the linearization at the boundary does not worsen the lower bound either (see the middle of Fig. 1).

In the case of nonsmooth $\breve{g}$ an element of the subgradient close to the optimum can still be used to obtain a lower bound. But the lower bound based on subgradients near sharp minima can be arbitrarily worse than the minimum of the McCormick relaxation (see the right of Fig. 1). In order to better deal with nonsmooth $\breve{g}$ we introduce a second method based on successive piecewise linearization.

2.2 *Successive Piecewise Linearization*

Operator-overloading [11] is used to obtain McCormick relaxations of factorable functions. Nonsmoothness of the convex McCormick relaxation $\breve{g}$ is the result of the composition with Lipschitz continuous elemental functions such as $\min(x, y) = -\max(-x, -y)$ and $\text{mid}(x, y, z) = x + y + z - \max(\max(x, y), z) - \min(\min(x, y), z)$ that can all be rewritten in terms of the absolute value function as

$$\max(x, y) = \frac{1}{2}(x + y + |x - y|)\ .$$

The McCormick relaxations themselves are abs-factorable and fall into the class of functions to which we can apply piecewise linearization as suggested in [3].

The piecewise linearization $\breve{g}^{\text{PL}}_{x_0}(x)$ of $\breve{g}$ at x_0 can be evaluated at a point x by a variation of tangent mode AD and satisfies a quadratic approximation property for nonsmooth functions

$$\breve{g}^{\text{PL}}_{x_0}(x) - \breve{g}(x) = O(\|x - x_0\|^2)\ .$$

Successive piecewise linearization (SPL) is an optimization method that repeatedly utilizes piecewise linearization models in a proximal point type method [6]. A simplified version of the SPL step for minimizing $\breve{g}$ is given as

$$x^{k+1} = \arg\min_{x \in [x]} \breve{g}^{\mathrm{PL}}_{x^k}(x) + \lambda \|x - x^k\|_2^2$$

where the choice of λ is similar to a step size in gradient descent in that its value should be chosen depending on curvature. The piecewise linearization plus the quadratic proximal term is a piecewise quadratic function that can be efficiently treated by combinatorial methods similar to e.g. the simplex method because the kinks are explicitly available through their linearizations. This explicit combinatorial treatment of the nonsmoothness is the advantage of considering the minimization of the above subproblem over the general nonlinear convex McCormick relaxation.

For sharp minima SPL converges super-linearly [5]. Super-linear convergence to $x^\star$ implies that within a small number of steps K we have $x^K = x^\star$ up to machine precision. Knowing $x^\star$ up to machine precision allows us to evaluate the true convex McCormick lower bound $\breve{g}(x^\star)$ up to machine precision as well. For sharp minima gradient descent with geometrically decaying step size only converges linearly and hence it cannot be used to quickly find $x^\star$ with sufficient accuracy. Figure 2 shows the super-linear convergence of SPL and linear convergence of gradient descent with decaying step size in the case of a sharp minimum.

For nonsharp minima the convergence of the above version of SPL is comparable to that of gradient descent for smooth problems. Convergence still depends on the condition number of the problem curvature and we cannot generally expect to get close enough to $x^\star$ without including curvature information.

To get a super-linearly convergent method in the nonsharp case we need to include the curvature of the space orthogonal to the nonsmoothness. One way to accomplish this is to use a curvature matrix B^k for the proximal term in the SPL step

$$x^{k+1} = \arg\min_{x \in [x]} \breve{g}^{\mathrm{PL}}_{x^k}(x) + \frac{1}{2}(x - x^k)^T B^k (x - x^k) .$$

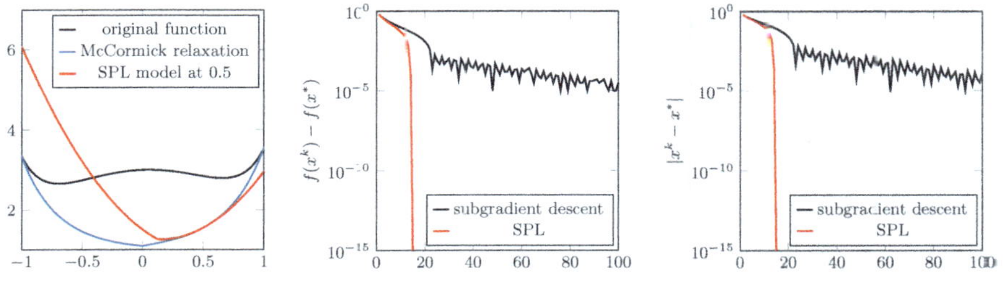

Fig. 2 Successive piecewise linearization to minimize convex McCormick relaxation with sharp minimum

The matrix B^k can for example be obtained via a quasi-Newton BFGS type iteration or as a convex combination of Hessians. Some additional measures such as trust regions may be required to guarantee robust global convergence. Similar ideas have previously been implemented for bundle methods [10] and the idea was hinted at in [5].

3 Numerical Results

To compare the tightness of bounds obtained by intervals and McCormick relaxations in the context of global optimization we consider five test functions in Table 1. The optima of the McCormick relaxations are obtained by the proposed projected gradient descent method. Since McCormick relaxations are at least as tight as the interval bounds, the corresponding branch and bound method requires at most as many branching steps. The benefit appears to grow with increasing dimension of the objective as indicated by the ratio of tasks generated by using intervals and McCormick relaxations in Table 1. Integration of multidimensional successive piecewise linearization is the subject of ongoing work.

4 Conclusions

Aspects of an enhanced branch and bound algorithm for finding a minimum among all local minima of a smooth nonconvex objective over a given domain were discussed. Smooth as well as nonsmooth optimization methods needed to be employed. A sometimes significant reduction in the number of branching steps could be observed. Translation of these savings into improved overall run time will require careful implementation of the method as a robust and computationally efficient software solution.

Acknowledgments This work was funded by Deutsche Forschungsgemeinschaft under grant number NA487/8-1. Further financial support was provided by the School of Simulation and Data Science at RWTH Aachen University.

Table 1 Number of branching steps performed in the context of the branch and bound method for interval bounds (I) and McCormick relaxations (R) with domain $[x]$ and problem dimension n

	6H	BO	RB	GW				ST			
$[x]$	$[-3, 3]$	$[-10, 10]$	$[-5, 10]$	$[-60, 80]$				$[-60, 80]$			
n	2	2	4	2	4	6	8	2	4	6	8
I	853	653	874,881	141	12,433	148,993	10,994,177	301	6785	285,121	15,190,785
R	829	505	873,137	129	10,577	105,985	8,194,817	285	4609	123,777	4,159,233
R/I	**0.97**	**0.77**	**1.00**	**0.91**	**0.85**	**0.71**	**0.75**	**0.95**	**0.68**	**0.43**	**0.27**

Bold values in table are ratio of R/I. Showing that this value decreases by increasing n was described in the result section
Test functions are Six-Hump Camel Back (6H), Booth (BO), Rosenbrock (RB), Griewank (GW) and Styblinski-Tang (ST) [8]

References

1. Beckers, M., Mosenkis, V., Naumann, U.: Adjoint mode computation of subgradients for McCormick relaxations. In: Recent Advances in Algorithmic Differentiation, Springer, Berlin (2012)
2. Deussen, J., Naumann, U.: Discrete Interval Adjoints in Unconstrained Global Optimization. In: Le Thi H., Le H., Pham Dinh T. (eds) Optimization of Complex Systems: Theory, Models, Algorithms and Applications. WCGO 2019. Advances in Intelligent Systems and Computing, vol 991. Springer, Cham (2020)
3. Griewank, A.: On stable piecewise linearization and generalized algorithmic differentiation. Optim. Methods Softw. **28**(6), 1139–1178 (2013)
4. Griewank, A., Walther, A.: Evaluating Derivatives: Principles and Techniques of Algorithmic Differentiation, 2nd edn. SIAM, Philadelphia (2008)
5. Griewank, A., Walther, A.: Relaxing Kink Qualifications and Proving Convergence Rates in Piecewise Smooth Optimization. SIAM J. Optim. **29**(1), 262–289 (2019)
6. Griewank, A., Walther, A., Fiege, S., Bosse, T.: On lipschitz optimization based on gray-box piecewise linearization. Math. Program. **158**(1–2), 383–415 (2016)
7. Hansen, E., Walster, G.W.: Global Optimization using Interval Analysis. Marcel Dekker, New York (2004)
8. Jamil, M., Yang, X.: A literature survey of benchmark functions for global optimization problems. Int. J. Math. Model. Numer. Optim. **4**(2), 150–194 (2013)
9. Lotz, J., Leppkes, K., Naumann, U.: dco/c++: Derivative Code by Overloading in C++. Aachener Informatik Berichte (AIB-2011-06) (2011)
10. Mifflin, R.: A quasi-second-order proximal bundle algorithm. Math. Program. **73**(1), 51–72 (1996)
11. Mitsos, A., Chachuat, B., Barton, P.: McCormick-based relaxation of algorithms. SIAM J. Optim. **20**(2), 573–601 (2009)
12. Naumann, U.: The art of Differentiating Computer Programs. In: An Introduction to Algorithmic Differentiation. SIAM, Philadelphia (2012). https://doi.org/10.1137/1.9781611972078

First Experiments with Structure-Aware Presolving for a Parallel Interior-Point Method

Ambros Gleixner, Nils-Christian Kempke, Thorsten Koch, Daniel Rehfeldt, and Svenja Uslu

Abstract In linear optimization, matrix structure can often be exploited algorithmically. However, beneficial presolving reductions sometimes destroy the special structure of a given problem. In this article, we discuss structure-aware implementations of presolving as part of a parallel interior-point method to solve linear programs with block-diagonal structure, including both linking variables and linking constraints. While presolving reductions are often mathematically simple, their implementation in a high-performance computing environment is a complex endeavor. We report results on impact, performance, and scalability of the resulting presolving routines on real-world energy system models with up to 700 million nonzero entries in the constraint matrix.

Keywords Block structure · Energy system models · HPC · Linear programming · Interior-point methods · Parallelization · Presolving

1 Introduction

Linear programs (LPs) from energy system modeling and from other applications based on time-indexed decision variables often exhibit a distinct block-diagonal structure. Our extension [1] of the parallel interior-point solver PIPS-IPM [2] exploits this structure even when both linking variables and linking constraints are present simultaneously. It was designed to run on high-performance computing (HPC) platforms to make use of their massive parallel capabilities. In this article, we present a set of highly parallel presolving techniques that improve PIPS-IPM's

---

A. Gleixner · N.-C. Kempke (✉) · S. Uslu
Zuse Institute Berlin, Berlin, Germany
e-mail: kempke@zib.de

T. Koch · D. Rehfeldt
Zuse Institute Berlin, Berlin, Germany
Technische Universität Berlin, Berlin, Germany

J. S. Neufeld et al. (eds.), *Operations Research Proceedings 2019*, Operations Research Proceedings, https://doi.org/10.1007/978-3-030-48439-2_13

$$
\begin{array}{rlllllll}
\min & \mathbf{c_0^T x_0} & + & c_1^T x_1 & + & \cdots & + & c_N^T x_N \\
\text{s.t.} & \mathbf{A_0 x_0} & & & & & & = \mathbf{b_0} \\
\mathbf{d_0} \leq & \mathbf{C_0 x_0} & & & & & & \leq \mathbf{f_0} \\
& A_1 \mathbf{x_0} & + & B_1 x_1 & & & & = b_1 \\
d_1 \leq & C_1 \mathbf{x_0} & + & D_1 x_1 & & & & \leq f_1 \\
& \vdots & & & & \ddots & & \vdots \\
& A_N \mathbf{x_0} & + & & & & + & B_N x_N = b_N \\
d_N \leq & C_N \mathbf{x_0} & + & & & & + & D_N x_N \leq f_N \\
& \mathbf{F_0 x_0} & + & F_1 x_1 & + & \cdots & + & F_N x_N = \mathbf{b_{N+1}} \\
\mathbf{d_{N+1}} \leq & \mathbf{G_0 x_0} & + & G_1 x_1 & + & \cdots & + & G_N x_N \leq \mathbf{f_{N+1}} \\
& & & & & & & \ell_i \leq x_i \leq u_i \quad \forall i = \mathbf{0}, 1, \ldots, N
\end{array}
$$

Fig. 1 LP with block-diagonal structure linked by variables and constraints. The block for $i = 0$, present on all processes, is marked in bold

performance while preserving the necessary structure of a given LP. We give insight into the implementation and the design of said routines and report results on their performance and scalability.

The mathematical structure of models handled by the current version of the solver are block-diagonal LPs as specified in Fig. 1. The $x_i \in \mathbb{R}^{n_i}$ are vectors of decision variables and $\ell_i, u_i \in (\mathbb{R} \cup \{\pm\infty\})^{n_i}$ are vectors of lower and upper bounds for $i = 0, 1, \ldots, N$. The extended version of PIPS-IPM applies a parallel interior-point method to the problem exploiting the given structure for parallelizing expensive linear system solves. It distributes the problem among several different processes and establishes communication between them via the Message Passing Interface (MPI). Distributing the LP data among these MPI processes as evenly as possible is an elementary feature of the solver. Each process only knows part of the entire problem, making it possible to store and process huge LPs that would otherwise be too large to be stored in main memory on a single desktop machine.

The LP is distributed in the following way: For each index $i = 1, \ldots, N$ only one designated process stores the matrices A_i, B_i, C_i, D_i, F_i, G_i, the vectors c_i, b_i, d_i, f_i, and the variable bounds ℓ_i, u_i. We call such a unit of distribution a *block* of the problem. Furthermore, each process holds a copy of the block with $i = 0$, containing the matrices A_0, C_0, F_0, G_0 and the corresponding vectors for bounds. All in all, N MPI processes are used. Blocks may be grouped to reduce N. The presolving techniques presented in this paper are tailored to this special distribution and structure of the matrix.

2 Structure-Specific Parallel Presolving

Currently, we have extended PIPS-IPM by four different presolving methods. Each incorporates one or more of the techniques described in [3–5]: singleton row elimination, bound tightening, parallel and nearly parallel row detection, and a few methods summarized under the term model cleanup. The latter includes the detection of redundant rows as well as the elimination of negligibly small entries from the constraint matrix.

The presolving methods are executed consecutively in the order listed above. Model cleanup is additionally called at the beginning of the presolving. A presolving routine can apply certain reductions to the LP: deletion of a row or column, deletion of a system entry, modification of variable bounds and the left- and right-hand side, and modification of objective function coefficients. We distinguish between local and global reductions. While *local reductions* happen exclusively on the data of a single block, *global reductions* affect more than one block and involve communication between the processes. Since MPI communication can be expensive, we reduced the amount of data sent and the frequency of communication to a minimum and introduced local data structures to support the synchronization whenever possible.

In the following, singleton row elimination is used as an example to outline necessary data structures and methods. Although singleton row elimination is conceptually simple, its description still covers many difficulties arising during the implementation of preprocessing in an HPC environment. A singleton row refers to a row in the constraint matrix only containing one variable with nonzero coefficient. Both for a singleton equality and a singleton inequality row, the bounds of the respective variable can be tightened. This tightening makes the corresponding singleton row redundant and thus removable from the problem. In the case of an equality row, the corresponding variable is fixed and removed from the system.

Checking whether a non-linking row is singleton is straightforward since a single process holds all necessary information. The detection of singleton linking rows requires communication between the processes. Instead of asking all processes whether a given row is singleton, we introduced auxiliary data structures. Let $g = (g_0, g_1, \ldots, g_N)$ denote the coefficient vector of a linking row. Every process i knows the number of nonzeros in block i, i.e., $||g_i||_0$, and in block 0, i.e., $||g_0||_0$, at all times. At each synchronization point, every process also stores the current number of nonzeros overall blocks, $||g||_0$. Whenever local changes in the number nonzeros of a linking row occur, the corresponding process stores these changes in a buffer, instead of directly modifying $||g_i||_0$ and $||g||_0$. From that point on the global nonzero counters for all other processes are outdated and provide only an upper bound. Whenever a new presolving method that makes use of these counters is entered, the accumulated changes of all processes get broadcast. The local counters $||g_i||_0$ and $||g||_0$ are updated and stored changes are reset to zero.

After a singleton row is detected, there are two cases to consider, both visualized in Fig. 2. A process might want to delete a singleton row that has its singleton

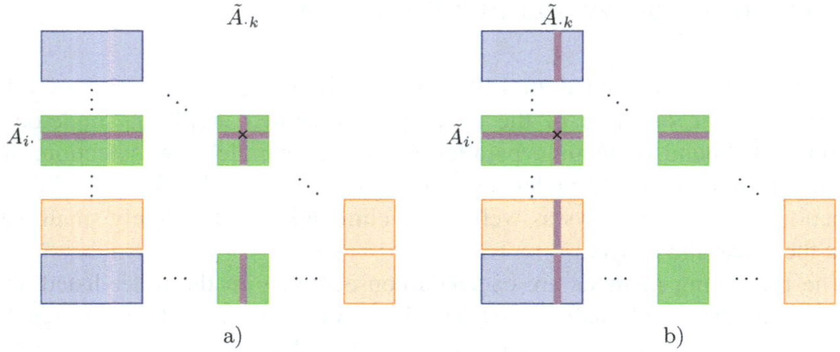

Fig. 2 LP data distributed on different processes and an entry $\tilde{a}_{ik}$ with corresponding singleton row $\tilde{A}_{i\cdot}$ and column $\tilde{A}_{\cdot k}$. Here $\tilde{A}$ is the complete system matrix from 1 and the coloring reflects the system blocks. (**a**) Singleton row leading to local changes. (**b**) Singleton row leading to global changes

entry in a non-linking part of the LP (Fig. 2a). This can be done immediately since none of the other processes is affected. By contrast, modifying the linking part of the problem is more difficult since all other processes have to be notified about the changes, e.g., when a process fixes a linking variable or when it wants to delete a singleton linking row (Fig. 2b). Again, communication is necessary and we implemented synchronization mechanisms for changes in variable bounds similar to the one implemented for changes in the nonzeros.

3 Computational Results

We conducted two types of experiments. First, we looked at the general performance and impact of our presolving routines compared with the ones offered by a different LP solver. For the second type of experiment, we investigated the scalability of our methods. The goal of the first experiment was to set the performance of our preprocessing into a more general context and show the efficiency of the structure-specific approach. To this end, we compared to the sequential, source-open solver SoPlex [6] and turned off all presolving routines that were not implemented in our preprocessing. With our scalability experiment, we wanted to further analyze the implementation and speed-up of our presolving. We thus ran several instances with different numbers of MPI processes.

The instances used for the computational results come from real-world energy system models found in the literature, see [7] (`elmod` instances) and [8] (`oms` and `yssp` instances). All tests with our parallel presolving were conducted on the JUWELS cluster at Jülich Supercomputing Centre (JSC). We used JUWELS' standard compute nodes running two Intel Xeon Skylake 8168 processors each with 24 cores 2.70 GHz and 96 GB memory. Since reading of the LP and presolving

Table 1 Runtimes and nonzero reductions for parallel presolving and sequential SoPlex presolving

	Input		PIPS-IPM			SoPlex	
Instance	N	nnzs	t_1 [s]	t_N [s]	nnzs	t_S [s]	nnzs
oms_1	120	2891 k	1.13	0.02	2362 k	1.51	2391 k
oms_2	120	11432 k	5.10	0.19	9015 k	11.09	9075 k
oms_3	120	1696 k	1.01	2.88	1639 k	0.54	1654 k
oms_4	120	131264 k	57.25	3.45	126242 k	206.31	127945 k
oms_5	120	216478 k	157.12	85.41	158630 k	>24 h	–
oms_6	120	277923 k	187.73	88.39	231796 k	>24 h	–
elmod_1	438	272602 k	125.62	0.48	208444 k	>24 h	–
elmod_2	876	716753 k	365.47	1.05	553144 k	>24 h	–
yssp_1	250	27927 k	13.01	0.44	22830 k	92.63	23758 k
yssp_2	250	68856 k	33.80	7.28	55883 k	1034.77	59334 k
yssp_3	250	32185 k	14.10	0.36	28874 k	95.08	29802 k
yssp_4	250	85255 k	39.71	7.25	76504 k	1930.16	80148 k

The number of nonzeros in columns "nnzs" are given in thousands

it with SoPlex was too time-consuming on JUWELS, we had to run the tests for SoPlex on a shared memory machine at Zuse institute Berlin with an Intel(R) Xeon(R) CPU E7-8880 v4, 2.2 GHz, and 2 TB of RAM.

The results of the performance experiment are shown in Table 1. We compared the times spent in presolving by SoPlex t_S, our routines running with one MPI process t_1 and running with the maximal possible number of MPI processes t_N. The nnzs columns report the number of nonzeros when read in (input) and after preprocessing. The key observations are:

- Except on the two smallest instances with less than 3 million nonzeros, already the sequential version of structure-specific presolving outperformed SoPlex significantly. The four largest instances with more than 200 million nonzeros could not be processed by SoPlex within 24 h.
- Reduction performed by both was very similar with an average deviation of less than 2%. Nonzero reduction overall instances was about 16% on average.
- Parallelization reduced presolving times on all instances except the smallest instance `oms_3`. On `oms_2`, `elmod_{1,2}`, and `yssp_{2,4}` the speed-ups were of one order of magnitude or more. However, on instances `oms_{4,5,6}` and `yssp_{2,4}` the parallel speed-up was limited, a fact that is further analyzed in the second experiment.

The results of our second experiment can be seen in Fig. 3. We plot times for parallel presolving, normalized by time needed by one MPI process. Let $S_n = t_1/t_n$ denote the speed-up obtained with n MPI processes versus one MPI process. Whereas for `elmod_2` we observe an almost linear speed-up $S_{146} \approx 114$, on `yssp_2` and `oms_4` the best speed-ups $S_{50} \approx 36$ and $S_{60} \approx 31$, respectively,

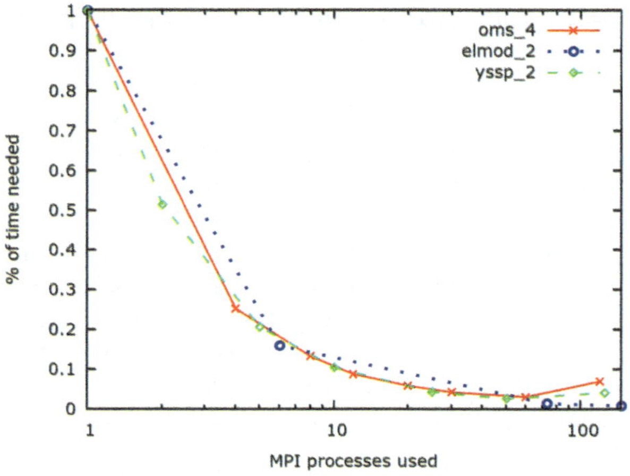

Fig. 3 Total presolving time for three instances of each type, relative to time for sequential presolving with one MPI process

are sublinear. For larger numbers of MPI processes, runtimes even start increasing again.

The limited scalability on these instances is due to a comparatively large amount of linking constraints. As explained in Sect. 2, performing global reductions within linking parts of the problem increases the synchronization effort. As a result, this phenomenon usually leads to a "sweet spot" for the number of MPI processes used, after which performance starts to deteriorate again. This effect was also responsible for the low speed-up on `oms_{5,6}` in Table 1. A larger speed-up can be achieved when running with fewer processes.

To conclude, we implemented a set of highly parallel structure-preserving presolving methods that proved to be as effective as sequential variants found in an out-of-the-box LP solver and outperformed them in terms of speed on truly large-scale problems. Beyond the improvements of the presolving phase, we want to emphasize that the reductions helped to accelerate the subsequent interior-point code significantly. On the instance `elmod_1`, the interior-point time could be reduced by more than half, from about 780 to about 380 s.

Acknowledgments This work is funded by the Federal Ministry for Economic Affairs and Energy within the BEAM-ME project (ID: 03ET4023A-F) and by the Federal Ministry of Education and Research within the Research Campus MODAL (ID: 05M14ZAM). The authors gratefully acknowledge the Gauss Centre for Supercomputing e.V. (www.gauss-centre.eu) for funding this project by providing computing time through the John von Neumann Institute for Computing (NIC) on the GCS Supercomputer JUWELS at Jülich Supercomputing Centre (JSC).

References

1. Breuer, T., et al.: Optimizing large-scale linear energy system problems with block diagonal structure by using parallel interior-point methods. In: Kliewer, N., Ehmke, J.F., Borndörfer, R. (eds.) Operations Research Proceedings 2017, pp. 641–647 (2018)
2. Petra, C.G., Schenk, O., Anitescu, M.: Real-time stochastic optimization of complex energy systems on high-performance computers. Comput. Sci. Eng. **16**, 32–42 (2014)
3. Achterberg, T., et al.: Presolve reductions in mixed integer programming. ZIB-Report 16–44, Zuse Institute, Berlin (2016)
4. Andersen, E.D., Andersen, K.D.: Presolving in linear programming. Math. Program. **71**, 221–245 (1995)
5. Gondzio, J.: Presolve analysis of linear programs prior to applying an interior point method. INFORMS J.Comput. **9**, 73–91 (1997)
6. Gleixner, A., et al.: The SCIP Optimization Suite 6.0. ZIB-Report 18–26, Zuse Institute, Berlin (2018)
7. Hinz, F.: Voltage Stability and Reactive Power Provision in a Decentralizing Energy System. PhD thesis, TU Dresden (2017)
8. Cao, K., Metzdorf, J., Birbalta, S.: Incorporating power transmission bottlenecks into aggregated energy system models. Sustainability **10**, 1–32 (2018)

A Steepest Feasible Direction Extension of the Simplex Method

Biressaw C. Wolde and Torbjörn Larsson

Abstract We present a feasible direction approach to general linear programming, which can be embedded in the simplex method although it works with non-edge feasible directions. The feasible direction used is the steepest in the space of all variables, or an approximation thereof. Given a basic feasible solution, the problem of finding a (near-)steepest feasible direction is stated as a strictly convex quadratic program in the space of the non-basic variables and with only non-negativity restrictions. The direction found is converted into an auxiliary non-basic column, known as an external column. Our feasible direction approach allows several computational strategies. First, one may choose how frequently external columns are created. Secondly, one may choose how accurately the direction-finding quadratic problem is solved. Thirdly, near-steepest directions can be obtained from low-dimensional restrictions of the direction-finding quadratic program or by the use of approximate algorithms for this program.

Keywords Linear program · Steepest-edge · Feasible direction · External pivoting

1 Derivation

Let $A \in \mathbb{R}^{m\times n}$, with $n > m$, have full rank, and let $b \in \mathbb{R}^m$ and $c \in \mathbb{R}^n$. Consider the Linear Program (LP)

$$z^* = \min z = c^{\mathrm{T}}x$$
$$\text{s.t. } Ax = b$$
$$x \geq 0,$$

B. C. Wolde · T. Larsson (✉)
Department of Mathematics, Linköping University, Linköping, Sweden
e-mail: biressaw.wolde@liu.se; torbjorn.larsson@liu.se,tolar@mai.liu.se

J. S. Neufeld et al. (eds.), *Operations Research Proceedings 2019*, Operations Research Proceedings, https://doi.org/10.1007/978-3-030-48439-2_14

whose feasible set is assumed to be non-empty. Let a non-optimal non-degenerate basic feasible solution be at hand. With a proper variable ordering, the solution corresponds to the partitioning $A = (B, N)$ and $c = (c_B^{\mathrm{T}}, c_N^{\mathrm{T}})^{\mathrm{T}}$, where $B \in \mathbb{R}^{m\times m}$ is the non-singular basis matrix and $N \in \mathbb{R}^{m\times(n-m)}$ is the matrix of non-basic columns. Let $x = (x_B^{\mathrm{T}}, x_N^{\mathrm{T}})^{\mathrm{T}}$. Introducing the reduced cost vector $\bar{c}_N^{\mathrm{T}} = c_N^{\mathrm{T}} - u^{\mathrm{T}}N$, where $u^{\mathrm{T}} = c_B^{\mathrm{T}}B^{-1}$ is the complementary dual solution, and letting I_m be the identity matrix of size m, problem LP is equivalent to

$$z^\star = \min z = c_B^{\mathrm{T}}B^{-1}b + \bar{c}_N^{\mathrm{T}}x_N$$
$$\text{s.t. } I_m x_B + B^{-1}Nx_N = B^{-1}b$$
$$x_B,\ x_N \geq 0.$$

The basic solution is $x_B = B^{-1}b$ and $x_N = 0$. By assumption, $B^{-1}b > 0$ and $\bar{c}_N \ngeq 0$ hold.

Let $\mathscr{N} = \{m+1, \ldots, n\}$, that is, the index set of the non-basic variables, and let a_j be column $j \in \mathscr{N}$ of the matrix A. Geometrically, the given basic feasible solution corresponds to an extreme point of the feasible polyhedron of problem LP, and a variable x_j, $j \in \mathscr{N}$, that enters the basis corresponds to the movement along a feasible direction that follows an edge from the extreme point. The edge direction is given by, see e.g. [4],

$$\eta_j = \begin{pmatrix} -B^{-1}a_j \\ e_{j-m} \end{pmatrix} \in \mathbb{R}^n,$$

where $e_{j-m} \in \mathbb{R}^{n-m}$ is a unit vector with a one entry in position $j-m$. The directional derivative of the objective function along an edge direction of unit length is $c^{\mathrm{T}}\eta_j/\|\eta_j\| = (c_B^{\mathrm{T}}, c_N^{\mathrm{T}})\eta_j/\|\eta_j\| = (-c_B^{\mathrm{T}}B^{-1}a_j + c_j)/\|\eta_j\| = \bar{c}_j/\|\eta_j\|$ (where $\|\cdot\|$ is the Euclidean norm). This is the rationale for the *steepest-edge criterion* [3], which in the simplex method finds a variable x_r, $r \in \mathscr{N}$, to enter the basis such that $\bar{c}_r/\|\eta_r\| = \min_{j\in\mathscr{N}} \{\bar{c}_j/\|\eta_j\|\}$.

We consider feasible directions that are constructed from non-negative linear combinations of the edge directions. To this extent, let $w \in \mathbb{R}_+^{n-m}$ and consider the direction

$$\eta(w) = \sum_{j\in\mathscr{N}} w_j\eta_j = \begin{pmatrix} -B^{-1}N \\ I_{n-m} \end{pmatrix} w,$$

where I_{n-m} is the identity matrix of size $n-m$. Note that any feasible solution to LP is reachable from the given basic feasible solution along some direction $\eta(w)$, and that $c^{\mathrm{T}}\eta(w) = \bar{c}_N^{\mathrm{T}}w$. Our development is founded on the problem and theorem below.

Define the Steepest-Edge Problem (SEP)

$$\begin{aligned} \min\ & \bar{c}_N^{\mathrm{T}} w \\ \text{s.t. } & \|\eta(w)\|^2 \leq 1 \\ & \operatorname{supp}(w) = 1 \\ & w \geq 0, \end{aligned}$$

where supp(·) is the support of a vector, that is, its number of nonzero components.

Theorem 1 *An index $r \in \mathcal{N}$ fulfils the steepest-edge criterion if and only if the solution*

$$w_j = \begin{cases} 1/\|\eta_j\| \ if\ j = r \\ 0\ otherwise \end{cases}, \quad j \in \mathcal{N},$$

solves SEP.

The theorem follows directly from an enumeration of all feasible solutions to problem SEP. Note that the optimal value of SEP is the steepest-edge slope $\bar{c}_r/\|\eta_r\|$.

To find a feasible direction that is steeper than the steepest edge, the support constraint in problem SEP is relaxed. The relaxed problem, called Direction-Finding Problem (DFP), can be stated as

$$\begin{aligned} \min\ & \bar{c}_N^{\mathrm{T}} w \\ \text{s.t. } & w^{\mathrm{T}} Q w \leq 1 \qquad (1) \\ & w \geq 0, \end{aligned}$$

where the matrix $Q = N^{\mathrm{T}}(B^{-1})^{\mathrm{T}} B^{-1} N + I_{n-m} \in \mathbb{R}^{(n-m)\times(n-m)}$, and it gives the *steepest* feasible direction. Because Q is symmetric and positive definite, and $\bar{c}_N \ngeq 0$ holds, the problem DFP has a unique, nonzero optimal solution, which fulfils the normalization constraint (1) with equality and has a negative objective value. Further, the optimal solution will in general yield a feasible direction that is a non-trivial non-negative linear combination of the edge directions, and has a directional derivative that is strictly better than that of the steepest edge.

As an example, we study the linear program

$$\min \{-x_1 - 2x_2 \mid 5x_1 - 2x_2 \leq 10;\ -2x_1 + 4x_2 \leq 8;\ 2x_1 + x_2 \leq 6;\ x_1, x_2 \geq 0\},$$

which is illustrated to the left in Fig. 1. For the extreme point at the origin, which has the slack basis with $B = I_3$, we have $\eta_1 = (-5, 2, -2, 1, 0)^{\mathrm{T}}$ and $\eta_2 = (2, -4, -1, 0, 1)^{\mathrm{T}}$, with $\bar{c}_1/\|\eta_1\| = -1/\sqrt{34}$ and $\bar{c}_2/\|\eta_2\| = -2/\sqrt{22}$. If using the steepest-edge criterion, the variable x_2 would therefore enter the basis. The feasible set of DFP is shown to the right in Fig. 1. The optimal

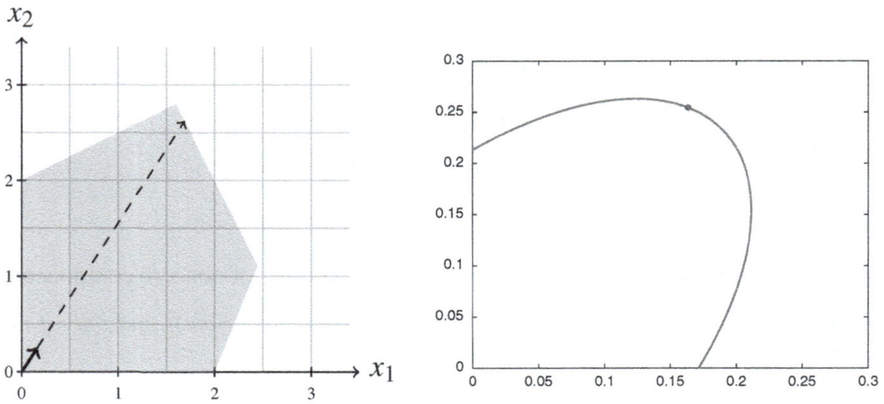

Fig. 1 Illustration of steepest feasible direction and problem DFP

solution of DFP is $w^* \approx (0.163, 0.254)^{\mathrm{T}}$ with $\bar{c}_N^{\mathrm{T}} w^* = c^{\mathrm{T}}\eta(w^*) \approx -0.672$, which should be compared with $-2/\sqrt{22} \approx -0.426$. The feasible direction is $\eta(w^*) \approx (-0.309, -0.690, -0.581, 0.163, 0.254)^{\mathrm{T}}$. The maximal feasible step in this direction yields the boundary point $(x_1, x_2) \approx (1.687, 2.625)$, whose objective value is better than those of the extreme points that are adjacent to the origin. (Further, this boundary point is close to the extreme point $(1.6, 2.8)^{\mathrm{T}}$, which is optimal.)

We next establish that problem DFP can be solved by means of a relaxed problem. Let $\mu/2 > 0$ be an arbitrary value of a Lagrangian multiplier for the constraint (1), and consider the Lagrangian Relaxation (LR)

$$\min_{w \geq 0} r(w) = \bar{c}_N^{\mathrm{T}} w + \frac{\mu}{2} w^{\mathrm{T}} Q w$$

of problem DFP (ignoring the constant $-\mu/2$). Since r is a strictly convex function, problem LR has a unique optimum, denoted $w^*(\mu)$. The following result can be shown.

Theorem 2 *If $w^*(\mu)$ is optimal in problem LR, then $w^* = w^*(\mu)/\|\eta(w^*(\mu))\|$ is the optimal solution to problem DFP.*

The proof is straightforward; both problems are convex and have interior points, and it can be verified that if $w^*(\mu)$ satisfies the Karush–Kuhn–Tucker conditions for problem LR then $w^* = w^*(\mu)/\|\eta(w^*(\mu))\|$ satisfies these conditions for problem DFP.

Hence, the steepest feasible direction can be found by solving the simply constrained quadratic program LR, for any choice of $\mu > 0$. The following result, which is easily verified, gives an interesting characterization of the gradient of the objective function r. It should be useful if LR is approached by an iterative descent method, such as for example a projected Newton method [1].

Proposition 1 *Let* $\eta_B(w) = -B^{-1}Nw$ *and* $\Delta u^{\mathrm{T}} = \mu\eta(w)^{\mathrm{T}}B^{-1}$. *Then*

$$\nabla r(w) = c_N - N^{\mathrm{T}}(u + \Delta u) + \mu w.$$

Note that the expression for Δu is similar to that of the dual solution, and that the pricing mechanism of the simplex method is used to compute $\nabla r(w)$, but with a modified dual solution. Further, $-\eta_B(w) = B^{-1}Nw = B^{-1}\sum_{j\in\mathcal{N}} w_j a_j$, that is a non-negative linear combination of the original columns $(a_j)_{j\in\mathcal{N}}$ expressed in the current basis.

In order to use a feasible direction within the simplex method, it is converted into an *external column* [2], which is a non-negative linear combination of the original columns in LP. Letting w^* be an (approximate) optimal solution to problem DFP, we define the external column as $c_{n+1} = c_N^{\mathrm{T}}w^*$ and $a_{n+1} = Nw^*$. Letting $\bar{c}_{n+1} = c_{n+1} - u^{\mathrm{T}}a_{n+1}$, the problem LP is augmented with the variable x_{n+1}, giving

$$\begin{aligned} z^\star = \min z = c_B^{\mathrm{T}}B^{-1}b + \bar{c}_N^{\mathrm{T}}x_N + \bar{c}_{n+1}x_{n+1} \\ \text{s.t. } I_m x_B + B^{-1}Nx_N + B^{-1}a_{n+1}x_{n+1} = B^{-1}b \\ x_B,\ x_N,\ x_{n+1} \geq 0, \end{aligned}$$

where $\bar{c}_{n+1} < 0$. By letting the external column enter the basis, the feasible direction will be followed. Note that the augmented problem has the same optimal value as the original one (If $B^{-1}a_{n+1} \leq 0$ holds, then $z^\star = -\infty$.) Further, if the external column is part of an optimal solution to the augmented problem, then it is easy to recover an optimal solution to the original problem [2].

The approach presented above is related to those in [5] and [2], which both use auxiliary primal variables for following a feasible direction. (The term external column is adopted from the latter reference.) These two works do however use *ad hoc* rules for constructing the feasible direction, for example based on only reduced costs, instead of solving a direction-finding problem with the purpose of finding a steep direction.

2 Numerical Illustration and Conclusions

It is in practice reasonable to use a version of LR that contains only a restricted number of edge directions. Letting $J \subseteq \mathcal{N}$, $w_J = (w_j)_{j\in J}$, $\bar{c}_J = (\bar{c}_j)_{j\in J}$, $N_J = (a_j)_{j\in J}$ and $Q_J = N_J^{\mathrm{T}}B^{-\mathrm{T}}B^{-1}N_J + I_{|J|}$, the restricted LR is given by $\min_{w_J\geq 0} r_J(w_J) = \bar{c}_J^{\mathrm{T}}w_J + \frac{\mu}{2}w_J^{\mathrm{T}}Q_J w_J$.

An external column should be advantageous to use compared to an ordinary non-basic column, but it is computationally more expensive. Further, pivots on external columns will lead to points that are *not* extreme points in the original polyhedron. It is therefore reasonable to combine pivots on external columns with ordinary pivots. Considering the choice of $|J|$, a high value makes the restricted LR computationally more expensive, but it then also has the potential to yield a better feasible direction. Hence, there is a trade-off between the computational burden of creating the external columns, both with respect to frequency of external columns and the choice of $|J|$, and the reduction in simplex iterations that they may lead to.

To make an initial assessment of the potential usefulness of generating external columns as described above, we made a simple implementation in MATLAB of the revised simplex method, using the Dantzig entering variable criterion but with the option to replace ordinary pivots with pivots on external column, which are found by solving restricted versions of LR. Letting k be the number of edge directions to be included in the restriction, we consider two ways of making the selection: by finding the k most negative values among $\{\bar{c}_j\}_{j \in \mathcal{N}}$ or among $\{\bar{c}_j / \|\eta_j\|\}_{j \in \mathcal{N}}$. These ways are called *Dantzig selection* and *steepest-edge selection*, respectively. The latter is computationally more expensive. The restricted problem LR is solved using the built-in solver `quadprog`.

We used test problem instances that are randomly generated according to the principle in [2], which allows the simplex method to start with the slack basis. The external columns are constructed from original columns only (although it is in principle possible to include also earlier external columns). We study the number of simplex iterations and running times required for reaching optimality when using the standard simplex method, and when using the simplex method with external columns with different numbers of edge directions used to generate the external columns and when generating external column only once or repeatedly. Table 1 shows the results. Figure 2 shows the convergence histories for the smallest problem instance when using $k = 200$.

Our results indicate that the use of approximate steepest feasible directions can considerably reduce both the number of simplex iterations and the total running times, if the directions are based on many edge directions and created seldomly; if the directions are based on few edge directions and created frequently, then the overall performance can instead worsen. These findings clearly demand for more investigations. Further, our feasible direction approach must be extended to properly handle degeneracy, and tailored algorithms for fast solution of the restricted problem LR should be developed.

Table 1 Simplex iterations and running times

Size		Std simplex		External columns: Dantzig selection				External columns: steepest-edge selection			
m	n	Iterations	Time	k	nmax	Iterations	Time	k	nmax	Iteration	Time
1000	2000	35,181	418.0	20	∞	28,198	310.6	20	∞	28,091	302.4
				20	50	43,999	720.4	20	50	27,261	552.8
				100	∞	20,200	211.3	100	∞	23,455	245.0
				100	500	21,237	231.5	100	500	20,133	225.9
				200	∞	20,561	217.7	200	∞	20,891	219.7
				200	500	18,926	210.2	200	500	19,168	220.9
1000	3000	47,647	715.2	20	∞	38,545	571.0	20	∞	36,324	511.2
				20	50	41,217	754.4	20	50	34,367	998.1
				100	∞	30,676	440.5	100	∞	22,886	305.8
				100	500	29,212	424.6	100	500	22,799	315.6
				200	∞	24,504	365.0	200	∞	26,150	342.0
				200	500	24,405	344.0	200	500	21,372	303.1
1000	4000	57,572	1096.1	20	∞	50,150	842.1	20	∞	41,857	713.9
				20	50	65,081	1593.8	20	50	61,379	2725.6
				100	∞	35,819	599.7	100	∞	33,346	574.5
				100	500	40,863	674.0	100	500	32,866	588.1
				200	∞	34,441	586.6	200	∞	31,678	536.8
				200	500	33,430	544.3	200	500	25,350	462.6

Here, (m, n) = problem size; k = number of edge directions used to generate the external columns; $nmax$ = number of simplex iterations between external columns, where "∞" means that an external column is generated only at the initial basis

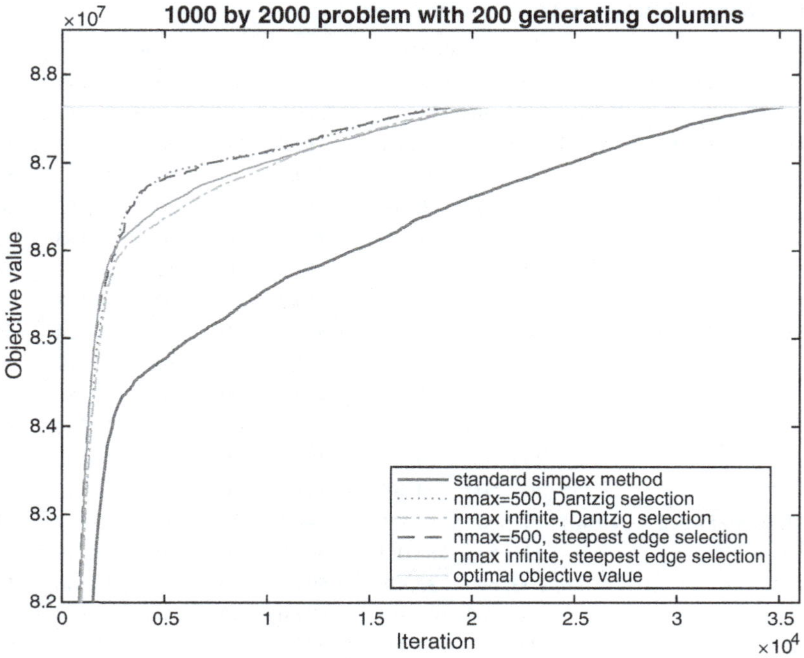

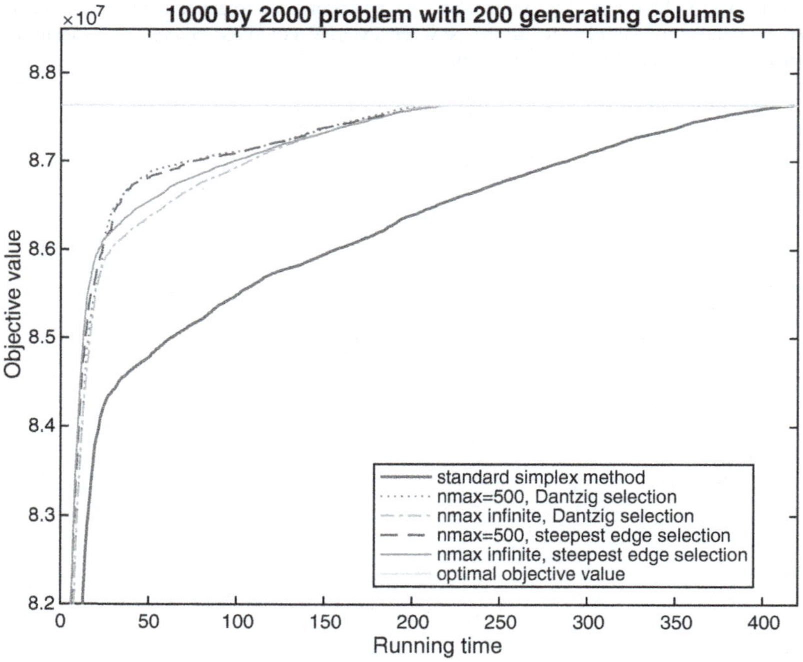

Fig. 2 Objective value versus iteration and running time, respectively

References

1. Bertsekas, D.P.: Projected Newton methods for optimization problems with simple constraints. SIAM J. Control Optim. **20**, 221–246 (1982)
2. Eiselt, H.A., Sandblom, C.-L.: Experiments with external pivoting. Comput. Oper. Res. **17**, 325–332 (1990)
3. Goldfarb, D., Reid, J.K.: A practicable steepest-edge simplex algorithm. Math. Program. **12**, 361–371 (1977)
4. Murty, K.G.: Linear Programming. Wiley, New York (1983)
5. Murty, K.G., Fathi, Y.: A feasible direction method for linear programming. Oper. Res. Lett. **3**, 121–127 (1984)

Convex Quadratic Mixed-Integer Problems with Quadratic Constraints

Simone Göttlich, Kathinka Hameister, and Michael Herty

Abstract The efficient numerical treatment of convex quadratic mixed-integer optimization poses a challenging problem. Therefore, we introduce a method based on the duality principle for convex problems to derive suitable lower bounds that can used to select the next node to be solved within the branch-and-bound tree. Numerical results indicate that the new bounds allow the tree search to be evaluated quite efficiently compared to benchmark solvers.

Keywords Mixed-integer nonlinear programming · Duality · Branch-and-bound

1 Introduction

Convex optimization problems with quadratic objective function and linear or quadratic constraints often appear in operations management, portfolio optimization or engineering science, see e.g. [1, 3, 11] and the references therein. To solve general convex mixed-integer nonlinear problems (MINLP), Dakin [6] proposed the branch-and-bound (B&B) method in 1965. This method extends the well-known method for solving linear mixed-integer problems [10]. The relation between primal and dual problems in the convex case has been used to obtain lower bounds and early branching decisions already in [5]. Therein, the authors suggested to use a quasi-Newton method to solve the Lagrangian saddle point problem appearing in the branching nodes. This nonlinear problem has only bound constraints. Fletcher and Leyffer [8] report on results of this approach for mixed-integer quadratic problems (MIQP) with linear constraints. In this work, we will extend those results by investigating the dual problem in more detail for an improved tree search within a

S. Göttlich (✉) · K. Hameister
Mannheim University, Department of Mathematics, Mannheim, Germany
e-mail: goettlich@uni-mannheim.de

M. Herty
RWTH Aachen University, Department of Mathematics, Aachen, Germany

J. S. Neufeld et al. (eds.), *Operations Research Proceedings 2019*, Operations Research Proceedings, https://doi.org/10.1007/978-3-030-48439-2_15

B&B algorithm. Although the approximate bounds we compute from theory are not always sharp, they can be used within the B&B algorithm as a node selection rule (or child selection method) to select the next node to be solved. The proposed heuristic is integrated into customized B&B algorithm and to the best of our knowledge, this idea has not been used before.

2 Lagrangian Duality Principle

MINLPs have been studied intensively over the last decades, see for example [3, 5–7, 12]. Here, we focus on convex linear-quadratic mixed-integer problems, where concepts from convex optimization [9] can be applied to compute the dual problem and hence suitable lower bounds to the primal problem. Within a tree search those bounds can be used to select the next node to be solved. However, the dual problem to convex minimization problems is usually hard to solve. Therefore, we provide approximations to the dual problem leading to a heuristic procedure that can be used to determine suitable lower bounds.

A MINLP for strictly convex quadratic functions $f : \mathbb{R}^n \to \mathbb{R}$, $g_j : \mathbb{R}^n \to \mathbb{R}$ and affine linear functions $h_1 : \mathbb{R}^n \to \mathbb{R}^m$, $h_2 : \mathbb{R}^n \to \mathbb{R}^m$ is given by

$$\begin{aligned} &\min_{x \in \mathbb{R}^n} f(x) \\ &\text{subject to } \; g_j(x) \leq 0, \; j = 1, \ldots, p, \; h_1(x) \leq 0, \; h_2(x) = 0, \\ &\qquad\qquad x_i \in \mathbb{Z}, \; \forall i \in I, \; \text{and } x_j \in \mathbb{R}, \; \forall j \in \{1, \ldots, n\} \backslash I \end{aligned} \tag{1}$$

including p quadratic constraints g_j. We assume $I = \{1, \ldots, \ell\}$ and $m < n$. The set I contains the indices of the integer components of $x \in \mathbb{R}^n$. In order to simplify the notation we introduce the subset $X \subset \mathbb{R}^n$ as $X = \{x \in \mathbb{R}^n \mid x_i \in \mathbb{Z}, \; \forall\, i \in I\}$.

Due to the given assumptions we may write $f(x) = \frac{1}{2}x^T Q_0 x + c_0^T x$, $g_j(x) = \frac{1}{2}x^T Q_j x + c_j^T x$, $h_1(x) = A_1 x - b_1$, $h_2(x) = A_2 x - b_2$. Here, Q_0 and Q_j, $j = 1, \ldots, p$, are positive definite and symmetric matrices. We also assume that the matrices $A_1 \in \mathbb{R}^{m_1 \times n}$ and $A_2 \in \mathbb{R}^{m_2 \times n}$ with $m = m_1 + m_2$ have maximal column rank to avoid technical difficulties, cf. [1].

In the algorithmic framework of the B&B method we need to solve relaxed problems, where X is replaced by $\mathbb{R}^n$. In order to select suitable nodes we require lower bounds on the relaxation problem. Those will be calculated using a dual formulation. The setting of the relaxed problem allows to derive a dual formulation due to the convexity assumptions [9]. Then, we have $x \in \mathbb{R}^n$ and the Lagrangian of the problem of (1) is given by

$$L(x, \alpha_1, \ldots, \alpha_p, \lambda, \mu) = f(x) + \sum_{j=1}^{p} \alpha_j g_j(x) + (\lambda, \mu)^T h(x),$$

where $\alpha \in \mathbb{R}^p_+$, $\lambda \in \mathbb{R}^{m_1}_+$ and $\mu \in \mathbb{R}^{m_2}$. Provided that the functions f and g are continuously differentiable, the dual problem can be stated by the first order optimality conditions.

Lemma 1 *Let $x \in \mathbb{R}^n$ be a feasible solution of the primal problem and let $(\alpha, \lambda, \mu) \in \mathbb{R}^p \times \mathbb{R}^m$ be a feasible solution of the dual problem. Then, for any $\alpha \in \mathbb{R}^p_+$, the value*

$$\tilde{L} = L(\hat{x}, \alpha_1, \ldots, \alpha_p, 0, \hat{\mu})$$

provides a lower bound for $f(x)$, $x \in X$, with

$$\hat{x}(\alpha_1, \ldots, \alpha_p, 0, \mu) = -M(\tilde{c} + A_2^T \mu),\ \hat{\mu} = -(ZA_2^T - 2BA_2^T)^{-1}(Z\tilde{c} - 2B\tilde{c} - b_2),$$

$$M^{-1} = (Q_0 + \sum_{j=1}^{p} \alpha_j Q_j),\ Z = A_2 M^T (Q_0 + \sum_{j=1}^{p} \alpha_j Q_j) M,$$

$$B = A_2 M^T,\ \tilde{c} = (c_0 + \sum_{j=1}^{p} \alpha_j c_j).$$

Proof Before we verify the Lagrangian multipliers, it is necessary to show, that $\hat{x}$ is the minimizer of the relaxed problem to (1). Therefore, we solve the following equation for x:

$$\nabla_x L(x, \alpha_1, \ldots, \alpha_p, \lambda, \mu) = \nabla f(x) + \sum_{j=1}^{p} \alpha_j \nabla g_j(x) + (\lambda^T, \mu^T) \nabla h(x) = 0.$$

This gives the unique minimizer $\hat{x}$ in dependence of the multipliers $\alpha_1, \ldots, \alpha_p, \lambda, \mu$

$$\hat{x}(\alpha_1, \ldots, \alpha_p, 0, \mu) = -(Q_0 + \sum_{j=1}^{p} \alpha_j Q_j)^{-1}(c_0 + \sum_{j=1}^{p} \alpha_j c_j + A_2^T \mu),$$

where the multiplier λ for the linear inequalities is zero. This allows to obtain the closed form of the dual function in terms of the multipliers. Calculating the gradient with respect to μ, we obtain $\hat{\mu}$ as the zero of the partial derivative of the dual function with respect to μ. We end up with

$$\hat{\mu} = -(A_2 M^T \tilde{Q} M A_2^T - 2A_2 M^T A_2^T)^{-1}(A_2 M^T \tilde{Q} M \tilde{c} - 2A_2 M^T \tilde{c} - b_2)$$

as a Lagrangian multiplier with $\tilde{Q} = Q_0 + \sum_{j=1}^{p} \alpha_j Q_j$, $M = (Q_0 + \sum_{j=1}^{p} \alpha_j Q_j)^{-1}$ and $\tilde{c} = (c_0 + \sum_{j=1}^{p} \alpha_j c_j)$.

Solution Procedure We present a new node selection rule based on the approximation of lower bounds on the optimal solutions of the relaxed problem [1, 5, 8]. This rule is integrated in a depth-first search strategy within the B&B algorithm and can be therefore seen as a *diving heuristic*. The B&B algorithm searches a tree structure to find feasible integer solutions, where the nodes correspond to the relaxed quadratic constraint problems. Due to the fact that we do not know the solution associated with a node in advance, we have to look for alternative strategies through the B&B tree. There are different strategies to build up the decision tree of the B&B algorithm, see [1] for more information. We focus on the following node selection strategy and refer to as B&B dual:

First, we compute a bound on the objective value of the arising subproblems. We intend to save computational costs by solving only one of the two subproblems. Thanks to Lemma 1, we are able to decide which branch should be solved first by comparing the values of the dual bound of the corresponding child nodes. We continue on the branch with the lower dual bound, i.e. the best-of-two node selection rule or best lower bound strategy [4].

3 Computational Results

The proposed strategy is implemented in MATLAB Release 2013b based on a software for solving MINLP[1] . Furthermore, we use IBM's ILOG CPLEX Interactive Optimizer 12.6.3. to solve the relaxed subproblems. All tests have been performed on a Unix PC equipped with 512 GB Ram, Intel® Xeon® CPU E5-2630 v2 @ 2.80 GHz.

Performance Measures We start with the introduction of performance measures inspired by Berthold [2] for the numerical comparison. Therefore, we compare the time needed to find the first integer solution $\tilde{x}_1$ and the optimal solution $\tilde{x}_{opt}$ as well as the time needed to prove optimality. To show the quality of the first found integer solution, we also record the objective function value of the first and the optimal solution, i.e., $f(\tilde{x}_1)$ and $f(\tilde{x}_{opt})$.

Let $\tilde{x}$ be again an integer optimal solution and $\tilde{x}_{opt}$ the optimum. We define $t_{max} \in \mathbb{R}_+$ as the time limit of the solution process. Then, the *primal gap* $\gamma \in [0, 1]$ of $\tilde{x}$ is given by

$$\gamma(\tilde{x}) := \begin{cases} 0, & \text{if } |f(\tilde{x}_{opt})| = |f(\tilde{x})| = 0, \\ 1, & \text{if } f(\tilde{x}_{opt}) \cdot f(\tilde{x}) < 0, \\ \frac{|f(\tilde{x}_{opt}) - f(\tilde{x})|}{\max\{|f(\tilde{x}_{opt})|, |f(\tilde{x})|\}}, & \text{else.} \end{cases}$$

[1] http://www.mathworks.com/matlabcentral/fileexchange/96-fminconset.

The monotone decreasing step function $p : [0, t_{max}] \to [0, 1]$ defined as

$$p(t) := \begin{cases} 1, & \text{if no increment until point } t, \\ \gamma(\tilde{x}(t)), & \text{with } \tilde{x}(t) \text{ increment at point } t \end{cases}$$

is called *primal gap function*. The latter changes its value whenever a new increment is found. Next, we define the measure $P(T)$ called the *primal integral* $P(T)$ for $T \in [0, t_{max}]$ as

$$P(T) := \int_0^T p(t)\,dt = \sum_{k=1}^{K} p(t_{k-1}) \cdot (t_k - t_{k-1}),$$

where $T \in \mathbb{R}_+$ is the total computation time of the procedure and $t_k \in [0, T], k \in 1, \ldots, K-1$ with $t_0 = 0, t_K = T$.

Tests for Data Sets from Academic Literature We take six instances from the MINLPLib library[2] to test the B&B dual method. All instances need to be convex and consist of at least one quadratic constraint. The first instances are the so-called CLay problems that are constrained layout problems. From literature we know that these problems are ill-posed in the sense that there is no feasible solution near the optimal solution of the continuous relaxation, see [3]. As a second application we consider portfolio optimization problems (called portfol_classical). Those problems arise by adding a cardinality constraint to the mean-variance portfolio optimization problem, see [13].

We compare the performance of the B&B dual algorithm with CPLEX and Bonmin-BB while focusing on the quality of the first solutions found. We consider the quality measures computing times t_1, t_{opt}, primal integral $P(t)$, primal gap $\gamma(\tilde{x}_1)$ and total number of integer solutions MIP Sols. Apparently, CPLEX and Bonmin-BB benefit from additional internal heuristics to prove optimality, cf. computing times in Table 1.

In particular, CPLEX is able to find good integer solutions reasonably fast, independent of the problem size and the number of quadratic constraints. However, the performance of the B&B dual algorithm depends on the choice of the Lagrangian multiplier α. For the CLay problems, it is reasonable to choose the multipliers close to zero, e.g. α has been chosen as $\alpha \in (0, \frac{1}{p})$, where p is the number of quadratic constraints. In contrast, in case of the portfolio optimization problems, α has been selected from the set $\{0.5, 0.6, 0.7, 0.8, 0.9\}$ and a small random number has been added to prevent symmetric solutions.

The first four columns in Table 1 describe the problems. QC and LC count the amount of quadratic and linear constraints of the initial problem. We document the solution times, the primal integral $P(T)$ and integer MIP solutions. It turns out that

[2]http://www.minlplib.org/instances.html.

Table 1 Examples taken from MINLPLib2

Instance	# var l/n	QC	LC	Solver	Time in sec. t_1	t_{opt}	T	$P(T)$	MIP Sols
CLay0204m	32/52	32	58	B&B dual	11	140	1197	15.66	4
				Bonmin-BB	<1	2.8	149	2.07	3
				CPLEX	1	22	41	2.92	8
CLay0205m	50/80	40	95	B&B dual	64	448	$54000^{\dagger}$	236.71	6
				Bonmin-BB	7	12,682	$54,000^{\dagger}$	166.22	6
				CPLEX	2	831	2693	15.87	9
CLay0303m	21/33	36	30	B&B dual	8	8	343	8.41	1
				Bonmin-BB	1.8	2.4	49	2.02	2
				CPLEX	1	15	15	1.67	5
CLay0305m	55/85	60	95	B&B dual	106	5557	$54000^{\dagger}$	652.35	9
				Bonmin-BB	8	–	$54000^{\dagger}$	(716.81)	(4)
				CPLEX	4	1669	4749	9.38	8
portfol_classical050_1	50/150	1	102	B&B dual	23	32,025	43,996	937.35	20
				Bonmin-BB	<1	$26^{\ddagger}$	30	1.77	4
				CPLEX	<1	4	4	0.31	3
portfol_classical200_2	200/600	1	402	B&B dual	398	–	$54000^{\dagger}$	1184.34	12
				Bonmin-BB	15	$15^{\ddagger}$	30	16.42	1
				CPLEX	7	5548	$54000^{\dagger}$	34.14	6

The symbol † states that the solution process has been stopped after 15 h. If the solution found by the solver is different from the one given by MINLPLib2, we marked t_{opt} with the symbol ‡

Table 2 Primal gap $\gamma(\tilde{x}_1)$ of the first integer feasible solution

Solver	CLay0204m	CLay0205m	CLay0303m	CLay0305m
B&B dual	**0.2839**	0.7171	**0**	0.6898
Bonmin-BB	0.9119	**0.3329**	0.3610	**0.0676**
CPLEX	0.9046	0.6832	0.3636	0.4730
Solver	portfol_classical050_1		portfol_classical200_2	
B&B dual	0.1254		0.2444	
Bonmin-BB	0.0154		0.0852	
CPLEX	**0.0070**		**0.0072**	

The values in bold indicate the solver (B&B dual, Bonmin-BB or CPLEX) with the best primal gap for a specific data set

CPLEX performs best for all instances. However, the primal gap of the first integer solution $\gamma(\tilde{x}_1)$ is fairly good for the B&B dual implementation, in particular in comparison with CPLEX, compare Table 2.

Acknowledgments This work has been supported by the BMBF project ENets and the DFG projects GO 1920/4-1, HE 5386/15-1.

References

1. Belotti, P., Kirches, C., Leyffer, S., Linderoth, J., Luedtke, J., Mahajan, A.: Mixed-integer nonlinear optimization. Acta Numer. **22**, 1–131 (2013). https://doi.org/10.1017/S0962492913000032
2. Berthold, T.: Measuring the impact of primal heuristics. Oper. Res. Lett. **41**(6), 611–614 (2013). https://doi.org/10.1016/S0167637713001181
3. Bonami, P., Biegler, L.T., Conn, A.R., Cornuéjols, G., Grossmann, I.E. Laird, C.D., Lee, J., Lodi, A., Margot, F., Sawaya, N., Wächter, A.: An algorithmic framework for convex mixed integer nonlinear programs. Discrete Optim. **5**(2), 186–204 (2008). https://doi.org/10.1016/j.disopt.2006.10.011
4. Bonami, P., Kilinç, M., Linderoth, J.: Algorithms and Software for Convex Mixed Integer Nonlinear Programs, pp. 1–39. Springer, New York (2012). https://doi.org/10.1007/978-1-4614-1927-3_1
5. Borchers, B., Mitchell, J.E.: An improved branch and bound algorithm for mixed integer nonlinear programs. Comput. Oper. Res. **21**(4), 359–367 (1994). https://doi.org/10.1016/0305-0548(94)90024-8
6. Dakin, R.J.: A tree-search algorithm for mixed integer programming problems. Comput. J. **8**(3), 250–255 (1965). https://doi.org/10.1093/comjnl/8.3.250
7. Fletcher, R., Leyffer, S.: Solving mixed integer nonlinear programs by outer approximation. Math. Program. **66**(1–3), 327–349 (1994). https://doi.org/10.1007/BF01581153
8. Fletcher, R., Leyffer, S.: Numerical experience with lower bounds for MIQP branch-and-bound. SIAM J. Control Optim. **8**(2), 604–616 (1998). https://doi.org/10.1137/S1052623494268455
9. Geiger, C., Kanzow, C.: Theorie und Numerik restringiererter Optimierungsaufgaben. Springer, Berlin (2002)
10. Land, A.H., Doig, A.G.: An automatic method of solving discrete programming problems. Econometrica **28**(3), 497 (1960). https://doi.org/10.2307/1910129
11. Misener, R., Floudas, C.A.: GloMIQO: Global mixed-integer quadratic optimizer. J. Global Optim. **57**, 3–30 (2013)
12. van Thoai, N.: Duality bound method for the general quadratic programming problem with quadratic constraints. J. Optim. Theory Appl. **107**(2), 331–354 (2000). https://doi.org/10.1023/A:1026437621223
13. Vielma, J.P., Ahmed, S., Nemhauser, G.L.: A lifted linear programming branch-and-bound algorithm for mixed-integer conic quadratic programs. INFORMS J. Comput. **20**(3), 438–450 (2008)

Part V
Decision Theory and Multiple Criteria Decision Making

The Bicriterion Maximum Flow Network Interdiction Problem in s-t-Planar Graphs

Luca E. Schäfer, Tobias Dietz, Marco V. Natale, Stefan Ruzika, Sven O. Krumke, and Carlos M. Fonseca

Abstract A biobjective extension of the maximum flow network interdiction problem is considered: Two maximum flows from source to sink are to be computed independently from each other while an interdictor aims to reduce the value of both maximum flows simultaneously by interdicting arcs. We show that this problem is intractable and propose two procedures to solve this problem on specific graph classes.

Keywords Network interdiction · Dynamic programming · Multicriteria optimization

1 Introduction

The maximum network flow problem is one of the best studied optimization problems in operations research with several applications (cf. [1, 3, 9]). A flow from a source vertex s to a sink vertex t in a directed network $G = (V, A)$ is a function $f : A \to \mathbb{R}_+$ that assigns a flow value to each arc and satisfies flow balance equations $\sum_{a \in A: a \in \delta^-(v)} f(a) - \sum_{a \in A: a \in \delta^+(v)} f(a) = 0$ for all vertices $v \in V \backslash \{s, t\}$, where $\delta^-(v)$ and $\delta^+(v)$ denotes the set of incoming and outgoing arcs of $v \in V$, respectively. Moreover, a flow f is called feasible with respect to a given capacity function $u : A \to \mathbb{N}$, if $f(a) \leq u(a)$ for all $a \in A$. The value of a flow f is equal to the excess at the sink vertex t, i.e., $val(f) = \sum_{a \in A: a \in \delta^-(t)} f(a) - \sum_{a \in A: a \in \delta^+(t)} f(a)$. The maximum flow problem asks for a feasible s-t-flow with maximum flow value $val^*(f)$ (cf. [1]).

L. E. Schäfer (✉) · T. Dietz · M. V. Natale · S. Ruzika · S. O. Krumke
Department of Mathematics, Technische Universität Kaiserslautern, Kaiserslautern, Germany
e-mail: luca.schaefer@mathematik.uni-kl.de

C. M. Fonseca
CISUC, Department of Informatics Engineering, University of Coimbra, Coimbra, Portugal

J. S. Neufeld et al. (eds.), *Operations Research Proceedings 2019*, Operations Research Proceedings, https://doi.org/10.1007/978-3-030-48439-2_16

In the network flow interdiction problem (NFI), an attacker is introduced who aims to reduce the maximum flow value in a network as much as possible by deleting arcs. Each arc is associated with an interdiction cost and the attacker is constrained to an interdiction budget. Several applications of this problem have been studied in the literature, see e.g., drug interdiction [15] and hospital infection control [2]. It has been shown that (NFI) is strongly $\mathcal{NP}$-complete, even when interdicting an arc requires exactly one unit of cost [12, 15]. Nevertheless, it can be solved in pseudopolynomial time on planar networks [12] and in polynomial time on source-sink-planar networks when restricting to unit interdiction costs. On the contrary, little is known about the approximability of (NFI) [6]. A fully polynomial time approximation scheme for planar graphs is presented in [12], whereas in [6] an approximation algorithm with an approximation ratio of $2(n-1)$ for general graphs is presented. In [4], the authors propose an algorithm that either returns a $(1, 1+\frac{1}{\varepsilon})$-approximation or a $(1+\varepsilon, 1)$-pseudoapproximation.

In this article, we consider a biobjective extension of (NFI), the biobjective maximum flow network interdiction problem (BMFNI). Two maximum flows are to be computed from source to sink constrained to some given arc capacities while an attacker aims at reducing both maximum flow values simultaneously. This leads to several incommensurable (Pareto-optimal) interdiction strategies. This could be of particular interest e.g. in the case of planning local heating networks or, more generally speaking, whenever two supply systems share a common network for which one aims to identify critical components or potential possibilities of savings.

The rest of the paper is organized as follows. In Sect. 2, we introduce the problem setting and then address the complexity of the problem. In Sect. 4, we propose solution procedures for some classes of graphs. Section 5 concludes the paper and proposes further research directions.

2 Problem Formulation

Let $G = (V, A)$ be a directed graph with vertex set V and arc set A. We set $n := |V|$ and $m := |A|$. A capacity function assigns two capacity values to every arc, i.e., $\mathbf{u} : A \to \mathbb{N}^2$, where $\mathbf{u}(a) = (u^1(a), u^2(a))^\top$, $a \in A$, with $u^i : A \to \mathbb{N}$, $i = 1, 2$. Let $f^1 : A \to \mathbb{R}_+$ and $f^2 : A \to \mathbb{R}_+$ be two functions assigning flow values to the arcs of G. Interdicting an arc $a \in A$ is associated with a cost $c(a) \in \mathbb{N}$, i.e., $c\colon A \to \mathbb{N}$, where interdicting arc a means that a is removed from the network. Further, we denote the set of all feasible interdiction strategies by Ω, i.e.,

$$\Omega := \left\{ \omega = (\omega_a)_{a\in A} \in \{0, 1\}^{|A|} \mid \sum_{a\in A} c(a) \cdot \omega_a \leq B \right\},$$

where $B \in \mathbb{N}$ denotes the interdiction budget of the attacker. The binary variable ω_a indicates whether arc a is interdicted or not, i.e., $\omega_a = 1$ or $\omega_r = 0$, respectively.

For a given strategy $\omega \in \Omega$ of the interdictor, $A(\omega)$ denotes the set of destroyed arcs, i.e., $A(\omega) := \{a \in A \mid \omega_a = 1\}$. Consequently, $G_\omega := G \backslash A(\omega)$ describes the graph G with node set V and arc set $A_\omega := A \backslash A(\omega)$. By $val^*(\mathbf{f}_{|G_\omega}) := \left(val^*(f^1_{|G_\omega}), val^*(f^2_{|G_\omega})\right)^\top$ we denote the vector of independent maximum flows in the interdicted network G_ω.

Remark 1 After fixing the interdiction strategy $\omega \in \Omega$, both maximum flows can be computed independently in G_ω with respect to u^1 and u^2.

We use the following order in $\mathbb{R}^2_+$, cf. [7]:

$$y^1 \leq y^2 \Leftrightarrow y^1_k \leq y^2_k \text{ for } k = 1, 2 \text{ and } y^1 \neq y^2$$

In (BMFNI), one asks for the feasible interdiction strategies ω that do not allow to improve the one maximum flow value without worsening the other.

Definition 1 A feasible interdiction strategy $\omega \in \Omega$ is called efficient, if there does not exist $\omega' \in \Omega$ such that

$$val^*(\mathbf{f}_{|G_{\omega'}}) \leq val^*(\mathbf{f}_{|G_\omega}).$$

In this case, we call $val^*(\mathbf{f}_{|G_\omega})$ a non-dominated solution.

3 Complexity Results

In this section, we consider the following decision version of (BMFNI): Given a directed graph $G = (V, A)$, an interdiction budget $B \in \mathbb{N}$, and a value $K = (K_1, K_2)^\top \in \mathbb{N}^2$, decide whether there exists an interdiction strategy $\omega \in \Omega$ with $val^*(\mathbf{f}_{|G_\omega}) \leq K$.

Theorem 1 *The decision version of (BMFNI) is $\mathcal{NP}$-complete, even for unit interdiction costs, i.e., $c(a) = 1$ for all $a \in A$ and even on series-parallel graphs.*

Proof The problem is clearly in $\mathcal{NP}$. To show $\mathcal{NP}$-completeness, we conduct a polynomial time reduction from the 0-1-knapsack problem, which is known to be $\mathcal{NP}$-complete, cf. [10]. The decision version of the 0-1-knapsack problem is as follows: Given a finite set $I = \{1, \ldots, n\}$ with profits $p_i \in \mathbb{N}$ and weights $w_i \in \mathbb{N}$ for all elements $i \in I$ and positive integers P and W, does there exist a solution $\mathbf{x} \in \{0, 1\}^n$ such that $\sum_{i=1}^n w_i \cdot x_i \leq W$ and $\sum_{i=1}^n p_i \cdot x_i \geq P$ with x_i being equal to 1 if item i is in the knapsack and 0 otherwise?

Given an instance of the 0-1-knapsack problem, we construct an instance of (BMFNI) as follows: Let $G = (V, A)$ with $V = \{s, t\}$ and $A = \{a_1, \ldots, a_{2n}\}$, where $a_i = (s, t)$ for $i = 1, \ldots, 2n$. Further, let two capacity functions $u^1 : A \to \mathbb{N}$ and $u^2 : A \to \mathbb{N}$ be defined as follows.

$$u^1(a_i) = \begin{cases} p_i, & \text{if } i = 1, \ldots, n \\ 0, & \text{if } i = n+1, \ldots, 2n \end{cases} \qquad u^2(a_i) = \begin{cases} U - w_i, & \text{if } i = 1, \ldots, n \\ U, & \text{if } i = n+1, \ldots, 2n, \end{cases}$$

where $U \geq \max_{i \in I}\{w_i\}$. Additionally, let $B = n$. Note that the following two statements are equivalent:

(1) Does there exist an interdiction strategy $\omega \in \Omega$ such that $val^*(\mathbf{f}_{|G_\omega}) \leq K'$?

(2) Does there exist an interdiction strategy $\omega \in \Omega$ such that $\sum_{i=1}^{2n} x_i \cdot \mathbf{u}(a_i) \geq K$, where $K_i' = val^*(f^i_{|G}) - K_i$, $i = 1, 2$ and x_i being equal to 1 if a_i is interdicted and 0 otherwise?

Suppose that ω is an interdiction strategy such that (2) holds. Then, we set $K_1 = P$ and $K_2 = nU - W$. It follows that

$$\sum_{i=1}^{2n} x_i p_i = \sum_{i=1}^{n} x_i p_i + \sum_{i=n+1}^{2n} x_i 0 \geq K_1 = P$$

$$nU - W = K_2 \leq \sum_{i=1}^{n} x_i(U - w_i) + \sum_{i=n+1}^{2n} x_i U = \ldots = nU - \sum_{i=1}^{n} x_i w_i$$

Consequently, it holds that $\sum_{i=1}^{n} x_i w_i \leq W$. This reduction can be done in polynomial time, which concludes the proof. □

Using the result shown in [5] we can conclude:

Theorem 2 *The problem (BMFNI) is intractable, i.e., the number of non-dominated solutions might be exponential in the number of vertices.* □

4 Solution Procedures

In this section, we propose two methods to solve (BMFNI) on two different classes of directed graphs. The first class of directed graphs was already informally introduced in the proof of Theorem 1. More specifically, we refer to graphs $G = (V, A)$ with $V = \{s, t\}$ and $A = \{a_1, \ldots, a_m\}$ with $a_i = (s, t)$ to TypeA-graphs. The second class of graphs under consideration are directed, series-parallel graphs. Due to [8], this class is defined as follows.

Definition 2 A directed graph $G = (V, A)$ is called series-parallel with source s and sink t, if G can be constructed by a sequence of the following operations.

(1) Construct $G' = (V', A')$ with $V' = (s, t)$ and $A' = \{(s, t)\}$
(2) (Parallel Composition) Given two directed, series-parallel graphs G_1 with source s_1 and sink t_1 and G_2 with source s_2 and sink t_2, form a new graph G by identifying $s = s_1 = s_2$ and $t = t_1 = t_2$.
(3) (Series Composition) Given two directed, series-parallel graphs G_1 with source s_1 and sink t_1 and G_2 with source s_2 and sink t_2, form a new graph G by identifying $s = s_1$, $t_1 = s_2$ and $t_2 = t$.

Note that a series-parallel graph G can be recognized in polynomial time along with the corresponding decomposition tree T_G. Further, the size of T_G is linear in the size of G and T_G can be computed in linear time, cf. [14].

It is easy to see that (BMFNI), when restricted to TypeA-graphs, can be solved by using a biobjective knapsack problem, where each element i of a set of elements I is associated with a weight $w_i \in \mathbb{N}$ and with a biobjective profit $(p_i^1, p_i^2)^\top$. One has to simply identify the set of items I with the set of arcs A. Then, let w_a be equal to the cost of interdicting arc $a \in A$, i.e., $w_a = c(a)$. Further, let the profit $(p_a^1, p_a^2)^\top$ be equal to the capacities of arc a. Although the biobjective knapsack problem is still hard to solve, there is a fully polynomial time approximation scheme (FPTAS), cf. [11, 13], yielding the following result.

Corollary 1 *There is a FPTAS for (BMFNI) constructing an ε-approximate Pareto-curve, when restricted to TypeA-graphs.*

Next, we propose a dynamic programming algorithm for (BMFNI) for the case of directed, series-parallel graphs. We assume a decomposition tree T_G for a given series-parallel graph to be given. For a graph H in T_G, we denote by $L(H, x)$ the set of non-dominated solutions in H with an interdiction budget of $x \in \mathbb{N}$. By s_H and t_H, we refer to the source and sink of H, respectively. Note that the size of $L(H, x)$ is bounded from above, i.e., $|L(H, x)| \leq (mU)^2$, where U denotes the maximal arc capacity. For the case of H being a primitive graph, i.e., a leaf of T_G, we can easily compute $L(H, x)$ for all $x \in \{0, 1, \ldots, B\}$, where $B \in \mathbb{N}$ denotes the given interdiction budget.

$$L(H, x) = \begin{cases} \{(u^1(s_H, t_H), u^2(s_H, t_H))^\top\}, & \text{if } x = 0, 1, \ldots, c(s_H, t_H) - 1 \\ \{(0, 0)^\top\}, & \text{if } x = c(s_H, t_H), \ldots, B \\ \{(u^1(s_H, t_H), u^2(s_H, t_H))^\top\}, & \text{if } c(s_H, t_H) > B \end{cases} \tag{1}$$

Now, let H be the parallel composition of H_1 and H_2. Then, $L(H, x)$ can be computed as follows:

$$L(H,x) = \text{" min "}\left\{\bigcup_{k=0}^{x} L(H_1,k) \oplus L(H_2, x-k)\right\} \text{ for all } x = 0, 1, \dots, B, \tag{2}$$

where $\oplus$ denotes the Minkowski sum and " min " denotes the Pareto dominance.

Let H be the series composition of H_1 and H_2. Then, $L(H, x)$ can be computed as follows:

$$L(H,x) = \text{" min "}\left\{\bigcup_{k=0}^{x} L(H_1,k) \odot L(H_2, x-k)\right\} \text{ for all } x = 0, 1, \dots, B, \tag{3}$$

where $x \odot y := \left(\min\{x^1, y^1\}, \min\{x^2, y^2\}\right)^\top$ for $x, y \in \mathbb{R}^2$ and $A \odot B := \{a \odot b \mid a \in A, b \in B\}$ for $A, B \subset \mathbb{R}^2$.

Theorem 3 *(BMFNI) can be solved on series-parallel graphs using dynamic programming by using formulas* (1), (2) *and* (3) *in pseudopolynomial time.*

5 Conclusion

We introduced the biobjective maximum flow network interdiction problem, called (BMFNI). We showed that (BMFNI) is $\mathcal{NP}$-complete and intractable. Further, we have seen that (BMFNI) on TypeA-graphs can be approximated within arbitrary precision due to its equivalence to the biobjective knapsack problem. We proposed a procedure to solve (BMFNI) on series-parallel graphs with dynamic programming over the decomposition tree.

Acknowledgments This work was partially supported by the Bundesministerium für Bildung und Forschung (BMBF) under Grant No. 13N14561.

References

1. Ahuja, R., Magnanti, T., Orlin, J.: Network Flows: Theory, Algorithms, and Applications. Prentice-Hall, Upper Saddle River (1993)
2. Assimakopoulos, N. (1987). A network interdiction model for hospital infection control. Comput. Biol. Med. **17**(6), 413–422

3. Boykov, Y., Kolmogorov, V.: An experimental comparison of min-cut/max-flow algorithms for energy minimization in vision. IEEE Trans. Pattern Anal. Mach. Intell. **26**(9), 1124–1137 (2004)
4. Burch, C., Carr, R., Krumke, S., Marathe, M., Phillips, C., Sundberg, E.: A decomposition-based pseudoapproximation algorithm for network flow inhibition. In: Network Interdiction and Stochastic Integer Programming, pp. 51–68. Springer, Berlin (2003)
5. Carstensen, P.J.: Complexity of some parametric integer and network programming problems. Math. Program. **26**(1), 64–75 (1983)
6. Chestnut, S.R., Zenklusen, R.: Hardness and approximation for network flow interdiction. Networks **69**(4), 378–387 (2017)
7. Ehrgott, M.: *Multicriteria Optimization*, 2nd edn. Springer, Berlin (2005)
8. Eppstein, D.: Parallel recognition of series-parallel graphs. Inf. Comput. **98**(1), 41–55 (1992)
9. Ford, L.R., Fulkerson, D.R.: Maximal flow through a network. Can. J. Math. **8**(3), 399–404 (1956)
10. Garey, M.R., Johnson, D.S.: Computers and Intractability, vol. 29. wh freeman, New York (2002)
11. Papadimitriou, C.H., Yannakakis, M.: On the approximability of trade-offs and optimal access of web sources. In: Proceedings 41st Annual Symposium on Foundations of Computer Science, pp. 86–92. IEEE, New York (2000)
12. Phillips, C.A.: The network inhibition problem. In: Proceedings of the Twenty-Fifth Annual ACM Symposium on Theory of Computing, pp. 776–785. ACM, New York (1993)
13. Safer, H.M., et al.: Fast approximation schemes for multi-criteria flow, knapsack, and scheduling problems. Technical report, Massachusetts Institute of Technology (MIT), New York. Sloan School of Management
14. Valdes, J., Tarjan, R.E., Lawler, E.L.: The recognition of series parallel digraphs. SIAM J. Comput. **11**(2), 298–313 (1982)
15. Wood, R.K.: Deterministic network interdiction. Math. Comput. Model. **17**(2), 1–18 (1993)

Assessment of Energy and Emission Reduction Measures in Container Terminals using PROMETHEE for Portfolio Selection

Erik Pohl, Christina Scharpenberg, and Jutta Geldermann

Abstract In this paper we present an approach to assess energy and emission reduction measures in container terminals. A modified PROMETHEE V approach is used to select suitable portfolios of these measures based on the decision makers preferences and further inter-dependencies.

Keywords Energy policy and planning · MCDA · PROMETHEE · Container terminal

1 Introduction

Environmental protection and energy efficiency have become major topics in seaport management, especially for container terminals. A reduction of emissions in container terminals can be achieved inter alia by using electrically-powered container handling equipment and renewable energy which, however, causes significant investments. Thus, the project "Simulation-based evaluation of measures for the improvement of energy sustainability in port operations" (SuStEnergyPort) [5], carried out by the Hamburg Port Consulting GmbH (HPC) and the University Duisburg-Essen (UDE) develops a structured, model-based methodology to identify suitable measures to improve the energy efficiency and utilization of renewable energy. The goal of the project is to suggest an optimal strategy consisting of several emission abatement measures as decision support for port operators and port authorities. A selection of promising measures for the abatement of CO_2 emissions is implemented in a simulation tool covering both logistic and energetic aspects. As some measures include using new container handling equipment, e.g. straddle carriers, a Life Cycle Assessment (LCA) is conducted to measure and compare

E. Pohl (✉) · C. Scharpenberg · J. Geldermann
Universität Duisburg-Essen, Chair of Business Administration and Production Management, Duisburg, Germany
e-mail: erik.pohl@uni-due.de

J. S. Neufeld et al. (eds.), *Operations Research Proceedings 2019*, Operations Research Proceedings, https://doi.org/10.1007/978-3-030-48439-2_17

the environmental impact [6]. The challenge, however, is a multicriteria portfolio selection problem of a bundle of emission abatement measures.

In the remainder of this paper, the assessment of energy consumption and emission reduction measures applicable to container terminals is described using PROMETHEE in Chap. 2. Chapter 3 briefly presents the outranking method PROMETHEE, before in Chap. 4 further developments and modifications of PROMETHEE V are introduced. To construct a portfolio of measures, we use a knapsack model with further constraints regarding dependencies among measures and the decision makers' preferences. This has been recently conducted using the ELECTRE method in [1]. In our case, the model is based on the PROMETHEE V method [2]. Based on this assessment, a road map towards a more profitable and ecological port operation can be developed.

2 Assessment of Emission Abatement Measures

Based on expert interviews with port authorities and port terminal operators, fifteen promising energy and emission reduction measures in container terminals are identified for further investigation within the research project. Thirteen relevant sustainability assessment-criteria within the ecological, economic, social and technical categories are elicited and organized as a criteria hierarchy, see Table 1. To set the weights, we conducted an expert workshop, where we applied SIMOS weighing method [8]. The energy and emission reduction measures contain, besides others, changes in operational processes (e.g. speed limitation), technical specifications of port operating equipment (e.g. start-stop routines of equipment) and alternative drive trains (e.g. hybrid and full-electric drive trains). In order to consider the current state of the terminal and to be able to evaluate the individual measures, the potential improvements with regard to each criterion are determined for each measure. E.g. for the measure *speed limitation* we calculate the global warming potential (GWP) of the terminal operation with and without implementing the measure. The performance of the measure is then calculated as the difference of both values, i.e. the GWP saving potential of this measure. To assess the measure according to the presented criteria, a simulation tool developed by HPC is used. A life cycle assessment is performed to determine the ecological impact of e.g. new equipment [9]. We further assume that equipment is only replaced at the end of its life time. Thus, at the end of the life time a port operator has to decide whether the status quo should be kept or replaced by an alternative energy or emission reduction measure. Therefore, we can compare the ecological, economic, technical and social impacts of a measure with those of the current situation.

Table 1 Criteria hierarchy for assessing the reduction measures

Category	Criteria	Unit	Min\Max	Weight [%]
Ecological	Energy saving potential	kWh/a	Max	13
	Global warming potential	kg CO_2e/a	Min	12
	Particulate matter	kg PM10e/ a	Min	3
	Acidification	kg SO_2e/a	Min	4
	Photochemical smog	kg NMVOCe/a	Min	2
Economic	Capital expenditure	Euro	Min	16
	Net Present value	Euro	Max	16
	Amortization	a	Min	9
	Operating costs	Euro/a	Max	8
Technical	Implementation effort	Points	Min	6
	Waterside productivity	%	Max	6
Social	Employee acceptance	Points	Max	7
	Image promotion	Points	Max	11

3 The PROMETHEE Method

The PROMETHEE method [3] like other outranking and multi-criteria methods has been applied in many environmental case studies [4]. It is applied to the performance values of a set of discrete alternatives $a \in A$ on a finite number of criteria $g \in G$. The performance matrix is given as the matrix of evaluation $g(a)$, $g \in G, a \in A$. PROMETHEE I ranks the alternatives according to a partial ranking based on two preference flows. The positive outranking flow of an alternative a is given by

$$\Phi^+(a) = \frac{1}{n-1} \sum_{b \in A} \Pi(a, b)$$

and the negative outranking flow by

$$\Phi^-(a) = \frac{1}{n-1} \sum_{b \in A} \Pi(b, a),$$

where n is the number of alternatives $n = |A|$ and Π is the aggregated preference index defined for each pair of actions [3]. This preference index is given by

$$\Pi(a, b) = \frac{\sum_{g=1}^{m} w_g P_g(a, b)}{\sum_{g=1}^{m} w_g},$$

where $P_g(a, b)$ represents the preference of alternative a over b for a given preference function of criterion $g \in G$. The preference function is a non-decreasing function and the shape is chosen according to the preference information given by the decision maker. In [2], six different functions are presented to model the preference. While the PROMETHEE I method gives a partial ranking, the net outranking flow is obtained using PROMETHEE II which yields a total ranking of the alternatives. The net outranking flow is given by

$$\Phi(a) = \Phi^{+}(a) - \Phi^{-}(a)$$

The net flow Φ lies in the interval $[-1, 1]$. For the following extension, we consider the net score $\Phi^{'}$ defined as

$$\Phi^{'}(a) = \frac{\Phi^{+}(a) + (1 - \Phi^{-}(a))}{2}.$$

The net score $\Phi^{'}$ lies in the interval [0, 1] [7]. The PROMETHEE V method is an extension of the PROMETHEE I and II method and identifies a subset of alternatives given a set of constraints e.g. number of measures, budget, investment. The goal is to select alternatives in such a way that the net outranking flow is maximized while all constraints are satisfied. This is modelled as a binary knapsack problem [2].

4 Building Portfolios

The PROMETHEE V method considers alternatives with a positive net outranking flow for selection while alternatives with a negative net outranking flow are therefore ignored. With regard to the project we wish to construct a portfolio of measures which may consist of every measure examined. While [7] argues that the PROMETHEE score is easier for decision makers to understand, we suggest to use it to associate each alternative with a positive rating for the selection step in PROMETHEE V. Additionally, while selecting measures, we have to consider dependencies and logical constraints between some measures as well as the decision makers' preferences. This can be modelled similarly to the PROMETHEE V selection step as a binary knapsack problem with additional requirements concerning logical constraints and further preferences of the decision maker. The goal is to select as many measures as possible with the highest score value. For this, we can associate a binary decision variable x_a, $a \in A$, such that $x_a = 1$ if measure $a \in A$ is selected and $x_a = 0$ otherwise. Then, the objective function to select measures with the highest total amount of PROMETHEE score is given by

$$\max \sum_{a \in A} \Phi^{'}(a) x_a$$

For the identification of suitable portfolios of energy and emission reduction measures in container terminals, a set of resources, constraints as well as further requirements and goals of the decision makers have to be taken into account. The most important constraint is the one related to the capital expenditure to implement the measures. Each measure is associated with an expenditure i_a. With the maximal available budget denoted as B, this constraint can be formulated as

$$\sum_{a \in A} x_a i_a \leq B. \tag{1}$$

There are two types of dependencies among the selected measure. First, the realization of one measure may require another measure to be implemented first. E.g. an adaptive lighting system for the terminal only works with LED lighting technology. The adaptive lighting system therefore requires the terminal operator to install LED lighting. This can be modeled as

$$x_b \geq x_a \tag{2}$$

where measure a depends on measure $b \in A$. Secondly, there are measures that are mutually exclusive, e.g. using hybrid VCs excludes using full electric VCs and vice versa. This can be modelled as

$$x_a + x_b \leq 1 \tag{3}$$

where a and b are the measures in question. If a decision maker wants to select measures to reduce the energy consumption during the next year by at least a given amount E, this can be modeled with the following constraint

$$\sum_{a \in A} e_a x_a \geq E \tag{4}$$

where e_a is the energy reduction potential of measure a. Since the terminal is in continuous operation, only a limited number of measures can be implemented simultaneously. The decision maker can restrict the number of measures which are selected. The following constraint models this.

$$\sum_{a \in A} x_a \leq N \tag{5}$$

where N is the number of allowed measures. Furthermore, the measures can be restricted by a total amount of implementation effort analogously if required. The resulting single-objective, binary linear programming model can be solved with any common optimization solver.

Table 2 shows the decision table with an abridged list of abatement measures. Due to trade secrets, the performance values are given in percentage of the highest value in each row. The presented net flow and Phi score correspond to the actual project results. To build the model, we proceed as follows. The objective function can stated with the presented Phi scores in Table 2. The capital expenditure i_a

Table 2 Abridged decision table

Criteria/Measures	Solar photovoltaics	Hybrid propulsion VC	...
Energy saving potential	0.35%	23.96%	...
Global warming potential	4.24%	30.50%	...
Particulate matter	1.06%	19.71%	...
Acidification	1.54%	16.88%	...
Photochemical smog	1.21%	18.27%	...
Amortization	4.37%	2.78%	...
Net present value	93.05%	100.00%	...
Capital expenditure	8.30%	11.20%	...
Operating costs	91.99%	100.00%	...
...	...	...	...
Φ net flow	−0.0417	0.4068	...
Φ score	0.47915	0.7034	...

in Eq. (1) and the energy reduction e_a in Eq. (4) are also stated in Table 2. The total amount of capital B, the required energy consumption reduction E as well as the number of measures N must be determined together with the decision maker. Solving the model will then yield an optimal set of emission abatement measures which can be presented to the decision maker. Since the presented model is a well-investigated bin-packing problem with a few items, it can be solved in less than a second by any modern Solver.

5 Conclusion

In this paper we presented an approach to assess energy and emission reduction measures in container terminals and how we use a modified PROMETHEE V approach using the PROMETHEE score allows to select optimal portfolios based on the decision makers preferences and logical constraints. In the project we consider measures that can be easily implemented and measures that need substantial capital expenditure and changes the operation. With the new presented knapsack formulation we can determine suitable portfolios for both, the terminal and the terminal operator and point out a road map towards a more sustainable port operation.

References

1. Barbati, M., Figueira, J.R., Greco, S., Ishizaka, A., Panaro, S.: A Multiple Criteria Methodology for Prioritizing and Selecting Portfolios of Urban Projects (13.12.2018). http://arxiv.org/pdf/1812.10410v2

2. Brans, J.P., Mareschal, B.: Promethee methods. In: Multiple Criteria Decision Analysis: State of the Art Surveys, pp. 163–186. Springer, Berlin (2005)
3. Brans, J.P., Vincke, P.: Note—a preference ranking organisation method: (the PROMETHEE method for multiple criteria decision-making). Manag. Sci. **31**(6), 647–656 (1985)
4. Greco, S., Ehrgott, M., Figueira, J. (eds.): Multiple Criteria Decision Analysis: State of the Art Surveys. International Series in Operations Research and Management Science, vol. 233, 2nd edn. Springer, New York (2016)
5. IHATEC—Innovative Hafentechnologien: Sustenergyport—simulationsbasierte bewertung von maßnahmen zur steigerung der energienachhaltigkeit im hafenbetrieb (2017). https://www.innovativehafentechnologien.de/wp-content/uploads/2017/08/Projektsteckbrief_SustEnergyPort.pdf
6. Klöpffer, W., Grahl, B.: Life cycle assessment (LCA): a guide to best practice. Wiley, New York (2014)
7. Mladineo, M., Veza, I., Gjeldum, N.: Solving partner selection problem in cyber-physical production networks using the humant algorithm. Int. J. Prod. Res. **55**(9), 2506–2521 (2017)
8. Pictet, J., Bollinger, D.: The silent negotiation or how to elicit collective information for group MCDA without excessive discussion. J. Multi-Criteria Decis. Anal. **13**(5–6), 199–211 (2005)
9. Scharpenberg, C., Pohl, E., Lauven, L.P., Geldermann, J.: Ecological assessment of port equipment for container terminals. In: Hamburg International Conference of Logistics (HICL) 2018, pp. 3–20. epubli (2018)

Decision-Making for Projects Realization/Support: Approach Based on Stochastic Dominance Rules Versus Multi-Actor Multi-Criteria Analysis

Dorota Górecka

Abstract The selection of projects for realization or co-financing is a complex process, often strategic in nature, which involves confronting trade-offs between multiple, frequently conflicting, factors. In this process the expectations of various stakeholders should be taken into account, while simultaneously ensuring the achievement of numerous objectives. This work presents and compares the application of two approaches—multi-criteria decision aiding (MCDA) methods for mixed evaluations (deterministic and stochastic ones, thus with stochastic dominance (SD) rules), e.g. BIPOLAR MIX, and multi-actor multi-criteria analysis (MAMCA)—to help decision-makers (DMs) in the project selection process. The problem is illustrated by ranking environmental infrastructure projects.

Keywords MCDA · MAMCA · BIPOLAR MIX

1 Introduction

The choice of projects for realization or financial support is a complicated, usually a group process, often strategic in nature, which involves confronting trade-offs between multiple, frequently opposing, factors. In this process the expectations and requirements of various stakeholders should be taken into consideration, while simultaneously ensuring the achievement of numerous objectives: economic, financial, technical, technological, political, legal, socio-cultural and ecological, as well as objectives connected with public relations, health, and safety. Thus, it is necessary to use structured, explicit approaches and carry out a thorough analysis of the advantages and disadvantages of individual projects from the point of view of many criteria, taking into account different, or even contradictory, priorities of decision-makers (DMs) as well as facing mixed evaluations, which often occur in

D. Górecka (✉)
Nicolaus Copernicus University in Toruń, Toruń, Poland
e-mail: dgorecka@umk.pl

J. S. Neufeld et al. (eds.), *Operations Research Proceedings 2019*, Operations Research Proceedings, https://doi.org/10.1007/978-3-030-48439-2_18

real-world applications, in order to make a decent, informed decision: to select good projects that will be implemented, and to reject bad projects—infeasible, impractical, unprofitable, unhelpful, unattractive.

The aim of this work is to propose a two-track group decision-making procedure for projects choice, consisting of two approaches, namely: multi-criteria decision aiding (MCDA) methods for mixed measurements of the alternatives' performance (deterministic and stochastic, hence with stochastic dominance (SD) rules), such as BIPOLAR MIX [1, 2], and multi-actor multi-criteria analysis (MAMCA) [3, 4]. The case study presents the real-life problem of ordering environmental infrastructure projects.

2 The Proposed Procedure of Appraising and Selecting Projects

Since selecting projects for implementation or funding is usually a group decision-making process, two different approaches can be used to solve it, namely: (1) MCDA methods for mixed data (based on SD rules) and (2) MAMCA [3, 4]. While within the first approach all stakeholders' (experts') opinions are analysed together, the second one assumes that the analysis is conducted for each stakeholder (expert) separately. Besides, within the second approach each stakeholder (expert) can have a different preference model, which is impossible within the first approach. Thus, in competitive procedures we recommend using the first approach, while in non-competitive—the second one (Fig. 1).

MCDA techniques accepting different types of the performances of the alternatives (e.g. ordinal and cardinal as well as deterministic, stochastic and/or fuzzy ones) are quite rare and not very well known [1]. The motivation for using the BIPOLAR

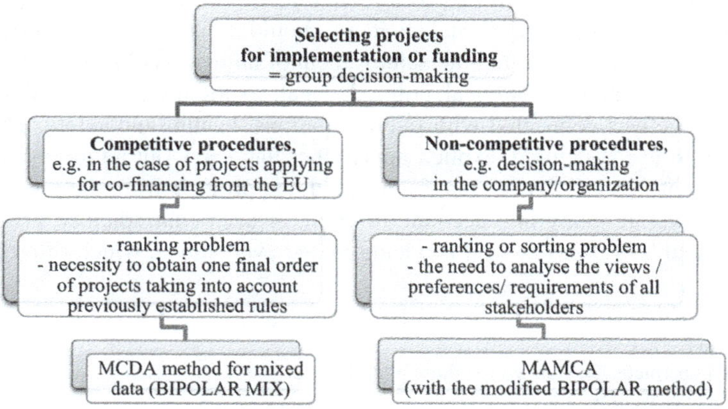

Fig. 1 The proposed procedure of appraising and selecting projects for realization/support

MIX method [1, 2] stems from the characteristics of the analysed decision-making problem and the general expectations of the contestants, which are as follows [5]:

- the decision-making problem should be formulated as a problem of ordering a finite number of alternatives—each alternative (project) should be classified in the ranking and obtain a numerical overall score;
- there is no place for the incomparability of the alternatives—ranking obtained should be complete, since otherwise it may be unconvincing for the applicants (contestants);
- there should be a possibility to employ both quantitative and qualitative criteria and to use mixed data (deterministic and stochastic information);
- decision-makers are able to present the information about their preferences, but they do not have much time for the interaction and cooperation with the analyst;
- the possibility of complete compensation occurrence should be removed—in the case of some criteria it may be risky and in the case of others projects should fulfil minimum quality standards;
- the occurrence of ties in the ranking should be limited;
- it is desired that the method allows us to determine whether the highly ranked projects are really good or just better than the weak ones.

Compared with other MCDA methods, BIPOLAR MIX responds to all these requirements. It is based on the classical BIPOLAR method proposed by Konarzewska-Gubała [6] and on its alteration, namely the modified BIPOLAR method with stochastic dominance (SD) rules, proposed by Górecka [7]. It uses a homogenous preference model and SD rules. It allows us: to obtain a complete order of the alternatives; to use mixed data; to determine whether highly ranked alternatives are really good or just not bad—it allows for ranking and sorting alternatives as well as for determining their quality, considering what is good and undesirable from the DM's point of view in the decision-making problem (reference system); to eliminate both the phenomenon of full compensation and the problem of the incomparability of the alternatives [1].

As far as non-competitive procedure is concerned, we suggest using MAMCA [3, 4]. Its steps are presented in Fig. 2 and briefly summarized below [4]:

1. Defining the problem and the alternatives.
2. Conducting stakeholder analysis.
3. Defining criteria and weights.

The choice of criteria is based on the stakeholders' objectives. The weights are determined by the importance the stakeholder is ascribing to his or her objectives.

4. Determining indicators and measurement methods for criteria.
5. Conducting overall analysis and ranking.

Every alternative is assessed against the different criteria for each stakeholder. Any MCDA method can be used to choose, order or sort the alternatives considered. Stakeholders are free to have their own criteria, weights and preference structure, and only at the end of the analysis their points of view are confronted.

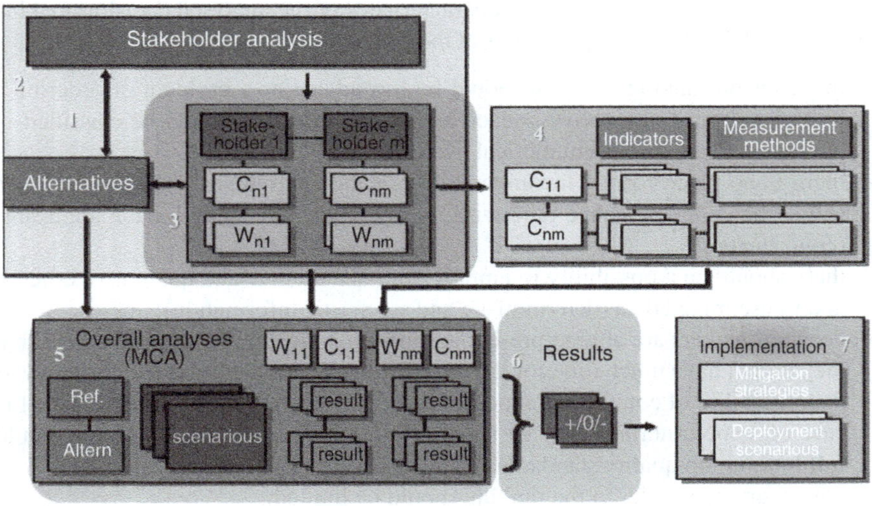

Fig. 2 The concept of the Multi-Actor Multi-Criteria Analysis (MAMCA) (Source: [3])

6. Obtaining results and conducting sensitivity analysis.

MAMCA aims at providing insight into what is significant for each stakeholder, not just at summing up different points of view and coming to a final conclusion.

7. Implementation.

In order to be consistent with the recommended approach, we suggest employing the modified BIPOLAR method [6, 7] within the MAMCA methodology [3, 4].

3 Case Study

In the analysis seven environmental infrastructure projects (construction and modernization of wastewater and rainwater collection networks and wastewater treatment plants) applying for co-financing from the European Union were considered. They were evaluated using 12 criteria: 2 deterministic ones (total cost—the only one minimized, and the number of people using project) and 10 stochastic ones. Regarding the latter, five experts – specialists in the field of environmental protection infrastructure – scored them from 0 (the lowest evaluation) to 10 (the highest evaluation). The model of preferences for the problem is presented in Table 1 while Table 2 provides the performance matrix for seven projects considered in the case study and four reference ones (two 'good' and two 'bad'). The results obtained by applying the BIPOLAR MIX method are given in Table 3.

Table 1 Model of preferences

f_k	Criterion	w_k	q_k	p_k	v_k
f_1	Total cost [PLN million]	0.12	1	3	30
f_2	People using project [number]	0.05	100	600	600
f_3	Efficiency	0.14	1	3	3
f_4	Influence on the environment	0.15	2	4	3
f_5	Influence on the employment	0.05	3	4	2
f_6	Influence on the inhabitants' health	0.14	3	5	2
f_7	Influence on the investment attractiveness	0.07	2	4	2
f_8	Influence on the tourist attractiveness	0.06	2	5	2
f_9	Validity of the technical solutions	0.08	1	3	2
f_{10}	Sustainability and institutional feasibility of the project	0.06	1	3	2
f_{11}	Complementarity with other projects	0.04	2	4	2
f_{12}	Comprehensiveness	0.04	2	4	2

Table 2 Performance matrix

f_k	a_1	a_2	a_3	a_4	a_5	a_6	a_7	d_1	d_2	z_1	z_2
	Performance $f_k(a_i)$							**$f_k(r_j)$**			
f_1	8.42	31.55	9.24	9.25	5.93	20.0	26.01	5.0	3.0	20.0	30.0
f_2	2000	4784	9128	1550	582	3900	2782	4500	6000	750	1000
	Average performance (expected value) $\mu_k(a_i)$							**$\mu_k(r_j)$**			
f_3	5.6	7.2	7.0	7.6	5.8	6.8	4.6	9.0	7.2	4.0	4.6
f_4	7.2	9.2	8.4	8.8	8.4	6.6	7.6	9.0	7.4	5.0	3.8
f_5	4.4	7.8	3.8	7.2	7.4	8.4	6.2	7.0	6.6	4.0	5.6
f_6	4.8	5.8	5.0	5.6	5.8	6.0	6.0	7.0	7.8	6.0	4.8
f_7	4.6	7.8	5.8	7.6	7.2	7.6	6.6	7.0	7.6	6.0	5.4
f_8	4.6	9.0	6.2	8.4	4.6	6.8	6.6	7.0	7.4	4.0	4.4
f_9	7.8	8.4	7.6	8.4	8.2	6.6	8.2	7.0	7.2	5.0	4.2
f_{10}	7.4	8.4	6.4	7.4	8.6	9.0	8.0	7.0	8.0	5.0	4.0
f_{11}	7.0	8.6	3.4	3.8	7.4	7.0	7.4	8.0	6.4	4.0	5.0
f_{12}	4.8	9.0	4.0	4.6	5.6	5.4	5.8	8.0	6.8	4.0	5.2

Table 3 Rankings of the projects obtained using the BIPOLAR MIX method

No.	Monorankings				Bipolar-ranking	
	a_i	Success indices d_{iS}	a_i	Anti-failure indices d_{iN}	a_i	Final scores d_{iSN}
1	a_2	−0.142	a_4	0.634	a_4	0.210
2	a_3	−0.177	a_3	0.518	a_3	0.170
3	a_5	−0.195	a_6	0.396	a_6	0.089
4	a_4	−0.214	a_1	0.367	a_1	0.029
5	a_6	−0.218	a_7	0.306	a_7	0.006
6	a_7	−0.294	a_2	0.000	a_2	−0.071
7	a_1	−0.310	a_5	0.000	a_5	−0.097

Figures 3 and 4 show the results received using the MAMCA model with the modified BIPOLAR method. Values in respective columns of Fig. 4 present the final scores for each project obtained for each expert. The higher the value is, the better is the performance of the project from the point of view of a given expert. MAMCA overall scores were established as the weighted sum of the individual scores. The results are also visualized on Fig. 3, using the multi-actor line chart representing particular projects (in general, the higher the line is, the more preferred the project is by a given expert).

In trying to find a satisfying solution from the point of view of all experts using the MAMCA model, we have come to the same conclusion as using the BIPOLAR MIX method, namely that the best project is a_4 (it takes first, second, first, fifth, and second place in subsequent expert rankings). However, its overall performance is only a little bit higher than the overall performance of project a_6 (which, in turn, wins seventh, first, second, first, and fourth place), occupying third spot according to the BIPOLAR MIX method.

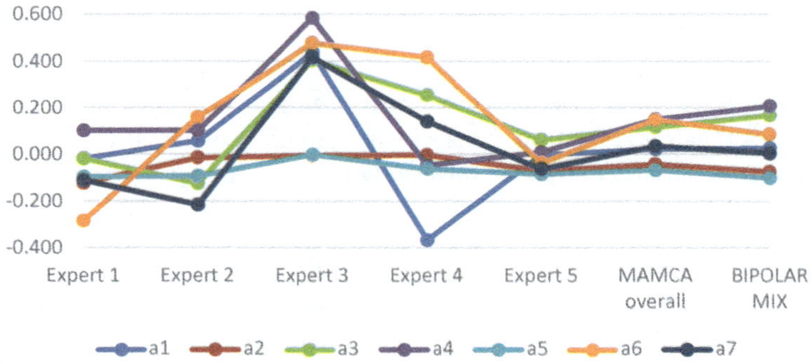

Fig. 3 Results of the MAMCA approach with the modified BIPOLAR method (different preference models—weights and thresholds—for each expert)

a_i	Expert 1	Expert 2	Expert 3	Expert 4	Expert 5	MAMCA overall Experts 1-5	BIPOLAR MIX Experts 1-5
a_1	-0.014	0.060	0.440	-0.364	0.000	0.025	0.029
a_2	-0.118	-0.012	-0.002	0.000	-0.066	-0.040	-0.071
a_3	-0.015	-0.120	0.408	0.259	0.065	0.119	0.170
a_4	0.104	0.104	0.586	-0.044	0.011	0.152	0.210
a_5	-0.093	-0.089	0.000	-0.059	-0.082	-0.065	-0.097
a_6	-0.280	0.164	0.481	0.422	-0.033	0.151	0.089
a_7	-0.109	-0.213	0.418	0.142	-0.061	0.036	0.006

Fig. 4 Results of the MAMCA approach and the BIPOLAR MIX method

4 Conclusions

In the paper we have compared the application of two approaches that can be used in the decision-making process regarding the challenging problem of project selection, namely: (1) MCDA methods for mixed data (BIPOLAR MIX, employing SD rules) and (2) MAMCA. While within the first approach a single hierarchical structure is built and consequently we receive one common ranking, then within the second approach different hierarchies can be used for each separate stakeholder (expert) and results take the form of different rankings. The first approach is appropriate when we need to get directly a global perspective (in competitive procedures), whereas MAMCA is suitable if we want to obtain a higher acceptance rate of the proposed solution, which might be more advantageous for the majority of stakeholders (in non-competitive procedures).

References

1. Górecka, D.: BIPOLAR MIX – a method for mixed evaluations and its application to the ranking of European projects. Mult. Crit. Decis. Making. **12**, 36–48 (2017)
2. Górecka, D.: Using Bipolar Mix in the process of selecting projects applying for co-financing from the European Union. In: Zadnik Stirn, L., Kljajić Borštnar, M., Žerovnik, J. Drobne, S. (eds.) Proceedings of the 14th International Symposium on Operational Research SOR'17, pp. 174–179. Slovenian Society Informatika, Section for Operational Research, Ljubljana (2017)
3. Macharis, C.: The importance of stakeholder analysis in freight transport: the MAMCA methodology. Eur. Transp./Transport. Europei. **25–26**, 114–126 (2004)
4. Macharis, C., Turcksin, L., Lebeau, K.: Multi actor multi criteria analysis (MAMCA) as a tool to support sustainable decisions: State of use. Decis. Support. Syst. **54**, 610–620 (2012)
5. Górecka, D.: On the choice of method in multi-criteria decision aiding process concerning European projects. In: Trzaskalik, T., Wachowicz, T. (eds.) Multiple Criteria Decision Making'10-11, pp. 81–103. Publisher of The University of Economics in Katowice, Katowice (2011)
6. Konarzewska-Gubała, E.: Wspomaganie decyzji wielokryterialnych: system BIPOLAR. Wydawnictwo Uczelniane Akademii Ekonomicznej we Wrocławiu, Wrocław (1991)
7. Górecka, D.: Wielokryterialne wspomaganie wyboru projektów europejskich. TNOiK „Dom Organizatora", Toruń (2009)

Part VI
Discrete and Integer Optimization

A Stochastic Bin Packing Approach for Server Consolidation with Conflicts

John Martinovic, Markus Hähnel, Waltenegus Dargie, and Guntram Scheithauer

Abstract The energy consumption of large-scale data centers or server clusters is expected to grow significantly in the next couple of years contributing to up to 13% of the worldwide energy demand in 2030. As the involved processing units require a disproportional amount of energy when they are idle, underutilized or overloaded, balancing the supply of and the demand for computing resources is a key issue to obtain energy-efficient server consolidations. Whereas traditional concepts mostly consider deterministic predictions of the future workloads or only aim at finding approximate solutions, here we propose an exact bin packing based approach to tackle the problem of assigning jobs with (not necessarily independent) stochastic characteristics to a minimal amount of servers subject to further practical constraints. Finally, this new approach is tested against real-world instances obtained from a Google data center.

Keywords Cutting and packing · Server consolidation · Bin packing problem · Normal distribution · HAEC

1 Introduction and Problem Statement

Nowadays, data centers are representing one of the most significant elements in the next stage of growth for the information and communication technology industry [5]. According to a forecast made by Cisco Systems [4], the global data center IP traffic is expected to grow threefold between 2016 and 2021, leading to a compound annual growth rate of 25% [4, Fig. 4]. Naturally, to cope with this huge traffic, a

J. Martinovic (✉) · G. Scheithauer
Institute of Numerical Mathematics, Technische Universität Dresden, Dresden, Germany
e-mail: john.martinovic@tu-dresden.de; guntram.scheithauer@tu-dresden.de

M. Hähnel · W. Dargie
Chair for Computer Networks, Technische Universität Dresden, Dresden, Germany
e-mail: markus.haehnel1@tu-dresden.de; waltenegus.dargie@tu-dresden.de

J. S. Neufeld et al. (eds.), *Operations Research Proceedings 2019*, Operations Research Proceedings, https://doi.org/10.1007/978-3-030-48439-2_19

very large number of processing and storage servers is required in the data centers. Already today these computational units inevitably consume a significant amount of energy, which is going to increase exponentially over the next couple of years [1, Fig. 4]. Independent studies revealed that existing servers are typically not used optimally for fear of providing too low availability at peak times [6]. Consequently, efficient server consolidation strategies and accurate predictions of future workloads are key elements to balance the demand for and the supply of computing resources. Whereas traditional strategies tend to allocate services based on deterministic estimates, see [15] and references therein, recent measurements and studies suggest that a considerable amount of data center workload for different applications is highly volatile [3]. Hence, characterizing the services in a probabilistic way turned out to be a more promising approach [9, 15, 16]. To this end, we consider the following general scenario: Given a fixed number n of jobs (services, tasks, etc.) whose resource demands are described as a stochastic process $\boldsymbol{X} : \Omega \times T \rightarrow \mathbb{R}^n$, where Ω is the sample space and $T := [0, \tau]$, $\tau > 0$, is a bounded time horizon. We aim at computing the lowest number of servers (processors, cores, etc.) of capacity $C > 0$ that is able to accomodate all jobs, so that overloading a single server is allowed (in a probabilistic sense) up to a maximal tolerable limit of $\varepsilon > 0$ at any instant of time $t \in T$. This basic problem can be referred to as a *stochastic bin packing problem (SBPP)*, where the items have nondeterministic lengths, while the bin capacity is fixed to some constant. In recent years, server consolidation or load balancing have already partly been addressed in connection with the SBPP. However, these approaches either assume stochastic independence [15, 16] or specific distributions (not being applicable to our real-world data) of the workloads [8, 11], or replace the random variables by deterministic *effective item sizes* [11, 15]. Moreover, the workloads are not treated as stochastic processes, and only heuristics (instead of exact formulations) are presented. For a large class of distributions, the first exact mathematical approaches for server consolidation have been introduced in [13] and were shown to achieve the best trade-off between resource consumption and performance when compared to other relevant consolidation strategies [9]. Here, this very promising approach is extended to (1) also treat stochastically dependent jobs, (2) use the concept of *overlap coefficients* to separate conflicting jobs, and (3) consider real data from a Google data center [14]; all in all leading to a *stochastic bin packing problem with conflicts (SBPP-C)*. Due to the limited space available, the full details can be found in the supplementary material [12].

2 Notation and Preliminaries

Let us consider $n \in \mathbb{N}$ *jobs*, indexed by $i \in I := \{1, \ldots, n\}$, whose *workloads* can be described by a stochastic process $\boldsymbol{X} : \Omega \times T \rightarrow \mathbb{R}^n$. Moreover, we assume $\boldsymbol{X}_t \sim \mathcal{N}_n(\boldsymbol{\mu}, \Sigma)$ for all $t \in T$, where $\boldsymbol{\mu} := (\mu_i)_{i \in I}$ and $\Sigma := (\sigma_{ij})_{i,j \in I}$ are a known mean vector and a known positive semi-definite, symmetric covariance matrix, respectively, of an n-dimensional multivariate normal distribution $\mathcal{N}_n$.

Hence, any individual workload $(X_t)_i$, $i \in I$, $t \in T$, follows the one-dimensional normal distribution $(X_t)_i \sim \mathcal{N}(\mu_i, \sigma_{ii})$. Considering normally distributed jobs is a common approach [15] or reasonable approximation, see [13, Remark 3] or [16, Fig. 4], and also warrantable for our real-world data [12, Fig. 4].

These jobs shall once be assigned to a minimal number of *servers* (or *machines, processors, cores*) with given capacity $C > 0$, i.e., it is not allowed to reschedule the jobs at any subsequent instant of time. Similar to the ordinary BPP [7], we use incidence vectors $\boldsymbol{a} \in \mathbb{B}^n$ to display possible item combinations.

Definition 1 Any vector $\boldsymbol{a} \in \mathbb{B}^n$ satisfying the

(A) (stochastic) capacity constraint: For a given threshold $\varepsilon > 0$, we demand $\mathbb{P}[\boldsymbol{X}_t^\top \boldsymbol{a} > C] \leq \varepsilon$ for all $t \in T$ to limit the probability of overloading the bin.
(B) non-conflict constraint: Let $F \subset I \times I$ describe a set of forbidden item combinations (to be specified later), then $a_i + a_j \leq 1$ has to hold for all $(i, j) \in F$.

is called (feasible) pattern/consolidation. The set of all patterns is denoted by P.

To state a more convenient and computationally favorable description of F, observe that we have $\boldsymbol{X}_t^\top \boldsymbol{a} \sim \mathcal{N}(\boldsymbol{\mu}^\top \boldsymbol{a}, \boldsymbol{a}^\top \Sigma \boldsymbol{a})$ for all $t \in T$, even if the individual components of $\boldsymbol{X}_t$ are not stochastically independent [2, Chapter 26]. Hence, we obviously have $\mathbb{P}[\boldsymbol{X}_t^\top \boldsymbol{a} > C] \leq \varepsilon$ for all $t \in T$ if and only if $\mathbb{P}[\boldsymbol{c}^\top \boldsymbol{a} > C] \leq \varepsilon$, where $\boldsymbol{c} \sim \mathcal{N}_n(\boldsymbol{\mu}, \Sigma)$ is a representative random vector (as to the distribution) for the workloads. These observations lead to an easier description of P:

Lemma 1 *Let* $0 < \varepsilon \leq 0.5$ *and* $F \subset I \times I$, *then* $\boldsymbol{a} = (a_i)_{i \in I} \in P$ *holds if and only if the following constraints are satisfied*

$$\sum_{i \in I} \mu_i a_i \leq C, \tag{1}$$

$$\sum_{i \in I} (2C\mu_i + q_{1-\varepsilon}^2 \sigma_{ii} - \mu_i^2) a_i + 2 \sum_{i \in I} \sum_{j > i} a_i a_j \left(q_{1-\varepsilon}^2 \sigma_{ij} - \mu_i \mu_j \right) \leq C^2, \tag{2}$$

$$\forall (i, j) \in F : a_i + a_j \leq 1, \tag{3}$$

where $q_{1-\varepsilon}$ *is the* $(1 - \varepsilon)$*-quantile of the standard normal distribution.*

The quadratic terms $a_i a_j$ in (2) will later be replaced by linearization techniques. To define a reasonable set F of forbidden item combinations, note that Condition (A) only states an upper bound for the overloading probability of a server. However, for a specific realization $\omega \in \Omega$ the consolidated workloads can still satisfy $\boldsymbol{c}(w)^\top \boldsymbol{a} > C$, which would then lead to some latency. To preferably "avoid" these performance-degrading situations, we introduce the concept of *overlap coefficients*.

Definition 2 For given random variables $Y, Z : \Omega \to \mathbb{R}$ with mean values $\mu_Y, \mu_Z \in \mathbb{R}$ and variances $\sigma_Y, \sigma_Z > 0$, the *overlap coefficient* Ω_{YZ} is defined by

$$\Omega_{YZ} := \mathbb{E}\left[(Y-\mu_Y)\cdot(Z-\mu_Z)\cdot S(Y-\mu_Y, Z-\mu_Z)\right]/(\sqrt{\sigma_Y}\cdot\sqrt{\sigma_Z})$$

with $S(y,z) = -1$ for $\max\{y,z\} < 0$ and $S(y,z) = 1$ otherwise.

Lemma 2 *For Y, Z as described above we have $\Omega_{YZ} \in [-1, 1]$.*

Contrary to the ordinary *correlation coefficient*, the new value Ω_{YZ} does not "penalize" the situation, where both workloads Y and Z are below their expectations μ_Y and μ_Z, since this scenario is less problematic in server consolidation. For a given threshold $S \in [-1, 1]$, we now define $F := F(S) := \left\{(i,j) \in I \times I \,\middle|\, i \neq j,\ \Omega_{ij} > S\right\}$, where Ω_{ij} represents the overlap coefficient between distinct jobs $i \neq j \in I$. In general, and specifically for our real-world data from [14], choosing $S \approx 0$ is reasonable to (1) preferably avoid those cases where both jobs operate above their expectations and (2) not to exclude too many feasible consolidations, see [12] for further explanations.

3 An Exact Solution Approach

To model the SBPP-C, we propose an *integer linear program (ILP)* with binary variables that is similar to the Kantorovich model [10] of the ordinary BPP [7]. Given an upper bound $u \in \mathbb{N}$ for the required number of servers (bins), we introduce decision variables $y_k \in \mathbb{B}$, $k \in K := \{1, \dots, u\}$, to indicate whether server k is used ($y_k = 1$) or not ($y_k = 0$). Moreover, we require assignment variables $x_{ik} \in \mathbb{B}$, $(i,k) \in Q$, to model whether job i is executed on server k ($x_{ik} = 1$) or not ($x_{ik} = 0$), where $Q := \{(i,k) \in I \times K \mid i \geq k\}$. As already pinpointed in the previous section, the quadratic terms in the pattern definition can be replaced by additional binary variables ξ_{ij}^k (and further constraints) with $k \in K$ and $(i,j) \in T_k := \{(i,j) \in I \times I \mid (i,k) \in Q, (j,k) \in Q, j > i\}$ in order to only consider those index tuples (i,j,k) that are compatible with the indices of the x-variables. Altogether, abbreviating the coefficients

$$\alpha_i := 2C\mu_i + q_{1-\varepsilon}^2\sigma_{ii} - \mu_i^2, \quad \beta_{ij} := q_{1-\varepsilon}^2\sigma_{ij} - \mu_i\mu_j$$

for $i, j \in I$ appearing in (2), the exact model for the SBPP-C results in:

Linear Assignment Model for SBPP-C

$$z = \sum_{k \in K} y_k \to \min$$

$$\text{s.t.} \quad \sum_{(i,k) \in Q} x_{ik} = 1, \qquad i \in I, \tag{4}$$

$$\sum_{(i,k)\in Q} \alpha_i x_{ik} + 2 \sum_{(i,j)\in T_k} \beta_{ij}\xi_{ij}^k \leq C^2 \cdot y_k, \qquad k \in K, \tag{5}$$

$$\sum_{(i,k)\in Q} \mu_i x_{ik} \leq C \cdot y_k, \qquad k \in K, \tag{6}$$

$$x_{ik} + x_{jk} \leq 1, \qquad k \in K, (i, j) \in F, \tag{7}$$

$$\xi_{ij}^k \leq x_{ik}, \qquad k \in K, (i, j) \in T_k, \tag{8}$$

$$\xi_{ij}^k \leq x_{jk}, \qquad k \in K, (i, j) \in T_k, \tag{9}$$

$$x_{ik} + x_{jk} - \xi_{ij}^k \leq 1, \qquad k \in K, (i, j) \in T_k, \tag{10}$$

$$y_k \in \mathbb{B},\ x_{ik} \in \mathbb{B}, \qquad k \in K, (i, k) \in Q, \tag{11}$$

$$\xi_{ij}^k \in \mathbb{B}, \qquad k \in K, (i, j) \in T_k. \tag{12}$$

The objective function minimizes the sum of all y-variables, that is the number of servers required to execute all jobs feasibly. Conditions (4) ensure that each job is assigned exactly once. According to Lemma 1, for any server $k \in K$, conditions (5)–(7) guarantee that the corresponding vector $\boldsymbol{x}^k = (x_{ik})$ represents a feasible pattern. Remember that, here, we replaced the quadratic terms $x_{ik} \cdot x_{jk}$ by new binary variables ξ_{ij}^k, so that conditions (8)–(10) have to be added to couple the x- and the ξ-variables. Moreover, we use the lower bound $\eta := \lceil \sum_{i\in I} u_i / C \rceil$ to fix some y-variables in advance. In order to obtain an upper bound u (to be used for the set K), an adapted first fit decreasing algorithm [12, Alg. 1] is applied.

4 Numerical Experiments

We implemented the above model in MATLAB R2015b and solved the obtained ILP by means of its CPLEX interface (version 12.6.1) on an Intel Core i7-8550U with 16 GB RAM. To this end, we consider a dataset containing 500 workloads from a Google data center [14] with $S = 0$. For given $n \in \mathbb{N}$, we always constructed 10 instances by randomly drawing n jobs from this dataset. Moreover, in accordance with [13], we chose $\varepsilon = 0.25$. A more detailed discussion of these parameters as well as further numerical tests are contained in [12, Sect.4.1].

In Table 1, we collect the average values: lower and upper bound η, u for the optimal value $z^\star$, numbers n_v, n_c and n_{it} of variables, constraints and iterations, and time t (in seconds) to solve the ILP. Obviously, for increasing values of n the instances become harder with respect to the numbers of variables, constraints, and iterations, so that more time is needed to solve the problems to optimality. However, any considered instance could be coped with in the given time limit ($\bar{t} = 100s$). Our main observations are: (1) As η does neither include the set F nor the covariances, its performance is rather poor. (2) The upper bound u is very close to the exact

Table 1 Average results for the SBPP-C based on 10 instances each (with $S = 0$)

n	25	30	35	40	45	50
η	4.0	5.0	5.3	6.0	6.9	7.6
$z^\star$	10.5	11.9	14.0	15.3	18.6	19.7
u	10.9	12.2	14.2	15.9	19.2	20.7
t	0.4	0.9	1.7	6.2	12.2	23.9
n_{it}	758.9	1126.8	2075.9	5192.9	7222.6	12453.1
n_v	2346.0	3820.5	6005.2	8777.6	12874.8	17342.5
n_c	8085.9	13032.3	20594.4	29890.3	44560.9	59371.1

optimal value. (3) Contrary to the less general approach from [13], it is possible to deal with much larger instance sizes in short times.

Consequently, this new approach does not only contribute to a more realistic description of the consolidation problem itself (since additional application-oriented properties are respected), but also to a wider range of instances that can be solved to optimality.

5 Conclusions

In this article, we developed an exact approach for server consolidation with (not necessarily stochastically independent) jobs whose workloads are given by stochastic characteristics. Moreover, the new concept of overlap coefficients contributes to separate mutually influencing jobs to avoid performance degradations such as latency. Based on numerical experiments with real-world data, this new approach was shown to outperform an earlier and less general method [13]. However, finding improved lower bounds (preferably using all of the problem-specific input data) or alternative (pseudo-polynomial) modeling frameworks are part of future research challenges.

Acknowledgments This work is supported in part by the German Research Foundation (DFG) within the Collaborative Research Center SFB 912 (HAEC).

References

1. Andrae, A.S.G., Edler, T.: On global electricity usage of communication technology: trends to 2030. Challenges **6**(1), 117–157 (2015)
2. Balakrishnan, N., Nevzorov, V.B.: A Primer on Statistical Distributions, 1st edn. Wiley, New York (2003)
3. Benson, T., Anand, A., Akella, A., Zhang, M.: Understanding data center traffic characteristics. Comput. Commun. Rev. **40**(1), 92–99 (2010)

4. Cisco: Cisco Global Cloud Index: Forecast and Methodology, 2016–2021. White Paper (2018). http://www.cisco.com/en/US/solutions/collateral/ns341/ns525/ns537/ns705/ns1175/Cloud_Index_White_Paper.html
5. Corcoran, P.M., Andrae, A.S.G.: Emerging Trends in Electricity Consumption for Consumer ICT. Technical Report (2013). http://aran.library.nuigalway.ie/xmlui/handle/10379/3563
6. Dargie, W.: A stochastic model for estimating the power consumption of a server. IEEE Trans. Comput. **64**(5), 1311–1322 (2015)
7. Delorme, M. Iori, M., Martello, S.: Bin packing and cutting stock problems: mathematical models and exact algorithms. Eur. J. Oper. Res. **255**, 1–20 (2016)
8. Goel, A., Indyk, P.: Stochastic Load Balancing and Related Problems. In: Proceeding of 40th Annual Symposium on Foundations of Computer Science, pp. 579–586 (1999)
9. Hähnel, M., Martinovic, J., Scheithauer, G., Fischer, A., Schill, A., Dargie, W.: Extending the cutting stock problem for consolidating services with stochastic workloads. IEEE Trans. Parallel Distrib. Syst. **29**(11), 2478–2488 (2018)
10. Kantorovich, L.V.: Mathematical methods of organising and planning production. Manag. Sci. **6**, 366–422 (1939 Russian, 1960 English)
11. Kleinberg, J., Rabani, Y., Tardos, E.: Allocating bandwidth for Bursty connections. SIAM J. Comput. **30**(1), 191–217 (2000)
12. Martinovic, J., Hähnel, M., Dargie, W., Scheithauer, G.: A Stochastic Bin Packing Approach for Server Consolidation with Conflicts. Preprint MATH-NM-02-2019, Technische Universität Dresden (2019). http://www.optimization-online.org/DB_HTML/2019/07/7274.html
13. Martinovic, J., Hähnel, M., Scheithauer, G., Dargie, W., Fischer, A.: Cutting stock problems with nondeterministic item lengths: a new approach to server consolidation. 4OR **17**(2), 173–200 (2019)
14. Reiss, C., Wilkes, J., Hellerstein, J.L.: Google cluster-usage traces: format + schema. Technical Report, Google Inc., Mountain View (2011)
15. Wang, M., Meng, X., Zhang, L.: Consolidating virtual machines with dynamic bandwidth demand in data centers. Proceedings of IEEE INFOCOM, pp. 71–75 (2011)
16. Yu, L., Chen, L., Cai, Z., Shen, H., Liang, Y., Pan, Y.: Stochastic load balancing for virtual resource management in datacenters. IEEE Trans. Cloud Comput. **8**(2), 459–472 (2020)

Optimal Student Sectioning at Niederrhein University of Applied Sciences

Steffen Goebbels and Timo Pfeiffer

Abstract Degree programs with a largely fixed timetable require centralized planning of student groups (sections). Typically, group sizes for exercises and practicals are small, and different groups are taught at the same time. To avoid late or weekend sessions, exercises and practicals of the same or of different subjects can be scheduled concurrently, and the duration of lessons can vary. By means of an integer linear program, an optimal group division is carried out. To this end, groups have to be assigned to time slots and students have to be divided into groups such that they do not have conflicting appointments. The optimization goal is to create homogeneous group sizes.

Keywords Timetabling · Integer linear programming

1 Introduction

A large number of articles deals with the "University Course Timetable Problem", see [1, 5] and the literature cited there. Here we are concerned with the subproblem "student sectioning", more precisely with "batch sectioning after a complete timetable is developed", see [3, 4, 6] for a theoretical discussion and overview. At our faculty for electrical engineering and computer science, we perform student sectioning on a fixed time table that already provides general time slots for groups, see Fig. 1. Based on enrollment data, also the number of groups per lecture, exercise and practical is known. The setting is typical for technical courses at universities of applied sciences. Groups for a given subject might be taught weekly or only in every second or every fourth week. This gives freedom to choose the starting week for such groups and to place up to four groups in the same time slot. These groups

S. Goebbels (✉) · T. Pfeiffer
iPattern Institute, Niederrhein University of Applied Sciences, Krefeld, Germany
e-mail: steffen.goebbels@hs-niederrhein.de; Steffen.Goebbels@hsnr.de; timo.pfeiffer@stud.hn.de

J. S. Neufeld et al. (eds.), *Operations Research Proceedings 2019*, Operations Research Proceedings, https://doi.org/10.1007/978-3-030-48439-2_20

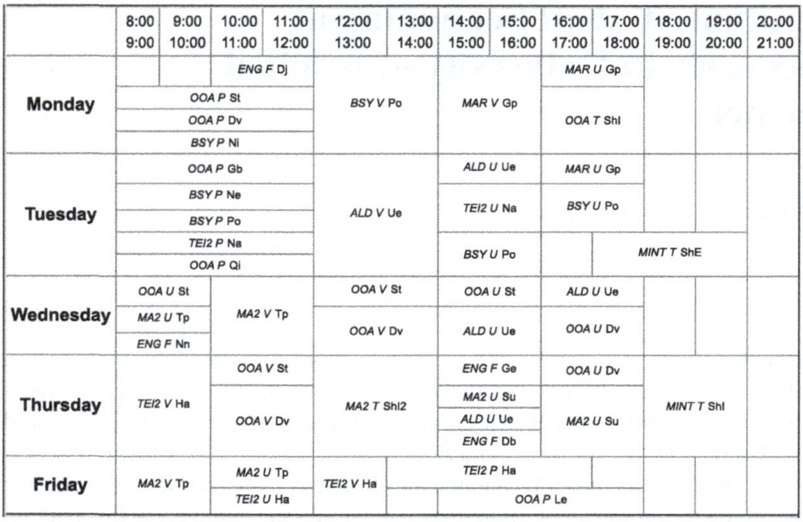

Fig. 1 Schedule of the bachelor course in computer science, second semester: the first acronym refers to the subject, then P denotes practical, U stands for exercise, V for lecture, T for tutorial and F for language course. Practicals, exercises (including language courses) and parallel lectures are subject to group planning. The last acronym denotes the lecturer. In this plan, all practicals but "OOA P" have a 4 week frequency. Groups for "OOA P" are taught every second week. Exercises "MA2 U", "ALD U", and "OOA U" have a weekly frequency, the other exercises are taught every second week. Lecture "OOA V" is split into two groups

can be taught in alternating weeks. Based on the time table's general time slots, our student sectioning problem has to select suitable slots and suitable starting weeks for groups. It also has to assign students to groups such that groups become nearly equal in size for each subject. A somewhat similar model with a different optimization goal is presented in [7]. For example, the often cited technique presented in [2] does not provide automated assignment of groups to time table slots. In the next section, we describe our model. Using IBM's solver ILOG CPLEX 12.8.0,[1] we applied the model with real enrollment data to our time table. Section 3 summarizes results.

2 Model

For simple terminology, lectures, exercises and practicals that require group division are considered to be separate modules which are numbered by $1, \dots, N$.

Every module $k \in [N] := \{1, \dots, n\}$ is taught for n_k groups $G_{k,j}$, $j \in [n_k]$. The number of groups is determined in advance, based on current enrollment

[1] See https://www.ibm.com/customer-engagement/commerce.

Fig. 2 Model and notation

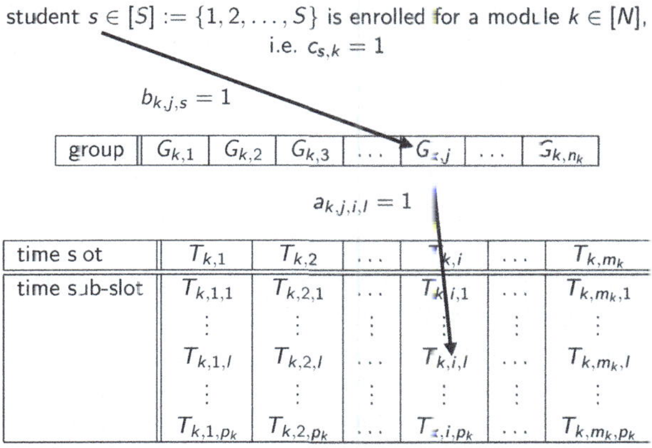

figures and teaching capacities. Each module k can be offered on at most m_k time slots $T_{k,1}, \ldots, T_{k,m_k}$ per week, cf. Fig. 2. Not all time slots might be needed, for example if $n_k < m_k$. For the participants of each group, a module takes place either weekly ($p_k := 1$), every second week ($p_k := 2$) or every fourth week ($p_k := 4$), see Table 1. If a module k is given in every second week, then at $T_{k,i}$ two groups can be planned in alternating weeks. This allows us to split $T_{k,i}$ into two simultaneous sub-slots $T_{k,i,1}$ and $T_{k,i,2}$. At most one group can be assigned to each of these sub-slots. The third index indicates whether the module is given for the assigned group in odd or even weeks. In the other week, participating students can take part in another bi-weekly module. If a module is given weekly, we only use a sub-slot $T_{k,i,1}$, for modules given every fourth week, time sub-slots $T_{k,i,l}$, $l \in [4]$, are considered. However, due to the workload of the instructors, there may be restrictions for assigning groups to time slots. The variables $1 \le q_{k,i} \le p_k$, $\sum_{i=1}^{m_k} q_{k,i} \ge n_k$, indicate the maximum number of groups that can be assigned to slots $T_{k,i}$, i.e., the maximum number of sub-slots with group assignment. For each group $G_{k,j}$ we define binary variables that assign a time sub-slot to the group. Let $a_{k,j,1,1}, \ldots, a_{k,j,1,p_k}, a_{k,j,2,1}, \ldots, a_{k,j,2,p_k}, \ldots, a_{k,j,m_k,1}, \ldots, a_{k,j,m_k,p_k} \in \{0, 1\}$ with

$$\sum_{i=1}^{m_k} \sum_{l=1}^{p_k} a_{k,j,i,l} = 1. \tag{1}$$

Table 1 Left: a group assigned to a sub time-slot $T_{k,i,l}$ is taught depending on frequency p_k in alternating weeks; right: a student can be member of groups of different modules that are taught in overlapping time slots but in different weeks ($p_{k_1} := 2$, $p_{k_2} =: 4$, and $p_{k_3} =: 4$)

Frequency p_k	Week 1	Week 2	Week 3	Week 4	Week 1	Week 2	Week 3	Week 4
1	$T_{k,i,1}$	$T_{k,i,1}$	$T_{k,i,1}$	$T_{k,i,1}$	$T_{k_1,i_1,1}$		$T_{k_1,i_1,1}$	
2	$T_{k,i,1}$	$T_{k,i,2}$	$T_{k,i,1}$	$T_{k,i,2}$		$T_{k_2,i_2,2}$		
4	$T_{k,i,1}$	$T_{k,i,2}$	$T_{k,i,3}$	$T_{k,i,4}$				$T_{k_3,i_3,4}$

If $a_{k,j,i,l} = 1$, the lesson for group $G_{k,j}$ takes place on time sub-slot $T_{k,i,l}$. Every sub-slot $T_{k,i,l}$ has to be assigned at most once, and only $q_{k,i}$ groups may be scheduled for a time slot $T_{k,i}$, i.e., for all $k \in [N]$ and $i \in [m_k]$

$$\sum_{j=1}^{n_k} a_{k,j,i,l} \leq 1 \text{ for all } l \in [p_k], \quad \sum_{j=1}^{n_k} \sum_{l=1}^{p_k} a_{k,j,i,l} \leq q_{k,i}. \tag{2}$$

Let $S \in \mathbb{N}$ be the number of all students. Each student can register for the modules individually or, as in the case of English courses, is automatically assigned based on her or his level. Students can select modules that belong to different semesters, as modules may have to be repeated. We use a matrix $C \in \{0, 1\}^{S \times N}$ to describe whether a student has enrolled for a module. Thereby, $c_{s,k} = 1$ indicates that student s has chosen module k. We have to assign exactly one group to each student $s \in [S]$ for each chosen module. To this end, we use binary variables $b_{k,j,s} \in \{0, 1\}$. Student s is in group j of module k iff $b_{k,j,s} = 1$. We get

$$\sum_{j=1}^{n_k} b_{k,j,s} = c_{s,k} \text{ for all } k \in [N] \text{ and } s \in [S]. \tag{3}$$

An external group assignment takes place for some modules (language courses). In this case, variables $b_{k,j,s}$ have to be set accordingly.

However, there must be no collision with simultaneous group assignments, cf. Table 1. Each time slot consists of 1 to 4 h in a fixed time grid covering one week. Per module, the duration of time slots is (approximately) the same. Each week can be modeled with the set [50] representing hours, i.e., $T_{k,i}, T_{k,i,l} \subset [50]$. It is allowed that the hours of time slots T_{k,i_1} and T_{k,i_2} (of the same module k) overlap, i.e., $T_{k,i_1} \cap T_{k,i_2} \neq \emptyset$ for $i_1 \neq i_2$, only if the module is given simultaneously by several instructors in different rooms.

- If a student is in a weekly group of one module, he may not be in a time-overlapping group of another module.
- If a student is in a bi-weekly group on a time sub-slot with a third index l, he may not be in another bi-weekly group on a time-overlapping sub-slot with the same third index l. He also must not be assigned to a group that belongs to an overlapping 4-weekly time sub-slot with a third index l or $l + 2$.
- If a student belongs to a group that is placed on a 4-weekly time sub-slot with a third index l, he must not be in another group belonging to an overlapping 4-weekly time sub-slot with the same third index l.

Conflicting time sub-slots are calculated in advance. Let T_{k_1,i_1,l_1} and T_{k_2,i_2,l_2}, $k_1 \neq k_2$, be two conflicting time slots for which, according to previous rules, two non-disjoint groups cannot be assigned. Group G_{k_1,j_1} is assigned to time sub-slot T_{k_1,i_1,l_1} iff $a_{k_1,j_1,i_1,l_1} = 1$, and a student s is assigned the group G_{k_1,j_1} iff $b_{k_1,j_1,s} = 1$. If also group G_{k_2,j_2} is assigned to time sub-slot T_{k_2,i_2,l_2} via $a_{k_2,j_2,i_2,l_2} = 1$ and if student s is assigned to this group via $b_{k_2,j_2,s} = 1$, then there is a collision. Thus,

$$a_{k_1,j_1,i_1,l_1} + b_{k_1,j_1,s} + a_{k_2,j_2,i_2,l_2} + b_{k_2,j_2,s} \leq 3 \tag{4}$$

has to be fulfilled for all colliding pairs $(T_{k_1,i_1,l_1}, T_{k_2,i_2,l_2})$ of time sub-slots, all groups $j_1 \in [n_{k_1}]$, $j_2 \in [n_{k_2}]$ and all students $s \in [S]$.

Collisions between group assignments are not defined independently of students by rule (4). This leads to a significant combinatorial complexity that has to be reduced. To speed-up the algorithm, certain groups can be assigned to sub-slots in a fixed manner. This can be done easily if, for a subject k, the number of sub-slots $m_k \cdot p_k$ equals the number of groups n_k. For such modules k we can set

$$a_{k,j,i,l} := \begin{cases} 1 : j = (i-1) \cdot p_k + l \\ 0 : \text{otherwise.} \end{cases} \tag{5}$$

By assigning groups to sub-slots in a chronologically sorted manner due to their group number, one can also avoid many permutations. Sorting can be established by following restrictions for all modules $k \in [N]$, all time slots $i_1 \in [m_k]$ and all sub-slots T_{k,i_1,l_1}, $l_1 \in [p_k]$, and all groups $j_1 \in [n_k]$:

$$\max \left\{ \sum_{i_2=i_1+1}^{m_k} \sum_{j_2=1}^{j_1-1} \sum_{l_2=1}^{p_k} a_{k,j_2,i_2,l_2}, \sum_{j_2=1}^{j_1-1} \sum_{l_2=l_1+1}^{p_k} a_{k,j_2,i_1,l_2} \right\} \leq n_k \quad (1 - a_{k,j_1,i_1,l_1}). \tag{6}$$

The inequality can be interpreted as follows. If group j_1 has been assigned to sub-slot T_{k,i_1,l_1} then no group with smaller index $j_2 < j_1$ must be assigned to a "later" sub-slot T_{k,i_2,l_2} in the sense that either $i_2 \geq i_1$ or $i_2 = i_1$ and $l_2 > l_1$.

Dual education and part-time students s may only be divided into those groups of their semester, that are assigned to time slots on certain days. This restriction does not apply to modules that do not belong to the students' semester (repetition of courses). For all time sub-slots $T_{k,i,l}$, at which s cannot participate, we require

$$a_{k,j,i,l} + b_{k,j,s} \leq 1 \text{ for all } j \in [n_k]. \tag{7}$$

Two (but no more) students s_1 and s_2 can choose to learn together. Then they have to be placed into the same groups of the modules that they have both chosen. This leads to constraints if $c_{s_1,k} = c_{s_2,k}$, $k \in [N]$. Then for all $j \in [n_k]$

$$b_{k,j,s_2} = b_{k,j,s_1}. \tag{8}$$

Students should be assigned to groups such that, for each module, groups should be of (nearly) equal size (cf. [2]). To implement this target we represent the difference of sizes of groups j_1 and j_2 of module k with the difference of two non-negative variables Δ^+_{k,j_1,j_2}, $\Delta^-_{k,j_1,j_2} \geq 0$:

$$\Delta^+_{k,j_1,j_2} - \Delta^-_{k,j_1,j_2} = \sum_{s=1}^{S}(b_{k,j_1,s} - b_{k,j_2,s}). \tag{9}$$

Thus, we have to minimize

$$\sum_{k=1}^{N}\sum_{j_1=1}^{n_k-1}\sum_{j_2=j_1+1}^{n_k}(\Delta^+_{k,j_1,j_2} + \Delta^-_{k,j_1,j_2} - \varepsilon_{k,j_1,j_2}) \text{ s.t. (1)–(4) and (7)–(9).} \tag{10}$$

We observed long running times if an uneven number of students have to be divided into an even number of groups, and vice versa. To further simplify complexity, we propose to subtract float or integer variables $0 \leq \varepsilon_{k,j_1,j_2} \leq \min\{D, \Delta^+_{k,j_1,j_2} + \Delta^-_{k,j_1,j_2}\}$ within the objective function (10). They serve as slack variables that allow absolute group size differences to vary between 0 and $D \in \mathbb{N}_0 = \{0, 1, 2, \ldots\}$ without penalty. Significant speedup already is obtained for $D = 1$, see Sect. 3. Thus, we consider a group sectioning as being optimal even if there exist slightly better solutions with fewer differences.

In certain groups j, a contingent of r places can be reserved (e.g., for participants of other faculties). This is done by adding or subtracting the number r on the right side of (9): If $j = j_1$, then r has to be added, if $j = j_2$, then r is subtracted.

3 Results

The program can be applied separately for each field of study. Presented results belong to our bachelor programs in computer science (second and fourth semester, 330 students including 59 dual education and part-time students, 30 modules, up to 8 groups per module) and electrical engineering (second, fourth, and sixth semester, 168 students including 40 dual education and part-time students, 27 modules, up to 4 groups per module). Table 2 summarizes running times with respect to combinations of speed-up measures $D \in \{1, 2\}$, sorting (6), and fixed assignment of certain groups to time-slots (5). Choosing $D = 0$ leads to memory overflow after 8 h in case of computer science (independent of speed-up measures), whereas group division for electrical engineering finishes in 420.67 s (without speed-up measures).

4 Enhancements

To assign additional students to groups by maintaining all previously done assignments, one can also use the integer linear program as an online algorithm.

Table 2 CPLEX 12.8.0 processor times measured in seconds on an Intel Core i5-6500 CPU, 3.20 GHz x4 with 16 GB RAM

Slack size D	Sorting (6)	Initialization (5)	Running time Computer science	Running time Electrical engineering
2	–	–	51.59	0.13
2	–	✓	3.35	0.1
2	✓	–	2.8	0.06
2	✓	✓	1.57	0.05
1	–	–	57.92	0.15
1	–	✓	6.32	0.08
1	✓	–	3.49	0.09
1	✓	✓	3.33	0.07
0	– / ✓	– / ✓	Memory overflow	≤420.67

If students choose modules from different semesters then the existence of a feasible solution is not guaranteed. However, such situations could be identified prior to group planning. Alternatively, one can deal with such students by applying the online version of the algorithm in order to individually identify conflicts.

As a secondary optimization goal, one could maximize the number of students that get commonly assigned to groups along all modules. Students who have to repeat modules could be distributed as evenly as possible among groups, since experience has shown that for such students the risk of non-appearance is high.

References

1. Bettinelli, A., Cacchiani, V., Roberti, R., Toth, P.: An overview of curriculum-based course timetabling. TOP **23**(2), 313–349 (2015)
2. Laporte, G., Desroches, S.: The problem of assigning students to course sections in a large engineering school. Comput. Oper. Res. **13**(4), 387–394 (1986)
3. Müller, T., Murray, K.: Comprehensive approach to student sectioning. Ann. Oper. Res. **181**(1), 249–269 (2010)
4. Schaerf, A.: A survey of automated timetabling. Artif. Intell. Rev. **13**(2), 87–127 (1999)
5. Schimmelpfeng, K., Helber, S.: Application of a real-world university-course timetabling model solved by integer programming. OR Spectr. **29**(4), 783–803 (2007)
6. Schindl, D.: Student sectioning for minimizing potential conflicts on multi-section courses. In: Proceedings of the 11th International Conference of the Practice and Theory of Automated Timetabling (PATAT 2016), Udine, pp. 327–337 (2016)
7. Sherali, H.D., Driscoll, P.J.: Course scheduling and timetabling at USMA. Mil. Oper. Res. **4**(2), 25–43 (1999)

A Dissection of the Duality Gap of Set Covering Problems

Uledi Ngulo, Torbjörn Larsson, and Nils-Hassan Quttineh

Abstract Set covering problems are well-studied and have many applications. Sometimes the duality gap is significant and the problem is computationally challenging. We dissect the duality gap with the purpose of better understanding its relationship to problem characteristics, such as problem shape and density. The means for doing this is a set of global optimality conditions for discrete optimization problems. These decompose the duality gap into two terms: near-optimality in a Lagrangian relaxation and near-complementarity in the relaxed constraints. We analyse these terms for numerous instances of large size, including some real-life instances. We conclude that when the duality gap is large, typically the near-complementarity term is large and the near-optimality term is small. The large violation of complementarity is due to extensive over-coverage. Our observations should have implications for the design of solution methods, and especially for the design of core problems.

Keywords Discrete optimization · Set covering problem · Duality gap

1 Theoretical Background

Consider the general primal problem $f^* := \min\{f(x) \mid g(x) \leq 0 \text{ and } x \in X\}$ where the set $X \subset \mathbb{R}^n$ is compact and the functions $f : X \to \mathbb{R}$ and $g : X \to \mathbb{R}^m$ are continuous. Letting $u \in \mathbb{R}^m_+$ be a vector of Lagrangian multipliers, the dual function $h : \mathbb{R}^m_+ \to \mathbb{R}$ is defined by the problem $h(u) = \min_{x \in X} \; f(x) + u^{\mathrm{T}} g(x)$, which is a Lagrangian relaxation. It is well known that the function h is finite, concave and continuous on $\mathbb{R}^m_+$, and that $h(u) \leq f^*$ holds for all $u \in \mathbb{R}^m_+$. The Lagrangian dual problem is defined as $h^* = \max_{u \in \mathbb{R}^m_+} \; h(u)$ and it provides the best

U. Ngulo · T. Larsson · N.-H. Quttineh (✉)
Department of Mathematics, Linköping University, Linköping, Sweden
e-mail: uledi.ngulo@liu.se; nils-hassan.quttineh@liu.se

J. S. Neufeld et al. (eds.), *Operations Research Proceedings 2019*, Operations Research Proceedings, https://doi.org/10.1007/978-3-030-48439-2_21

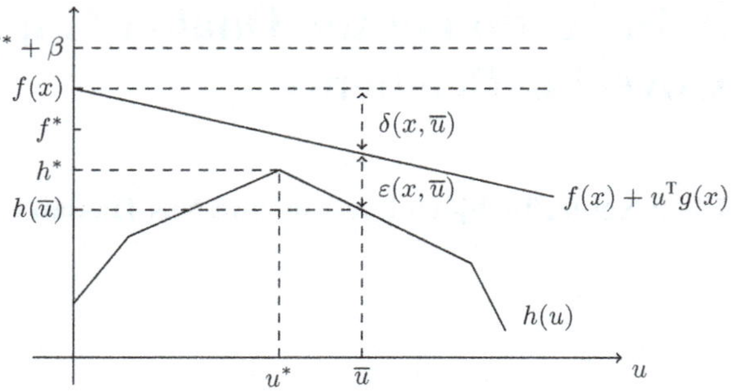

Fig. 1 Illustration of the optimality condition

lower bound for the primal optimal value f^*. The duality gap for this primal-dual pair is $\Gamma := f^* - h^*$.

The following result is a known Lagrangian dual characterization of optimal or near-optimal primal solutions [5, Prop. 5]. Define the functions $\varepsilon : X \times \mathbb{R}^m_+ \to \mathbb{R}_+$ and $\delta : X \times \mathbb{R}^m_+ \to \mathbb{R}$ with $\varepsilon(x, u) := f(x) + u^{\mathrm{T}}g(x) - h(u)$ and $\delta(x, u) := -u^{\mathrm{T}}g(x)$, respectively. The quantity $\varepsilon(x, u)$ is the degree of near-optimality of an $x \in X$ in the Lagrangian relaxation obtained with the dual solution u, and $\delta(x, u)$ is the degree of near-complementarity of an $x \in X$ with respect to u. Clearly, if x is primal feasible then $\delta(x, u) \geq 0$. Theorem 1 states a global optimality condition based on these quantities.

Theorem 1 *Let $u \in \mathbb{R}^m_+$. Then a primal feasible solution x is β−optimal if and only if $\varepsilon(x, u) + \delta(x, u) \leq f^* + \beta - h(u)$ holds.*

Figure 1 illustrates this result for a given $\overline{u}$. In particular, a primal feasible solution x^* and a $u^* \in \mathbb{R}^m_+$ are both optimal if and only if $\varepsilon(x^*, u^*) + \delta(x^*, u^*) = \Gamma$ holds, where $\varepsilon(x^*, u^*) \geq 0$ and $\delta(x^*, u^*) \geq 0$.

2 Set Covering Problem

Consider the Set Covering Problem (SCP)

$$z^*_{IP} := \min \left\{ \sum_{j=1}^{n} c_j x_j \;\middle|\; \sum_{j=1}^{n} a_{ij} x_j \geq 1, \; i = 1, \ldots, m, \text{ and } x \in \{0, 1\}^n \right\}$$

and the linear programming relaxation

$$z^*_{LP} := \min \left\{ \sum_{j=1}^{n} c_j x_j \;\middle|\; \sum_{j=1}^{n} a_{ij} x_j \geq 1, \; i = 1, \ldots, m, \text{ and } x \in \mathbb{R}^n_+ \right\}.$$

Note that the upper bounds on the variables are not needed in the latter problem.

We apply the optimality condition above to the SCP with the purpose of dissecting the duality gap. Given Lagrangian multipliers $u \in \mathbb{R}^m_+$ the dual function is $h : \mathbb{R}^m_+ \to \mathbb{R}$ with $h(u) = \sum_{i=1}^m u_i + \sum_{j=1}^n h_j(u)$, where $h_j : \mathbb{R}^m_+ \to \mathbb{R}$ with $h_j(u) = \min_{x_j \in \{0,1\}}(c_j - \sum_{i=1}^m u_i a_{ij})x_j = \min\{0, c_j - \sum_{i=1}^m u_i a_{ij}\}$. The dual problem is $h^* = \max_{u \in \mathbb{R}^m_+} h(u)$. The Lagrangian relaxation made has the integrality property [6, p. 177]. Hence, $h^* = z^*_{LP}$ and $\Gamma = z^*_{IP} - z^*_{LP}$. Further, any optimal solution to the dual of the linear programming relaxation, u^*, is an optimal solution to the Lagrangian dual problem. Since $\bar{c}_j = c_j - \sum_{i=1}^m u^*_i a_{ij} \geq 0$ holds, it follows that $h_j(u^*) = 0$.

The function $\varepsilon : \{0, 1\}^n \times \mathbb{R}^m_+ \to \mathbb{R}$ can be separated over the primal variables into $\varepsilon(x, u) = \sum_{j=1}^n \varepsilon_j(x_j, u)$ where $\varepsilon_j(x_j, u) = (c_j - \sum_{i=1}^m u_i a_{ij})x_j - h_j(u)$, and the function $\delta : \{0, 1\}^n \times \mathbb{R}^m_+ \to \mathbb{R}$ can be separated over the dual variables into $\delta(x, u) = \sum_{i=1}^m \delta_i(x, u_i)$ where $\delta_i(x, u_i) = -u_i(1 - \sum_{j=1}^n a_{ij}x_j)$. Further, $\varepsilon_j(x_j, u^*) = \bar{c}_j x_j$. Clearly, $\varepsilon(x, u) + \delta(x, u) = \sum_{j=1}^n c_j x_j - h(u)$ holds for any $x \in \{0, 1\}^n$ and $u \in \mathbb{R}^m_+$, and in particular it holds for any x that is feasible in SCP and the dual optimum u^*.

Finally, for a primal optimal solution x^* we obtain that $\Gamma = \varepsilon(x^*, u^*) + \delta(x^*, u^*)$. Hence, the duality gap can be dissected into the near-optimality term $\varepsilon(x^*, u^*)$ and the near-complementarity term $\delta(x^*, u^*)$.

3 Numerical Study

The SCP instances studied are taken from the OR-Library and originate from [1–4]. Some of these instances are real-life problems. Out of 87 instances, 11 have been removed because $\Gamma = 0$. We investigate the duality gap of each of the 76 remaining instances in terms of the near-optimality and near-complementarity terms. In our investigation we calculate the following quantities: density $\rho := (\sum_{i=1}^m \sum_{j=1}^n a_{ij})/(m \times n)$, Average Excess Coverage (AEC) $:= \frac{1}{m}\sum_{i=1}^m(\sum_{j=1}^n a_{ij}x^*_j - 1)$, relative duality gap $\Gamma_{\text{rel}} := (z^*_{IP} - z^*_{LP})/z^*_{LP}$, relative near-optimality $\varepsilon_{\text{rel}} := \varepsilon(x^*, u^*)/(z^*_{IP} - z^*_{LP})$, relative near-complementarity $\delta_{\text{rel}} := \delta(x^*, u^*)/(z^*_{IP} - z^*_{LP})$, matrix shape m/n, and relative cardinality $\kappa_{\text{rel}} := \kappa_{LP}/\kappa_{IP}$, where κ_{LP} and κ_{IP} are the cardinalities of linear programming and integer programming optimal solutions, respectively. We study the relationship between these quantities as illustrated in Figs. 2 and 3. The quantity ε_{rel} is not shown since $\varepsilon_{\text{rel}} = 1 - \delta_{\text{rel}}$.

Figure 2 shows that the relative near-complementarity quantity, δ_{rel}, is always close to one whenever the relative duality gap is large. Hence, in such situations the relative near-optimality quantity, ε_{rel}, is always close to zero. For small gaps, both quantities can contribute to the gap. On the other hand, whenever ε_{rel} is large, then the relative duality gap is always small. Further, a large δ_{rel} is due to excess coverages of constraints, which typical occur for instances where $m/n > 1$.

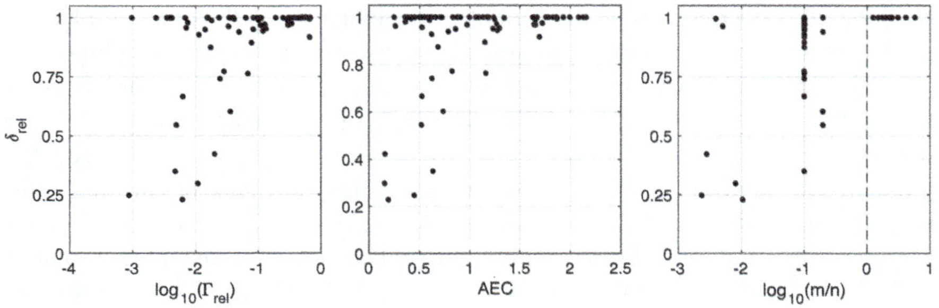

Fig. 2 Illustration of δ_{rel} versus $\Gamma_{\rm rel}$, AEC and m/n, respectively, for the 76 instances

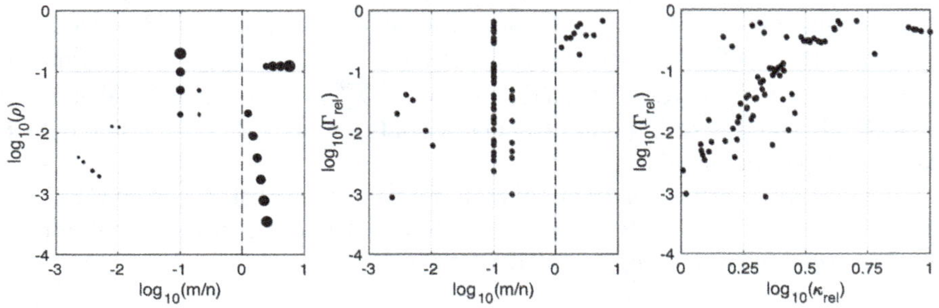

Fig. 3 Illustration of $\Gamma_{\rm rel}$ versus m/n, ρ and $\kappa_{\rm rel}$, respectively, for the 76 instances. To the left, the area of a circle is proportional to $\Gamma_{\rm rel}$

Figure 3 shows that the relative duality gap is large whenever $m/n > 1$, while the gaps are moderate or small whenever m/n is small. Further, the gap tends to increase with increasing density. (Also, all instances with $\Gamma = 0$ have $m/n \leq 0.2$ and are sparse with $\rho \leq 2\%$.) Furthermore, the relative duality gap increases whenever $\kappa_{\rm rel}$ increases. Detailed results are given in Table 1.

Table 1 Source of problem, name of problem, problem size (m, n), density (ρ), LP optimal value (z^*_{LP}), cardinality of x^*_{LP} (κ_{LP}), IP optimal value (z^*_{IP}), cardinality of x^*_{IP} (κ_{IP}), relative cardinality $\kappa_{\text{rel}} = \kappa_{LP}/\kappa_{IP}$, Average Excess Cover (AEC), relative duality gap (Γ_{rel}), relative near-complementarity (δ_{rel})

Ref.	Name	m	n	ρ	z^*_{LP}	κ_{LP}	z^*_{IP}	κ_{IP}	κ_{rel}	AEC	Γ_{rel}	δ_{rel}
[1]	scpa1	300	3000	2.01	246.84	123	253	67	1.84	0.62	2.5	74.2
	scpa2	.	.	2.01	247.50	132	252	68	1.94	0.60	1.8	98.8
	scpa3	.	.	2.01	228	119	232	70	1.70	0.68	1.8	87.5
	scpa4	.	.	2.01	231.40	108	234	67	1.61	0.62	1.1	92.9
	scpa5	.	.	2.01	234.89	93	236	72	1.29	0.63	0.5	35.0
	scpb1	300	3000	4.99	64.54	83	69	39	2.13	1.16	6.9	76.4
	scpb2	.	.	4.99	69.30	95	76	40	2.38	1.09	9.7	100
	scpb3	.	.	4.99	74.16	81	80	40	2.03	1.15	7.9	89.5
	scpb4	.	.	4.99	71.22	90	79	39	2.31	0.97	10.9	96.8
	scpb5	.	.	4.99	67.67	79	72	38	2.08	1.06	6.4	100
	scpc1	400	4000	2.00	223.80	138	227	82	1.68	0.86	1.4	95.0
	scpc2	.	.	2.00	212.85	140	219	81	1.73	0.82	2.9	77.3
	scpc3	.	.	2.00	234.53	148	243	74	2.00	0.72	3.6	100
	scpc4	.	.	2.00	213.85	139	219	76	1.83	0.83	2.4	100
	scpc5	.	.	2.00	211.64	147	215	77	1.91	0.71	1.6	100
	scpd1	400	4000	5.01	55.31	93	60	40	2.33	1.19	8.5	100
	scpd2	.	.	5.01	59.35	98	66	40	2.45	1.13	11.2	100
	scpd3	.	.	5.01	65.07	99	72	41	2.41	1.17	10.7	100
	scpd4	.	.	5.00	55.84	104	62	46	2.26	1.34	11.0	99.7
	scpd5	.	.	5.01	58.62	95	61	44	2.16	1.24	4.1	100
	scpe1	50	500	19.66	3.48	50	5	5	10.00	0.50	43.7	100
	scpe2	.	.	20.05	3.38	43	5	5	8.60	0.58	47.9	100
	scpe3	.	.	20.16	3.30	41	5	5	8.20	0.66	51.6	100
	scpe4	.	.	19.81	3.45	46	5	5	9.20	0.62	44.8	100
	scpe5	.	.	20.07	3.39	44	5	5	8.80	0.52	47.5	100
[1]	scp46	200	1000	2.04	557.25	77	560	64	1.20	0.52	0.5	54.5
	scp48	.	.	2.01	488.67	78	492	59	1.32	0.46	0.7	100
	scp49	.	.	1.98	638.54	100	641	61	1.64	0.39	0.4	100
	scp410	.	.	1.95	513.5	67	514	64	1.05	0.45	0.1	100
	scp51	200	2000	2.00	251.23	93	253	62	1.50	0.53	0.7	95.8
	scp52	.	.	2.00	299.76	97	302	58	1.67	0.36	0.8	98.0
	scp54	.	.	1.98	240.5	79	242	66	1.20	0.53	0.6	66.7
	scp56	.	.	2.00	212.5	59	213	58	1.02	0.42	0.2	100
	scp57	.	.	2.02	291.78	78	293	64	1.22	0.46	0.4	100
	scp58	.	.	1.98	287	77	288	62	1.24	0.50	0.4	100

(continued)

Table 1 (continued)

Ref	Name	m	n	ρ	z^*_{LP}	κ_{LP}	z^*_{IP}	κ_{IP}	κ_{rel}	AEC	Γ_{rel}	δ_{rel}
[1]	scp61	200	1000	4.93	133.14	62	138	34	1.82	0.74	3.7	60.3
	scp62	.	.	5.00	140.46	67	146	36	1.86	0.91	4.0	100
	scp63	.	.	4.96	140.13	69	145	35	1.97	0.88	3.5	100
	scp64	.	.	4.93	129	49	131	38	1.29	1.00	1.6	100
	scp65	.	.	4.97	153.35	76	161	36	2.11	0.79	5.0	94.0
[2]	scpnre1	500	5000	9.98	21.38	88	29	29	3.03	2.16	35.6	100
	scpnre2	.	.	9.98	22.36	89	30	26	3.42	1.72	34.2	97.6
	scpnre3	.	.	9.98	20.49	81	27	25	3.24	1.64	31.8	100
	scpnre4	.	.	9.97	21.35	84	28	27	3.11	1.76	31.1	100
	scpnre5	.	.	9.98	21.32	83	28	26	3.19	1.78	31.3	100
	scpnrf1	500	5000	19.97	8.80	61	14	14	4.36	1.93	59.1	100
	scpnrf2	.	.	19.97	9.99	62	15	15	4.13	2.13	50.1	100
	scpnrf3	.	.	19.97	9.49	58	14	14	4.14	1.83	47.5	100
	scpnrf4	.	.	19.97	8.47	60	14	14	4.29	1.87	65.3	99.4
	scpnrf5	.	.	19.97	7.84	66	13	13	5.08	1.69	65.9	91.8
	scpnrg1	1000	10,000	2.00	159.89	268	176	104	2.58	1.25	10.1	99.1
	scpnrg2	.	.	2.00	142.07	253	154	102	2.48	1.23	8.4	95.1
	scpnrg3	.	.	2.00	148.27	263	166	105	2.50	1.27	12.0	94.5
	scpnrg4	.	.	2.00	148.95	263	168	103	2.55	1.27	12.8	97.2
	scpnrg5	.	.	2.00	148.23	264	168	103	2.56	1.30	13.3	95.5
	scpnrh1	1000	10,000	5.00	48.13	185	***63***	52	3.56	1.64	30.9	97.0
	scpnrh2	.	.	5.00	48.64	190	***63***	52	3.65	1.65	29.5	98.7
	scpnrh3	.	.	5.00	45.20	197	***59***	52	3.79	1.67	30.5	100
	scpnrh4	.	.	5.00	44.04	170	***58***	52	3.27	1.63	31.7	100
	scpnrh5	.	.	5.00	42.37	171	***55***	52	3.29	1.65	29.8	96.9
[3]	rail507	507	63,009	1.28	172.15	307	174	114	2.69	0.16	1.1	29.8
	rail582	582	55,515	1.24	209.71	362	211	156	2.32	0.19	0.6	22.9
	rail2536	2536	1,081,841	0.40	688.40	939	689	431	2.18	0.45	0.1	24.7
	rail2586	2586	920,683	0.34	935.22	1692	***956***	591	2.86	0.16	2.2	38.9
	rail4284	4284	1,092,610	0.24	1054.06	2036	***1098***	732	2.78	0.42	4.2	99.5
	rail4872	4872	968,672	0.20	1509.64	2689	***1561***	1051	2.56	0.26	3.4	96.3
[4]	scpclr10	511	210	12.33	21	150	25	25	6.00	2.08	19.1	100
	scpclr11	1023	330	12.42	16.5	330	23	23	14.35	1.86	39.4	100
	scpclr12	2047	495	12.46	16.5	400	***23***	23	17.39	1.87	39.4	100
	scpclr13	4095	715	12.48	14.3	715	***24***	24	29.79	1.99	67.8	100
[4]	scpcyc06	240	192	2.08	48	96	***60***	60	1.60	0.25	25.0	100
	scpcyc07	672	448	0.89	112	224	***152***	152	1.47	0.36	35.7	100
	scpcyc08	1792	1024	0.39	256	920	***348***	348	2.64	0.36	35.9	100
	scpcyc09	4608	2304	0.17	576	1763	***820***	820	2.15	0.42	42.4	100
	scpcyc10	11,520	5120	0.08	1280	3820	***1984***	1984	1.93	0.55	55.0	100
	scpcyc11	28,160	11,264	0.04	2816	9374	***4524***	4524	2.07	0.61	60.7	100

Objective values z^*_{IP} in bolded italics are not proven to be optimal. Here, ρ, Γ_{rel}, ε_{rel} and δ_{rel} are in percentage. Recall that $\varepsilon_{\text{rel}} = 1 - \delta_{\text{rel}}$

4 Conclusions

The duality gap for a non-convex problem can be dissected into two terms: degree of near-optimality in a Lagrangian relaxation and degree of near-complementarity in the relaxed constraints. We have empirically studied these terms for a large collection of large-scale set covering problems, and their relationship to problem characteristics. A main conclusion is that large duality gaps are consistently caused solely by violation of complementarity, due to extensive excess coverage of constraints. As expected, the relative duality gap is largely affected by the density and shape of the problem.

Our observations should be exploited when designing heuristic approaches for large-scale set covering problems. In particular, our observations can be utilized when designing core problems for set covering problems. Core problems are restricted but feasible versions of full problems; such a problem should be of a manageable size and is constructed by selecting a subset of the original columns, see for example [3]. Our results indicate that if Γ is expected to be large then it can also be expected that $\varepsilon_{\text{rel}} = 0$. Since $\varepsilon(x^*, u^*) = \sum_{j=1}^{n} \bar{c}_j x_j^* \geq 0$, it is then likely that $\bar{c}_j > 0$ implies that $x_j^* = 0$ holds, and therefore columns with $\bar{c}_j > 0$ can most likely be excluded from the core problem. Otherwise, if Γ is expected to be moderate, the core problem must also contain variables with small non-zero reduced costs. This conclusion gives a theoretical justification of the core problem construction used in [3].

References

1. Beasley, J.E.: An algorithm for set covering problem. Eur. J. Oper. Res. **31**(1), 85–93 (1987)
2. Beasley, J.E.: A Lagrangian heuristic for set-covering problems. Nav. Res. Logist. **37**(1), 151–164 (1990)
3. Ceria, S., Nobili, P., Sassano, A.: A Lagrangian-based heuristic for large-scale set covering problems. Math. Program. **81**(2), 215–228 (1998)
4. Grossman, T., Wool, A.: Computational experience with approximation algorithms for the set covering problem. Eur. J. Oper. Res. **101**(1). 81–92 (1997)
5. Larsson, T., Patriksson, M.: Global optimality conditions for discrete and nonconvex optimization – With applications to Lagrangian heuristics and column generation. Oper. Res. **54**(3), 436–453 (2006)
6. Wolsey, L.A.: Integer Programming. Wiley, Hoboken (1998)

Layout Problems with Reachability Constraint

Michael Stiglmayr

Abstract Many design/layout processes of warehouses, depots or parking lots are subject to reachability constraints, i.e., each individual storage/parking space must be directly reachable without moving any other item/car. Since every storage/parking space must be adjacent to a corridor/street one can alternatively consider this type of layout problem as a network design problem of the corridors/streets.

More specifically, we consider the problem of placing quadratic parking spaces on a rectangular shaped parking lot such that each of it is connected to the exit by a street. We investigate the optimal design of parking lot as a combinatorial puzzle, which has—as it turns out—many relations to classical combinatorial optimization problems.

Keywords Combinatorial optimization · Network design problem · Maximum leaf spanning tree · Connected dominating set

1 Introduction

In contrast to articles [1, 2] in the area of civil engineering and architecture investigating the (optimal) design of a parking lot, we focus on the topological layout of the parking lot rather than on issues like the optimal width of parking spaces and streets, one- or two-way traffic, traffic flow, angle of parking spaces to the streets, or irregular shaped parking lots. These modeling assumptions and restrictions allow to formulate a complex practical problem as a combinatorial optimization problem. This model was proposed in the didactic textbook [3] as a combinatorial puzzle. In the Bachelor thesis [4] integer programming formulations were presented and solved using a constructive heuristic.

M. Stiglmayr (✉)
University of Wuppertal, Wuppertal, Germany
e-mail: stiglmayr@math.uni-wuppertal.de
http://www.uni-w.de/u9

J. S. Neufeld et al. (eds.), *Operations Research Proceedings 2019*, Operations Research Proceedings, https://doi.org/10.1007/978-3-030-48439-2_22

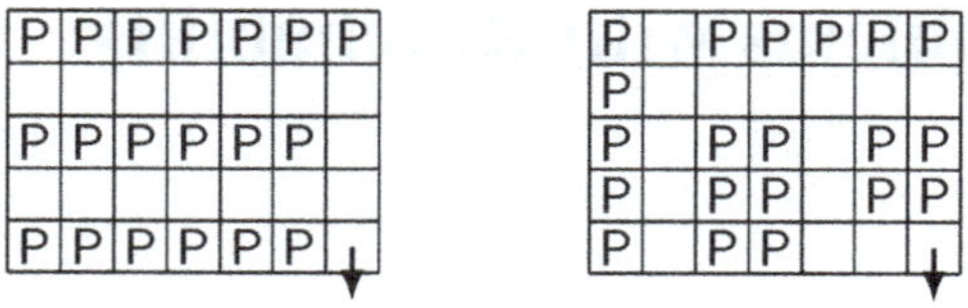

Fig. 1 Sketch of a parking lot layout, individual parking spaces are marked with P. Feasible solution with 19 parking spaces (on the left) and optimal solution with 20 parking spaces (right)

We search for the optimal layout of a parking lot on a given a rectangular shaped area maximizing the number of individual parking spaces. The rectangular area is subdivided by a square grid such that every cell of the grid represents either an individual parking space or a part of a street. For the reachability constraint we assume that cars can move on that grid only vertically or horizontally. This implies that every individual parking space must be connected by a street to the exit, where "connected" is defined based on the 4-neighborhood N_4 of a grid cell (see Fig. 1 for an example). One of the major modeling aspects is thereby the connectivity of street fields to the exit, for which we will present two different formulations. In general there may be cells not neighboring a street, which can not be used as parking space and are not streets fields connected to the exit. However, such *blocked* cells can be neglected since there is always a solution without blocked cells having the same number of individual parking spaces.

2 Model Formulations

2.1 Formulation Based on the Distance to the Exit

Let (i, j) denote a cell in the grid with $i \in \mathcal{I} = \{1, \ldots, m\}$ being its row index, $j \in \mathcal{J} = \{1, \ldots, n\}$ its column index. Then, we introduce a binary variable x_{ij} which is equal to one if (i, j) serves as a parking space and zero if (i, j) is part of a street. A method to model the reachability constraint based on the connectivity of street fields is to measure the discrete distance to the exit by a binary variable z_{ij}^k, with $k \in \mathcal{K} = \{1, \ldots, n \cdot m\}$ denoting the number of street cells to the exit. Thereby $z_{ij}^k = 1$, if (ij) represents a street field, which is k steps away from the exit, and zero otherwise. Note that many of these variables can be set to zero in advance, if k is smaller than the shortest path to the exit.

Then, the parking lot problem can be written as:

$$\max \sum_{i=1}^{n} \sum_{j=1}^{m} x_{ij} \tag{1}$$

$$\text{s. t.} \quad z_{ij}^k + x_{ij} \leq 1 \qquad \forall i, j, k \tag{1a}$$

$$z_{ij}^{k} \leq z_{i-1,j}^{k-1} + z_{i,j-1}^{k-1} + z_{i+1,j}^{k-1} + z_{i,j+1}^{k-1} \qquad \forall i,j,k \qquad \text{(1b)}$$

$$x_{ij} \leq \sum_{k=1}^{mn} \left(z_{i-1,j}^{k} + z_{i,j-1}^{k} + z_{i+1,j}^{k} + z_{i,j+1}^{k} \right) \qquad \forall i,j \qquad \text{(1c)}$$

$$\sum_{k} z_{ij}^{k} \leq 1 \qquad \forall i,j \qquad \text{(1d)}$$

$$z_{m,n}^{0} = 1 \qquad \text{(1e)}$$

$$x_{ij}, z_{ij}^{k} \in \{0,1\} \qquad \forall i,j,k \qquad \text{(1f)}$$

The constraint (1a) ensures that the distance values $z_{ij}^{k} = 0$ for all k if the cell i, j is a parking space, i.e., the distance to the exit is only measured on and along the streets. A street cell can only have a distance of k to the exit if one of its neighboring cells has a distance of $k - 1$ to the exit (constraint (1b)) and one cell can not have two different distances to the exit (constraint (1d)). Note that (1a) and (1d) can be merged to $\sum_k z_{ij}^{k} + x_{ij} \leq 1 \ \forall i, j$. Constraint (1c) states that any parking space requires one neighboring street field, i.e., a cell with a distance to the exit. The large number of binary variables to formulate the reachability constraint, $\mathcal{O}(n^2m^2)$, makes this problem difficult for branch and bound solvers.

2.2 *Network Flow Based Formulation*

Consider the movement of the cars to leave the parking lot as a network flow. To model this approach we identify each cell of the grid by one node and connect a node (i, j) with a node (rs) if (rs) is in the 4-neighborhood of (i, j), i.e., $(rs) \in N_4(ij) := \{(i-1, j-1), (i-1, j+1), (i+1, j-1), (i+1, j+1)\}$. Since every cell is the potential location of a parking space, we set the supply of every node to one unit of flow, and associate with the exit node (in our instances node (mn)) a demand of $m\,n - 1$. Then, nodes without inflow represent parking spaces, all other nodes represent street fields. For this network flow based formulation we need two types of variables: continuous flow variables and binary decision variables.

$$x_{(ij),(rs)} \in \mathbb{R}_{+} \quad \text{flow between node } (ij) \text{ and node } (rs)$$

$$z_{(ij)} = \begin{cases} 1 & \text{if node } (ij) \text{ has not zero inflow} \\ 0 & \text{otherwise} \end{cases}$$

$$\min \sum_{i=1}^{n} \sum_{j=1}^{m} z_{(ij)} \tag{2}$$

$$\text{s.t.} \quad \sum_{(rs)\in N_4(ij)} x_{(ij),(rs)} - \sum_{(rs)\in N_4(ij)} x_{(rs),(ij)} = 1 \qquad \forall (ij) \setminus (mn) \tag{2a}$$

$$\sum_{(rs)\in N_4(mn)} x_{(mn),(rs)} - \sum_{(rs)\in N_4(mn)} x_{(rs),(mn)} = -m\,n + 1 \tag{2b}$$

$$\sum_{(rs)\in N_4(ij)} x_{(rs),(ij)} \leq M \cdot z_{(ij)} \quad \forall (ij) \tag{2c}$$

$$x_{(ij),(rs)} \in \mathbb{R}_+ \qquad (rs) \in N_4(ij) \tag{2d}$$

$$z_{(ij)} \in \{0, 1\} \tag{2e}$$

Equations (2a)–(2b) are classical flow conservation constraints. The big-M constraint (2c) couples the inflow to the decision variable z_{ij}: If the sum of incoming flow of a node i, j is zero, $z_{(ij)}$ can be set to zero. Otherwise, if there is a non zero inflow, $z_{(ij)} = 1$. Setting $M := m \cdot n - 1$ does not restrict the amount of inflow if $z_{(ij)} = 1$.

3 Properties of the Parking Lot Problem

3.1 Upper Bound and Geometrically Implied Constraints

We will in the following investigate theoretical properties of the parking lot problem, which hold for both problem formulations (1) and (2). Based on the 4-neighborhood obviously every street field serves directly at most three parking spaces, if it is end point of a street, two parking spaces, if it is in the middle of a straight street and one parking space, if it is a T-junction. Since every additional end point of a street is associated with one T-junction we obtain the following upper bound on the number of parking spaces:

$$\#\,\text{parking spaces} \leq \frac{2}{3}\, n\, m$$

Besides this global upper bound on the number of parking spaces, the grid structure of the problem allows to state several geometrically implied constraints which are all satisfied by at least one optimal solution.

no streets along the border A street along the border of the parking lot can be moved in parallel one step to the interior of the parking lot, where each street field can serve directly more than one parking space.

street field next to the exit The exit field has only two neighboring fields, only one of which can be a street field in a optimal solution.

street field on the border of every rectangle In the border of every rectangular shaped region of size larger than 3×3 is at least one street field.

street field in every row/column cut Every row $1 < i \leq m$ and every column $1 < j \leq n$ contains at least one street field.

3.2 Tree Structure

Theorem 1 *For any instance of* (2) *there is an optimal solution in which the edges with positive flow value form a spanning tree.*

Proof Consider a basic feasible solution of the minimum cost flow problem given by the supply and demand values of (2). The edges with strictly positive flow form a tree of the grid graph, since there are no capacity constraints and consequently all non-basic edges have a flow value of zero. Furthermore, the tree is spanning (every basic edge has positive flow), since every node has a supply of one flow unit.

Definition 1 (See e.g., [5]) Let $G = (V, E)$ be a graph. A *connected dominating set* in G is a subset $U \subseteq V$ for which the following two properties hold

- connected: $\forall u_1, u_2 \in U$ there is a path $P = (u_1, \ldots, u_2) \subset U$ from u_1 to u_2
- dominating: $\forall v \in V \setminus U \; \exists u \in U$ such that $(u, v) \in E$

Theorem 2 ([6]) *Let $n := |V|$ and $d := |U|$ be the cardinality of a minimum connected dominating set U, then $\ell = n - d$ is the maximal number of leafs of a spanning tree in $G = (V, E)$.*

Proof Let U be a minimum connected dominating set. Then there exists a tree T in U, and all nodes in $V \setminus U$ can be connected as leafs to T, consequently $\ell \geq n - d$. Contrary, let $T = (V, E')$ be a spanning tree in $G = (V, E)$ and L the set of leafs of T. Then $V \setminus L$ is a connected dominating set. Thus, $\ell = n - d$

Identifying the leafs with the individual parking spaces and the street fields with a connected dominating set, the maximum leaf spanning tree problem maximizes the number of individual parking spaces, while the minimum connected dominating set minimizes the number of street fields. Independently, Reis et al. [7] proposed a flow based formulation of the maximum leaf spanning tree problem which is equivalent to (2). Alternative formulation of the maximum leaf spanning tree problems are presented in [8].

Theorem 3 ([9]) *The maximum leaf spanning tree problem is $\mathcal{NP}$-complete even for planar graphs with maximum node degree 4.*

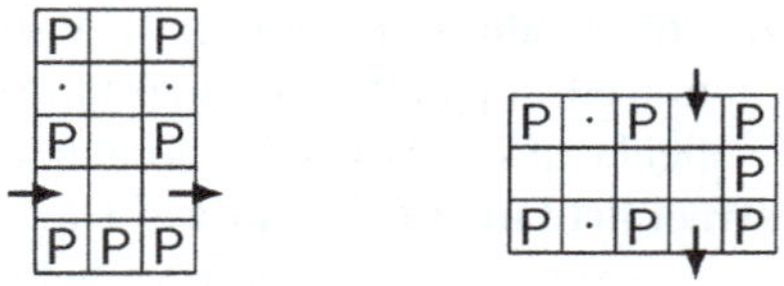

Fig. 2 Illustration of column- and row-wise building blocks

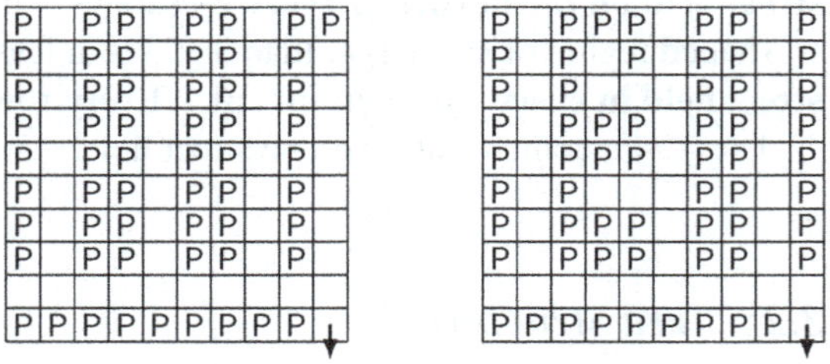

Fig. 3 Example of a suboptimal heuristic solution consisting of column building blocks for 10×10 with 58 parking spaces (left), optimal solution with 60 parking spaces (right)

Proof By reduction from *dominating set*, which can be reduced form *vertex cover*.

The parking lot problem is, thus, a special case of an $\mathcal{NP}$-complete optimization problem. In contrast to general maximum leaf spanning tree problems the parking lot problem has a very regular structure, such that this complexity result does not directly transfer.

4 Heuristic Solution Approach

In [4] a constructive heuristic is proposed, which is based on the use of building blocks of three row or columns, respectively (see Fig. 2). The parking lot is filled row- or column-wise with blocks of three rows/columns, where the last block of rows/columns has one additional street field at the exit. If the number of rows n is not a multiple of three, one or two rows remain, which can be used for one or $m-1$ additional parking spaces. Analogously, in the case of column building blocks. Based on the number of rows and columns the performance of the row- and column-wise building blocks differs. This constructive heuristic works best if the number of rows/columns is a multiple of three, since the building blocks achieve the theoretical upper bound of $\frac{2}{3}$. See Fig. 3 for a suboptimal heuristic solution in comparison to the optimal solution.

5 Conclusion

We presented two integer programming formulations of the parking lot problem and focused thereby in particular on the reachability constraint. The first model (1) based on a distance along the streets to the exit is intuitive but requires many binary variables. However, this formulation allows to limit the distance to the exit which could be relevant, e.g., in evacuation scenarios. In the second model (2), which is based on network flows, the distances to the exit are not encoded. Possible extensions of it could, e.g., balance the flow on streets and thus avoid congestions.

References

1. Bingle, R., Meindertsma, D., Oostendorp, W., Klaasen, G.: Designing the optimal placement of spaces in a parking lot. Math. Modell. **9**(10), 765–776 (1987)
2. Abdelfatah, A.S., Taha, M.A.: Parking capacity optimization using linear programming. J. Traffic Logist. Eng., 2(3), 2014.
3. Verhulst, F., Walcher, S. (eds.): Das Zebra-Buch zur Geometrie. Springer, Berlin (2010)
4. Kleinhans, J.: Ganzzahlige Optimierung zur Bestimmung optimaler Parkplatz-Layouts. Bachelor Thesis, Bergische Universität Wuppertal (2013)
5. Du, D.-Z., Wan, P.-J.: Connected Dominating Set: Theory and Applications, vol. 77. Springer Optimization and Its Applications. Springer, Berlin (2013)
6. Douglas, R.J.: NP-completeness and degree restricted spanning trees. Discret. Math. **105**(1), 41–47 (1992)
7. Reis, M.F., Lee, O., Usberti, F.L.: Flow-based formulation for the maximum leaf spanning tree problem. Electron Notes Discrete Math. **50**, 205–210 (2015). LAGOS'15 – VIII Latin-American Algorithms, Graphs and Optimization Symposium
8. Fujie, T.: The maximum-leaf spanning tree problem: Formulations and facets. Networks **43**(4), 212–223 (2004)
9. Garey, M.R., Johnson, D.S.: Computers and Intractability: A Guide to the Theory of NP-Completeness. W. H. Freeman, San Francisco (1979)

Modeling of a Rich Bin Packing Problem from Industry

Nils-Hassan Quttineh

Abstract We present and share the experience of modeling a real-life optimization problem. This exercise in modeling is a text book example of how a naive, straightforward mixed-integer modeling approach leads to a highly intractable model, while a deeper problem analysis leads to a non-standard, much stronger model. Our development process went from a weak model with burdensome run times, via meta-heuristics and column generation, to end up with a strong model which solves the problem within seconds. The problem in question deals with the challenges of planning the order-driven continuous casting production at the Swedish steel producer SSAB. We study the cast planning problem, where the objective is to minimize production waste which unavoidably occurs as orders of different steel grades are cast in sequence. This application can be categorised as a rich bin packing problem.

Keywords Mixed-integer programming · Cutting and packing · Industrial optimization

1 The Cast Planning Problem

We present the challenges of planning the order-driven continuous casting production at the Swedish steel producer SSAB. Customers place orders on slabs of a certain steel grade and specified width. Currently more than 200 steel grades are available, and possible slab widths are within [800, 1600] millimeters. Slabs are produced in a continuous caster, and steel grades are produced in batches of 130 tonnes. A single order might be as small as 10–15 tonnes, hence orders of the same (or similar) steel grade are identified and cast simultaneously. (This is the continuous

N.-H. Quttineh (✉)
Department of Mathematics, Linköping University, Linköping, Sweden
e-mail: nils-hassan.quttineh@liu.se
http://users.mai.liu.se/nilqu94

J. S. Neufeld et al. (eds.), *Operations Research Proceedings 2019*, Operations Research Proceedings, https://doi.org/10.1007/978-3-030-48439-2_23

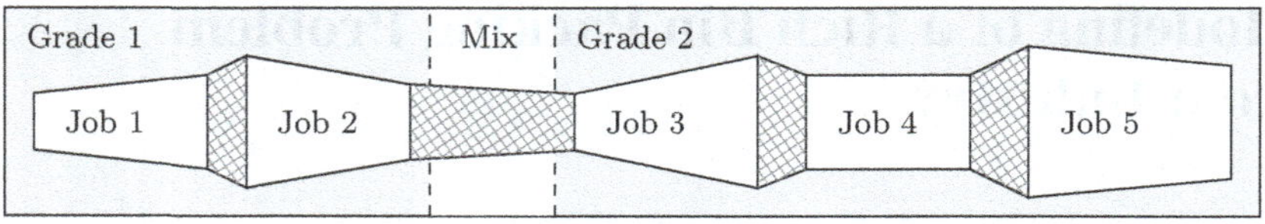

Fig. 1 A job sequence for a single tundish, and the waste (gray areas) generated

casting problem, see e.g. [1, 2].) Each batch of melted metal of a certain steel grade is emptied into a *tundish* which due to the extreme heat wears down over time. Hence the tundish needs to be replaced regularly.

Assume we are given a set of jobs $i \in \mathcal{I}$ that should be cast, where each job represents one or more batches of the same steel grade. Each job has a size (number of batches), a wear on the tundish, and start and end cast widths inherited from the first and last slab in the job. The objective is to assign and sequence all jobs into identical tundishes (bins) so that production waste, which occurs as jobs of different steel grades are cast in sequence, is minimized.

There are many complicating factors. A single tundish can withstand at most $\overline{P} = 5$ batches and must be replaced whenever the accumulated wear is above a certain threshold. When casting jobs of different steel grades in sequence, a so called mix-zone (waste) is created. Certain grades are more alike than others, hence the ordering of the jobs in a tundish affects the amount of waste created. Further, start and end widths of jobs usually differ, and this also causes waste (since the casting is a continuous process). All this is illustrated in Fig. 1.

2 Mixed-Integer Programming Model

We define the Cast Planning Problem (CPP), where the degrees of freedom are (1) the assignment of jobs to tundishes, (2) the sequence of the jobs in a tundish, and (3) a job can either be cast from wide to narrow or the other way around. All sets, parameters and variables are defined in Table 1. Many variants of this problem have been studied, see for example [3] and [4].

The objective function (1) strives to minimize the total waste generated and the penalty costs for not scheduling jobs.

$$[\text{CPP}] \qquad \min \sum_{i\in\mathcal{I}}\sum_{j\in\mathcal{I}}\sum_{m\in\mathcal{M}}\sum_{n\in\mathcal{M}} c_{ij}^{mn} y_{ij}^{mn} + \sum_{i\in\mathcal{I}} f_i u_i \tag{1}$$

subject to

$$\sum_{m\in\mathcal{M}}\sum_{p\in\mathcal{P}}\sum_{t\in\mathcal{T}} x_{ipt}^{m} = 1 - u_i, \qquad i \in \mathcal{I} \tag{2}$$

Table 1 Sets, parameters and variables

Sets and indices	
$\mathcal{I}$, $\mathcal{T}$	Set of jobs i to be cast, and set of tundishes t available
$\mathcal{P}$, $\mathcal{P}^*$	Set of sequence numbers $p = 1, \ldots, \overline{P}$ in a tundish, where $\mathcal{P}^* = \mathcal{P} \setminus \{\overline{P}\}$
$\mathcal{M}$	Set of modes m for a job: either `wide-to-narrow` or `narrow-to-wide`
Parameters	
c_{ij}^{mn}	Waste created if job i, in mode m, preceeds job j, in mode n
f_i	Penalty cost for *not* scheduling job i
q_i, w_i	Job size (number of batches) and wear on a tundish for job i
$\underline{P}$, $\overline{P}$	Minimal and maximal number of batches in a tundish
$\overline{W}$	Maximal allowed wear on a tundish
Variables	
x_{ipt}^m	1 if job i in mode m is given sequence number p in tundish t, 0 otherwise
y_{ij}^{mn}	1 if job i in mode m preceeds job j in mode n, 0 otherwise
z_{pt}	Equals 1 if sequence number p of tundish t is used, and 0 otherwise
$\overline{z}_t$	Equals 1 if tundish t is used, and 0 otherwise
u_i	Equals 1 if job i is *not* scheduled, and 0 otherwise

$$\sum_{i\in\mathcal{I}}\sum_{m\in\mathcal{M}} x_{ipt}^m = z_{pt}, \qquad p \in \mathcal{P},\ t \in \mathcal{T} \tag{3}$$

$$z_{p+1,t} \le z_{pt}, \qquad p \in \mathcal{P}^*,\ t \in \mathcal{T} \tag{4}$$

$$z_{pt} \le \overline{z}_t, \qquad p \in \mathcal{P},\ t \in \mathcal{T} \tag{5}$$

$$\underline{P}\cdot\overline{z}_t \le \sum_{i\in\mathcal{I}}\sum_{m\in\mathcal{M}}\sum_{p\in\mathcal{P}} q_i x_{ipt}^m \le \overline{P}\cdot\overline{z}_t, \qquad t \in \mathcal{T} \tag{6}$$

$$\sum_{i\in\mathcal{I}}\sum_{m\in\mathcal{M}}\sum_{p\in\mathcal{P}} w_i x_{ipt}^m \le \overline{W}\cdot\overline{z}_t, \qquad t \in \mathcal{T} \tag{7}$$

$$x_{ipt}^m + x_{j,p+1,t}^n - 1 \le y_{ij}^{mn}, \qquad i,j \in \mathcal{I},\ m,n \in \mathcal{M},\ p \in \mathcal{P}^*,\ t \in \mathcal{T} \tag{8}$$

and

$$x_{ipt}^m,\ y_{ij}^{mn},\ z_{pt},\ \overline{z}_t,\ u_i \in \{0,1\}, \qquad \forall\, i,j,m,n,p,t \tag{9}$$

Constraint (2) assigns each job i, if scheduled, to exactly one sequence number in one tundish. Constraint (3) states that if a sequence number in a tundish is used, exactly one job must be assigned to it. Constraint (4) states that if a sequence number in a tundish is used, the previous sequence number must also be used. Constraint (5) states that if any sequence number in a tundish is used, the tundish is used. Constraints (6) and (7) are tundish knapsack constraints for number of batches and allowed wear. Constraint (8) is used to decide if a job i preceeds another job j or not. Since tundishes are identical the model contains a lot of symmetry, therefore standard symmetry-breaking constraints are added.

This MIP model is very weak, and it is difficult to prove optimality even for small problem instances. It is therefore necessary to seek another approach.

3 Column Generation Approach

In a column generation approach for the CPP, a relaxation of a Master Problem (MP) and a Column Generation Problem (CGP) are solved alternatingly. A column represents an optimal setup (with respect to sequence and mode) for a subset of jobs to be cast in a single tundish.

Let $\mathcal{N}$ be the set of all feasible columns, and assume we currently consider a subset $\mathcal{K}$ of those columns. Further, we introduce a binary variable v_k which is 1 if a certain column $k \in \mathcal{K}$ is used and 0 otherwise, while variable u_i is the same as before. Parameter $T = |\mathcal{T}|$ is the number of available tundishes.

3.1 Master Problem

$$[\text{MP}] \qquad \min \sum_{k \in \mathcal{K}} c_k v_k + \sum_{i \in \mathcal{I}} f_i u_i \tag{10}$$

$$\text{s.t.} \quad \sum_{k \in \mathcal{K}} a_{ik} v_k = 1 - u_i, \qquad i \in \mathcal{I} \qquad | \; \lambda_i \tag{11}$$

$$\sum_{k \in \mathcal{K}} v_k \leq T, \qquad | \; \pi \tag{12}$$

$$v_k \in \{0, 1\}, \qquad k \in \mathcal{K} \tag{13}$$

$$u_i \in \{0, 1\}, \qquad i \in \mathcal{I} \tag{14}$$

Here the binary parameter a_{ik} specifies if a certain job i is part of column k or not, and it is defined according to

$$a_{ik} = \sum_{m \in \mathcal{M}} \sum_{p \in \mathcal{P}} x_{ip}^{mk}, \qquad i \in \mathcal{I}, \tag{15}$$

where the binary parameters x_{ip}^{mk} come from the CGP solution k and they specify whether a certain job i is scheduled at sequence number p in mode m or not. Further, the cost of a column k (the waste it generates) is defined by

$$c_k = \sum_{i\in\mathcal{I}} \sum_{j\in\mathcal{I}} \sum_{m\in\mathcal{M}} \sum_{n\in\mathcal{M}} c_{ij}^{mn} y_{ij}^{mnk}, \tag{16}$$

where parameters y_{ij}^{mnk} also come from the CGP solution k. In order to get dual information, λ_i and π, to CGP, the linear programming relaxation of the MP is solved. Constraints (13) and (14) are replaced by $v_k \geq 0$ and $u_i \geq 0$. The relaxation of the upper bounds on these variables is valid due to the nature of the objective function.

3.2 Column Generation Problem

The column generation problem is solved for a single tundish, since the tundishes are identical, and therefore constraints and variables are modified accordingly.

$$[\text{CGP}] \qquad \min \sum_{i\in\mathcal{I}} \sum_{j\in\mathcal{I}} \sum_{m\in\mathcal{M}} \sum_{n\in\mathcal{M}} c_{ij}^{mn} y_{ij}^{mn} - \sum_{i\in\mathcal{I}} \sum_{m\in\mathcal{M}} \sum_{p\in\mathcal{P}} \lambda_i x_{ip}^{m} - \pi \tag{17}$$

$$\text{s.t.} \quad \sum_{m\in\mathcal{M}} \sum_{p\in\mathcal{P}} x_{ip}^{m} \leq 1, \qquad i \in \mathcal{I} \tag{18}$$

and

$$(3), (4), (6), (7), (8), (9).$$

Columns are added to the MP as long as the reduced cost of the generated column (the objective value) is negative. When no more favourable column is found, the integer version of MP is solved once to produce a feasible solution.

3.3 Column Enumeration

The column generation approach improves both solution times and lower bounds, but is not always able to find an optimal solution. Since the final integer version of MP is always solved within a second, we also investigate the possibility to generate all columns a priori, that is, complete column enumeration.

4 Numerical Results

For a set of problem instances derived from company order data, we provide numerical results in Table 2. The instances span different combinations of jobs and steel grades, and since it is difficult to compare the cost of increased waste and the use of an additional tundish, different values of parameter T are used. Note that the number of jobs does not alone affect the difficulty of the instances.

We have implemented the models using `AMPL` and utilized the general purpose solver `cplex`, both for the full model as well as for the column generation and enumeration models. Parameter values $\underline{P} = 3$, $\overline{P} = 5$, and $\overline{W} = 650$ have been used

Table 2 Problem ID, number of jobs and number of tundishes available

Problem			CPP/`cplex`			Column Generation (CG)					Enumeration			
ID	$\|I\|$	T	10 m	1 h	Gap	#	Time	z^*_{LP}	z^*_{IP}	Gap	$\|\mathcal{N}\|$	Time	z^*	Time
1	19	7	162	162	67%	65	88	161.00	**161**	–	2317	3	161	0.02
		8	147	**145**	72%	56	70	145.00	147	1%			145	0.02
		9	138	**132**	78%	61	70	132.00	**132**	–			132	0.02
2	22	9	183	**182**	80%	61	63	182.00	189	4%	2014	2	182	0.03
		10	165	163	91%	65	67	161.00	**161**	–			161	0.02
		11	149	**142**	92%	64	65	142.00	**142**	–			142	0.01
3	23	9	196	**174**	91%	78	234	173.00	175	1%	2900	2.5	174	0.03
		10	162	160	91%	77	190	157.00	**157**	–			157	0.02
		11	152	144	94%	78	173	143.00	**143**	–			143	0.02
4	24	15	**119**	**119**	86%	31	32	119.00	**119**	–	158	0.1	119	0.01
		16	**96**	**96**	89%	29	30	96.00	**96**	–			96	0.01
		17	**81**	**81**	79%	36	37	81.00	**81**	–			81	0.01
5	25	15	123	**121**	100%	47	48	121.00	**121**	–	284	0.1	121	0.01
		16	102	**99**	97%	42	43	99.00	**99**	–			99	0.01
		17	**84**	**84**	95%	45	46	84.00	**84**	–			84	0.01
6	28	14	203	197	100%	83	159	183.00	**183**	–	1913	2	183	0.02
		15	182	173	100%	80	131	168.50	**169**	–			169	0.02
		16	171	157	100%	84	121	155.00	**155**	–			155	0.01
7	28	15	235	209	100%	96	496	184.00	190	3%	7940	6.5	185	0.07
		16	212	195	100%	103	563	171.00	175	2%			172	0.07
		17	212	179	100%	102	434	159.50	162	1%			160	0.07
8	35	20	219	196	100%	89	177	187.00	**187**	–	1829	2	187	0.03
		21	197	174	100%	88	139	167.00	178	6%			167	0.02
		22	187	156	100%	94	148	152.50	154	1%			153	0.02

Best found solution in CPP after 10 min and 1 h by `cplex`, and its gap after 1 h. For CG, the number of columns generated, solution time (in seconds), the LP and IP optimal values, and the gap. For Enumeration, the total number of feasible columns and the time needed to generate them, optimal objective value, and solution time (in seconds) for all problem instances. The penalty costs used, $f_i = 1000$, were high enough to always schedule all jobs (i.e. $u_i^* = 0$ for all i).

The CPP model produces feasible solutions of good quality for all instances, sometimes actually optimal (in bold), but even after 1 h of computing time the gap from `cplex` is huge. Comparing with the optimal solutions found by the enumeration approach, we see that the huge gaps are not only caused by weak lower bounds from CPP; best found solutions are in some cases far from optimal.

The column generation approach successfully generates high quality solutions within minutes, most of them optimal, and the lower bound quality is excellent. Finally, a complete enumeration of all feasible columns only takes a few seconds, and solving the corresponding "complete" master problem is instant.

5 Conclusions

We have presented the Cast Planning Problem where a set of jobs should be assigned to tundishes so that total waste is minimized. A straightforward MIP model is not ideal; it produces weak lower bounds and is not always able to find the optimal solution within 1 h. The use of stronger column variables, which contain more information, improves both solution times and solution quality considerably. Complete enumeration of columns turned out to be possible, and this model approach can be generated and solved to optimality within seconds.

The technique of complete enumeration yields a strong model (when feasible, of course). Further, it allows for taking all kinds of complicated constraints into account, such as non-linear expressions and logical tests that could be challenging to model. For example, assume there are restrictions that forbid certain jobs to be sequenced last in a tundish. Although possible to incorporate in a mathematical model, it requires additional variables and constraints since one does not know in advance how many jobs that will be assigned to a specific tundish. In an enumeration scheme, such restrictions are trivial to incorporate.

References

1. Bellabdaoui, A., Teghem, J.: A mixed-integer linear programming model for the continuous casting planning. Int. J. Prod. Econ. **104**, 260–270 (2006)
2. Chang, S.Y., Chang, M.-R., Hong, Y.: A lot grouping algorithm for a continuous slab caster in an integrated steel mill. Prod. Plan. Control **11**(4), 363–368 (2000)
3. Tang, L., Luo, J.: A new ILS algorithm for cast planning problem in steel industry. ISIJ Int. **47**(3), 443–452 (2007)
4. Yang, F., Wang, G., Li, Q.: Self-organizing optimization algorithm for cast planning of steel making — Continuous casting. In: 2014 IEEE International Conference on System Science and Engineering (ICSSE), pp. 210–214. https://doi.org/10.1109/ICSSE.2014.6887936

Optimized Resource Allocation and Task Offload Orchestration for Service-Oriented Networks

Betül Ahat, Necati Aras, Kuban Altınel, Ahmet Cihat Baktır, and Cem Ersoy

Abstract With the expansion of mobile devices and new trends in mobile communication technologies, there is an increasing demand for diversified services. Thus, it becomes crucial for a service provider to optimize resource allocation decisions to satisfy the service requirements. In this paper, we propose a stochastic programming model to determine server placement and service deployment decisions given a budget restriction when certain service parameters are random. Our computational tests show that the Sample Average Approximation method can effectively find good solutions for different network topologies.

Keywords Stochastic programming · Network optimization

1 Introduction

The popularity of mobile devices has led to a rapid evolution on mobile communication industry since it enables offering new customized services with various attributes in terms of resource consumption and latency tolerance. Although such devices are getting more powerful, they are still restricted in terms of battery life and storage, which makes it hard to process various complicated services locally. Extending their capabilities by offloading the applications to central cloud cannot solve the problem as it imposes additional load on the network and introduces a significant Wide Area Network (WAN) delay. An emerging concept called edge computing brings computational resources closer to the end-users, enabling to run highly demanding services at the edge of the network to meet strict delay requirements defined in their Service Level Agreements (SLAs). Within this

B. Ahat (✉) · N. Aras · K. Altınel
Boğaziçi University, Department of Industrial Engineering, Istanbul, Turkey

A. C. Baktır · C. Ersoy
Boğaziçi University, Department of Computer Engineering, Istanbul, Turkey

J. S. Neufeld et al. (eds.), *Operations Research Proceedings 2019*, Operations Research Proceedings, https://doi.org/10.1007/978-3-030-48439-2_24

context, optimal resource allocation and task offloading scheme is crucial to handle the service requests within the specified delay limit [1].

Server placement and service deployment decisions are among key factors having an effect on the profitability of a service provider. In this study, we investigate a multi-tier computation architecture where dynamic changes in the number of user requests are taken into account and the resource consumption of different services is uncertain. The aim is to maximize the expected profit of successfully handled requests by optimally allocating computational resources for a limited budget. We formulate a mixed-integer linear stochastic programming model and employ Sample Average Approximation (SAA) method suggested by Kleywegt et al. [3] for its solution. The performance of the solution method is assessed on realistic test instances.

The remainder of the paper is organized as follows. In Sect. 2, we describe the problem and give the two-stage stochastic integer programming formulation. Computational results are presented in Sect. 3. A brief discussion on the possible future research directions is given in the last section.

2 Problem Definition

In this problem, the network topology is assumed to be given in advance, where N represents the nodes of the network. The set U denotes the end-user locations, which are considered as aggregated demand points. There is a set of services, denoted by Q, with different characteristics including unit revenue, computation load on servers, and network load on nodes. The number of service requests from a user location may vary in time. Similarly, two different requests from the same service type may require different load on the network and computational resources. We let the number of user requests and service load requirements to be stochastic, but assume that their distribution is known. Since most user-centric services, such as augmented reality and healthcare applications, are latency intolerant, services have a maximum acceptable delay requirement.

The set S denotes the potential server locations on the network and there are discrete capacity levels for the servers. For each level, capital costs and server capacities are specified. To operate effectively, the maximum number of service instances provided by a server is restricted depending on the server capacity level. To eliminate the possibility of excessive delay, the maximum utilization for networking and computational resources is set to ϕ, where $0 < \phi < 1$. Finally, the total budget that can be spent by the service provider for server deployment decisions is also given. All index sets, parameters, and decision variables utilized in the model are summarized below:

Sets:

N:	Nodes
U:	User locations

S:	Potential server locations
L:	Capacity levels of the server
Q:	Services
$N_{us} \in N$:	Nodes on the shortest path between user location u and potential server location s

Parameters:

a_n:	Capacity of node n
e_l:	Capacity of server at capacity level l
c_l:	Capital cost of server at capacity level l
n_l:	Max. number of service deployments on a server at capacity level l
d_{uq}:	Total number of requests from user location u for service type q
r_q:	Unit revenue obtained by satisfying a service type q request
m_q:	Computation load on server for service type q
h_q:	Network load on nodes for service type q
α_q:	Max. allowed delay of service type q
ϕ:	Max. allowed utilization for networking and computational resources
b:	Total budget for server placement decisions

Decision Variables:

X_{sl}:	1 if level l server is placed at server location s; 0 otherwise
Y_{qs}:	1 if service type q is deployed at server location s; 0 otherwise
θ_{uqs}:	Fraction of service type q requests from user location u that is assigned to server location s
F_s:	Total flow on server location s
F_n:	Total flow on node n
Z_{uqs}:	1 if type q service requests from user location u are ever assigned to server location s; 0 otherwise

Let $\boldsymbol{\xi} = (\boldsymbol{d}, \boldsymbol{m}, \boldsymbol{h})$ represent the random data vector corresponding to the number of user requests, computation and network load requirements with known distribution. Also let the parameters $\xi = (d, m, h)$ be actual realizations of the random data. Using this notation, two-stage stochastic integer programming formulation of the problem can be written as follows

$$\max \mathbb{E}[Q(\boldsymbol{x}, \boldsymbol{y}, \boldsymbol{\xi})] \tag{1}$$

$$\text{s.t.} \sum_{l} X_{sl} \leq 1 \qquad \forall s \tag{2}$$

$$\sum_{s} \sum_{l} c_l X_{sl} \leq b \tag{3}$$

$$\sum_{q} Y_{qs} \leq \sum_{l} n_l X_{sl} \qquad \forall s \tag{4}$$

$$X_{sl}, Y_{qs} \in \{0, 1\}, \tag{5}$$

where $Q(\boldsymbol{x}, \boldsymbol{y}, \xi)$ is the optimal value of the second-stage problem

$$\max \sum_u \sum_q \sum_s r_q d_{uq} \theta_{uqs} \tag{6}$$

$$\text{s.t.} \sum_s \theta_{uqs} \le 1 \qquad \forall u, q \tag{7}$$

$$\theta_{uqs} \le Y_{qs} \qquad \forall u, q, s \tag{8}$$

$$F_s = \sum_u \sum_q m_q d_{uq} \theta_{uqs} \qquad \forall s \tag{9}$$

$$F_s \le \phi \sum_l e_l X_{sl} \qquad \forall s \tag{10}$$

$$F_n = \sum_q \sum_{(u,s):n \in N_{us}} h_q d_{uq} \theta_{uqs} \qquad \forall n \tag{11}$$

$$F_n \le \phi \, a_n \qquad \forall n \tag{12}$$

$$\theta_{uqs} \ge 0 \qquad \forall u, q, s \tag{13}$$

Note that $Q(\boldsymbol{x}, \boldsymbol{y}, \xi)$ is a function of the first-stage decision variables $\boldsymbol{x}$ and $\boldsymbol{y}$, and a realization $\xi = (d, m, h)$ of the random parameters. $\mathbb{E}[Q(\boldsymbol{x}, \boldsymbol{y}, \boldsymbol{\xi})]$ denotes the expected revenue obtained by the satisfaction of the user requests. In the first stage, we determine the server placement and service deployment decisions before the realization of the uncertain data. In the second stage, after a realization of ξ becomes available, we optimize the task assignments for the given server placement and service deployment decisions.

The objective function of the first stage problem (1) tries to maximize the expected revenue. Constraints (2) enforce that at most one server can be placed at every potential server location. Constraint (3) guarantees that the total capital cost for server placement decisions cannot exceed the total budget. Constraints (4) state that the total number of service deployments cannot exceed the maximum number depending on the server level decisions at each server location.

In the second stage, when the server placement and service deployment decisions are made and the uncertain data is revealed, the model determines the task assignments to maximize the revenue while satisfying the delay requirements. The objective function (6) aims to maximize the total revenue of successful task assignments. Constraints (7) state that a service request by a user can be assigned to at most one server. Task assignment is valid only if the corresponding service is deployed on the server. This is guaranteed by constraints (8). The total flow on computational and network resources are calculated in constraints (9) and (11), respectively. To prevent excessive delay on both resources, the maximum utilization is bounded by the parameter ϕ, which is ensured by constraints (10) and (12).

Finally, most services in such an environment are latency-intolerant and their SLA definitions may impose maximum latency values to enhance user experience. Therefore, the end-to-end delay for each successful service request should not exceed the maximum delay limit of the service. The total end-to-end delay includes code execution on the server, and routing the request and the response between user location and server through nodes and links. In this study, we assume that a service request and its corresponding response follow the shortest path in terms of the number of hops between user and server locations. On this path, the contribution of each node and link to the routing delay is assumed to be constant for each service. Therefore, the routing delay between each user and possible server location can be calculated by a preprocessing step. Moreover, the execution delay on servers is determined by the formulations of $M/M/1$ queuing model [2]. Let β_{uqs} denote the routing delay between user location u and server location s for service type q. Then, the delay requirement can be written as

$$\frac{m_q}{\sum_l e_l X_{sl} - F_s} \leq (\alpha_q - \beta_{uqs}) \qquad \text{if } \theta_{uqs} > 0 \tag{14}$$

Let Z_{uqs} be a binary decision variable that takes value 1 if service type q requests from user location u are ever assigned to server location s, and 0 otherwise. Then condition (14) can be rewritten as

$$m_q Z_{uqs} \leq (\alpha_q - \beta_{uqs})(\sum_l e_l X_{sl} - F_s) \qquad \forall u, q, s \tag{15}$$

$$\theta_{uqs} \leq Z_{uqs} \qquad \forall u, q, s \tag{16}$$

3 Computational Results

It is difficult to solve the stochastic program (1)–(5) since $\mathbb{E}[Q(\boldsymbol{x}, \boldsymbol{y}, \boldsymbol{\xi})]$ cannot be written in a closed-form expression. Therefore, we implement SAA scheme described by Kleywegt et al. [3] that consists of three phases. In the first phase, we generate $M = 20$ independent samples of size $N = 20$ and solve the SAA problem. Then, we calculate an upper statistical bound (UB) for the optimal value of the true problem. In the second phase, by fixing each optimal solution $(\boldsymbol{x}, \boldsymbol{y})$ obtained in the first phase of the SAA method, we solve the same problem with a sample size of $N' = 50$. The solution providing the largest estimated objective value is used to obtain an estimate of a lower bound (LB) for the true optimal value by solving $N'' = 1000$ independent second stage problems in the third phase. Finally, we compute an estimate of the optimality gap and its estimated standard deviation. In our preliminary analysis, we observe that these parameters are sufficient to obtain good results.

To evaluate the performance of the proposed method, three different topology instances from Topology Zoo [4] are utilized with varying number of nodes, users and server locations. Three service types are generated along with their diversified characteristics and requirements. Similarly, we allow three capacity levels for servers. For each topology, instances with low and high number of user requests are generated. The number of requests from each user location for any service type follows $U(10, 50)$ and $U(10, 100)$ for low and high user requests cases, respectively. Finally, we use three different budget levels to see their effect on the optimal objective value, optimality gap, and solution time. The proposed method is then implemented in C++ with CPLEX 12.8 running on a computer with Intel Xeon E5-2690 2.6 GHz CPU and 64 GB main memory.

The results of the computational study are provided in Table 1. First of all, the overall performance of the method seems to be satisfactory since small optimality gaps are obtained. It can be observed that the SAA method requires more time as the topology size increases. For each topology, as the expected number of user requests increases, the LB and UB for the optimal objective value increase. However, for higher number of the user requests, the standard deviation of the optimality gap is also large since the uncertainty in the input parameters gets higher. In addition, the increase in the budget that can be spent on server placement decisions also increases the LB and UB for the optimal objective value. Since the problem becomes tighter at low budget values, the SAA method takes longer as the budget decreases.

Table 1 The results of the SAA method

(N, U, S)	User requests	Budget	LB	UB	Optimality gap (%)	St. Dev. of the gap	Time (s)
(11, 10, 10)	Low	70k	2615.7	2628.6	0.49	13.3	202.8
		80k	2647.1	2658.3	0.42	13.8	157.9
		90*k*	2652.1	2664.8	0.48	14.1	133.6
	High	70k	3526.4	3586.8	1.71	44.5	156.4
		80k	3880.3	3939.1	1.51	43.5	155.5
		90k	4181.2	4237.7	1.35	39.4	142.1
(18, 15, 15)	Low	80*k*	3623.0	3651.5	0.79	24.7	1542.2
		100k	3907.6	3924.9	0.44	17.5	530.3
		120k	3973.8	3989.3	0.39	18.0	354.9
	High	80k	4357.1	4361.6	0.11	45.4	1357.8
		100k	5203.5	5214.1	0.20	50.9	403.7
		120k	5926.5	5962.3	0.60	50.4	409.1
(25, 20, 20)	Low	100*k*	4705.0	4710.8	0.12	34.0	4711.9
		120k	5085.8	5108.6	0.45	22.0	1804.5
		150k	5282.6	5318.3	0.68	15.7	727.4
	High	100k	5408.0	5491.2	1.54	60.9	6662.2
		120k	6321.1	6376.4	0.87	69.6	1337.5
		150k	7504.3	7561.3	0.76	72.5	848.0

4 Conclusions and Future Research

In this paper, we have studied the resource allocation and task assignment problem under uncertainty. The SAA method is utilized to provide an efficient framework to decide on server placement and service deployment decisions. Our computational results reveal the efficacy of the method on realistic use-cases. As the increase in the network size increases the solution time significantly, one can focus on acceleration techniques on the SAA method to solve larger instances and enhance the solution quality.

Acknowledgments The first two authors was partially supported by Boğaziçi University Scientific Research Project under the Grant number: BAP 14522.

References

1. Baktır, A.C., Özgövde, A., Ersoy, C.: How can edge computing benefit from software-defined networking: A survey, use cases, and future directions. IEEE Commun. Surv. Tutorials **19**(4), 2359–2391 (2017)
2. Jia, M., Cao, J., Liang, W.: Optimal cloudlet placement and user to cloudlet allocation in wireless metropolitan area networks. IEEE Trans. Cloud Comput. **5**(4), 725–737 (2015)
3. Kleywegt, A.J., Shapiro, A., Homem-de Mello, T.: The sample average approximation method for stochastic discrete optimization. SIAM J. Optim. **12**(2), 479–502 (2002)
4. Knight, S., Nguyen, H.X., Falkner, N., Bowden, R., Roughan, M.: The internet topology zoo. IEEE J. Sel. Areas Commun. **29**(9), 1765–1775 (2011)

Job Shop Scheduling with Flexible Energy Prices and Time Windows

Andreas Bley and Andreas Linß

Abstract We consider a variant of the job shop scheduling problem, which considers different operational states of the machines (such as off, ramp up, setup, processing, standby and ramp down) and time-dependent energy prices and aims at minimizing the energy consumption of the machines. We propose an integer programming formulation that uses binary variables to explicitly describe the non-operational periods of the machines and present a branch-and-price approach for its solution. Our computational experiments show that this approach outperforms the natural time-indexed formulation.

1 Introduction

Energy efficiency in scheduling problems is a relevant topic in modern economy. Instead of only optimizing the makespan, Weinert et al. [8] introduced the energy blocks methodology, including energy efficiency into scheduling. Dai et al. [2] improved a genetic simulated annealing algorithm to model and solve energy efficient scheduling problems with makespan and energy consumption. Shrouf et al. [6] proposed a formulation of the single machine scheduling problem including the machine states processing, off, idling and the transition states of turning on and off. Selmair et al. [5] extend this approach to multiple machines proposing a time-indexed IP formulation for the job shop scheduling problem with flexible energy prices and time windows. It is known that time-indexed formulations often provide better dual bounds than time-continuous models. However, they lead to huge formulations that are hard to solve. An alternative and computational successful approach is to use column generation and branch-and-price techniques to solve scheduling problems, c.f. van den Akker [7], for example.

A. Bley · A. Linß (✉)
Universität Kassel, Institut für Mathematik, Kassel, Germany
e-mail: andreas.bley@uni-kassel.de; andreas.linss@uni-kassel.de

J. S. Neufeld et al. (eds.), *Operations Research Proceedings 2019*, Operations Research Proceedings, https://doi.org/10.1007/978-3-030-48439-2_25

2 Problem Description

In this paper, we consider the job shop scheduling problem with flexible energy prices and time windows introduced in [5]. We are given a set of (non-uniform) machines $M := \{1, \dots, n_M\}$ that have to process a set of jobs $J := \{1, \dots, n_J\}$. A job $j \in J$ consists of a finite list of operations O_j, which we call tasks (j, k) with $j \in J$, $k \in O_j$. The tasks of each job have to obey the precedence order given by the sequence of the operations, i.e., task (j, k) can start only if $(j, k-1)$ is finished. We let $O := \{(j, k) : j \in J,\ k \in O_j\}$.

The planning horizon is a discretized time window $[T] := \{0, \dots, T-1\}$ consisting of T (uniform) time periods. For each $t \in [T]$, we are given the energy price $C_t \in \mathbb{R}_{\geq 0}$ at time t (that is assumed to be valid for the period t to $t+1$). For each task $(j, k) \in O$, we are given its setup time $d^{se}_{j,k} \in \mathbb{N}$ and its processing time $d^{pr}_{j,k} \in \mathbb{N}$ on its associated machine $m_{j,k} \in M$. In addition, we are given a release time a_j and due date f_j for each job $j \in J$, that apply to the first and the last task of the job respectively.

In each period, a machine can be in one of the operating states: *off*, *processing*, *setup*, *standby*, *ramp-up* or *ramp-down*, summarized as $\mathcal{S} = \{\mathit{off}, \mathit{pr}, \mathit{se}, \mathit{st}, \mathit{ru}, \mathit{rd}\}$, with the canonical switches and implications. For each machine $i \in M$ and state $s \in \mathcal{S}$, P^s_i denotes the energy demand of machine i in state s. The duration of the *ramp-up* phase, changing from state *off* to any state $s \in \{\mathit{se}, \mathit{pr}, \mathit{st}, \mathit{rd}\}$, is d^{ru}_i. The duration of the *ramp-down* phase is d^{rd}_i.

In our problem setting we assume that each machine is in exactly one operational state in each period. Also, it is not allowed to preempt any task and the setup for a task must be followed immediately by the processing of the task. Clearly, the processing of each task can start only after its predecessor task did complete its processing. However, the setup for a task can be performed already during its predecessor is still processed (on some other machine). Also, the setup may be performed before the release date of the job, the processing may not.

From the given release and due dates and the precedence constraints, we can easily derive the implied release and due dates for each task $(j, k) \in O$, with the placeholder $a_{j,0} = a_j$, as $a_{j,k} = \max\left(a_{j,k-1} + d^{pr}_{j,k-1}, d^{ru}_{m_{j,k}} + d^{se}_{j,k}\right)$ and $f_{j,k} = \min(f_j, T - d^{rd}_{m_{j,k}}) + 1 - \sum_{q=k}^{|O_j|} d^{pr}_{j,q}$, leading to the set $A_{j,k} := \{a_{j,k}, \dots, f_{j,k}\}$ of all valid start times of (j, k).

3 Problem Formulation

In contrast to [5], we propose a time-indexed formulation that uses variables to explicitly describe the non-operational periods of the machines. Our formulation consists of three different types of variables.

For each task $(j, k) \in O$ and period $t \in [T]$, a binary variable $x_{j,k,t}$ is used to indicate if the processing of task (j, k) on machine $m_{j,k}$ starts exactly in period t. Note that a value $x_{j,k,t} = 1$ implies that the setup for task (j, k) starts in period $t - d^{se}_{j,k}$, so machine $m_{j,k}$ is in state *setup* from $t - d^{se}_{j,k}$ until $t - 1$ and in state *processing* from t until $t + d^{pr}_{j,k} - 1$.

In order to model the other machine states, we introduce two types of variables. For each period $t \in [T]$ and machine $i \in M$, a binary variable $z^{st}_{i,t}$ indicates if i is in *standby* in period t. Additionally, so-called *break variables* are used to describe possible sequences of the transition states {*rd*, *off*, *ru*} of the machines. For each machine $i \in M$ and each pair of periods $t_0, t_1 \in [T]$ with $t_1 - t_0 \geq d^{ru}_i + d^{rd}_i$, a binary variable $z^{rd,ru}_{i,t_0,t_1}$ is used to indicate that machine m is starting to *ramp-down* at t_0, is *off* from $t_0 + d^{rd}_i$ until $t_1 - d^{ru}_i$, and in *ramp-up* from $t_1 - d^{ru}_i + 1$ to t_1. The energy costs induced by setting a task start variable $x_{j,k,t}$ or a break variable $z^{rd,ru}_{i,t_0,t_1}$ to 1 are

$$\hat{c}_{j,k,t} = \sum_{q=t-d^{se}_{j,k}}^{t-1} C_q P^{se}_{m_{j,k}} + \sum_{q=t}^{t+d^{pr}_{j,k}-1} C_q P^{pr}_{m_{j,k}} \tag{1}$$

$$\hat{d}_{t_0,t_1,i} = \sum_{q=t_0}^{t_0+d^{rd}_i-1} C_q P^{rd}_i + \sum_{q=t_1-d^{ru}_i}^{t_1} C_q P^{ru}_i \,. \tag{2}$$

In order to simplify the formulation and to also model the initial *ramp-up* and the final *ramp-down* of a machine using *break variables* (that start with a *ramp-down* and end with a *ramp-up*), we extend the time window for each machine $i \in M$ to $T_+{}^i := \{-d^{rd}_i, \ldots, T + d^{ru}_i - 1\}$ and enforce that the machine is in state *off* at time 0 and at time T. Setting $C_t = 0$ for $t \in T_+ \setminus T$, the costs of the corresponding break variable are set correctly in (2). The set of all feasible pairs (t_0, t_1), with $t_0, t_1 \in [T_+]$ and $t_1 - t_0 \geq d^{ru}_i + d^{rd}_i$, of these *ramp-down–off–ramp-up* phases of machine i is denoted by B_i. We obtain the following integer programming formulation of the problem:

$$\min \sum_{t\in[T]} \Big(C_t P^{st}_i z^{st}_{i,t} + \sum_{j\in J} \sum_{k\in O_j} \hat{c}_{j,k,t} x_{j,k,t} \Big) + \sum_{i\in M} \sum_{(t_0,t_1)\in B_i} \hat{d}_{t_0,t_1,i} z^{rd,ru}_{i,t_0,t_1} \tag{3}$$

$$\sum_{t\in[T]} x_{j,k,t} = 1 \qquad j \in J,\ k \in O_j \tag{4}$$

$$\sum_{q=0}^{t-d^{pr}_{j,k}} x_{j,k,q} - \sum_{q=0}^{t} x_{j,k+1,q} \geq 0 \quad j \in J,\ k \in [|O_j| - 1],\ t \in [T] \tag{5}$$

$$\sum_{\substack{(j,k)\in O:\\ m_{j,k}=i}} \sum_{q=\max(t-d^{pr}_{j,k},0)}^{\min(t+d^{se}_{j,k},T-1)} x_{j,k,q} + z^{st}_{i,t} + \sum_{\substack{(t_0,t_1)\in B_i:\\ t\in\{t_0,\dots,t_1\}}} z^{rd,ru}_{i,t_0,t_1} = 1 \quad i \in M,\ t \in [T] \tag{6}$$

$$\sum_{\substack{(t_0,t_1)\in B_i:\\ t\in\{t_0,\dots,t_1\}}} z^{rd,ru}_{i,t_0,t_1} = 1 \quad i \in M,\ t \in \{-d^{rd}_i, -1\} \tag{7}$$

$$\sum_{\substack{(t_0,t_1)\in B_i:\\ t\in\{t_0,\dots,t_1\}}} z^{rd,ru}_{i,t_0,t_1} = 1 \quad i \in M,\ t \in \{T, T + d^{ru}_i - 1\} \tag{8}$$

$$x_{j,k,t} \in \{0,1\} \quad (j,k) \in O,\ t \in [T] \tag{9}$$

$$z^{rd,ru}_{i,t_0,t_1} \in \{0,1\} \quad i \in M,\ (t_0,t_1) \in B_i \tag{10}$$

$$z^{st}_{i,t} \in \{0,1\} \quad i \in M,\ t \in [T] \tag{11}$$

The objective (3) describes the minimization of the costs of energy consumption caused by the standby, setup and processing, and by the break phases of the machines. Constraints (4) ensure the execution of each task and (5) the precedence order of the tasks. Constraints (6) ensure that each machine is in the proper state at each time. Constraints (7) and (8) ensure that all machines are *off* at the beginning and at the end of the planning horizon, while (9), (10) and (11) describe the domains of the variables.

4 Branch-and-Price Approach

We use a branch-and-price approach [1] to solve our model. Starting with a restricted master problem (RMP) using a subset $\hat{B}_i \subseteq B_i$ of the break variables of each $i \in M$, we iteratively (re-)solve the RMP and add missing variables $z^{rd,ru}_{i,t_0,t_1}$ with $(t_0, t_1) \in B_i \setminus \hat{B}_i$ with negative reduced costs when solving the LP-relaxation of (3)–(11). We choose $\hat{B}_i := \{(-d^{rd}_i, d^{ru}_i - 1), (T - d^{rd}_i, T + d^{ru}_i - 1)\}$ for the initial set of break variables, as this guarantees the feasibility of the restricted model if there exists any feasible solution for the given instance. The pricing problem for the break variables $z^{rd,ru}_{i,t_0,t_1}$, $(t_0, t_1) \in B_i$, can be formulated and solved as a shortest path problem with node weights and a limited number of offline periods in the time-expanded machine state network for each machine $i \in M$ individually: For each pair $s \in \{\textit{off}, ru, rd\}$ and $t \in T_+$, we introduce a node (s, t), and two nodes (s_1, t_1), (s_2, t_2) are connected by a directed arc if and only if one can switch from state s_1 to s_2 in exactly $t_2 - t_1$ periods. We also add artificial *start*- and *end*-nodes connected to the *ramp-down* and *ramp-up*-nodes as illustrated in Fig. 1 (with an additionally row to describe the time period of the columns).

Combining the dual variables $\pi_{i,t}$ of constraints (6),(7) and (8) to node weights ℓ as follows

Fig. 1 Time-expanded machine state network used in pricing problem

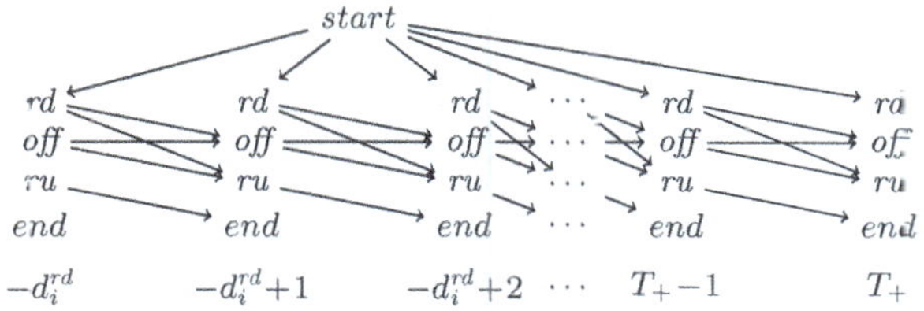

$$\ell_{start} = \ell_{end} = 0,\ \ell_{(off,t)} = -\pi_{i,t} \text{ and } \ell_{(s,t)} = \sum_{q=t}^{t+d_i^s-1} C_q P_i^s - \pi_{i,q} \text{ for } s \in \{ru, rd\},$$

it is easy to see that start-end-paths of negative weight correspond to break variables with negative reduced costs. So, the pricing problem for these variables reduces to a hop-constrained node-weighted shortest path problem.

To branch on fractional solutions, we create child nodes by imposing the linear branching constraints

$$\sum_{t\in\{q\in[T]:\ q\le t'\}} x_{\bar{j},\bar{k},t} = 0 \quad \text{and} \quad \sum_{t\in\{q\in[T]:\ q\ge t'+1\}} x_{\bar{j},\bar{k},t} = 0,$$

where $(\bar{j}, \bar{k}, t')$ is an appropriately chosen branch candidate. The restrictions imposed by this branching scheme do not disturb the combinatorial structure of the pricing problem and, furthermore, have a significant impact on both children.

In order to determine the task $(\bar{j}, \bar{k})$ of the branching candidate, we denote $l(j,k) := \min\{t \in [T] : x_{j,k,t} > 0\}$ and $r(j,k) := \max\{t \in [T] : x_{j,k,t} > 0\}$, $m(j,k) := \lceil(\frac{r(j,k)+l(j,k)}{2})\rceil$ and $N(j,k) := |\{t \in [T] : x_{j,k,t} > 0\}|$. At branch-and-bound nodes with an LP bound near the global lower bound, auspicious tasks $(j,k) \in O$ are obtained by choosing for each machine $i \in M$ the task maximizing the score function $\rho(j,k) := (r(j,k) - l(j,k)) \cdot \frac{1}{N(j,k)}$. Note that if $\rho(j,k) > 1$, then even the fractional processing of task (j,k) is preempted. For each candidate task we determine the average over all nonzero start variables of this task of the pseudocosts of branching these variables to zero, and then choose the candidate $(\bar{j}, \bar{k})$ that maximizes this average. At branch-and-bound nodes with *nodeobjective* $> 0.9 \cdot$ *cutoffbound* $+ 0.1 \cdot$ *lowerbound*, when we do not expect substantially better solutions in this branch, we simply choose the task $(\bar{j}, \bar{k}) = \arg\max\{r(j,k) - l(j,k) : (j,k) \in O\}$ to branch.

The time t' of the branching candidate is chosen from a candidate set: Let $O|_{\bar{i}} := \{(j,k) \in O : m_{j,k} = \bar{i}\}$ be the set of all tasks, that have to run on machine $m_{\bar{j},\bar{k}} = \bar{i} \in M$. Let $t_l = \frac{1}{2}(l(\bar{j},\bar{k}) + m(\bar{j},\bar{k}))$, $t_r = \frac{1}{2}(r(\bar{j},\bar{k}) + m(\bar{j},\bar{k}))$ and $t_m = m(\bar{j},\bar{k})$. The number of all nonzero variables $x_{j,k,q}$ on machine $\bar{i}$ in the interval $[l(\bar{j},\bar{k}), r(\bar{j},\bar{k})]$ is denoted by $K = \sum_{(j_1,k_1)\in O|_{\bar{i}}} \sum_{q=l(\bar{j},\bar{k})}^{r(\bar{j},\bar{k})} \lceil x_{j_1,k_1,q} \rceil$. We choose

Table 1 Computational results

$algo(\|O\|, \|M\|, \|T\|, obj)$	Cols	Rows	Nodes	Time (s)	Opt in node	Gap (%)
$bpa(20, 5, 72, const)$	1767	1122	1089	15	24	0
$cILP(20, 5, 72, const)$	2515	2695	7267	239	368	0
$bpa(20, 5, 72, sin)$	1890	1121	547	6	531	0
$cILP(20, 5, 72, sin)$	2515	2695	4280	51	2560	0
$bpa(20, 5, 720, const)$	12.9k	8986	1487	3142	1274	0
$cILP(20, 5, 720, const)$	17.3k	237k	2107	40k	–	5.18
$bpa(20, 5, 720, sin)$	13.2k	8984	805	3060	151	0
$cILP(20, 5, 720, sin)$	17.3k	227k	1508	22.5k	1385	0

$$t' = \arg\min\left\{ \left| \frac{1}{2} - \frac{K - \sum_{(j_1,k_1)\in O|_{\bar{t}}} \sum_{q=l(\bar{j},\bar{k})}^{t} \lceil x_{j_1,k_1,q} \rceil}{K} \right| \ : \ t \in \{t_l, t_m, t_r\} \right\},$$

which leads to a branching period t' that divides $[l(\bar{j}, \bar{k}), r(\bar{j}, \bar{k})]$ into two nearly balanced intervals, with nearly as many conflicts before as after t'.

The first task selection strategy often leads to integer solutions, because we partition the solution space into equally good subspaces. The second strategy aims at pruning the branches, where we do not expect better solutions.

5 Implementation and Results

We implemented the presented approach in `C++` using the branch-and-price framework of `SCIP 6.0.2` [3] and `gurobi 8.1.1` [4] to solve the linear relaxations. A pricing iteration is stopped if for $\frac{|n_M|}{2}$ subproblems variables with negative reduced costs have been found. Per subproblem, at most 10 such variables are generated per iteration. Depth-first-search is used as node selection, as (near) optimal solutions are found very quickly. SCIP's cut generation and presolving is disabled. We also use a time limit of 40k s.

In our computational experiments, we compared our branch-and-price algorithm (bpa) with the compact ILP formulation (cILP) proposed in [5] solved by `gurobi` with default settings and the presented variable domain reduction. Our branch-and-price algorithm runs entirely single threaded, while gurobi can use up to eight threads. In the benchmark instances used in these test, we have $C_t = 1$ and $C_t = \lfloor (\sin(\pi \cdot t/T) + 1) \cdot 10 \rfloor$, for all $t \in [T]$.

The results reported in Table 1 show significant performance improvements. Not only the solution times are reduced, also the sizes of the resulting models and of the branch-and-bound tree are reduced.

Our goal for the future is to attack much bigger problem instances using our approach. For this, we plan to further reduce the size of the models, to investigate cuts to strengthen the formulation, and to further analyse and improve the branching schemes and the presolving.

References

1. Barnhart, C., Johnson, E.L., Nemhauser, G.L., Savelsbergh, M.W.P., Vance, P.H.: Branch-and-price: column generation for solving huge integer programs. Oper. Res. **46**(3), 316–329 (1998)
2. Dai, M., Tang, D., Giret, A., Salido, M.A., Li, W.: Energy-efficient scheduling for a flexible flow shop using an improved genetic-simulated annealing algorithm. Robot. Comput. Integr. Manuf. **29**(5), 418–429 (2013)
3. Gleixner, A., Bastubbe, M., Eifler, L., Gally, T., Gamrath, G., Gottwald, R.L., Hendel, G., Hojny, C., Koch, T., Lübbecke, M.E., Maher, S.J., Miltenberger, M., Müller, B., Pfetsch, M.E., Puchert, C., Rehfeldt, D., Schlösser, F., Schubert, C., Serrano, F., Shinano, Y., Viernickel, J.M., Walter, M., Wegscheider, F., Witt, J.T., Witzig, J.: The SCIP Optimization Suite 6.0. ZIB-Report 18–26, Zuse Institute Berlin, Berlin (2018)
4. Gurobi Optimization, I.: Gurobi optimizer reference manual (2019)
5. Selmair, M., Claus, T., Herrmann, F., Bley, A., Trost, M.: Job shop scheduling with flexible energy prices. In: Proceedings of ECMS 2016, pp. 488–494 (2016)
6. Shrouf, F., Ordieres-Meré, J., García-Sánchez, A., Ortega-Mier, M.: Optimizing the production scheduling of a single machine to minimize total energy consumption costs. J. Clean. Prod. **67**, 197–207 (2014)
7. van den Akker, J.M., Hoogeveen, J.A., van de Velde, S.L.: Parallel Machine Scheduling by Column Generation. Memorandum COSOR. Technische Universiteit Eindhoven, Eindhoven (1997)
8. Weinert, N., Chiotellis, S., Seliger, G.: Methodology for planning and operating energy-efficient production systems. CIRP Ann. Manuf. Technol. **60**, 41–44 (2011)

Solving the Multiple Traveling Salesperson Problem on Regular Grids in Linear Time

Philipp Hungerländer, Anna Jellen, Stefan Jessenitschnig, Lisa Knoblinger, Manuel Lackenbucher, and Kerstin Maier

Abstract In this work we analyze the multiple Traveling Salesperson Problem (mTSP) on regular grids. While the general mTSP is known to be NP-hard, the special structure of regular grids can be exploited to explicitly determine optimal solutions in linear time.

Our research is motivated by several real-world applications, for example delivering goods with swarms of unmanned aerial vehicles (UAV) or search and rescue operations. In order to obtain regular grid structures, we divide large search areas in several equal-sized squares, where we choose the square size as large as the sensor range of a UAV.

First, we use an Integer Linear Program (ILP) to formally describe our considered mTSP variant on regular grids that aims to minimize the total tour length of all salespersons, which corresponds to minimizing the average search time for a missing person.

With the help of combinatorial counting arguments and the establishment of explicit construction schemes, we are able to determine optimal mTSP solutions for specific grid sizes with two salespersons, where the depot is located in one of the four corners.

Keywords Combinatorial optimization · Mixed-integer programming · Routing

1 Introduction

The multiple Traveling Salesperson Problem (mTSP), also known as the Vehicle Routing Problem, is a generalization of the NP-hard Traveling Salesperson Problem (TSP). Given p points, including a depot, a feasible mTSP solution consists of m

P. Hungerländer · A. Jellen · S. Jessenitschnig · L. Knoblinger · M. Lackenbucher · K. Maier (✉)
Department of Mathematics, University of Klagenfurt, Klagenfurt, Austria
e-mail: philipp.hungerlaender@aau.at; anna.jellen@aau.at; stefanje@edu.aau.at; lisakn@edu.aau.at; malackenbuch@edu.aau.at; kerstin.maier@aau.at

J. S. Neufeld et al. (eds.), *Operations Research Proceedings 2019*, Operations Research Proceedings, https://doi.org/10.1007/978-3-030-48439-2_26

shortest Hamiltonian cycles, such that the depot is visited by all salespersons and the remaining $p - 1$ points are visited by exactly one salesperson.

In this paper we consider the mTSP on regular $\ell \times n$ grid graphs in the Euclidean plane, where the number of grid points is ℓn. The special structure of the grid is exploited to find lower bounds, explicit construction schemes and hence, optimal mTSP solutions.

There are dozens of variants and related applications of the TSP, since it is one of the most famous and important (combinatorial) optimization problems. An important special case is the metric TSP, for which the costs between points are symmetric and satisfy the triangle inequality. The Euclidean TSP, which is still NP-hard, see [3, 5] for details, is a special metric TSP, where the grid points lie in $\mathcal{R}^d$ and the distances are measured by the ℓ_d-norm. Although the mTSP on grid graphs is related to these problems, we succeed in providing optimal solutions in linear time for several special cases.

Explicit construction schemes and corresponding optimal solutions are also known for another related problem, namely the TSP with Forbidden Neighborhoods (TSPFN), where consecutive points along the Hamiltonian cycle must have a minimal distance. The TSPFN was studied on regular 2D and 3D grids, see [1, 2] for details. To the best of our knowledge, this extended abstract provides the first lower bounds and explicit construction schemes, and hence, optimal solutions for the mTSP with more than one salesperson.

Our research is motivated by several real-world applications, like search and rescue operations or delivering goods with swarms of unmanned aerial vehicles (UAV), see, e.g., [4]. Large search areas can be divided into several equal-sized squares, resulting in a regular grid structure. The size of a square is chosen as large as the sensor or camera range of a UAV.

The remainder of this paper is structured as follows. In Sect. 2 we formally describe our considered variant of the mTSP with the help of an Integer Linear Program (ILP). In Sect. 3 we provide optimal tour lengths and linear time construction schemes for the mTSP with two salespersons for different grid dimensions and the depot located in one of the four corners. Finally, future research directions are pointed out.

2 Problem Description

In this paper we consider the mTSP on a regular $\ell \times n$ grid and use the grid numbering depicted in Fig. 2a). A regular grid can be viewed as a chessboard with black and white squares. We assume w.l.o.g. that the depot is located at (1, 1) and colored black.

The mTSP can be modeled as an adapted version of the ILP discussed in [6]. We use the Euclidean norm to measure the distances between the grid points. Furthermore, we relax the constraint responsible for balancing the lengths of separate routes of the salespersons, by limiting the number of visited grid points (including the depot) for each salesperson to

$$\left\lceil \frac{\ell n}{m} \right\rceil + 1. \quad (1)$$

Like in [6], we aim to minimize the total tour length of all salespersons, which corresponds to minimizing the average search time for a missing person.

With our ILP we are able to solve mTSP instances with small grid sizes to optimality. These results for specific grids are often quite helpful for coming up with hypotheses on more general results like the ones stated in Theorems 1–3.

We close this section by proving a simple proposition that is used repeatedly throughout the remainder of this paper.

Proposition 1 *For a Hamiltonian path on a regular grid, viewed as a chessboard, that uses only steps of length 1 and visits ℓ grid points, the following properties hold: The path starts and ends on grid points with different (same) colors, if and only if ℓ is even (odd).*

Proof W.l.o.g. the start grid point is black. A step of length 1 always connects grid points of different colors.

For $\ell = 2$ ($\ell = 3$) we can only reach a white grid point with a step of length 1 (and a black grid point with a further step of length 1). We assume that the proposition is true for a particular even (odd) value ℓ. The statement is also true for $\ell + 2$, because according to our assumption we reach a white (black) grid point with ℓ steps of length 1 and with 2 further steps of length 1 we reach again a white (black) grid point. □

3 Linear Time mTSP Solutions

Lemma 1 *The length of the lower bound of an mTSP solution on an $\ell \times n$ grid with $m = 2$ and the depot located in a corner is $(\ell n - 1) + \sqrt{2} + \sqrt{5}$*

Proof As the depot is located at $(1, 1)$, there must be two moves of length > 1, such that both salespersons are able to leave and return to the depot. The shortest such moves have lengths $\sqrt{2}$ and 2 and thus a trivial lower bound for the value of an mTSP solution on an $\ell \times n$ grid is $(\ell n - 1) + \sqrt{2} + 2$. The steps of length 1 and $\sqrt{2}$ are unique, i.e., they end at $(1, 2)$, $(2, 1)$, and $(2, 2)$, respectively. There are two possibilities for the step of length 2, i.e., $(1, 3)$ or $(3, 1)$. No matter which step of length 2 is chosen, either $(1, 2)$ or $(2, 1)$ can not be connected to any unvisited grid point by a step of length 1. Hence, we derive a stronger lower bound of length $(\ell n - 1) + \sqrt{2} + \sqrt{5}$. □

Theorem 1 *The length of an optimal mTSP solution on an $\ell \times n$ grid with $m = 2$, $\ell, n > 4$, n even, and the depot located in a corner is $(\ell n - 1) + \sqrt{2} + \sqrt{5}$.*

Proof As proven in Lemma 1, the lower bound for the defined mTSP tours has length $(\ell n - 1) + \sqrt{2} + \sqrt{5}$. In order to finish the proof a construction scheme for mTSP tours with length $(\ell n - 1) + \sqrt{2} + \sqrt{5}$ is needed. The construction is given in Algorithm 1, where the number of grid points visited by the first salesperson, see (1) for the respective upper bound, is set to $x := \frac{\ell n}{2} - \left((\frac{\ell n}{2} + 1) \mod 2\right)$. Figure 1a) depicts a representative example of an optimal solution. □

Algorithm 1 Construction scheme for optimal mTSP solutions with $m = 2$, the depot located in a corner, and the grid dimensions given in Theorems 1 and 2, respectively

```
Input: ℓ × n grid, number x of points visited by the first salesperson
Output: Optimal mTSP solution
connect (1,1) with (2,3) and (1,n) with (2,n)
for i = 1 to n − 1 do
  | connect (1,i) with (1,i + 1)
end
for i = 0 to n/2 − 3 do
  | connect (2,4 + 2i) with (2,5 + 2i)
end
▷ Now 2n − 2 grid points located in the first two rows are added to the cycle.
for j = 0 to n/2 − 2 do
    for i = 1 to ℓ − 2 do
        connect (1 + i, n − 2j) with (2 + i, n − 2j) and
        (1 + i, n − 2j − 1) with (2 + i, n − 2j − 1)
        ▷ Draw two parallel edges in the two most right columns.
        if x gridpoints are visited then
            if 2 + i = ℓ − 1 then
                remove edges ((1 + i, n − 2j), (2 + i, n − 2j)) and
                  ((1 + i, n − 2j − 1), (2 + i, n − 2j − 1))
                connect (2, n − 2j − 2) to (3, n − 2j − 2) to
                (3, n − 2j − 3) to (2, n − 2j − 3)
                ▷ If the edges end in row ℓ − 1, remove them and draw two
                  parallel edges in the columns to the left.
            end
            Connect all remaining subpaths to a cycle by edges of length 1.
            The second cycle can easily be drawn on the remaining grid points.
             return mTSP solution.
        end
    end
end
```

Theorem 2 *The length of an optimal mTSP solution on an $\ell \times n$ grid with $m = 2$, $\ell, n > 3$ odd, and the depot located in a corner is $(\ell n - 2) + 2\sqrt{2} + \sqrt{5}$.*

Proof As shown in Lemma 1, an optimal mTSP solution has a minimal length of $(\ell n - 1) + \sqrt{2} + \sqrt{5}$. Steps of length 1 and $\sqrt{2}$ are fixed, i.e., steps to (1, 2), (2, 1) and (2, 2) are unique. There are two possibilities for the step of length $\sqrt{5}$, i.e., to

(2, 3) or to (3, 2). In both cases we have three points adjacent to the depot with the same color. Now, to obtain the lower bound, there must be one salesperson visiting an even and the other salesperson visiting an odd number of grid points.

Since ℓn is odd and by excluding the depot, it remains an even number of grid points for the two salespersons. Each salesperson visits at most $\frac{\ell n+1}{2}$ grid points (excluding the depot), see (1) for the respective upper bound (including the depot), i.e., the optimal solution has either two salespersons who visit $\frac{\ell n-1}{2}$ grid points each or salespersons visiting $\frac{\ell n+1}{2}$ and $\frac{\ell n-3}{2}$ grid points, respectively. This means in each case both salespersons visit either an even or odd number of grid points. Therefore, due to Proposition 1, one salesperson needs an additional step of length > 1, which results in an mTSP solution with at least length $(\ell n - 2) + 2\sqrt{2} + \sqrt{5}$.

There is one case left that has to be considered: $(\ell n - 2) + 2\sqrt{2} + 2$. Steps to (1, 2), (2, 1) and (2, 2) are fixed again and there are two possibilities for the step of length 2. In both cases each salesperson has one black and one white point adjacent to the depot. Taking into account the additional step of length $\sqrt{2}$ one salesperson visits an even and the other an odd number of grid points. In summary, this yields a contradiction due to Proposition 1. Therefore $(\ell n - 2) + 2\sqrt{2} + \sqrt{5}$ is a valid lower bound.

To complete the proof, it remains to find a general construction scheme. We can apply Algorithm 1, where the number of grid points visited by the first salesperson is set to $x := \lfloor \frac{\ell n}{2} \rfloor - ((\lfloor \frac{\ell n}{2} \rfloor + 1) \mod 2)$. Figure 1b depicts a representative example of an optimal solution for a 9 × 13 grid. □

Theorem 3 *The length of an optimal mTSP solution on an* $\ell \times n$ *grid with* $m = 2$, $n = 4$, *and the depot located in a corner is* $(\ell n - 1) + 2 + \sqrt{5}$.

Proof As shown in Lemma 1, an optimal mTSP solution has a minimal length of $(\ell n - 1) + \sqrt{2} + \sqrt{5}$. As above the steps to (1, 2), (2, 1), and (2, 2) are fixed and w.l.o.g. as the step of length $\sqrt{5}$ we go to (3, 2).

Due to (1) the salespersons must visit $2\ell - 1$ and $2\ell - 2$ grid points (excluding the depot), respectively, i.e., one salesperson visits an even and the other an odd number

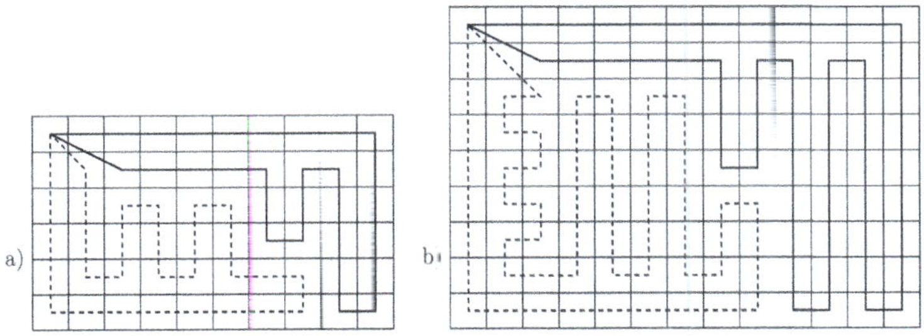

Fig. 1 Construction scheme for an $\ell \times n$ grid with $m = 2$ and (**a**) $\ell, n > 4$, ℓ even, depicted by the example of a 6 × 10 grid. (**b**) $\ell, n > 3$ odd, is displayed by the example of a 9 × 13 grid. The first salesperson is depicted by the drawn-through line in both (**a**) and (**b**)

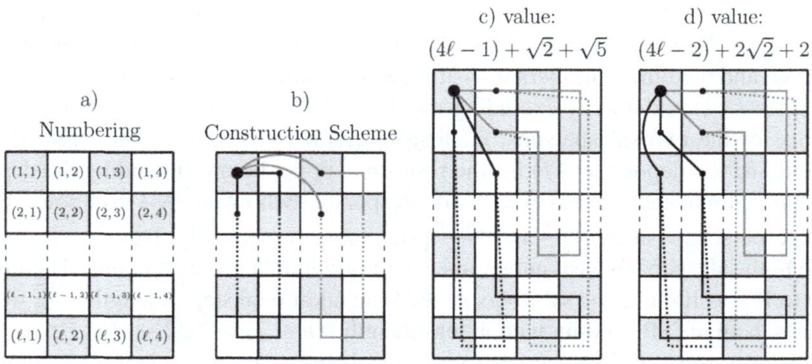

Fig. 2 (**a**) Numbering of a regular grid. (**b**) Optimal construction scheme for $\ell \times 4$ grids. (**c**) and (**d**) Visualizations indicates why mTSP solutions if the given lengths with $m = 2$ and the depot located at (1, 1) cannot exist on the $\ell \times 4$ grid

of grid points. However, starting at the given points and respecting the length of the given lower bounds results in the salespersons visiting $2\ell - 4$ and $2\ell + 1$ or 2ℓ and $2\ell - 3$ grid points, which contradicts (1) (see Fig. 2c) for a visualization).

Now let us consider further relevant lower bounds, where similar arguments as above can be applied:

- $(4\ell - 1) + 2 \cdot 2$: The steps of length 1 and 2 are unique, they end at (1, 2), (1, 3), (2, 1), and (3, 1). Either from (1, 2) or (1, 3) a step of length > 1 is necessary, due to Proposition 1.
- $(4\ell - 2) + 2\sqrt{2} + 2$: The steps to (1, 2), (2, 2), and (2, 1) are unique. For the step of length 2 we have two possibilities: W.l.o.g. we go to (3, 1). A step from (2, 1) to (3, 2) is implied. As in the first case it is not possible the tours of both salespersons fulfill (1) (see Fig. 2d) for a visualization).
- $(4\ell - 2) + 2\sqrt{2} + \sqrt{5}$: The steps to (1, 2), (2, 2), and (2, 1) are unique. We have two analogous possibilities for the step of length $\sqrt{5}$, either to (3, 2) or (2, 3). In both cases we have three points adjacent to the depot with the same color and an additional step of length $\sqrt{2}$. Due to Proposition 1 both salespersons need to visit an even or an odd number of grid points, which yields to a contradiction of 1.

In summary we derive a lower bound of $(4\ell - 1) + 2 + \sqrt{5}$. The related construction scheme is shown in Fig. 2b. □

For future work it remains to determine lower bounds, explicit construction schemes, and hence, optimal mTSP solutions for further grid sizes, different locations of the depot, more than two salespersons, and other distance measures.

References

1. Fischer, A., Hungerländer, P.: The traveling salesman problem on grids with forbidden neighborhoods. J. Comb. Optim. **34**(3), 891–915 (2017)
2. Fischer, A., Hungerländer, P., Jellen, A.: The traveling salesperson problem with forbidden neighborhoods on regular 3D grids. In: Operations Research Proceedings, 2017. Springer, Cham (2018)
3. Garey, M.R., Graham, R.L., Johnson, D.S.: Some NP-complete geometric problems. Proceedings of the Eighth Annual ACM Symposium on Theory of Computing, pp. 10–22 (1976)
4. Hayat, S., Yanmaz, E., Muzaffar, R.: Survey on unmanned aerial vehicle networks for civil applications: a communications viewpoint. IEEE Commun. Surv. Tutorials **18**, 2624–2661 (2016)
5. Papadimitriou, C.H.: The Euclidean travelling salesman problem is NP-complete. Theor Comput. Sci. **4**(3), 237–244 (1977)
6. Xu, X., Yuan, H., Liptrott, M., Trovati, M.: Two phase heuristic algorithm for the multiple-travelling salesman problem. Soft Comput. **22**, 6567–6581 (2018)

References

The Weighted Linear Ordering Problem

Jessica Hautz, Philipp Hungerländer, Tobias Lechner, Kerstin Maier, and Peter Rescher

Abstract In this work, we introduce and analyze an extension of the Linear Ordering Problem (LOP). The LOP aims to find a simultaneous permutation of rows and columns of a given weight matrix such that the sum of the weights in the upper triangle is maximized. We propose the weighted Linear Ordering Problem (wLOP) that additionally considers individual node weights.

First, we argue that in several applications of the LOP the optimal ordering obtained by the wLOP is a worthwhile alternative to the optimal solution of the LOP. Additionally, we show that the wLOP constitutes a generalization of the well-known Single Row Facility Layout Problem.

We introduce an Integer Linear Programming formulation as well as a Variable Neighborhood Search for solving the wLOP. Finally, we provide a benchmark library and examine the efficiency of our exact and heuristic approaches on the proposed instances in a computational study.

Keywords Integer linear programming · Variable neighborhood search · Ordering problem

1 Introduction

In this paper we introduce and analyze the *weighted Linear Ordering Problem* (wLOP) that constitutes an extension of the well-known *Linear Ordering Problem* (LOP).

Ordering problems associate to each ordering (or permutation) of a set of nodes $[n] = \{1, \dots, n\}$ a profit and the aim is to determine an ordering with maximum profit. For the LOP, this profit is calculated by those pairs $(u, v) \in [n] \times [n]$, where

J. Hautz · P. Hungerländer · T. Lechner · K. Maier (✉) · P. Rescher
Department of Mathematics, Alpen-Adria-Universität Klagenfurt, Klagenfurt, Austria
e-mail: jehautz@edu.aau.at; philipp.hungerlaender@aau.at; toblechner@edu.aau.at; kerstin.maier@aau.at; perescher@edu.aau.at

J. S. Neufeld et al. (eds.), *Operations Research Proceedings 2019*, Operations Research Proceedings, https://doi.org/10.1007/978-3-030-48439-2_27

u is arranged before v in the ordering. Thus, in its matrix version the LOP can be formulated as follows. A matrix $W = (w_{ij}) \in \mathbb{Z}^{n \times n}$ is given and the task is to find a simultaneous permutation π of rows and columns in W such that the following sum is maximized: $\sum_{\substack{i,j \in [n] \\ \pi(i) < \pi(j)}} w_{ij}$.

The LOP occurs in a large variety of applications, like assessment of corruption perception [1], scheduling (with precedences [2]), graph drawing (one sided crossing minimization [5]), and ranking in sports tournaments. We refer to [6] for further applications and different solution approaches for the LOP.

A wLOP instance consists like an LOP instance of n nodes and pairwise node weights w_{ij}, $i, j \in [n]$. In addition, we are given individual node weights w_i, $i \in [n]$. Then, the wLOP can be defined as the following optimization problem

$$\max_{\pi \in \Pi_n} \sum_{\substack{i,j \in [n] \\ i<j}} w_{ij} d_{ij}^{\pi}, \tag{1}$$

where Π_n is the set of permutations of the nodes $[n]$ and d_{ij}^{π} is the sum of the weights of the nodes located between node i and node j in the permutation $\pi \in \Pi_n$, plus half of the weights of i and j:

$$d_{ij}^{\pi} = \begin{cases} \frac{w_i + w_j}{2} + \sum\limits_{\substack{k \in [n] \\ \pi(i) < \pi(k) < \pi(j)}} w_k, & \pi(i) < \pi(j), \\ 0, & \text{otherwise.} \end{cases}$$

The LOP can be expressed as a special case of the wLOP, where we do not consider individual node weights and therefore d_{ij}^{π} is set to 1, if $\pi(i) < \pi(j)$, and to 0, otherwise. Hence, the wLOP can be applied to the same problems as the LOP and provides an alternative way to determine meaningful orderings and rankings.

In various applications the orderings and rankings determined by the wLOP might even be preferable to corresponding LOP solutions. Often, we prefer that nodes with high pairwise weights are positioned far away from each other and nodes with no or few weights are placed in the center of the ordering. In the LOP we have no means for modeling this aspect, but such orderings can be obtained by taking the pairwise LOP node weights, setting the individual weights of all nodes to 1 and solving the wLOP, as in this case the distances between nodes are considered in the objective function.

For further clarifying the workings of the wLOP, let us consider a toy example of scheduling a set of exams $\mathcal{E}$, where we include preferences in which order the exams should be taken. Additionally, we want to ensure that dependent exams with high credits are as far apart from each other as possible. To obtain an ordering that considers these properties, the pairwise weights for two exams $i, j \in \mathcal{E}$, $i < j$, are set to $w_{ij} = 2$, if exam i is preferred to be taken before exam j, $w_{ij} = 1$, if there is no preference and $w_{ij} = 0$, otherwise. Furthermore, the individual node

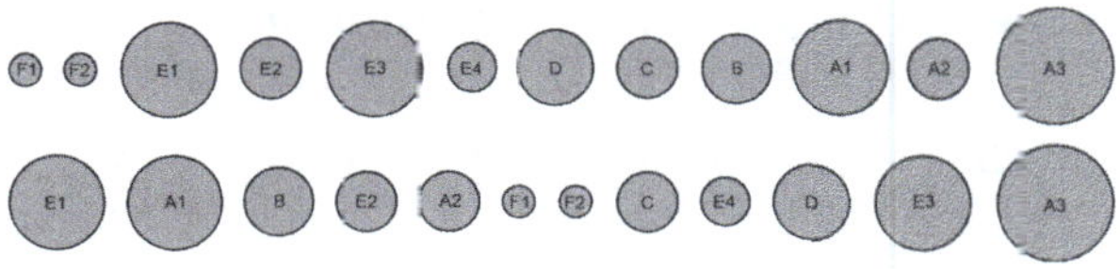

Fig. 1 Exams with individual credits and optimal LOP and wLOP solutions

weights are set to $w_i = \frac{1}{c_i}$, $i \in \mathcal{E}$, such that large distances between exams with high credits are more likely in the optimal wLOP solution. In our example, exams starting with the same letter are preferred to be scheduled in ascending order. There are no preferences among the remaining pairwise combinations. Figure 1 depicts optimal LOP and wLOP solutions of our toy example as well as the credits of the exams that are displayed by their respective circle sizes.

Moreover, the wLOP constitutes a generalization of the well-known *Single Row Facility Layout Problem* (SRFLP). The aim of the SRFLP is to determine an optimal arrangement with respect to material-flow of n machines on a straight line without gaps. The machines have given lengths $\ell_1, \dots, \ell_n$ as well as pairwise weights c_{ij}, $i, j \in [n]$, $i < j$, indicating the amount of material transported between them. Then the NP-hard optimization problem can be defined as follows

$$\min_{\pi \in \Pi_n} \sum_{\substack{i,j \in [n] \\ i<j}} c_{ij} z_{ij}^{\pi}, \tag{2}$$

where Π_n is the set of permutations of the machines and z_{ij}^{π} is the center-to-center distance between machines i and j with respect to a particular permutation $\pi \in \Pi_n$.

We refer to [4] for a survey on applications and solution approaches for the SRFLP. Note that assuming $c_{ij} \geq 0$ in row layout problems guarantees boundedness of the objective value of the optimal layout. For the SRFLP the non-negativity of the weights further ensures that all machines are placed next to each other without spacing. For the wLOP we do not need this restriction on the pairwise weights as space between the nodes is not allowed. If we use minimization instead of maximization and set $w_i := \ell_i$, $i \in [n]$, and $w_{ij} = w_{ji} := c_{ij}$, $i, j \in [n]$, $i < j$, the optimization problem defined in (1) reduces to (2). Hence, the wLOP constitutes a generalization of the SRFLP by additionally allowing for asymmetric costs.

The remaining paper is structured as follows. We propose an Integer Linear Programming (ILP) formulation and a Variable Neighborhood Search (VNS) for solving the wLOP in Sects. 2 and 3, respectively. In Sect. 4 we examine the efficiency of our exact and heuristic approaches and point out several future research directions.

2 Integer Linear Programming Formulation

For modeling the wLOP as ILP, we use binary variables

$$x_{ikj} \in \{0, 1\}, \; i, j, k \in [n], \; i \neq k \neq j \neq i, \tag{3}$$

which are set to 1, if node i is located before node k and node k is located before node j, and to 0, otherwise.

Any feasible ordering of the nodes has to fulfill the following constraints (we assume $i, j, k, l \in [n]$ pairwise different):

$$x_{ikj} + x_{jki} + x_{ijk} + x_{kji} + x_{jik} + x_{kij} = 1, \quad i, j, k \in [n], \tag{4}$$

$$x_{ijk} + x_{ikj} + x_{kij} - x_{ijl} - x_{ilj} - x_{lij} = 0, \quad i, j, k, l \in [n]. \tag{5}$$

Equations (4) ensure that exactly one of the variables associated to the six possible permutations of three nodes $i, j, k \in [n]$ is set to one. Equations (5) guarantee that for each node $m \in [n]$ the same number of variables associated to the three permutations of the nodes i, j and m, where node i is before node j, is set to one. Due to (4), at most one of these three variables can be set to one. Hence, in summary for each set of nodes $i, j, m \in [n]$, $i \neq j \neq m \neq i$, the following holds:

$$x_{mij} + x_{imj} + x_{ijm} = \begin{cases} 1, & \text{if node } i \text{ is before node } j, \\ 0, & \text{otherwise.} \end{cases} \tag{6}$$

Next, let us assume that node i is before node j and node j is before node k. Due to (6), $x_{mij} + x_{imj} + x_{ijm} = 1$, $m \in [n]$, $i \neq m \neq j$ and $x_{mjk} + x_{jmk} + x_{jkm} = 1$, $m \in [n]$, $i \neq m \neq j$ hold. Choosing $m := k$ and $m := i$, respectively, yields $x_{kij} + x_{ikj} + x_{ijk} = 1$ and $x_{ijk} + x_{jik} + x_{jki} = 1$. Subtracting the second from the first equation results in $x_{kij} + x_{ikj} - x_{jik} - x_{jki} = 0$. Due to (4) $x_{kij} = x_{ikj} = x_{jik} = x_{jki} = 0$ holds, and thus, $x_{ijk} = 1$. Therefore, also the equation $x_{ijk} + x_{ikj} + x_{kij} = 1$ holds, and due to (6), it is guaranteed that node i is before node k. In summary, Eqs. (4) together with Eqs. (5) ensure transitivity and hence, suffice to guarantee feasible orderings of the n nodes.

Using (6), we can model the distances d_{ij}, $i, j \in [n]$, $i \neq j$, between nodes i and j as follows:

$$d_{ij} = \left(\frac{w_i + w_j}{2}\right) \frac{\sum_{\substack{k \in [n] \\ i \neq k \neq j}} \left(x_{kij} + x_{ikj} + x_{ijk}\right)}{n - 2} + \sum_{\substack{k \in [n] \\ i \neq k \neq j}} x_{ikj} w_k.$$

Note, that for each pair of nodes i and j, $i, j \in [n]$, $i < j$, either d_{ij} (if j is before i) or d_{ji} (if i is before j) is 0.

Finally, we can formulate the following ILP for solving the wLOP:

$$\max \sum_{i,j\in[n]} w_{ij} d_{ij} \qquad \text{s. t. (3)–(5).}$$

3 Variable Neighborhood Search

We adapted the basic VNS for the LOP proposed in [3] in order to obtain a VNS for the wLOP. Our heuristic uses up to k_{max} neighborhoods. A k-neighborhood, $k > 1$, of a solution is reached by applying insertion moves $k-1$ times. The 1-neighborhood consists of changing the current solution by swapping two consecutive nodes.

An iteration of the basic VNS consists of three steps: *shaking*, *improving* and *updating*. First, the *shaking* step randomly generates a new solution based on the initial or current best solution within the considered neighborhood. After that, a local search method is performed to *improve* this solution. Finally, the current neighborhood is *updated*: k is set to 1, if it was possible to improve the current best solution, and otherwise k is increased by one (if $k > k_{max}$, then $k := k_{max}$). These three steps are repeated until no improvement of the current solution is found within *maxIter* consecutive iterations.

As local search method we apply a descent procedure within the 2-neighborhood, i.e., we use insertion moves as long as it is possible to improve the current solution. The difference between the basic VNS for the LOP and for the wLOP lies in the determination of the best insertion move. When calculating the move value of a node i for the LOP, just the pairwise weights in the $i-th$ column/row of the weight matrix need to be considered. In contrast, for the wLOP we have to take into account the whole weight matrix for determining the move value.

4 Computational Results

All experiments were performed on a Linux 64-bit machine equipped with Intel(R) Xeon(R) CPU e5-2630 v3@2.40 GHz and 128 GB RAM. As ILP-solver, we use Gurobi 8.0.1 in single thread mode, restricted to a time limit of 6 h. We generated appropriate benchmark instances for the wLOP, by adapting well-known LOP and SRFLP instances from literature. For the LOP instances, the weights were adapted by setting all the individual node weights $w_i := 1$, $i \in [n]$. For the SRFLP instances: we set the individual node weights $w_i := \ell_i$ and decided randomly, with equal probability, whether $w_{ij} := c_{ij}$ and $w_{ji} := 0$ or $w_{ij} = 0$ and $w_{ji} := c_{ij}$, $i, j \in [n]$, $i \neq j$. All instances and best known solutions can be downloaded from https://tinyurl.com/wLOP-Instances.

The VNS uses up to 5 neighborhoods ($k_{max} = 5$) and $maxIter$ is set to 15. For each instance we apply our heuristic 10 times with randomly generated initial

Table 1 Results for the wLOP generated by our ILP and VNS on adapted LOP and SRFLP instances with up to 76 nodes

Instance	ILP			VNS		
	Objective value	Gap [%]	Time [hh:mm:ss]	Objective value	Gap [%]	Time [mm:ss.ms]
N-pal13	230.0	0.00	00:00:01	230.0	0.00	00:00.082
N-pal23	1012.0	0.00	00:00:07	1012.0	0.00	00:00.310
N-econ36	11,665,494.0	0.00	00:01:11	11,665,494.0	0.00	00:11.093
N-econ43	17,695,359.0	0.00	00:04:17	17,695,359.0	0.00	00:23.782
N-econ47	21,150,538.0	0.00	00:05:06	21,150,538.0	0.00	00:39.784
N-econ58	37,186,607.0	0.00	00:20:24	37,186,607.0	0.00	01:54.423
N-econ59	37,391,397.0	0.00	00:31:22	37,391,397.0	0.00	01:57.969
N-econ61	37,417,458.0	0.00	01:18:42	37,417,458.0	0.00	02:20.407
N-econ68	49,614,951.0	0.00	03:00:36	49,614,951.0	0.00	03:31.742
N-econ72	53,085,343.0	19.42	06:00:00	62,710,120.0	1.09	05:16.809
N-econ73	75,008,624.0	0.00	01:45:02	75,008,624.0	0.00	05:01.723
N-econ76	99,029,608.0	3.93	06:00:00	102,311,935.0	0.60	06:10.363
Am-2006-15dept-set1	9577.5	0.00	00:00:02	9577.5	0.00	00:00.966
Am-2008-17dept-set1	14,193.0	0.00	00:00:05	14,193.0	0.00	00:00.769
Am-2008-18dept-set1	17,680.5	0.00	00:00:02	17,680.5	0.00	00:00.976
Am-2009-33dept-set1	61,614.0	10.71	06:00:00	65,440.5	4.24	00:08.807
Am-2009-35dept-set1	69,539.0	16.40	06:00:00	76,429.5	5.90	00:11.752
QAP-sko42-01	19,715.0	39.14	06:00:00	24,161.0	13.53	00:28.946
QAP-sko49-01	29,669.0	55.47	06:00:00	39,688.0	16.22	00:51.715

The VNS Gap ($= \frac{\text{Upper Bound} - \text{Objective}}{\text{Objective}}$) indicates the relative gap between the objective value determined by VNS and the best upper bound computed by our ILP

solutions and state the best overall solution found. The corresponding time indicated is the total solving time for all 10 runs.

In Table 1 we report the results obtained by our proposed solution approaches for all instances considered. The first set (from *N-pal13* to *N-econ76*) is composed of the LOP instances and the second set (from *Am-2006-15dept-set1* to *QAP-sko49-01*) consists of SRFLP instances. Our ILP computes optimal and close to optimal solutions for all LOP instances except for *N-econ72*. Additionally, the SRFLP instances with up to 18 nodes are solved to optimality. However, we obtain large gaps for SRFLP instances with 33 to 49 nodes. The VNS either finds the optimal solution or improves the best ILP solution within less than 7 min.

In summary, we introduced and studied the wLOP, which constitutes a generalization of two well-known combinatorial optimization problems, namely the LOP and the SRFLP. For future work, it would be interesting to investigate, if Eqs. (4) and (5) describe the smallest linear subspace that contains the wLOP polytope for the variables defined in (3). Furthermore, it would be interesting to determine additional valid inequalities and facets for this polytope and also to examine their impact in computational results. Finally, it would be worthwhile to extend further heuristics for the LOP to the wLOP.

References

1. Achatz, H., Kleinschmidt, P., Lambsdorff, J.: Der corruption perceptions index und das linear ordering problem. ORNews **26**, 10–12 (2006)
2. Boenchendorf, K.: Reihenfolgenprobleme/Mean-flow-time sequencing. Mathematical Systems in Economics 74. Verlagsgruppe AthenÄaum, Hain, Scriptor (1982)
3. García-González, C., Pérez-Brito, D., Campos, V., Marti, R.: Variable neighborhood search for the linear ordering problem. Comput. Oper. Res. **33**(12), 3549–3565 (2006)
4. Hungerländer, P., Rendl, F.: A computational study and survey of methods for the single-row facility layout problem. Comput. Optim. Appl. **55**(1), 1–20 (2013)
5. Jünger, M., Mutzel, P.: 2-layer straightline crossing minimization: performance of exact and heuristic algorithms. J. Graph Algorithms Appl. **1**, 1–25 (1997)
6. Martí, R., Reinelt, G.: The Linear Ordering Problem: Exact and Heuristic Methods in Combinatorial Optimization. Applied Mathematical Sciences. Springer, Berlin (2011)

Adaptation of a Branching Algorithm to Solve the Multi-Objective Hamiltonian Cycle Problem

Maialen Murua, Diego Galar, and Roberto Santana

Abstract The Hamiltonian cycle problem (HCP) consists of finding a cycle of length N in an N-vertices graph. In this investigation, a graph G is considered with an associated set of matrices, in which each cell in the matrix corresponds to the weight of an arc. Thus, a multi-objective variant of the HCP is addressed and a Pareto set of solutions that minimizes the weights of the arcs for each objective is computed. To solve the HCP problem, the Branch-and-Fix algorithm is employed, a specific branching algorithm that uses the embedding of the problem in a particular stochastic process. To address the multi-objective HCP, the Branch-and-Fix algorithm is extended by computing different Hamiltonian cycles and fathoming the branches of the tree at earlier stages. The introduced anytime algorithm can produce a valid solution at any time of the execution, improving the quality of the Pareto Set as time increases.

Keywords Multi-objective optimization · Discrete optimization problems · Hamiltonian cycle problem · Branching algorithm

---

M. Murua (✉)
TECNALIA, Donostia-San Sebastián, Spain
e-mail: maialen.murua@tecnalia.com

D. Galar
TECNALIA, Donostia-San Sebastián, Spain

Division of Operation and Maintenance Engineering, Luleå University of Technology, Luleå, Sweden

R. Santana
Department of Computer Science and Artificial Intelligence, University of the Basque Country (UPV/EHU), Donostia-San Sebastián, Spain

J. S. Neufeld et al. (eds.), *Operations Research Proceedings 2019*, Operations Research Proceedings, https://doi.org/10.1007/978-3-030-48439-2_28

1 Introduction

The Hamiltonian cycle problem (HCP) consists of finding a cycle of length N in an N-vertices graph [4]. These cycles are called Hamiltonian cycles (HC)s. If the graph G contains at least one HC, G is said to be a Hamiltonian graph. The traveling salesman problem (TSP) is the problem of finding the shortest route to go to N different cities visiting each city once and returning to the city of origin [3]. The distances between all pairs of cities are known. Let G be a directed graph of N vertices with a weight c_{ij} for each arc (i, j), then a solution of the TSP corresponds to a HC of minimum total weight [5]. The main difference between these two problems is that finding a HC in a graph might be complex.

In this investigation, the HCP is considered for a graph G of size N, in which the Branch-and-Fix (BF) algorithm [1] has been employed to solve the problem. G has an associated set of matrices $\boldsymbol{W}^1, \ldots, \boldsymbol{W}^k$, where the cell $w^l_{i,j}$ represents the lth weight associated to an arc (i, j) for $l \in \{1, \ldots, k\}$. Each matrix represents a different way to evaluate HCs in G. We address a multi-objective variant of the HCP, considering only graphs that are Hamiltonian. The optimal Pareto set (PS) will comprise HCs that minimize the sum of weights for the different matrices. Notice that we simultaneously address the problem of finding a HC and minimizing the weights associated to the arcs. Our approach consists of modifying and extending the BF to deal with the multi-objective variant of the HCP.

The BF algorithm uses the embedding of the HCP in a discounted Markov Decision Process (MDP), a particular stochastic process that provides a mathematical framework for modeling decision making [2]. The problem addressed in this investigation consists of finding the set of non-dominated HCs (HCs that have associated non-dominated objective functions), where HCs will be found using the BF algorithm. The algorithm is modified in order to stop exploring a branch of the tree that will lead to a HC that is dominated by other previously found HC.

2 Branch-and-Fix Method

The BF was proposed in 2009 by Ejov et al. [1] to solve the HCP in a given graph. It is based on the idea that the HCP can be embedded in a MDP, and uses the polytope $\boldsymbol{X}_\beta$ defined by Feinberg [2] in 2000. In that investigation it was shown that the extreme points of the polytope correspond to HCs, though in 2009 it was found [6] that the extreme points could also be 1-randomized policies. The BF avoids arriving to those undesirable extreme points that induce 1-randomized policies. It solves sequences of linear programs (LP)s, two at each branching point of the logical tree.

1. **Initialization**. The original LP is solved to find $x \in \mathbf{X}_\beta$. If the feasible solution x_0 induces a deterministic policy $\boldsymbol{\zeta}_0$, the solution is found, if not, $\boldsymbol{\zeta}_0$ is a 1-randomized policy.

2. **Branching**. The 1-randomized policy ζ_0 serves to identify the splitting node i where the randomization occurs. If d is the number of arcs $\{(i, a_1), \ldots, (i, a_d)\}$ that emanate from i, d subgraphs $G_1, G_2, \ldots, G_d$ are constructed. In each G_k where $k = 1, \ldots, d$, (i, a_k) is fixed. These graphs are identical to G in all other vertices and arcs.
3. **Fixing**. In this step some other arcs are fixed, as in some graphs fixing one arc implies fixing other arcs.
4. **Iteration**. A second LP is solved to check the feasibility of the current fixed arcs. If that is the case, the algorithm is called recursively with the updated graph. If it is not feasible, the algorithm returns to step 2 fixing the following arc that emanates from i. The algorithm terminates when a HC is found or after exploring all the branches (in that case the graph is not Hamiltonian).

3 Multi-Objective HCP

Let $G = (V, \mathcal{A})$ be a graph of size N, where V is the set of nodes $|V| = N$ and $\mathcal{A}$ the set of the undirected arcs. We define k asymmetric weight-matrices, $\mathbf{W}^l$, $l \in \{1, \ldots, k\}$, where each arc $(i, j) \in \mathcal{A}$ has associated a weight $w^l_{ij} > 0$ in each of the k matrices. At the same time, we denote δ^{HC}_l as the objective function associated to a matrix $\mathbf{W}^l$ and specific HC, defined in Eq. (1).

$$\delta^{HC}_l = \sum_{m=1}^{|\mathcal{A}|} x_m \times w^l_m \qquad x_m \in \{0, 1\} \tag{1}$$

Where $x_m = 1$ if the mth arc belongs to the HC and $x_m = 0$ otherwise. w^l_m represents the entry of $\mathbf{W}^l$ corresponding to arc m. Finally, $\boldsymbol{\delta}^{HC}$ is the multiple-objective function associated to a HC defined in Eq. (2).

$$\boldsymbol{\delta}^{HC} = \{\delta^{HC}_1, \ldots, \delta^{HC}_k\} \tag{2}$$

An important constraint in this framework is that G is not a complete graph, therefore finding a HC is not trivial. A straightforward approach for finding the non-dominated HCs is by enumerating all possible solutions and selecting the optimal ones. However, complete enumeration is not a feasible approach even for small values of N. BF searches the space of HCs by pruning branches that are guaranteed not to lead to a valid HC. We extend the same rational but adding the possible dominance of a point as a new criterion for fathoming.

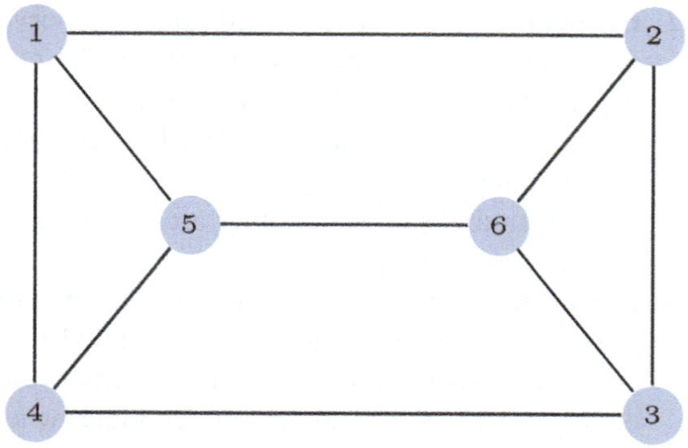

Fig. 1 An undirected graph of 6 vertices

Example 1 Consider the undirected graph of size 6 shown in Fig. 1 and two asymmetric weight-matrices. There are a total of 6 HCs in the graph:

$$\{(4,6,2,3,1,5),(2,6,4,5,1,3),(2,3,6,1,4,5),(4,1,2,5,6,3),(5,3,4,1,6,2)$$
$$(5,1,6,3,4,2)\}$$

For instance, for the first $HC_1 = (4,6,2,3,1,5)$, the set of arcs belonging to the HC is $\{(1,4),(2,6),(3,2),(4,3),(5,1),(6,5)\}$. To calculate $\delta_1^{HC_1}$, the w_{ij}^1 of the first weight matrix $\boldsymbol{W}^1$ that corresponds to each arc in the set are summed up.

$$\delta_1^{HC_1} = w_{14}^1 + w_{26}^1 + w_{32}^1 + w_{43}^1 + w_{51}^1 + w_{65}^1$$

In the same way $\delta_2^{HC_1}$ is calculated using $\boldsymbol{W}^2$. The bi-objective function $\boldsymbol{\delta}^{HC_1}$, is compound by $\delta_1^{HC_1}$ and $\delta_2^{HC_1}$. The following bi-objective functions are considered:

$$\boldsymbol{\delta}^{HC_1} = \{5309, 613.4\}\ \boldsymbol{\delta}^{HC_2} = \{6416, 514.3\}\ \boldsymbol{\delta}^{HC_3} = \{5347, 516.98\}$$

$$\boldsymbol{\delta}^{HC_4} = \{4338, 516.98\}\ \boldsymbol{\delta}^{HC_5} = \{5256, 613.4\}\ \boldsymbol{\delta}^{HC_6} = \{5422, 514.3\}.$$

Based on this, the PS and non-dominated HCs can be calculated. The non-dominated HCs are HC_4 and HC_6 since the vectors of objectives $\boldsymbol{\delta}^{HC_4}$ and $\boldsymbol{\delta}^{HC_6}$ are non-dominated.

$$\text{PS} = \{(4338, 516.98), (5422, 514.3)\}$$

3.1 Adaptation of the BF to Compute a PS

The BF algorithm is extended to deal with the multi-objective HCP computing a PS of a given graph. In principle, the algorithm allows more than two objectives. The user may define an objective using the corresponding matrix as input. For that purpose the following changes were made to the original algorithm:

1. **Initialization:** Save a HC whenever it is found without stopping the search and continuing exploring the tree.
2. **Initialization:** Compute the objective functions for the HCs and calculate the PS.
3. **Iteration:** Stop exploring a branch that will return a HC which is dominated by any solution already in the PS. A branch will be fathomed if the $\delta^{\mathcal{U}}$ is dominated by a δ^{HC}, being $\mathcal{U}$ the set of fixed arcs at the current stage. This check is performed in the fourth stage of the method, called iteration, after computing the second LP.

An anytime algorithm is an algorithm whose quality of results enhances progressively as computation time increases. One of the main properties of these algorithms is the interruptibility, as the algorithm can be stopped at any time and provide a solution. Provided that at least one HC has been found, our approach can be understood as an anytime algorithm, as a PS is returned at any time and it is improved as computation time increases.

4 Experiments

The algorithm was implemented in Python (version 3.5.2) and CPLEX (version 12.7.1) was used to solve the LPs. Twenty random Hamiltonian graphs of size 20 were generated to perform the experiments, using the following procedure: (1) Generate random symmetric matrices of 20 × 20 dimension of zeros and ones; (2) Use a random permutation of length 20 to assure that at least there is one HC, connecting the vertices indicated by the permutation; (3) Fill the diagonal of the matrices with zeros.

For each of the graphs, two weight matrices were also generated randomly. To generate the first matrix, the samples were extracted randomly from a normal distribution, $w_{ij}^1 \sim \mathcal{N}(\mu, \sigma) \times c$, where $\mu = 10$, $\sigma = 3$ and $c = 10$. The values of the second matrix were generated using the following formula $w_{ij}^2 = \alpha \times w_{ij}^1 + |\epsilon|$, where $\alpha = 0.01$ and $\epsilon \sim \mathcal{N}(\mu, \sigma)$, $\mu = 0$ and $\sigma = 100$.

Table 1 shows the number of HCs and the cardinality of the PS computed by the BF in 10, 20 and 30 min of time execution. It can be observed that the number of HCs increases with the time as expected, though the size of the increase differs from graph to graph. The reason for this is that the difficulty of finding a HC depends on the graph, as in some cases the algorithm has to explore deeper levels of the tree.

Table 1 Number of HCs found by the BF and the cardinality of the PS for the 20 Hamiltonian graphs in three time periods

	10 min		20 min		30 min	
	HCs	PS	HCs	PS	HCs	PS
G_1	175	8	347	11	417	11
G_2	337	8	558	7	768	8
G_3	758	15	1468	17	2227	18
G_4	983	12	1622	13	2025	13
G_5	487	7	1085	5	1573	7
G_6	830	16	1324	15	1617	16
G_7	397	6	614	6	691	6
G_8	788	6	1430	6	2027	6
G_9	399	13	713	10	974	9
G_{10}	1073	14	1631	17	2234	18
G_{11}	538	7	877	8	1278	8
G_{12}	287	6	604	6	905	12
G_{13}	501	9	841	9	1277	15
G_{14}	767	13	1290	13	1915	11
G_{15}	622	9	1297	13	1817	11
G_{16}	606	14	1436	13	1768	14
G_{17}	443	4	881	7	1188	12
G_{18}	576	14	898	15	1240	17
G_{19}	1056	15	1811	11	2592	11
G_{20}	515	3	862	5	1102	5

However, this phenomenon is not repeated for the cardinality of the PS since more HCs do not necessarily imply more elements in the PS.

5 Conclusions

We have introduced an extension of the BF algorithm to solve a multi-objective variant of the HCP. It was shown that the extended BF algorithm can simultaneously find several HCs and compute continuously an improving PS. The research presented can be improved executing the BF simultaneously by changing the labels of the graph vertices, as different HCs will be found for each of the representations. As a consequence, the obtained PSs will also be different and the algorithm could be considered as a parallel algorithm.

Acknowledgments This project was funded by the ELKARTEK Research Programme of the Basque Government (project KK-2019/00068). This work has been possible thanks to the support of the computing infrastructure of the i2BASQUE academic network. The work of Roberto Santana was funded by the Basque Government (project IT-1244-19), and Spanish Ministry of Economy and Competitiveness MINECO (project TIN2016-78365-R).

References

1. Ejov, V., Filar, J., Haythorpe, M., Nguyen, G.: Refined MDP-based branch-and-fix algorithm for the Hamiltonian cycle problem. Math. Oper. Res. **34**, 758–768 (2009)
2. Feinberg, E.: Constrained discounted Markov decision process and Hamiltonian cycles. Math. Oper. Res. **25**, 130–140 (2000)
3. Goldberg, D., Lingle, R.: Alleles, loci and the traveling salesman problem. In: Proceedings of the 1st International Conference on Genetic Algorithms and Their Applications, pp. 154–159 (1985)
4. Haythorpe, M.: Markov chain based algorithms for the hamiltonian cycle problem. PhD. Thesis, University of South Australia (2011)
5. Lenstra, J., Kan, A.R.: Some simple applications of the travelling salesman problem. J. Oper. Res. Soc. **26**, 717–733 (1975)
6. Nguyen, G.: Hamiltonian cycle problem, Markov decision processes and graph spectra. Ph.D. Thesis, University of South Australia (2009)

Part VII
Energy and Environment

Combinatorial Reverse Auction to Coordinate Transmission and Generation Assets in Brazil: Conceptual Proposal Based on Integer Programming

Laura S. Granada, Fernanda N. Kazama, and Paulo B. Correia

Abstract In Brazil, reverse auctions are the main mechanism for contracting generation and transmission assets. However, mismatches between their start-up dates have depleted supply on approximately 17,000 GWh. So, we propose a new Combinatorial Reverse Auction format, that jointly auctions' transmission and generation assets. The conceptual proposal was divided into two phases: descending clock rounds and sealed bid round. Activity rules as eligibility points and preference revealed were implemented to restrict the packages allowed for bidding in each round. The Winner Determination Problem was modeled as an Integer Linear Program adjusted to the Brazilian electric market. Computational experiments will be developed based on real data.

Keywords Combinatorial auction · Winner determination problem · Integer programming · Brazil electric sector

1 Introduction

The Brazilian electric sector until the 1990s was formed by vertically integrated companies that held generation, transmission, and distribution activities in a given concession area. In 2001, the energy crisis suffered in the country had significant economic and political impacts that were reflected in a new model for the electric sector. This model, implemented from 2003 onward, established the segregation of the activities of the electric sector to stimulate competition in each segment. Since then, simultaneous and sequential reverse auctions have been used to promote the expansion of both the generation facilities and the transmission capacity, respectively.

L. S. Granada (✉) · F. N. Kazama · P. B. Correia
University of Campinas - UNICAMP, Campinas, Brazil
e-mail: lauragranada@fem.unicamp.br

J. S. Neufeld et al. (eds.), *Operations Research Proceedings 2019*, Operations Research Proceedings, https://doi.org/10.1007/978-3-030-48439-2_29

The coordination of these auctions are key to prevent mismatches between generation and transmission assets. Minor distortions can have major long-term impacts, raising the cost of energy or compromising the system's security. However, transmission assets without any bid in auctions became a common practice since 2013, resulting in generation assets, already built and authorized to operate, but disconnected from the basic network. The negative impacts of mismatches were circa of 17,000 GWh, near 3% of the total energy generated in the country, overcharging the final consumers on US$802 million.

Simultaneous and sequential reverse auctions could fit substitute assets, but generation and transmission assets are essentially complementaries. So the bidders face the exposure problem.[1] Combinatorial Reverse Auctions (CRAs) allow participants to bid on an individual product, or on packages of them to explore synergies among products, mitigating the exposure problem. CRAs are difficult to implement because the number of packages grows significantly with the number of products. Given n products, a participant could submit $(2^n - 1)$ bids. In this way, finding the set of bids that fit the requirements of the auctioneer at a reasonable time is the main obstacle of this auction. This problem is known as the Winner Determination Problem (WDP) and is classified as NP-hard. The WDP can be formulated as a classic Integer Linear Program: set packing and knapsack. Several algorithms have been implemented to solve the WDP. Kameshwaran and Benyoucef [4] and Sandholm [5] used Branch and Bound (BB) search trees; Dynamic Programming (DP) was used by Terence [6]; Lagrangian Relaxation was used by Fu-Shiung [2], among others.

In this context, this study aims to propose a new CRA format adjusted to the Brazilian electricity market necessities.

2 Overview of Brazil's Auctions

One of the main challenges for the expansion of the electric sector is the coordination between generation and transmission auction. Nowadays, transmission assets are auctioned after generation assets, presenting a gap that can vary between 6 months to a year.

2.1 *Energy Assets Auction*

The Energy Assets Auctions (EAA) are multi-technology simultaneuos reverse auctions, that is, two or more generation technologies are competing simultaneously. In these auctions, the energy demand is pooled and procure jointly.

[1]For example, looking for a couple of products in a sequential auction, someone overbids on the first but do not get the second.

EAAs are divided into A-4 for quick-build technologies like wind, solar and small-hydro, and A-6 for conventional technologies with longer build time [3]. The National Electricity Regulatory Agency (ANEEL) is responsible for defining the auction rules, and the Chamber of Electric Energy Commercialization Trading (CCEE) conducts the auctioning process, acting as the auctioneer.

A-4 auctions are divided into an Initial and a Continuous phase. In the former, bidders, regardless of generation technology, submit sealed bids with a price (R$/MWh) and a capacity (MWh). Bidders are organized from lower to higher price and then, classified as temporary winners those, whose accumulated sum of their capacities does not exceed the remaining transmission capacity.

The Continuous phase is a clock descending auction. Generation assets are divided by technology, and then, they are organized and classified as in the previous phase. In this stage, any temporarily disqualified bidder can replace a temporary winner by submitting a bid lower than the marginal price minus a decrement. If this happens, a new list of temporary winners is created and the process restarts. This mechanism continues until no temporarily disqualified bidders submit a bid capable of ousting a temporary winner [3].

2.2 *Transmission Assets Auction*

Transmission Assets Auctions (TAAs) are sequential sealed bid reverse auctions for multi-items. The products offered in TAAs are lots comprised of transmission lines and electrical substations. For each lot to be auctioned, ANEEL defines the delivery term and the maximum value of the Allowed Annual Revenue (RAP).

For each lot, bidders submit a RAP-bid (R$/year). If the difference between the lowest bid and the others is more than 5%, the bidder with the lowest RAP-bid is declared the winner of the auction. Otherwise, only bidders with a bid difference less than or equal to 5% of the lowest bid are classified to an Open Outcry Session. If there are lots without receiving any bid, they are classified as Empty Lots.

In Open Outcry Session, the bidding process continues until no further bids are submitted. The bids in this session must be less than or equal to the lowest classified bid. Finally, the winner is the one with the lowest RAP-bid.

3 Conceptual Proposal: Methodology

The proposed CRA intends to sell both generation and transmission assets at the same time. The format is hybrid, composed of two phases: descending combinatorial clock rounds and combinatorial sealed bid round.

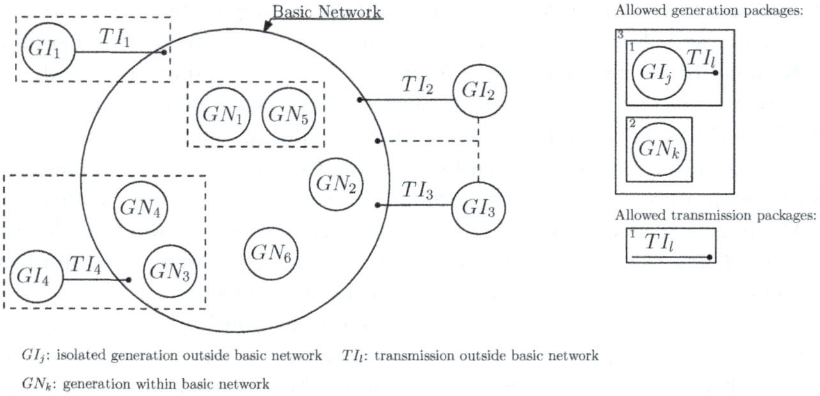

Fig. 1 Products and packages

3.1 Products and Packages

The products offered in our auction are generation and transmission assets. As shown in Fig. 1, these products are classified according to their location: within the basic network (GN_k) and outside it (($GI_j + TI_l$) and TI_l). Each generation asset has associated a generation capacity CI_j and CN_k, respectively.

If a bidder i, submits a bid on GI_j, he has to submit also on TI_l, guaranteeing the connection to the basic network. For this reason, these assets are always offered jointly. However, products TI_l are also offered individually, allowing other agents to submit more competitive bids on them.

In order to guarantee an unbundled electric market, the generating agents will only be able to submit bids on packages composed by generation assets, as shown in Fig. 1. The same consideration is valid for transmission agents.

3.2 Descending Combinatorial Clock Rounds

This phase is composed of multiple rounds. The auctioneer defines a reserve price at which bidders submit a single bid for each package of items of their interest. For items with an excess of demand, the price decrease and the bidders are allowed to submit new bids. These rounds end when there is no longer excess demand for any item.

Activity Rules The eligibility points and preference revealed presented by Ausubel and Cramton [1] were adapted for our CRA and considered jointly as activity rules.

Each item n offered in the auction is assigned a number of eligibility points e. Each package S_m has a number of eligibility points $E(S_m)$, which is calculated by adding the e of the items that conformer it. In the first round, each bidder i

will be assigned with a number of points associated with their financial situation or the deposit done. In the following rounds t, the bidder's eligibility E_i^t will be equal to the points of the previous round E_i^{t-1} or the package points submitted in the previous round $E(S_m)^{t-1}$, the lowest. In each round, while the price decrease, bidders can bid on packages with the same or smaller eligibility points than their own. Bidder's eligibility points can never increase in later rounds.

The revealed preference rule will be applied only in rounds in which bidders' eligibility is reduced. A bidder i will be allowed to bid on a certain package S_m in round t, regardless the eligibility points rule, if package S_m^t is relatively more advantageous than S_m^r, that is, $(p^t - p^r) \cdot (S_m^t - S_m^r) \geq 0, \forall r < t \mid r \in \mathcal{E}_i(S)$, where p is the clock price in each round and $\mathcal{E}_i(S)$ denotes set of packages in the rounds where their eligibility points were reduced.

3.3 Combinatorial Sealed Round

In this phase, bidders may submit bids on their classified packages. When all bids are received, the auctioneer will solve the WDP considering the bids of both descending clock rounds and sealed round.

The WDP was modeled as a set packing problem. For its resolution, an algorithm based on exact methods will be developed.

Activity Rule Our CRA activity rule was adapted from [1]. The set of packages in which the bidder i may bid are $\mathcal{E}_i(S)$, and the package of the final clock round is f. The bid b_{im} in the sealed round might satisfy that $b_{im} \geq b_{im}^t + p^t \cdot (S_m - S_m^t), \forall S_m \neq S_m^t, \forall t \in \mathcal{E}_i(S)$, where S_m represents the packages submitted in the sealed round. Following this rule, the items allocated in the final clock round will be unchanged after the sealed-bid round [1].

3.4 Mathematical Formulation

Each bidder $i, \forall i \in \mathcal{I}$, may bid[2] on any of the allowed packages $S_m, \forall m \in \mathcal{M}$. The bidder i's bid on a package m is represented as b_{im}. The packages can be formed by the products $GI_j, \forall j \in \mathcal{J}, TI_l, \forall l \in \mathcal{L}$, and $GR_k, \forall k \in \mathcal{K}$ as mentioned above.

Three matrices ($\mathbf{GI}_{m \times j}$, $\mathbf{GN}_{m \times k}$ and $\mathbf{TI}_{m \times l}$) were used to represent the formation of packages. The components of these matrices have values of 1 if the product belongs to a package, or zero otherwise. AI_{lj} is a Boolean matrix that indicate the association between GI_j and TI_l assets.

[2]The bid (R$/year) is the yearly revenue for the supplied energy, typically for 15–30 years forward contracts.

To identify which packages are winners, the model uses a decision variable $x_{im} \in \{0, 1\}$, $\forall i \in \mathcal{I}$ e $\forall m \in \mathcal{M}$. In this way, the objective function minimizes the value paid by final consumers according to:

$$z = \min \sum_{i \in \mathcal{I}} \sum_{m \in \mathcal{M}} b_{im} \cdot x_{im} \quad (1)$$

where $x_{im} = 1$ indicates the package m belonging to the bidder i is winning, or $x_{im} = 0$ otherwise.

EAA aims to increase generation capacity, through the energy coming from the winning assets. Equation 2 formalizes a demand constraint as:

$$\sum_{i \in \mathcal{I}} \sum_{m \in \mathcal{M}} \sum_{j \in \mathcal{J}} x_{im} \cdot GI_{mj} \cdot CI_j + \sum_{i \in \mathcal{I}} \sum_{m \in \mathcal{M}} \sum_{k \in \mathcal{K}} x_{im} \cdot GN_{mk} \cdot CN_k \geq D \quad (2)$$

Generation assets outside the basic network must have an associated transmission asset that connects them to the network. Equation 3 formalized this constraint:

$$\sum_{i \in \mathcal{I}} \sum_{m \in \mathcal{M}} x_{im} \cdot GI_{mj} - \sum_{i \in \mathcal{I}} \sum_{m \in \mathcal{M}} \sum_{l \in \mathcal{L}} x_{im} \cdot TI_{ml} \cdot AI_{lj} \leq 0 \qquad \forall j \in \mathcal{J} \quad (3)$$

Set-Packing constraints guarantee that a product is not allocated more than once, as shown in Eqs. 4.

$$\begin{aligned} \sum_{i \in \mathcal{I}} \sum_{m \in \mathcal{M}} x_{im} \cdot GI_{mj} &\leq 1 \qquad \forall j \in \mathcal{J} \\ \sum_{i \in \mathcal{I}} \sum_{m \in \mathcal{M}} x_{im} \cdot GN_{mk} &\leq 1 \qquad \forall k \in \mathcal{K} \\ \sum_{i \in \mathcal{I}} \sum_{m \in \mathcal{M}} x_{im} \cdot TI_{mj} &\leq 1 \qquad \forall j \in \mathcal{J} \end{aligned} \quad (4)$$

Other constraints such as the number of packages submitted per bidders, bid value on sealed bid round, market investment and financial guarantee values, can be treated as input validations in the computational interface to reduce the variables of the model.

4 Conclusions and Future Works

The Brazilian electricity market can take advantage of the CRA to overcome its difficulties. The current methodology makes the sector inefficient, decreases competition, and raises the prices paid by final consumers.

This paper aimed to present a hybrid CRA format for generation and transmission assets in Brazil. The model was formulated from the point of view of the auctioneer, minimizing the total contracted value.

A computational interface will be developed to simulate the proposal. Input data, the demand to be contracted and the characteristics of the assets to be simulated, will be based on real data. WDP will be solved using algorithms based on exact methods as Lagrangian Relaxation or Branch and Bound search trees.

Acknowledgments The authors acknowledge the financial support of the companies CPFL Paulista, Foz do Chapecó Energia, Campos Novos Energia and Transmissora Morro Agudo. This project was developed through the ANEEL research and development program, with number PD 00063-3042.

References

1. Ausubel, L.M., Cramton, P.: Activity rules for the combinatorial clock auction. University of Maryland, Department of Economics, Maryland (2011)
2. Fu-Shiung, H.: Combinatorial reverse auction based on revelation of Lagrangian multipliers. In: Decision Support Systems. Elsevier, Taiwan (2008)
3. Hochberg, M., Poudineh, R.: Renewables auction design in theory and practice: lessons from the experiences of Brazil an Mexico. The Oxford Institute for Energy Studies (2018). https://doi.org/10.26889/9781784671068
4. Kameshwaran, S., Benyoucef, L.: A branch and bound framework for optimal winner determination in combinatorial auction with interval bids, Metz (2007)
5. Sandholm, T.: Improved algorithms for optimal winner determination in combinatorial auctions and generalizations. In: Proceedings of the Seventeenth National Conference on Artificial Intelligence and Twelfth Conference on Innovative Applications of Artificial Intelligence, Texas (2000)
6. Terence, K.: Generalized knapsack solver for multi-unit combinatorial auction: analysis and application to computational resource allocation. In: 6th International Workshop on Agent-Mediated Electronic Commerce. Springer, Berlin (2005)

References

A Lagrangian Decomposition Approach to Solve Large Scale Multi-Sector Energy System Optimization Problems

Andreas Bley, Angela Pape, and Frank Fischer

Abstract We consider the capacity and operations planning of a European energy supply system with a high share of renewable energy. Our model includes the energy sectors electricity, heat, and transportation and it considers numerous types of consumers and power generation, storage, and transformation technologies, which participate in these energy sectors. Given time series for the regional demands in each sector and the potential renewable production, the goal is to simultaneously optimize the strategic dimensioning and the hourly operation of all components in the system such that the overall costs are minimized.

In this paper, we propose a Lagrangian solution approach that decomposes the model into many independent unit-commitment-type problems by relaxing several coupling constrains. This allows us to compute high quality lower bounds quickly and, in combination with some problem tailored heuristics, globally valid solutions with less computational effort.

1 Problem Description

We consider the problem of optimizing the design of several interconnected national or regional energy systems with a high share of renewable energy. Our goal is to simultaneously determine the dimensioning and the operation of a power plant fleet involving several energy sectors at minimum overall costs. Given a scenario

A. Bley
Institut für Mathematik, Universität Kassel, Kassel, Germany
e-mail: andreas.bley@uni-kassel.de

A. Pape (✉)
Fraunhofer IEE, Kassel, Germany
e-mail: angela.pape@iee.fraunhofer.de

F. Fischer
Institut für Informatik, Universität Mainz, Mainz, Germany
e-mail: frank.fischer@uni-mainz.de

J. S. Neufeld et al. (eds.), *Operations Research Proceedings 2019*, Operations Research Proceedings, https://doi.org/10.1007/978-3-030-48439-2_30

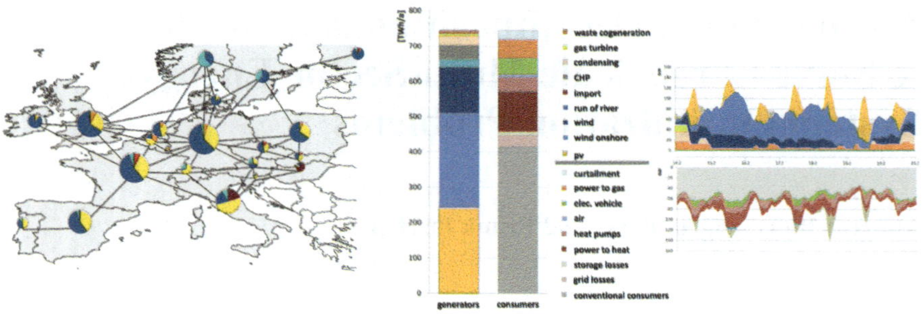

Fig. 1 Left: Primary energy sources and exchange capacities defined by technology installation. Right: Electricity production and consumption per technology defined by operational schedules

year, this fleet must cover regional demands for electricity, several types of heat, and transportation, which are given by time series at an hourly resolution, while adhering to the technical constraints of all components and a CO_2 emission budget. Figure 1 illustrates the general decisions we have to take, i.e., the dimensioning of all technologies in each region and the hourly operation of all components producing, consuming, or transforming energy.

Integer and linear programming formulations, which are commonly used for combined capacity and operations planning problems of this type, often lead to models that are way too large to be solved efficiently in practice. The main contribution of this paper is to show the efficiency of a Lagrangian solution approach that decomposes the problem into several single unit-commitment-type problems and, thus, allows us to solve problems with much less computational resources. For the sake of brevity, we neglect data uncertainty, the multi-period nature of the long-term energy system development and non-convexities stemming from discrete decisions in this paper and only consider the setting of a deterministic single-stage planning problem formulated as a continuous linear model. Possible extensions are discussed briefly in the outlook chapter.

The problem described above is modeled as a linear program using the SCOPE toolset developed at Fraunhofer IEE. SCOPE is a comprehensive and flexible energy systems modeling and planning tool implemented in MATLAB [12]. A detailed description of SCOPE and its modules can be found in [3–5, 9, 13] and the references therein.

In the problem variant addressed in this paper, we consider pure power generators (i.e. condensing power plants, gas turbines, wind turbines and PV systems) as well as storage technologies (i.e. batteries, hydrogen storages, and hydroelectric storages) in the electricity sector. These technologies contribute to regional electricity markets and a global CO_2 budget. The regional electricity markets are interconnected via so-called NTC links, which form the underlying electrical power network. In the heat sector, we distinguish between local and district heating and industrial process heating at different temperature levels, whose respective demand time series must be met. In our model, we consider several types of renewable or conventionally operated combined heat and power plants as well as a wide variety of pure heat producers (e.g. condensing boilers, solarthermal and geothermal energy) and heat pumps, which participate in the electricity market as flexible consumers. In the transport sector, industrial and personal transportation needs and driving characteristics are described by different time series of transportation demands. The optimization can draw on pure electric vehicles, catenary trucks, hydrogen vehicles or plug-in hybrids, which can use both electricity and liquid (fossil or renewable) fuels. In addition to the NTC links, some regions and energy sectors are also linked via the exchange of fossil or renewable gas, which can be produced in power-to-gas plants or electrolysers and used in gas turbines, for example. The task is to optimize the dimensioning and the operation of all components in the system such that the sum of investment and operational costs is minimized. For this, the model has to simultaneously decide how much capacity of which type to install in which region of the network and how to operate these over the planning horizon.

2 Lagrangian Approach and Implementation

Using the LP-tool set of SCOPE, the above problem is modelled as a linear programming model. In principle, the overall model contains a sub-model for each technology and for each region where this technology can be installed. These sub-models contain the variables and the 'local' constraints describing the dimensioning and the operation of a single technology without paying attention to the other technologies, similar to a unit commitment model for a single power plant. Additional 'global' energy balance, demand, and budget constraints then link the sub-models of all technologies that contribute to the respective regional energy balances, markets, and budgets. For each regional thermal and transport market, a single variable per participating technology is used to describe its market share and a single equality enforces a proper market split. This allows to optimize the mix of technologies in each region, but each technology's share remains fix for the entire planning horizon. Once the shares are decided, the power production of each technology must exactly meet its share of the corresponding demand time series over the entire planning horizon. For electricity more flexibility among the technologies' shares is permitted. Here, the model contains an individual energy balance equality per region and per hour of the planning horizon to ensure that power production and import meet consumption, export, and base demand. Finally, there are a few global constraints describing the CO_2 emission budget and a simplified gas network.

We illustrate this for the combined heat and power plant (CHP) technology. The overall LP contains one sub-model per region and fuel type to (cumulatively) model all corresponding CHP units. Each sub-model contains a variable P_{inst} to describe the installed capacity and a variable $x_{chp} \in [0, 1]$ to describe the thermal demand share of the CHP units in the corresponding region. To model the operation of CHP, each sub-model uses numerous time-indexed variables: $P(t)$ and $Q(t)$ with $t \in T := \{1, \ldots, 8760\}$ describe the electric and thermal power generation of the main unit per hour, $P_{pth}(t)$ and $Q_{bk}(t)$ the corresponding backup units' thermal generation, and $QS(t)$, $QS_{in}(t)$, $QS_{out}(t)$ the level, the addition, and the withdrawal from the integrated thermal storage unit, etc. Numerous local constraints ensure the valid operation of the units: Inequalities $P(t) \leq P_{inst}$ and $Q(t) \leq \gamma P_{inst}$ for all $t \in T$ bound the power and heat generation by the installed capacity and equalities $(1 - loss) \cdot QS(t-1) + QS_{in}(t) - QS_{out}(t) = QS(t)$ for all $t \in T$ model the dynamics of the thermal storage, for example. To ensure that the CHP units satisfy their share of the regional thermal demand time series $D_{th}(t)$, each sub-model contains the equalities $Q(t)+Q_{bk}(t)+QS_{out}-QS_{in}(t) = x_{chp} \cdot D_{th}(t)$ for all $t \in T$. Global constraints linking the CHP sub-models with other sub-models, are, for example, thermal market share and electric energy balance equalities. For each region, a single equality $\sum_{u \in ThT} x_u = 1$ ensures that the shares of all technologies in ThT that contribute to the thermal demand (with $chp \in ThT$) sum up to 1. Similarly, but permitting to vary the technologies' shares over time, equalities $\sum_{u \in ElT} P_u(t) = D_{el}(t)$ for all $t \in T$ (with $P_{chp}(t) := P(t) - P_{pth}(t)$ and $chp \in ElT$) ensure that electricity production and consumption are balanced in each region and hour.

For the large European scenarios we are interested in, even solving the continuous linear models is extremely challenging: The presolved LPs contain up to 50 mio variables, 50 mio constraints, and 200 mio non-zeros and solving them as monolithic LPs using CPLEX on a machine with 64 CPUs and 264 GB RAM takes up to 30 days.

In order to compute (near-optimal) solutions for these LPs faster, we developed a Lagrangian relaxation approach that actively exploits the special structure of the model. For an overview on Lagrangian relaxation and decomposition we refer the reader to the survey articles [10, 11]. A description of its application to unit commitment and network design problems can be found in [1, 2]. As mentioned above, the coefficient matrix of the overall LP has a near-block structure, with blocks corresponding to the technologies' sub-models linked only by few energy balance and budget constraints. The idea of the proposed approach is to relax these linking constraints, which disturb the block structure, and penalize their violation in the objective function by a linear term. Thus, the remaining relaxation decomposes into independent sub-problems, where each sub-problem only models the dimensioning and the operation of a single technology in one region. As these sub-problems are much smaller and easier to solve, this allows a more efficient solution of the overall problem by solving the sub-problems independently.

For the CHP sub-model described above, the thermal demand share and electric energy balance equalities are relaxed. This affects the objective coefficients of the variables x_{chp}, $P(t)$, $P_{pth}(t)$ via the Lagrangian multipliers, but breaks the interdependency with other sub-models contributing to the same region's heat and electric balances.

In our implementation we use the general purpose bundle method CONICBUNDLE [6, 7] to solve the Lagrangian dual problem. As the model is generated by the SCOPE toolset, a special MATLAB-to-C++ interface has been implemented to use CONICBUNDLE with MATLAB and to solve the independent sub-problems of the relaxation in parallel. The sub-problems are (re-)solved using the dual simplex of

CPLEX 12.7.1 [8], which performed most efficient when iteratively resolving the subproblems.

A disadvantage of the Lagrangian relaxation approach is that solutions obtained by solving the subproblems are in general not fully feasible for the original problem. However, these partial solutions are useful in heuristics to construct feasible near-optimal solutions. Aggregating the subproblems' solutions, the ConicBundle solver used in our implementation computes fractional solution candidates for the original model, for which the violation of the relaxed constraints converges towards 0. Thus, solution candidates obtained in later iterations of the algorithms are nearly feasible, which is often sufficient in our application. In order to compute fully feasible solutions, we also implemented a heuristic that fixes the variables of all but the most flexible and controllable technologies (i.e. gas units, boilers, and emergency units) to the candidate's values and then solves the original model for the few remaining variables of those technologies.

In order to guarantee a stable and efficient operation of the method, all subproblems must be reasonably bounded and numerically well-conditioned. To ensure this, we had to implement extra bound setting/strengthening and preprocessing routines for the sub-models within Scope. To cope with numerical difficulties, the sub-models for several technologies had to be revised. Still, especially problems with large operational time-horizons are numerically notoriously difficult. Finally, we had to ensure that the constraints that are relaxed in the approach are scaled (before relaxing) in such a way that the corresponding optimal Lagrangian multipliers do not differ by too many orders of magnitude and that the resulting penalties are in the same order of magnitude as the original objective function. Otherwise relaxed constraints might be fully ignored by the method, as they appear numerically insignificant, or convergence takes too long.

3 Results

Using our approach, we can solve the considered problems with less computational resources. Solving the subproblems sequentially substantially reduces the required RAM, so even large problems can be handled on a standard workstation. Instead of 128 GB and more required for the barrier algorithm to run efficiently, the Lagrangian approach only needed 12 GB RAM to solve models with operational schedules covering all 8760 h of a 1 year time-horizon. On the other hand, solving the subproblems in parallel, we can exploit multiprocessor machines very effectively. In contrast to the barrier algorithm, the value of the Lagrangian relaxation progresses rather fast in the first iterations before stalling near the optimal value. Thus, the method allows us to efficiently compute bounds and solutions near the optimum, which is sufficient for parameter studies. Table 1 shows the progression of both methods on a machine with two E5-2690v4 CPUs (28 cores at 2.6 GHz) and 256 GB RAM for two typical benchmark instances with 902 units in 21 regions and time-horizons of 200 and 8760 h, respectively.

Table 1 Progression of Lagrangian decomposition and barrier algorithm (without crossover) for two small benchmark instances with 200 h (left) and a 8760 h (right) planning horizon

Time	Gap lagrange	Status barrier	Time	Gap lagrange	Status barrier
6 s	6.8%	Start first iteration	848 s	51.3%	Start first iteration
14 s	3.5%	Dual bound < 0	5930 s	2.3%	First dual bound > 0
63 s	0.5%	First dual bound > 0	8420 s	1.8%	Gap <1%
94 s	0.4%	Optimal	9452 s	1.5%	Optimal
184 s	0.1%	–	14,459 s	1.0%	–

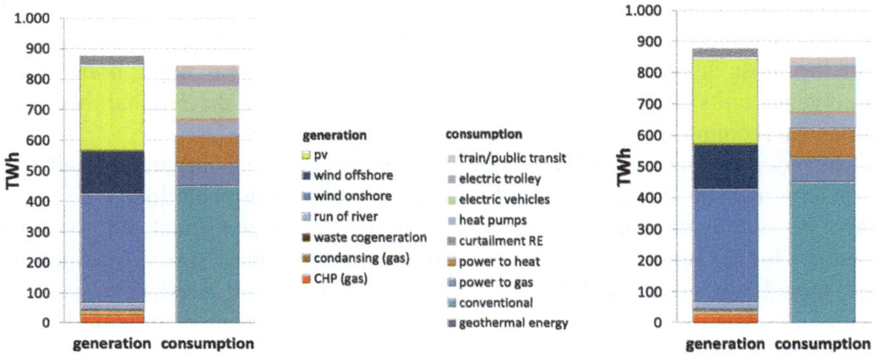

Fig. 2 Aggregated electricity production and consumption for the Lagrangian relaxation's solution candidate and the solution of the complete LP model for the 8760 h benchmark instance

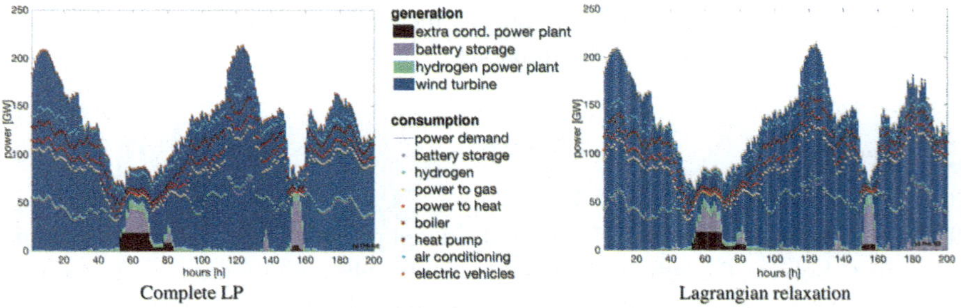

Fig. 3 Operational schedules for the Lagrangian relaxation's solution candidate and the solution of the complete LP model for the 200 h benchmark instance

The solution candidates obtained from the Lagrangian relaxation in later iterations are nearly feasible and converge towards the original LP solution. Figures 2 and 3 show how well the solution candidates obtained from the Lagrangian approach at the end resemble both the aggregated energy production and consumption values as well as the operational schedule of the optimal solution of the original complete LP model.

4 Outlook

In the future, we plan to apply this approach to mixed-integer versions of the model. These permit a more accurate modeling of operational limits, downtimes, and start-up times of certain technologies, which may have a significant impact on the cost and operation of these components. Furthermore, we plan to address the multi-period version of the problem, where several incremental expansion stages of the energy system are considered. This will enable us to analyze so-called development paths of energy systems, to quantify the path's impact on the final system, and to identify so-called bridging technologies. In both cases, the generalization of the approach is straightforward.

Acknowledgments This work has been supported by the German Federal Ministry for Economic Affairs and Energy (BMWi), Grant 0325879, Project *PfadE*^{3}.

References

1. Belloni, A., Diniz Souto Lima, A.L., Piñeiro Maceira, M.E., Sagastizábal, C.A.: Bundle relaxation and primal recovery in unit commitment problems. The Brazilian case. Ann. Oper. Res. **120**, 21–44 (2003)
2. Bley, A., Fischer, F., Hahn, P.: Decomposition techniques for large scale optimisation problems. In: Proceedings of the 1st WindAc Africa, Cape Town (2016)
3. Gerhardt, N., Sandau, F., Scholz, A., Hahn, H.: Interaktion EE-Strom, Wärme und Verkehr. Tech. Rep., Fraunhofer IEE (2015)
4. Gerhardt, N., Böttger, D., Trost, T., Scholz, A., Pape, C., Gerlach, A.K., Härtel, P., Ganal, I.: Analyse eines europäischen 95-Prozent-Klimaschutzszenarios über mehrere Wetterjahre. Tech. Rep., Fraunhofer IEE (2017)
5. Gerhardt, N., Ganal, I., Jentsch, M., Rodriguez, J., Stroh, K., Buchmann, E.K.: Entwicklung der Gebäudewärme und Rückkopplung mit dem Energiesystem in 95-Prozent THG-Klimazielszenarien. Tech. Rep., Fraunhofer IEE (2019)
6. Helmberg, C.: ConicBundle 0.3.11 (2011)
7. Helmberg, C., Kiwiel, K.: A spectral bundle method with bounds. Math. Program. **93**, 173–194 (2002)
8. IBM ILOG CPLEX: IBM ILOG CPLEX 12.7, User's Manual for CPLEX (2015)
9. Jentsch, M.: Potenziale von Power-to-Gas Energiespeichern. Ph.D. Thesis, University of Kassel (2014)
10. Lemaréchal, C.: Lagrangian relaxation. In: Computational Combinatorial Optimization, pp. 112–156. Springer, Berlin (2001)
11. Lemaréchal, C.: The omnipresence of Lagrange. Ann. Oper. Res. **153**, 9–27 (2007)
12. MathWorks MATLAB: MATLAB R2017a (2017)
13. Pape, C., Gerhardt, N., Härtel, P., Scholz, A., Schwinn, R., Drees, T., Maaz, A., Sprey, J., Breuer, C., Moser, A., Sailer, F., Reuter, S., Müller, T.: Roadmap Speicher. Speicherbedarf für Erneuerbare Energien - Speicheralternativen - Speicheranreiz - Überwindung rechtlicher Hemmnisse. Tech. Rep., Fraunhofer IEE (2014)

Operational Plan for the Energy Plants Considering the Fluctuations in the Spot Price of Electricity

Masato Dei, Tomoki Fukuba, Takayuki Shiina, and K. Tokoro

Abstract Electricity is now traded in various ways in the market. Planning is necessary to buy or sell electricity in the market, because the fluctuations in the market price may incur significant costs. Regardless of the irregularities in power generation due to plant failures or inspections, or inconsistent weather conditions, the electricity demand of the market must be delivered. Therefore, it is necessary to have an operational plan that takes into account the uncertainty of the market price. Such uncertainty problems are usually solved by using the expected value minimization model. However, this model is risk-neutral, and does not consider fluctuations in the costs of different constituents. Accordingly, we propose a conditional value at risk minimization model that avoids the risk of fluctuating costs. In this study, we formulate a stochastic programming model for the operational plan (considering uncertainty), for factories Further, we show the effectiveness of the proposed model by comparing it with the expected value minimization model.

Keywords Stochastic programming · Energy policy and planning · Integer programming

1 Introduction

Recently, environmental problems such as global warming have become serious Therefore, efforts are being made to realize sustainable local communities that have a low environmental impact, and the development of smart communities is one of these efforts. Tokoro and Fukuyama [1] presented a basic model for the quantitative evaluation of a smart community, which considers seven developmental

M. Dei · T. Fukuba · T. Shiina (✉)
Waseda University, Tokyo, Japan
e-mail: tshiina@waseda.jp

K. Tokoro
Central Research Institute of Electric Power Industry, Yokosuka, Kanagawa, Japan

J. S. Neufeld et al. (eds.), *Operations Research Proceedings 2019*, Operations Research Proceedings, https://doi.org/10.1007/978-3-030-48439-2_31

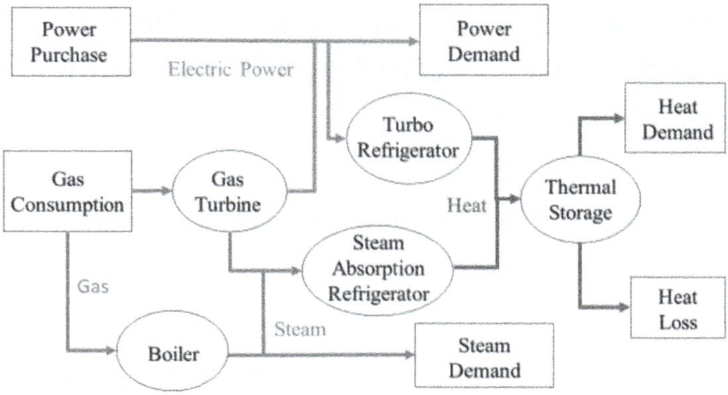

Fig. 1 Constituents of the energy plant

areas electricity: gas, water treatment, industry, business, home, and railroad. The basic model facilitates the quantitative evaluation of smart communities, which considers the interaction among the seven areas of development. In their industrial model, Suzuki and Okamoto [2] proposed a benchmark model for energy plant planning in the context of industries. The model applies to factories. Fukuba et al. [3] extended the benchmark model by introducing PV (photovoltaic) generation. The benchmark problem is a model of an energy plant that assumes the purchase of electricity and natural gas, and uses the equipment to meet the demand for electricity, steam, and heat. Figure 1 indicates the constituents of the system: a gas turbine, boiler, two types of refrigerators, and a thermal storage tank. Here, the objective is to design an operational plan that minimizes the total cost incurred in purchasing power and natural gas under the constraints of operation and energy flow.

The Japan electric power exchange (JEPX) maintains a stable supply of electricity by balancing the power generation plan submitted by the power producer with the demand plan submitted by the consumer. The power generation plan and the demand plan for every 30 min of the next day are determined 1 day before. Accordingly, the plan for the next day is fixed in the day ahead market (spot market) [4]. Depending on the shortage or excess of power according to the plan, the electricity supplier buys or sells power in the market. In the case of a shortage of power in the market, the price may rise, and the supplier has to pay a substantial amount to buy the power. Thus, the decision has to be taken under uncertain circumstances. In this study, we consider trading only in the spot market, and plans are made 1 day ahead. Though there is an adjusting market where adjustments in the power plans are made after the planning, the trading in the adjusting market is integrated in the spot market (assuming that the day ahead market and the adjusting market are highly correlated).

2 Problem Formulation

The following nomenclature is used while formulating the problem.

Parameters

C_{Er}^{i} Purchase cost of electricity at time i
C_{Fr}^{i} Purchase cost of gas at time i
$\overline{E}_{L}^{i}$ Demand of electricity at time i
E_{rm}^{i} Unutilized electric power at time i
K Number of scenarios
I Number of planning periods
P_k Occurrence probability of scenario k
$C_{Er}^{i,k}$ Purchase cost of electricity at time i under scenario k
α Risk parameter

Decision variables

x_g^i Gas consumption of a gas turbine at time i
x_t^i Heat production of a turbo refrigerator at time i
x_b^i Gas consumption of a boiler at time i
z VaR (value at risk; α quantile of random cost)

Functions

$E_r^i\left(x_g^i, x_t^i; \overline{E}_L^i, E_{rm}^i\right)$ Purchase amount of electric power at time i

Here, our aim is to minimize the total cost incurred in the purchase of electricity. Accordingly, the objective function is given by,

$$\min \sum\nolimits_{i=1}^{I} \left\{ C_{Er}^{i} E_r^i\left(x_g^i, x_t^i; \overline{E}_L^i, E_{rm}^i\right) + C_{Fr}^{i}\left(x_g^i + x_b^i\right)\right\} \tag{1}$$

The first term in (1) represents the purchase cost of electricity, and the second term represents the purchase cost of gas. Usually, the expected value minimization model is used to address the problems of price fluctuations. We also use the same model to minimize the expected purchase cost of electricity. The corresponding objective function can be shown as follows.

$$\min \sum\nolimits_{k=1}^{K} P_k \sum\nolimits_{i=1}^{I} C_{Er}^{i,k} E_r^i\left(x_g^i, x_t^i; \overline{E}_L^i, E_{rm}^i\right) + \sum\nolimits_{i=1}^{I} C_{Fr}^{i}\left(x_g^i + x_b^i\right) \tag{2}$$

The first term in (2) represents the expected purchase cost of electricity. However, the model is based on a risk-neutral framework and does not take into account the risk of fluctuations in the cost. Therefore, to reduce the cost fluctuations in the worst-case scenario, we use the conditional value at risk (CVaR) indicator proposed by Rockafellar and Uryasev [5]. CVaR is the expected cost exceeding VaR. The optimization problems that include VaR do not appear under the category of convex programming problems, and an efficient approach is required to solve them. On the other hand, the optimization problems that include CVaR are convex programming problems and can be solved easily. Accordingly, we propose a CVaR minimization model to minimize the risk of increase in the purchase cost caused by the fluctuations in the spot price. The corresponding objective function is shown below.

$$\min z + \frac{1}{1-\alpha}\sum_{k=1}^{K} P_k \left\{ \sum_{i=1}^{I} C_{Er}^{i,k} E_r^i \left(x_g^i, \boldsymbol{x}_t^i; \overline{E}_L^i, E_{rm}^i \right) - z \right\}^+ + \sum_{i=1}^{I} C_{Fr}^i \left(x_g^i + x_b^i \right) \tag{3}$$

3 Numerical Experiments

Numerical experiments were performed using the AMPL modeling language and Gurobi 7.5.0 solver in Windows 10 (64 bit architecture) operating system installed on a computer with core i7-4790S (3.20 GHz) CPU and 16.0 GB of RAM. A scenario representing the uncertainty of the power spot price is generated using monthly historical data of the power spot price of the JEPX [4]. Therefore, the number of scenarios is the number of days in each month. We can observe that the power spot price fluctuates throughout the day. The power spot price for 31 days in January 2017 is shown in Fig. 2. In the figure, the horizontal axis corresponds to the number of planning periods, and the time interval between two successive planning periods is 30 min. It is evident from Fig. 2 that the maximum power spot price is five times the minimum power spot price. Under the consideration of fluctuations in the power spot prices, we evaluate the cost of operation using the expected value minimization and CVaR minimization models. Figure 3 shows the demand of electric power for a particular day [2]. It can be observed that the demand of electric power is approximately 20 MWh during daytime.

The calculations were performed using both the expected value minimization model and the CVaR minimization model for January–December, 2017. Risk parameters $\alpha = 0.95, 0.9, 0.85$ were used in CVaR minimization model. Figure 4 shows the expected power purchase cost and gas purchase cost as a function of month in 2017. The figure indicates large fluctuation in power purchase costs throughout the year if the expected value minimization model is used. In contrast, cost fluctuations are small in the CVaR minimization model. Although the power purchase cost is lower, the gas purchase cost is higher in the CVaR minimization model as compared to the expected value minimization model. A comparison between the outcomes of both models indicates that trading volumes of power and gas have been swapped. Because gas is not traded in the market in Japan, it is effective to buy a bulk amount of the natural gas at a stable price to avoid the risk of high cost of electricity due to fluctuating spot prices. Figure 5 shows the distribution of costs and cumulative probabilities for 31 scenarios of January 2017 using both models. Here, the total costs for each scenario are listed in ascending order.

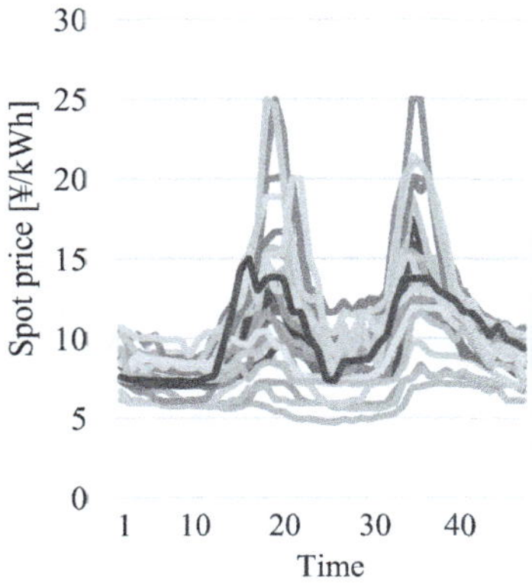

Fig. 2 Fluctuation of power spot price

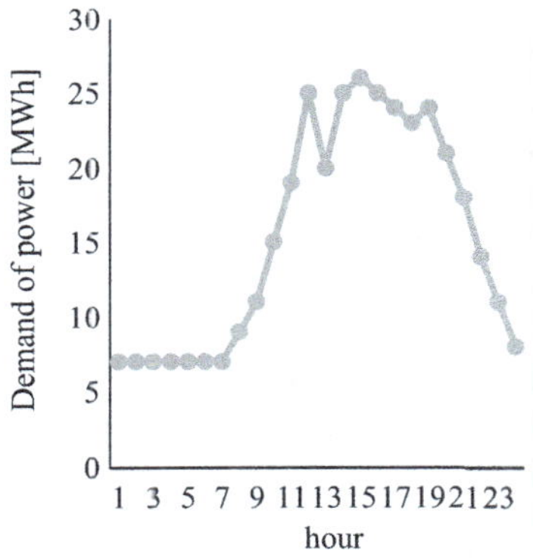

Fig. 3 Demand of power

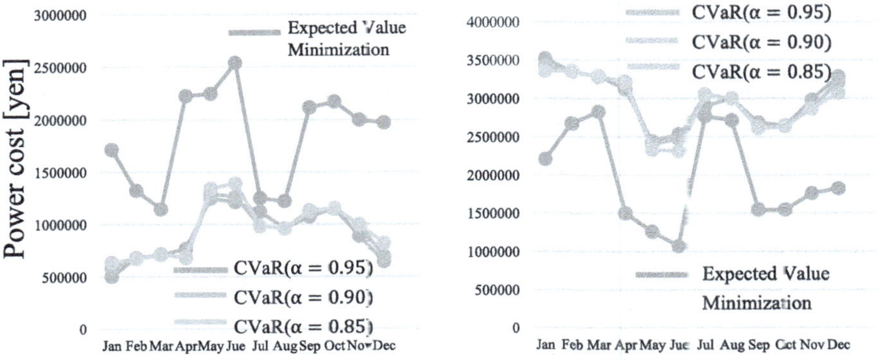

Fig. 4 Monthly expected cost of power and gas in 2017

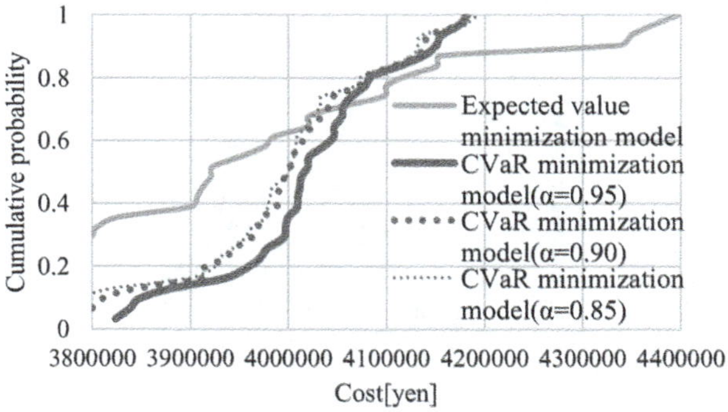

Fig. 5 Comparison between the expected value minimization model and the CVaR minimization model

Table 1 Comparison of the cost in the worst-case scenario ($\alpha = 0.90$)

Cumulative probability [%]	Quantile of expected value minimization	Quantile of CVaR minimization ($\alpha = 0.90$)	GAP [%]
100 = 31/31 days	4,397,657	4,188,711	4.75
96.77 = 30/31 days	4,371,748	4,174,012	4.52
93.55 = 29/31 days	4,349,400	4,142,154	4.76

In addition, Table 1 shows the comparison of the costs in the worst-case scenario between the expected value minimization model and the CVaR minimization model ($\alpha = 0.90$). It shows the amount of worst cost that can be avoided by the CVaR minimization model compared with the expected value minimization model. The criteria for evaluation of GAP is defined as follows.

$$\text{GAP} = \frac{\text{Expected value minimization} - \text{CVaR minimization}}{\text{Expected value minimization}} \times 100\,[\%]$$

Results show that the CVaR minimization model reduces the cost in the worst-case scenario compared to the expected value minimization model. Accordingly, a reduction in the fluctuations of the costs is observed. Further, the risk-neutral expected value minimization model minimizes the expected and resultant costs, as expected, which are smaller than the CVaR minimization model. However, the costs fluctuations and the cost incurred in the worst scenario are considerably large in the case of expected value minimization model as compared to the CVaR minimization model.

We note that although CVaR minimization model resulted in higher expected cost compared to the expected value minimization model, the cost in the worst-case scenario is reduced. Moreover, the effect is considerably large, as the cost in the worst-case scenario is reduced by 4.75% compared to the expected value minimization model.

4 Concluding Remarks

In this paper, we proposed a CVaR minimization model that avoids the risk of fluctuating power spot prices, and compared it with the expected value minimization model. The CVaR minimization model suppresses the worst cost that occurs on the upper side of the cumulative probability, and it has been shown that the cost variation is reduced. In addition, by introducing uncertainty into the original operational model, this could contribute to smart communities. In the future, we will extend the model by considering plans when other facilities are introduced, or plans for selling the generated power to the market.

References

1. Tokoro, K., Fukuyama, Y.: Development and extension of smart community model for energy efficiency. Oper. Res. **62**(1), 44–49 (2017)
2. Suzuki, R., Okamoto, T.: An Introduction of the Energy Plant Operational Planning Problem: A Formulation and Solutions. IEEJ International Workshop on Sensing, Acutuation, and Motion Control (2015)
3. Fukuba, T., Shiina, T., Tokoro, K., Sato, T.: Optimization of Power Plant Operation via Stochastic Programming with Recourse, pp. 1–12. Springer (2020)
4. JEPX Japan Electric power Exchange. http://www.jepx.org/market/index.html. Accessed 07 Nov 2017
5. Rockafellar, R.T., Uryasev, S.: Optimization of conditional value-at-risk. J. Risk. **2**, 21–41 (2000)

Design of an Electric Bus Fleet and Determination of Economic Break-Even

Marius Madsen and Marc Gennat

Abstract In this contribution the planning and optimization of a mixed electric and diesel bus fleet for a municipal transportation company is presented. The goal and objective is to maximize the driven electric mileage without expanding the total number of buses in the fleet. Boundary conditions are driving schedules and timetable planning must be hold. Moreover, buses are to be charged exclusively in the depot. No opportunity charging outside the depot is allowed due to costs of additional installations. Based on an existing timetable, cycles are identified which can be driven with a parametrizable battery capacity. Furthermore, climatic conditions and weight of batteries, which are incorporated into the specific energy requirement per kilometer, are modeled.

Keywords Mobility · Vehicle scheduling · Economic modeling

1 Introduction and Background

Public transport in Krefeld is provided by the local transport company SWK MOBIL GmbH using buses and trams. The SWK is discussing a possible purchase of electric buses. Economic analyses come to the conclusion that electric buses are currently not economical compared to diesel buses, although the mileage-related costs are lower, the higher capital expenditures are not compensated [9].

In this work, a heuristic model will be used to check under which conditions electric standard buses (12 m) and electric articulated buses (18 m) can be operated economically in Krefeld. For this purpose, it is investigated how a combination of given cycles can be used to maximize electrical operating kilometers in order to benefit from lower mileage-dependent costs.

M. Madsen (✉) · M. Gennat
SWK E^2 Institute of Energy Technology and Energy Management, Hochschule Niederrhein University of Applied Science, Krefeld, Germany
e-mail: marius.madsen@hs-niederrhein.de; marc.gennat@hs-niederrhein.de

J. S. Neufeld et al. (eds.), *Operations Research Proceedings 2019*, Operations Research Proceedings, https://doi.org/10.1007/978-3-030-48439-2_32

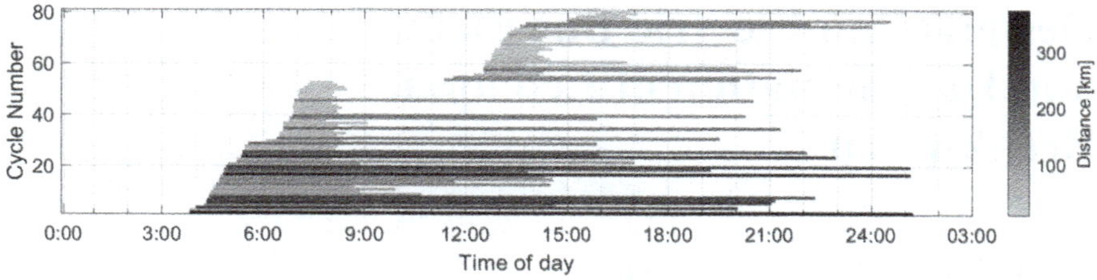

Fig. 1 18-meter bus cycles on day type *Monday*

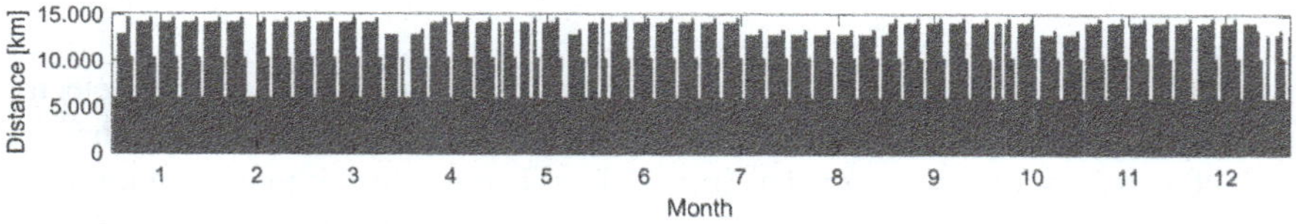

Fig. 2 Resulting daily mileage

2 Modeling of Cycles

The scheduling software used by SWK organizes and stores cycle information according to the standard interface *ÖPNV-Datenmodell 5.0* [12]. Matrices are generated for all different day types. Travel times, stops, routes and line numbers are available for each cycle. A cycle starts with depot exit and ends with a depot entrance [7].

As an example, Fig. 1 shows 18-meter-bus cycles on a Monday. Linking the calendar to the cycle database results in the mileage shown in Fig. 2, depending on the day of the year.

3 Heuristic Maximization of Electric Operated Mileage

Mileage-related operating costs of electric buses are lower than those of diesel buses [2, 9]. For this reason, it is examined under which conditions electric busses can be used economically if the electrically driven mileage is maximized. Investments in infrastructure are already kept low using existing DC systems of tram infrastructure (see [5]), because they are operated with the same voltage [10]. The capital expenditures are therefore significantly affected by the selected traction battery capacity [3]. Therefore, an heuristic algorithm is used, which maximizes the electrically driven mileage with the smallest possible battery capacity.

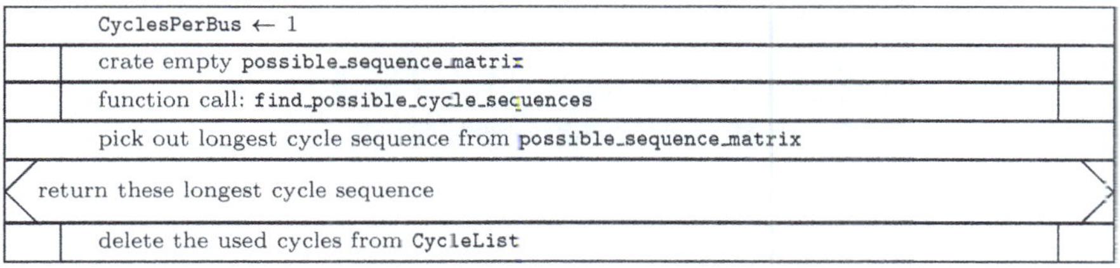

Fig. 3 Function: `longest_cycle_sequence`

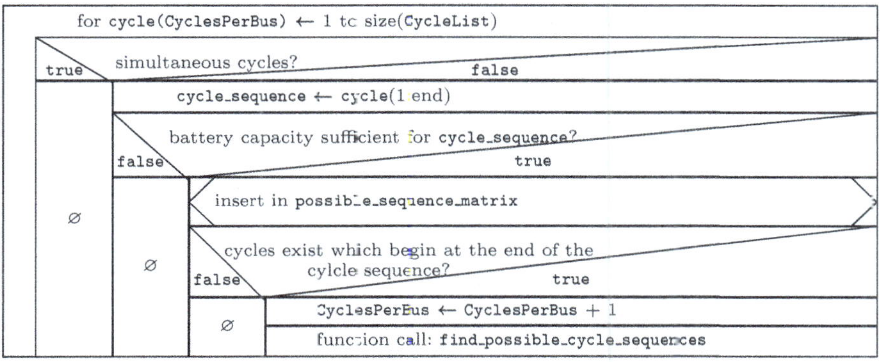

Fig. 4 Function: `find_possible_cycle_sequences`

3.1 Combination of Suitable Cycles

A bus can travel several short cycles in a row in one day instead of a long one (see Fig. 1). If an electric bus travels several short cycles in a row and is recharged in the depot between the cycles, a longer total daily mileage is achievable. This results in a combined system of opportunity and overnight charging [6]. Energy requirements of the individual cycles and days vary depending on the bus type, weather, battery design and cycle specifications. Depending on the day, there are about 80 cycles. To find the combination of cycles with highest mileage per day a heuristic was developed. The heuristic is shown below as a Nassi–Shneiderman diagram in Fig. 3.

In a first step the heuristic computes with function *find_possible_cycle_sequences* (Fig. 4) all possible combinations for each day and bus. This function checks in advance, whether the battery capacity is sufficient to drive the cycle. Required heating and cooling capacity and possible charging time between cycles in the depot are considered. In a second step the combination with the highest mileage is selected.

A possible result is shown in Fig. 5 for the combination of most suitable cycles for ten 18-meter-buses as an example for 1 day.

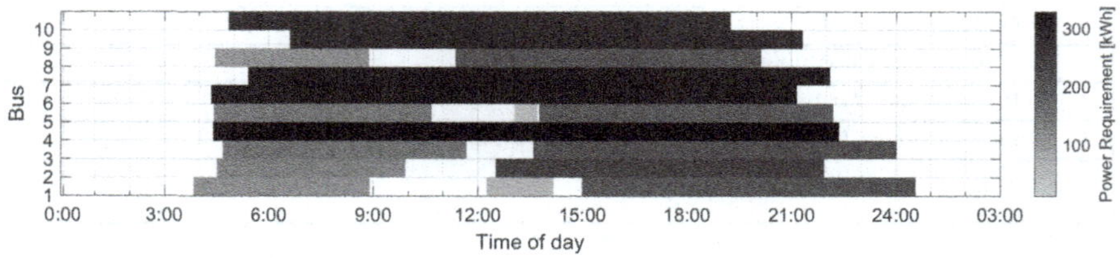

Fig. 5 Resulting cycle sequence for ten 18-meter-buses on a monday

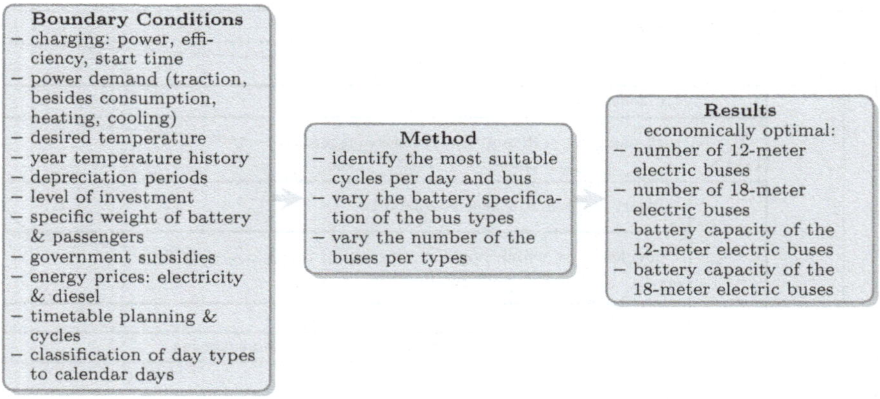

Fig. 6 Central parameters in the algorithm

3.2 *Embedding the Heuristic in an Expanded Algorithm*

Until now, the mileage of individual days and buses have been maximized. For consideration of a whole year, an expansion is needed. The central parameters of the expanded algorithm are shown in Fig. 6.

To determine the cost-effectiveness according to different boundary conditions, the illustrated heuristic is implemented in an algorithm (Fig. 7). This algorithm determined the electric bus fleet with minimum costs and maximum of kilometers. For a single bus, this procedure calculates the best solution. For multiple buses, an optimal solution cannot be guaranteed, because a heuristic is used. However, it is believed that the solution is near optimal or optimal.

Figure 8 shows the daily mileage of separate buses and for each day per year as an example of the algorithm for 40 articulated buses with a battery capacity of 500 kWh each.

If Fig. 8 is compared with Fig. 2, the high impact of the seasons becomes clear, as the range is reduced, depending on the heating or cooling requirements. The possible daily mileage decreases with the number of buses. In summer holidays, a maximum of 16 electric 18-meter buses can be operated on weekdays, and only five

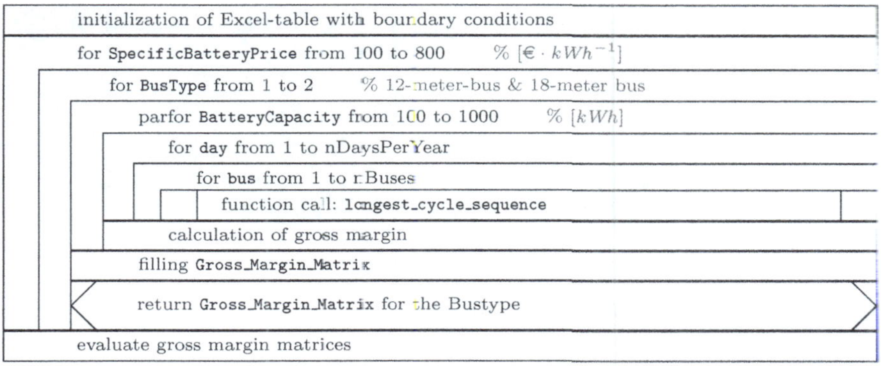

Fig. 7 Structure of the expanded algorithm

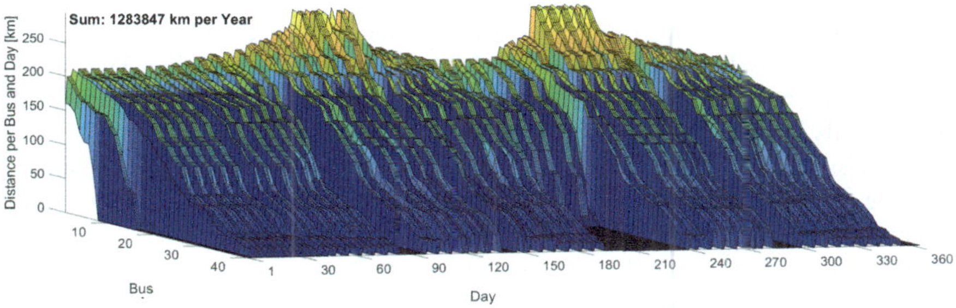

Fig. 8 Driven mileage per day and 18-meter-bus

on Sundays, depending on the weather. Thus, an economic optimum is expected with a small number of electric buses.

3.3 *Results of Maximizing the Electric Operating Mileage*

Even with the maximization of electric driven kilometers, an economical viable purchase of electric buses is not feasible, even with government subsidies of 40% at today's market prices. Figure 9 shows the gross-margin matrix as a function of the specific battery price and the battery capacity for three 18-meter-buses. An break-even is archieved, if specific prices of traction batteries fall below 370 €/kWh. When iterating the algorithm with different number of buses, the economic optimum results for a quantity of three electric buses (at above mentioned battery prices of 370 €/kWh).

There are more cycles for 18-meter-buses in Krefeld than 12-meter-bus cycles, so there is a higher potential in selecting suitable combinations. For 12-meter-buses, the battery price may not exceed 350 €/kWh.

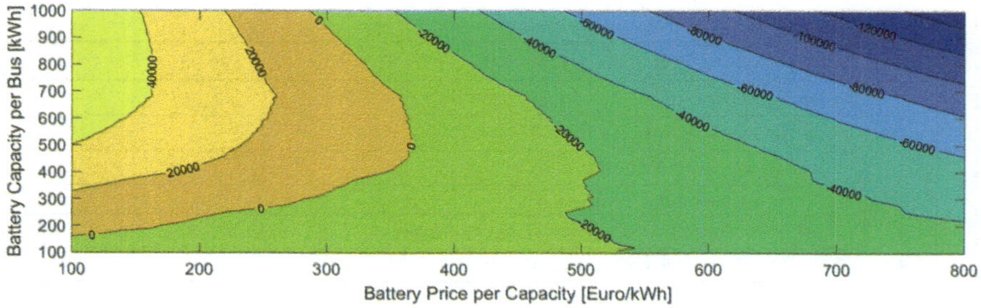

Fig. 9 Gross margin results for diffenrent battery capacities and prices

4 Results and Conclusions

It has been shown that electric buses in Krefeld can not be used economically under current conditions. Break even is achieved at a battery price of less than 370 €/kWh. Although the prices of battery cells have dropped [1], the prices for battery packs are currently at 600–1000 €/kWh [3, 4, 9].

It is assumed that these results can be transferred to cities with a similar number of cycles. Thus, electric buses can not be used economically in such cities under current conditions (see [8, 11]).

If it is politically desired to internalize external effects, such as CO_2 or noise emissions, the framework conditions must change. These could be a taxation of fossil fuels or higher subsidies of electric buses. These changes would have a high impact on the initial conditions of this study.

Neither the actual timetables in the scheduling software, nor the cycles are changed in this study. Only existing cycles were assigned daily to electric buses. In continuing investigations, further potential can be identified in breaking up and optimizing the cycles.

Acknowledgement This work was supported by a grant from the `SWK Mobil GmbH` in a joined R&D Project.

References

1. Ahlswede, A.: Weltweite Preisentwicklung für Lithium-Ionen-Batterien von 2013 bis 2020, Statista (2019). https://de.statista.com/statistik/daten/studie/534429/umfrage/weltweite-preise-fuer-lithium-ionen-akkus
2. Hondius, H.: Was sparen Elektrobusse wirklich? Wirkungsgrade von Elektrobussen im Vergleich zu Dieselbussen vor dem Hintergrund der Konsequenzen der Energiewende. Der Nahverkehr **32**(10), 38–42 (2014)
3. Kunith, A.: Elektrifizierung des urbanen öffentlichen Busverkehrs. Springer, Berlin (2017). https://doi.org/10.1007/978-3-658-19347-8

4. Lehner, S., Rogge, M., Becker, J., Sauer, D.: Battery design for successful electrification in public transport. Energies **8**(7), 6715–6737 (2015). https://doi.org/10.3390/en8076715
5. Müller-Hellmann, A.: Stadtbahnen und Elektromobilität - Ideen zu einer Symbiose. Der Nahverkehr **30**(3), 10–14 (2012)
6. Müller-Hellmann, A.: Vielfalt der Ladestrategien. Variationsmöglichkeiten von Betriebsweisen, Ladeverfahren und -strategien für den Batteriebuseinsatz im ÖPNV. Der Nahverkehr, vol. 36, pp. 9–12. Sonderheft Elektrobusse (2018)
7. Schnieder, L.: Betriebsplanung im öffentlichen Personennahverkehr. VDI & Springer Vieweg, Berlin (2018). https://doi.org/10.1007/978-3-662-57318-1
8. Schwarze, J., Schroeteler, K., Hasler, P., Anemüller, S.: Köln setzt auf E-Mobilität auch im Busverkehr. Der Nahverkehr **35**(6), 39–44 (2017)
9. Seeliger, A., Jeschull S., Krönauer B., Limberg S., Schreiner C., Albuquerque C., de Souza M., Verza M.: Elektrobusse im ÖPNV, HS Niederrhein (2016)
10. Tschöke, H.: Die Elektrifizierung des Antriebsstrangs. Springer, Berlin (2015). https://doi.org/10.1007/978-3-658-04644-6
11. van Liemt, C.: Stadt Bingen – Referenzstandort für Elektromobilität. Diplomica Verlag (2018)
12. Verband Deutscher Verkehrsunternehmen (VDV): VDV Standardschnittstelle Liniennetz/-Fahrplan. VDV-Schriften (2013)

Tradeoffs Between Battery Degradation and Profit from Market Participation of Solar-Storage Plants

Leopold Kuttner

Abstract Large scale energy storage installations have seen a rise in recent years, especially in combination with renewable power plants. Their increasing integration is contingent on successful market participation. So far, a common objective in energy resource scheduling is the maximization of profit from market participation, which requires high storage utilization and, consequently, leads to their degradation. This contribution presents a multi-objective mixed-integer program for operational planning of solar-storage plants that maximizes profit and minimizes the number of charge cycles. In a case study, solar-storage plants are studied in a short-term market environment. Pareto optimal solutions are determined using the epsilon constraint method. Analysis of the pareto-front shows that even minor sacrifices in profit allow to significantly reduce the number of charge cycles and therefore increase the lifetime of the battery.

Keywords Distributed energy resource scheduling · Multi-objective · Battery storage · Solar plant

1 Introduction

In recent years, large-scale energy storage plant projects have been successfully integrated into the power grid. Continuing advances in battery technology drive down costs and make participation of battery plants in energy markets viable. Especially in combination with renewable power plants, such as wind or photovoltaic plants, storage plants allow to use the full potential of renewables, thereby accelerating the transition to a carbon free energy supply.

Plant operators must determine power schedules ahead of time in order to operate efficiently. These scheduling decisions are in large parts driven by the goal of prof-

L. Kuttner (✉)
TU Dresden, Faculty of Business and Economics, Dresden, Germany
e-mail: leopold.kuttner@tu-dresden.de

J. S. Neufeld et al. (eds.), *Operations Research Proceedings 2019*, Operations Research Proceedings, https://doi.org/10.1007/978-3-030-48439-2_33

itable market participation. But, because batteries suffer from gradual degradation over time, which decreases their profitability in the long-term, extending the lifetime of batteries must be considered as well. A key determining factor to battery lifetime and performance is the cycle rate [1, 2]. By minimizing the number of charge cycles, the lifetime of batteries can be significantly extended. Therefore, two conflicting objectives emerge: maximizing profit from market participation in the short-term versus prolonging battery lifetime in the long-term through minimization of charge cycles.

This contribution presents a multi-objective mixed-integer scheduling model that respects both objectives. In a case study, solar-storage plants are studied in a short-term market environment. Pareto optimal solutions are determined using the epsilon constraint method. We further analyze the tradeoff between profit and battery degradation by selecting solutions from the pareto front using the utopian point method and by comparing them against those obtained by a simple penalty approach.

2 Battery Degradation and Market Participation

2.1 Battery Degradation

There are two components to battery lifetime [1]: calendar life and cycle life. The calendar life describes the natural lifetime, inasmuch as the technical capabilities of the battery will degrade over time even if it is not utilized. The cycle life describes degradation as a function of battery utilization, especially w.r.t. the number of charges cycles. Although there are multiple factors impacting the lifetime of batteries, this paper focuses on charge cycle degradation only.

In the next chapter, a model for operational planning of a solar-storage plant is presented. One objective of that model is to reduce battery degradation by minimizing charge cycles. One charge cycle can be defined as discharging and charging the full capacity of a battery once. More sophisticated calculations of charge cycle degradation that account for state-of-charge and depth-of-discharge will not be studied [2].

2.2 Short-Term Energy Markets

Energy is traded in different markets with different time horizons. Short-term markets with day-ahead or intra-day focus are prevalent throughout the world. In those markets, participants place bids to purchase or sell electricity. After the gate closes, prices and traded volumes are determined from the market equilibrium and contracts are awarded accordingly. To determine bid size and quantity, schedules of power plants are a key determining factor. Therefore, an integrated approach to scheduling and bidding is common in energy resource scheduling literature.

2.3 *Conflicting Incentives*

In short-term scheduling, a common objective is to maximize revenue from market participation. On the one hand, the utilization of storage plants can greatly increase the profitability of a market participant. On the other hand, it is beneficial to not overuse batteries in order to minimize lifetime degradation. Rather, the focus should be on periods with a high potential for profit such as transitions to and from peak hours or periods with high price volatility. Optimization models that only take revenue from market participation into consideration cannot account for the storage utilization tradeoff. Therefore, both objectives must be considered simultaneously.

3 Multi-Objective Solar-Storage Plant Scheduling Model

3.1 *Literature Review*

There is a vast literature on energy planning and scheduling problems. Multi-objective optimization problems and solution methods are well studied [3, 4]. Storage degradation is studied in different contexts. In a case study of a big data center, [5] finds a considerable tradeoff in favor of battery lifetime. Reference [6] studies the effects of provision of frequency control on battery lifetime. Overall, tradeoffs of storage degradation in market environments are not well studied. This contribution aims to be a first step in this direction.

3.2 *Mathematical Formulation*

The following model considers $t \in \mathcal{T}$ time periods. Parameters are given in capital letters, variables are lower case. The two conflicting objectives are the maximization of profit, i.e.

$$\max \sum_{t \in \mathcal{T}} \left(x_t^+ - x_t^-\right) P_t^{\mathrm{x}}, \tag{1}$$

and the minimization of charge cycles, i.e.

$$\min \sum_{t \in \mathcal{T}} \left(y_t^{\mathrm{c}} + y_t^{\mathrm{d}}\right). \tag{2}$$

Objective (1) describes the revenue from selling x_t^+ or purchasing x_t^- energy from the market at price P_t^{x}, whereas objective (2) is a measure for the number of charge cycles, given as the sum of charged y_t^{c} and discharged y_t^{d} power. The storage capacity need not be included in the objective function, as it would only act as a scaling factor.

Storage continuity constraints are given by (3)–(7). The state of charge s_t, as well as the initial state of charge S_0, cannot exceed the maximum capacity $\overline{S}$. Charge y_t^{c} and discharge y_t^{d} rates are limited by P^{c} and P^{d}, respectively, and are subject to charge E^{c} and discharge E^{d} efficiencies. Within a time period, the storage plant can either charge or discharge, which is indicated by $q_t^{\mathrm{c}} \in \mathbb{B}$.

$$s_t = s_{t-1} + E^{\mathrm{c}} y_t^{\mathrm{c}} - y_t^{\mathrm{d}}, \qquad \forall t \in \mathcal{T}, \tag{3}$$

$$s_0 = S_0, \tag{4}$$

$$s_t \leq \overline{S}, \qquad \forall t \in \mathcal{T}, \tag{5}$$

$$y_t^{\mathrm{c}} \leq P^{\mathrm{c}} q_t^{\mathrm{c}}, \qquad \forall t \in \mathcal{T}, \tag{6}$$

$$y_t^{\mathrm{d}} \leq P^{\mathrm{d}}(1 - q_t^{\mathrm{c}}), \qquad \forall t \in \mathcal{T}. \tag{7}$$

Power balance constraints state that power generation and power consumption must be in balance at all times:

$$Y_t^{\mathrm{s}} + E^{\mathrm{d}} y_t^{\mathrm{d}} + x_t^- = x_t^+ + y_t^{\mathrm{c}}, \qquad \forall t \in \mathcal{T}. \tag{8}$$

Solar output Y_t^{s}, discharged power $E^{\mathrm{d}} y_t^{\mathrm{d}}$, and purchased power x_t^- must be equal to power sold x_t^+ and charged power y_t^{c}. In each period, power can either be purchased or sold, indicated by $q_t^{\mathrm{x}} \in \mathbb{B}$, s.t.

$$x_t^- \leq M^- q_t^{\mathrm{x}}, \qquad \forall t \in \mathcal{T}, \tag{9}$$

$$x_t^+ \leq M_t^+(1 - q_t^{\mathrm{x}}), \qquad \forall t \in \mathcal{T}, \tag{10}$$

where $M^- = P^{\mathrm{c}}$ and $M_t^+ = Y_t^{\mathrm{s}} + E^{\mathrm{d}} P^{\mathrm{d}}$. Given decision variables

$$q_t^{\mathrm{x}}, q_t^{\mathrm{c}} \in \mathbb{B}, \qquad \forall t \in \mathcal{T}, \tag{11}$$

$$s_t, x_t^+, x_t^-, y_t^{\mathrm{c}}, y_t^{\mathrm{d}} \in \mathbb{R}_{\geq 0}, \qquad \forall t \in \mathcal{T}, \tag{12}$$

the optimization problem can be formulated as (1) and (2) subject to (3)–(12).

4 Analysis of Multi-Objective Solution Frontiers

4.1 Methodology

The numerical analysis comprises 100 instances (`I1–I100`) that are synthesized from multiple real-world data sets, and are available on request.

First, approximations of the continuous pareto-fronts are determined using the equidistant epsilon constraint method. These solutions are then compared to those obtained by a penalty approach using a linear weighted sum:

$$\max\left(\sum_{t\in\mathcal{T}} A\left(x_t^+ - x_t^-\right) P_t^{\mathrm{x}} - \sum_{t\in\mathcal{T}}\left((1-A)\left(y_t^{\mathrm{c}} + y_t^{\mathrm{d}}\right)\right)\right). \tag{13}$$

Solutions are determined for equidistant values of $A \in [0, 1]$. Additionally, (13) is normalized to its single objective optima to avoid scaling issues. Finally, for each instance, the pareto-optimal solution with minimal distance to the utopia point is determined.

4.2 Results and Discussion

In Fig. 1a, three characteristic segments of the solution front with different slopes can be identified. The almost linear segments represent solutions with similar marginal profit as a function of storage utilization. Depending on the preference of the decision maker, two solutions stand out: I82ϵS23 and I82ϵS27. The former has coordinates (0.91, 0.24), meaning that 91% of profits can be obtained while requiring only 0.24% of charge cycles compared to the single objective optimum for profit (1). Solution I82ϵS27 requires 14% charge cycles while retaining 79%

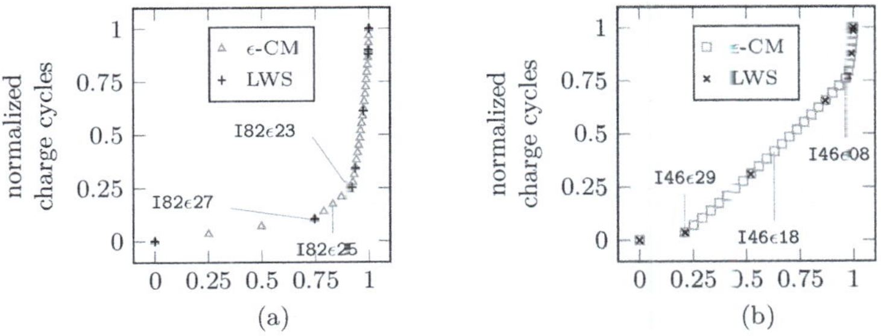

Fig. 1 Tradeoff between profit from market participation (1) and charge cycles (2). Objective function values are normalized to their single objective optima. Two representative instances are plotted, compare Fig. 2. The epsilon constraint method (ϵ-CM) and the weighted objective method (LWS) each show 30 solutions. **(a)** I82, $E^{\mathrm{c}}E^{\mathrm{d}} = 0.92$. **(b)** I46, $E^{\mathrm{c}}E^{\mathrm{d}} = 0.86$

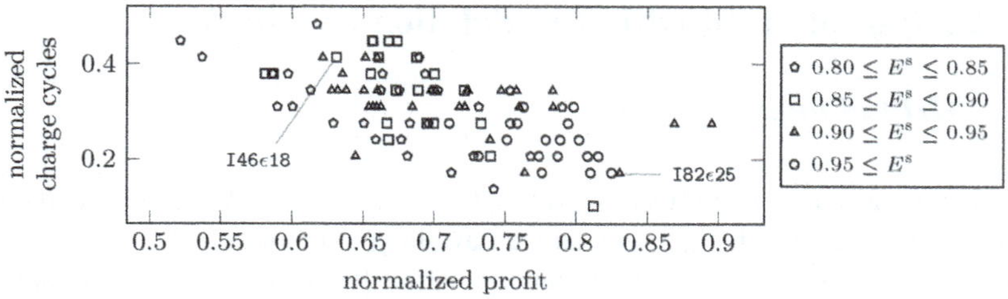

Fig. 2 Pareto-optimal solutions with minimal distance to the utopia point of instances `I1–I100`, categorized according to the round-trip system efficiency $E^{s} = E^{c}E^{d}$

of profits. Solution `I82ϵS25` has the minimum distance to the utopia point $(1, 0)$. The same interpretation applies to Fig. 1b.

The solution quality of both methods are on par. The epsilon constraint method shows a better distributed solution front compared to the weighted objective method. Even though the latter method often does not find the solutions with minimal distance to the utopia point, it does reliably find solutions at the intersections of solution frontier segments. Both methods show that a minor sacrifice in profit allows to significantly reduce the number of charge cycles, and therefore extend battery lifetime.

Figure 2 suggests that high system efficiencies benefit the most from a multi-objective approach, i.e. they exhibit only a small decrease in profit while requiring substantially less charge cycles. In part, that is because load shifting is a defining characteristic of storage systems and (a) profit mainly results from taking advantage of price volatility and shifting power from one period to another, e.g. purchase at a low price and sell at a higher one, and (b) storage efficiency can be interpreted as a penalty, or a spread, on the market price. As a consequence, load shifting is less lucrative the lower the system efficiency is, and it becomes unprofitable if efficiency losses exceed the price difference.

Recall that the plotted solutions in Fig. 2 are those with minimum distance to the utopia point. These might not be the preferred solutions of the decision maker. Instead, solutions at the intersection of two segments might be more desirable, compare Fig. 1.

5 Critical Evaluation and Future Research

Analysis of the pareto-front shows that accepting minor losses in revenue allows to significantly prolong battery lifetime. Instead of the epsilon constraint method, weighted objectives can be used to reduce the computational burden with the drawback of generating more fragmented solution fronts.

The charge cycle model of this contribution is very simple. Future research could focus on studying more sophisticated degradation metrics [2] or the tradeoff in stochastic optimization models. As a more general critique, operational scheduling models are heavily driven by forecast data quality, especially market prices and power forecasts, and such data might not always be available. But, recent advances in time series forecasting increasingly legitimate data driven scheduling approaches [7, 8].

Acknowledgement This work was supported by the European Union and the Free State of Saxony under SAB-Nr. 100331224.

References

1. Vetter, J., Novák, P., Wagner, M.R., Veit, C., Möller, K.-C., Besenhard, J., Winter, M., Wohlfahrt-Mehrens, M., Vogler, C., Hammouche, A.: Ageing mechanisms in lithium-ion batteries. J. Power Sources **147**(1–2), 269–281 (2005)
2. Koller, M., Borsche, T., Ulbig, A., Andersson, G.: Defining a degradation cost function for optimal control of a battery energy storage system. In 2013 IEEE Grenoble Conference, pp. 1–6. IEEE, Piscataway (2013)
3. Alarcon-Rodriguez, A., Ault, G., Galloway, S.: Multi-objective planning of distributed energy resources: a review of the state-of-the-art. Renew. Sustain. Energy Rev. **14**(5), 1353–1366 (2010)
4. Fadaee, M., Radzi, M.: Multi-objective optimization of a stand-alone hybrid renewable energy system by using evolutionary algorithms: a review. Renew. Sustain. Energy Rev. **16**(5), 3364–3369 (2012)
5. Mamun, A.-A., Narayanan, I., Wang, D., Sivasubramaniam, A., Fathy, H.K.: Multi-objective optimization to minimize battery degradation and electricity cost for demand response in datacenters. In: ASME 2015 Dynamic Systems and Control Conference, pp. 1–10. American Society of Mechanical Engineers, New York (2015)
6. Stroe, D.-I., Knap, V., Swierczynski, M., Stroe, A.-I., Teodorescu, R.: Operation of a grid-connected lithium-ion battery energy storage system for primary frequency regulation: a battery lifetime perspective. IEEE Trans. Ind. Appl. **53**(1), 430–438 (2016)
7. Ziel, F., Steinert, R.: Electricity price forecasting using sale and purchase curves: the x-model. Energy Econ. **59**, 435–454 (2016)
8. Chen, Y., Wang, Y., Kirschen, D., Zhang, B.: Model-free renewable scenario generation using generative adversarial networks. IEEE Trans. Power Syst. **33**(3), 3265–3275 (2018)

On the Observability of Smart Grids and Related Optimization Methods

Claudia D'Ambrosio, Leo Liberti, Pierre-Louis Poirion, and Sonia Toubaline

Abstract Management of energy systems is one of the biggest challenges of our time. The daily demand for energy increases constantly for many reasons. Moreover, the wide use of renewable energies, aimed at limiting polluting emissions, can create instability in the networks and uncertainty in energy production. In this context Operational Research is a crucial tool that allows optimizing strategic, tactical, and operational decisions to be taken. We focus on a strategic problem in smart grids, the phasor measurement units placement. The aim is to make the grid observable, i.e., to install devices that can communicate the status of the grid. Being able to observe it allows the grid manager to improve efficiency. We introduce and compare several mathematical optimization formulations proposed in the literature.

Keywords Smart grids · Observability · Optimal PMU placement

1 Introduction

Smart Grids caught an increasing interest in the last few decades. Several reasons explain why smart grids are so important and, to understand them, we should first answer the question: what is a smart grid? Different answers have been proposed, we quote here a Deliverable of the European Union (EU) Commission Task Force for Smart Grids [4]:

C. D'Ambrosio (✉) · L. Liberti
LIX, CNRS, École Polytechnique, Institut Polytechnique de Paris, Palaiseau, France
e-mail: dambrosio@lix.polytechnique.fr

P.-L. Poirion
RIKEN Center for Advanced Intelligence Project, Tokyo, Japan

S. Toubaline
Université Paris–Dauphine, Université PSL, CNRS, LAMSADE, Paris, France

J. S. Neufeld et al. (eds.), *Operations Research Proceedings 2019*, Operations Research Proceedings, https://doi.org/10.1007/978-3-030-48439-2_34

> A Smart Grid is an electricity network that can cost efficiently integrate the behaviour and actions of all users connected to it—generators, consumers and those that do both—in order to ensure economically efficient, sustainable power system with low losses and high levels of quality and security of supply and safety. Though elements of smartness also exist in many parts of existing grids, the difference between a today's grid and a smart grid of the future is mainly the grid's capability to handle more complexity than today in an efficient and effective way.

Thus, the capability of dealing with complexity could be considered a measure of smartness. The document continues:

> A smart grid employs innovative products and services together with intelligent monitoring, control, communication, and self-healing technologies in order to:
>
> - Better facilitate the connection and operation of generators of all sizes and technologies.
> - Allow consumers to play a part in optimising the operation of the system.
> - Provide consumers with greater information and options for how they use their supply.
> - Significantly reduce the environmental impact of the whole electricity supply system.
> - Maintain or even improve the existing high levels of system reliability, quality and security of supply.
> - Maintain and improve the existing services efficiently.
> - ...

Making a grid smart is a challenge. In the following, we discuss how optimization could help progress in this direction.

2 Smart Grids Observability

A first step to make a grid smart is making it observable, i.e., the status of the grid should be accessible at any instant. To do so, devices measuring the phasors (both angle and magnitude) should be installed. These devices, called Phasor Measurement Units (PMUs), were created in the eighties by Phadke and Thorp (VirginiaTech University, U.S.A.). Their idea was to ease the state estimation of a grid by considering the measurements performed periodically by the PMUs. In order to synchronize the PMUs, a timestamp is added to the measures, thanks to a GPS. The reader is referred to [5] for more details on PMUs technicalities.

Phadke et al. [6] proposed the installation of PMUs at every bus. However, Baldwin et al. [1] observed that installing them only on a subset of the buses might be sufficient to guarantee full observability, i.e., the knowledge of voltage at each bus and current at each line. In [1] the authors proposed for the first time to minimize the number of installed PMUs via a Simulated Annealing algorithm which identified a heuristic solution.

3 Mathematical Optimization Approaches

In the next two sections, we report about some Mathematical Optimization (MO) approaches proposed in the literature to solve the PMU installation problem by guaranteeing full observability of the smart grid. In particular, in Sect. 3.1 we consider the case of PMUs that can observe all the lines out of the bus where they are installed (infinite line observability capacity). In Sect. 3.2, we focus on the case of PMUs that can observe only one of the lines out of the bus where they are installed (unit line observability capacity).

In both cases, we consider an undirected graph $G = (V, E)$, representing an electrical grid, where V is the set of busses and E is the set of lines. For ease of notation, we introduce also $\Gamma(i) = \{j | \{i, j\} \in E\}$ for each $i \in V$.

3.1 Infinite Line Observability Capacity

The first Integer Programming formulation for the PMU placement problem was introduced in [9]. The decision variables x_i ($\forall i \in V$) are equal to 1 iff a PMU is installed at node i. We report the model (P^∞) in the following:

$$\min_x \sum_{i \in V} x_i \tag{1}$$

$$\forall i \in V \quad f_i(x) \geq 1 \tag{2}$$

$$\forall i \in V \quad x_i \in \{0, 1\} \tag{3}$$

where the objective function (1) aims at minimizing the number of installed PMUs while the constraints (2) guarantee the full observability. The constraints are general enough to model several variants of the PMU placement problem. In this paper we focus on the zero injections variant, i.e., when the nodes of the grid are all buses (no loads or generators are considered). In the case, $f(x)$ can be represented by a matrix whose entry i, j is equal to 1 if $i = j$ or $\{i, j\} \in E$. Formulation (P^∞) becomes a classical set covering problem.

Baldwin et al. [1] introduced the concept of observability propagation rules. More recently, Brueni and Heath [2] provided a simplified version of these rules restricted to observing nodes (buses). We report here the two simplified rules:

R1 if a node $v \in V$ has a PMU installed, then all the nodes in $\Gamma(v)$ are observed too

R2 if a node $v \in V$ is observed and all the nodes in $\Gamma(v)$ are observed, save one, then all nodes in $\Gamma(v)$ are observed.

These rules are derived from Ohm's law and Kirchoff's law and can be used to deduce the voltage of some nodes from some PMU measurements. Brueni and Heath

[2] did not propose a MO formulation, but analyzed the problem from a theoretical viewpoint.

3.2 Unit Line Observability Capacity

When the PMUs can observe only one line out of the bus where they are installed, rule R1 can be modified as follow:

R1' if a PMU is installed on node $i \in V$ and observes edge $\{i, j\} \in E$, node j is also observed.

This variant can be modeled by considering that the PMU is installed on an edge (line) instead of a vertex (bus). If a PMU is installed on an edge $\{i, j\}$, from an optimization viewpoint it does not matter at which vertex the PMU is installed in practice: both nodes i and j are observed thanks to rule R1'.

In the case of zero injections, formulation (P^{∞}) becomes (P^1) (see [3]):

$$\min_x \sum_{\{i,j\}\in E} x_{ij} \tag{4}$$

$$\forall i \in V \qquad \sum_{\substack{j\in V\\ \{i,j\}\in E}} x_{ij} \geq 1 \tag{5}$$

$$\forall \{i, j\} \in E \qquad x_{ij} \in \{0, 1\} \tag{6}$$

where x_{ij} is equal to 1 iff a PMU is installed on edge $\{i, j\}$ for each $\{i, j\} \in E$. As for (P^{∞}), objective function (4) aims at minimizing the number of installed PMUs while the constraints (5) guarantee the full observability. Moreover, the authors mention that the model could be modified to propagate the observability thanks to R2. However, they only consider a local propagation.

More recently Toubaline et al. [8] and Poirion et al. [7] studied the same problem and proposed three MO approaches modeling the global propagation of R1' and R2. For lack of space, we present only the first one, i.e., a MO formulation based on a propagation iteration index. Note that the iteration does not correspond to a time step, it only models the iterations needed to deduce the value of the voltage at the nodes. The (binary) variables are:

- $\forall i \in V, j \in \Gamma(i)$, s_{ij} is equal to 1 iff a PMU is installed on edge $\{i, j\}$
- $\forall i \in V, t \in \{0, \ldots, T\}$, ω_{it} is equal to 1 iff node i is observed at iteration t. Note that once a node is observed at iteration t, it will be observed at iterations $\{t+1, t+2, \ldots, T\}$
- $\forall i \in V, j \in \Gamma(i), t \in \{0, \ldots, T\}$, y_{ijt} is equal to 1 iff rule R2 is used at iteration t to observe node j using node i as already observed

We report here the complete MO formulation (P_T^1) proposed in [7, 8]:

$$\min_{s,\omega,y} \sum_{i\in V} \sum_{\substack{j\in\Gamma(i)\\ j>i}} s_{ij} \tag{7}$$

$$\forall i \in V, j \in \Gamma(i) \qquad s_{ij} = s_{ji} \tag{8}$$

$$\forall i \in V \qquad \omega_{iT} = 1 \tag{9}$$

$$\forall i \in V \qquad \sum_{j\in\Gamma(i)} s_{ij} \geq \omega_{i0} \tag{10}$$

$$\forall j \in V, t \in \tau' \qquad \omega_{j,t+1} - \omega_{jt} \leq \sum_{\substack{i\in V\\ j\in\Gamma(i)}} y_{ijt} \tag{11}$$

$$\forall i \in V, j \in \Gamma(i), t \in \tau' \qquad \omega_{j,t+1} - \omega_{jt} \leq \omega_{it} - y_{ijt} + 1 \tag{12}$$

$$\forall i \in V, j, k \in \Gamma(i), j \neq k, t \in \tau' \qquad \omega_{j,t+1} - \omega_{jt} \leq \omega_{kt} - y_{ijt} + 1 \tag{13}$$

$$\forall i \in V, t \in \tau' \qquad \omega_{it} \leq \omega_{i,t-1} \tag{14}$$

$$\forall i \in V, j \in \Gamma(i) \qquad s_{ij} \in \{0, 1\} \tag{15}$$

$$\forall i \in V, t \in \tau \qquad \omega_{it} \in \{0, 1\} \tag{16}$$

$$\forall i \in V, j \in \Gamma(i), t \in \tau' \qquad y_{ijt} \in \{0, 1\} \tag{17}$$

where $\tau = \{0, \dots, T\}$ and $\tau' = \{0, \dots, T - 1\}$. Similarly to (1) and (4), the objective (7) minimizes the number of installed PMUs. Constraints (8) enforce the undirectedness of the graph. Full observability is guaranteed by constraints (9). Rule R1' is modeled by constraints (10): if a node is observed at the iteration 0, there must be a PMU installed at some incident edge. Constraints (11)–(13) describe the dynamics of rule R2. Constraints (14) guarantee that, once a node i is observed, it stays observed for the following iterations. Finally, Constraints (15)–(17) force the variables to be binary.

Note that T is not an input of the problem. However, the authors showed that it can be bounded by $n - 2$. In fact, to observe a grid at least a PMU should be installed. At iteration 0, a PMU observes the two nodes of the edge on which it is installed (see rule R1). Thus, there are at most $n - 2$ nodes left to be observed and, in the worse case, we observe one of them at each iteration.

4 Computational Results

In this section, we compare the three MO formulations, (P^∞), (P^1), and (P_T^1). Note that the different formulations model different problems, thus the objective function value of the best solution is different.

Table 1 Objective function value of the best feasible solution

n	(P^{∞})	(P^1)	(P^1_1)	(P^1_{n-2})
5	2	3	2	1
7	2	4	3	2
14	4	7	4	2
24	7	12	7	3
30	10	15	10	6
39	13	21	12	6
57	18	29	18	8
118	32	61	39	23

The tests were performed on a MacBook Pro mid-2015 running Darwin 15.5.0 on a (virtual) quad-core i7 at 3.1 GHz with 16 GB RAM. The MO formulations are implemented in AMPL environment and solved with CPLEX v. 12.8. A time limit of 2 h was set. Instances were created using networks with topologies of standard IEEE n-bus systems like in [7, 8].

For formulation (P^1_T), we tested different values of T:

- $T = 0$: corresponds to consider just R1' and not R2
- $T = 1$: local propagation
- $T = n - 2$: global propagation.

Formulation (P^1_0) is equivalent to P^1, thus we do not report the corresponding results. In Table 1, we report: in the first column, the number of nodes of the instances; in the following columns, the objective function value of the best solution found with the different MO formulations. All the instances were solved to optimality besides for $n \geq 24$ with formulation (P^1_{n-2}).

First, we can observe that the lines observability capacity highly influence the minimum number of PMUs needed: in (P^{∞}) a smaller number of PMUs is needed with respect to (P^1) (between 10 and 36% less). The problem solved by (P^{∞}) might be unrealistic and its optimal solution can be far from the real number of PMUs needed.

Focusing on (P^1) and (P^1_T), it is clear that the use of propagation rules influences largely the number of PMUs needed to guarantee full observability. Let us consider the case of no propagation (P^1) and the case of local propagation (P^1_1): the number of installed PMUS in their optimal solution is much lower in the second case (between 25% and 43% less). As expected, the difference is even more striking if we compare no propagation and the global propagation (P^1_{n-2}) with a decrease of number of PMUs between 50 and 75%. If we compare (P^1_1) and (P^1_{n-2}), the difference is between 33 and 57%. However, from a computational viewpoint, the former formulation scales better than the latter.

Finally, note that the size of the grids is limited: to scale up to large grids, decomposition methods can be considered.

References

1. Baldwin, T.L., Mili, L., Boisen, M.B. Jr., Adapa, R.: Power system observability with minimal phasor measurement placement. IEEE Trans. Power Syst. **8**, 707–715 (1993)
2. Brueni, D.J., Heath, L.: The PMU placement problem. SIAM J. Discrete Math. **19**(3), 744–761 (2005)
3. Emami, R., Abur, A., Galvan, F.: Optimal placement of phasor measurements for enhanced state estimation: a case study. In: Proceedings of the 16th IEEE Power Systems Computation Conference, pp. 923–928 (2008)
4. EU Commission Task Force for Smart Grids – Expert Group 1. Functionalities of smart grids and smart meter (2010)
5. Phadke, A.G., Thorp, J.S.: Synchronized Phasor Measurements and Their Applications. Power Electronics and Power Systems Book Series. Springer, Cham (2008)
6. Phadke, A.G., Thorp, J.S., Karimi, K.J.: State estimation with phasor measurements. IEEE Trans. Power Syst. **1**, 233–241 (1986)
7. Poirion, P.-L., Toubaline, S., D'Ambrosio, C., Liberti, L.: The power edge set problem. Networks **68**(2), 104–120 (2016)
8. Toubaline, S., Poirion, P.-L., D'Ambrosio, C., Liberti, L.: Observing the state of a smart grid using bilevel programming. In: Lu, Z. et al. (eds.) Combinatorial Optimization and Applications. COCOA 2015. Lecture Notes in Computer Science, vol. 9486, pp. 364–376. Springer, Berlin (2015)
9. Xu, B., Abur, A.: Observability analysis and measurement placement for systems with PMUs. In: Proceedings of 2004 IEEE PES Power Systems Conference and Exposition, New York, October 10–13, vol. 2, pp. 943–946 (2004)

Part VIII
Finance

A Real Options Approach to Determine the Optimal Choice Between Lifetime Extension and Repowering of Wind Turbines

Chris Stetter, Maximilian Heumann, Martin Westbomke, Malte Stonis, and Michael H. Breitner

Abstract The imminent end-of-funding for an enormous number of wind energy turbines in Germany until 2025 is confronting affected operators with the challenge of deciding whether to further extend the lifetime of old turbines or to repower and replace it with new and more efficient ones. By means of a real options approach, we combine two methods to address the question if extending the operational life of a turbine is economically viable, and if so, for how long until it is replaced by a new turbine. It may even be the case that repowering before leaving the renewable energy funding regime is more viable. The first method, which is the net present repowering value, determines whether replacing a turbine before the end of its useful life is financially worthwhile. The second method, which follows a real options approach, determines the optimal time to invest in the irreversible investment (i.e., replacing the turbine) under uncertainty. The combination allows for continuously evaluating the two options of lifetime extension and repowering in order to choose the most profitable end-of-funding strategy and timing. We finally demonstrate the relevance of our approach by applying it to an onshore wind farm in a case study.

Keywords Real options · Net present repowering value · Onshore wind · Repowering · Lifetime extension

C. Stetter · M. Heumann (✉) · M. H. Breitner
Information Systems Institute, Leibniz University Hannover, Hannover, Germany
e-mail: stetter@iwi.uni-hannover.de; heumann@iwi.uni-hannover.de; breitner@iwi.uni-hannover.de

M. Westbomke · M. Stonis
Institut für Integrierte Produktion Hannover gGmbH, Hannover, Germany
e-mail: westbomke@iph-hannover.de; stonis@iph-hannover.de

J. S. Neufeld et al. (eds.), *Operations Research Proceedings 2019*, Operations Research Proceedings, https://doi.org/10.1007/978-3-030-48439-2_35

1 Introduction

More than one third of the installed wind energy capacity in Germany will leave the renewable energy funding regime between 2020 and 2025. As a result, operators of affected turbines are increasingly facing the decision of choosing the most profitable end-of-funding strategy. One option is lifetime extension at the level of electricity spot market prices or power purchase agreements (PPA), whereas the other option is repowering, which is the replacement by a new and more efficient turbine. This development features the opportunity to reduce the number of turbines while increasing the installed capacity nationwide. In the German case, repowered turbines correspond to the renewable energy funding regime for another 20 years. However, restrictive regulations regarding required minimum distances to settlements and other areas may impede a repowering in some countries, which is particularly relevant in Germany. The resulting challenge is to determine the optimal lifetime of the old turbine and corresponding repowering timing as previous research has shown [4]. While the decision criteria for finding the best option follow economic principles incorporating viability analyses, its methodological determination is subject to the problem of replacement and the optimal stopping problem.

Like any capital investment, the decision of repowering a wind turbine is subject to three important characteristics. Investing in a repowering project is irreversible once conducted, there is uncertainty about future income due to multiple factors such as feed-in tariff uncertainty, and investors have flexibility in the timing of the investment [1]. Given these characteristics, a real options approach is the most appropriate method to adequately account for managerial flexibility and uncertainty in the realm of repowering projects [2].

Current research tends to investigate lifetime extension and repowering separately and is limited to basic net present value analysis, which do not account for uncertainty. We utilize the net present repowering value (NPRV) of Silvosa et al. [5] to holistically capture the technical and economic key variables of onshore wind projects, for both the old and new wind farm of the repowering project. The operators' possibility of delaying the investment is considered on a sequential basis combining the NPRV and a real options approach. The first method, the NPRV, determines whether replacing a turbine before the end of its useful life is financially worthwhile. The second method, which follows a real options approach, determines the optimal time to invest in the irreversible investment (i.e., replacing the turbine) under uncertain conditions.

We focus on deriving a methodology that allows for continuously evaluating the two options of lifetime extension and repowering to choose the most profitable end-of-funding strategy and timing. Section 2 therefore presents the developed methodology where the revenues of the new wind farm are considered uncertain and are modeled using a stochastic process. To evaluate our methodology, we conduct a case study of a wind farm in Lower Saxony, Germany in Sect. 3, where we simulate the decision whether to extend the lifetime of the old wind farm or to repower by means of our modeling approach. Finally, conclusions are drawn in Sect. 4.

2 Methodology

The subsequent methodology to determine the optimal choice between lifetime extension and repowering of wind turbines is based on the real options approach of Himpler and Madlener [2]. For the sake of simplicity, we derive the model with regard to only one uncertain variable, the price per unit of electricity of the new wind turbines. To address the economics of the problem of replacement, we adjust the NPRV of Silvosa et al. [5] for a consistent application in the real options approach to accurately solve the optimal stopping problem. The NPRV compares the annuity value k of the new farm and the constant payment δ that accrues from selling the old turbines with the annuity value Q of the old wind farm:

$$NPRV = k + \delta - Q. \tag{1}$$

A value of the NPRV greater than zero indicates that the old wind farm should be replaced by the new one as a higher payoff is expected. The monthly constant payment of an annuity Q is derived from the old turbines' cash flow for each month of the project life cycle of the old wind farm $t^0 \in (1, \ldots, T^0)$ as follows:

$$Q_{t^0} = \frac{\sum_{t^0}^{T^0} \frac{(P^0 - O^0) \cdot C^0 \cdot N^0}{(1+r_m^0)^{t^0}}}{a_{T^0 - t^0 | r_m^0}} + \frac{R^0_{T^0} - D^0_{T^0}}{s_{T^0 - t^0 | r_m^0}}, \tag{2}$$

where we denote the price per unit of electricity as P, the operational expenditures as O, the installed capacity as C, the annual full load hours per turbine as N, the residual value as R and dismantling cost as D. The present annuity factor a for T months is quantified as:

$$a_{T|r_m} = \sum_{t=1}^{T} \frac{1}{(1 + r_m)^t}. \tag{3}$$

The future annuity factor s for T months is determined as:

$$s_{T|r_m} = \frac{(1 + r_m)^T - 1}{r_m}. \tag{4}$$

The monthly discount rate r_m is derived from the annual discount rate r:

$$r_m = (1 + r)^{\frac{1}{12}} - 1. \tag{5}$$

The monthly constant payment of an annuity k related to the new wind farm at each month of the project life cycle of the old wind farm is determined respectively:

$$k_{t^0} = \frac{-I^n \cdot C^n}{a_{T^0 - t^0 | r_m^n}} + \frac{\sum_{t^n}^{T^n} \frac{(P_{t^0}^n - O^n) \cdot C^n \cdot N^n}{(1+r_m^n)^{t^n}}}{a_{T^0 - t^0 | r_m^n}} + \frac{R_{T_n}^n - D_{T_n}^n}{s_{T^n | r_m^n}}, \tag{6}$$

where the capital expenditures for the new turbines are denoted as I and each month of the project life cycle of the new wind farm is $t^n \in (1, \ldots, T^n)$. For the constant payment δ that accrues from selling the old turbine, a linear depreciation is considered for the residual value assumed for the first period of valuation and the last period of the lifetime extension:

$$\delta_{t^0} = \frac{R_{t^0=0}^0 - \frac{R_{t^0=0}^0 - R_{T_o}^0}{T^0} \cdot t^0 - D_{T_o}^0}{a_{T^0 - t^0 | r_m^n}}. \tag{7}$$

An increasing number of countries have adopted auctions for the allocation of permits and financial support for new onshore wind projects. Thus, at the time of valuation, the feed-in tariff for the new project is uncertain as it will be determined in a future auction. We model the price uncertainty as a stochastic process with a geometric Brownian motion (GBM) [2]:

$$dP_t^n = \mu P_t^n dt + \sigma P_t^n d W_t, \tag{8}$$

where μ represents the monthly drift of the auction results, σ the volatility and W_t is a Wiener process. However, the presented model is discrete in time whereas the GBM is a continuous-time stochastic process. Utilizing Itô's lemma, the discretized process can be expressed by the closed-form solution of the GBM:

$$P_t^n = P_0^n e^{(\mu - \frac{1}{2}\sigma^2)t + \sigma W_t}. \tag{9}$$

A rational investor always wants to maximize the return of an investment. Here, the investor has the managerial flexibility of deferring the investment by extending the lifetime of the old project or exercising the option to invest in the new project. This is analogous to an American option [2] as the investor can decide to realize the new project during the investment period (i.e. project life cycle of the old wind farm) or not. The resulting problem is to find the stopping time τ^* which maximizes the expected NPRV:

$$V_t = \sup_{\tau \in \gamma} \mathbb{E}\big[NPRV_\tau\big], \tag{10}$$

where V_t is the value function of the optimal stopping problem, γ is the set of all stopping times $t^0 \leq \tau \leq T^0$ and $\mathbb{E}[\ldots]$ denotes the expectations operator.

The investor should invest in the irreversible asset, the new wind farm, if it has the advantage of an expected annuity payment $\mathbb{E}\left[k_t + \delta_t\right]$ that is greater than or equal to the constant payment from the annuity of the old wind farm Q_t. Hence, the optimal stopping time τ^* for the problem of Eq. 10 is defined as the first time the constant payment of the new project reaches this threshold. We can express the optimal investment time mathematically as [8]:

$$\tau^* = \inf\left\{t^0 \leq \tau \leq T^0 \middle| \mathbb{E}\left[k_\tau + \delta_\tau\right] \geq Q_\tau\right\} \tag{11}$$

The presented methodology is derived for a monthly evaluation but can be adjusted to any time dimension. Note that it may even be the case that replacing the old turbine before leaving the renewable energy funding regime is more viable. This could occur if the effect of the efficiency gain of the new wind farm exceeds the additional revenues of the old wind farm resulting from a lifetime extension.

3 Case-Study and Results

To demonstrate the applicability of our methodology, we determine the optimal choice between lifetime extension and repowering in a case study of an exemplary project in Lower Saxony, Germany. We assume that repowering is spatially feasible from a regulatory point of view for all turbine sites. The case study features an existing wind farm with ten Vestas V82-1.65 wind turbines commissioned in January 2000 and three potential repowering turbines of type Vestas V150-4.2.

It is assumed that the considered location exhibits 100% site quality such that it is equal to the reference site legally specified in the latest amendment to Germany's Renewable Energy Sources Act (EEG). Based on this assumption, average wind speeds of 6.45 m/s result for the considered site at an assumed hub height of 100 meter from which the full load hours were estimated. For this site quality, the operational and capital expenditures were derived with regard to current cost studies [3, 6, 7]. The maximum lifetime extension of an old turbine is presumed to be 5 years. For this period, the residual value of the old turbine remains unchanged. All project characteristics are summarized in Table 1.

We simulate the decision if extending the operational life of the old wind farm is economically viable, and if so, for how long until repowering. The time of assessment is chosen to be January 2020, where the old turbines leave the funding regime. The price per unit of electricity is assumed to be at the level of an agreed PPA of 4.90 € ct/kWh and, thus, is certain. The uncertain price for the repowering turbines, which will be determined in a future auction are modeled with the GBM of Eq. 9. We utilize a Monte Carlo simulation to model the uncertainty of the price which is expected to decrease from $p_0 = 6.07$ with $\mu = -0.01$ and $\sigma = 0.15$ annually. An excerpt from the resulting simulation paths at every month of the investment period is shown in Fig. 1, where the red line represents the expected value

Table 1 Project characteristics

	Unit	Old	New
C	MW	10×1.65	3×4.2
N	h/a/turb.	2784	3435
P	€ ct/kWh	4.90	6.07
O	€ ct/kWh	1.96	2.26
I	€/kW	–	1308
r	%	3.5	4.75
R	€/turb.	28,700	60,000
D	€/turb.	11,000	20,000

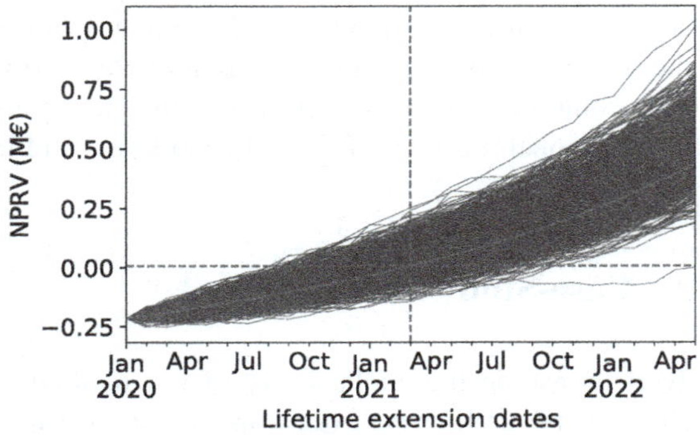

Fig. 1 Simulated NPRV

of the NPRV. The NPRV increases as the progress in time predominates the effect of a lower price for the new wind farm. The first passage time that the expected annuity payment of the new project hits the threshold defined by the annuity payment of the old project is the optimal time of repowering (see Eq. 11), which is March 2021. The investor should extend the lifetime of the old turbines and then exercise the option to repower.

4 Conclusion

Determining the optimal choice between lifetime extension and repowering of wind turbines is a major challenge for operators as more than one third of the installed wind energy capacity in Germany will leave the renewable energy funding regime between 2020 and 2025. Operators need to examine whether extending the operational life of the old turbine is economically viable, and if so, for how long until it is repowered. The resulting problem is to find the optimal time to invest in an irreversible investment under uncertain conditions.

We have proposed a real options based modeling approach for continuously evaluating the two options of lifetime extension and repowering to choose the most profitable end-of-funding strategy and timing. On this account, we utilize the NPRV that estimates whether replacing a turbine before the end of its useful life is financially worthwhile. The uncertain revenues of the new wind farm resulting from unknown future auction results and the managerial flexibility to defer the investment are considered by means of a real options approach to determine the optimal time to invest. For the analyzed project, which represents a mid-sized potential repowering project in Germany, it was demonstrated that extending the lifetime of the old turbines at the level of a PPA is an economically feasible strategy. The level of remuneration for the old turbines is a significant variable driving the optimal repowering timing in the proposed model.

As our modeling approach permits to examine the options of lifetime extension and repowering simultaneously, we contribute to comprehensive methodological support for operators of old wind turbines to find the best end-of-funding strategy and timing. Most importantly, our introduced modeling approach aims at accounting for uncertainty of the irreversible investment of repowering.

References

1. Dixit, A.K., Dixit, R.K., Pindyck, R.S.: Investment Under Uncertainty. Princeton University Press, Princeton (1994)
2. Himpler, S., Madlener, R.: Optimal timing of wind farm repowering: a two-factor real options analysis. J. Energy Markets **7**(3), (2014)
3. Lüers, S., Wallasch, A.K., Rehfeldt, K.: Kostensituation der Windenergie an land in Deutschland–update. Deutsche WindGuard Technical Report (2015)
4. Piel, J., Stetter, C., Heumann, M., Westbomke, M., Breitner, M.: Lifetime extension, repowering or decommissioning? Decision support for operators of ageing wind turbines. In: Journal of Physics: Conference Series, vol. 1222 (2019)
5. Silvosa, A.C., Gòmez, G.I., del Rìo, P.: Analyzing the techno-economic determinants for the repowering of wind farms. Eng. Econ. **58**(4), 282–303 (2013)
6. Wallasch, A.K., Lüers, S., Rehfeldt, K.: Weiterbetrieb von Windenergieanlagen nach 2020. Deutsche WindGuard, Technical Report (2016)
7. Wallasch, A.K., Lüers, S., Rehfeldt, K., Vogelsang, K.: Perspektiven für den Weiterbetrieb von Windenergieanlagen nach 2020. Deutsche WindGuard Technical Report (2017)
8. Welling, A., Lukas, E., Kupfer, S.: Investment timing under political ambiguity. J. Business Econ. **85**(9), 977–1010 (2015)

Measuring Changes in Russian Monetary Policy: An Indexed-Based Approach

Nikolay Nenovsky and Cornelia Sahling

Abstract Russia's transition to a market economy was accompanied by several monetary regime changes of the Bank of Russia (BoR) and even different policy goals. In this context we should mention the transformation of the exchange rate regime from managed floating to free floating (since November 2014) and several changes of the monetary regimes (exchange rate targeting, monetary targeting, and inflation targeting). As a measurement of changes in Russian monetary policy in 2008–2018 we develop a Monetary policy index (MPI). We focus on key monetary policy instruments: interest rates (key rate, liquidity standing facilities and standing deposit facilities rates), amount of REPO operations, BoR foreign exchange operations and required reserve ratio on credit institutions' liabilities. Our investigation provides a practical contribution to the discussion of Russian monetary regimes by creating a new MPI adopted to the conditions in Russia and enlarges the discussion of appropriate monetary policy regimes in transition and emerging countries.

Keywords Russia · Monetary policy instruments · Monetary policy index · Exchange rate regime · Inflation targeting

1 Introduction: The Russian Monetary Policy Framework

The transition of central planning to a market economy in former Soviet states caused a broad range of modifications of the economic and political system. This

N. Nenovsky
University of Picardie Jules Verne, CRIISEA, Amiens, France

National Research University Higher School of Economics, Moscow, Russia

C. Sahling (✉)
Peoples' Friendship University of Russia (RUDN University), Moscow, Russian Federation

J. S. Neufeld et al. (eds.), *Operations Research Proceedings 2019*, Operations Research Proceedings, https://doi.org/10.1007/978-3-030-48439-2_36

paper concentrates on changes in the monetary policy framework of the Bank of Russia (BoR) and its implications for monetary policy instruments.

In recent years most significant changes in Russian monetary policy are related to the switch to free floating (since Nov 2014—instead of managed floating with a dual-currency basket) and to an inflation targeting regime (since 2015—instead of monetary targeting). In this context, several changes in the monetary policy framework happened. A part of this transition was the development of a new interest rate corridor (since Sep 2013) with the newly introduced key rate as a core element of this system. In addition, since 2007 the BoR has declared price stability as primary goal of monetary policy (currently an inflation rate of 4% is intended). For a more detailed explanation of frequent monetary policy changes see e.g. Johnson (2018) [1] and Gurvich (2016) [2].

These changes in monetary policy patterns seem to complicate the estimation of policy in the case of Russia. Evidence from other investigations confirms this thesis. Vdovichenko/Voronina (2006) [3] conclude that for the period 1999–2003 a money-based rule is more suitable to explain Russian Monetary Policy. Korhonen, Nuutilainen (2016) [4] consider the Taylor rule appropriate only for 2003–2015.

Our new MPI is motivated by Girardin et al. (2017) [5] who developed an index for measuring Chinese monetary policy. An overview on different policy indexes for the People's Bank of China is provided in Sun (2018) [6]. To the best of our knowledge, there is no suitable MPI developed for Russia for the recent years and the aim of this paper is to develop an appropriate MPI for the BoR.

2 A Monetary Policy Index for Russia

In computing our MPI, we should consider the underlying indicators, their theoretical framework and policy implications. Our index is based on several indicators: key rate, required reserve ratio, liquidity provision standing facility, standing deposit facility, amount of REPO operations (fixed rate) and BoR foreign exchange operations. The meanings of these variables are taken from the BoR statistical database [7]. The mentioned indicators and its main characteristics are summarized in Table 1. The data for the MPI is on a monthly basis for the period of 2008–2018. The choice of key determinants is based on literature investigation considering Russian monetary policy and emerging countries' experience and, of course, official statements of the BoR.

We define our MPI as a weighted average of the above-mentioned monetary policy instruments. These different instruments include the key rate, liquidity standing facilities and standing deposit facilities rates, amount of REPO operations, BoR foreign exchange operations and required reserve ratio on credit institutions' liabilities. As a first step we have normalized all policy variables series on a scale from 1 to 10 (where 1 is the minimum value of the index, while 10 is the maximum value of the index).

Table 1 BoR monetary policy tools

Indicator	Description/purpose	Technical explanations (usage for MPI)
Some elements of the interest rate policy	1. Key rate (core element for the new interest rate corridor)	Before Feb 2014—main refinancing rate and then key rate (introduced in Sep 2013 and since Jan 2016—refinancing rate equals to key rate)
	2. Liquidity provision standing facility	Overnight credit rates; lombard loans (1 calendar day); fixed-term rates (long-term)
	3. Standing deposit facility	Overnight deposit rate (until Feb 2010: Tomorrow next)
Required reserve ratio	Amount of money that a credit institute should keep in an account with the BoR	In rubles and in foreign currency on the example of ratios to liabilities to individuals (from Dec 2017 for banks holding a general licence + non-bank credit institutions)
Fixed rate REPO	Steering money market interest rates; liquidity providing	Amount allotted (term: 1 and 7 days)
BoR foreign exchange operations	Influence the ruble exchange rate	Amount of BoR FX operations on the domestic market for 1 month (in Nov 2014 switch to a floating rate)

Following the literature (for ex. on China) and our personal interpretations, we gave the following weights to the variables in the index: 0.3 to the interest rates (REPO rate, standing liquidity providing facilities and standing deposit facilities rates), 0.25 to the amount of REPO operations, 0.25 to BoR foreign exchange operations and 0.2 to required reserve ratio on credit institutions' liabilities (both in rubles and in foreign currency), see Eq. (1).

$$\begin{aligned} \text{MPI} = {} & 0.3\,(\text{REPO rate, liquidity and deposit standing facilities}) \\ & + 0.25\,(\text{REPO volume}) - 0.25\,(\text{Forex operation volume}) \\ & - 0.2\,(\text{required reserves ratio}) \end{aligned} \tag{1}$$

The MPI is designed to be interpreted as follows: when it grows, we have a restriction (reduction of liquidity in rubles), and when it diminishes, we observe a loosening of monetary policy (liquidity injection in rubles). Therefore, operations in rubles and forex are taken with a negative sign (considering the BoR data definition, and our MPI construction, see Eq. (1)). We want to clarify that no theoretical approach has been developed in determining the weights in the literature (this is also a task for future research). So far, we have been guided by similar research as well as by our observations on the conduct and official statements of the representatives of the Bank of Russia.

3 Results and Discussion

The calculated results of the MPI (normalized 1–10) are presented in Fig. 1.[1] The results are compared with the development of the key rate (normalized 1–10). As the figure shows, the two indicators have moved in the same direction for a long time. But since the end of 2013 and especially beginning of 2014 (geopolitical tensions) greater differences between MPI and the key rate can be indicated. This observation requires deeper explanation. As mentioned above, since September 2013 the BoR announced a gradually switch to an inflation targeting regime; at that time a new system with an interest rate corridor was introduced.

Based on this observation, we should consider other components of our MPI. Therefore, Fig. 2 presents the comparison with the volumes of REPO operations and foreign exchange intervention. The amounts of these indicators are reported on the left-hand side of the chart (normalized). When integrating the chart with the BoR forex interventions, the currency crisis of 2014/2015 is of main importance. In November 2014 the BoR employed a floating exchange regime. As a consequence, a sharp depreciation of the Russian ruble occurred (from 45.89 RUB/USD on the 11th of November 2014 to 65.45 RUB/USD on the 11th of February 2015, [7]). High exchange rate fluctuations negatively influence trade and investment patterns, the depreciation of the national currency decreases the purchasing power of local residents (see Broz/Frieden (2008) [8]). The BoR was forced to stop the ruble devaluation (for further explanation of BoR anti-crisis measures in literature see e.g. Golovnin (2016) [9]) by increasing the key rate (from 9.5% on the 11th of November 2014 to 17% on the 13th of January 2015, [7]).

In contrary to declining foreign exchange interventions after switching to a floating rate, the volume of REPO operations increased since 2012 (as you can see in Fig. 2). This BoR policy behavior is related to liquidity providing (via REPO in Rubles). During the crisis in 2014/2015, the two monetary policy instruments (REPO auctions and FX interventions) moved in opposite directions: increasing REPO volume and US dollar sales (FX interventions with a sign "-"). Therefore, the empirical data shows two opposite effects: liquidity providing (REPO) and

[1]The components of the MPI (i.e. the individual variables) are normalized individually in a scale from 1 to 10.

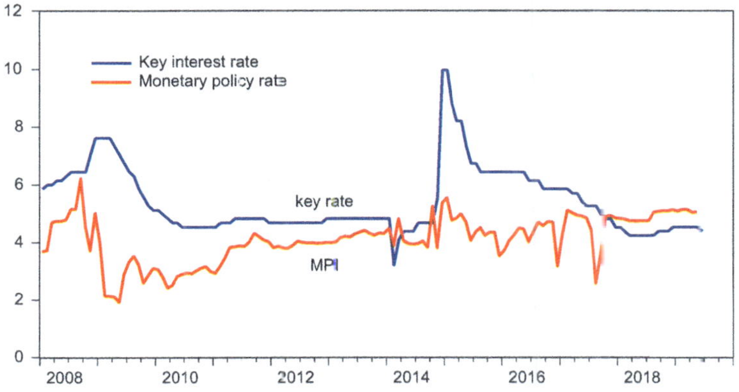

Fig. 1 Key policy interest rate and monetary policy index (normalized 1–10)

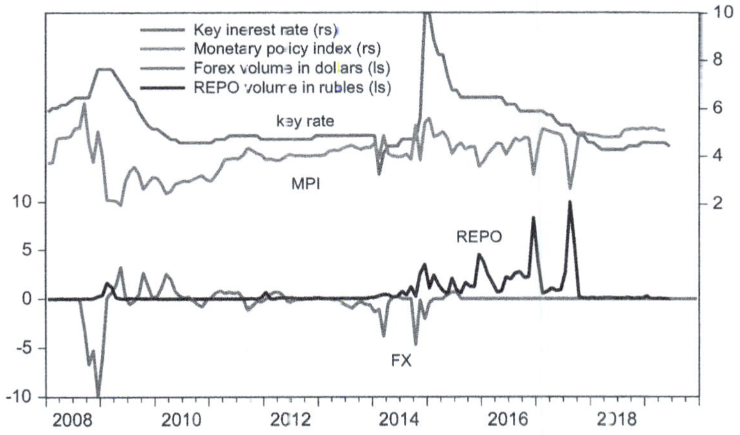

Fig. 2 Key rate, monetary policy index and REPO and forex intervention (normalized 1–10)

absorbing (FX). This BoR crisis behavior could be explained by sterilization considerations. We assume that the BoR has sterilized and targeted the monetary base. The restructure of the official BoR balance sheet into a simplified form with three positions (net foreign assets, domestic assets and monetary base) confirms this thesis (see Fig. 3). The basic idea of these calculations and the theoretical background are shown in Nenovsky/Sahling [10].

For the post-crisis period (since 2016), the estimated MPI got closer to the key rate. Based on this idea, the proposed MPI represents an indicator of crises in the financial sector. The opposite movement of the MPI and the key rate are signs of a crisis (2008/2009 and 2014/2015); the subsequent convergence could be treated as a sign of recovery.

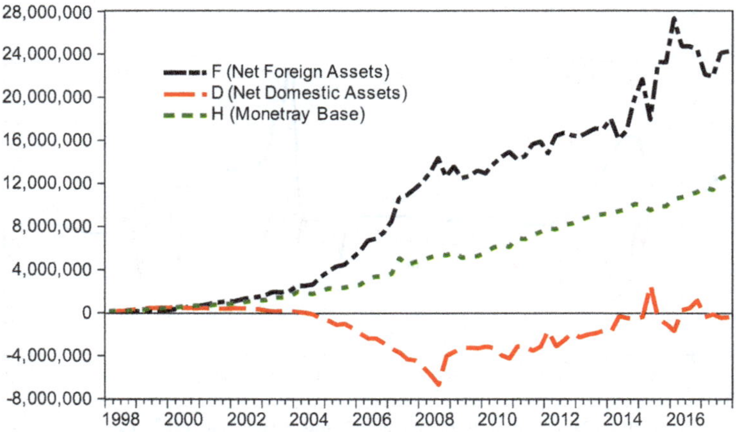

Fig. 3 Dynamics of the monetary base and its sources in Russia for the 1998–2017 period (millions of rubles)

4 Conclusions

The recent changes in Russian monetary policy are difficult to measure considering opposite trends for some policy instruments (e.g. REPO operations and FX interventions). To solve this problem, we have proposed a MPI with several normalized policy instruments.

Promising future research fields with this MPI could be related to some index improvements. First, an interesting point would be the usage of other variables/BoR policy instruments for the index. Further, the weights to the composing variables of the MPI could be modified and theoretically motivated. Due to the interconnection between the Central Bank and the Ministry of Finance (for a theoretical discussion about monetary and fiscal macroeconomic policies see Jordan (2017) [11]), the exploration of possible influence channels of the Ministry of Finance on monetary policy would be desirable for the case of the BoR.

References

1. Johnson, J.: The Bank of Russia: from central planning to inflation targeting. In: Conti-Brown, P., Lastra, R.M. (eds.) Research Handbook on Central Banking, pp. 94–116. Edward Elgar, Cheltenham (2018)
2. Gurvich, E.T.: Evolution of Russian macroeconomic policy in three crises. J. New Econ. Assoc. **29**(1), 174–181 (2016). (in Russian)
3. Vdovichenko, A., Voronina, V.: Monetary policy rules and their application for Russia. Res. Int. Bus. Financ. **20**(2), 145–162 (2006)
4. Korhonen, I., Nuutilainen, R.: A monetary policy rule for Russia, or is it rules? BOFIT Discussion Papers No. 2 (2016)

5. Girardin, E., Lunven, S., Ma, G.: China's evolving monetary policy rule: from inflation-accommodating to anti-inflation policy. BIS Working Papers No. 641 (May 2017)
6. Sun, R.: A narrative indicator of monetary conditions in China. Int. J. Cent. Bank. **14**(4), 1–42 (2018)
7. The Central Bank of Russia: Statistics. https://www.cbr.ru/eng/statistics/default.aspx. Accessed 14 Aug 2019
8. Broz, J.L., Frieden, J.A.: The political economy of exchange rates. In: Wittman, D.A., Weingast, B.R. (eds.) The Oxford Handbook of Political Economy, pp. 587–598. Oxford University Press (2008)
9. Golovnin, M.Y.: Monetary policy in Russia during the crisis. J. New Econ. Assoc. **29**(1), 168–174 (2016). (in Russian)
10. Nenovsky, N., Sahling, C.: Monetary targeting versus inflation targeting: empirical evidence from Russian Monetary Policy (1998–2017). Forthcoming. Paper presented at the XX April International Academic Conference "On Economic and social development", National Research University Higher School of Economics, Moscow, Russia (2019)
11. Jordan, J.L.: Rethinking the monetary transmission mechanism. Cato J. **37**(2), 361–384 (2017)

Part IX
Graphs and Networks

Usage of Uniform Deployment for Heuristic Design of Emergency System

Marek Kvet and Jaroslav Janáček

Abstract In this contribution, we deal with an emergency service system design, in which the average disutility is minimized. Optimization of the average user disutility is related to the large weighted p-median problem. The necessity to solve large instances of the problem has led to the development of many heuristic and approximate approaches. Due to complexity of the integer programming problems, the exact methods are often abandoned for their unpredictable computational time in the case, when a large instance of a location problem has to be solved. For practical use, various kinds of metaheuristics and heuristics are used to obtain a good solution. We focus on usage of uniform deployment of p-median solutions in heuristic tools for emergency service system design. We make use of the fact that the uniformly deployed set of solutions represents a partial mapping of the "terrain" and enables to determine areas of great interest. We study here the synergy of the uniformly deployed set and heuristics based on neighborhood search, where the solution neighborhood is set of all p-median solutions, Hamming distance of which from the current solution is 2.

1 Introduction

The family of discrete location problems belongs to the hard solvable combinatorial problems with plethora real-life applications. That is why solving methods of these problems attract attention of many researchers and practitioners [2, 6, 9]. A special class of the location problems has been formulated and studied with the purpose to design efficient public service systems [1, 8]. This special class of the location problems is characterized by a given number p of service centers, which can be deployed across a serviced region. Within this paper, we concentrate our effort on this kind of location problems and we call them briefly the p-location

M. Kvet (✉) · J. Janáček
University of Žilina, Faculty of Management Science and Informatics, Žilina, Slovakia
e-mail: marek.kvet@fri.uniza.sk; jaroslav.janacek@fri.uniza.sk

J. S. Neufeld et al. (eds.), *Operations Research Proceedings 2019*, Operations Research Proceedings, https://doi.org/10.1007/978-3-030-48439-2_37

problems. The solving methods associated with p-location problems can be divided into two main classes, which distinguish exact and heuristic algorithms. The exact and approximate algorithms yield either optimal or near-to-optimal solution with guaranteed deviation from optimum, but their necessary computational time is almost unpredictable [3, 10]. The heuristic algorithms are not able to ensure that resulting solution will differ from the optimal one in a given limit, but they are able to keep a given amount of computational time [4, 11]. Nevertheless, the Achilles heel of each heuristic and metaheuristic is being trapped at a local extreme, which is far from the optimal solutions [5]. To face the effect of trapping, various tools have been developed. Evolutionary metaheuristics have been equipped with diversity maintenance mechanism and the simple incrementing heuristic are started several times from randomly generated starting solutions.

Within the paper, we will study a memetic approach to a simple incrementing heuristic performance improvement. The approach is based on exploitation of a maximal or near-to-maximal uniformly deployed set of p-location problem solutions [7]. Elements of the set represent vertices of a unit hypercube, where minimal mutual Hamming distance must be greater than or equal to a given threshold. The set of vertices resembles a set of triangular points in a terrain, where the shape of the surface is estimated according to altitude at the individual triangular points. We use the partial mapping of the p-location problem solutions in a range of objective function values to identify an area of the greatest interest and we submit the area to proper exploration employing the simple heuristic.

2 Exchange Heuristics for Emergency Medical System Design Problem

The emergency medical system design problem (EMSP) can be considered as a task of determination of p service center locations from a set I of m possible service center locations so that the sum of weighted distances from users' locations to the nearest located center is minimal. Let symbol J denote the set of the users' locations and b_j denote the weight associated with user location $j \in J$. If d_{ij} denotes the distance between locations i and j, then the studied problem known also as the weighted p-median problem can be described by (1).

$$min\left\{\sum_{j \in J} b_j min\left\{d_{ij} : i \in I_1\right\} : I_1 \subseteq I, |I_1| = p\right\} \tag{1}$$

The problem (1) is also known as the weighted p-median problem and each feasible solution can be represented by a subset I_1, cardinality $|I_1|$ of which equals to p. The objective function $F(I_1)$ of the solution I_1 is described by (2).

$$F(I_1) = \sum_{j \in J} b_j min\left\{d_{ij} : i \in I_1\right\} \tag{2}$$

The simple incrementing heuristics (exchange heuristics) are based on a search across a neighborhood of a current solution. The neighborhood is formed by all feasible solutions, which differ from the current solution only in one service center location. The discussed exchange heuristics make use either of the best admissible strategy or the first admissible strategy.

In the algorithms, a solution I_1 is represented by a list of p selected locations, i.e. it contains p subscripts of service center locations.

The algorithm *getBA* based on the best admissible strategy starts with an initial solution, which inputs as a list I_1 and it is described below by the following five steps.

Algorithm $getBA(I_1)$

0. Initialize $F^{**} = F(I_1)$, $F^* = F(I_1)$, $C = I - I_1$.
1. For each pair (i, j) for $i \in I_1$, $j \in C$ subsequently perform the following operations.
   - Set $\underline{I_1} = (I_1 - \{i\}) \cup \{j\}$ and compute $F(\underline{I_1})$.
   - If $F(\underline{I_1}) < F^*$, then set $F^* = F(\underline{I_1})$ and $I_1^* = \underline{I_1}$.
2. If $F^{**} \leq F^*$ then terminate, the resulting solution is I_1^* and its objective function value is F^*, else set $F^{**} = F^*$, $I_1 = I_1^*$, $C = I - I_1$ and go to 1.

The algorithm *getFA* based on the first admissible strategy starts also with an initial solution I_1 and is described below by the following steps.

Algorithm $getFA(I_1)$

0. Initialize $F^* = F(I_1)$, $C = I - I_1$, $cont = true$ and create a set P of all pairs (i, j) for $i \in I_1$, $j \in C$.
1. While $((cont)$ and $(P \neq \emptyset))$ subsequently perform the following operations.
   - Withdraw a pair (i, j) from P, i.e. set $P = P - \{(i, j)\}$.
   - Set $\underline{I_1} = (I_1 - \{i\}) \cup \{j\}$ and compute $F(\underline{I_1})$.
   - If $F(\underline{I_1}) < F^*$, then set $F^* = F(\underline{I_1})$ and $I_1^* = \underline{I_1}$ and $cont = false$.
2. If $cont = true$ then terminate, the resulting solution is I_1^* and its objective function value is F^*, else set $I_1 = I_1^*$, $C = I - I_1$, $cont = true$ and create a new set P of all pairs (i, j) for $i \in I_1$, $j \in C$. Then go to 1.

Obviously, the algorithm $getBA$ is able to find a better solution in the first neighborhood processing than the algorithm $getFA$, when starting from the same solution. Nevertheless, a better solution is paid for by longer computing time and, in addition, it is questionable, whether the move to the better solution in the first neighborhood processing will influence the quality of the final solution. Therefore, it is worth studying both strategies.

3 Uniformly Deployed Set Usage for EMSP Optimization

The above p-location problem can be reformulated as a zero-one programming problem introducing zero-one variable y_i for each $i \in I$, where the variable y_i gets the value of one if a center is located at location i and it gets the value of zero otherwise. Then, the problem (1) can be studied as a search in a sub-set of m-dimensional hypercube vertices, when m denotes the cardinality of the set I. The distance between two feasible solutions $\boldsymbol{y}$ and $\boldsymbol{x}$ can be measured by so called Hamming distance or Manhattan distance defined by (3).

$$H(\boldsymbol{y}, \ \boldsymbol{x}) = \sum_{i=1}^{m} |y_i - x_i| \qquad (3)$$

We have to realize that the distance of two different feasible solutions of (3) is only even integer number ranging from 2 to $2p$. The expression $p - H(\boldsymbol{y}, \boldsymbol{x})/2$ gives the number of common centers in the both solutions. We note that the neighborhood of a current solution I_1 is formed by all feasible solutions, Hamming distance of which from the solution I_1 is equal to 2, what means that they differ in one center location from the current solution.

The notion of Hamming distance enables to define maximal uniformly deployed set of p-location problem solutions. The maximal uniformly deployed set is defined for given distance d as a maximal set of p-location problem solutions, where every two solutions have minimal Hamming distance d.

If a uniformly deployed set S of the p-location problem solutions obtained in advance by arbitrary process is known, we can employ it for improvement of the incrementing algorithms in the following way.

1. Compute $F(s)$ for each $s \in S$ and determine the s^* with the lowest value of (2).
2. Set $I_1 = s^*$ and perform $getBA(I_1)$ or $getFA(I_1)$.

The above scheme can be generalized by applying the procedures *getBA*(I_1) or *getFA*(I_1) to a given portion of the good solutions of S or the whole algorithm can be applied on several uniformly deployed sets, which can be obtained by permutations of the original numbering of the possible service center locations.

4 Computational Study

The main goal of performed computational study was to study the efficiency of suggested approximate solving methods, which are based on making use of the uniformly deployed set of p-location problem solutions and on applying the algorithms *getFA* and *getBA* respectively. The used benchmarks were obtained from the road network of Slovak self-governing regions. The mentioned instances are further denoted by the names of capitals of the individual regions followed by triples

Table 1 Results of numerical experiments for the self-governing regions of Slovakia

Region	Optimal solution		$\|S\|$	F^{inp}	Algorithm $getFA$			Algorithm $getBA$		
	F^{opt}	CT [s]			F^*	gap [%]	CT [s]	F^*	gap [%]	CT [s]
BA	19,325	0.28	23	30,643	19,325	0.00	0.03	19,325	0.00	0.02
BB	29,873	2.22	172	54,652	29,888	0.05	5.04	29,873	0.00	1.32
KE	31,200	1.44	60	52,951	31,252	0.17	2.23	31,451	0.80	0.79
NR	34,041	2.84	83	59,207	34,041	0.00	0.95	34,075	0.10	0.32
PO	39,073	35.01	232	75118	39153	0.20	7.89	39,117	0.11	1.74
TN	25,099	1.38	137	46,035	25,125	0.10	0.35	25,125	0.10	0.10
TT	28,206	0.87	212	45,771	28,206	0.00	0.22	28,372	0.59	0.07
ZA	28,967	0.71	112	501,38	28,989	0.08	0.83	28,967	0.00	0.29

(XX, m, p), where XX is commonly used abbreviation of the region denotation, m stands for the number of possible center locations and p is the number of service centers, which are to be located in the mentioned region. The list of instances follows: Bratislava (BA, 87, 14), Banská Bystrica (BB, 515, 36), Košice (KE, 460, 32), Nitra (NR, 350, 27), Prešov (PO, 664, 32), Trenčín (TN, 276, 21), Trnava (TT, 249, 18) and Žilina (ZA, 315, 29).

An individual experiment was organized in such a way that the optimal solution of the min-sum location problem was obtained using the radial approach described in [10], first. The objective function value of the exact solution denoted by F^{opt} together with the computational time in seconds denoted by "CT [s]" are reported in the left part of Table 1. The next two columns are used for the uniformly deployed set characteristics. The symbol $|S|$ denotes the cardinality of the set S and F^{inp} denotes the minimal objective function value computed according to (2) for each solution from the set S. The right part of Table 1 contains the comparison of suggested approaches based on algorithms *getFA* and *getBA* respectively. For each method, three different values are reported. The resulting objective function value is reported in the column denoted by F^*. The solution accuracy is evaluated also by gap, which expresses a relative difference of the obtained result from the optimal solution. The value of gap is expressed in percentage, where the optimal objective function value of the problem is taken as the base. Finally, the computational time in seconds is reported in columns denoted by "CT [s]".

Comparison of the results reported in parts "Algorithm *getFA*" and "Algorithm *getBA*" of Table 1 showed that there is almost no winner in the competition of resulting solution objective function. As far as the computational time is concerned, the algorithm *getBA* demonstrates better stability. It is worth to note that the both variants reached almost the optimal solution of the solved problems.

5 Conclusions

The main goal of this paper was to explore efficiency of simple incrementing exchange algorithms in combination with a uniformly deployed set. Efficiency was studied in the case when emergency medical service system is to be designed. Presented results of performed numerical experiments confirm that the both variants of heuristics give satisfactory solution accuracy in acceptable computational time. Therefore, we can conclude that we have constructed a very successful heuristic method for efficient and fast emergency medical service system design.

Future research in this field may be aimed at other forms of uniformly deployed set employment. Another interesting topic for next research could be focused on other ways of processing elements of the uniformly deployed set different from the neighborhood search.

Acknowledgments This work was supported by the research grants VEGA 1/0342/18 "Optimal dimensioning of service systems", VEGA1/0089/19 "Data analysis methods and decisions support tools for service systems supporting electric vehicles", and VEGA 1/0689/19 "Optimal design and economically efficient charging infrastructure deployment for electric buses in public transportation of smart cities" and APVV-15-0179 "Reliability of emergency systems on infrastructure with uncertain functionality of critical elements".

References

1. Doerner, K.F., et al.: Heuristic solution of an extended double-coverage ambulance location problem for Austria. Central Eur. J. Oper. Res. **13**(4), 325–340 (2005)
2. Erlenkotter, D.: A dual-based procedure for uncapacitated facility location. Oper. Res. **26**(6), 992–1009 (1978)
3. García, S., Labbé, M., Marín, A.: Solving large p-median problems with a radius formulation. INFORMS J. Comput. **23**(4), 546–556 (2011)
4. Gendreau, M., Potvin, J.: Handbook of Metaheuristics, 3rd edn, 610pp. Springer, Berlin (2019)
5. Gupta, A., Ong, Y.S.: Memetic Computation, 104pp. Springer, Berlin (2019)
6. Janáček, J., Buzna, Ľ.: An acceleration of Erlenkotter-K¨rkel's algoriths for uncapacitated facility location problem. Ann. Oper. Res. **164**, 97–109 (2008)
7. Janáček, J., Kvet, M.: Uniform deployment of the p-location problem solutions. In: OR 2019 Proceedings. Springer, Berlin (2020)
8. Jánošíková, Ľ., Žarnay, M.: Location of emergency stations as the capacitated p-median problem. In: International Scientific Conference: Quantitative Methods in Economics-Multiple Criteria Decision Making XVII, Virt, Slovakia, pp. 116–122 (2014)
9. Korkel, M.: On the exact solution of large-scale simple plant location problem. Eur. J. Oper. Res. **39**, 157–173 (1989)
10. Kvet, M.: Advanced radial approach to resource location problems. In: Studies in Computational Intelligence: Developments and Advances in Intelligent Systems and Applications, pp. 29–48. Springer, Berlin (2015)
11. Rybičková, A., Burketová, A., Mocková, D.: Solution to the location-routing problem using a genetic algorithm. In: Smart Cities Symposium Prague, pp. 1–6 (2016)

Uniform Deployment of the p-Location Problem Solutions

Jaroslav Janáček and Marek Kvet

Abstract The uniform deployment has emerged from the need to inspect the enormously large set of feasible solutions of an optimization problem and due to inability of the exact methods to terminate the computation in an acceptable time. The objective function values of the solutions of the uniformly deployed set enable to determine areas of great interest. The uniformly deployed set can also represent population with maximal diversity for evolutionary metaheuristics. The paper deals with a notion of uniformity based on minimal Hamming distance between each pair of solutions. The set of selected solutions is considered to be uniformly deployed if the minimal Hamming distance across the set of all pairs of selected solutions is greater than or equal to a given threshold and if there is no possibility to add any other solution to the set. The paper contains a way of suggesting an initial uniformly deployed set of solutions and an iterative approach to the set enlargement.

1 Introduction

The family of the p-location problems includes such standard problems as p-median and p-center problems and their special versions used in emergency service system designing [2, 6, 8]. Due to complexity of the problems, the exact methods [1, 3, 5] are often abandoned for their unpredictable computational time. For practical use, various kinds of metaheuristics are used. Evolutionary metaheuristics as the genetic algorithm or scatter search method hold an important position in the family of solving tools [4, 9]. These metaheuristics start from an initial set of solutions called the population and they transform the solutions of the current population into members of the new population. These metaheuristics search through the large set of all solution trying to find a good resulting solution. This searching process may prematurely collapse, if the population becomes homogenous, i.e. if the solutions

J. Janáček (✉) · M. Kvet
University of Žilina, Faculty of Management Science and Informatics, Žilina, Slovakia
e-mail: jaroslav.janacek@fri.uniza.sk; marek.kvet@fri.uniza.sk

J. S. Neufeld et al. (eds.), *Operations Research Proceedings 2019*, Operations Research Proceedings, https://doi.org/10.1007/978-3-030-48439-2_38

of the current population are close to each other in the term of some metric. To face the loss of population diversity, various approaches have been suggested starting from simple diversification processes and going on to means of machine learning including exploitation of orthogonal arrays [10]. The uniformly deployed set of solutions can be also used as a preliminary face of an incrementing algorithm [7]. In this paper, we focus on construction of a uniformly deployed set of solutions, which can represent maximally diversified population and may play an important role in all metaheuristics, in which population diversity has to be maintained.

2 The p-Location Problem

The family of p-location problems includes a series of network problems, which can be generally defined as a task of locating p centers at some of possible center locations so that an associated objective function value is minimal. The possible center locations correspond to some network nodes. Similarly, user or customer locations also coincide with network nodes. Formulation of the problems uses denotations I and J for the set of all possible center locations and the set of user locations respectively. The p-location problem can be defined by (1), where the decision on locating a center at i is modelled by a zero-one variable y_i for $i \in I$. The variable y_i gets the value of one if a center is located at i and it gets the value of zero otherwise.

$$min\left\{f(\boldsymbol{y}) : y_i \in \{0, 1\},\ \ i \in I,\ \ \sum_{i \in I} y_i = p\right\} \tag{1}$$

The associated min-sum objective function f^s gets the form of (2).

$$f^s(\boldsymbol{y}) = \sum_{j \in J} b_j min\left\{d_{ij} : i \in I,\ \ y_i = 1\right\} \tag{2}$$

The problem (1) can be studied as a search in a sub-set of m-dimensional hypercube vertices, where m denotes the cardinality of the set I. The distance between two solutions $\boldsymbol{y}$ and $\boldsymbol{x}$ can be measured by Hamming distance defined by (3).

$$H(\boldsymbol{y},\ \boldsymbol{x}) = \sum_{i=1}^{m} |y_i - x_i| \tag{3}$$

The distance of two feasible solutions is only even integer number ranging from 0 to $2p$. The expression $p - H(\boldsymbol{y},\ \boldsymbol{x})/2$ gives the number of possible center locations occupied by centers in both solutions. The uniform deployment problem can be established as a task to find such maximal sub-set S of feasible solutions of the problem (1) so that the inequality $H(\mathbf{y}, \mathbf{x}) \geq h$ holds for each $\mathbf{x}, \mathbf{y} \in S$.

3 Generation of the Set with Minimal Mutual Distance

We present two trivial cases of uniform deployment problem for $h = 2$ and $h = 2p$. In the first case, we assume that the set I of possible locations is numbered by integers from 1 to m. The case of $h = 2$ corresponds to enumeration of all p-tuples of the m locations. In the case $h = 2p$, the i-th solution for $i = 1, \ldots, \lfloor m/p \rfloor$ consists of locations, subscripts of which can be obtained for $k = 1, \ldots, p$ as $p * i + k$.

For medium sized instances of the p-location problem, the first case of maximal set S is usually too large to be used as a starting population in any evolutionary algorithm. Contrary to the first case, the second case defines the set S, which is too small to reach a demanded cardinality of the starting population. That is why we focus on uniformly deployed p-location solution sets, which guarantee the minimal Hamming distance $h = 2(p - 1)$, or $2(p - 2)$ or $2(p - 3)$ etc.

The suggested approach to the uniformly deployed set is based on the following algorithm, which forms $4q$ sub-sets of cardinality q of q^2 different items for an odd integer q so that any pair of the sub-sets has at most one common item. The algorithm produces four groups G^1, G^2, G^3, and G^4 of the sub-sets so that any pair of sub-sets of the same group has no common item.

Input of the algorithm is a matrix $\{a_{ij}\}_{i=1,\ldots,q,j=1,\ldots,q}$, in which each element corresponds to exactly one of the q^2 items.

The first group creation: for $i = 1, \ldots, q$ perform initialization $G^1(i) = \emptyset$ and for $j = 1, \ldots, q$ do $G^1(i) = G^1(i) \cup \{a_{ij}\}$. The second group creation: for $i = 1, \ldots, q$ perform initialization $G^2(i) = \emptyset$ and for $j = 1, \ldots, q$ do $G^2(i) = G^2(i) \cup \{a_{ji}\}$. The third group creation: for $i = 1, \ldots, q$ perform initialization $G^3(i) = \emptyset$ and for $j = 1, \ldots, q$ do $G^3(i) = G^3(i) \cup \{a_{k(i,j),j}\}$. If $j > i$ then $k(i, j) = q + 1 + i - j$, else $k(i, j) = 1 + i - j$. The fourth group creation: for $i = 1, \ldots, q$ perform initialization $G^4(i) = \emptyset$ and for $j = 1, \ldots, q$ do $G^4(i) = G^4(i) \cup \{a_{s(i,j),j}\}$. If $j > q - i$ then $s(i, j) = i + j - q$, else $s(i, j) = i + j$.

The above algorithm yields a uniformly deployed set for $h = 2(p - 1)$ for the special case that p is odd and $p^2 \leq m$ holds. If p is odd and $(p + 1)^2 \leq m$, then the algorithm can be also applied and the $4p$ sub-sets of cardinality p each can be obtained by simple removal of one item from each sub-set.

The suggested algorithm can also be used as a building block for obtaining some starting uniformly deployed set of p-location problem solutions for cases, when $p^2 > m$ holds, but a lower minimal mutual distance must be accepted. An application to the case $p^2 > m$ needs determination of integers r and odd q so that $rq = p$ and $rq^2 \leq m$.

After the integers r and q are determined, r disjoint portions of q^2 possible items are selected from the original set of m possible locations, the algorithm is used to produce G_k^1, G_k^2, G_k^3, and G_k^4 groups of q tuples of locations for $k = 1, \ldots, r$.

Now, the four groups G^1, G^2, G^3, and G^4 of q sub-sets of cardinality p, where each pair of the sub-sets has at most r common items can be created according to the prescription (4):

$$For\ \ t = 1, \ldots, 4 \text{ perform } for\ \ i = 1, \ldots, q\ \ G^t(i) = \bigcup_{k=1}^{r} G_k^t(i) \tag{4}$$

4 Completion of a Uniform Set of p-Location Solutions

The initial uniformly deployed set of p-location solutions can be constructed according to rules formulated in the previous section. Here, we formulate the problem of identification whether the uniformly deployed set is maximal.

Let S be a studied set of r p-location problem solutions with minimal mutual distance $h = 2(p - d)$. Let the s-th solution from S be described by a series of zero-one constants e_{si} for $i = 1, \ldots, m$, where e_{si} equals to one, if the solution includes location i and it equals to zero otherwise. We introduce zero-one variables $y_i \in \{0, 1\}$ for $i = 1, \ldots, m$ to described hypothetical additional solution $\boldsymbol{y}$, which could be used for extension of the set S. We also introduce auxiliary variables z_s for $s \in S$ to identify exceeding number of common locations in solution s and $\boldsymbol{y}$. Then, the identification problem, whether the set is or is not maximal, can be modelled as follows.

$$Minimize \quad \sum_{s \in S} z_s \tag{5}$$

$$Subject\ \ to: \quad \sum_{i=1}^{m} y_i = p \tag{6}$$

$$\sum_{i=1}^{m} e_{si} y_i \le d + z_s \quad for\ \ s \in S \tag{7}$$

$$y_i \in \{0, 1\} \quad for\ \ i = 1, \ldots, m \tag{8}$$

$$z_s \ge 0 \quad for\ \ s \in S \tag{9}$$

The objective function (5) expresses the sum of surpluses of common locations of the suggested solution $\boldsymbol{y}$ and s over all solutions from S. If the objective function value of the optimal solution equals to zero, then the solution $\boldsymbol{y}$ has at most d

common locations with each solution from S and it follows that the set S is not maximal and the solution $\boldsymbol{y}$ can be added to the set S. Constraint (6) is a feasibility constraint imposed upon solution $\boldsymbol{y}$. Constraints (7) links up suggested solution $\boldsymbol{y}$ to the variables z_s for $\in S$. A procedure solving the problem (5)–(9) can be used for extension of an initial set of p-median problem according to the following process.

0. Start with an initial uniformly deployed set S.
1. Solve the problem (5)–(9) to optimality and obtain the associated solution $\boldsymbol{y}$.
2. If the objective function (5) of the solution $\boldsymbol{y}$ equals to zero, update $S = S \cup \{\boldsymbol{y}\}$ and go to 1, otherwise terminate, the current set S is maximal.

5 Numerical Experiments

The numerical experiments were performed to demonstrate the process of providing the maximal or near-to-maximal uniformly deployed sets in real cases, where $p^2 > m$. The instances were obtained from the road network of self-governing regions of Slovakia. The mentioned instances are further denoted by the names of capitals of the individual regions followed by triples (XX, m, p), where XX is commonly used abbreviation of the region denotation, m stands for the number of possible center locations and p is the number of service centers, which are to be located in the mentioned region. The list of instances follows: Bratislava (BA, 87, 14), Banská Bystrica (BB, 515, 36), Košice (KE, 460, 32), Nitra (NR, 350, 27), Prešov (PO, 664, 32), Trenčín (TN, 276, 21), Trnava (TT, 249, 18) and Žilina (ZA, 315, 29).

As the rules for generating $4p$ solutions of the p-location problem cannot be used, we employed the rules for generating $4q$ solutions of a q-location problem formed from q^2 locations with the minimal Hamming distance $2(q-1)$. The set of $4q$ solutions can be easily extended to the set of $2q$-location problem solutions using a next portion of q^2 locations. Each pair of the new solutions has minimal distance $2(2q-2)$, i.e. the new solutions have at most two common locations. This process of extension can be performed r-times using rq^2 locations and solutions of each pair have at most r common locations. Odd integer q and integer r are to be chosen so that $rq^2 \leq m$ holds and the number rq approximates the value p. In the case, when $rq > p$, the associated solutions of rq-location problem can be adjusted to p-location problem solutions by removing $p - rq$ locations from each solution. In the opposite case, when $rq < p$, the solutions of rq-location problem can be extended by addition of some unused locations. We performed these operations with the above mentioned instances and obtained the initial uniformly deployed sets, parameters of which are reported in Table 1. The symbol $|S|$ denotes the cardinality of the set S and d denotes the maximal number of common locations for any pair of solutions. Time characteristics of the computational process are plotted in Table 1. The column denoted by "First" contains the time of solution of the problem (5)–(9) for the first extension of the original set. The column denoted by "Last" contains

Table 1 Time characteristics of the uniformly deployed set extension

Region	r	q	$\|S\|$	d	Computational time in seconds			
					First	Last	Avg	StDev
BA	2	5	23	2	0.1	0.03	0.05	0.02
BB	3	13	172	3	0.1	17.87	15.28	140.02
KE	2	15	60	2	–	–	–	–
NR	2	13	83	2	0.1	20,070.57	10,035.33	14,191.97
PO	2	17	232	2	0.2	116.83	58.51	82.48
TN	2	9	137	2	0.1	4433.59	2216.84	3134.96
TT	2	9	212	2	0.1	99.66	49.88	70.41
ZA	3	9	112	3	0.1	3466.16	1733.12	2450.89

the time of solution of the problem (5)–(9) for the last extension of the original set. The denotation "Avg" stands for the average time and "StDev" denotes the standard deviation of the times. To solve the problems, the software FICO Xpress 7.3 was used on a PC equipped with the Intel® Core™ i7 5500U 2.4 GHz processor and 16 GB RAM.

6 Conclusions

The paper dealt with the uniformly deployed set of the p-location problem solution. We presented a method of obtaining the starting uniformly deployed set and integer programming model of its enlargement. The resulting uniformly deployed set can be employed in approximate solving techniques for the p-location problem. It can be also used either as a partial mapping of the "terrain" to determine areas of great interest or it can be considered as a population with maximal diversity for evolutionary algorithm. Future research could be focused on more general ways of obtaining the maximal uniformly deployed set and on their usage in heuristics approaches.

Acknowledgments This work was supported by the research grants VEGA 1/0342/18 "Optimal dimensioning of service systems", VEGA1/0089/19 "Data analysis methods and decisions support tools for service systems supporting electric vehicles", and VEGA 1/0689/19 "Optimal design and economically efficient charging infrastructure deployment for electric buses in public transportation of smart cities" and APVV-15-0179 "Reliability of emergency systems on infrastructure with uncertain functionality of critical elements".

References

1. Avella, P., Sassano, A., Vasil'ev, I.: Computational study of large scale p-median problems. Math. Program. **109**, 89–114 (2007)

2. Doerner, K.F., et al.: Heuristic solution of an extended double-coverage ambulance location problem for Austria. Cent. Eur. J. Oper. Res. **13**(4), 325–340 (2005)
3. García, S., Labbé, M., Marín, A.: Solving large p-median problems with a radius formulation INFORMS J. Comput. **23**(4), 546–556 (2011)
4. Gendreau, M., Potvin, J.: Handbook of Metaheuristics, 3rd edn., 610 pp. Springer, Berlin (2019)
5. Janáček, J., Kvet, M.: Min-max optimization and the radial approach to the public service system design with generalized utility. Croat. Oper. Res. Rev. **7**(4), 49–61 (2016)
6. Jánošíková, Ľ., Jankovič, P., Márton, P.: Models for relocation of emergency medical stations. In: The Rise of Big Spatial Data. Lecture Notes in Geoinformation and Cartography, pp. 225–239. Springer, Berlin (2016)
7. Kvet, M., Janáček, J.: Usage of uniform deployment for heuristic design of emergency system. In: Neufeld, J.S., Buscher, U., Lasch, R., Möst, D., Schönberger, J. (eds.) Operations Research Proceedings 2019: Selected Papers of the Annual International Conference of the German Operations Research Society (GOR), Dresden, Germany, September 4–6, 2019
8. Marianov, V., Serra, D.: Location problems in the public sector. In: Drezner, Z., et al. (eds.) Facility Location: Applications and Theory, pp. 119–150. Springer, Berlin (2002)
9. Rybičková, A., Burketová, A., Mocková, D.: Solution to the location-routing problem using a genetic algorithm. In: Smart Cities Symposium Prague, pp. 1–6 (2016)
10. Zhang, O., Leung, Y.: An orthogonal genetic algorithm for multimedia multicast routing. IEEE Trans. Evol. Comput. **3**(1), 53–62 (1999)

Algorithms and Complexity for the Almost Equal Maximum Flow Problem

R. Haese, T. Heller, and S. O. Krumke

Abstract In the Equal Maximum Flow Problem (EMFP), we aim for a maximum flow where we require the same flow value on all arcs in some given subsets of the arc set. We study the related Almost Equal Maximum Flow Problems (AEMFP) where the flow values on arcs of one homologous arc set differ at most by the valuation of a so called homologous function Δ. We prove that the integer AEMFP is in general $\mathcal{NP}$-complete, and that even finding a fractional maximum flow in the case of convex homologous functions is also $\mathcal{NP}$-complete. This is in contrast to the EMFP, which is polynomial time solvable in the fractional case. We also provide inapproximability results for the integral AEMFP. For the integer AEMFP we state a polynomial algorithm for the constant deviation and concave case for a fixed number of homologous sets.

1 Introduction

The Maximum Flow Problem (MFP) is a well studied problem in the area of network flow problems. Given a graph $G = (V, A)$ with arc capacities $u: A \mapsto \mathbb{R}_+$, a source $s \in V$, a sink $t \in V\backslash\{s\}$ one searches for a s-t-flow $f: A \mapsto \mathbb{R}_{\geq 0}$ such that $0 \leq f \leq u$ (capacity constraints), for all $v \neq s, t$ we have $f(\delta^+(v)) - f(\delta^-(v)) = 0$ (flow conservation) and such that the total amount of flow reaching the sink $\text{val}(f) := f(\delta^-(t)) - f(\delta^+(t))$ is maximized. Like in standard notation from the literature, we denote by $\delta^-(v)$ the set of ingoing arcs of some node v and the set of outgoing arcs by $\delta^+(v)$, and for $S \subseteq A$ abbreviate $f(S) := \sum_{a \in S} f(a)$.

In this paper, we study a variant of the family of equal flow problems, which we call the Almost Equal Flow Problems (AEMFP). In addition to the data for

---

R. Haese · S. O. Krumke
University of Kaiserslautern, Kaiserslautern, Germany

T. Heller (✉)
Fraunhofer ITWM, Kaiserslautern, Germany
e-mail: till.heller@itwm.fraunhofer.de

J. S. Neufeld et al. (eds.), *Operations Research Proceedings 2019*, Operations Research Proceedings, https://doi.org/10.1007/978-3-030-48439-2_39

the MFP one is given (not necessarily disjoint) homologous subsets $R_i \subseteq A$ for $i = 1, \ldots, k$, functions Δ_i and one requires for the flow f the condition that $f(a) \in [f_i, \Delta_i(f_i)]$ for all $a \in R_i$, $i = 1, \ldots, k$ (homologous arc set condition), where $f_i := \min_{a \in R_i} f(a)$ denotes the smallest flow value of an arc in R_i. In the special case that all Δ_i are the identity, all arcs in a homologous set are required to have the same flow value. This problem is known as the Equal MFP (EMFP).

The AEMFP is motivated by application in the distribution of energy where it is undesirable to have strongly differing amounts of energy between different time periods. The energy distribution can be modeled as a flow problem in a time expanded network and the homologous arcs correspond to subsequent time periods. The simplest case there is a constant deviation allowed between periods which leads to the AEMFP with constant deviation (see Sect. 4). However, the relation between different periods might be more complex, which motivates the study of non-linear deviation functions such as concave and convex functions.

Ali et al. [2] studied a variant of the Minimum Cost Flow Problem, where K pairs of arcs are required to have the same flow value, which they called *equal flow problem*. An integer version of this problem, where flow on the arcs required to be integer, was studied by Ali et al. [2] and was shown to be $\mathcal{NP}$-complete. Further, they obtained a heuristic algorithm based a Lagrangian relaxation technique. Meyers and Schulz [7] showed that the integer equal flow problem is not approximable in polynomial time (unless $\mathcal{P} = \mathcal{NP}$), even if the arc sets are of size two. Ahuja et al. [1] considered the *simple equal flow problem*, where the flow value on arcs of a single subset of the arc set has to be equal. Using Megiddo's parametric search technique [5, 6], they present a strongly polynomial algorithm which has a running time of $\mathcal{O}(\{m(m + n \log n) \log n\}^2)$.

1.1 Our Contribution

We provide the first complexity results for the AEMFP. The first three columns in Table 1 denote the complexity classes of the different problem variants while the entries of the fourth column contain an upper bound for the best approximation factor for a polynomial algorithm (unless $\mathcal{P} = \mathcal{NP}$). If a function Δ is of the form $x \mapsto x + c$ for a fixed constant $c \geq 0$ we call Δ a *constant deviation function*. For the AEFMP with k homologous arc sets and constant deviation functions, we obtain a running time of $\mathcal{O}(n^k m^k \log(\log(n))^k T_{mf}(n, n + m))$ where $T_{mf}(n, m)$ denotes the running time of a maximum flow algorithm on a graph G with n nodes and m arcs.

Note that general polynomial time solvability of the AEMFP in the case of constant deviation functions also follows from Tardos' Algorithm, see e.g. [8]. Our main algorithmic contribution is a combinatorial method which not only works in the constant deviation case but also for concave functions.

Table 1 Overview of results

Function Δ	Fractional	Integer	Fixed k	Lower bound for approximation
Const. deviation	$\mathcal{P}$	$\mathcal{NP}$	$\mathcal{P}$	$2-\epsilon$
Concave	$\mathcal{P}$	$\mathcal{NP}$	$\mathcal{P}$	No constant
Convex	$\mathcal{NP}$	$\mathcal{NP}$	$\mathcal{NP}$	No constant

2 Problem Definition

The (AEMFP) can be formulated as the following optimization problem in the variables $f(a)$ $(a \in A)$ and $f_i, i = 1, \ldots, k$:

$$\text{(AEMFP)} \quad \max \quad \text{val}(f) \tag{1a}$$

$$\text{s.t.}\, f(\delta^+(s)) - f(\delta^-(s)) \geq 0 \tag{1b}$$

$$f(\delta^+(t)) - f(\delta^-(t)) \leq 0 \tag{1c}$$

$$f(\delta^+(v)) - f(\delta^-(v)) = 0 \qquad \forall v \in V\backslash\{s,t\} \tag{1d}$$

$$0 \leq f(r) \leq u(r) \qquad \forall r \in A \tag{1e}$$

$$f_i \leq f(r_i) \leq \Delta_i(f_i) \qquad \forall r_i \in R_{\angle_i}, \forall R_{\Delta_i} \tag{1f}$$

In the integral version, we additionally require f to attain only integral values. Note that, in general, problem AEMFP (1a)–(1f) is nonlinear due to the nonlinearity of the deviation functions Δ_i and condition (1f). However, if each Δ_i is a constant deviation function, then (1f) becomes $f_i \leq f(r_i) \leq f_i + c_i$ and AEMFP is a Linear Program.

The simple AEMFP is the AEMFP with just one homologous arc set R_Δ. Note that by subdividing arcs that are contained in several homologous arc sets, we can assume w.l.o.g. that the homologous arc sets are disjoint.

3 Complexity and Approximation

While for standard flow problem there is always an optimal solution which is also integral, simple examples show that this is not the case for the AEMFP. In fact, the following results show that finding even approximately optimal integral solutions is hard in general for AEMFP.

Theorem 1 *The integer AEMFP is $\mathcal{NP}$-complete, even if all homologous functions are the same constant deviation function or the same concave function, the homologous sets are disjoint, the capacities are integral, and the graph is bipartite. Unless $\mathcal{P} = \mathcal{NP}$, for any $\varepsilon > 0$, there is no polynomial time $(2-\varepsilon)$-approximation* *algorithm for the integer AEMFP, even if we consider disjoint sets and a constant deviation* $x \mapsto x + 1$. □

Theorem 2 *In the case of concave homologous functions, the integer AEMFP is $\mathcal{NP}$-hard to solve and there is no polynomial time constant approximation algorithm (unless $\mathcal{P} = \mathcal{NP}$). In the case of convex homologous functions, even the fractional AEMFP is $\mathcal{NP}$-hard to solve and there is no polynomial time constant approximation algorithm (unless $\mathcal{P} = \mathcal{NP}$).* □

4 The Constant Deviation Case

We start with the simple AEMFP. Let $G = (V, A)$ be a graph with a single homologous arc set R and constant deviation function $\Delta_R : x \mapsto x + c$. For easier notation, we define $Q := A \backslash R$ as the set of all arcs that are not contained in the homologous arc set R. By the homologous arc set condition (1f), we know that the flow value on each of the corresponding arcs must lie in an interval $[\lambda^*, \Delta(\lambda^*)] = [\lambda^*, \lambda^* + c]$, where λ^* is unknown.

For a guess value λ consider the modified network G_λ, where we modify the upper capacity of every arc in R to $\lambda + c$ and its lower capacity from 0 to λ. All arcs in Q keep their upper capacities and have lower capacity of 0. By f_λ we denote a traditional s-t-flow which is feasible in G_λ.

For an (s, t)-cut (S, T) let us denote by

$$g_S(\lambda) := u(\delta^+(S \cap Q)) + \sum_{r \in \delta^+(S \cap R)} \min\{u(r), \Delta_R(\lambda)\} - \sum_{r \in \delta^-(S \cap R)} \lambda$$

its capacity in G_λ. By the Max-Flow Min-Cut Theorem we get

$$\max_{f_\lambda} \mathrm{val}(f_\lambda) = \min_{(S,T) \text{ is a } (s,t)\text{-cut}} g_S(\lambda)$$

We summarize some structural results in the following observation:

Observation 3 *The function $F(\lambda) := \min_{(S,T) \text{ is a } (s,t)\text{-cut}} g_S(\lambda)$ is a piecewise linear concave function. AEMFP can be solved by solving*

$$\max \left\{ F(\lambda) : 0 \leq \lambda \leq \min_{r \in R_\Delta} u(r) \right\}.$$

The function $F(\lambda)$ has at most $2m$ breakpoints, The minimum distance between two of these breakpoints is $\frac{1}{m^2}$. □

Observe that the optimal value λ^* is attained at a breakpoint of F. At this point the slope to the left is positive or the slope to the right is negative. If there exists a cut such that the slope is 0, we simply take the breakpoint to the left or right of the current value λ. Now we apply the parametric search technique by Megiddo [5, 6] to search for the optimal value λ^* on the interval $[0, u_R]$, where $u_R := \min_{r \in R_\Delta} u(r)$. We simulate an appropriate maximum flow algorithm (the Edmonds-Karp algorithm) for symbolic lower capacities λ^* and upper capacities $\lambda^* + c$ on the arcs in R.

Observation 4 *If we run the Edmonds-Karp algorithm with a symbolic input parameter λ, all flow values and residual capacities which are calculated during the algorithm steps are of the form $a + b\lambda$ for $a, b \in \mathbb{Z}$.* □

Lemma 1 *The algorithm computes a maximum almost equal flow in time $\mathcal{O}(n^3 m\, T_{MF}(n, n+m))$, where $T_{MF}(n, n+m)$ denotes the time needed to compute a maximum flow on a graph with n nodes and $n + m$ arcs.* □

By exploiting implicit parallelism [6] one can improve the running time to $\mathcal{O}(nm(n \log n + m \log m) T_{MF}(n, n+m))$. Interestingly, using one of the known faster maximum flow algorithms instead of the Edmonds-Karp algorithm does not yield an improved running time.

To solve the integer version of the maximum AEMFP, we simply use the optimal value λ^* of the non-integer version and compute two maximum flow on the graphs $G_{\lfloor \lambda^* \rfloor}$ and $G_{\lceil \lambda^* \rceil}$. By taking the $\text{argmax}\{val(f_{\lfloor \lambda^* \rfloor}), val(f_{\lceil \lambda^* \rceil})\}$ we get the optimal parameter λ^*_{int} for the integer version.

In the general constant deviation AEMFP we consider more than one homologous arc set. By iteratively using the algorithm for the simple constant deviation AEMFP, we obtain a combinatorial algorithm for the general constant deviation AEMFP. We present the algorithm for the case of two homologous arc sets, but it can be generalized to an arbitrary number of homologous arc sets. The idea behind our algorithm is to fix some λ_1 and use then the algorithm for the simple case to find the optimal corresponding λ_2. Once we found $\lambda_2^*(\lambda_1)$, we check if λ_1 is to the left, right or equal to λ_1^*. Note that the objective function is still a concave function in λ_1 and λ_2 since it is the sum of concave functions. Also, like in the simple case, all flow values and capacities both in the network G and the residual network G_f during the algorithm are of the form $a + b\lambda_1 + c\lambda_2$.

Note that the running time of the algorithm for the general constant deviation AEMFP increases for every additional homologous arc set roughly by a factor of the running time of the algorithm for the simple constant deviation AEMFP.

Theorem 5 *Let $T_{mf}(n, m)$ denote the running time of a maximum flow algorithm on a graph G with n nodes and m arcs. The AEMFP with k homologous sets can be solved in time $\mathcal{O}\left(\min\left\{n^k m^k \log(\log(n))^k, n^{3k}\right\} \cdot T_{mf}(n, n+m)\right)$.*

Here we see that the running time for an arbitrary number of homologous arc sets becomes exponential.

5 Convex and Concave Deviations

If we deviation function is a convex function $\Delta_{conv}: R \mapsto \mathbb{R}_{\geq 0}$, we get the convex AEMFP. Note that this problem is neither a convex nor a concave program due to the constraint (1f). Hence, standard methods of convex optimization can not be applied. In fact, the next theorem says that (unless $\mathcal{P} = \mathcal{NP}$) one cannot hope to find a polynomial time algorithm that solves this problem:

Theorem 6 *The AEMFP with a convex homologous function Δ is $\mathcal{NP}$-complete, even if all homologous functions are given as $\Delta_R(x) = 2x^2 + 1$ for all homologous sets R, the homologous sets disjoint, the capacities are integral, and the graph is bipartite.* □

Theorem 7 *Unless $\mathcal{P} = \mathcal{NP}$, there is no polynomial time constant factor approximation algorithm for the integer convex AEMFP.* □

In contrast to the convex case, which is $\mathcal{NP}$-complete even for the fractional case, the concave case is polynomially solvable since in this case (1) becomes a concave program. We provide a combinatorial algorithm by combining results of the previous section together with techniques of Toledo [9]. This enables us to prove the following result:

Theorem 8 *The AEMFP with a piecewise polynomial concave homologous function Δ with maximum degree p can be solved in polynomial time for one homologous arc set in time $\mathcal{O}(mp \cdot (nm \cdot (n + 2m + n^2)(T_{MF}(n, n + m))))$ under the assumption that we can compute the roots of a polynomial p of maximum degree q in constant time $\mathcal{O}(1)$.* □

Our algorithm yields in the worst case a better running time than a direct implementation of the Megiddo-Toledo algorithm for maximizing non-linear concave function in k dimensions, which runs in $\mathcal{O}((T_{mf}(n, m))^{2^k})$ [9].

The integral version of the concave AEMFP turns out to be still hard and hard to approximate.

Theorem 9 *The concave integer AEMFP is $\mathcal{NP}$-complete. Moreover, unless $\mathcal{P} = \mathcal{NP}$, there is no constant approximation factor for the integer concave AEMFP.* □

6 Outlook

We have provided the first complexity results for the Almost Equal Maximum Flow Problems (AEMFP). Our results can be extended to the *almost equal minimum cost flow* problem where one searches for a minimum cost flow subject homologous arc set constraints (1f).

Whereas the complexity for the fractional versions of the AEMFP is essentially settled, an interesting open questions raised by our work is the existence of polynomial approximation algorithms for the integer AEMFP with constant deviation functions.

References

1. Ahuja, R.K., Orlin, J.B., Sechi, G.M., Zuddas, P.: Algorithms for the simple equal flow problem. Manag. Sci. **45**(10), 1440–1455 (1999)
2. Ali, A.I., Kennington, J., Shetty, B.: The equal flow problem. Eur. J. Oper. Res. **36**(1), 107–115 (1988)
3. Cohen, E., Megiddo, N.: Maximizing concave functions in fixed dimension, In: Panos M. Pardalos (ed.) Complex. Numer. Optim., pp. 74–87 (1993)
4. Haese, R.: Almost equal flow problems. Master thesis, University of Kaiserslautern (2019)
5. Megiddo, N.: Combinatorial optimization with rational objective functions. In: Proceedings ACM Symposium on Theory of Computing, pp. 1–12 (1978)
6. Megiddo, N.: Applying parallel computation algorithms in the design of serial algorithms. In: Symposium on Foundations of Computer Science, pp. 399–408 (1981)
7. Meyers, C.A., Schulz, A.S.: Integer equal flows. Oper. Res. Lett. **37**(4), 245–249 (2009)
8. Tardos, E.: A strongly polynomial algorithm to solve combinatorial linear programs. Oper. Res. **34**(2), 250–256 (1986)
9. Toledo, S.: Maximizing non-linear concave functions in fixed dimension, In: Panos M. Pardalos (ed.) Complex. Numer. Optim., pp. 429–447 (1993)

Exact Solutions for the Steiner Path Cover Problem on Special Graph Classes

Frank Gurski, Stefan Hoffmann, Dominique Komander, Carolin Rehs, Jochen Rethmann, and Egon Wanke

Abstract The Steiner path problem is a restriction of the well known Steiner tree problem such that the required terminal vertices lie on a path of minimum cost. While a Steiner tree always exists within connected graphs, it is not always possible to find a Steiner path. Despite this, one can ask for the Steiner path cover, i.e. a set of vertex disjoint simple paths which contains all terminal vertices and possibly some of the non-terminal vertices. We show how a Steiner path cover of minimum cardinality for the disjoint union and join composition of two graphs can be computed in linear time from the corresponding values of the involved graphs. The cost of an optimal Steiner path cover is the minimum number of Steiner vertices in a Steiner path cover of minimum cardinality. We compute recursively in linear time the cost within a Steiner path cover for the disjoint union and join composition of two graphs by the costs of the involved graphs. This leads us to a linear time computation of an optimal Steiner path, if it exists, for special co-graphs.

1 Introduction

For the well known Steiner tree problem there are several solutions on special graph classes like series-parallel graphs [10], outerplanar graphs [9] and graphs of bounded tree-width [2]. The class Steiner tree problem (CSP) is a generalization of the well known Steiner tree problem in which the vertices are partitioned into classes of terminal vertices [8]. The unit-cost version of CSP can be solved in linear time on co-graphs [11].

F. Gurski · S. Hoffmann · D. Komander · C. Rehs · E. Wanke
University of Düsseldorf, Institute of Computer Science, Düsseldorf, Germany

J. Rethmann (✉)
Niederrhein University of Applied Sciences, Faculty of Electrical Engineering and Computer Science, Krefeld, Germany
e-mail: jochen.rethmann@hsnr.de

J. S. Neufeld et al. (eds.), *Operations Research Proceedings 2019*, Operations Research Proceedings, https://doi.org/10.1007/978-3-030-48439-2_40

The Steiner path problem is a restriction of the Steiner tree problem such that the required terminal vertices lie on a path. The Euclidean bottleneck Steiner path problem was considered in [1] and a linear time solution for the Steiner path problem on trees was given in [7]. The Steiner path cover problem on interval graphs was considered in [6].

In this article we consider the Steiner path cover problem. Let G be a given undirected graph on vertex set $V(G)$ and edge set $E(G)$ and let $T \subseteq V(G)$ be a set of terminal vertices. The problem is to find a set of vertex-disjoint simple paths in G that contain all terminal vertices of T and possibly also some of the non-terminal (Steiner) vertices of $S := V(G) - T$. The size of a Steiner path cover is the number of its paths, the cost is defined as the minimum number of Steiner vertices in a Steiner path cover of minimum size.

2 Co-graphs

Co-graphs (short for complement reducible graphs) have been introduced in the 1970s by a number of authors with different notation. Co-graphs can be characterized as the set of graphs without an induced path P_4 with four vertices [3].

Definition 1 The class of *co-graphs* is recursively defined as follows.

(i) Every graph on a single vertex $(\{v\}, \emptyset)$, denoted by $\bullet_v$, is a *co-graph*.
(ii) If A, B are vertex-disjoint co-graphs, then

  (a) the disjoint union $A \oplus B$, which is defined as the graph with vertex set $V(A) \cup V(B)$ and edge set $E(A) \cup E(B)$, and
  (b) the join composition $A \otimes B$, defined by their disjoint union plus all possible edges between vertices of A and B, are *co-graphs*.

For every co-graph G one can define a tree structure, called *co-tree*. The leaves of the co-tree represent the vertices of the graph and the inner vertices of the co-tree correspond to the operations applied on the subgraphs of G defined by the subtrees. For every co-graph one can construct a co-tree in linear time, see [4].

3 Solution for the Steiner Path Cover Problem

Let G be a co-graph and $T \subseteq V(G)$ be a set of terminal vertices. We define $p(G, T)$ as the minimum number of paths within a Steiner path cover for G with respect to T. Further we define $s(G, T)$ as the minimum number of Steiner vertices in a Steiner path cover of size $p(G, T)$ with respect to T. We do not specify set T if it is clear from the context which set is meant.

Lemma 1 *Let A and B be two vertex-disjoint co-graphs and let $T_A \subseteq V(A)$ and $T_B \subseteq V(B)$ be two sets of terminal vertices. Without loss of generality, let $|T_A| \leq |T_B|$. Then the following equations hold true:*

1. $p(\bullet_v, \emptyset) = 0$ *and* $p(\bullet_v, \{v\}) = 1$
2. $p(A \oplus B, T_A \cup T_B) = p(A, T_A) + p(B, T_B)$
3. $p(A \otimes B, \emptyset) = 0$
4. $p(A \otimes B, T_A \cup T_B) = \max\{1, p(B, T_B) - |V(A)|\}$ *if* $1 \leq |T_B|$

Proof

1. Obvious.
2. Since the disjoint union does not create any new edges, the minimum size of a Steiner path cover for the disjoint union of A and B is equal to the sum of the sizes of Steiner path covers of minimum size for A and B.
3. If $A \otimes B$ does not contain any terminal vertices, there is no path in a Steiner path cover of minimum size.
4. We show that $p(A \otimes B) \geq \max\{1, p(B) - |V(A)|\}$ applies by an indirect proof. Assume that a Steiner path cover C for $A \otimes B$ contains less than $\max\{1, p(B) - |V(A)|\}$ paths. The removal of all vertices of A from all paths in C gives a Steiner path cover of size at most $|C| - |V(A)| < p(B)$ for B. ↯

   To see that $p(A \otimes B) \leq \max\{1, p(B) - |V(A)|\}$ applies, consider that we can use any vertex of A to combine two paths of the cover of B to one path, since the join composition of A and B creates all edges between A and B. If there are more terminal vertices in T_A than there are paths in the cover of B, i.e. $p(B) < |T_A|$, then we have to split paths of B and reconnect them by terminal vertices of T_A. This can always be done since $|T_A| \leq |T_B|$. □

Let C be a Steiner path cover for a co-graph G with respect to a set $T \subseteq V(G)$ of terminal vertices. Then $\mathfrak{p}(C)$ denotes the number of paths in cover C, and $\mathfrak{s}(C)$ denotes the number of Steiner vertices in the paths of cover C.

Lemma 2 *Let C be a Steiner path cover for some co-graph $G = A \otimes B$ with respect to a set T of terminal vertices. Then there is a Steiner path cover C' with respect to T which does not contain paths p and p' satisfying one of the structures (1)–(7), such that $\mathfrak{p}(C) \geq \mathfrak{p}(C')$ and $\mathfrak{s}(C) \geq \mathfrak{s}(C')$ holds true. Let $q_1, \ldots, q_4$ denote subpaths which may be empty.*

1. $p = (x, q_1)$ *where* $x \notin T$. *Comment: No path starts with a Steiner vertex.*
2. $p = (q_1, u, x, v, q_2)$ *where* $u, x \in V(A)$, $v \in V(B)$, *and* $x \notin T$. *Comment: On a path, the neighbors u, v of a Steiner vertex x are both contained in the same graph.*
3. $p = (q_1, x, y, q_2)$, $p' = (q_3, u, v, q_4)$ *where* $x, y \in V(A)$, $u, v \in V(B)$. *Comment: The cover only contains edges of one of the graphs.*
4. $p = (x, q_1)$, $p' = (q_2, u, y, v, q_3)$, *where* $x, y \in V(A)$, $u, v \in V(B)$, *and* $y \notin T$. *Comment: If a path starts in A then there is no Steiner vertex in A with two neighbors on the path in B.*

5. $p = (x, q_1)$, $p' = (q_2, u, v, q_3)$, *where* $x \in V(A)$ *and* $u, v \in V(B)$. *Comment: If a path starts in A, then no edge of B is contained in the cover.*
6. $p = (x, q_1)$, $p' = (q_2, u)$, *where* $x \in V(A)$, $u \in V(B)$, $p \neq p'$. *Comment: All paths start in the same graph.*
7. $p = (\ldots, x, u, v, y, \ldots)$ *where* $u, v \notin T$. *Comment: The paths contain no edge between two Steiner vertices.*

Proof If one of the forbidden configurations is present, it is removed by an operation described next.

1. If x is removed from p, we get a cover with one Steiner vertex less than C.
2. If x is removed from p, we get a cover with one Steiner vertex less than C.
3. If $p \neq p'$, then (q_1, x, v, q_4) and (q_3, u, y, q_2) are the paths in cover C'. If $p = p'$, then we have to distinguish whether $\{u, v\} \in q_1$, $\{u, v\} \in q_2$, $\{x, y\} \in q_3$, or $\{x, y\} \in q_4$. We show the first case, the other three cases can be handled similar. Let $p = (q_3, u, v, q_5, b, a, q_6, x, y, q_2)$, where $b \in V(B)$ and $a \in V(A)$. Then the new path in cover C' is $(q_3, u, a, q_6, x, v, q_5, b, y, q_2)$. In any case cover C' is as good as C.
4. If $p \neq p'$, then q_1 and (q_2, u, x, v, q_3) are the new paths in cover C'. If $p = p'$, i.e. $q_1 = (q_2', u, y, v, q_3)$, where q_2' is obtained from q_2 by removing x, then (q_2', u, x, v, q_3) is the new path in cover C'. The cover C' contains one Steiner vertex less than C.
5. If $p \neq p'$, then q_1 and (q_2, u, x, v, q_3) are the new paths in cover C'. If $p = p'$, i.e. $q_1 = (q_2', u, v, q_3)$, where q_2' is obtained from q_2 by removing x, then (q_2', u, x, v, q_3) is the new path in cover C'. The cover C' is as good as C.
6. We combine the paths to only one path (q_2, u, x, q_1) and we get a cover C' with one path less than C.
7. Since a co-graph contains no P_4 as an induced subgraph, there has to be one of the edges $\{x, v\}$, $\{u, y\}$, or $\{x, y\}$ in G. In the first case we remove vertex u from p, in the second case we remove vertex v, in the last case we remove vertices u and v. In any case we get a cover C' with at least one Steiner vertex less than C.

The number of Steiner vertices in C is decreased by one each time using Modifications 1, 2, 4, and 7, and remains the same when using Modifications 3, 5, and 6. Thus, Modifications 1, 2, 4, and 7 can be applied at most $|V(G)| - |T|$ times. The number of edges between vertices of A and vertices of B decreases when using the Modifications 3, 5, and 6 and remains the same when using the Modifications 1, 2, and 4. Only Modification 7 can reduce the number of edges between vertices of A and vertices of B by a maximum of two. Since Modification 7 reduces the number of Steiner vertices, a maximum of $3(|V(G)| - |T|) + |V(G)| - 1$ modifications can be made until the process stabilizes. □

Since the hypothesis of Lemma 2 is symmetric in A and B, the statement of Lemma 2 is also valid for co-graphs $G = A \otimes B$ if A and B are switched.

Definition 2 A Steiner path cover C for some co-graph $G = A \otimes B$ is said to be in normal form if none of the operations described in the proof of Lemma 2 is applicable.

Theorem 1 *For each co-graph $G = A \otimes B$ and set of terminal vertices T any Steiner path cover C with respect to T can be transformed into a Steiner path cover C' in normal form such that C' does not contain an edge of graph A, no path in C' starts or ends in graph A, $\mathfrak{p}(C') \leq \mathfrak{p}(C)$ and $\mathfrak{s}(C') \leq \mathfrak{s}(C)$ if $|T_A| < |T_B|$.*

Proof *(By Contradiction)* Assume, cover C contains an edge of graph A. Then by Lemma 2(5), all paths starts in graph A. By Lemma 2(4), it holds that no Steiner vertex v of $V(A)$ is contained in C, where the neighbors of v are both of graph B. By Lemma 2 (1), (2), and (5), it holds that all vertices of $V(B)$ from C are connected with a terminal vertex of $V(A)$, thus $|T_A| > |T_B|$. ↯

Second, we have to show that no path in C starts or ends in graph A. Assume on the contrary, that there is one path that starts in A. By Lemma 2(6), it holds that all paths start in A. Continuing as in the first case this leads to a contradiction. □

Remark 1 For two vertex-disjoint co-graphs A, B and two sets of terminal vertices $T_A \subseteq V(A)$, $T_B \subseteq V(B)$ it holds that $s(A \oplus B, T_A \cup T_B) = s(A, T_A) + s(B, T_B)$, since the disjoint union does not create any new edges.

What follows is the central lemma of our work, the proof is by induction on the structure of the co-graph.

Lemma 3 *For every co-graph G and every Steiner path cover C for G with respect to a set T of terminal vertices it holds that $p(G) + s(G) \leq \mathfrak{p}(C) + \mathfrak{s}(C)$*

Proof *(By Induction)* The statement is obviously valid for all co-graphs which consist of only one vertex. Let us assume that the statement is valid for co-graphs of n vertices.

Let $G = A \otimes B$ be a co-graph that consists of more than n vertices, where A and B are vertex-disjoint co-graphs of at most n vertices each. Without loss of generality let $|T_A| \leq |T_B|$.

1. Let $X(A)$ denote the vertices of A used in Cover C, and let D denote the cover for B that we obtain by removing the vertices of $X(A)$ from cover C. By the induction hypothesis, it holds that $p(B) + s(B) \leq \mathfrak{p}(D) + \mathfrak{s}(D)$.
2. Let $nt(X(A))$ denote the number of non-terminal vertices of $X(A)$. By Theorem 1 it holds that $\mathfrak{s}(C) = \mathfrak{s}(D) + nt(X(A))$ and $\mathfrak{p}(C) = \mathfrak{p}(D) - |T_A| - nt(X(A))$. Thus, we get $\mathfrak{p}(C) + \mathfrak{s}(C) = \mathfrak{p}(D) + \mathfrak{s}(D) - |T_A|$.

We put these two results together and obtain:

$$p(B) + s(B) - |T_A| \leq \mathfrak{p}(D) + \mathfrak{s}(D) - |T_A| = \mathfrak{p}(C) + \mathfrak{s}(C)$$

To show the statement of the lemma, we first consider the case $p(B) - 1 \leq |V(A)|$. Then it holds that $p(A \otimes B) = 1$. If $|T_A| \geq p(B) - 1$, then $d := |T_A| - (p(B) - 1)$ many Steiner vertices from B can be replaced by terminal vertices from A

Otherwise if $|T_A| < p(B) - 1$, then $-d = (p(B) - 1) - |T_A|$ many Steiner vertices from A are used to combine the paths. Thus, it holds that $s(A \otimes B) \leq s(B) - d$ since the number of Steiner vertices in an optimal cover is at most the number of Steiner vertices in a certain cover. Thus, for $p(A \otimes B) = 1$ we get:

$$\begin{aligned} p(A \otimes B) + s(A \otimes B) \leq 1 + s(B) - d \ &= \ 1 + s(B) - (|T_A| - (p(B) - 1)) \\ &= 1 + s(B) - |T_A| + p(B) - 1 \ \leq \ \mathfrak{p}(C) + \mathfrak{s}(C) \end{aligned}$$

Consider now the case where $p(B) - 1 > |V(A)|$ holds, i.e. not all paths in an optimal cover for B can be combined by vertices of A. By Lemma 1, it holds that $p(A \otimes B) = \max\{1, p(B) - |V(A)|\}$. Thus, for $p(A \otimes B) > 1$ we get:

$$\begin{aligned} p(A \otimes B) + s(A \otimes B) &\leq p(B) - |V(A)| + s(B) + nt(A) \\ &= p(B) + s(B) - |T_A| \ \leq \ \mathfrak{p}(C) + \mathfrak{s}(C) \end{aligned}$$

The non-terminal vertices of A must be used to combine paths of the cover, thus the non-terminal vertices of A become Steiner vertices. □

Obviously, it holds that $s(\bullet_v, \emptyset) = 0$ since a single non-terminal vertex does not define a path in a Steiner path cover of minimum size. It holds that $s(\bullet_v, \{v\}) = 0$ since a single terminal vertex leads to one path of length 0 in a Steiner path cover of minimum size.

Lemma 4 *Let A and B be two vertex-disjoint co-graphs, and let $T_A \subseteq V(A)$, $T_B \subseteq V(A)$ be sets of terminal vertices. Then the following equation holds true:*

$$s(A \otimes B) = s(B) + p(B) - p(A \otimes B) - |T_A| \ \textit{if} \ |T_A| \leq |T_B|$$

Proof First, we show $s(A \otimes B) \leq s(B) + p(B) - p(A \otimes B) - |T_A|$.

By Lemma 3, we know that $s(G) + p(G) \leq \mathfrak{s}(C) + \mathfrak{p}(C)$ holds true for any cover C for co-graph G and set of terminal vertices T. Consider cover C for $A \otimes B$ obtained by an optimal cover D for B in the following way: Use the terminal vertices of A to either combine paths of D or to remove a Steiner vertex of D by replacing $v \notin T$ by some terminal vertex of A in a path like $(\ldots, u, v, w, \ldots) \in D$, where $u, w \in T$. Then, we get $\mathfrak{s}(C) + \mathfrak{p}(C) = s(B) + p(B) - |T_A|$, and by Lemma 3, we get the statement.

$$\begin{aligned} & s(A \otimes B) + p(A \otimes B) \leq s(B) + p(B) - |T_A| \ = \ \mathfrak{s}(C) + \mathfrak{p}(C) \\ \iff \quad & s(A \otimes B) \leq s(B) + p(B) - p(A \otimes B) - |T_A| \end{aligned}$$

We prove now that $s(A \otimes B) \geq s(B) + p(B) - p(A \otimes B) - |T_A|$.

Let $X(A)$ be the vertices of $V(A)$ that are contained in the paths of an optimal cover C for $A \otimes B$. Let D be the cover for B obtained by removing the vertices of

$X(A)$ from C. Then by Theorem 1, the following holds true:

$$
\begin{aligned}
& |X(A)| = nt(X(A)) + |T_A| = \mathfrak{p}(D) - p(A \otimes B) \\
\iff \quad & nt(X(A)) = \mathfrak{p}(D) - p(A \otimes B) - |T_A|
\end{aligned}
$$

Thus, we get:

$$
\begin{aligned}
& \mathfrak{s}(D) = s(A \otimes B) - nt(X(A)) = s(A \otimes B) - \mathfrak{p}(D) + p(A \otimes B) + |T_A| \\
\iff \quad & s(A \otimes B) = \mathfrak{s}(D) + \mathfrak{p}(D) - p(A \otimes B) - |T_A| \\
\Rightarrow \quad & s(A \otimes B) \geq s(B) + p(B) - p(A \otimes B) - |T_A|
\end{aligned}
$$

The implication follows by Lemma 3. □

By Lemmas 1 and 4, and since a co-tree can be computed in linear time from the input co-graph [4], we have shown the following result.

Theorem 2 *The value of a Steiner path cover of minimum cost for a co-graph can be computed in linear time.*

4 Conclusions

In this paper we have shown how to compute recursively the value of a Steiner path cover of minimum cost for a co-graph in linear time. This can be extended to an algorithm which constructs a cover of minimum cost. Since trivially perfect graphs, threshold graphs, and weakly quasi threshold graphs are all co-graphs, our results hold for these graph classes, too. In our future work we want to extend the results to directed co-graphs as defined in [5].

References

1. Abu-Affash, A.K., Carmi, P., Katz, M.J., Segal, M.: The Euclidean bottleneck Steiner path problem and other applications of (α,β)-pair decomposition. Discrete Comput. Geom. **51**(1), 1–23 (2014)
2. Bodlaender, H., Cygan, M., Kratsch, S., Nederlof, J.: Deterministic single exponential time algorithms for connectivity problems parameterized by treewidth. Inf. Comput. **243**, 86–111 (2015)
3. Corneil, D., Lerchs, H., Stewart-Burlingham, L.: Complement reducible graphs. Discrete Appl. Math. **3**, 163–174 (1981)
4. Corneil, D., Perl, Y., Stewart, L.: A linear recognition algorithm for cographs. SIAM J. Comput. **14**(4), 926–934 (1985)
5. Crespelle, C., Paul, C.: Fully dynamic recognition algorithm and certificate for directed cographs. Discrete Appl. Math. **154**(12), 1722–1741 (2006)

6. Custic, A., Lendl, S.: On streaming algorithms for the Steiner cycle and path cover problem on interval graphs and falling platforms in video games. ACM Comput. Res. Repository abs/1802.08577, 9 pp. (2018)
7. Moharana, S.S., Joshi, A., Vijay, S.: Steiner path for trees. Int. J. Comput. Appl. **76**(5), 11–14 (2013)
8. Reich, G., Widmayer, P.: Beyond Steiner's problem: a VLSI oriented generalization. In: Proceedings of Graph-Theoretical Concepts in Computer Science (WG). Lecture Notes in Computer Science, vol. 411, pp. 196–210. Springer, Berlin (1990)
9. Wald, J., Colbourn, C.: Steiner trees in outerplanar graphs. In: Thirteenth Southeastern Conference on Combinatorics, Graph Theory, and Computing, pp. 15–22 (1982)
10. Wald, J., Colbourn, C.: Steiner trees, partial 2-trees, and minimum IFI networks. Networks **13**, 159–167 (1983)
11. Westbrook, J., Yan, D.: Approximation algorithms for the class Steiner tree problem (1995). Research Report

Subset Sum Problems with Special Digraph Constraints

Frank Gurski, Dominique Komander, and Carolin Rehs

Abstract The subset sum problem is one of the simplest and most fundamental NP-hard problems in combinatorial optimization. We consider two extensions of this problem: The subset sum problem with digraph constraint (SSG) and subset sum problem with weak digraph constraint (SSGW). In both problems there is given a digraph with sizes assigned to the vertices. Within SSG we want to find a subset of vertices whose total size does not exceed a given capacity and which contains a vertex if at least one of its predecessors is part of the solution. Within SSGW we want to find a subset of vertices whose total size does not exceed a given capacity and which contains a vertex if all its predecessors are part of the solution. SSG and SSGW have been introduced by Gourvès et al. who studied their complexity for directed acyclic graphs and oriented trees. We show that both problems are NP-hard even on oriented co-graphs and minimal series-parallel digraphs. Further, we provide pseudo-polynomial solutions for SSG and SSGW with digraph constraints given by directed co-graphs and series-parallel digraphs.

1 Introduction

Within the *subset sum problem (SSP)* there is given a set $A = \{a_1, \ldots, a_n\}$ of n items. Every item a_j has a size s_j and there is a capacity c. All values are assumed to be positive integers and $s_j \leq c$ for every $j \in \{1, \ldots, n\}$. The task is to choose a subset A' of A, such that $s(A') := \sum_{a_j \in A'} s_j$ is maximized and the *capacity constraint* holds, i.e.

$$s(A') \leq c. \tag{1}$$

F. Gurski (✉) · D. Komander · C. Rehs
University of Düsseldorf, Institute of Computer Science, Algorithmics for Hard Problems Group, Düsseldorf, Germany
e-mail: frank.gurski@hhu.de

J. S. Neufeld et al. (eds.), *Operations Research Proceedings 2019*, Operations Research Proceedings, https://doi.org/10.1007/978-3-030-48439-2_41

In order to consider generalizations of the subset sum problem we will consider the following constraints for some digraph $G = (A, E)$. For some vertex $y \in A$ we define its predecessors by $N^-(y) = \{x \in A \mid (x, y) \in E\}$. The *digraph constraint* ensures that $A' \subseteq A$ contains y, if it contains at least one predecessor of y, i.e.

$$\forall y \in A \left(N^-(y) \cap A' \neq \emptyset\right) \Rightarrow y \in A'. \tag{2}$$

The *weak digraph constraint* ensures that A' contains y, if it contains every predecessor of y, i.e.

$$\forall y \in A \left(N^-(y) \subseteq A' \wedge N^-(y) \neq \emptyset\right) \Rightarrow y \in A'. \tag{3}$$

This allows us to state the following optimization problems given in [5].

Name Subset sum with digraph constraint (SSG)
Instance A set $A = \{a_1, \ldots, a_n\}$ of n items and a digraph $G = (A, E)$. Every item a_j has a size s_j and there is a capacity c.
Task Find a subset A' of A that maximizes $s(A')$ subject to (1) and (2).

Name Subset sum with weak digraph constraint (SSGW)
Instance A set $A = \{a_1, \ldots, a_n\}$ of n items and a digraph $G = (A, E)$. Every item a_j has a size s_j and there is a capacity c.
Task Find a subset A' of A that maximizes $s(A')$ subject to (1) and (3).

For both problems a subset A' of A is called *feasible*, if it satisfies the prescribed constraints of the problem. Further by $OPT(I)$ we denote the value of an optimal solution on input I.

The complexity for SSG and SSGW restricted to DAGs and oriented trees was considered in [5].

2 SSG and SSGW on Directed Co-graphs

2.1 *Directed Co-graphs*

Definition 1 (Directed Co-graphs, [3]) The class of *directed co-graphs* is recursively defined as follows.

(i) Every digraph on a single vertex $(\{v\}, \emptyset)$, denoted by v, is a *directed co-graph*.
(ii) If $G_1 = (V_1, E_1)$ and $G_2 = (V_2, E_2)$ are two directed co-graphs, then

  (a) the *disjoint union* $G_1 \oplus G_2$, which is defined as the digraph with vertex set $V_1 \cup V_2$ and edge set $E_1 \cup E_2$,
  (b) the *order composition* $G_1 \oslash G_2$, defined by their disjoint union plus all possible edges only directed from V_1 to V_2, and

(c) the *series composition* $G_1 \otimes G_2$, defined by their disjoint union plus all possible edges between V_1 and V_2 in both directions,

are *directed co-graphs*.

Every expression X using these four operations is called a *di-co-expression* and digraph(X) the defined graph. Obviously, for every directed co-graph we can define a tree structure, denoted as *di-co-tree*. The leaves of the di-co-tree represent the vertices of the graph and the inner nodes of the di-co-tree correspond to the operations applied on the subexpressions defined by the subtrees. For every directed co-graph one can construct a di-co-tree in linear time, see [3].

By omitting the series composition within Definition 1 we obtain the class of all *oriented co-graphs*. The class of oriented co-graphs was already analyzed by Lawler in [7] using the notation of *transitive series-parallel (TSP) digraphs*.

Theorem 1 (★[1]) *SSG and SSGW are NP-hard on oriented co-graphs.*

Next, we will show pseudo-polynomial solutions for SSG and SSGW restricted to directed co-graphs.

2.2 Subset Sum with Digraph Constraint (SSG)

Given some instance of SSG such that $G = (A, E)$ is a directed co-graph which is given by some binary di-co-expression X. For some subexpression X' of X let $F(X', s) = 1$ if there is a solution A' in the graph defined by X' such that $s(A') = s$, otherwise let $F(X', s) = 0$. We use the notation $s(X') = \sum_{a_j \in X'} s_j$.

Lemma 1 (★) *Let* $0 \le s \le c$.

1. $F(a_j, s) = 1$ *if and only if* $s = 0$ *or* $s_j = s$.
   In all other cases $F(a_j, s) = 0$.
2. $F(X_1 \oplus X_2, s) = 1$, *if and only if there are some* $0 \le s' \le s$ *and* $0 \le s'' \le s$ *such that* $s' + s'' = s$ *and* $F(X_1, s') = 1$ *and* $F(X_2, s'') = 1$.
   In all other cases $F(X_1 \oplus X_2, s) = 0$.
3. $F(X_1 \oslash X_2, s) = 1$, *if and only if*
   - $F(X_2, s) = 1$ *for* $0 \le s \le s(X_2)$[2] *or*
   - *there is an* $s' > 0$, *such that* $s = s' + s(X_2)$ *and* $F(X_1, s') = 1$.

   In all other cases $F(X_1 \oslash X_2, s) = 0$.
4. $F(X_1 \otimes X_2, s) = 1$, *if and only if* $s = 0$ *or* $s = s(X_1) + s(X_2)$.
   In all other cases $F(X_1 \otimes X_2, s) = 0$.

[1] The proofs of the results marked with a ★ are omitted due to space restrictions.

[2] The value $s = 0$ is for choosing an empty solution in digraph($X_1 \oslash X_2$).

Corollary 1 *There is a solution with sum s for some instance of SSG such that G is a directed co-graph which is given by some binary di-co-expression X if and only if* $F(X, s) = 1$, *i.e.* $OPT(I) = \max\{s \mid F(X, s) = 1\}$.

Theorem 2 (★) *SSG can be solved in directed co-graphs on n vertices and m edges in* $\mathcal{O}(n \cdot c^2 + m)$ *time and* $\mathcal{O}(n \cdot c)$ *space.*

2.3 Subset Sum with Weak Digraph Constraint (SSGW)

In order to get useful informations about the sources within a solution, we use an extended data structure. We consider an instance of SSGW such that $G = (A, E)$ is a directed co-graph which is given by some di-co-expression X. For some subexpression X' of X let $H(X', s, s') = 1$ if there is a solution A' in the graph defined by X' satisfying (1) and (3) such that $s(A') = s$ and the sum of sizes of the sources in A' is s', otherwise let $H(X', s, s') = 0$. We denote by $o(X)$ the sum of the sizes of all sources in digraph(X).

A remarkable difference between SSGW and SSG w.r.t. co-graph operations is the following. When considering $X_1 \oslash X_2$ we can combine solutions A_1 of X_1 satisfying (1) and (3) which do not contain all items of X_1 with solutions A_2 of X_2 satisfying only (1) to obtain solution $A_1 \cup A_2$ of $X_1 \oslash X_2$ satisfying (1) and (3), if $s(A_1) + s(A_2) \leq c$. Furthermore, within $X_1 \otimes X_2$ we can combine solutions A_1 of X_1 satisfying (1) which do not contain all items and solutions A_2 of X_2 satisfying (1) which do not contain all items to obtain solution $A_1 \cup A_2$ of $X_1 \otimes X_2$ satisfying (1) and (3), if $s(A_1) + s(A_2) \leq c$. For a subexpression X' of X let $H'(X', s) = 1$ if there is a solution A' in the digraph defined by X' satisfying (1) such that $s(A') = s$, otherwise let $H'(X', s) = 0$. This allows us to compute the values $H(X', s, s')$ as follows.

Lemma 2 (★) *Let* $0 \leq s, s' \leq c$.

1. $H(a_j, s, s') = 1$ *if and only if* $s = s' = 0$ *or* $s_j = s = s'$.
   In all other cases $H(a_j, s, s') = 0$.
2. $H(X_1 \oplus X_2, s, s') = 1$, *if and only if there are* $0 \leq s_1 \leq s$, $0 \leq s_2 \leq s$, $0 \leq s'_1 \leq s'$, $0 \leq s'_2 \leq s'$, *such that* $s_1 + s_2 = s$, $s'_1 + s'_2 = s'$, $H(X_1, s_1, s'_1) = 1$, *and* $H(X_2, s_2, s'_2) = 1$.
   In all other cases $H(X_1 \oplus X_2, s, s') = 0$.
3. $H(X_1 \oslash X_2, s, s') = 1$, *if and only if*
   - $H(X_1, s, s') = 1$ *for* $1 \leq s < s(X_1)$ *or*
   - $H'(X_2, s) = 1$ *for* $0 \leq s \leq s(X_2)$[3] *and* $s' = 0$ *or*
   - *there are* $1 \leq s_2 \leq s(X_2)$, *such that* $s(X_1) + s_2 = s$, $o(X_1) = s'$, *and* $H(X_2, s_2, o(X_2)) = 1$, *or*

[3]The value $s = 0$ is for choosing an empty solution in digraph($X_1 \oslash X_2$).

- $s = s(X_1) + s(X_2)$ *and* $s' = o(X_1)$, *or*
- *there are* $0 \leq s_1 < s(X_1)$, $0 \leq s_2 \leq s(X_2)$, *such that* $s_1 + s_2 = s$, $H(X_1, s_1, s') = 1$, *and* $H'(X_2, s_2) = 1$.

In all other cases $H(X_1 \oslash X_2, s, s') = 0$.

4. $H(X_1 \otimes X_2, s, 0) = 1$, *if and only if*

- $H'(X_1, s) = 1$ *for* $1 \leq s < s(X_1)$ *or*
- $H'(X_2, s) = 1$ *for* $0 \leq s < s(X_2)$[4] *or*
- *there are* $1 \leq s_2 \leq s(X_2)$, *such that* $s(X_1) + s_2 = s$, *and* $H(X_2, s_2, o(X_2)) = 1$, *or*
- *there are* $1 \leq s_1 \leq s(X_1)$, *such that* $s_1 + s(X_2) = s$, *and* $H(X_1, s_1, o(X_1)) = 1$, *or*
- $s = s(X_1) + s(X_2)$, *or*
- *there exist* $1 \leq s_1 < s(X_1)$ *and* $1 \leq s_2 < s(X_2)$ *such that* $s_1 + s_2 = s$, $H'(X_1, s_1) = 1$, *and* $H'(X_2, s_2) = 1$.

In all other cases $H(X_1 \otimes X_2, s, s') = 0$.

Corollary 2 *There is a solution with sum s for some instance of SSGW such that G is a directed co-graph which is given by some binary di-co-expression X if and only if* $H(X, s, s') = 1$, *i.e.* $OPT(I) = \max\{s \mid H(X, s, s') = 1\}$.

Theorem 3 *SSGW can be solved in directed co-graphs on n vertices and m edges in* $\mathcal{O}(n \cdot c^4 + m)$ *time and* $\mathcal{O}(n \cdot c)$ *space.*

3 SSG and SSGW on Series-Parallel Digraphs

3.1 Series-Parallel Digraphs

We recall the definitions of from [2] which are based on [8].

Definition 2 (Minimal Series-Parallel Digraphs) The class of *minimal series-parallel digraphs*, *msp-digraphs* for short, is recursively defined as follows.

(i) Every digraph on a single vertex $(\{v\}, \emptyset)$, denoted by v, is a *minimal series-parallel digraph*.
(ii) If $G_1 = (V_1, A_1)$ and $G_2 = (V_2, A_2)$ are vertex-disjoint minimal series-parallel digraphs, then the parallel composition $G_1 \cup G_2 = (V_1 \cup V_2, A_1 \cup A_2)$ is a *minimal series-parallel digraph*.
(iii) If G_1 and G_2 are vertex-disjoint minimal series-parallel digraphs and O_1 is the set of vertex of outdegree 0 (set of sinks) in G_1 and I_2 is the set of vertices

[4]The value $s = 0$ is for choosing an empty solution in digraph$(X_1 \otimes X_2)$.

of indegree 0 (set of sources) in G_2, then series composition $G_1 \times G_2 = (V_1 \cup V_2, A_1 \cup A_2 \cup (O_1 \times I_2))$ is a *minimal series-parallel digraph.*

An expression X using these three operations is called an *msp-expression* and digraph(X) the defined graph. For every minimal series-parallel digraph we can define a tree structure, denoted as *msp-tree*. The leaves of the msp-tree represent the vertices of the graph and the inner nodes of the msp-tree correspond to the operations applied on the subexpressions defined by the subtrees. For every minimal series-parallel digraph one can construct a msp-tree in linear time [8].

Theorem 4 (★) *SSG and SSGW are NP-hard on minimal series-parallel digraphs.*

The transitive closure $td(G)$ of a digraph G has the same vertex set as G and for two distinct vertices u, v there is an edge (u, v) in $td(G)$ if and only if there is a path from u to v in G. The transitive reduction $tr(G)$ of a digraph G has the same vertex set as G and as few edges of G as possible, such that G and $tr(G)$ have the same transitive closure. The transitive closure is unique for every digraph. The transitive reduction is not unique for arbitrary digraphs, but for acyclic digraphs. The time complexity of the best algorithm for finding the transitive reduction of a graph is the same as the time to compute the transitive closure of a graph or to perform Boolean matrix multiplication [1]. The best known algorithm to perform Boolean matrix multiplication has running time $\mathcal{O}(n^{2.3729})$ by [4].

Lemma 3 (★) *Given some instance of SSG on acyclic digraph G, then the set of feasible (optimal) solutions of SSG for G and for $tr(G)$ are equal.*

Definition 3 (Series-Parallel Digraphs) Series-parallel digraphs are exactly the digraphs whose transitive closure equals the transitive closure of some minimal series-parallel digraph.

3.2 Subset Sum with Digraph Constraint (SSG)

Given some instance of SSG such that $G = (A, E)$ is a minimal series-parallel digraph which is given by some binary msp-expression X. For some subexpression X' of X let $F(X', s) = 1$ if there is a solution A' in the graph defined by X' such that $s(A') = s$, otherwise let $F(X', s) = 0$. We use the notation $s(X') = \sum_{a_j \in X'} s_j$.

Lemma 4 (★) *Let $0 \leq s \leq c$.*

1. *$F(a_j, s) = 1$ if and only if $s = 0$ or $s_j = s$.*
   In all other cases $F(a_j, s) = 0$.
2. *$F(X_1 \cup X_2, s) = 1$, if and only if there are some $0 \leq s' \leq s$ and $0 \leq s'' \leq s$ such that $s' + s'' = s$ and $F(X_1, s') = 1$ and $F(X_2, s'') = 1$.*
   In all other cases $F(X_1 \cup X_2, s) = 0$.

3. $F(X_1 \times X_2, s) = 1$, *if and only if*

   - $F(X_2, s) = 1$ *for* $0 \leq s \leq s(X_2)$[5] *or*
   - *there is some* $1 \leq s' \leq s(X_1)$ *such that* $s = s' + s(X_2)$ *and* $F(X_1, s') = 1$.

   In all other cases $F(X_1 \times X_2, s) = 0$.

Corollary 3 *There is a solution with sum s for some instance of SSG such that G is a minimal series-parallel digraph which is given by some binary msp-expression X if and only if* $F(X, s) = 1$, *i.e.* $OPT(I) = \max\{s \mid F(X, s) = 1\}$.

Theorem 5 (★) *SSG can be solved in minimal series-parallel digraphs on n vertices and m edges in* $\mathcal{O}(n \cdot c^2 + m)$ *time and* $\mathcal{O}(n \cdot c)$ *space.*

Theorem 6 (★) *SSG can be solved in series-parallel digraphs on n vertices and m edges in* $\mathcal{O}(n \cdot c^2 + n^{2.3729})$ *time and* $\mathcal{O}(n \cdot c)$ *space.*

3.3 Subset Sum with Weak Digraph Constraint (SSGW)

In order to get information on the sinks within a solution, we use an extended data structure. Given some instance of SSGW such that $G = (A, E)$ is an msp-digraph which is given by some binary msp-expression X. For some subexpression X' of X let $H(X', s, s') = 1$ if there is a solution A' in the graph defined by X' such that $s(A') = s$ and the sum of sizes of the sinks in A' is s' otherwise let $H(X', s, s') = 0$. We denote by $i(X)$ the sum of the sizes of all sinks in X.

Lemma 5 (★) *Let* $0 \leq s, s' \leq c$.

1. $H(a_j, s, s') = 1$ *if and only if* $s = s' = 0$ *or* $s_j = s = s'$.
   In all other cases $H(a_j, s, s') = 0$.
2. $H(X_1 \cup X_2, s, s') = 1$, *if and only if there are* $0 \leq s_1 \leq s$, $0 \leq s_2 \leq s$, $0 \leq s_1' \leq s'$, $0 \leq s_2' \leq s'$, *such that* $s_1 + s_2 = s$, $s_1' + s_2' = s'$, $H(X_1, s_1, s_1') = 1$, *and* $H(X_2, s_2, s_2') = 1$.
   In all other cases $H(X_1 \cup X_2, s, s') = 0$.
3. $H(X_1 \times X_2, s, s') = 1$, *if and only if*

   - $0 \leq s \leq s(X_2)$[6] *and* $0 \leq s' \leq s(X_2)$, *such that* $H(X_2, s, s') = 1$ *or*
   - *there are* $1 \leq s_1 \leq s(X_1)$ *and* $1 \leq s_1' < i(X_1)$, *such that* $s_1 = s$, $0 = s'$, *and* $H(X_1, s_1, s_1') = 1$, *or*
   - *there are* $1 \leq s_1 \leq s(X_1)$, *such that* $s_1 + s(X_2) = s$, $i(X_2) = s'$, *and* $H(X_1, s_1, i(X_1)) = 1$, *or*

[5]The value $s = 0$ is for choosing an empty solution in digraph$(X_1 \times X_2)$.

[6]The value $s = s' = 0$ is for choosing an empty solution in digraph$(X_1 \times X_2)$. The values $s > s' = 0$ are for choosing a solution without sinks in digraph$(X_1 \times X_2)$

- *there are* $1 \leq s_1 \leq s(X_1)$, $1 \leq s'_1 < i(X_1)$, $1 \leq s_2 \leq s(X_2)$, *and* $1 \leq s'_2 \leq s(X_2)$, *such that* $s_1 + s_2 = s$, $s'_2 = s'$, $H(X_1, s_1, s'_1) = 1$, *and* $H(X_2, s_2, s'_2) = 1$.

In all other cases $H(X_1 \times X_2, s, s') = 0$.

Corollary 4 *There is a solution with sum s for some instance of SSGW such that G is a minimal series-parallel digraph which is given by some binary msp-expression X if and only if* $H(X, s, s') = 1$, *i.e.* $OPT(I) = \max\{s \mid H(X, s, s') = 1\}$.

Theorem 7 (★) *SSGW can be solved in minimal series-parallel digraphs on n vertices and m edges in* $\mathcal{O}(n \cdot c^4 + m)$ *time and* $\mathcal{O}(n \cdot c)$ *space.*

4 Conclusions and Outlook

The presented methods allow us to solve SSG and SSGW with digraph constraints given by directed co-graphs and (minimal) series-parallel digraphs in pseudo-polynomial time. It remains to find a solution for SSGW in general series-parallel digraphs. By simple counter examples we cannot use Lemma 3 and the recursive structure of minimal series-parallel digraphs. In our future work we want to analyze whether the shown results also hold for other graph classes. Therefore, we want to consider edge series-parallel digraphs from [8]. Furthermore, we intend to consider related problems. These include the two minimization problems which are introduced in [5] by adding a maximality constraint to SSG and SSGW. Moreover, we want to generalize the results for SSG to the partially ordered knapsack problem [6].

References

1. Aho, A., Garey, M., Ullman, J.: The transitive reduction of a directed graph. SIAM J. Comput. **1**(2), 131–137 (1972)
2. Bang-Jensen, J., Gutin, G. (eds.): Classes of Directed Graphs. Springer, Berlin (2018)
3. Crespelle, C., Paul, C.: Fully dynamic recognition algorithm and certificate for directed cographs. Discrete Appl. Math. **154**(12), 1722–1741 (2006)
4. Gall, F.L.: Powers of tensors and fast matrix multiplication. In: Proceedings of the International Symposium on Symbolic and Algebraic Computation (ISSAC), pp. 296–303. ACM, New York (2014)
5. Gourvès, L., Monnot, J., Tlilane, L.: Subset sum problems with digraph constraints. J. Comb. Optim. **36**(3), 937–964 (2018)
6. Johnson, D., Niemi, K.: On knapsacks, partitions, and a new dynamic programming technique for trees. Math. Oper. Res. **8**(1), 1–14 (1983)
7. Lawler, E.: Graphical algorithms and their complexity. Math. Centre Tracts **81**, 3–32 (1976)
8. Valdes, J., Tarjan, R., Lawler, E.: The recognition of series-parallel digraphs. SIAM J. Comput. **11**, 298–313 (1982)

Part X
Health Care Management

A Capacitated EMS Location Model with Site Interdependencies

Matthias Grot, Tristan Becker, Pia Mareike Steenweg, and Brigitte Werners

Abstract A rapid response to emergencies is particularly important. When an emergency call arrives at a site that is currently busy, the call is forwarded to a different site. Thus, the busy fraction of each site depends not only on the assigned area but also on the interactions with other sites. Typically, the frequency of emergency calls differs throughout the city area. The assumption made by existing standard models for ambulance location of an average server busy fraction may over- or underestimate the actual coverage. Thus, we introduce a new mixed-integer linear programming formulation with an upper bound for the busy fraction of each site to explicitly model site interdependencies. We apply our mathematical model to a realistic case of a local EMS provider and evaluate the optimal results provided by the model using a discrete event simulation. The performance of the emergency network is improved compared to existing standard ambulance location models.

1 Introduction

In medical emergencies, patients require fast and qualified assistance. The response time of emergency services until the patient is reached depends primarily on the location of the nearest ambulance. In urban areas, there is a large difference in the number of calls between the inner city and suburbs. As a result, the busy fraction may greatly differ, since vehicles close to the city center are usually faced with calls at a higher frequency in comparison to vehicles located at the suburbs. Thus, the assumption of a uniform system-wide busy fraction may under- or overestimate the actual busyness. A typical policy is that an incoming emergency call is served by the closest site if an ambulance is available. In case all ambulances at that site are busy,

M. Grot (✉) · P. M. Steenweg · B. Werners
Ruhr University Bochum, Faculty of Management and Economics, Bochum, Germany
e-mail: matthias.grot@rub.de

T. Becker
RWTH Aachen University, School of Business and Economics, Aachen, Germany

J. S. Neufeld et al. (eds.), *Operations Research Proceedings 2019*, Operations Research Proceedings, https://doi.org/10.1007/978-3-030-48439-2_42

the call is forwarded to the second closest site and so on [3]. Therefore, different sites have to support each other in case of unavailability. To efficiently use the resources of the emergency medical service (EMS) system, an optimal distribution of ambulances over the urban area should explicitly take interdependencies into account.

In contrast to [2] and [6] who explicitly model vehicle interdependencies, we present an innovative mixed-integer linear programming formulation that prescribes site locations and ambulance allocations taking into account site interdependencies (Sect. 2). By introducing a capacity constraint for each site, the maximum busy fraction is limited. Section 3 presents a computational experiment with real-world data that evaluates the performance of the mathematical formulation on the basis of a discrete event simulation. Finally, a brief conclusion and directions for further research are given in Sect. 4.

2 Mathematical Formulation

Due to the practical relevance, there is a rich body of literature on EMS location problems. Two recent literature reviews provide an extensive overview of many important aspects for EMS systems. Aringhieri et al. [3] discuss the complete care pathway of emergency medical care from emergency call to recovery in the hospital while [1] focus on emergency and non-emergency healthcare facility location problems. Well known models in the EMS context include the deterministic Maximal Covering Location Problem (*MCLP*) [4] and the probabilistic Maximum Expected Covering Location Problem (*MEXCLP*) [5]. In modeling reality, these standard models rely on several simplifying assumptions, e.g. independence between all ambulances (servers). These assumptions are used to deal with the challenges that arise from the probabilistic nature of EMS systems and to obtain tractable linear mathematical formulations. As a result, some of the system dynamics (Fig. 1) are neglected and lead to errors in determining the level of coverage.

To address this issue, we focus on an extension of the *MEXCLP* that includes capacity constraints. Capacity is interpreted as the expected rate of calls that can be covered while an upper bound on the busy fraction is not violated. Constraints on

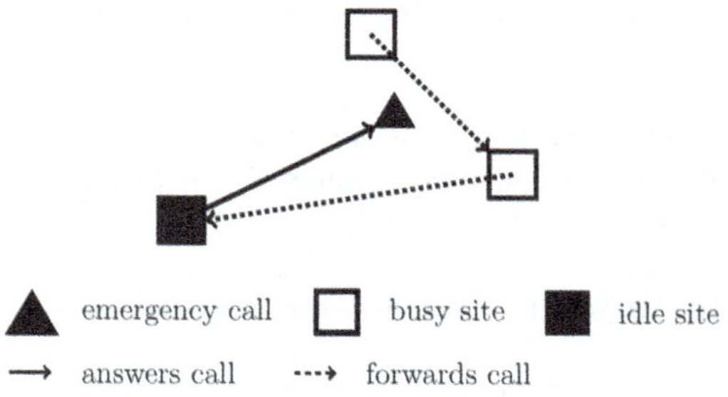

Fig. 1 Scheme of site interdependencies

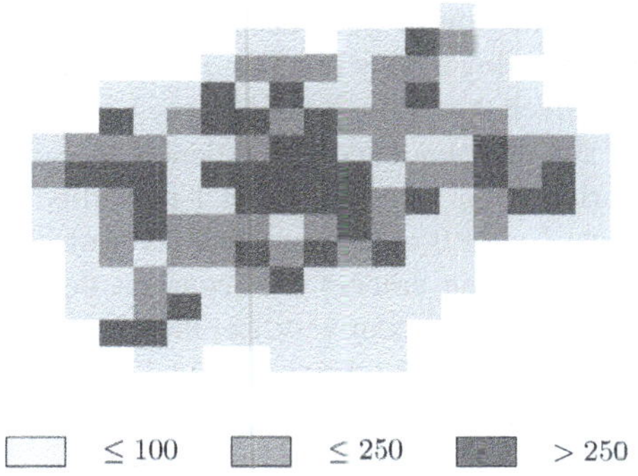

Fig. 2 Spatial distribution of emergency calls over 1 year

the busy fraction of each vehicle were proposed by Ansari et al. [2] and Shariat-Mohaymany et al. [6]. These models limit the busy fraction by introducing a constraint that limits the number of direct calls plus forwarded calls that each vehicle has to serve. Instead, we propose constraints ensuring that the solution of the strategic optimization model respects a maximum busy fraction for each site.

To maintain a linear formulation, models often assume values for the busy fraction that are based only on system-wide demand. However, the frequency of emergency calls at the city center is higher than in the suburbs. Thus, the simplifying assumption of the *MEXCLP* tends to overestimate the expected coverage at the city center and underestimate it in the suburbs (Fig. 2). By introducing an upper bound for the utilization of each site, the error can be limited. Furthermore, it is possible to explicitly model site interdependencies in the capacity constraints, while maintaining a linear model.

The average busy fraction of a server can be derived by the total duration of emergency operations in hours per day, divided by the total available capacity in ambulance hours per day [5]. Then, the average busy fraction q is calculated taking into account that there are a number of V servers and 24 h per day:

$$q = \frac{\bar{t} \cdot \sum_{j \in J} d_j}{24 \cdot V} \tag{1}$$

Let $\bar{t}$ be the average duration (in hours) of an emergency operation and let d_j denote the number of calls per day at demand node $j \in J$. To ensure that the site busy fraction does not exceed an upper bound q^{ub}, the following inequality obtained from Eq. (1) must hold:

$$\bar{t} \cdot \sum_{j \in J} d_j \leq 24 \cdot V \sqrt[V]{q^{ub}} \tag{2}$$

The left-hand side of Eq. (2) denotes the expected time consumed by answering calls during a day if the set of demand nodes J is served. The right-hand side states the maximum utilization of a site given a number of V vehicles and a maximum busy fraction of q^{ub}. The maximum amount of demand that can be served while maintaining a busy fraction q^{ub} thus depends only on the number of vehicles V.

Given this upper bound for the busy fraction, it is possible to relax the assumption of site independence. Let L denote the set of possible levels of coverage. The binary variable y_{ijl} takes the value 1 if facility i provides coverage to demand node j on level l, and 0 otherwise. The level of coverage specifies which priority a facility has when an emergency occurs at demand node j. The servers of each site are numbered consecutively. Let the binary variable x_{ik} take the value 1 if facility i has k servers, and 0 otherwise. Finally, we propose the following capacity constraints under consideration of site interdependencies to ensure a maximum busy fraction of q^{ub}:

$$\sum_{j\in J}\sum_{l\in L} q^{ub^{(l-1)}}(1-q^{ub})c_{ij}d_j y_{ijl} \leq$$
$$\sum_{k\in K}(24\cdot k\sqrt[k]{q^{ub}} - 24\cdot(k-1)\sqrt[(k-1)]{q^{ub}})x_{ik} \quad \forall i \in I \tag{3}$$

The left-hand side of constraints (3) captures the load of site i. Given a first level assignment, a site has to serve a fraction of $1 - q^{ub}$ of the demand (d_j). On each successive level, the probability that a call is forwarded is the product of the busy fractions of all preceding facilities and the probability that the facility is available. The maximum amount of demand that is forwarded is thus defined by $q^{ub^{(l-1)}}(1-q^{ub})$. The average time required to serve an emergency at demand node j when served by facility i is given by c_{ij}. The right-hand side specifies the maximum amount of demand that can be served given k vehicles while a busy fraction of q^{ub} is maintained.

$$\max \quad \sum_{i\in I}\sum_{j\in J}\sum_{l\in L}(1-q^{ub})q^{ub^{(l-1)}}d_j y_{ijl} \tag{4}$$

$$s.t. \quad (3)$$

$$\sum_{i\in I}\sum_{k\in K} x_{ik} = V \tag{5}$$

$$x_{ik} \leq x_{i(k-1)} \quad \forall i \in I, k \in K \mid k > 1 \tag{6}$$

$$\sum_{i\in I} y_{ijl} = 1 \quad \forall j \in J, l \in L \tag{7}$$

$$y_{ijl} \leq x_{i1} \quad \forall i \in I, j \in J, l \in L \tag{8}$$

$$\sum_{l\in L} y_{ijl} \leq 1 \quad \forall i \in I, j \in J \tag{9}$$

$$x_{ik}, y_{ijl} \in \{0,1\} \quad \forall i \in I, j \in J, k \in K, l \in L \tag{10}$$

This *cap*acitated *e*xtension (*CAPE*) of the *MEXCLP* maximizes the expected number of calls (4) reached within the time standard across all demand nodes. Constraint (5) ensures that a number of V servers are distributed among all facilities. Constraints (6) state that, except for the first server, a facility may only be equipped with an additional server k if it has server $k - 1$. Constraints (7) require that every demand node is covered on each level. A facility i can only provide coverage to a demand node j if it is activated, i. e. $x_{i1} = 1$ (Constraints (8)) Additionally, each facility i can provide coverage to demand node j only on a single level l (Constraints (9)). Finally, to model site interdependencies and to ensure a maximum busy fraction q^{ub}, Constraints (3) are added according to the previous discussion.

3 Computational Experiments

To evaluate the computational characteristics and solution quality of the proposed mathematical formulation, extensive computational experiments were conducted. Our computational experiments are based on the real historical call data of an EMS provider in Germany. The performance of the *MEXCLP* (see [5] for the mathematical formulation of the *MEXCLP*) and its capacitated extension *CAPE* as stated in Sect. 2 is compared across a large set of test instances, which has been obtained by systematically varying the number of sites from 4 to 10 and ambulances from 6 to 12. Both model formulations were implemented in Python 3.7 and solved with Gurobi 8.1. In order to determine the upper bound q^{ub} the *CAPE* formulation is solved for busy fractions from 0.05 to 0.65 in 0.025 increments for each instance. The respective q^{ub} value was selected that provides the best solution in terms of simulated coverage. For the *MEXCLP* the system-wide busy fraction q was determined from the data according to the typical assumptions [5]. To evaluate the quality of the solutions obtained by the two mathematical formulations for our test instances, a discrete event simulation (DES) is used. The *Open Street Map Routing Machine* and *Nominatim* were used to estimate the actual street distances between planning squares. Each simulation run then determines the time required to reach the scene of the emergency for all calls over the course of 1 year. The coverage achieved by the two mathematical formulations is shown in Table 1 for each test instance. It is calculated as the mean simulation coverage of 10 runs. In our experiments, an emergency call is considered covered if it is reached within a time threshold of 10 min.

Looking at the results in Table 1, the *CAPE* formulation improves the simulated coverage by 0.84% on average and up to 2.52% at maximum. In 3 out of 39 cases, the *MEXCLP* finds a slightly better solution. One reason might be that other exogenous factors like deviations in travel times also have an impact on the solution. In all other instances, ambulances are located more efficiently towards the city center by the *CAPE* formulation compared to the *MEXCLP*. As a result, the spatially heterogeneous distributed demand can be covered more efficiently. Except for 3 instances with optimality gaps of at most 0.14%, all instances were solved to

Table 1 Comparison of the simulation results for the solutions of the different mathematical formulations

# Sites	# Vehicles	Coverage[a]		
		MEXCLP	CAPE	Δ
4	6	67.03%	**67.85%**	0.81%
4	7	72.23%	**74.18%**	1.94%
4	8	76.89%	**79.42%**	2.52%
4	9	81.52%	**82.89%**	1.38%
4	10	83.30%	**85.57%**	2.26%
4	11	86.14%	**87.02%**	0.88%
4	12	88.61%	**88.86%**	0.25%
5	6	67.84%	**68.79%**	0.94%
5	7	74.01%	**74.43%**	0.42%
5	8	79.01%	**79.58%**	0.57%
5	9	82.26%	**83.26%**	1.00%
5	10	84.39%	**85.33%**	0.94%
5	11	87.60%	**89.09%**	1.48%
5	12	90.04%	**90.13%**	0.09%
6	6	68.40%	**68.53%**	0.12%
6	7	74.51%	**75.38%**	0.87%
6	8	79.62%	**80.47%**	0.85%
6	9	82.80%	**84.28%**	1.48%
6	10	86.18%	**87.22%**	1.04%
6	11	**89.25%**	89.04%	−0.21%
6	12	**90.73%**	90.59%	−0.15%
7	7	74.42%	**74.89%**	0.47%
7	8	**80.24%**	80.21%	−0.03%
7	9	83.82%	**84.20%**	0.38%
7	10	86.46%	**87.17%**	0.71%
7	11	88.68%	**89.50%**	0.82%
7	12	89.87%	**91.34%**	1.47%
8	8	80.05%	**80.32%**	0.27%
8	9	84.09%	**85.23%**	1.14%
8	10	87.42%	**88.32%**	0.90%
8	11	89.50%	**90.06%**	0.55%
8	12	90.85%	**91.48%**	0.62%
9	9	83.62%	**84.63%**	1.01%
9	10	87.11%	**88.08%**	0.97%
9	11	89.70%	**90.33%**	0.63%
9	12	91.05%	**91.24%**	0.19%
10	10	87.72%	**89.25%**	1.53%
10	11	90.31%	**91.49%**	1.18%
10	12	90.97%	**92.34%**	1.37%
			∅	0.84%

[a] Fraction of calls reached within time threshold
Bold values highlight the better result obtained

optimality for the *CAPE* formulation within the time limit of 1800 s. The optimality gap is 0.01% on average and the mean runtime amounts to 187.38 s. The *MEXCLP* could be solved to optimality for all instances with a very low mean run time. The complexity of the *CAPE* compared to the *MEXCLP* stems from the combinatorial nature of the multi-level assignments. As a result, our computational experiments have illustrated the importance of considering site interdependencies. The solution is improved for almost all cases while the runtime increases moderately.

4 Conclusion and Outlook

This paper has introduced the *CAPE* formulation, a capacitated extension of the *MEXCLP* that explicitly models site interdependencies. It represents a mixed-integer linear programming formulation that includes new upper bound chance constraints on the utilization of each site leading to a more accurate representation of the real EMS system. Results of our computational experiments indicate that the choice of sites and their capacities is improved by modeling site interdependencies. Site interdependencies are therefore an important feature in optimizing an EMS system with heterogeneous demand distribution, as their consideration may lead to a more efficient use of resources and thus to better coverage. In comparison to the MEXCLP, the computational effort required for the *CAPE* is higher, since the complexity of the mathematical formulation increases and multiple runs are required to determine sensible values for the upper bound of the site busy fraction. However, our computational results indicate that the computational effort for this type of long-term strategic problem is worthwhile. In further research, the impact of including probabilistic response times on the solution should be evaluated in more detail.

References

1. Ahmadi-Javid, A., Seyedi, P., Syam, S.S.: A survey of healthcare facility location. Comput. Oper. Res. **79**, 223–263 (2017)
2. Ansari, S., Mclay, L.A., Mayorga, M.E.: A maximum expected covering problem for district design. Transport. Sci. **51**(1), 376–390 (2017)
3. Aringhieri, R., Bruni, M.E., Khodaparasti, S., van Essen, J.T.: Emergency medical services and beyond: addressing new challenges through a wide literature review. Comput. Oper. Res. **78**, 349–368 (2017)
4. Church, R., ReVelle, C.: The maximal covering location problem. Pap. Reg. Sci. **32**(1), 101–118 (1974)
5. Daskin, M.: A maximum expected covering location model: formulation, properties and heuristic solution. Transport. Sci. **17**(1), 48–70 (1983)
6. Shariat-Mohaymany, A., Babaei, M., Moadi, S., Amiripour, S.M.: Linear upper-bound unavailability set covering models for locating ambulances: application to Tehran rural roads. Eur. J. Oper. Res. **221**(1), 263–272 (2012)

Online Optimization in Health Care Delivery: Overview and Possible Applications

Roberto Aringhieri

Abstract Health Care Delivery is the process in charge of providing a certain health service addressing different questions (equity, rising cost, ...) in such a way to find a balance between service quality for patients and efficiency for health care providers. The intrinsic uncertainty and the dynamic nature of the processes in health care delivery are among the most challenging issues to deal with. This paper illustrates how online optimization could be a suitable methodology to address such challenges.

Keywords Online optimization · Health care delivery · Radiotherapy · Operating room · Emergency care

1 Introduction

Health Care Delivery is the process in charge of providing a certain health service addressing different questions (equity, rising cost, ...) in such a way to find a balance between service quality for patients and efficiency for health care providers. Such processes or *care pathway* or *patient flow* are defined as "health-care structured multidisciplinary plans that describe spatial and temporal sequences of activities to be performed, based on the scientific and technical knowledge and the organizational, professional and technological available resources" [1]. A care pathway can be conceived as an algorithm based on a flowchart that details all decisions, treatments, and reports related to a patient with a given pathology, with a logic based on sequential stages [2].

The intrinsic uncertainty and the dynamic nature of the health care delivery processes are among the most challenging issues to deal with. Despite its intuitive potential to address such issues, online optimization was little applied to solve health

R. Aringhieri (✉)
Dipartimento di Informatica, Università degli Studi di Torino, Torino, Italy
e-mail: roberto.aringhieri@unito.it; http://di.unito.it/aringhieri

J. S. Neufeld et al. (eds.), *Operations Research Proceedings 2019*, Operations Research Proceedings, https://doi.org/10.1007/978-3-030-48439-2_43

care delivery problems: most of the applications are in the field of the appointment scheduling [3] and in the real-time management of ambulances [4–7]. This paper illustrates how online optimization could be a suitable methodology to address such challenges through a brief overview of the more recent applications.

2 Radiotherapy Patient Scheduling

The Radiotherapy Patient Scheduling (RPS) problem falls into the broader class of multi-appointment scheduling problems in which patients need to visit sequentially multiple or single resource types in order to receive treatment or be diagnosed [8]. A radiotherapy treatment consists in a given number of radiation sessions, one for each (working) day, to be delivered in one of the available time slots. Waiting time is the main critical issue when delivering this treatment to patients suffering from malignant tumors.

In [9] a hybrid method combining stochastic optimization and online optimization has been proposed to deal with the fact that the patients have different characteristics that are not known in advance: their release date (day of arrival), the due date (last day for starting the treatment), their priority, and their treatment duration. Tested on real data, the proposed approach outperforms the policies typically used in treatment centers.

From a combinatorial point of view, the most critical feature of the RPS is the fact that the first treatment usually requires (at least) two slots instead of the only one required by the remaining treatments [10]: as a matter of fact, the first slot in the first day is needed to setup the linear accelerator. Such a feature determines a sort of hook shape (on the right in Fig. 1), which complicates the scheduling with respect to the case without hooks (on the left in Fig. 1).

After proposing a general patient-centered formulation of the problem, several new but simple online algorithms are proposed in [10] to decide the treatment schedule of each new patient as soon as she/he arrives. Such algorithms exploit a pattern whose shape recall that of a water fountain, which was discovered visualizing the exact solution on (really) small instances. Among them, the best algorithm is capable to schedule all the treatments before the due date in both

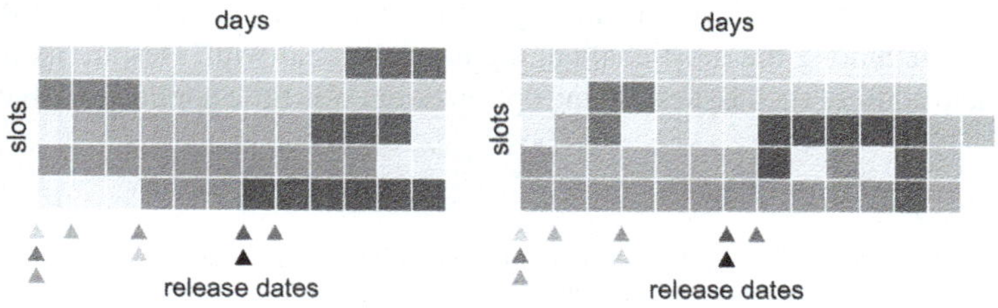

Fig. 1 Hook shape as complicating constraints (from [10])

scenarios considered, whose workload are set to the 80% and 100% of the available slots over a time horizon of 1 year, respectively.

3 Operating Rooms

Operating Room (OR) is probably one of the most treated topics in the health care literature. Topics in this area are usually classified into three phases corresponding to three decision levels, namely strategic (long term), tactical (medium term) and operational (short term). At the operational decision level, the problem is also called "surgery process scheduling" and concern all the decisions regarding the scheduling and the allocation of resources for elective and non-elective surgeries.

In [11], a new online decision problem has been introduced and called Real Time Management (RTM) of ORs, which arises during the fulfillment of the surgery process scheduling. The RTM addresses the problem of supervising the execution of the schedule when uncertainty factors occur, which are mainly two: (1) actual surgery duration exceeds the expected time, and (2) non-elective patients need to be inserted in the OR sessions within a short time. In the former, the more rational decision should be taken regarding the surgery cancellation or the overtime assignment to end he surgery. In the latter, the decision concerns the OR and the instant in which the non-elective patient should be inserted, taking into account the impact on the elective patients previously scheduled. The proposed approach in [11, 12] is a hybrid model for the simulation of the elective and non-elective patient flow (in accordance with [13]) in which the embedded offline and online optimization algorithms address the above decision problems. Further, the simulation allows to evaluate their impact over a time horizon of 2 years considering both patient-centered and facility-centered indices.

The quantitative analysis confirms the well-known trade-off between the number of cancellations and the number of operated patients (or, equivalently, the OR session utilization) showing also that the overtime could be interpreted as a really flexible resources that can be used to bring under control several challenging situations and policy evaluation [14]. Further, an approximated competitive analysis obtained by comparing the online solutions with the offline ones (obtained solving the corresponding mathematical model with Cplex) showed really good competitive ratios confirming also the challenging of dealing with a flow of non-elective patients sharing the ORs with a flow of elective patients.

4 Emergency Care Delivery System

The Emergency Care Delivery System (ECDS) is usually composed of an Emergency Medical Service (EMS) serving a network of Emergency Departments (EDs). ECDS plays a significant role as it constitutes an important access point to the national health system. Overcrowding affects EDs through an excessive number of

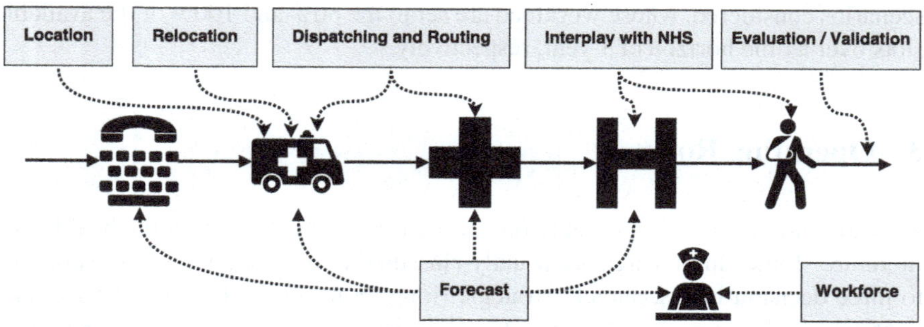

Fig. 2 Emergency Care Pathway (from [16])

patients in the ED, long patient waiting times and patients leaving without being visited, and also imposing to treat patients in hallways and to divert ambulances. From a medical point of view, when the crowding level raises, the rate of medical errors increases and there are delays in treatments, that is a risk to patient safety. Furthermore, overcrowding decreases productivity, causes stress among the ED staff, patient dissatisfaction and episodes of violence [15]. The Emergency Care Pathway (ECP) was introduced in [16] formalizing the idea of the ECDS from an Operations Research perspective. The ED overcrowding can be directly addressed in the following two stages of the ECP: (1) the ambulance rescue performed by the EMS and (2) the management of the ED patient flow (Fig. 2).

As discussed in the introduction, online optimization has been applied to relocation, dispatching and routing of ambulances [4–7]. A particular application of online optimization in dispatching is proposed in [17] in which the ED network of Piedmont Region has been considered. To this end, simple online dispatching rules has been embedded within a simulation model powered by the health care big data of Piedmont Region: exploiting the big data, one can replicate the behavior of the health system modeling how each single patient flows within her/his care pathway. The basic idea is to exploit clusters of EDs (EDs enough close to each other) in such a way to fairly distribute the workload: as soon as a non urgent patient is in charge of the EMS, the ambulance will transport her/him to the ED with minimal workload belonging to the cluster. The results prove a general improvements of the waiting times and crowding reduction, which further improves as soon as the percentage of the patients transported by the EMS increases. This remark has an evident implication in terms of health policy making, which would not have been possible without an analysis of the entire ED network.

Online optimization has been applied to the management of ED patient flow only in [19] even if the ED seems the perfect operative context for its application. A possible reason for this could be probably the wide variety of different patient paths within the ED pathway (Fig. 3) and the missing of data or tools to mine them. This implies strong assumptions and simplifications usually neglecting fundamental aspects, such as the interdependence between activities and, by consequence, the

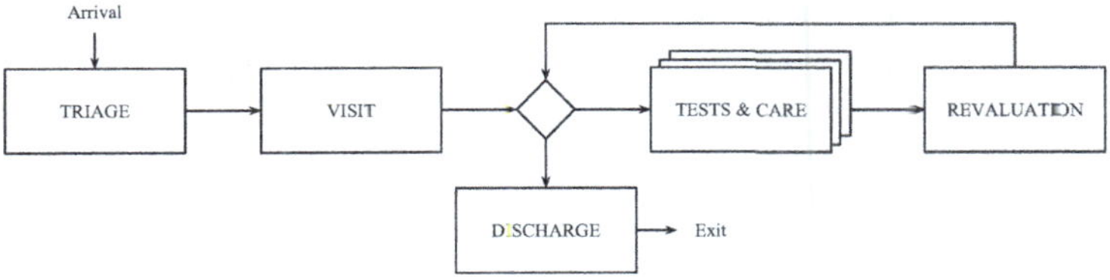

Fig. 3 A generic care pathway for a patient within the ED (from [18])

access to the ED resources. The access to the usually limited ED resources is a challenge issue: as a matter of fact, the resources needed by each patient are known only after the visit (Fig. 3) while are unknown for all the triaged patients, which could be the majority in a overcrowded situation.

The proposed approach in [18] is based on an online optimization approach with lookahead [20] embedded in a simulation model: exploiting the prediction based on ad hoc process mining model [21], the proposed online algorithm is capable to pursue different policies to manage the access of the patients to the critical resources. The quantitative analysis—based on a real case study—proves the feasibility of the proposed approach showing also a consistent crowding reduction on average, during both the whole day and the peak time. The most effective policies are those that tend to promote patients (1) needing specialized visits or exams that are not competence of the ED staff, or (2) waiting for their hospitalization. In both cases the simulation reports a reduction of the waiting times of more than 40% with respect to the actual case study under consideration.

5 Conclusions

The applications briefly reported here support the claim that online optimization could be a suitable methodology to cope with the intrinsic uncertainty and the dynamic nature of the problems arising in the management of health care delivery processes. Summing up our experience in online optimization applied to health care delivery, we can report that a sub optimal decision at the right time could be enough and sometimes better than the best *a priori* stochastic-based decision. And often, this is the only (computational) option we have.

Even if this claim requires more investigation, this seems due to the difficulty of incorporating an efficient model of the stochasticity and the dynamics of the usually complex care pathway in the optimization process. On the other side, online optimization can take advantage of the complete knowledge of the past, and of the reason(s) determining an unattended situation (e.g., a delay of a certain surgery) and its possible (usually limited in number) effects in the nearest future to deal with.

As reported in [20], the effect of the lookahead can be decomposed into an informational and a processual component. The main challenge is therefore to structurally include the lookahead exploiting the (mined) knowledge of the health care process in the online optimization for health care delivery.

Acknowledgments I would like to acknowledge Davide Duma from the University of Turin for his precious collaboration in this area of research.

References

1. Campbell, H., Bradshaw, N., Porteous, M.: Integrated care pathways. Br. Med. J. **316**, 133–144 (1998)
2. De Bleser, L., Depreitere, R., De Waele, K., Vanhaecht, K., Vlayen, J., Sermeus, W.: Defining pathways. J. Nurs. Manag. **14**, 553–563 (2006)
3. van de Vrugt, M.: Efficient healthcare logistics with a human touch. Ph.D. thesis, University of Twente (2016)
4. Jagtenberg, C., van den Berg, P., van der Mei, R.: Benchmarking online dispatch algorithms for emergency medical services. Eur. J. Oper. Res. **258**(2), 715–725 (2017)
5. van Barneveld, T., Jagtenberg, C., Bhulai, S., van der Mei, R.: Real-time ambulance relocation: assessing real-time redeployment strategies for ambulance relocation. Socio-Econ. Plan. Sci. **62**, 129–142 (2018)
6. Nasrollahzadeh, A., Khademi, A., Mayorga, M.: Real-time ambulance dispatching and relocation. Manuf. Serv. Oper. Manag. (2018). https://doi.org/10.1287/msom.2017.0649. Published Online: April 11, 2018
7. Aringhieri, R., Bocca, S., Casciaro, L., Duma, D.: A simulation and online optimization approach for the real-time management of ambulances. In: 2018 Winter Simulation Conference (WSC), vol. 2018, December, pp. 2554–2565. IEEE, Piscataway (2019)
8. Marynissen, J., Demeulemeester, E.: Literature review on multi-appointment scheduling problems in hospitals. Eur. J. Oper. Res. **272**(2), 407–419 (2019)
9. Legrain, A., Fortin, M.A., Lahrichi, N., Rousseau, L.M.: Online stochastic optimization of radiotherapy patient scheduling. Health Care Manag. Sci. **18**(2), 110–123 (2015)
10. Aringhieri, R., Duma, D., Squillace, G.: Pattern-based online algorithms for a general patient-centred radiotherapy scheduling problem. In Health Care Systems Engineering. HCSE 2019, volume 316 of Springer Proceedings in Mathematics and Statistics, pp. 251–262. Springer, Cham (2020)
11. Duma, D., Aringhieri, R.: An online optimization approach for the Real Time Management of operating rooms. Oper. Res. Health Care **7**, 40–51 (2015)
12. Duma, D., Aringhieri, R.: The real time management of operating rooms. In: Operations Research Applications in Health Care Management. International Series in Operations Research & Management Science, vol. 262, pp. 55–79. Springer, Berlin (2018)
13. Dunke, F., Nickel, S.: Evaluating the quality of online optimization algorithms by discrete event simulation. Cent. Eur. J. Oper. Res. **25**(4), 831–858 (2017)
14. Duma, D., Aringhieri, R.: The management of non-elective patients: shared vs. dedicated policies. Omega **83**, 199–212 (2019)
15. George, F., Evridiki, K.: The effect of emergency department crowding on patient outcomes. Health Sci. J. **9**(1), 1–6 (2015)
16. Aringhieri, R., Dell'Anna, D., Duma, D., Sonnessa, M.: Evaluating the dispatching policies for a regional network of emergency departments exploiting health care big data. In: International Conference on Machine Learning, Optimization, and Big Data. Lecture Notes in Computer Science, vol. 10710, pp. 549–561. Springer International Publishing, Berlin (2018)

17. Aringhieri, R., Bruni, M., Khodaparasti, S., van Essen, J.: Emergency medical services and beyond: addressing new challenges through a wide literature review. Comput. Oper. Res. **78**, 349–368 (2017)
18. Duma, D.: Online optimization methods applied to the management of health services. Ph.D. thesis, School of Sciences and Innovative Technologies (2018)
19. Luscombe, R., Kozan, E.: Dynamic resource allocation to improve emergency department efficiency in real time. Eur. J. Oper. Res. **255**(2), 593–603 (2016)
20. Dunke, F., Nickel, S.: A general modeling approach to online optimization with lookahead. Omega (United Kingdom) **63**, 134–153 (2016)
21. Duma, D., Aringhieri, R.: An ad hoc process mining approach to discover patient paths of an emergency department. Flex. Serv. Manuf. J. **32**, 6–34 (2020). https://doi.org/10.1007/s10696-018-9330-1

Part XI
Logistics and Freight Transportation

On a Supply-Driven Location Planning Problem

Hannes Hahne and Thorsten Schmidt

Abstract In this article, a generalized model in the context of logistics optimization for renewable energy from biomass is presented and analyzed. It leads us to the conclusion that demand-driven location planning approaches have to be expanded by a supply-driven one.

Keywords Location planning · Convex optimization · Mixed-integer programming

1 Preliminaries: Location Planning

So far, most of the location planning problems discussed in literature are demand-driven. It means that a set of facilities must be assigned to a set of allowed locations (e. g. modeled as a network) in order to completely satisfy the demands of a set of customer sites (see [1] and [2]). In order to differentiate between feasible solutions, an objective function must be taken into account in most cases. Therefore, these problems are considered as optimization problems (see [3]).

1.1 Use Case: Biogas Plants

Biogas plants are facilities that convert biomass substrates into electricity and by-products, such as waste heat and different types of digestate. The production of substrates in agricultural and urban environments is restricted to specific areas, likewise is the demand for all by-products. In general, these areas and the potential

H. Hahne (✉) · T. Schmidt
Technische Universität Dresden, Dresden, Germany
e-mail: hannes.hahne@tu-dresden.de; thorsten.schmidt@tu-dresden.de; https://tu-dresden.de/ing/maschinenwesen/itla/tl

J. S. Neufeld et al. (eds.), *Operations Research Proceedings 2019*, Operations Research Proceedings, https://doi.org/10.1007/978-3-030-48439-2_44

location of the biogas plants differ. Hence, transport distances are existing (see [4]). The transport of biomass leads to costs, which are increasing (partwise) linear with the transport distance (see [5]). It can be assumed that the biomass transport is realized considering the shortest path. The amount of substrate supply in an area is always limited. Moreover, digestate can be delivered only under legal regulations and its demand is therefore limited as well. If the supply with digestate exceeds the demand it has to be stored costly (see [6] and [7]). The waste heat is partly used during the biogas-genesis itself but to a much greater extent it is available for reuse in heat sinks. The transport of heat is always lossy. Facility types are distinguished by their power capacity and therefore by their substrate demands in one period (see [8]). In addition each type of biogas plant produces a specific amount of by-products during that time. Moreover, each one has its own periodical occurring operating and maintenance costs. A distinction is made between fix and variable costs as well as between fix and variable returns. Fix costs are periodically appearing costs such as capital depreciation, maintenance or substrate costs. Fix returns arise through the sale of electricity in one period. All variable costs and variable returns are depending on the transport distance. The underlying metrics differ between energy transport (heat) and goods transport (substrate and digestate). The main objective is to maximize the profit (see [9]).

2 A Generalized Model

Based on the preliminaries of Sect. 1.1 the following formulation of a generalized model gives a reliable answer to the question: "Which number and types of biogas plants should be opened at potential locations under given demand and supply restrictions, so that a configuration results in maximum profit?"

2.1 Formalization

The model developed in this section is called Biogas Graph Analyzer (BGA). It is defined on a complete, edge-weighted and undirected multigraph G with $G := \{V, E, L\}$. G can also be called a multilayer-network. Let V denote a set of nodes with $V = \{1, \dots, n\}$ and $i, j \in V$. V corresponds to all (potential) locations of substrate suppliers, by-product costumers and biogas plants. Let $E = V \times V$ denote a set of edges with $(i, j) \in E$ and L a set of layers with $L = \{\alpha, \beta\}$ and $l \in L$. Let T be a set of types of biogas plants with $t \in T$ representing a specific type. P denotes a set of products, including all substrates and by-products. Thus, $p \in P$ represents a specific product. For each type of a biogas plant t there is a profit function with $e_t := e(t)$ and a cost function with $f_t := f(t)$ which describes all fixed costs and profits. Furthermore, there is a function for transport costs $c_{T_p} := t(p)$, for storage costs $c_{S_p} := s(p)$ and for profits $r_p := r(p)$. Because each product is transported on a specific layer l, we set $l = \alpha$ if $p =$ heat and $l = \beta$ otherwise. Taking into account that each layer needs its specific distance function m

the following formulation is proposed:

$$m_{ijl_p} := m((i,j),l_p) \begin{cases} 1 & \text{if } i = j \wedge l_p = \alpha \\ [0,1) & \text{if } i \neq j \wedge l_p = \alpha \\ 0 & \text{if } i = j \wedge l_p = \beta \\ \bar{p}_{ij} & \text{shortest path from } i \text{ to } j\ (i \neq j) \wedge l_p = \beta \end{cases}.$$

For a specific biogas plant type t the function $\delta_{tp} := \delta(t,p)$ define its product demand and $\sigma_{tp} := \sigma(t,p)$ its product supply. For a specific location i the function $d_{ip} := d(i,p)$ define its product demand d and $s_{ic} := s(i,p)$ its product supply s. Let y_{jt} denote the number of opened biogas plants of type t at node j with $y_{jt} \in \mathbb{Z}_0^+$. Let χ_{ijp} be a variable that describes the transportation amount of product p from a node i to an opened biogas plant at node j with $\chi_{ijp} \in \mathbb{R}_0^+$. Let x_{ijp} denote the transportation amount of a product p from an opened biogas plant at node j to a customer at node i with $x_{ijp} \in \mathbb{R}_0^+$.

$$\begin{aligned} \varphi(y_{jt},\chi_{ijp},x_{ijp}) = & \sum_{j\in V}\sum_{t\in T} y_{jt}\cdot(e_t - f_t) \\ & + \sum_{i\in V}\sum_{j\in V}\sum_{p\in P \wedge l_p=\alpha} m_{ijp}\cdot x_{ijp}\cdot r_p \\ & - \sum_{i\in V}\sum_{j\in V}\sum_{p\in P \wedge l_p=\beta} m_{ijp}\cdot x_{ijp}\cdot c_{T_p} \\ & - \sum_{i\in V}\sum_{j\in V}\sum_{p\in P \wedge l_p=\beta} m_{ijp}\cdot \chi_{ijp}\cdot c_{T_p} \\ & - \sum_{j\in V}\sum_{t\in T}\sum_{p\in P}\left(y_{jt}\cdot\sigma_{tp} - \sum_{i\in V} x_{ijp}\right)\cdot c_{S_p} \\ \rightarrow \max \end{aligned} \tag{1}$$

$$\sum_{i\in V}\chi_{ijp} = \sum_{t\in T} y_{jt}\cdot\delta_{tp} \qquad \forall j \in J, \forall p \in P \tag{2}$$

$$\sum_{j\in V}\chi_{ijp} \leq s_{ip} \qquad \forall i \in I, \forall p \in P \tag{3}$$

$$\sum_{j\in V} x_{ijp} \leq d_{ip} \qquad \forall i \in I, \forall p \in P \tag{4}$$

$$\sum_{i\in V} x_{ijp} \leq \sum_{t\in T} y_{jt}\cdot\sigma_{tp} \qquad \forall j \in J, \forall p \in P \tag{5}$$

An objective function consisting of five terms is suggested (see Eq. 1): The first term considers the fixed returns and the losses in one period. The second term describes the returns for waste heat. Terms three and four take care of the transportation costs for substrates and by-products. The last term expresses the storage costs for products, in particular by-products. The first restriction (see Eq. 2) expresses that the product demand of all opened plants at one location has to be fulfilled completely and therefore matches the amount of products transported to these plants. Furthermore, the transported amount of products from a specific supplier to all opened plants cannot exceed its inventories. This is expressed through the second restriction (see Eq. 3). The demand of a customer for by-products may not be exceeded, which is stated in the third restriction (see Eq. 4). Finally, the amount of by-products provided by some opened plants at one location must be less or equal than the amount of by-products that are transported away from there (see Eq. 5). The planning problem described by BGA depends heavily on the available amount of substrates in an examined area. It defines the highest number of plants that can be opened. Unlike other location planning problems there are no costumer demands that needs to be satisfied by products physically transported from these plants (e. g. biogas plants are no main sources for digestate/ fertilizer or heat). Rather, there is a potential for biomass of biomass-energy-conversion. It is *not* necessarily stated that this potential of biomass has to be completely used in the converting process. Taking that into account, the resulting location planning problem should be considered as *supply-driven*.

3 Computational Studies

The future energy concept of the Federal Republic of Germany is based on decentralization and renewability. In order to get statistical reliable statements how the currently applicable legal regulations, in interaction with the analyzed logistics constraints, effects the opening of biogas plants an extensive computational study was performed.

3.1 Instance-Generation

To create the needed instances a generator tool was written in Python 2.7. It allows to freely specify all relevant parameters for the networks (e.g. connectivity, distance intervals), customer demands and biomass supplies as well as facility types and their properties. The parameter values have been taken from a wide literature study, including related legal regulations. The generator tool mainly consists of three parts: a network generator that creates the connected and undirected graphs with the needed randomization, functions to allocate supplies and demands over the generated networks and a representation of the model given in Sect. 2.1.

3.2 *Runtime-Analysis*

We used a python library called PuLP (ver. 1.6.10) to communicate with solvers and evaluate the results. Starting with runtime experiments, the time efforts which have to be considered while solving "real world sized" instances were obtained. These tests were done under different solver settings available in GLPK (LP/MIP Solver, ver. 4.65) on an Intel i5 2.2 GHz processor and 8 GB RAM. Four settings have been varied: the backtracking technique, the branching technique, the generation of cuts and the presolving technique. Hence, a 2^4 full factorial design was carried out. Thus, although bigger networks are not usual in the agricultural context, it was analyzed in detail how the network size influences the runtime. Choosing a combination of LP-presolving and branching on the most fractional variable during integer optimization leads to runtimes less than 10 s for all study-relevant problem sizes.

3.3 *Results*

Figure 1 shows the development of the objective function and the ratio of opened systems for two typical network sizes in an agricultural environment with comparatively low demand for heat. The experiments were repeated three times on a 20 node network, each time with 1000 generated instances. Although four different plant types were considered (75, 150, 250, 500 and 750 kW), only 75 and 500 kW plants were opened.

The objective value obviously depends on the transport distances. Our results show, that the median decreases to less then one-third if the average distances are doubled. Furthermore, the ratio of opened 75 to 500 kW plants changes significantly. It turns out, that the opening of decentralized, smaller facilities in more spacious areas seems to be more profitable. This partly contradicts the argument, that larger biogas plants should be built into areas with wide sourcing distances for biomass.

4 Summary and Outlook

Starting from the literature discussion, we presented an use-case in location planning theory, i.e. locating biogas plants. Our proposition of a generalized model leads to the conclusion, that the widely focused demand-driven problems should be extended by a supply-driven one. In order to derive statements from the model, we developed an instance generator tool. Based on multiple instances, we have shown that the calculation efforts of the formulated model are acceptable for all "real world size" instances. Thus, we were able to perform a more extensive computational study. This led to statistical reliable results about the relationship between network size, plant distribution and objective value. So far, we identified some similar problems

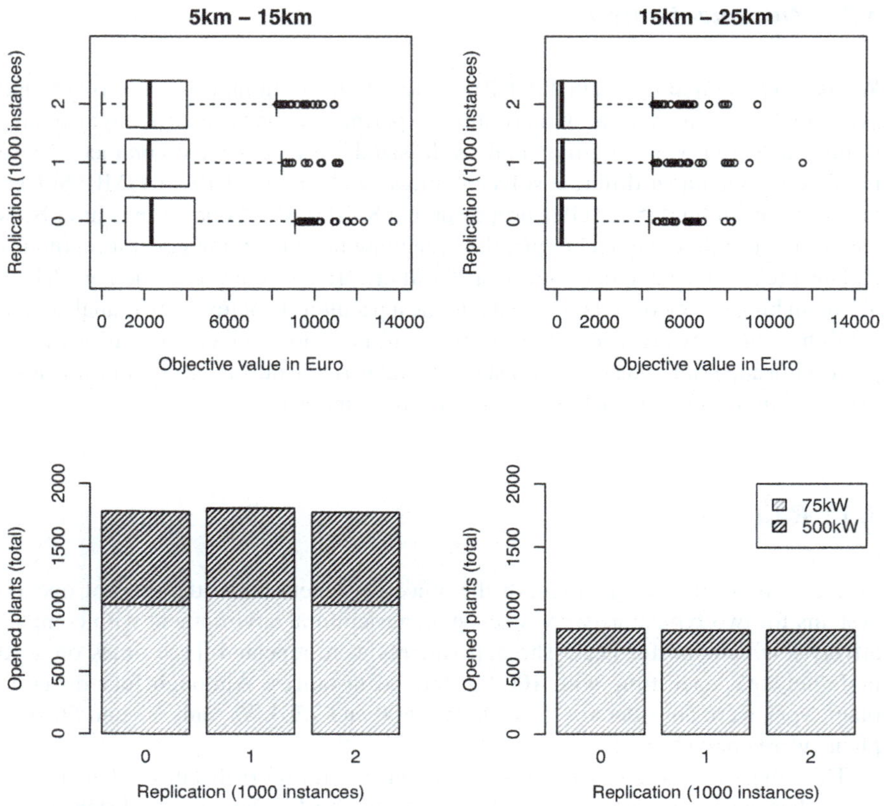

Fig. 1 Computational study results for two different network distance intervals.

(e.g. locations of sawmills or recycling centers) that are solvable with the given approach here, even though the instance generator tool would need to be adapted to these problems.

References

1. Laporte, G., Nickel, S., Saldanha de Gama, F.: Location Science. Springer International Publishing, Berlin (2015)
2. Hanne, T., Dornberger, R.: Location planning and network design. In: Computational Intelligence in Logistics and Supply Chain Management. International Series in Operations Research & Management Science, vol. 244. Springer, Cham (2017)
3. Domschke, W., Drexl, A., Mayer, G., Tadumadze, G.: Betriebliche Standortplanung. In: Tempelmeier, H. (ed.) Planung logistischer Systeme. Fachwissen Logistik. Springer Vieweg, Berlin (2018)

4. Silva, S., Alçada-Almeida, L., Dias, L.C.: Multiobjective programming for sizing and locating biogas plants. Comput. Oper. Res. **83**, 189–198 (2017)
5. Jensen, I.G, Münster, M., Pisinger, D.: Optimizing the supply chain of biomass and biogas for a single plant considering mass and energy losses. Eur. J. Oper. Res. **262**, 744–758 (2017)
6. De Meyer, A., Cattrysse, D., Van Orshoven, J.: A generic mathematical model to optimise strategic and tactical decisions in biomass-based supply chains (OPTIMASS). Eur. J. Oper. Res. **245**, 247–264 (2015)
7. Yuruk, F., Erdogmus, P.: Finding an optimum location for biogas plant. Neural Comput. Appl. **29**, 157–165 (2018)
8. Unal, H.B., Yilmaz, H.I., Miran, B.: Optimal planning of central biogas plants and evaluation of their environmental impacts. Ekoloji **20**, 21–28 (2011)
9. Delzeit, R.: Choice of location for biogas plants in germany - description of a location module. Working Papers Series of the Institute for Food and Resource Economics (2008)

Dispatching of Multiple Load Automated Guided Vehicles Based on Adaptive Large Neighborhood Search

Patrick Boden, Hannes Hahne, Sebastian Rank, and Thorsten Schmidt

Abstract This article describes a dispatching approach for Automated Guided Vehicles with a capacity of greater than one load (referred as Multiple Load Automated Guided Vehicles). The approach is based on modelling the dispatching task as a Dial-a-Ride Problem. An Adaptive Large Neighborhood Search heuristic was employed to find solutions for small vehicle fleets online. To investigate the performance of this heuristic the generated solutions are compared to results of an exact solution method and well established rule-based dispatching policies. The comparison is based on test instances of a use case in semiconductor industry.

Keywords Dispatching · Multiple Load AGV · Dial-a-Ride Problem · Adaptive Large Neighborhood Search

1 Introduction

Starting in the 1950s Automated Guided Vehicles (AGVs) became well established for the automation of transport jobs within logistics and production sites like in automated warehouses and manufacturing plants. Due to the ongoing technical development of AGV system components new aspects for the planning and control of AGV systems are emerging. This article is dedicated to the aspect of dispatching within the control process of Multiple Load Automated Guided Vehicles (MLAGV). The dispatching process requires the assignment of transport jobs to vehicles and the determination of the processing sequence.

In contrast to more common single load AGV, MLAGV are able to transport more than one load. This is accompanied with a significant increase of possibilities to assign transportation jobs to the vehicles. Against the background of high transportation system performance this results in the challenge of efficiently assign-

P. Boden (✉) · H. Hahne · S. Rank · T. Schmidt
Technische Universität Dresden, Dresden, Germany
e-mail: patrick.boden@tu-dresden.de; http://tu-dresden.de/mw/tla

J. S. Neufeld et al. (eds.), *Operations Research Proceedings 2019*, Operations Research Proceedings, https://doi.org/10.1007/978-3-030-48439-2_45

ing/dispatching pending load transport requests to available transport resources. Since this is an online problem, dispatching decisions must be made within seconds (online). In general, AGV dispatching is considered as a challenging algorithmic problem (see Sinriech and Kotlarski [10]).

2 Related Work

Following Egbelu and Tanchoco [2], AGV dispatching strategies can be categorized in work center and vehicle initiated approaches. While in the first case a sufficient amount of vehicles is assumed, in the second case vehicles are a limited resource. Depending on the related category dispatching concepts differ.

This article focuses on the vehicle initiated approach. Thus, a limited amount of vehicles is under investigation. A vehicle needs to select the next transportation job from a list of pending requests. For MLAGV systems the common way to solve the outlined problem is to select the next transportation job for the MLAGV by predefined, static dispatching rules. According to Ho and Chien [3], there are four different subtasks:

Task assignment	Selection of the next task (pickup or drop off),
Pickup selection	Selection of the next pickup location,
Drop off selection	Selection of the next drop off location and
Load selection	Selection of the next load at location.

In case of MLAGV, the rules for the pickup and the drop off selection are similar to the dispatching rules for single load AGVs. Quite common are the *shortest location* and *longest time in system* rules (see Ho and Chien [3]). These static rules typically only consider the next activity to be executed. This, on the one hand, leads to transparent dispatching decisions and short solution computation times. On the other hand, results from literature indicate that high performance reserves for the MLAGV system are still remaining by using these simple approaches (see Sinriech and Kotlarski [10]). Nevertheless recent publications by Ndiaye et al. [6] or Li and Kuhl [4] demonstrate that these rule-based approach is still common for MLAGV dispatching.

In contrast to the rule-based approaches the dispatching problem can be modeled as a Pickup and Delivery Problem (PDP) or as a Dial-a-Ride Problem (DARP) (see Schrecker [9]). Both are variants of the Traveling Salesman Problem with some additional constraints which makes both NP-hard. The PDP and DARP allow the consideration of multiple vehicles, sequence restrictions for the transport loads (to ensure that a load is picked up before it will be dropped off) and time windows for pickup and drop off tasks.

Since the DARP also takes constraints relating to driving times for loads and vehicles into account (important in the logistics context), this article focuses on that approach. We follow the Mixed Integer Programming formulation of Cordeau [1]. The survey of Molenbruch et al. [5] provides a comprehensive overview on aspects of modelling and solution generation. Following them, a common approach to determine exact solutions for PDP and DARP is the Branch and Cut (B&C) algorithm. This approach leads to optimal solutions but is only applicable for small problem sizes due to extensive computational times.

3 Heuristic Approach

In order to be able to dispatch an MLAGV fleet under real world conditions, a heuristical, non optimal solution approach was identified and implemented. Our decision bases on a literature survey of Molenbruch et al. [5]. They showed that metaheuristics based on local search techniques are common to solve such routing problems. We choose the Adaptive Large Neighborhood Search (ALNS) approach presented by Ropke and Pisinger [8] since they are able to investigate offline problem instances with several hundred transport jobs in a reasonable time. Furthermore, Ropke [7] shows an adequate speed up of the ALNS heuristic by parallelisation.

Because the ALNS was originally developed for the PDP with time windows, further constraints like a maximum ride time for the vehicles were considered to generate feasible solutions for the DARP. These additional conditions were implemented by adjusting the cost function. They were considered as hard-constraints, which lead to ∞ costs in case of a violation. To make the heuristic applicable for online planning, jobs that are already running on the vehicles are considered, too.

The heuristic needs to compute a valid schedule s for a set of vehicles K, containing all requests (already running RR and initially unplanned UR). Minimizing the sum of all request delivery times is the objective. The heuristic can be subdivided into two steps. In a first step, for generating the initial solution we start with $|K|$ empty routes. Next we insert iteratively the remaining transport jobs from UR to the schedule, each at its minimum cost position. This is done always selecting the request that increase the cost of S at least. In a second step (see Algorithm 1), we apply ALNS presented by Ropke and Pisinger [8] to improve the solution.

The algorithm iteratively revises the current solution by removing q transport jobs from the schedule. Transport jobs that are physically already running at a vehicle are prohibited to remove (running requests RR). Afterwards the schedule is reconstructed by reinserting these requests. For removing and reinserting, different neighborhood search heuristics like *remove the worst request* are employed. These sub-heuristics are randomly chosen by adaptive weights that are adjusted by their performance. Performance is measured by whether they contribute to improvement or to finding new solutions. To counter the risk of getting trapped in an local optimum, the ALNS is combined with a Simulated Annealing approach like in Ropke and Pisinger [8]. The algorithm terminates when the specified time limit is reached.

Algorithm 1 Improve initial solution

```
function IMPROVE(initial solution s, running requests RR, amount to remove q)
    solution : s_best = s;
    while stop-criterium not met do
        s' = s;
        remove q requests (not in RR) from s';
        reinsert removed requests into s';
        if evaluation(s') < evaluation(s_best) then
            s_best = s';
        end if
        if accept(s', s) == true then
            s = s'
        end if
    end while
end function
```

4 Use Case

The use case is based on a application of MLAGV in the semiconductor industry. Two MLAGV, each of them with the capacity of two loads, are employed to serve 25 production tools. From a previous material flow simulation study it is known, that up to 45 jobs per hour need to be performed in the area of investigation.

To dispatch the MLAGV fleet several dispatching approaches are employed. Besides three simple rule based dispatching strategies (see Table 1), a B&C algorithm (based on CPLEX 12.8.0) and the ALNS (implemented in Python) are applied. The B&C and the ALNS are terminated by time (3 and 30 s) to test them for online decision making. Both algorithms are also tested with a time limit of 15 min to investigate their behaviour in offline planning situations.

For comparison static test instances based on the use case application are created. The scenarios are catagorized by the number of transport jobs to be done, ranging from 1 to 20. Each scenario needs to be dispatched separately. For each category, 10 different scenarios are considered and appropriate solutions are calculated. The objective was to compute a dispatching decision that minimizes the sum of the load's delivery times. In addition to the calculation of the sequence of pickup and drop off for each load, time constraints need to be considered. It was assumed that each load was ready to transport at the moment when the dispatching decision was made and

Table 1 Investigated rule based dispatching approaches

Abb.	Task determination	Pickup selection	Drop off selection
PSS	Pickup first	Shortest distance	Shortest distance
PLL	Pickup first	Longest time in system	Longest time in system
RRR	Random	Random	Random

that all jobs need to be performed within 30 min. For the rule based dispatching approaches (see Table 1) these time window constraints are neglected. Since they are not considered by these approaches.

5 Computational Results

The results of the computational experiment are summarized in Fig. 1 by illustrating the average deviation to the minimum cost result for the investigated problem sizes. Finding optimal solutions with the B&C algorithm was just possible for scenarios with a small number of transport jobs. Within 3 s problems with up to 3 transport jobs could be solved optimally. With up to 15 min optimal solutions with up to 5 jobs are solved optimally. Generating valid non optimal solutions for larger problem instances was limited by 8 transport jobs.

Compared to B&C, the ALNS heuristic performed better. All optimal solutions found by the B&C algorithm are also detected by the ALNS with a time limit of 3 s. In comparison of all approaches, the ALNS with a 15 min time limit was able to find the minimum solution in each case. The difference between the 3 s and the 15 min variant was quite small with around 1% in the 20 transport jobs scenarios. This indicates that the ALNS is suitable for online decision making for dispatching transport jobs for MLAGV. The solution quality of the rule based dispatching approaches were worse compared to B&C and ALNS. Dispatching based on the

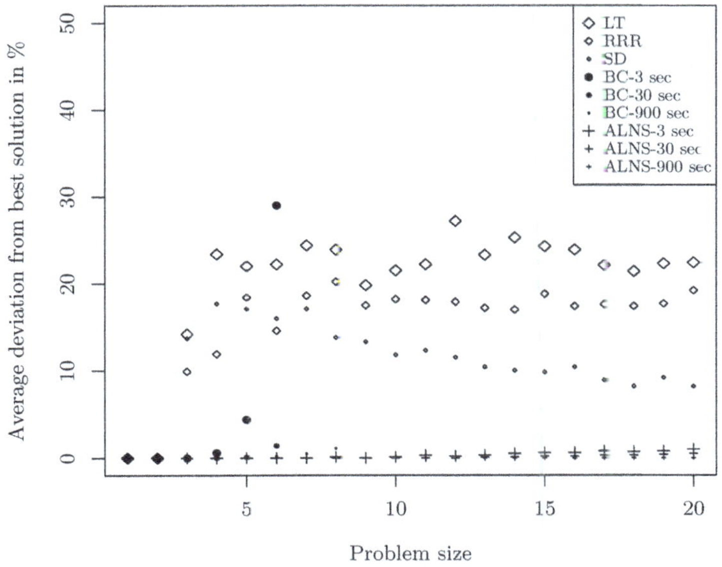

Fig. 1 Comparison of the average deviation from the best found solution by different dispatching approaches. Note: overlapping datapoints around 0% result

PSS rule provided the best results with a decreasing gap to the results from the ALNS with increasing problem size.

6 Conclusion and Outlook

Within this article several approaches for the generation of dispatching decisions for a small MLAGV fleet in a use case from semiconductor industry are compared. We demonstrate that applying heuristics to dispatching based on a tour generation approach (DARP) can improve the solution quality, measured by the sum of delivery times, of classic rule based dispatching approaches. Calculating such a tour by the B&C was limited in terms of the problem size. The ALNS heuristic found in each case the best known solution. Even within a time limit of 3 s the solution quality was near to the results of the ALNS with a time limit of 15 min.

However, the investigation was based on static test instances. For further research a material flow simulation study should be performed to investigate the influence of these dispatching approaches in a dynamic environment.

Acknowledgments The work was performed in the project Responsive Fab, co-funded by grants from the European Union (EFRE) and the Free State of Saxony (SAB).

References

1. Cordeau, J.-F.: A branch-and-cut algorithm for the dial-a-ride problem. Oper. Res. **54**(3), 573–586 (2006)
2. Egbelu, P.J., Tanchoco, J.M.A.: Characterization of automatic guided vehicle dispatching rules. Int. J. Prod. Res. **22**(3), 359–374 (1984)
3. Ho, Y.-C., Chien, S.-H.: A simulation study on the performance of task determination rules and delivery-dispatching rules for multiple-load AGVs. Int. J. Prod. Res. **44**(20), S. 4193–4222 (2006)
4. Li, M.P., Kuhl, M.E.: Design and simulation analysis of PDER: a multiple-load automated guided vehicle dispatching algorithm. In: Proceedings of the 2017 Winter Simulation Conference, pp. 3311–3322 (2017)
5. Molenbruch, Y., et al.: Typology and literature review for dial-a-ride problems. Ann. Oper. Res. **259**, 295–325 (2017)
6. Ndiaye, M.A., et al.: Automated transportation of auxiliary resources in a semiconductor manufacturing facility. In: Proceedings of the 2016 Winter Simulation Conference, pp. 2587–2597 (2016)
7. Ropke, S.: PALNS - a software framework for parallel large neighborhood search. In: 8th Metaheuristic International Conference CDROM (2009)
8. Ropke, S., Pisinger, D.: An adaptive large neighborhood search heuristic for the pickup and delivery problem with time windows. Transp. Sci. **40**, 455–472 (2006)
9. Schrecker, A.: Planung und Steuerung Fahrerloser Transportsysteme: Ansätze zur Unterstützung der Systemgestaltung. In: Gabler Edition Wissenschaft. Produktion und Logistik. Wiesbaden und s.l.: Deutscher Universitätsverlag (2000)
10. Sinriech, D., Kotlarski, J.: A dynamic scheduling algorithm for a multiple load multiple-carrier system. Int. J. Prod. Res. **40**(5), 1065–1080 (2006)

Freight Pickup and Delivery with Time Windows, Heterogeneous Fleet and Alternative Delivery Points

Jérémy Decerle and Francesco Corman

Abstract Several alternatives to home delivery have recently appeared to give customers greater choice on where to securely pickup goods. Among them, the click-and-collect option has risen through the development of locker points for unattended goods pickup. Hence, transportation requests consist of picking up goods from a specific location and dropping them to one of the selected delivery locations. Also, transfer points allow the exchange of goods between heterogeneous vehicles. In this regard, we propose a novel three-index mixed-integer programming formulation of this problem. Experiments are performed on various instances to estimate the benefits of taking into account several transfer points and alternative delivery points instead of the traditional home delivery.

Keywords Freight transport · Alternative delivery · Mixed fleet

1 Introduction

Recently, the rapid and constant growth of online sales has raised new challenges for freight delivery. While the volume of goods to deliver increases, last-mile delivery should improve to meet customers' expectations such as same-day delivery. Nowadays, customers expect to receive their order at a precise time with perfect reliability. However, traditional home delivery is showing its limits. Last-mile secured delivery to home requires the presence of the customer at his home, mostly during office hours when there is most likely no one at home. To deal with that, several alternatives have recently appeared to give customer greater choice on where to pickup goods. Among them, the click-and-collect option has risen through the development of locker points where goods can be delivered and later picked up at any time [3]. To solve this problem, we study a novel variant of the pickup

J. Decerle (✉) · F. Corman
Institute for Transport Planning and Systems, ETH Zürich, Zurich, Switzerland
e-mail: jeremy.decerle@ivt.baug.ethz.ch; francesco.corman@ivt.baug.ethz.ch

J. S. Neufeld et al. (eds.), *Operations Research Proceedings 2019*, Operations Research Proceedings, https://doi.org/10.1007/978-3-030-48439-2_46

and delivery problem. The aim is to determine the optimal way to construct a set of routes in order to satisfy transportation requests. Each transportation request consists of picking up goods from a location and dropping them to one of the selected delivery locations. The main contribution of this planning approach lies in its flexibility, which offers several options for the delivery location, including the customer's home, parcel lockers or stores. An assignment of requests to potential delivery locations is performed prior to the resolution according to the wishes of the customers. Finally, various types of vehicles of different capacity, speed and characteristics are considered for the planning. Also, transfer points allow the exchange of goods between heterogeneous vehicles. Benefits of transfer operations are intuitively guessed over cost reduction and distance savings [2]. Hence, it is also important to determine which vehicle best fits each transportation request. In this regard, we propose a novel three-index mixed-integer programming formulation on a novel variant of the pickup and delivery problem with time windows, heterogeneous fleet, transfer, and alternative delivery points in comparison with an existing four-index formulation of a similar problem without transfer [4]. To estimate the potential gains in terms of traveling time and cost, some experiments are performed using the optimization solver Gurobi.

2 Model Description

2.1 Basic Notation

The freight pickup and delivery problem is modeled on a graph $G = (N, A)$ where N is the set of nodes and A the set of arcs. The sets P and D denote the pickup/delivery locations of each transportation request. In addition, the set C denotes the alternative delivery points. Goods can be transferred between vehicles at a defined set T of transfer points. The set of depot nodes W contains the departure and arrival nodes of the vehicles respectively noted w^- and w^+. As a result, $W = \{w^-, w^+\}$. Each transfer point $r \in T$ is composed of two nodes t_r^- and t_r^+ respectively representing the loading and unloading of goods between vehicles. Thus, $N = W \cup P \cup D \cup C \cup T$.

Concerning the set K of vehicles, the maximal capacity of a vehicle k is defined by the parameter c_k. The types of vehicles (bike, car, ...) are represented by the set V. The assignment of a vehicle k to a type is indicated by the parameter $v_k \in V$. Consequently, the arc $(i, j) \in A$ has a duration of $t_{i,j}^m$ and a cost $\lambda_{i,j}^m$ using the vehicle type $m \in V$.

Each pickup and delivery request i is associated to a load q_i. Pickup nodes are associated with a positive value and delivery nodes with a negative value. The service duration at a node i is represented by the parameter d_i. Each transportation request and each depot is restricted by the parameters e_i and l_i respectively representing the earliest and latest time to service the request i, or in the case of depots their opening hours. To couple the pickup and delivery alternatives, the binary parameter $\omega_{i,j}$ is true if the pickup request $i \in P$ may be delivered to $j \in D \cup C$.

In order to formulate the freight pickup and delivery problem, we use several decision variables. The binary variable $x^k_{i,j} = 1$ if the vehicle k travels from i to j, 0 otherwise. Concerning the delivery location, the variable $y^k_{i,j} = 1$ if the goods picked at the node i are delivered at the location j by the vehicle k, 0 otherwise. Regarding the transfer points, an explicit tracking of each request is required. To do so, the variable $z^k_{i,j} = 1$ if the goods picked at the node i are in the vehicle k when it arrives to the node j. The load of the vehicle k when it arrives at the node i is determined by the variable Q^k_i. Finally, the variable τ_i represents the beginning time of service at a pickup or delivery node i.

2.2 Problem Formulation

The three-index freight pickup and delivery problem with time windows, heterogeneous fleet and alternative delivery points is formulated as follows :

$$min \sum_{k \in K} \sum_{(i,j) \in A} t^{v_k}_{i,j} \cdot \lambda^{v_k}_{i,j} \cdot x^k_{i,j} \tag{1}$$

subject to:

$$\sum_{k \in K} \sum_{\substack{j \in N \\ (i,j) \in A}} x^k_{i,j} = 1 \qquad \forall i \in P \tag{2}$$

$$\sum_{\substack{j \in N \\ (w^-,j) \in A}} x^k_{w^-,j} = 1 \qquad \forall k \in K \tag{3}$$

$$\sum_{\substack{i \in N \\ (i,w^+) \in A}} x^k_{i,w^+} = 1 \qquad \forall k \in K \tag{4}$$

$$\sum_{\substack{j \in N \\ (i,j) \in A}} x^k_{i,j} = \sum_{\substack{j \in N \\ (j,i) \in A}} x^k_{j,i} \qquad \forall i \in N \backslash \{W\}, k \in K \tag{5}$$

$$if \quad x^k_{i,j} = 1 \quad \Rightarrow \quad \tau_j \geq \tau_i + d_i + t^{v_k}_{i,j}$$
$$\forall i \in N \backslash \{T\}, j \in N \backslash \{T\}, k \in K, (i,j) \in A \tag{6}$$

$$if \quad x^k_{i,j} = 1 \quad \Rightarrow \quad Q^k_j \geq Q^k_i + q^k_j \qquad \forall (i,j) \in A, k \in K \tag{7}$$

$$max(0, q_i) \leq Q^k_i \leq min(c_k, c_k + q_i) \qquad \forall i \in N, k \in K \tag{8}$$

$$Q^k_0 = 0 \qquad \forall k \in K \tag{9}$$

$$e_i \leq \tau_i \leq l_i \qquad \forall i \in N \backslash \{T\}, k \in K \tag{10}$$

$$\sum_{k \in K} \sum_{j \in D \cup C} y^k_{i,j} \cdot \omega_{i,j} = 1 \qquad \forall i \in P \tag{11}$$

$$\sum_{k \in K} y^k_{i,j} \leq \omega_{i,j} \qquad \forall i \in P, j \in D \cup C \tag{12}$$

$$y^k_{i,j} \leq \sum_{p \in N} x^k_{p,j} \qquad \forall i \in P, j \in D \cup C, k \in K \tag{13}$$

$$\sum_{i \in N} \sum_{k \in K} x^k_{i,j} \leq \sum_{i \in P} \sum_{k \in K} y^k_{i,j} \qquad \forall j \in D \cup C \tag{14}$$

$$if \quad \sum_{i \in P} \sum_{k \in K} y^k_{i,j} = 0 \quad \Rightarrow \quad \sum_{i \in N} \sum_{k \in K} x^k_{i,j} = 0 \qquad \forall j \in D \cup C \tag{15}$$

$$\sum_{\substack{i \in N \\ (i,t^-_r) \in A}} x^k_{i,t^-_r} = x^k_{t^-_r,t^+_r} \qquad \forall r \in T, k \in K \tag{16}$$

$$\sum_{\substack{i \in N \\ (t^+_r,i) \in A}} x^k_{t^+_r,i} = x^k_{t^-_r,t^+_r} \qquad \forall r \in T, k \in K \tag{17}$$

$$z^k_{i,0} = 0 \qquad \forall i \in P, k \in K \tag{18}$$

$$z^k_{i,n} = 0 \qquad \forall i \in P, k \in K \tag{19}$$

$$if \quad x^k_{p,q} = 1 \quad \Rightarrow \quad z^k_{i,p} = z^k_{i,q}$$
$$\forall i \in P, p \in N, q \in N, k \in K, (p,q) \in A, p \notin T^- \tag{20}$$

$$if \quad x^k_{i,j} = 1 \quad \Rightarrow \quad z^k_{i,j} = 1 \qquad \forall i \in P, j \in N, k \in K, (i,j) \in A \tag{21}$$

$$if \quad x^k_{j,p} = 1 \quad \Rightarrow \quad z^k_{i,p} = 0 \qquad \forall i \in P, j \in D \cup C, k \in K, p \in N, (j,p) \in A \tag{22}$$

$$\sum_{p \in K} z^p_{i,t_r^-} - \sum_{q \in K} z^q_{i,t_r^+} = 0 \qquad \forall r \in T, i \in P \tag{23}$$

$$if \quad \sum_{\substack{p \in N \\ (p,j) \in A}} x^k_{p,j} = 0 \quad \Rightarrow \quad \sum_{i \in P} z^k_{i,j} \leq 0 \qquad \forall k \in K, j \in N \setminus \{D\} \tag{24}$$

Objective function (1) minimizes the total travel time and cost of the vehicles.

Constraint (2) makes sure that each request is picked only once. Departure and arrival of the delivery vehicles to their associated depot are guaranteed by constraints ((3) and (4)). Flow conservation is ensured by constraint (5) while constraint (6) ensures that no subtours occur. Constraints ((7)–(9)) guarantee that a vehicles capacity are not violated throughout their tours. Finally, time window compliance of the transportation requests is verified by constraint (10).

Constraints ((11)–(15)) are related to the delivery alternatives. Constraints ((11) and (12)) verify that each request is delivered only once. The variable y^c_{ij} tracking where each request is delivered is determined by constraint (13) Lastly, constraints ((14) and (15)) ensure that a delivery node is not visited if no goods are delivered to it.

Furthermore, constraints ((16)–(24)) concern the moves of goods between vehicles at transfer points. Flow conservation at transfer points is ensured by constraints ((16) and (17)). Constraints ((18) and (19)) assure that vehicles start and finish their routes empty. The loading and unloading of goods to/from vehicles at the relevant node is verified by constraints ((20)–(22)). Constraint (23) verifies that goods which arrive at a transfer on any vehicle must then leave the transfer on another vehicle. Finally, constraint (24) establishes that if a vehicle do not travel to a specific location, then he cannot transport any goods to this same location.

Finally, the model can be extended by adding several time-related constraints derived from the formulation of [1] specific to the transfer points that are not reported here due to space limits. In addition, non-linear constraints can be easily reformulated as a linear program by means of the usual big-M formulation.

3 Numerical Experiments

3.1 *Instances and Settings of Experiments*

The benchmark contains 10 instances whose pickup and delivery locations are randomly selected places in Zurich city. Each instance contains 10 pickup locations that are paired with 10 delivery locations. Moreover, instances may contain some delivery alternatives, either 0, 1 or 7. Similarly, either 0,1 or 5 transfer points are also considered. Each instance contains only one depot. As a result, each of the 10 initial instances is replicated into 9 different configurations to cover all configurations. One car and one bike are also available to perform the pickup and delivery requests. The cost per km of traveling by car is defined to 0.71 CHF/km and 0.19 CHF/km by bike. In addition, all travel times are vehicle-type-specific and computed using the open-source routing library GraphHopper.

The planning is solved using the commercial solver Gurobi (version 8.1.1) with a time limit of 60 min. Finally, experiments have been performed on a desktop with an Intel® Core™ i7-6700 CPU @3.40Ghz with 16 GB of RAM.

3.2 Computational Results

In this part, the results obtained by the optimization solver Gurobi on the pickup and delivery problem are presented. Based on the results presented in Table 1, the introduction of transfer points for heterogeneous vehicles and alternative delivery points shows promising results. When considering at least one delivery alternative and no transfer points, the objective value decreases by 19.31% from 216.47 to 167.03. Moreover, the cost and time of traveling decrease by 37.13% when considering 7 delivery alternatives compared with none. Indeed, delivery at a mutual location of several orders reduce trips for single-good delivery. In addition, the introduction of one transfer point without any delivery alternative allows decreasing the objective function by 6.95%. Consequently, the results highlight an immediate decrease of the objective value as soon as either one delivery alternative or transfer point is considered. However, the simultaneous introduction of several delivery alternatives and transfer points tend to complicate the resolution of the problem. In such a situation, the solver may not find the optimal or even a feasible solution within the 1-h time limit. Finally, the consideration of delivery alternatives provides the best balance between the solution's quality and the computational time required.

Table 1 Computational results on the pickup and delivery problem

Delivery alternatives	Transfer points	# Feasible solutions	# Optimal solutions	Objective value	Average time (s)
0	0	10	8	216.47	800
	1	10	4	201.26	2486
	5	9	0	221.49	3600
1	0	10	9	167.03	938
	1	9	0	233.26	3600
	5	6	0	214.02	3600
7	0	10	5	127.86	1940
	1	1	0	323.58	3600
	5	0	0	–	3600

4 Conclusion

In this paper, a mixed-integer programming model on a novel variant of the pickup and delivery problem with time windows, heterogeneous fleet, transfer, and alternative delivery points is presented. The problem is modeled in order to decrease the traveling time and cost of the vehicles. The results highlight the decrease of the objective function by taking into account transfer points between heterogeneous vehicles and alternative delivery points instead of the traditional home delivery. Indeed, goods of different customers can be delivered at a mutual location, and thence reduce trips for single-good delivery. The computational results show the decrease of the objective function by more than 19.3% including at least one delivery alternative. In addition, the transfer of goods between heterogeneous vehicles at transfer points is also effective to decrease the cost and time of travel. However, the addition of transfer points to the mathematical model tends to complicate the resolution, in particular by giving more opportunities to transfer goods between heterogeneous vehicles, but also having to track down the delivery goods to know which vehicle is transporting them at every moment. In future works, we would like to integrate public transport as a possible transportation mode to fulfill the requests. In addition, we aim to speed up the solving method in order to solve larger instances in a dynamic context.

References

1. Cortés, C.E., Matamala, M., Contardo, C.: The pickup and delivery problem with transfers: formulation and a branch-and-cut solution method. Eur. J. Oper. Res. **200**(3), 711–724 (2010)
2. Godart, A., Manier, H., Bloch, C., Manier, M.: Milp for a variant of pickup delivery problem for both passengers and goods transportation. In: 2018 IEEE International Conference on Systems, Man, and Cybernetics (SMC), Oct 2018, pp. 2692–2698
3. Morganti, E., Seidel, S., Blanquart, C., Dablanc, L., Lenz, B.: The impact of e-commerce on final deliveries: alternative parcel delivery services in France and Germany. Transp. Res. Procedia **4**, 178–190 (2014)
4. Sitek, P., Wikarek, J.: Capacitated vehicle routing problem with pick-up and alternative delivery (CVRPPAD): model and implementation using hybrid approach. Ann. Oper. Res. **273**(1-2), 257–277 (2019)

Can Autonomous Ships Help Short-Sea Shipping Become More Cost-Efficient?

Mohamed Kais Msakni, Abeera Akbar, Anna K. A. Aasen, Kjetil Fagerholt, Frank Meisel, and Elizabeth Lindstad

Abstract There is a strong political focus on moving cargo transportation from trucks to ships to reduce environmental emissions and road congestion. We study how the introduction of a future generation of autonomous ships can be utilized in maritime transportation systems to become more cost-efficient, and as such contribute in the shift from land to sea. Specifically, we consider a case study for a Norwegian shipping company and solve a combined liner shipping network design and fleet size and mix problem to analyze the economic impact of introducing autonomous ships. The computational study carried out on a problem with 13 ports shows that a cost reduction up to 13% could be obtained compared to a similar network with conventional ships.

Keywords Maritime transportation · Liner shipping network design · Hub-and-spoke · Autonomous ships

1 Introduction

The maritime shipping industry is experiencing a development towards the utilization of autonomous ships that will have a different design than conventional ships. With no crew on-board, it will not be necessary to have a deckhouse nor accommodation. The resulting saved space and weight can be used to carry more

M. K. Msakni · A. Akbar · A. K. A. Aasen · K. Fagerholt (✉)
Norwegian University of Science and Technology, Trondheim, Norway
e-mail: kais.msakni@ntnu.no; kjetil.fagerholt@ntnu.no

F. Meisel
School of Economics and Business, Kiel University, Kiel, Germany
e-mail: meisel@bwl.uni-kiel.de

E. Lindstad
Sintef Ocean AS, Marintek, Trondheim, Norway
e-mail: lindstad@sintef.no

J. S. Neufeld et al. (eds.), *Operations Research Proceedings 2019*, Operations Research Proceedings, https://doi.org/10.1007/978-3-030-48439-2_47

cargoes. In addition to the operational cost reduction, autonomous ships offer ecological advantages as they reduce fuel consumption and carbon dioxide emission. However, international regulations per today lead to challenges in introducing fully autonomous ships because traditionally the captain has the responsibility to ensure the safety of the ship at sea. Conversely, it is expected that Norwegian regulations will be adapted quickly to allow the utilization of autonomous ships nationally. For instance, Norway can be seen as a leading country within autonomous ship technology. Two Norwegian companies, Kongsberg Maritime and Yara, developed one of the world's first commercial autonomous ship, Yara Birkeland. This motivates the development of a shipping network based on mother and daughter ships that utilize advantages of hub and feeder networks, where conventional mother ships sail in international waters, while autonomous daughter ships sail in national waters and tranship the cargoes with the mother ships.

This concept of conventional mother and autonomous daughter ships is in this paper applied on a case study for a Norwegian shipping company that transports containers between Europe and several ports along the Norwegian coast. The aim is to determine an optimal liner shipping network design (LSND) and the optimal fleet of vessels to be deployed in terms of number and size, as well as the route to be sailed for each vessel so that ports demand and weekly services are satisfied. To study the economic impact of introducing autonomous ships, we first solve this problem with only conventional ships and compare this to the solution where daughter ships are autonomous.

Several research studies have been conducted for developing different versions of LSND problems, see for example [1–3]. More recently, Brouer et al. [4] develop a base integer programming model and benchmark suite for the LSND problem. This model is extended by Karsten et al. [5] to include transit time. Karsten et al. [6] base their article on the contribution of [5] and propose the first algorithm that explicitly handles transshipment time limits for all demands. Wang and Meng [7] consider an LSND with transit time restrictions, but the model does not consider transshipment costs. Holm et al. [8] study a LSND problem for a novel concept for short sea shipping where transshipment of daughter and mother ships is performed at suitable locations at sea.

In the following, we describe the LSND problem for the Norwegian case study considered in this paper in more detail, followed by a description of the proposed solution methodology, the computational study and conclusions.

2 Problem Description

The problem is considered from a shipping company's point of view, operating a fleet of mother and daughter ships in a transportation system. The mother ships sail on a main route between ports in Europe and the Norwegian coastline. The daughter ships are autonomous and sail along the Norwegian coastline, serving smaller ports.

The objective is to study the economic impact of introducing autonomous ships in the liner shipping network.

Ports are classified into *main* and *small* ports according to their size and location. Main ports are large ports and placed along the Norwegian coastline and can be served by mother ships and daughter ships. Small ports are not capable of docking a mother ship, and hence, can only be served by daughter ships. The small ports act as feeder ports, and the main ports can act as hubs. Furthermore, the *continental port* is the departure port of the mother ships and is located on the European continent.

The mother ships sail between the European continent and selected main ports, referred to as *main routes*. It is assumed that there is only one mother route. To maintain a weekly service frequency, several mother ships can be deployed on the main route. Also, mother ships sail on southbound journeys, i.e. they start by serving the northernmost Norwegian port, then serve other main ports located further south. The daughter ships sail between ports located along the Norwegian coastline and the routes are referred to as *daughter routes*. One daughter ship can be deployed on a daughter route, and hence, the duration of a daughter route cannot exceed 1 week. It is possible that a main port is not visited by a main route. In such a case, a daughter route must serve this port.

The fleet size and mix are to be determined in the problem. The fleet of daughter ships is heterogeneous, meaning the ships can differ in capacity and corresponding cost components. The fleet of mother ships is homogeneous, and their size is determined a priori so that all cargoes of the system can be transported. The aim is to create a network of main and daughter routes such that the total costs of operating the container transportation system are minimized.

3 Solution Methodology

To solve the underlying problem, a solution methodology is proposed based on two steps. A label setting algorithm is developed to generate candidate routes. These are taken as input in a mathematical model to find the best combination of routes while minimizing costs.

1st Step: Route Generation A label setting algorithm is used to generate all feasible and non-dominated mother and daughter routes.

The main routes are deployed by mother ships and start and end at the main continental port. A stage of mother route generation corresponds to a partial route that starts with the continental port and visits main Norwegian ports. This partial route is extended to visit another Norwegian port not previously visited or to return to the continental port. However, since main Norwegian ports are visited during the southbound journeys, the extension is only allowed to visit a port located further south. Furthermore, there is no time restriction on the main routes, meaning that a partial route can be extended to any main port located further south. To guarantee a weekly service, the number of mother ships that are deployed on a main route is proportional to the number of weeks the completion of the route takes.

A daughter route differs from a mother route and can include both main ports and small ports. The starting and ending port of a daughter route is called transshipment port. A partial daughter route can be extended to any main or small port. The extension is limited by the route time, which must be completed within 1 week. Also, the number of containers on board a daughter ship cannot exceed its capacity. In a case where two partial routes have the same visited ports but in a different order, the partial route with a lower number of containers on board and fuel cost dominates. By doing so, a set of feasible and non-dominated daughter routes is generated.

2nd Step: Path-flow Based Formulation Due to the space limitation of this manuscript, only a verbal description of the mathematical model is given. The input is a set of mother and daughter routes with their corresponding weekly operational costs. The binary decision variables decide on the mother and daughter routes to take up in the solution. The objective function of the 0-1 integer programming model minimizes the total costs of using a fleet of mother and daughter ships of the network. One set of constraints enforce that each main port must be visited by either a daughter or a main route. Another set of constraints ensure that each small port is served by a daughter route. A further constraint is used to select only one main route. A final set of constraints establishes the transshipment relation between main and daughter routes.

4 Computational Results

The test data consists of served ports by the liner shipping company in Norway. In total, there are 13 ports located at the Norwegian coastline and one main continental port located in Maasvlakte, Rotterdam. The cargo demand to and from Rotterdam for each port is provided by the company. This constitutes the normal demand scenario from which two additional scenarios are derived. The second (high) scenario reflects a 40% increase in demand. The third (very high) scenario represents an increase of 100% of demand.

The capacity of the mother ships is an input of the model and can be determined a priori by taking the maximum of the total number of containers going either from or to Rotterdam. For this case study, the capacity equals to 1000 TEU (twenty-foot equivalent unit) is considered for normal and high demand scenarios. A mother ship with this capacity requires an average fuel consumption of 0.61 tonnes/h when sailing at 12 knots, and its weekly charter cost is estimated to 53,000 USD. Conversely, the very high demand scenario requires a mother ship with a higher capacity, 1350 TEU. Such a mother ship requires 0.69 tonnes/h for fuel consumption and has a charter cost of 54,000 USD.

Three different ship types are selected for daughter ships with capacities of 86 TEU, 158 TEU, and 190 TEU, and referred to as small S, medium M and

Table 1 Conventional and autonomous daughter ship types with corresponding capacity, fuel consumption and weekly time charter cost

	Conventional			Autonomous		
	S	M	L	S	M	L
Capacity [TEU]	86	158	190	86	158	190
Fuel consumption [tonnes/h]	0.101	0.114	0.123	0.085	0.097	0.107
Weekly time charter cost [USD]	25	30	35	9.7	15	20.2

Table 2 Results of the liner shipping network design solved with conventional and autonomous daughter ships for the case study with 13 ports

	Normal		High		Very High	
	Conv.	Aut.	Conv.	Aut.	Conv.	Aut.
Total op. costs [k USD]	329	297	399	353	591	515
Fuel cost [k USD]	45	45	46	45	50	48
Time charter costs [k USD]	171	140	196	151	317	242
Cargo handl. Costs [k USD]	111	111	155	155	222	222
Port costs [k USD]	2	1	2	2	2	3
Fleet of daughter ships	1 M, 1 L	2 S, 1 M	3 M	3 M	4 M, 1 L	4 M, 1L
Fleet of mother ships	2 I	2 I	2 I	2 I	3 II	3 II
Nr. main routes	15	15	15	15	15	15
Nr. daughter routes	1553	1553	585	585	221	221
Total solution time [s]	6	6	6	4	4	4
Route generation [s]	5	5	5	4	4	4
Master problem [s]	1	1	1	< 1	< 1	< 1

large L daughter ships, respectively. Autonomous ships have no deckhouse nor accommodation, making these ships lighter with reduced air resistance. This has an immediate impact on the fuel consumption, which is lower than conventional ships. In addition, the time charter costs for autonomous ships exclude the crew cost as we assume full autonomy and estimate an increase of around 5% in construction costs (see [9]). In our experiment, we used the estimation provided by naval architects at SINTEF Ocean for autonomous and conventional ships with a sailing speed set to 8 knots. The different cost parameters for autonomous and conventional daughter ships are given in Table 1.

The route generation procedure has been implemented using MATLAB version R2018a, 64-bit. The mathematical model has been implemented with the commercial optimization software Xpress-IVE version 1.24.24 64-bit, with Xpress Model Version 4.8.3 and Xpress Optimizer Version 33.01.02.

Table 2 presents the results of solving the LSND problem with conventional and autonomous daughter ships. The total operating costs, with the corresponding cost components, are listed, together with the daughter fleet composition, the number of routes generated, and the solutions times. For the fleet composition, small daughter ships are denoted by S, medium by M and large by L. The mother ships deployed for normal and high scenarios are with 1000-TEU capacity and denoted by I, whereas very high scenario requires a larger vessel with 1350-TEU capacity denoted by II. The route generation time is the sum of the route generation time for both main and daughter ships.

As can be seen from Table 2, the number of daughter routes decreases as the demand increases, which is due to the ship capacity restriction of the generated routes. On the other hand, the normal demand scenario shows that the fleet of autonomous daughter ships is larger with smaller ships than the corresponding conventional daughter fleets. This effect is caused by the decrease in cost components for the autonomous ships, and in particular the time charter costs, which motivates the use of autonomous daughter ships in short sea shipping. Indeed, the reduction of the total operational costs is 9.7% for normal demand and reaches up to 12.9% for very high demand scenario.

5 Conclusion

This paper studies the introduction of autonomous ships to a liner shipping network design problem on a Norwegian case study. The liner shipping network consists of a single mother route that sails from a European port to main ports at Norway and daughter routes that do transshipment at main ports to serve small ports (and also main ports not visited by the main route). The problem is solved for a liner shipping company using a path-flow mathematical model. The results show that a reduction between 9 and 13% in operating costs can be obtained by utilizing autonomous daughter ships.

References

1. Agarwal, R., Ergun, H.: Ship scheduling and network design for cargo routing in liner shipping. Transp. Sci. **42**, 175–196 (2008). https://doi.org/10.1287/trsc.1070.0205
2. Álvarez, J.F.: Joint routing and deployment of a fleet of container vessels. Marit. Econ. Logist. **11**, 186–208 (2009). https://doi.org/10.1057/mel.2009.5
3. Reinhardt, L.B., Pisinger, D.: A branch and cut algorithm for the container shipping network design problem. Flex. Serv. Manuf. J. **24**, 349–374 (2011). https://doi.org/10.1007/s10696-011-9105-4
4. Brouer, B.D., Alvarez, J.F., Plum, C.E.M., Pisinger, D., Sigurd, M.M.: A base integer programming model and benchmark suite for liner-shipping network design. Transp. Sci. **48**, 281–312 (2013). https://doi.org/10.1287/trsc.2013.0471
5. Karsten, C.V., Brouer, B.D., Desaulniers, G., Pisinger, D.: Time constrained liner shipping network design. Transp. Res. E Logist. Transp. Rev. **105**, 152–162 (2017). https://doi.org/10.1016/j.tre.2016.03.010
6. Vad Karsten, C., Brouer, B.D., Pisinger, D.: Competitive liner shipping network design. Comput. Oper. Res. **87**, 125–136 (2017). https://doi.org/10.1016/j.cor.2017.05.018
7. Wang, S., Meng, Q.: Liner shipping network design with deadlines. Comput. Oper. Res. **41**, 140–149 (2014). https://doi.org/10.1016/j.cor.2013.08.014

8. Holm, M.B., Medbøen, C.A.B., Fagerholt, K., Schütz, P.: Shortsea liner network design with transhipments at sea: a case study from Western Norway. Flex. Serv. Manuf. J. (2018). https://doi.org/10.1007/s10696-018-9317-y
9. Kretschmann, L., Burmeister, H.-C., Jahn, C.: Analyzing the economic benefit of unmanned autonomous ships: an exploratory cost-comparison between an autonomous and a conventional bulk carrier. Res. Transp. Bus. Manag. **25**, 76–86 (2017). https://doi.org/10.1016/j.rtbm.2017.06.002

Identification of Defective Railway Wheels from Highly Imbalanced Wheel Impact Load Detector Sensor Data

Sanjeev Sabnis, Shravana Kumar Yadav, and Shripad Salsingikar

Abstract The problem solving competition organized by the Railway Application Section of the Institute of Operations Research and Management Sciences (INFORMS) in 2017 was to predict the values of load exerted by wheels on the track, when a currently empty rail car would be loaded in the next trip. The organizers provided Wheel Impact Load Detector (WILD) data i.e. value of peak force along with other input variables such as train number, car number, axle side, wheel age, loaded or empty status etc.

In this work, the original prediction problem is converted into a classification problem on the basis of peak force values in order to detect defects in railroad wheels. Peak force values greater than or equal to threshold value ($\geq$ 90 Kilo Pound Force (kips)) define one class, while its values less than threshold value ($<$ 90 kips) define its complement. Given data set is highly imbalanced as about 99.23% of the peak force values fall below threshold values and the remaining 0.76% peak force values fall into its complement. The statistical methodologies that have been attempted to come up with a classification rule include (1) Zero-Inflated Binomial (ZIB) regression model, (2) ZIB regression with L_1 norm regularization model, and (3) ZIB regression with L_2 norm regularization model. Out of these three methods, ZIB regression with L_2 norm model yielded satisfactory results with False Positive Rate reduced to 13.06% and False Negative Rate to 07.75% with accuracy of 87%.

Keywords Railway · Wheel impact load detectors · Zero-inflated binomial regression · Regularization

S. Sabnis (✉) · S. K. Yadav
Dept. of Mathematics, Indian Institute of Technology Bombay, Mumbai, India

S. Salsingikar
Ind. Engg. & Ops. Res., Indian Institute of Technology Bombay, Mumbai, India

TCS Research, TATA Consultancy Services, Thane, India

J. S. Neufeld et al. (eds.), *Operations Research Proceedings 2019*, Operations Research Proceedings, https://doi.org/10.1007/978-3-030-48439-2_48

1 Introduction

Wheel Load Impact Detectors are track side sensors used in railroads to measure the vertical force imposed by a wheel on the rail. The unit of the force is Kilo Pound Force. Higher the force, more damage will be imposed to wheels and tracks. Railroads use this measure to detect whether there are any defects in wheels. An alarm is triggered if the force value is above a given threshold value and involved cars need to be set out for inspection and repair (if necessary). Although immediate action is not required for wheels with peak force values close to the threshold value, the damage to tracks and wheels might deteriorate (and the force imposed by a defective wheel may further increase) if the defective wheels are not fixed in time. However, setting out a car for inspection when it is loaded will cause loss to railroad company. The shipment might get delayed and the network traffic might suffer from unnecessary disruptions. Therefore, the railroad has an incentive to pro-actively identify problematic wheels/cars and exclude them from being used for future shipments.

In order to monitor movement of goods trains in a complex railway network in any country, it becomes imperative to mine the relevant data and use adequate statistical methodology to unearth hidden information/patterns about the data. Given the importance of the problem, Railway Application Section of INFORMS floated a problem solving competition in 2017 [1]. The problem posed was to predict the values of load exerted by wheels on the track, when a currently empty rail car would be loaded in the next trip using historical WILD data i.e. value of peak force along with other input variables such as train number, axle side, wheel age, loaded or empty status etc. In this data the condition of a particular wheel in a goods train is monitored by keeping track of whether peak force value is above or below certain threshold value.

Very unusual feature of this data set is that its dichotomous response variable Y is abnormally imbalanced in the sense that the percentage of values $Y = 0$ (Peak force < 90 kips) is about 99.2% as opposed to the percentage of values $Y = 1$ (Peak force ≥ 90 kips) is only 0.8% as far as the training data set is concerned, while the respective percentages in testing sample are 99.0% and 1.0%. This peculiar nature of the data at hand renders the use of standard statistical classification procedure, namely, logistic regression model inefficient, and this calls for the use of the Zero-Inflated Binomial (ZIB) model for classification purpose. Further, use of ZIB in conjunction with L_1 and L_2 regularization is expected to give better results. In fact, ZIB with L_2 helps meet the objective of obtaining a reasonably good classification rule.

The organization of this paper is as follows. Section 2 contains detailed discussion about all the aspects of the data set that is being used in this case study. Section 3 discusses ZIB model. Section 3.1 discusses maximum likelihood procedure along with Expectation-Maximization algorithm (EM algorithm) for ZIB model, while Sect. 3.2 covers L_1 and L_2 regularization versions of ZIB model. Section 4 presents summary of numerical results obtained using ZIB model and its L_1 and L_2 regularization versions. Lastly, Sect. 5 contains conclusions.

Table 1 Partitioned data set

Sample	Count	*Actual Y=0* (Peak force < 90 kips)	*Actual Y=1* (Peak force ≥ 90 kips)
Training	541,116	537,301 (99.2%)	3815 (0.8%)
Testing	148,489	147,050 (99.0%)	1439 (1.0%)

2 Data Description

In this data set the binary random variable obtained using peak force threshold value is used as the response variable. Peak force involves two components average force and dynamic force and they are linearly related to the peak force,

$$\text{Peak force} = \text{Average force} + \text{Dynamic force}$$

where average force is the static part of the peak force associated with the weight of the car (when the car is empty, average force should be low; when the car is loaded with heavy commodities, the average force will be fairly high), while dynamic force is the dynamic part of the peak force caused by wheel defects, speed of the cars etc.

The data is comprised of 22 covariates and some of the important ones are (1) the time of peak force readings, (2) the location of the car in the train, (3) the axle number within the car where the wheel is located at, (4) the binary variable that indicates the side of the car that the wheel is attached to (L = left, R = right), (5) the age of the wheel in days, (6) the weight of the car when empty in tons, (7) average force value in kips, (8) peak force value in kips, (9) dynamic force value in kips, (10) the speed of the wheel, (11) the tonnage of the car that wheel is a part of, and (12) the binary variable indicating whether the car is loaded (L) or empty (E).

The given data set is partitioned into training sample and testing sample as shown in 'count' column of Table 1 above. The data in this table is further partitioned according to whether peak force value is less than the threshold value or greater than or equal to it, and they are appearing in columns 3 and 4. This highly imbalanced nature of the data is quite evident from the entries in this table, and this very nature renders standard methods of classification unusable and ineffective. This calls for the use of ZIB models.

3 Zero-Inflated Binomial (ZIB) Regression Model

Zero-inflated data are commonly modeled by mixtures of distributions, the count response variable is assumed to be distributed as a mixture of a count distribution and a distribution with point mass of one at zero [2, 3].

For a vector of responses $\underline{Y} = (Y_1, Y_2, \ldots, Y_N)^{'}$ in ZIB model,

$$Y_i = \begin{cases} 0 & \text{with probability}\,(p_i + (1-p_i)(1-\pi_i)^{n_i}) \\ y_i & \text{with probability}\,(1-p_i)\binom{n_i}{y_i}\pi_i{}^{y_i}(1-\pi_i)^{(n_i-y_i)}, \text{where} y_i = 1, 2, \ldots, n_i \end{cases}$$

In addition, the parameters $\underline{p} = (p_1, p_2, \ldots, p_N)^{'}$ and $\underline{\pi} = (\pi_1, \pi_2, \ldots, \pi_N)^{'}$ are modeled using Generalized Linear Models (GLMs) as $logit(\underline{\pi}) = B\underline{\beta}$ and $logit(\underline{p}) = G\underline{\gamma}$ where B and G denote design matrices. Although this model consists of two distinct parts, the model components must be fitted simultaneously.

Mean and variance of ZIB model are given by $E(Y_i) = n_i\pi_i(1-p_i)$ and $Var(Y_i) = n_i\pi_i(1-p_i)\{n_i\pi_i p_i - \pi_i + 1\}$.

The Expectation Maximization (EM) algorithm is a broadly applicable approach to the iterative computation of Maximum Likelihood (ML) estimates and is particularly useful in a variety of missing data problems. The missing values are replaced by their expected values and then the likelihood is maximized to find the estimates. It is an iterative procedure and each iteration consists of two steps called the Expectation step (E-step) and the Maximization step (M-step).

3.1 Likelihood Function of ZIB Model

Let u_i be the random variable defined as

$$u_i = \begin{cases} 1 & Y_i = 0 \\ 0 & Y_i > 0 \end{cases}$$

As $logit(\pi_i) = B_i\underline{\beta}$, it follows that $\pi_i = \frac{e^{(B_i\underline{\beta})}}{1+e^{(B_i\underline{\beta})}}$.

Also, $logit(p_i) = G_i\underline{\gamma}$, it follows that $p_i = \frac{e^{(G_i\underline{\gamma})}}{1+e^{(G_i\underline{\gamma})}}$ where B_i and G_i are the ith rows of design matrices B and G respectively.

The log-likelihood equation of ZIB is given by

$$\ln\left(L(\underline{\beta}, \underline{\gamma}/\underline{y})\right) = \sum_{i=1}^{N}[u_i \ln\left(e^{(G_i\underline{\gamma})} + (1+e^{(B_i\underline{\beta})})^{-n_i}\right) + (1-u_i)(y_i B_i\underline{\beta} - n_i(1+e^{(B_i\underline{\beta})}) + \ln\binom{n_i}{y_i} - \ln\left(1+e^{(G_i\underline{\gamma})}\right))] \tag{1}$$

The missing data in this problem is a vector of indicator variables $\underline{z} = (z_1, z_2, \ldots, z_N)^{'}$ where $z_i = 1$ when Y_i is from the zero state and $z_i = 0$ when Y_i is from the binomial state. The complete-data log-likelihood is given by

$$l_c(\underline{\beta}, \underline{\gamma}; \underline{y}, \underline{z}) = l_c(\underline{\gamma}; \underline{y}, \underline{z}) + l(\underline{\beta}; \underline{y}, \underline{z}) \tag{2}$$

where

$$l_c(\underline{\gamma};\, \underline{y}, \underline{z}) = \ln\Big(L_c(\underline{\gamma};\, \underline{y}, \underline{z})\Big)$$
$$= \sum_{i=1}^{N}\left[z_i \ln\left(\frac{e^{(G_i\underline{\gamma})}}{1+e^{(G_i\underline{\gamma})}}\right) + (1-z_i)\ln\left(\frac{1}{1-e^{(G_i\underline{\gamma})}}\right)\right] \quad (3)$$

and

$$l(\underline{\beta};\, \underline{y}, \underline{z}) = \ln\Big(L(\underline{\beta};\, \underline{y}, \underline{z})\Big)$$
$$= \sum_{i=1}^{N}(1-z_i)\ln\left(\binom{n_i}{y_i}\left(\frac{e^{(B_i\underline{\beta})}}{1+e^{(B_i\underline{\beta})}}\right)^{y_i}\left(1-\frac{e^{(B_i\underline{\beta})}}{1+e^{(B_i\underline{\beta})}}\right)^{(n_i-y_i)}\right) \quad (4)$$

It is clear that $l_c(\underline{\gamma};\, \underline{y}, \underline{z})$ is a linear function of z_i, $j = 1, 2, \cdots, N$. It is a convenient form for the EM algorithm [4]. Using the EM algorithm, at $(r+1)^{th}$ iteration the E step consists of replacing each z_i by its conditional expectation given $y_i, \underline{\gamma}^{(r)}, \underline{\beta}^{(r)}$ which is $E(z_i/y_i, \underline{\beta}, \underline{\gamma}) = [1 + e^{-G_i\underline{\gamma}}((1+e^{B_i\underline{\beta}})^{-n_i})]^{-1}$.

With this substitution, $l_c(\underline{\beta}, \underline{\gamma};\, \underline{y}, \underline{z})$ is easily maximized with respect to $\underline{\gamma}$ and $\underline{\beta}$ because it is equal to sum of the log-likelihood for an unweighted binomial logistic regression of $\underline{z}^{(r)}$ on G (a term not involving $\underline{\beta}$) and the log-likelihood for a weighted logistic regression of $\underline{y}$ on B, with weight $(\underline{1}_{N\times 1} - \underline{z}^{(r)})$ (a term not involving $\underline{\gamma}$). In this way, the parameters $\underline{\beta}$ and $\underline{\gamma}$ are estimated.

3.2 ZIB Model with L_1 and L_2 Regularizations

Overfitting is a phenomenon which occurs when a model learns the details and noise in the training data to an extent that it negatively impacts the performance of the model on new data. Regularization is a technique which is used to solve the overfitting problem of the statistical and machine learning models. L_1 regularization or Lasso regularization adds a penalty to the error function. The penalty is the sum of the absolute values of parameters in the logistic regression model under consideration.

$$\min_{\underline{\beta}}\left[\sum_{i=1}^{N}(1-z_i)\Big(y_i B_i\underline{\beta} - n_i(1+e^{(B_i\underline{\beta})}) + \ln\binom{n_i}{y_i}\Big) + \lambda\sum_{i=1}^{p}|\beta_i|\right] \quad (5)$$

L_2 regularization or Ridge regularization adds a penalty to the error function. The penalty is the sum of the squares of parameters in the logistic regression model under consideration.

$$\min_{\underline{\beta}}\Big[\sum_{i=1}^{N}(1-z_i)\Big(y_i B_i\underline{\beta}-n_i(1+e^{(B_i\underline{\beta})})+\ln\binom{n_i}{y_i}\Big)+\lambda\sum_{i=1}^{p}{\beta_i}^2\Big] \quad (6)$$

In the above two optimization models λ is the tuning parameter which determines how much to penalize the model.

4 Results

ZIB model and its two variants with L_1 and L_2 norms were trained using the training data set and they were tested on the test data set. The corresponding confusion matrix is given in Table 2 and the comparison of results between these three models is given in Table 3.

The independent use of ZIB model and that of L_1 or L_2 regularization in the GLMs context have been reported extensively in the literature. However, using ZIB model along with L_1 or L_2 regularization may be thought to be a clever methodological step in this paper, and its spin-off is quite evident from the numerical results that have been obtained and given below in Table 3.

Table 2 Confusion Matrix based on ZIB model and its variants

	Actual Y = 0 (Peak force < 90 kips)	*Actual Y = 1* (Peak force ≥ 90 kips)
Predicted Y = 0 (Peak force < 90 kips)	True Negative (TN)	False Negative (FN)
	146,867 (ZIB)	1217 (ZIB)
	146,510 (ZIB + L_1)	1435 (ZIB + L_1)
	127,851 (ZIB + L_2)	109 (ZIB + L_2)
Predicted Y = 1 (Peak force ≥ 90 kips)	False Positive (FP)	True Positive (TP)
	183 (ZIB)	222 (ZIB)
	540 (ZIB + L_1)	4 (ZIB + L_1)
	19,199 (ZIB + L_2)	1330 (ZIB + L_2)

Table 3 Comparison of ZIB model and its variants

Rate	Formula	ZIB	ZIB + L_1	ZIB + L_2
False Positive	FP/FP+TN	00.12%	00.37%	13.06%
True Negative	TN/TN+FP	99.87%	99.63%	86.94%
False Negative	FN/FN+TP	84.57%	99.72%	07.57%
True positive	TP/TP+FN	15.42%	00.28%	92.42%
Specificity	TN/TN+FP	99.87%	99.63%	86.94%
Precision	TP/TP+FP	54.81%	00.74%	06.47%
Accuracy	TP+TN/TP+FN+FP	99.05%	98.67%	87.00%

5 Conclusions

The use of ZIB model, for addressing highly imbalanced nature of the binary data considered here, along with L_2 regularization proves to be extremely efficient in coming up with a classification model having low FNR (7.57%) and FPR (13.06%) with high accuracy (87.00%).

References

1. INFORMS, Railway Application, Problem Solving Competition: Data Analytics for Railroad Empty-to-Load Peak Kips Prediction. https://connect.informs.org/railway-applications/new-item3/problem-solving-competition681/new-item11 (2017). Accessed 08 Nov 2019
2. Hall, B.D.: Zero-inflated Poisson and Binomial regression with random effects: a case study. Biometrics **56**(4), 1030–1039 (2000)
3. Diop, A., Diop, A., Dupuy, J.-F.: Simulation-based inference in a zero-inflated Bernoulli regression model. Commun. Stat. Simul. Comput. **45**(10), 3597–3614 (2016)
4. Dempster, A.P., Laird, N.M., Rubin, D.B.: Maximum likelihood from incomplete data via the EM Algorithm. J. R. Stat. Soc. Ser. B Methodol. **39**(1), 1–38 (1977)

Exact Approach for Last Mile Delivery with Autonomous Robots

Stefan Schaudt and Uwe Clausen

Abstract The Courier, Express and Parcel industry is facing great challenges due to constantly growing parcel volumes combined with increasing customer expectations. A concept to overcome these challenges bases on delivery robots. These robots have a unit capacity and are capable of transporting small goods autonomously on sidewalks. One field of application is the last mile of parcel delivery, where a small fleet of robots could deliver time-critical parcels over small distances. Compared to traditional trucking personnel costs can be saved and due to their autonomy, the robots can operate all day long. In this study, we model the arising optimization problem of scheduling robots and customers. Moreover, we compare two different assignment approaches between robots and micro-depots and evaluate the concept based on test instances.

Keywords Last mile · Delivery robots · Scheduling · Transportation

1 Introduction

A strong growing e-commerce market increases the parcel volumes to be delivered. Besides, cities in all over Western Europe take actions to become more environmentally friendly and limit emissions from transport. Therefore, some cities already established low emission zones. Thus, the Courier, Express and Parcel industry is seeking new innovative and environmentally friendly concepts for last mile deliveries. One concepts for time dependent deliveries are autonomous robots. These robots have been successfully tested in a large number of field tests. Among

S. Schaudt (✉) · U. Clausen
Technical University Dortmund, Dortmund, NRW, Germany
e-mail: stefan.schaudt@tu-dortmund.de; schaudt@itl.tu-dortmund.de
http://www.itl.tu-dortmund.de

J. S. Neufeld et al. (eds.), *Operations Research Proceedings 2019*, Operations Research Proceedings, https://doi.org/10.1007/978-3-030-48439-2_49

others, the manufacturer Starship Technologies tested three robots in cooperation with the parcel delivery company Hermes in Hamburg. Customers were able to choose a 15-min arrival time slot [1]. However, there are also other companies exploring automated delivery solutions including Amazon, Deutsche Post DHL, Domino's Pizza, or Swiss Post.

The research field of delivery robots combined with optimization has only received little attention in the literature. Poeting et al. [4] present the first simulation model of a last mile network involving delivery robots. Besides traditional trucking, a small percentage of all parcels is delivered by autonomous robots. Boysen et al. [2] present an optimization model for a delivery truck that is capable of transporting delivery robots. The truck can load robots at micro-depots and unload them at certain drop-off points. The goal is to minimize the number of late deliveries, which are transported by robots. An exact solution approach for the location routing problem with delivery robots is presented by Sonneberg et al.[5]. In a case study different numbers of compartments for robots are evaluated with focus on the total covered distance and the daily costs.

2 Problem Description

Delivery robots are small autonomous driving vehicles that have a single unit capacity and a speed of 6 km/h. The delivery procedure with an autonomous robot can be described as follows: First a customer, which has ordered a good, selects a delivery time window. When a cut-off time is reached a robot is loaded at a micro-depot and starts its delivery to the customer. After arriving, the robot waits on the sidewalk. Simultaneously, the customer receives an arrival notification together with an opening code. After removing the good, the robot returns to its micro-depot. When arrived the battery is replaced and the next delivery can be started. A schematic representation of this problem is shown in Fig. 1.

Formally described, let $G = (C \cup D, A)$ denote a directed graph, where $C = \{1, 2 \ldots, n\}$ denotes the set of customers, $D = \{n+1, n+2, \ldots, n+m\}$ the micro-depot set, and $A = \{(i, j), (j, i) : i \in C, j \in D\}$ the arc set. Set $R = \{1, 2 \ldots, r\}$ denotes the set of robots. All robots are identical, have a capacity of one and a fixed speed. Each customer $i \in C$ is associated with a due window $[a_i, b_i]$ and a weight w_i. The start time of service at customer $i \in C$ is denoted by s_i. If this time s_i is not within the due window of customer i, the weighted deviation is captured,

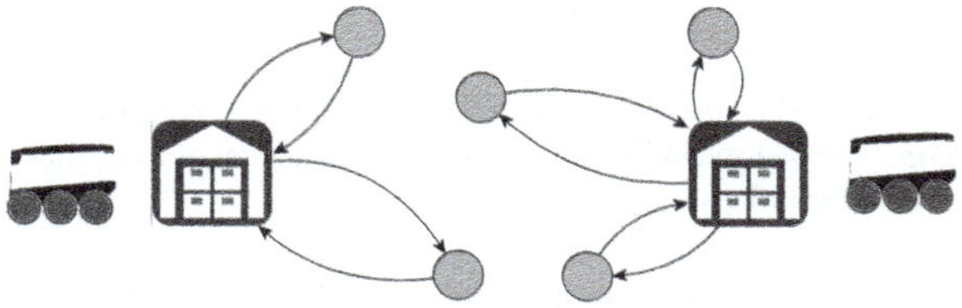

Fig. 1 Problem structure with two robots, five customers, and two micro-depots

$\max\{0, t_i - b_i, a_i - t_i\} \cdot w_i$. Based on this penalty term we are able to define the objective function, which is given by $\sum_{i \in C} \max\{0, t_i - b_i, a_i - t_i\} \cdot w_i$. The travel time from customer $i \in C$ to customer $j \in C$ via the micro-depot $\ell \in D$ is denoted by t_{ij}^{ℓ}, with $t_{ij}^{\ell} = t_{ji}^{\ell}$, and from customer i to micro-depot ℓ by t_i^{ℓ}. The service time τ consists of loading, unloading, and replacing the battery of a robot. All operations take place within the interval $[0, T]$, whereas T denotes the time horizon. Function $d(k)$ maps a robot $k \in R$ to a micro-depot. We distinguish between the following two cases: (1) the robots are equally pre-assigned to the micro-depots, or (2) the assignment between robots and micro-depots is optimized. Note, in both cases the assignment between customers and micro-depots is not given. All times are discrete and deterministic.

A solution π to this problem consists of r customer sequences S_k, $k \in R$, and a vector $\mathbf{s} \in Z_+^n$, where $\mathbf{s}_i$ is the start time of service at customer $i \in C$. If a customer cannot be served within the time horizon T, we set $\mathbf{s}_i = \infty$. A given solution π is feasible, if (1) every customer is contained in exactly one sequence S_k, $k \in R$, (2) for any two consecutive customers $i, i+1 \in S_k$, $k \in R$, the start time of service at customer $i+1$ is at least the start time of service at customer i plus the service time τ and the travel time $t_{ij}^{d(k)}$, and (3) the time $\mathbf{s}_i$ is within the interval $[0, T]$, for all $i \in C$.

In the first case, where robots are pre-assigned to the micro-depots, the problem consists of finding a feasible solution π, such that the penalty term is minimized. This problem can be transformed into an total weighted earliness and tardiness scheduling problem with unrelated parallel machines and machine dependent set up times. This problem and thus, the first case is NP-hard [3]. The processing time of a job $i \in C$ executed on machine $k \in R$, corresponds to the transporting time from micro-depot $d(k)$ to customer $i \in C$. The set up time on machine $k \in R$ after executing job $i \in C$, is equally to the time for unloading, loading, replacing the battery and the return time from $i \in C$ to micro-depot $d(k)$. Hence, the machines can be seen as unrelated. To solve the second case the problem is split into two steps. Firstly, the robots are assigned to the micro-depots and then the scheduling problem from the first case is solved.

3 Mixed Integer Linear Programming Formulation

The first case can be solved with a mixed integer linear programming (MIP) formulation. A MIP formulation is chosen to check which instance size can be solved optimal. Let x_{ij}^k be a binary variable indicating whether or not customer $i \in C \cup \{0\}$ precedes customer $j \in C \cup \{0\}$ supplied by robot $k \in R$ in an optimal solution. Customer 0 represents a dummy customer, which can be seen as the first and the last customer of each sequence. Let u_i and s_i be continuous variables indicating the earliness or tardiness/start time of service at customer $i \in C$. The MIP formulation is formulated as follows:

$$\min \quad \sum_{i\in C} w_i u_i \tag{1}$$

$$\text{s.t.} \quad \sum_{\substack{i\in C\cup\{0\},\\ k\in R}} x_{ij}^k = 1 \qquad \forall\, j \in C \tag{2}$$

$$\sum_{\substack{i\in C\cup\{0\},\\ k\in R}} x_{ji}^k = 1 \qquad \forall\, j \in C \tag{3}$$

$$\sum_{i\in C\cup\{0\}} x_{ij}^k - \sum_{i\in C\cup\{0\}} x_{ji}^k = 0 \qquad \forall\, j \in C, k \in R \tag{4}$$

$$\sum_{i\in C} x_{0i}^k \leq 1 \qquad \forall\, k \in R \tag{5}$$

$$s_i + \tau + \sum_{k\in R} (t_{ij}^{d(k)} x_{ij}^k) \leq s_j + (1 - \sum_{k\in R} x_{ij}^k) M_1 \qquad \forall\, i, j \in C \tag{6}$$

$$\sum_{k\in R} (t_i^{d(k)} x_{0i}^k) \leq s_i \qquad \forall\, i \in C \tag{7}$$

$$s_i + \tau + \sum_{k\in R} (x_{ij}^k t_i^{d(k)}) \leq T \qquad \forall\, i \in C \tag{8}$$

$$a_i - s_i \leq u_i \qquad \forall\, i \in C \tag{9}$$

$$s_i - b_i \leq u_i \qquad \forall\, i \in C \tag{10}$$

$$u_i, s_i \geq 0 \qquad \forall\, i \in C \tag{11}$$

$$x_{ij}^k \in \{0, 1\} \qquad \forall\, i, j \in C, k \in R \tag{12}$$

Objective function (1) minimizes the weighted sum of earliness and tardiness. With (2) and (3) it is ensured that each customer $j \in C$ has a single predecessor/successor. Constraints (4) are flow conservation constraints. Equations (5) ensure that each robot only executes at most one customer sequence. Inequalities (6) guarantee that if robot $k \in R$ first serve customer $j \in C$ and then customer $i \in C$, the start time of service at customer j is at least the start time of service at customer i plus the service time τ and the travel time from i to j via micro-depot $d(k)$, with $M_1 = T - t_i^{\ell_1} - t_j^{\ell_2} + t_{ij}^{\ell_3}$, whereas ℓ_1, ℓ_2 is the micro-depots with minimal distance to customer i and j and ℓ_3 to the micro-depot that maximizes to distance from i to j via ℓ_3. With (7) it is ensured that a robot cannot start before time zero. Constraints

(8) satisfy that the start time of service at a customer plus the service- and return time back to the micro-depot is not greater than the time horizon. With constraints (9) and (10) earliness and tardiness is measured. Finally, the domains of the variables are set by (11) and (12).

3.1 Assignment Problem

For the second case, where robots and micro-depots are not pre-assigned, we adapt the presented MIP formulation. The binary variable y_ℓ^k, $\ell \in D, k \in R$, is 1 if robot k is assigned to micro-depot ℓ and 0 otherwise. We replace constraints (6)–(8) by (13)–(15) and add constraints (16) and domains (17).

$$s_i + \tau + \sum_{\ell \in D} (t_{ij}^\ell y_\ell^k) \leq s_j + (1 - x_{ij}^k) M_2 \qquad \forall\, i, j \in C, k \in R \qquad (13)$$

$$\sum_{\ell \in D} (t_i^\ell y_\ell^k) \leq s_i + (1 - x_{0i}^k) M_3 \qquad \forall\, i \in C, k \in R \qquad (14)$$

$$s_i + \tau + \sum_{\ell \in D} (t_i^\ell y_\ell^k) \leq T + (1 - x_{i0}^k) M_3 \qquad \forall\, i \in C, k \in R \qquad (15)$$

$$\sum_{\ell \in D} (y_\ell^k) = 1 \qquad \forall\, k \in R \qquad (16)$$

$$y_\ell^k \in \{0, 1\} \qquad \forall\, \ell \in D, k \in R \qquad (17)$$

Constraints (13) has the same explanation as constraints (6) except that the micro-depot of robot $k \in R$ is determined by decision variable y_l^k, with $M_2 = T - t_i^{\ell_1} - t_j^{\ell_2} + t_{ij}^{\ell_3}$, where $\ell_1 = \text{argmin}_{\ell \in D}\{t_i^\ell\}$, $\ell_2 = \text{argmin}_{\ell \in D}\{t_j^\ell\}$, and $\ell_3 = \text{argmax}_{\ell \in D}\{t_{ij}^\ell\}$. With inequalities (14) it is ensured that the start time of service at customer $i \in C$ is a least the travel time from the chosen micro-depot to the customer, with $M_3 = t_i^{\ell_1} - t_i^{\ell_2}$, where $\ell_1 = \text{argmax}_{\ell \in D}\{t_i^\ell\}$ and $\ell_2 = \text{argmin}_{\ell \in D}\{t_i^\ell\}$. Constraints (15) guarantee that non of the robots returns after the time horizon T. Equations (16) ensures that each robot is assigned to exactly one of the micro-depot. Finally, the domains of the variables are set by (17).

4 Numerical Experiments

In this section, we describe test instances to evaluate the concept and measure the performance of the MIP formulation. Tests are conducted based on a square layout, with randomly distributed customers. Parameters of the test sets are displayed in Table 1. For a given instance value q is the number of customers divided by five.

Table 1 Parameter setting for the test instances

Square side length	4 km	Due windows distribution	[15,360]
Micro-depots	2	Due window length	15 min
Customers	10,15,20	Time horizon	420 min
Robots	q-1, q, q+1	Earliness/tardiness weight	1
Robot speed	6 km/h	Service time	10 min

Table 2 Computational results

		Pre-assignment			Optimized assignment		
c	r	inf	⌀ time	t/o	inf	⌀ time	t/o
10	1	20	73.45	0	19	117.9	0
	2	0	2.6	0	0	11.0	0
	3	0	0.1	0	0	0.05	0
15	2	0	1465.75	6	0	891.25	5
	3	0	10.8	0	0	3.6	0
	4	0	0.15	0	0	0.1	0
20	3	0	1598.8	8	0	1042.5	5
	4	0	374.05	2	0	135.75	0
	5	0	167.9	0	0	3.7	0

Tests are conducted with two micro-depots, which are located in position [1,1] and [3,3]. For each setting we calculate 20 instances, resulting in a total of 180 instances.

Both formulations are implemented in Java and tested on an Intel Xeon CPU E5-1630 (3.70 GHz) using mixed integer programming solver Gurobi 8.10. The computation time limit is 3600 s. Table 2 shows the results of the computational experiments. The first two columns correspond to the number of customers and robots. Columns three to five are the results from the first case and columns six to eight are the results obtained from the second case. In both cases, the first column represents the number of infeasible instances, the second column shows the average computation time, and the last column indicate the number of instances that reached the computational time limit. The results indicate that a standard MIP approach can handle instances of up to 20 customers if the number of robots is sufficient. A reduction of robots lead to an increase in computational time. Furthermore, for the instances with 15 and 20 customers the computational time of the second case is on average $\sim$2 times faster compared to the first one. Finally, the objective value between the first and the second case decreases by 22.34% on average when using an optimized assignment.

5 Conclusion

We presented a MIP formulation for the problem of delivering goods with a fleet of autonomous robots to a given set of customer with due windows. Moreover, two different assignment rules have been compared. The experiments showed that a small number of robots is sufficient to supply a given set of customers within 7 h. Furthermore, even for a small instances an optimized assignment results in significant savings. However, the size of the test instances is still unsatisfying and further research need to be made. to handle large instances.

References

1. Bertram, I.: Hermes startet Test mit Starship-Robotern in London. https://newsroom.hermesworld.com/hermes-startet-test-mit-starship-robotern-in-london-12325/
2. Boysen, N., Schwerdfeger, S., Weidinger, F.: Scheduling last-mile deliveries with truck-based autonomous robots. Eur. J. Oper. Res. **271**(3), 1085–1099 (2018). https://doi.org/10.1016/j.ejor.2018.05.058
3. Pinedo, M.: Scheduling: Theory, Algorithms, and Systems. Springer Science+Business Media, New York (2008). https://doi.org/10.1007/978-0-387-78935-4
4. Poeting, M., Schaudt, S., Clausen, C.: Simulation of an optimized last-mile parcel delivery network involving delivery Robots. In: Advances in Production, Logistics, and Traffic - Proceedings of the 4th Interdisciplinary Conference on Production Logistics and Traffic 2019, pp. 1–19. Springer Nature Switzerland AG, Basel (2019). https://doi.org/10.1007/978-3-030-13535-5
5. Sonneberg, M.-O., Leyerer, M., Kleinschmidt, A., Knigge, F., Breitner, M.H.: Autonomous unmanned ground vehicles for urban logistics: optimization of last mile delivery operations. In: Intelligent Decision Support and Big Data for Logistics and Supply Chain Management. Hawaii International Conference on System Sciences (2019). https://doi.org/10.24251/HICSS.2019.186

A Solution Approach to the Vehicle Routing Problem with Perishable Goods

Boris Grimm, Ralf Borndörfer, and Mats Olthoff

Abstract This paper focuses on a special case of vehicle routing problem where perishable goods are considered. Deliveries have to be performed until a due date, which may vary for different products. Storing products is prohibited. Since late deliveries have a direct impact on the revenues for these products, a precise demand prediction is important. In our practical case the product demands and vehicle driving times for the product delivery are dependent on weather conditions, i.e., temperatures, wind, and precipitation. In this paper the definition and a solution approach to the Vehicle Routing Problem with Perishable Goods is presented. The approach includes a procedure how historical weather data is used to predict demands and driving times. Its run time and solution quality is evaluated on different data sets given by the MOPTA Competition 2018.

Keywords Vehicle routing problem · Mixed integer programming

1 Introduction

Companies that sell goods to customers at multiple locations have to come up with a dedicated plan how to distribute their products over the different locations to perfectly fit the customers demands. Detailed vehicle routing problems arise naturally when planning these deliveries. The Vehicle Routing Problem and its variants are very well studied in the literature. We refer to [5] for a review. These problems become even more complex by adding constraints modeling characteristics of goods, vehicles, or strategic decisions. The authors of [2] call these problems rich vehicle routing problems. This paper focuses on a special case of these problems where perishable goods are considered, i.e. each product has a

B. Grimm (✉) · R. Borndörfer · M. Olthoff
Zuse Institute Berlin, Berlin, Germany
e-mail: grimm@zib.de; borndoerfer@zib.de; olthoff@zib.de
http://www.zib.de

J. S. Neufeld et al. (eds.), *Operations Research Proceedings 2019*, Operations Research Proceedings, https://doi.org/10.1007/978-3-030-48439-2_50

due date. Deliveries have to be performed before this date and it is prohibited to store the products for a long time at the individual locations. Since late deliveries have a direct impact on the revenues for these products, a precise demand prediction is even more important. Moreover, the demands and vehicle driving times required to transport the goods are dependent on the actual weather conditions. The problem we are dealing with was defined for the AIMMS-MOPTA Optimization Modeling Competition 2018 [3]. It came with a detailed problem description and a set of historical weather features, which have an influence on the product demands and the vehicle driving times to deliver the products. The data should be used to predict the demands and weather conditions on a fixed date and to create an optimal plan to deliver the perishable products to a set of customers. In this paper we present the definition and a solution approach to the Vehicle Routing Problem with Perishable Goods (VRPPG). This includes a procedure how historical weather data is used to predict the demands and driving times. The VRPPG can be seen as a mixture of the Capacitated Vehicle Routing Problem, first defined by Dantzig and Ramser [1] and the Vehicle Routing Problem with Time Windows presented in [4]. We test the approach which let our team finish second place in the MOPTA Competition 2018, on different data sets given by the competition. Finally, we discuss the quality of the approach in terms of run time and quality of the computed solutions.

2 The Vehicle Routing Problem with Perishable Goods

We consider a set of locations $I \subset \mathbb{N}$ including warehouse 0 and set of retailers $J \subset I$. Each retailer $j \in J$ has a demand of $d_j \in \mathbb{N}$ of perishable goods. Deliveries of goods are in time if they arrive before a fixed due date $h \in [0, H[$. Each good delivered later is penalized by c_m independently of the actual duration of the delay. A maximum number of $A \in \mathbb{N}$ vehicles are available to perform deliveries. Each vehicle has a capacity of $C \in \mathbb{N}$ items.

If a vehicle is used it induces the fixed cost $M_F \in \mathbb{R}$ and, additionally, the variable costs of $m_t, m_v \in \mathbb{R}$ for each hour and mile it is operated, respectively.

The possible trips between locations are given by the set $E \subseteq I \times I$. The resulting graph $G := (I, E)$ is used to compute feasible tours beginning and ending at the warehouse. It is assumed that unloading goods at a retailer takes no time. The duration of a trip between two locations and its length is given by $\tau, l : E \mapsto \mathbb{R}^+$, respectively.

Definition 1 The Vehicle Routing Problem with Perishable Goods is the task to find the following items: a set of cycles R in G, for each cycle $r \in R$ an assignment of amounts of goods to retailers j, and an assignment of each cycle $r \in R$ to a vehicle in $\{1, ..., A\}$. These assignments have to minimize penalties caused by late deliveries and the costs for operating the vehicles while satisfying the following conditions: each cycle visits 0; the sum of the goods assigned to a retailer of a cycle does not exceed C; the sum of all goods assigned to a fixed retailer j over all cycles satisfies d_j.

For the MOPTA Competition 2018 the duration of a trip between two locations was only given by historical data including the following three weather features: temperature $f \in \mathbb{R}^+$ in degrees Fahrenheit, precipitation $p \in \mathbb{R}^+$ in cubic centimeters per hour, and wind $w \in \mathbb{R}^+$ in kilometers per hour. Based on the historical data, the map $\tau : E \times \mathbb{R}^3 \mapsto \mathbb{R}^+$ for travel times between two locations in dependence of the features had to be generated.

3 A Solution Approach to the VRPPG

We present an approach to tackle the problem in this section. It is based on a mixed integer linear programming (MILP) model and optimizes the sequences of tours that have to be performed by the vehicles. We define a tour as a single vehicle movement along a cycle beginning at the warehouse, visiting one or more retailers $j \in J$, before returning to the warehouse. Let T denote the set of tours. The approach uses minimum and maximum numbers of tours given by $m := \lceil \frac{\sum_{j \in J} d_j}{C} \rceil$ and $M := \sum_{j \in J} \lceil \frac{d_j}{C} \rceil$, respectively.

Our solution approach can be summarized as follows. We assume that all travel times $\tau(e, f, p, w)$ are given for a fixed tuple of weather features. We connect all retailers in G to the warehouse 0. For those retailers originally not directly connected to 0, the length of the edges is set to the length of the shortest path with respect to travel time. We compute the set P of all possible cycles that visit at most $n + 1$ retailers with minimal travel time with respect to cycles that visit precisely the same retailers. Here, n is defined as $n := \max\{|S| \mid \sum_{j \in S} d_j \leq C, S \subseteq J\}$ and for a cycle p, let c_p and τ_p denote its length and duration. Further, let P_j denote the set of cycles that visit j and let $\tau_{p,j}$ denote the time to reach retailer j from the warehouse along some $p \in P_j$. The model given in Sect. 3.1 is then solved by an commercial state of the art MILP solver.

3.1 MILP Model for the VRPPG

The MILP model formulation contains binary decision variables $x_{t,p}$ for all $t \in T, p \in P$ that indicate if tour t performs cycle p, $p_{t,j}$ for all $t \in T, j \in J$ that indicate if retailer j is delivered late on tour t, v_t for all $t \in T$ that indicate if tour t is operated by an additional vehicle, and r_t for all $t \in T$ that indicate if tour t is operated by the same vehicle as tour $t - 1$. Let $l_{t,j}$ be continuous variables that define the load of goods delivered in time for retailer j on tour t and $s_{t,j}$ the respective load delivered late. The continuous variables d_t, a_t define the departure and arrival time of a tour t at the warehouse. Finally, let H denote the time horizon. To shorten notation $\tilde{d}_j$ defines $\tilde{d}_j := \min\{C, d_j\}$.

Reloading a vehicle for a tour requires that the preceding tour has already arrived at the warehouse, which is controlled by constraints (3) and (4). The set of equalities (5) decides which cycle is assigned to an operated tour. The correct arrival and departure times are set by the constraints (6) with respect to the cycle operated on a tour. To satisfy the demand of each retailer and to prevent exceeding the vehicle capacity, the amount of transported goods for each retailer and each tour are summed up in (7) and (8), respectively. The next two sets of inequalities restrict the individual loads of each retailer on each tour. In set (9) the upper bound of the (in time) load variable for a retailer is set to the minimum of the vehicle capacity and the retailer's demand, if a cycle containing the retailer is selected. It is set to zero if the retailer is visited late on the selected cycle. The upper bounds for the (late) load variables are enforced by (10). The set of constraints (11) takes care that the decision variable $p_{t,j}$ becomes 1 if retailer j will be handled late, i.e., tour t starts too late.

$$\min \sum_{t \in T} c_p x_{t,p} + \sum_{t \in T} M_F v_t + \sum_{t \in T} \sum_{j \in J} c_m s_{t,j},$$

$$\text{s.t.}: \quad \sum_{t \in T} v_t \leq A, \tag{1}$$

$$v_t + r_t \leq 1, \qquad \forall t \in T \tag{2}$$

$$d_{t+1} + H(1 - r_t) \geq a_t \qquad \forall t \in T, \tag{3}$$

$$\sum_{p \in P} x_{t,p} \geq r_{t+1} \qquad \forall t \in T, \tag{4}$$

$$\sum_{p \in P} x_{t,p} = v_t + r_t \qquad \forall t \in T, \tag{5}$$

$$d_t + \sum_{p \in P} \tau_p x_{t,p} \leq a_t \qquad \forall t \in T, \tag{6}$$

$$\sum_{t \in T} l_{t,j} + s_{t,j} \geq d_j \qquad \forall j \in J, \tag{7}$$

$$\sum_{j \in J} l_{t,j} + s_{t,j} \leq C(v_t + r_t) \qquad \forall t \in T, \tag{8}$$

$$\tilde{d}_j \sum_{p \in P_j} x_{t,p} - \tilde{d}_j p_{t,j} \geq l_{t,j} \qquad \forall t \in T, j \in J, \tag{9}$$

$$\tilde{d}_j \sum_{p \in P_j} x_{t,p} \geq s_{t,j} \qquad \forall t \in T, j \in J, \tag{10}$$

$$\sum_{p \in P_j} \tau_{p,j} x_{t,p} + d_t \leq h + H p_{t,j} \qquad \forall t \in T, j \in J, \qquad (11)$$

$$x_{t,p} \in \{0, 1\} \qquad \forall t \in T, p \in P, \qquad (12)$$

$$p_{t,j} \in \{0, 1\}, \ s_{t,l}, l_{t,l} \in \{0, \tilde{d}_j\}, \ c_t, d_t \in \{0, H\} \qquad \forall t \in T, j \in J, \qquad (13)$$

$$v_t, r_t \in \{0, 1\} \qquad \forall t \in T. \qquad (14)$$

3.2 *Weather Features*

So far we assumed that the travel times for each edge are given. As mentioned before this was not the case for the MOPTA Competition 2018. Instead a set of historical travel time data linked to weather features was given. To compute travel times for the edges of G for a fixed tuple of weather features we implemented two different methods.

The first method, which we call Closest Candidate Method (CCM), simply searches for the 'closest' weather feature tuple in the historical data set and applies these travel times to the edges of G. In this sense closest refers to the distance according to the following metric d. Let $w_1 := (f_1, p_1, w_1)$ and $w_2 := (f_2, p_2, w_2)$ be two weather tuples. The distance $d(w_1, w_2)$ is a normalized Euclidean distance given by $d(w_1, w_2) := \sqrt{\left(\frac{f_1 - f_2}{f_\Delta}\right)^2 + \left(\frac{p_1 - p_2}{p_\Delta}\right)^2 + \left(\frac{w_1 - w_2}{w_\Delta}\right)^2}$.

Where $f_\Delta, p_\Delta, w_\Delta$ are the ranges between minimum and maximum value of the respective weather feature included in the historical data set.

The second method, which we call Closest Pair Method (CPM), looks for two points $w_1 := (f_1, p_1, w_1)$ and $w_2 := (f_2, p_2, w_2)$ that enclose a given weather feature $w := (f, p, w)$, i.e., $f \in [\min\{f_1, f_2\}, \max\{f_1, f_2\}]$, $p \in [\min\{p_1, p_2\}, \max\{p_1, p_2\}]$, $w \in [\min\{w_1, w_2\}, \max\{f_1, f_2\}]$. CPM iterates through all pairs of historical weather w_1 and w_2 that enclose w to find the pair with a minimum cumulative distance $d(w, w_1) + d(w, w_2)$. The found pair is then used to interpolate the searched point as follows. Each edge $e \in E$ is assigned a travel time according to the weighted sum $\tau(e) := \left(\frac{\tau_1(e)}{d(w,w_1)} + \frac{\tau_2(e)}{d(w,w_2)}\right) / \left(d(u, w_1) + d(w, w_2)\right)$ where $\tau_i(e)$ is the travel time on edge e and weather feature i. If no suitable pair of features is found, CCM is used to assign travel times.

4 Computational Results

In this section we present the characteristics of the solution approach both for the CCM and CPM method to estimate the driving times. Table 1 contains an individual row for two data sets containing 50 respectively 100 VRPPG instances as shown

Table 1 Computational results for CCM and CPM

		CCM					CPM					Diff	
		CPU				Gap	CPU(s)				Gap		
Test set	#	Ø(s)	2m	15m	1 h	(%)	Ø(s)	2m	15m	1 h	(%)	#	%
Fixed	50	23.5	47	3	0	0.0	23.9	48	2	0	0.0	2	1.9
Dependent	100	650	48	28	7	1.01	535	49	33	6	1.12	1	1.0

in Column #. The "dependent" data set contains instances where the retailers' demands are dependent from the weather conditions. Therefore, the demands were determined from a historical data set in a similar way like it was done for the driving times. The next two blocks of four columns show the characteristics of the CCM and CPM approach, respectively. The column $\emptyset(s)$ marks the arithmetic average of the run times for the instances of each set. The columns 2m and 5m give the number of instances that were solved to optimality within 2 and 15 min, respectively, whereas the column 1h shows the number of instances that hit the time limit before proven to be optimal. The two columns headlined with *gap* mark the maximum gap between the upper and lower bound of instances that were not proven to be solved to optimality. Finally, the columns # and % in the *Diff* block compare the solutions found by each of the approaches. The column # gives the number of instances where a different number of vehicles is used in the solutions of CCM and CPM while Column % shows $|c(CCM) - c(CPM)|/c(CCM)$, i.e., the absolute value between the two objective function values divided by the CCM objective function value. All computations were done with AIMMS 4.53.5.7 with CPLEX 12.7.1 as internal MILP solver on a Intel(R) Core(TM) i7-4600U CPU with 2.10 GHz, four cores, 8 GB RAM and a maximum run time limit of 1 h.

Table 1 shows that the presented approaches generate high quality solutions for all instances of both test sets in a short time, about 24 s on average per instance, in case of the instances with weather independent demands. None of these instances takes more than 4 min to solve, which is a promising result. The *Diff* block shows that both approaches lead to very similar solutions for most of all instances.

5 Conclusion

We presented two approaches for the VRPPG based on a MILP problem formulation with path variables. The chosen approach leads to high quality solutions combined with fast running times for the instances subject to the competition, which justifies the presented approach. To further improve the model and solve larger instances a dynamic cycle construction, i.e., variable generation should be used.

Acknowledgments This work was done in the Research Campus MODAL funded by the German Ministry of Education and Research (BMBF).

References

1. Dantzig, G.B., Ramser, J.H.: The truck dispatching problem. Manag. Sci. **6**, 80–91 (1959)
2. Lahyani, R., Khemakhem, M., Semet, F.: Rich vehicle routing problems: from a taxonomy to a definition. Eur. J. Oper. Res. **241**, 1–14 (2015)
3. Problem Description for the 10th AIMMS-MOPTA Optimization Modeling Competition. http://coral.ie.lehigh.edu/~mopta/AIMMS_MOPTA_case_2018.pdf
4. Solomon, M.M.: Algorithms for the vehicle routing and scheduling problems with time window constraints. Oper. Res. **35**, 254–265 (1987)
5. Toth, P., Vigo, D. (eds.): The Vehicle Routing Problem. Society for Industrial and Applied Mathematics, Philadelphia (2002)

Part XII
Optimization Under Uncertainty

Solving Robust Two-Stage Combinatorial Optimization Problems Under Convex Uncertainty

Marc Goerigk, Adam Kasperski, and Paweł Zieliński

Abstract In this paper a class of robust two-stage combinatorial optimization problems with uncertain costs is discussed. It is assumed that the uncertainty is modeled by using a convex uncertainty set, for example of polyhedral or ellipsoidal shape. Several methods to compute exact and approximate solutions are introduced. Experimental results for robust two-stage version of the weighted set cover problem are presented.

Keywords Robust optimization · Two-stage approach · Convex uncertainty · Combinatorial optimization

1 Problem Formulation

Let $\mathcal{P}$ be a combinatorial optimization problem of the form: $\min\{\boldsymbol{C}^T\boldsymbol{x} : \boldsymbol{x} \in \mathcal{X} \subseteq \{0, 1\}^n\}$, where $\boldsymbol{C} = [C_1, \ldots, C_n]^T$ is a vector of nonnegative costs and $\mathcal{X}$ is a set of feasible solutions. Given $\boldsymbol{x} \in \{0, 1\}^n$, let $\mathcal{R}(\boldsymbol{x}) = \{\boldsymbol{y} \in \{0, 1\}^n : \boldsymbol{x} + \boldsymbol{y} \in \mathcal{X}\}$ be the set of *recourse actions* for $\boldsymbol{x}$. Then $\mathcal{X}' = \{\boldsymbol{x} : \mathcal{R}(\boldsymbol{x}) \neq \emptyset\}$ is called the set of *partial solutions*. Let us define $\mathcal{Z} = \{(\boldsymbol{x}, \boldsymbol{y}) : \boldsymbol{x} \in \mathcal{X}', \boldsymbol{y} \in \mathcal{R}(\boldsymbol{x})\}$. We assume that a feasible solution can be built in two stages. In the first stage we choose a

M. Goerigk
Network and Data Science Management, University of Siegen, Siegen, Germany
e-mail: marc.goerigk@uni-siegen.de

A. Kasperski (✉)
Faculty of Computer Science and Management, Wrocław University of Science and Technology, Wrocław, Poland
e-mail: adam.kasperski@pwr.edu.pl

P. Zieliński
Faculty of Fundamental Problems of Technology, Wrocław University of Science and Technology, Wrocław, Poland
e-mail: pawel.zielinski@pwr.edu.pl

J. S. Neufeld et al. (eds.), *Operations Research Proceedings 2019*, Operations Research Proceedings, https://doi.org/10.1007/978-3-030-48439-2_51

partial solution $\boldsymbol{x} \in \mathcal{X}'$ and complete it to $\boldsymbol{x}+\boldsymbol{y} \in \mathcal{X}$, by choosing a recourse action $\boldsymbol{y} \in \mathcal{R}(\boldsymbol{x})$. The overall solution cost is then $\boldsymbol{C}^T\boldsymbol{x}+\boldsymbol{c}^T\boldsymbol{y}$, where $\boldsymbol{c}=[c_1,\ldots,c_n]^T$ is a vector of second-stage costs. Suppose that $\boldsymbol{c}$ is only known to belong to a convex *uncertainty (scenario) set* $\mathcal{U}$. For example, we can use a *polyhedral uncertainty set* $\mathcal{U}^{\mathcal{HP}} = \{\underline{\boldsymbol{c}}+\boldsymbol{\delta} : \boldsymbol{A}\boldsymbol{\delta} \leq \boldsymbol{b}, \boldsymbol{\delta} \geq \boldsymbol{0}\}$, or an *ellipsoidal uncertainty set* $\mathcal{U}^{\mathcal{E}} = \{\underline{\boldsymbol{c}}+\boldsymbol{\delta} : ||\boldsymbol{A}\boldsymbol{\delta}||_2 \leq \Omega\}$, which are popular methods of uncertainty representation in robust optimization (see, e.g., [1]). The fixed vector $\underline{\boldsymbol{c}}$ represents the nominal second-stage costs and $\boldsymbol{\delta}$ represents deviations of these costs from their nominal values. Consider the following *incremental* and *two-stage* problems, respectively:

$$\textsc{Inc}(\boldsymbol{c},\boldsymbol{x}) = \min_{\boldsymbol{y}\in\mathcal{R}(\boldsymbol{x})} \boldsymbol{c}^T\boldsymbol{y}, \quad \textsc{TSt}(\boldsymbol{c}) = \min_{(\boldsymbol{x},\boldsymbol{y})\in\mathcal{Z}} \boldsymbol{C}^T\boldsymbol{x}+\boldsymbol{c}^T\boldsymbol{y}.$$

In the incremental problem we seek a best recourse action for a fixed partial solution $\boldsymbol{x} \in \mathcal{X}'$ and a fixed second-stage cost vector $\boldsymbol{c}$, while in the two-stage problem we seek an optimal solution (built in two stages) for a fixed second-stage cost vector $\boldsymbol{c} \in \mathcal{U}$. We will also consider the following *evaluation problem*:

$$\textsc{Eval}(\boldsymbol{x}) = \boldsymbol{C}^T\boldsymbol{x} + \max_{\boldsymbol{c}\in\mathcal{U}} \textsc{Inc}(\boldsymbol{c},\boldsymbol{x}),$$

where the inner maximization problem is called the *adversarial problem* and scenario $\boldsymbol{c} \in \mathcal{U}$ maximizing $\textsc{Inc}(\boldsymbol{c},\boldsymbol{x})$ is called a *worst scenario* for $\boldsymbol{x}$. Finally, we discuss the following *robust two-stage* problem:

$$\textsc{RTSt} = \min_{\boldsymbol{x}\in\mathcal{X}'} \textsc{Eval}(\boldsymbol{x}) = \min_{\boldsymbol{x}\in\mathcal{X}'} \max_{\boldsymbol{c}\in\mathcal{U}} \min_{\boldsymbol{y}\in\mathcal{R}(\boldsymbol{x})} (\boldsymbol{C}^T\boldsymbol{x}+\boldsymbol{c}^T\boldsymbol{y}). \tag{1}$$

The two-stage approach has a long tradition in stochastic optimization (see, e.g., [2, 3, 5]). First robust two-stage models for combinatorial problems were discussed in [8] and [7]. Some complexity results for convex uncertainty representation have been recently established in [4], while [6] develops lower bounds for a branch-and-bound method. In this paper we consider exact and approximate methods for solving and bounding (1).

2 Exact Methods of Solving the Problem

The incremental and two-stage problems have the same complexity as the underlying deterministic problem $\mathcal{P}$. Indeed, one can compute $\textsc{Inc}(\boldsymbol{x},\boldsymbol{c})$ by minimizing $\boldsymbol{c}^T\boldsymbol{y}$ over $\boldsymbol{y} \in \mathcal{X}$ after fixing $y_i = 1$ if $x_i = 1$ for $i \in [n]$. On the other hand, $\textsc{TSt}(\boldsymbol{c})$ can be computed by solving $\mathcal{P}$ for the costs $\hat{c}_i = \min\{C_i, c_i\}$, $i \in [n]$ (see [4]).

The evaluation and robust two-stage problems can be represented as the following mathematical programming formulations, respectively:

$$a)\quad \begin{aligned} \max\ & \boldsymbol{C}^T\boldsymbol{x}+\theta \\ \text{s.t.}\ & \theta \le \boldsymbol{c}^T\boldsymbol{y} \quad \forall \boldsymbol{y}\in\mathcal{R}(\boldsymbol{x}) \\ & \boldsymbol{c}\in\mathcal{U} \end{aligned} \qquad b)\quad \begin{aligned} \min\ & \boldsymbol{C}^T\boldsymbol{x}+\theta \\ \text{s.t.}\ & \theta \ge \boldsymbol{c}^T\boldsymbol{y}_{\boldsymbol{c}} && \forall \boldsymbol{c}\in\mathcal{U} \\ & \boldsymbol{x}+\boldsymbol{y}_{\boldsymbol{c}}\in\mathcal{X} && \forall \boldsymbol{c}\in\mathcal{U} \\ & \boldsymbol{x},\boldsymbol{y}_{\boldsymbol{c}}\in\{0,1\}^n && \forall \boldsymbol{c}\in\mathcal{U} \end{aligned} \tag{2}$$

Formulation (2*a*) is a convex optimization problem with possibly an exponential number of constraints. It can be solved by applying a standard row (constraint) generation technique. Observe that (2*a*) for a small subset $\mathcal{R}'(\boldsymbol{x}) \subseteq \mathcal{R}(\boldsymbol{x})$ is a tractable problem, which provides an upper bound on $\text{EVAL}(\boldsymbol{x})$. The cuts (rows) can be generated by solving the incremental problem. Problem (2*b*) has infinite number of variables and constraints. In general, no method of replacing $\mathcal{U}$ with a finite representation is known. However, after replacing $\mathcal{U}$ with a finite set $\mathcal{U}' = \{\boldsymbol{c}_1, \ldots, \boldsymbol{c}_K\} \subset \mathcal{U}$ we get a compact formulation which provides a lower bound on RTST. Unfortunately, the problem with $\mathcal{U}'$ can still be hard to solve, as it becomes the robust two stage problem with discrete uncertainty representation (see, e.g., [8] and [7]). If $\boldsymbol{x}'$ is an optimal solution to (2*b*), after replacing $\mathcal{U}$ with $\mathcal{U}' \subset \mathcal{U}$, then $\text{EVAL}(\boldsymbol{x}')$ provides an upper bound on RTST and a worst scenario for $\boldsymbol{x}'$ (computed by solving (2*a*)) gives a cut that can be added to $\mathcal{U}'$. Summarizing, for small problems, (2*b*) can be solved by applying a row and column generation technique (see, e.g., [9]). For larger problems, we can use (2*b*) to compute approximate solutions by limiting the running time or the number of cuts generated, i.e. the size of $\mathcal{U}'$.

3 Computing Lower Bounds

For any scenario $\boldsymbol{c} \in \mathcal{U}$, the value of $\text{TST}(\boldsymbol{c})$ gives us a lower bound on RTST. We get the best such bound by computing $LB_{Adv} := \max_{\boldsymbol{c}\in\mathcal{U}} \text{TST}(\boldsymbol{c})$. The value of LB_{Adv} can be computed by solving the following formulation:

$$\max\{\theta \mid \theta \le \boldsymbol{C}^T\boldsymbol{x}+\boldsymbol{c}^T\boldsymbol{y}\ \forall(\boldsymbol{x},\boldsymbol{y})\in\mathcal{Z},\ \boldsymbol{c}\in\mathcal{U}\}. \tag{3}$$

Problem (3) is similar to (2*a*). It is a convex optimization problem with an exponential number of constraints, which can be solved by using a row generation technique. Now the cuts are generated by solving the two-stage problem. If $\boldsymbol{c}^*$ is an optimal solution to (3) for some $\mathcal{Z}' \subseteq \mathcal{Z}$, then $\text{TST}(\boldsymbol{c}^*) \le LB_{Adv}$ gives us a lower bound on LB_{Adv}, which is also a lower bound on RTST. We can thus initialize $\mathcal{Z}'$ by solving the two-stage problem for a good heuristic scenario $\boldsymbol{c}_0 \in \mathcal{U}$ and try to improve it by applying the row generation algorithm to (3). We now propose a heuristic method of computing $\boldsymbol{c}_0$. For the simplicity of presentation, we assume that $\mathcal{U} = \mathcal{U}^{\mathcal{HP}}$. Let $\Gamma^* = \max\{\sum_{i\in[n]} \delta_i : \boldsymbol{A\delta} \leq \boldsymbol{b}, \boldsymbol{\delta} \geq \boldsymbol{0}\}$ be the maximal amount of uncertainty that can be allocated to the second stage costs. Consider the following problem:

$$\begin{aligned} &\min\ t \\ &\text{s.t. } \delta_i \leq \max\{0, t - \underline{c}_i\},\ \forall i \in [n] \\ &\qquad \textstyle\sum_{i\in[n]} \delta_i = \Gamma^* \\ &\qquad \boldsymbol{A\delta} \leq \boldsymbol{b}, \boldsymbol{\delta} \geq \boldsymbol{0} \end{aligned} \tag{4}$$

Note that (4) is always feasible by construction. The idea of (4) is to allocate uniformly the uncertainty to the variables with the smallest nominal second-stage costs. If $\boldsymbol{\delta}^*$ is an optimal solution to (4), then $\boldsymbol{c}_0 := \underline{\boldsymbol{c}} + \boldsymbol{\delta}^*$ is the heuristic scenario. Problem (4) can be solved efficiently by using a binary search on $t \in [0, \Gamma^* + \max_{i\in[n]} \underline{c}_i]$. Notice that for a fixed t, it reduces to an LP problem. Let $LB_0 := \text{TST}(\boldsymbol{c}_0)$ be the lower bound corresponding to $\boldsymbol{c}_0$.

We now show another lower bound, which can be computed by solving one compact, convex formulation. Assume that $\mathcal{X} = \{\boldsymbol{x} \in \{0,1\}^n : \boldsymbol{Hx} \geq \boldsymbol{g}\}$. Then $\mathcal{Z} = \{(\boldsymbol{x}, \boldsymbol{y}) \in \{0,1\}^{2n} : \boldsymbol{H}(\boldsymbol{x}+\boldsymbol{y}) \geq \boldsymbol{g}, \boldsymbol{x}+\boldsymbol{y} \leq \boldsymbol{1}\}$. By relaxing $(\boldsymbol{x}, \boldsymbol{y}) \in [0,1]^{2n}$ in $\mathcal{Z}$, we get a convex set $\mathcal{Z}^r$. Hence $\text{RTST} \geq \max_{\boldsymbol{c}\in\mathcal{U}} \min_{(\boldsymbol{x},\boldsymbol{y})\in\mathcal{Z}} (\boldsymbol{C}^T\boldsymbol{x} + \boldsymbol{c}^T\boldsymbol{y}) \geq \max_{\boldsymbol{c}\in\mathcal{U}} \min_{(\boldsymbol{x},\boldsymbol{y})\in\mathcal{Z}^r} (\boldsymbol{C}^T\boldsymbol{x} + \boldsymbol{c}^T\boldsymbol{y}) = \min_{(\boldsymbol{x},\boldsymbol{y})\in\mathcal{Z}^r} \max_{\boldsymbol{c}\in\mathcal{U}} (\boldsymbol{C}^T\boldsymbol{x} + \boldsymbol{c}^T\boldsymbol{y}) =: LB_M$. The first inequality follows from the well-known max-min inequalities. The second inequality is true, because $\mathcal{Z}^r$ is a relaxation of $\mathcal{Z}$. Finally, the last equality results from the fact that the sets $\mathcal{Z}^r$ and $\mathcal{U}$ are convex. The value of LB_M can be computed efficiently, for polyhedral and ellipsoidal uncertainty, by solving one convex problem, after dualizing the inner maximization problem [4]. The bound LB_0 requires solving (4) to identify $\boldsymbol{c}_0$ and one two-stage problem, which can be done in polynomial time if $\mathcal{P}$ is polynomially solvable. Finally, LB_{Adv} can be computed or approximated by solving a sequence of two-stage problems.

4 Computing Approximate Solutions

In this section we propose some methods of computing an approximate first stage solution $\boldsymbol{x} \in \mathcal{X}'$. Notice that evaluating $\boldsymbol{x}$ by using $(2a)$ can be time-consuming for large problems. We will show how to estimate efficiently the distance of $\boldsymbol{x}$ to the optimum, without evaluating it. In practice, it is not necessary to provide the exact value of $\text{EVAL}(\boldsymbol{x})$, as the first stage solution is evaluated a posteriori, i.e. after the true scenario $\boldsymbol{c} \in \mathcal{U}$ occurs. This task can be done by solving the much easier incremental problem. By exchanging the inner max-min operators in (1), we get

$$\text{RTST} \leq \min_{(\boldsymbol{x},\boldsymbol{y})\in\mathcal{Z}} \max_{\boldsymbol{c}\in\mathcal{U}} (\boldsymbol{C}^T\boldsymbol{x} + \boldsymbol{c}^T\boldsymbol{y}) =: UB. \tag{5}$$

We can compute the value of UB by solving the single-stage min-max robust problem. By dualizing the inner maximization problem for polyhedral $\mathcal{U}$, we get a compact MIP formulation for computing UB. Let $\boldsymbol{x}_M$ be an optimal solution to (5). It can be shown (see [4]) that $\text{EVAL}(\boldsymbol{x}_M) \leq UB$, and thus $\text{EVAL}(\boldsymbol{x}_M) \leq \frac{UB}{LB_M}\text{RTST} =: \rho_{MM} \cdot \text{RTST}$. So, the first-stage solution $\boldsymbol{x}_M$ has an approximation ratio ρ_{MM} equal to the integrality gap of the relaxation of the min-max problem.

Let $\boldsymbol{y} \in \{0, 1\}^n$ and define $\overline{c}(\boldsymbol{y}) = \max_{\boldsymbol{c}\in\mathcal{U}} \boldsymbol{c}^T\boldsymbol{y}$. The value of $\overline{c}(\boldsymbol{y})$ can be computed efficiently by solving a convex problem.

Proposition 1 *Fix any $\tilde{\boldsymbol{c}} \in \mathcal{U}$ and let $(\tilde{\boldsymbol{x}}, \tilde{\boldsymbol{y}}) \in \mathcal{Z}$ be an optimal solution to the two-stage problem under $\tilde{\boldsymbol{c}}$. Then for $k \in \{Adv, M, 0\}$ it holds that*

$$\text{EVAL}(\tilde{\boldsymbol{x}}) \leq \left(1 + \frac{\overline{c}(\tilde{\boldsymbol{y}}) - \tilde{\boldsymbol{c}}^T\tilde{\boldsymbol{y}}}{LB_k}\right)\text{RTST} = \rho_k \cdot \text{RTST}. \tag{6}$$

Proof $\text{EVAL}(\tilde{\boldsymbol{x}}) = \boldsymbol{C}^T\tilde{\boldsymbol{x}} + \max_{\boldsymbol{c}\in\mathcal{U}} \min_{\boldsymbol{y}\in\mathcal{R}(\tilde{\boldsymbol{x}})} \boldsymbol{c}^T\boldsymbol{y} \leq \boldsymbol{C}^T\tilde{\boldsymbol{x}} + \max_{\boldsymbol{c}\in\mathcal{U}} \boldsymbol{c}^T\tilde{\boldsymbol{y}} = \boldsymbol{C}^T\tilde{\boldsymbol{x}} + \overline{c}(\tilde{\boldsymbol{y}}) = \boldsymbol{C}^T\tilde{\boldsymbol{x}} + \tilde{\boldsymbol{c}}^T\tilde{\boldsymbol{y}} + \overline{c}(\tilde{\boldsymbol{y}}) - \tilde{\boldsymbol{c}}^T\tilde{\boldsymbol{y}} = \text{TST}(\tilde{\boldsymbol{c}}) + \overline{c}(\tilde{\boldsymbol{y}}) - \tilde{\boldsymbol{c}}^T\tilde{\boldsymbol{y}}$. Since $\text{RTST} \geq \text{TST}(\tilde{\boldsymbol{c}})$, we obtain $\text{EVAL}(\tilde{\boldsymbol{x}}) \leq \text{RTST} + \overline{c}(\tilde{\boldsymbol{y}}) - \tilde{\boldsymbol{c}}^T\tilde{\boldsymbol{y}}$, which implies (6). □

In the following, we use $\tilde{\boldsymbol{c}} = \boldsymbol{c}_0$. The corresponding first-stage solution $\boldsymbol{x}_0$ can be computed by solving one two-stage problem for $\boldsymbol{c}_0$, which has the same complexity as the underlying problem $\mathcal{P}$. Notice that we get the ratio ρ_0 as a byproduct, since $LB_0 = \text{TST}(\boldsymbol{c}_0)$. The ratio ρ_M can be computed in polynomial time and the ratio ρ_{Adv} can be computed or estimated by using (3).

5 Computational Tests

In this section we will show some results of computational tests. We used `IBM ILOG CPLEX 12.8.0.0` optimizer to solve the MIP formulations. We consider the case when $\mathcal{P}$ is the weighted set cover problem defined as follows. An instance (n, m, k) of this problem is specified as $\mathcal{X} = \{\boldsymbol{x} \in \{0, 1\}^n : \boldsymbol{H}\boldsymbol{x} \geq \boldsymbol{1}\}$, where $\boldsymbol{H}$ is an $m \times n$ binary matrix, having exactly k ones in each row, and $\boldsymbol{1}$ is a vector of ones (note that $(n, m, 2)$ is a vertex cover problem). Matrix $\boldsymbol{H}$ is generated randomly, i.e. the k ones are placed randomly in each row of $\boldsymbol{H}$. The first stage costs C_i, $i \in [n]$, are random integers from $[1, 60]$ and the nominal second-stage costs $\underline{c}_i$, $i \in [n]$, are random integers from $[1, 10]$. In the tests, we used the following special case of polyhedral uncertainty representation: $\mathcal{U}_1^{\mathcal{HP}} = \{\underline{\boldsymbol{c}} + \boldsymbol{\delta} : \boldsymbol{0} \leq \boldsymbol{\delta} \leq \boldsymbol{d}, \sum_{i\in[n]} \delta_i \leq \Gamma, \sum_{i\in U_j} \delta_i \leq \Gamma_j, j \in [K]\}$. So, it is assumed that the second stage costs are such that $c_i \in [\underline{c}_i + d_i]$ for $i \in [n]$. The total amount of uncertainty that can be allocated to the second stage costs is bounded by a global budget Γ. Additionally, each subset of variables $U_j \subseteq [n]$ has its own budget Γ_j. In our tests d_i is a random integer from $[1, 120]$, K is a random integer from $[0, 10]$ and the budgets are set to 20% of the maximum total deviation for each U_j, $j \in [K]$. The sets U_j were generated randomly.

We fix the running time of solving the master problem of (2b) to 120 s and the number of cuts was limited to 30. Hence the maximum allowed running time of solving (2b) was set to 1 h. The maximum running time of solving (3) was set to 60 s. After this time either the exact value of LB_{Adv} or its estimation from below has been returned. For each instance we computed the solutions $\boldsymbol{x}_0$ and $\boldsymbol{x}_M$ with the corresponding ratios (see the previous section). Solving (2) was only possible for small problems.

The first test was performed for $n = 30$. For each combination of $m \in \{50, 100, 500\}$ and $k \in \{2, 3, 4\}$ we generated 50 instances. The results are shown in Table 1. Problem (2b) can be hard to solve exactly for $m \geq 100$ and $k \geq 3$. Hence, applying the row and column generation technique to obtain an optimal solution can be inefficient, even for relatively small number of variables. The average ratios are better for smaller k. For $n = 30$ all obtained solutions were evaluated and the improved distance to the optimality of $\boldsymbol{x}_0$ and $\boldsymbol{x}_M$ (see the column "best") is quite small (the largest ratio observed was 1.65).

The second test was performed for larger instances, having $n = 100$ variables. For each assumed combination of m and k we generated 30 random instances. The results are shown in Table 2. In this case no problem was solved exactly within

Table 1 The (average, maximum) ratios for instances with $n = 30$; #opt is the number of instances solved to the optimality; "best" shows $\textsc{Eval}(\boldsymbol{x}_i)$ divided by the best lower bound found

$n = 30$		$\boldsymbol{x}_0$				$\boldsymbol{x}_M$	
(m, k)	#opt	ρ_0	ρ_{Adv}	ρ_M	Best	ρ_{MM}	Best
(50,2)	50	(1.94, 3.06)	(1.70, 2.51)	(1.77, 2.62)	(1.11, 1.45)	(1.21, 1.57)	(1.08, 1.39)
(50,3)	42	(2.35, 3.63)	(2.01, 2.60)	(2.11, 2.80)	(1.20, 1.65)	(1.15, 1.67)	(1.02, 1.30)
(50,4)	36	(2.81, 4.33)	(2.43, 3.87)	(2.60, 3.80)	(1.24, 1.63)	(1.20, 1.60)	(1.01, 1.18)
(100,2)	46	(1.84, 2.31)	(1.65, 1.99)	(1.85, 2.32)	(1.08, 1.18)	(1.51, 1.76)	(1.08, 1.26)
(100,3)	9	(2.10, 2.62)	(1.87, 2.19)	(2.19, 2.58)	(1.11, 1.37)	(1.55, 2.29)	(1.12, 1.56)
(100,4)	15	(2.62, 4.16)	(2.30, 3.26)	(2.81, 3.82)	(1.11, 1.25)	(1.57, 2.25)	(1.10, 1.41)
(500,2)	11	(1.50, 1.88)	(1.41, 1.75)	(1.71, 2.26)	(1.08, 1.19)	(1.86, 1.91)	(1.03, 1.11)
(500,3)	0	(1.72, 2.14)	(1.61, 1.92)	(2.15, 2.69)	(1.09, 1.17)	(2.24, 2.65)	(1.14, 1.36)
(500,4)	2	(1.94, 2.44)	(1.78, 2.24)	(2.56, 3.26)	(1.08, 1.19)	(2.37, 3.29)	(1.16, 1.46)

Table 2 The (average, maximum) ratios for instances with $n = 100$; "best" shows the best upper bound on $\boldsymbol{x}_i$ divided by the best lower bound found

$n = 100$	$\boldsymbol{x}_0$				$\boldsymbol{x}_M$	
(m, k)	ρ_0	ρ_{Adv}	ρ_M	Best	ρ_{MM}	Best
(500,2)	(1.67, 1.88)	(1.56, 1.69)	(1.77, 1.92)	(1.14, 1.21)	(1.73, 1.80)	(1.17, 1.31)
(500,3)	(2.18, 2.47)	(2.09, 2.32)	(2.45, 2.69)	(1.25, 1.31)	(1.84, 2.15)	(1.38, 1.58)
(500,4)	(2.46, 2.94)	(2,61, 2.94)	(3.06, 3.33)	(1.40, 1.48)	(1.77, 2.10)	(1.39, 1.55)
(1000,2)	(1.52, 1.73)	(1.46, 1.60)	(1.69, 1.88)	(1.16, 1.22)	(1.84, 1,87)	(1.11, 1.22)
(1000,3)	(1.91, 2.16)	(1.90, 2.15)	(2.33, 2.56)	(1.32, 1.42)	(1.74, 2.18)	(1.15, 1.48)
(1000,4)	(2.30, 2.39)	(2.30, 2.39)	(3.03, 3.11)	(1.36, 1.42)	(2.16, 2.24)	(1.42, 1.48)

the assumed time. Also, not all solutions x_0 and x_M were evaluated exactly, in which case we got an upper bound on the evaluation. Hence, for each solution we computed the ratio between the best upper bound and the best lower bound (see the column "best"). The average ratios are greater than those for $n = 30$. However, the improved ratios for x_0 and x_M suggest that the solutions can be still attractive.

Acknowledgments Adam Kasperski and Paweł Zieliński were supported by the National Science Centre, Poland, grant 2017/25/B/ST6/00486.

References

1. Ben-Tal, A., El Ghaoui, L., Nemirovski, A.: Robust Optimization. Princeton Series in Applied Mathematics. Princeton University Press, Princeton (2009)
2. Dhamdhere, K., Ravi, R., Singh, M.: On two-stage stochastic minimum spanning trees. In: Jünger, M., Kaibel, V. (eds.) IPCO 2005. Lecture Notes in Computer Science, vol. 3509, pp. 321–334. Springer, Berlin (2005)
3. Flaxman, A., Frieze, A., Krivelevich, M.: On the random 2-stage minimum spanning tree. Random Struct. Algoritm. **28**, 24–36 (2006)
4. Goerigk, M., Kasperski, A., Zieliński, P.: Robust two-stage combinatorial optimization problems under convex uncertainty. arXiv:1905.02469 (2019)
5. Kall, P., Mayer, J.: Stochastic Linear Programming. Models, Theory and Computation. Springer, New York (2005)
6. Kämmerling, N., Kurtz, J.: Oracle-based algorithms for binary two-stage robust optimization. Preprint, arXiv:1905.05257 (2019)
7. Kasperski, A., Zieliński, P.: On the approximability of robust spanning problems. Theor. Comput. Sci. **412**, 365–374 (2011)
8. Katriel, I., Kenyon-Mathieu, C., Upfal, E.: Commitment under uncertainty: two-stage matching problems. Theor. Comput. Sci. **408**, 213–223 (2008)
9. Zeng, B., Zhao, L.: Solving two-stage robust optimization problems using a column and constraint generation method. Oper. Res. Lett. **41**, 457–461 (2013)

Production Planning Under Demand Uncertainty: A Budgeted Uncertainty Approach

Romain Guillaume, Adam Kasperski, and Paweł Zieliński

Abstract The paper deals with a version of the capacitated single-item lot sizing problem with backordering under uncertainty. The interval representation of uncertain cumulative demands is considered. Two its variants: the discrete budgeted uncertainty and the continuous budgeted uncertainty, are examined. In order to choose a robust production plan, for the problem under consideration, that hedges against the uncertainty the well-known minmax criterion is adopted. Polynomial and pseudopolynomial methods for finding a robust production plan, respectively, under the discrete and continuous budgeted uncertainty are proposed.

Keywords Optimization under uncertainty · Supply chain · Production planning

1 Introduction

In this paper we focus on the tactical level of a capacitated production planning for the master production scheduling (MPS) under uncertainty. Namely, we deal with a version of the capacitated single-item lot sizing problem with backordering under uncertainty in the demand. Its deterministic counterpart is defined as follows: given T periods, let $d_t \geq 0$ be the *demand* and $x_t \geq 0$ be the *production amount* in period t, $t \in [T]$ ($[T]$ denotes the set $\{1, \ldots, T\}$). We assume that the production x_t in each period $t \in [T]$ can be under some constraints, namely, let $\mathbb{X} \subseteq \mathbb{R}^T_+$ be a set of production plans assumed, here, to be given by linear constraints. The *inventory*, *backordering costs*, and the *selling price*, denoted by c^I, c^B and b^P, respectively, do not depend on the period t. Accordingly, we wish to find a feasible production plan

R. Guillaume
Université de Toulouse-IRIT, Toulouse, France
e-mail: Romain.Guillaume@irit.fr

A. Kasperski · P. Zieliński (✉)
Wrocław University of Science and Technology, Wrocław, Poland
e-mail: AdamKasperski@pwr.edu.pl; Pawel.Zielinski@pwr.edu.pl

J. S. Neufeld et al. (eds.), *Operations Research Proceedings 2019*, Operations Research Proceedings, https://doi.org/10.1007/978-3-030-48439-2_52

$x \in \mathbb{X}$, subject to the conditions of satisfying each demand, which minimizes the total cost of storage and backordering minus the benefit from selling the product. Set $D_t = \sum_{i=1}^{t} d_i$ and $X_t = \sum_{i=1}^{t} x_i$, hence D_t and X_t stand for the cumulative demand up to period t and the cumulative production up to period t, respectively. Thus the problem under consideration can be given by the following linear program:

$$\min \sum_{t\in[T]} (c^I I_t + c^B B_t - b^P s_t) \tag{1}$$

$$\text{s.t. } B_t - I_t = D_t - X_t \qquad t \in [T], \tag{2}$$

$$\sum_{i=1}^{t} s_i = D_t - B_t \qquad t \in [T], \tag{3}$$

$$B_t, I_t, s_t \geq 0 \qquad t \in [T], \tag{4}$$

$$x \in \mathbb{X} \subseteq \mathbb{R}_+^T, \tag{5}$$

where I_t, B_t and s_t are inventory level, backordering level and sales of the product at the end of period $t \in [T]$, respectively. We can write (1)–(5) in the equivalent form: $\min_{x\in\mathbb{X}} \sum_{t\in[T]} \max\{c^I(X_t - D_t), c^B(D_t - X_t)\} - b^P \min\{X_T, D_T\}$.

Consider now the case when the cumulative demands are uncertain. The uncertainty is modeled by a *scenario set* $\mathcal{U}$ that contains all possible realizations (vectors) of the cumulative demands, called *scenarios*. Several methods of defining $\mathcal{U}$ have been proposed in the literature (see, e.g., [3, 4, 6]), among others, the interval uncertainty representation, in which the value of each cumulative demand in period $t \in [T]$ belongs to a closed interval $[\underline{D}_t, \overline{D}_t]$ and a scenario set is the Cartesian product of all the intervals (see, e.g., [6]). We assume that the intervals have a symmetric form, i.e. $[\hat{D}_t - \Delta_t, \hat{D}_t + \Delta_t]$, $t \in [T]$, where Δ_t, $\hat{D}_t \geq \Delta_t \geq 0$, is the maximal deviation of the value of cumulative demand in period t from its *nominal value* $\hat{D}_t$.

In this paper we study the following two special cases of the interval, symmetric, uncertainty representation (see, e.g., [1, 2, 9]):

$$\mathcal{U}^d = \{\boldsymbol{D} = (D_t)_{t\in[T]} : D_t \leq D_{t+1}, D_t \in [\hat{D}_t - \Delta_t, \hat{D}_t + \Delta_t], |\{t : D_t \neq \hat{D}_t\}| \leq \Gamma\},$$

$$\mathcal{U}^c = \{\boldsymbol{D} = (D_t)_{t\in[T]} : D_t \leq D_{t+1}, D_t \in [\hat{D}_t - \Delta_t, \hat{D}_t + \Delta_t], \| \boldsymbol{D} - \hat{\boldsymbol{D}} \|_1 \leq \Gamma\},$$

where Γ is called a *budget*. It controls the amount of uncertainty allocated by an adversary to the cumulative demands. The first representation $\mathcal{U}^d$ is *discrete budgeted uncertainty*, where $\Gamma \in \{0\} \cup [T]$, and the second one $\mathcal{U}^d$ is *continuous budgeted uncertainty*, where $\Gamma \in \mathbb{R}_+$. Moreover, we assume that the demand in a period could be equal to zero only at the worst case, i.e. when the cumulative demand in period t is at its upper bound and in period $t+1$ it is at its lower bound. Formally, we assume that $\hat{D}_t + \Delta_t \leq \hat{D}_{t+1} - \Delta_{t+1}$, $t \in [T-1]$.

In order to compute a robust production plan we adopt a *minmax* approach, in which we seek a plan minimizing its total cost over scenario set $\mathcal{U} \in \{\mathcal{U}^d, \mathcal{U}^c\}$. This leads to the following *minmax* problem:

$$MinMax : \min_{\boldsymbol{x}\in\mathbb{X}} \max_{\boldsymbol{D}\in\mathcal{U}} \sum_{t\in[T]} \max\{c^I(X_t - D_t), c^B(D_t - X_t)\} - b^P \min\{X_T, D_T\}. \tag{6}$$

The MINMAX problem contains the *adversarial problem* for a fixed $\boldsymbol{x} \in \mathbb{X}$, i.e.

$$Adv : \max_{\boldsymbol{D}\in\mathcal{U}} \sum_{t\in[T]} \max\{c^I(X_t - D_t), c^B(D_t - X_t)\} - b^P \min\{X_T, D_T\}. \tag{7}$$

2 Discrete Budgeted Uncertainty

In this section we propose a polynomial method for finding a robust production plan under the discrete budgeted uncertainty. Consider first the ADV problem.

Lemma 1 *The* ADV *problem for* $\mathcal{U}^d$ *boils down to the following problem:*

$$\max \sum_{t\in[T-1]} \max\{c^I(X_t - (\hat{D}_t - \delta_t\Delta_t)), c^B(\hat{D}_t + \delta_t\Delta_t - X_t)\} \tag{8}$$

$$+ \max\{c^I(X_T - (\hat{D}_T - \delta_T\Delta_T)) - b^P(\hat{D}_T - \delta_T\Delta_T),$$

$$c^B(\hat{D}_T + \delta_T\Delta_T - X_T) - b^P X_T\}$$

$$\text{s.t.} \sum_{t\in[T]} \delta_t \le \Gamma, \tag{9}$$

$$0 \le \delta_t \le 1, \quad t \in [T]. \tag{10}$$

Moreover (8)–(10) has an integral optimal solution $\boldsymbol{\delta}^*$ *such that* $\sum_{t\in[T]} \delta^*_t = \Gamma$ *and the value of (8) for* $\boldsymbol{\delta}^*$ *equals the optimal value of objective function of (7).*

Proof Since (8) is a convex function with respect to $\boldsymbol{\delta} \in [0, 1]^T$, it attains its maximum value at a vertex of the convex polytope (9) and (10) (see, e.g., [8]). Moreover, the matrix of constraints (9) and (10) is totally unimodular, hence an optimal solution $\boldsymbol{\delta}^*$ to (8)–(10) is integral. Let $\boldsymbol{D}' \in \mathcal{U}^d$ be an optimal solution of the ADV problem. It can be expressed by a feasible solution $\boldsymbol{\delta}'$ to (8)–(10). Indeed, $D'_t = \hat{D}_t \pm \delta'_t\Delta_t$, $t \in [T]$, $\boldsymbol{\delta}' \in [0, 1]^T$. Obviously $\sum_{t\in[T]} \delta'_t \le \Gamma$, since $|\{t : \Delta_t > 0\}| \le \Gamma$. The value of the objective function of (7) for $\boldsymbol{D}'$ can be bounded from above by the value of (8) for $\boldsymbol{\delta}'$ and, in consequence, by the value of (8) for $\boldsymbol{\delta}^*$. Let us form the cumulative demand scenario $\boldsymbol{D}^* \in \mathcal{U}^d$, that corresponds to $\boldsymbol{\delta}^*$, as follows:

$$D_t^* = \begin{cases} \hat{D}_t - \delta_t^* \Delta_t & \text{if} c^I(X_t - (\hat{D}_t - \delta_t^* \Delta_t)) > c^B(\hat{D}_t + \delta_t^* \Delta_t - X_t), t \in [T-1], \\ \hat{D}_t - \delta_t^* \Delta_t & \text{if} c^I(X_t - (\hat{D}_t - \delta_t^* \Delta_t)) - b^P(\hat{D}_t - \delta_t^* \Delta_t) \\ & > c^B(\hat{D}_t + \delta_t^* \Delta_t - X_t) - b^P X_t, t = T, \\ \hat{D}_t + \delta_t^* \Delta_t & \text{otherwise.} \end{cases}$$

It easily seen that the optimal value of (8) for $\boldsymbol{\delta}^*$ is equal to the value of the objective function of (7) for $\boldsymbol{D}^*$. By the optimality of $\boldsymbol{D}'$, this value is bounded from above by the value of the objective function of (7) for $\boldsymbol{D}'$. Hence $\boldsymbol{D}^*$ is an optimal solution to (7) as well, which proves the lemma. □

From Lemma 1, it follows that an optimal solution $\boldsymbol{D}^*$ to (7) is such that $D_t^* \in \{\hat{D}_t - \Delta_t, \hat{D}_t, \hat{D}_t + \Delta_t\}$ for every $t \in [T]$. Thus we can rewrite (8) as follows:

$$\sum_{t\in[T-1]} \max\{c^I(X_t - \hat{D}_t), c^B(\hat{D}_t - X_t)\}$$
$$+ \max\{c^I(X_T - \hat{D}_T) - b^P \hat{D}_T, c^B(\hat{D}_T - X_T) - b^P X_T\} + \sum_{t\in[T]} c_t \delta_t, \qquad (11)$$

where $c_t = \max\{c^I(X_t-(\hat{D}_t-\Delta_t)), c^B(\hat{D}_t+\Delta_t-X_t)\}-\max\{c^I(X_t-\hat{D}_t), c^B(\hat{D}_t-X_t)\}$ for $t \in [T-1]$; $c_T = \max\{c^I(X_T-(\hat{D}_T-\Delta_T)) - b^P(\hat{D}_T-\Delta_T), c^B(\hat{D}_T+\Delta_T - X_T) - b^P X_T\} - \max\{c^I(X_T - \hat{D}_T) - b^P \hat{D}_T, c^B(\hat{D}_T - X_T) - b^P X_T\}$. Therefore, we need to solve (11) with constraints (9) and (10). We first find, in $O(T)$ time (see, e.g., [5]), the Γth largest coefficient, denoted by $c_{\sigma(\Gamma)}$, such that $c_{\sigma(1)} \geq \cdots \geq c_{\sigma(\Gamma)} \geq \cdots \geq c_{\sigma(T)}$, where σ is a permutation of $[T]$. Then having $c_{\sigma(\Gamma)}$ we can choose Γ coefficients $c_{\sigma(i)}$, $i \in [\Gamma]$, and set $\delta^*_{\sigma(i)} = 1$.

Theorem 1 *The* ADV *problem for* $\mathcal{U}^d$ *can be solved in* $O(T)$ *time.*

We now examine the MINMAX problem. Writing the dual to (11) with (9) and (10) and taking into account the forms of coefficients c_t, we have:

$$\begin{aligned} \min \sum_{t\in[T]} \pi_t + \Gamma\alpha + \sum_{t\in[T]} \gamma_t & \\ \text{s.t.}\ \pi_t \geq c^I(X_t - \hat{D}_t), & \quad t \in [T-1], \\ \pi_t \geq c^B(\hat{D}_t - X_t), & \quad t \in [T-1], \\ \pi_T \geq c^I(X_T - \hat{D}_T) - b^P \hat{D}_T, & \\ \pi_T \geq c^B(\hat{D}_T - X_T) - b^P X_T, & \\ \alpha + \gamma_t \geq c^I(X_t - (\hat{D}_t - \Delta_t)) - \pi_t, & \quad t \in [T-1], \end{aligned}$$

$$
\begin{aligned}
&\alpha + \gamma_t \geq c^B(\hat{D}_t + \Delta_t - X_t) - \pi_t, && t \in [T-1],\\
&\alpha + \gamma_T \geq c^I(X_T - (\hat{D}_T - \Delta_T)) - b^P(\hat{D}_T - \Delta_T) - \pi_T, &&\\
&\alpha + \gamma_T \geq c^B(\hat{D}_T + \Delta_T - X_T) - b^P X_T - \pi_T, &&\\
&\alpha, \gamma_t \geq 0, \pi_t \text{unrestricted}, && t \in [T].
\end{aligned}
$$

Adding linear constraints $\boldsymbol{x} \in \mathbb{X}$ to the above model yields a linear program for the MINMAX problem.

Theorem 2 *The* MINMAX *problem for $\mathcal{U}^d$ can be solved in a polynomial time.*

3 Continuous Budgeted Uncertainty

In this section we provide a pseudopolynomial method for finding a robust production plan under the continuous budgeted uncertainty. We start with the ADV problem.

Lemma 2 *The* ADV *problem for $\mathcal{U}^c$ boils down to the following problem:*

$$
\begin{aligned}
\max \sum_{t\in[T-1]} & \max\{c^I(X_t - (\hat{D}_t - \delta_t)), c^B(\hat{D}_t + \delta_t - X_t)\} \qquad (12)\\
& + \max\{c^I(X_T - (\hat{D}_T - \delta_T)) - b^P(\hat{D}_T - \delta_T), c^B(\hat{D}_T + \delta_T - X_T) - b^P X_T\}\\
\text{s.t.} \sum_{t\in[T]} & \delta_t \leq \Gamma, \qquad (13)\\
& 0 \leq \delta_t \leq \Delta_t, \quad t \in [T]. \qquad (14)
\end{aligned}
$$

The optimal value of (12) equals the optimal value of objective function of (7).

Proof A proof is similar in spirit to that of Lemma 1. □

Lemma 2 shows that solving ADV is equivalent to solving (12)–(14). Rewriting (12) yields:

$$
\begin{aligned}
&\sum_{t\in[T-1]} \max\{c^I(X_t - \hat{D}_t), c^B(\hat{D}_t - X_t)\} \qquad (15)\\
&+ \max\{c^I(X_T - \hat{D}_T) - b^P \hat{D}_T, c^B(\hat{D}_T - X_T) - b^P X_T\} + \sum_{t\in[T]} c_t(\delta_t),
\end{aligned}
$$

where $c_t(\delta) = \max\{c^I(X_t-(\hat{D}_t-\delta)), c^B(\hat{D}_t+\delta-X_t)\} - \max\{c^I(X_t-\hat{D}_t), c^B(\hat{D}_t - X_t)\}$ for $t \in [T-1]$; $c_T(\delta) = \max\{c^I(X_T - (\hat{D}_T - \delta)) - b^P(\hat{D}_T - \delta), c^B(\hat{D}_T + \delta - X_T) - b^P X_T\} - \max\{c^I(X_T - \hat{D}_T) - b^P \hat{D}_T, c^B(\hat{D}_T - X_T) - b^P X_T\}$ are linear or piecewise linear convex functions in $[0, \Delta_t]$, $t \in [T]$. Therefore (15) with constraints (13) and (14), in particular the inner problem, is a special case of a continuous knapsack problem with separable convex utilities, which is weakly NP-hard (see [7]). It turns out that this inner problem is weakly NP-hard as well—a proof of this fact is a modification of that in [7]. Hence we get the following theorem.

Theorem 3 *The* ADV *problem for $\mathcal{U}^c$ is weakly NP-hard.*

We now propose an algorithm for the ADV problem, in which we reduce solving the problem to finding a longest path in a layered weighted graph. It can be shown that if $\Gamma, \Delta_t \in \mathbb{Z}_+$, $t \in [T]$, then there exists an optimal solution $\boldsymbol{\delta}^*$ to (15) with (13), (14) such that $\delta_t^* \in \{0, 1, \ldots, \Delta_t\}$, $t \in [T]$. Hence, we can build a layered graph $G = (V, A)$. The set V is partitioned into $T + 2$ disjoint layers $V_0, V_1, \ldots, V_T, V_{T+1}$ in which $V_0 = \{\mathfrak{s} = 0_0\}$ and $V_{T+1} = \{\mathfrak{t}\}$ contain two distinguished nodes, $\mathfrak{s}$ and $\mathfrak{t}$, and each layer V_t corresponding to period t, $t \in [T]$, has $\Gamma + 1$ nodes labeled in the following way: $t_0, \ldots, t_\Gamma$, where the notation t_i, $i = 0, \ldots, \Gamma$, means that i units of the available uncertainty Γ have been allocated by an adversary to the cumulative demands in periods from 1 to t. Each node $(t-1)_i \in V_{t-1}, t \in [T]$ (including the source node $\mathfrak{s} = 0_0$ in V_0) has at most $\Delta_t + 1$ arcs that go to nodes in layer V_t, namely arc $((t-1)_i, t_{i+\delta_t})$ exists if $i + \delta_t \le \Gamma$, where $\delta_t = 0, \ldots, \Delta_t$. Moreover, we associate with each arc $((t-1)_i, t_j) \in A$, $(t-1)_i \in V_{t-1}, t_j \in V_t$ the cost $c_{(t-1)_i t_j}$ in the following way:

$$c_{(t-1)_i t_j} = \begin{cases} \max\{c^I(X_t - (\hat{D}_t - (j-i))), c^B(\hat{D}_t + (j-i) - X_t)\} & \text{if t} \in [\text{T} - 1], \\ \max\{c^I(X_t - (\hat{D}_t - (j-i))) - b^P(\hat{D}_t - (j-i)), \\ c^B(\hat{D}_t + (j-i) - X_t) - b^P X_t\} & \text{if t} = \text{T}. \end{cases}$$

We finish with connecting each node from V_T with the sink node $\mathfrak{t}$ by the arc of zero cost. A trivial verification shows that each path from $\mathfrak{s}$ to $\mathfrak{t}$ models an integral feasible solution to (12)–(14) and in consequence its optimal solution. Hence solving the ADV problem boils down to finding a longest path from $\mathfrak{s}$ to $\mathfrak{t}$ in G, which can be done in $O(|A| + |V|)$ time in directed acyclic graphs (see, e.g., [5]).

Theorem 4 *Suppose that $\Gamma, \Delta_t \in \mathbb{Z}_+$, $t \in [T]$. Then the* ADV *problem for $\mathcal{U}^c$ can be solved in $O(T\Gamma\Delta_{\max})$ time, where $\Delta_{\max} = \max_{t\in[T]} \Delta_t$.*

Consider now the MINMAX problem. Its inner problem corresponds to the ADV one for a fixed $\boldsymbol{x} \in \mathbb{X}$, which can be reduced to finding a longest path in layered weighted graph G built. A linear program for the latter problem, with pseudopolynomial numbers of constraints and variables, is as follows:

$$\min \pi_{\mathfrak{t}}$$

$$\text{s.t.} \ \pi_{t_j} - \pi_{(t-1)_i} \ge c^I(X_t - (\hat{D}_t - (j-i))) \qquad ((t-1)_i, t_j) \in A, t \in [T-1],$$

$$\pi_{t_j} - \pi_{(t-1)_i} \geq c^B(\hat{D}_t + (j-i) - X_t) \qquad ((t-1)_i, t_j) \in A, t \in [T-1],$$

$$\pi_{t_j} - \pi_{(t-1)_i} \geq c^I(X_t - (\hat{D}_t - (j-i))) \qquad ((t-1)_i, t_j) \in A, t = T,$$

$$- b^P(\hat{D}_t - (j-i))$$

$$\pi_{t_j} - \pi_{(t-1)_i} \geq c^B(\hat{D}_t + (j-i) - X_t) - b^P X_t \qquad ((t-1)_i, t_j) \in A, t = T,$$

$$\pi_{\mathfrak{t}} - \pi_u \geq 0 \qquad u \in V_T,$$

$$\pi_{\mathfrak{s}} = 0, \pi_u \text{unrestricted}, \qquad u \in V.$$

The optimal value of $\pi_{\mathfrak{t}}$ is equal to the worst-case cost of a production plan $\boldsymbol{x}$. Adding linear constraints $\boldsymbol{x} \in \mathbb{X}$ gives a linear program for the MINMAX problem.

Theorem 5 *Suppose that Γ, $\Delta_t \in \mathbb{Z}_+$, $t \in [T]$. Then the* MINMAX *problem for $\mathcal{U}^c$ is pseudopolynomially solvable.*

Acknowledgments Romain Guillaume was partially supported by the project caasc ANR-18-CE10-0012 of the French National Agency for Research, Adam Kasperski and Paweł Zieliński were supported by the National Science Centre, Poland, grant 2017/25/B/ST6/00486.

References

1. Bertsimas, D., Sim, M.: Robust discrete optimization and network flows. Math. Program. **98**, 49–71 (2003)
2. Bertsimas, D., Sim, M.: The price of robustness. Oper. Res. **52**, 35–53 (2004)
3. Dolgui, A., Prodhon, C.: Supply planning under uncertainties in MRP environments: a state of the art. Annu. Rev. Control **31**, 269–279 (2007)
4. Guillaume, R., Thierry, C., Zieliński, P.: Robust material requirement planning with cumulative demand under uncertainty. Int. J. Prod. Res. **55**, 6824–6845 (2017)
5. Korte, B., Vygen, J.: Combinatorial Optimization: Theory and Algorithms, Algorithms and Combinatorics. Springer, Berlin (2012)
6. Kouvelis, P., Yu, G.: Robust Discrete Optimization and Its Applications. Kluwer, Dordrecht (1997)
7. Levi, R., Perakis, G., Romero, G.: A continuous knapsack problem with separable convex utilities: approximation algorithms and applications. Oper. Res. Lett. **42**, 367–373 (2014)
8. Martos, B.: Nonlinear Programming Theory and Methods. Akadémiai Kiadó, Budapest (1975)
9. Nasrabadi, E., Orlin, J.B.: Robust optimization with incremental recourse. CoRR abs/1312.4075 (2013)

Robust Multistage Optimization with Decision-Dependent Uncertainty

Michael Hartisch and Ulf Lorenz

Abstract Quantified integer (linear) programs (QIP) are integer linear programs with variables being either existentially or universally quantified. They can be interpreted as two-person zero-sum games between an existential and a universal player on the one side, or multistage optimization problems under uncertainty on the other side. Solutions are so called winning strategies for the existential player that specify how to react on moves—certain fixations of universally quantified variables—of the universal player to certainly win the game. In this setting the existential player must ensure the fulfillment of a system of linear constraints, while the universal variables can range within given intervals, trying to make the fulfillment impossible. Recently, this approach was extended by adding a linear constraint system the universal player must obey. Consequently, existential and universal variable assignments in early decision stages now can restrain possible universal variable assignments later on and vice versa resulting in a multistage optimization problem with decision-dependent uncertainty. We present an attenuated variant, which instead of an NP-complete decision problem allows a polynomial-time decision on the legality of a move. Its usability is motivated by several examples.

Keywords Robust optimization · Multistage optimization · Decision-dependent uncertainty · Variable uncertainty

1 Introduction

Optimization under uncertainty often pushes the complexity of problems that are in the complexity class P or NP, to PSPACE [14]. Nevertheless, dealing with uncertainty is an important aspect of planning and various solution paradigms for optimization under uncertainty exist, e.g. Stochastic Programming [3] and Robust

M. Hartisch (✉) · U. Lorenz
University of Siegen, Siegen, Germany
e-mail: michael.hartisch@uni-siegen.de

J. S. Neufeld et al. (eds.), *Operations Research Proceedings 2019*, Operations Research Proceedings, https://doi.org/10.1007/978-3-030-48439-2_53

Optimization [2]. In most settings it is assumed that the occurring uncertainty is embedded in a predetermined uncertainty set or that it obeys a fixed random distribution. In particular, planning decisions have no influence on uncertainty. Decision-dependent uncertainty has recently gained importance in both stochastic programming [1, 5, 8, 9] and robust optimization [11, 13, 15, 16]. We focus on quantified integer programming (QIP) [12], which is a robust multistage optimization problem. Only recently, an extension for QIP was presented such that existential and universal variable assignments in early decision stages now can restrain possible universal variable assignments later on and vice versa resulting in a multistage optimization problem with decision-dependent uncertainty [6]. The aim of this paper is to investigate the implications and possibilities of this extension for operations research.

2 Quantified Integer Programs with Interdependent Domains

Quantified Integer Programs (QIP) are Integer Programs (IP) extended by an explicit variable order and a quantification vector that binds each variable to a universal or existential quantifier. Existentially quantified variables depict decisions made by a planner, whereas universally quantified variables represent uncertain events the planner must cope with. In particular, a QIP can be interpreted as a zero-sum game between a player assigning existentially quantified variables against the player fixing the universally quantified variables. The first priority of the so-called existential player is the fulfillment of the existential constraint system $A^{\exists}x \leq b^{\exists}$ when all variables x are fixed. A solution of a QIP is a strategy for assigning existentially quantified variables, that specifies how to react on moves of the universal player—i.e. assignments of universally quantified variables—to certainly fulfill $A^{\exists}x \leq b^{\exists}$. By adding a min-max objective function the aim is to find the best strategy [12].

Definition 1 (Quantified Integer Program) Let $A^{\exists} \in \mathbb{Q}^{m_{\exists} \times n}$ and $b^{\exists} \in \mathbb{Q}^{m_{\exists}}$ for $n, m_{\exists} \in \mathbb{N}$ and let $\mathcal{L} = \{x \in \mathbb{Z}^n \mid x \in [l, u]\}$ with $l, u \in \mathbb{Z}^n$. Let $Q \in \{\exists, \forall\}^n$ be a vector of quantifiers. We call each maximal consecutive subsequence in Q consisting of identical quantifiers a *quantifier block* and denote the i-th block as $B_i \subseteq \{1, \ldots, n\}$ and the corresponding quantifier by $Q^{(i)} \in \{\exists, \forall\}$, the corresponding variables by $x^{(i)}$ and its domain by $\mathcal{L}^{(i)}$. Let $\beta \in \mathbb{N}, \beta \leq n$, denote the number of blocks. Let $c \in \mathbb{Q}^n$ be the vector of objective coefficients and let $c^{(i)}$ denote the vector of coefficients belonging to block B_i. Let $Q \circ x \in \mathcal{L}$ with the component wise binding operator $\circ$ denote the *quantification vector* $(Q^{(1)}x^{(1)} \in$

$\mathcal{L}^{(1)}, \ldots, Q^{(\beta)}x^{(\beta)} \in \mathcal{L}^{(\beta)})$ such that every quantifier $Q^{(i)}$ binds the variables $x^{(i)}$ of block i to its domain $\mathcal{L}^{(i)}$. We call

$$\min_{x^{(1)}\in\mathcal{L}^{(1)}} \left(c^{(1)}x^{(1)} + \max_{x^{(2)}\in\mathcal{L}^{(2)}} \left(c^{(2)}x^{(2)} + \ldots \min_{x^{(\beta)}\in\mathcal{L}^{(\beta)}} c^{(\beta)}x^{(\beta)}\right)\right)$$

$$\text{s.t.} \quad Q \circ x \in \mathcal{L}:\ A^{\exists}x \leq b^{\exists}$$

a QIP with objective function (for a minimizing existential player).

In the above setting the universally quantified variables only must obey the hypercube $\mathcal{L}$ given by the variable bounds. Hence, QIPs are rather unsymmetric as—even though the min-max semantics is symmetrical—only the existential player has to deal with a polytope (given by $A^{\exists}x \leq b^{\exists}$) the universal player can modify. In [7] this setting was extended to allow a polyhedral domain for the universal variables given by a second constraint system $A^{\forall}x \leq b^{\forall}$. However, still the existential player's variables had no influence on this system. Only recently, a further extension was presented allowing the interdependence of both variable domains [6]. The presented Quantified Integer Program with Interdependent Domains (QIP^{ID}) required the definition of a legal variable assignment, since now the case that both constraint systems are violated could occur and the player who made the first illegal move loses (we refer to [6] for more details).

Definition 2 (Legal Variable Assignment) For variable block $i \in \{1, \ldots, \beta\}$ the set of legal variable assignments $\mathcal{F}^{(i)}(\tilde{x}^{(1)}, \ldots, \tilde{x}^{(i-1)})$ depends on the assignment of previous variable blocks $\tilde{x}^{(1)}, \ldots, \tilde{x}^{(i-1)}$ and is given by

$$\mathcal{F}^{(i)} = \left\{\hat{x}^{(i)} \in \mathcal{L}^{(i)} \mid \exists x = (\tilde{x}^{(1)}, \ldots, \tilde{x}^{(i-1)}, \hat{x}^{(i)}, x^{(i+1)}, \ldots, x^{(\beta)}) \in \mathcal{L} : A^{Q^{(i)}}x \leq b^{Q^{(i)}}\right\}$$

i.e. after assigning the variables of block i there still must exist an assignment of x such that the system of $Q^{(i)} \in \{\exists, \forall\}$ is fulfilled. The dependence on the previous variables $\tilde{x}^{(1)}, \ldots, \tilde{x}^{(i-1)}$ will be omitted when clear.

Hence, even a local information—whether a variable is allowed to be set to a specific value—demands the solution of an NP-complete problem. Just like QIP, QIP^{ID} is PSPACE-complete [6].

Definition 3 (QIP with Interdependent Domains (QIP^{ID})) For given $A^{\forall}$, $A^{\exists}$, $b^{\forall}$, $b^{\exists}$, c, $\mathcal{L}$ and Q with $\{x \in \mathcal{L} \mid A^{\forall}x \leq b^{\forall}\} \neq \emptyset$ we call

$$\min_{x^{(1)}\in\mathcal{F}^{(1)}} \left(c^{(1)}x^{(1)} + \max_{x^{(2)}\in\mathcal{F}^{(2)}} \left(c^{(2)}x^{(2)} + \ldots \max_{x^{(\beta)}\in\mathcal{F}^{(\beta)}} c^{(\beta)}x^{(\beta)}\right)\right)$$

$$\text{s.t.} \quad \exists x^{(1)} \in \mathcal{F}^{(1)}\ \forall x^{(2)} \in \mathcal{F}^{(2)} \ldots \forall x^{(\beta)} \in \mathcal{F}^{(\beta)} :\ A^{\exists}x \leq b^{\exists}$$

a *Quantified Integer Program with Interdependent Domains* (QIP^{ID}).

We say a player $q \in \{\exists, \forall\}$ loses, if either a) $A^q \tilde{x} \not\leq b^q$ for a fully assigned $\tilde{x}$ or b) if there exists no legal move for this player at some point during the game, i.e. $\mathcal{F}^{(i)} = \emptyset$. As we will see in the following section, a general QIP^{ID} is too comprehensive for most problems of the OR-world and a few restrictions are sufficient in order to simplify the solution process.

3 Addition Structural Requirements for $A^{\forall}x \leq b^{\forall}$

The recurring NP-complete evaluation of $\mathcal{F}^{(i)}$ constitutes a massive overload when solving a QIP^{ID} via game-tree search [4]. In a general QIP^{ID} it can occur that the universal player has no strategy in order to ensure the fulfillment of $A^{\forall}x \leq b^{\forall}$. This makes sense in an actual two-person game where both players could lose. In an OR-setting, however, the universal player can be considered to be the one who decides which uncertain event will occur, the scope of which depends on previous planning decisions. But obviously there exists no planning decision that obliterates uncertainty in the sense that there is no further legal assignment for universal variables such that uncertainty "loses". Therefore, we make the following assumptions:

a) For each universal variable block $i \in \{1, \ldots, \beta\}$ we demand $\forall \left(\hat{x}^{(1)} \in \mathcal{L}^{(1)}, \hat{x}^{(2)} \in \mathcal{F}^{(2)}, \ldots, \hat{x}^{(i-2)} \in \mathcal{F}^{(i-2)}, \hat{x}^{(i-1)} \in \mathcal{L}^{(i-1)}\right) : \mathcal{F}^{(i)} \neq \emptyset$.
b) Let $i \in \{1, \ldots, \beta\}$ with $Q^{(i)} = \forall$ and let $\hat{x}^{(1)} \in \mathcal{L}^{(1)}, \ldots, \hat{x}^{(i-1)} \in \mathcal{L}^{(i-1)}$ be a partial variable assignment up to block i. If $\tilde{x}^{(i)} \notin \mathcal{F}^{(i)}(\hat{x}^{(1)}, \ldots, \hat{x}^{(i-1)})$ then

$$\exists k \in \{1, \ldots, m_{\forall}\} : \sum_{j<i} A^{\forall}_{k,(j)} \hat{x}^{(j)} + A^{\forall}_{k,(i)} \tilde{x}^{(i)} + \sum_{j>i} \min_{x^{(j)} \in \mathcal{L}^{(j)}} A^{\forall}_{k,(j)} x^{(j)} \not\leq b^{\forall} .$$

Restriction a) requests, that there always exists a legal move for the universal player, even if the existential player does not play legally. In particular, previous variable assignments—although they can restrict the set of legal moves— can never make $A^{\forall}x \leq b^{\forall}$ unfulfillable. In b) we demand that a universal variable assignment $\tilde{x}^{(i)} \in \mathcal{L}^{(i)}$ is illegal, if there is a universal constraint that cannot be fulfilled, even in the best case. Therefore, it is sufficient to check separately the constraints in which $\tilde{x}^{(i)}$ is present in order to ensure $\tilde{x}^{(i)} \in \mathcal{F}^{(i)}$. Hence, there always exists a strategy for the universal player to fulfill $A^{\forall}x \leq b^{\forall}$ (due to a)) and further checking $x^{(i)} \in \mathcal{F}^{(i)}$ can be done in polynomial time (due to a) and b)) for universal variables. The legality of existential variable assignments does not have to be checked immediately (due to a)) and can be left to the search.

4 Application Examples

In this section we briefly describe a few examples where QIP^{ID} can be used in order to grasp the relevance of multistage robust optimization with decision-dependent uncertainty. We will not explicitly specify how the described problems can be translated into linear constraints, but note, that all the upcoming examples can be modeled as QIP^{ID} while meeting the requirements described in Sect. 3. Further, keep in mind that QIP^{ID} is a multistage optimization framework. Therefore, rather than adhering to adjustable robust programming with only a single response stage, planning problems with multiple decision stages are realizable.

Maintenance Reduces Downtime Consider a job shop problem with several tasks and machines. One is interested in a robust schedule as machines can fail for a certain amount of time (universal variables indicate which machines fail and how long they fail). The basic problem can be enhanced by adding maintenance work to the set of jobs: the maintenance of a machine prevents its failure for a certain amount of time at the expense of the time required for maintenance and the maintenance costs. Therefore, the universal constraint system contains constraints describing the relationship between maintenance and failure: With existential variable $m_{i,t}$ indicating the maintenance of machine i at time t and universal variable $f_{i,t}$ indicating the failure of machine i at time t the (universal) constraint $f_{i,t+j} \leq 1 - m_{i,t}$ prohibits the failure of machine i for each of the $j \in \{0, \ldots, K\}$ subsequent time periods. The universal constraint system also could contain further restrictions regarding the number of machines allowed to fail at the same time, analogous to budget constraints common in robust optimization [15]. This budget amount also can depend on previous planning decisions, e.g. the overall machine utilization. Further, reduced operation speed can reduce wear and therefore increase durability and lessen the risk of failure.

Workers' Skills Affect Sources of Uncertainty The assignment of employees to various tasks may have significant impact on potential occurring failures, processing times and the quality of the product. For example, it might be cheaper to have a trainee carry out a task, but the risk of error is higher and the processing time might increase. Further, some worker might work slower but with more diligence—resulting in a long processing-time but a high quality output—than other faster, but sloppier, workers. Hence, the decision which worker performs a particular task has an impact on the anticipated uncertain events. In a more global perspective staff training and health-promoting measures affect the skills and availability of a worker, and thereby affecting potential risks.

Road Maintenance for Disaster Control In order to mitigate the impact of a disaster, road rehabilitation can improve traveling time as the damage of such roads can be reduced (see [13]). Again, a budget for the deterioration of travel times for all roads could be implemented, whereat the budget amount could be influenced by the number of emergency personal, emergency vehicles and technical equipment made available.

Time-Dependent Factors in Process Scheduling In [10] the authors present a process scheduling approach with uncertain processing-time of the jobs, whereat the range of the uncertain processing-time parameters depend on the scheduling time of the job itself. The selection of specific scheduling times therefore actively determines the range in which the uncertain processing-times are expected. For a QIP this influence on uncertainty could be achieved by adding universal constraints as follows: Let $x_{i,t}$ be the (existential) binary indicator whether task i is scheduled to start at time t and let y_i be the (universal) variable indicating the occurring processing-time of task i. Let $l_{i,t}$ and $u_{i,t}$ indicate the range of the processing-time of task i if scheduled at t. Adding $\sum_t l_{i,t}x_{i,t} \leq y_i \leq \sum_t u_{i,t}x_{i,t}$ to the universal constraint system would establish the intended interdependence.

5 Conclusion

We addressed the largely neglected potential of optimization under decision-dependent uncertainty. In the scope of quantified integer programming with decision-dependent uncertainty we presented reasonable restrictions such that a game-tree search algorithm must not cope with recurring NP-complete subproblems but rather polynomial evaluations. Further, we provided several examples where such a framework is applicable.

Acknowledgments This research is partially supported by the German Research Foundation (DFG) project "Advanced algorithms and heuristics for solving quantified mixed - integer linear programs".

References

1. Apap, R., Grossmann, I.: Models and computational strategies for multistage stochastic programming under endogenous and exogenous uncertainties. Comput. Chem. Eng. **103**, 233–274 (2017)
2. Ben-Tal, A., Ghaoui, L.E., Nemirovski, A.: Robust Optimization. Princeton University Press, Princeton (2009)
3. Birge, J., Louveaux, F.: Introduction to Stochastic Programming, 2nd edn. Springer, New York (2011)
4. Ederer, T., Hartisch, M., Lorenz, U., Opfer, T., Wolf, J.: Yasol: an open source solver for quantified mixed integer programs. In: Advances in Computer Games - 15th International Conferences, ACG 2017, pp. 224–233 (2017)
5. Gupta, V., Grossmann, I.: A new decomposition algorithm for multistage stochastic programs with endogenous uncertainties. Comput. Chem. Eng. **62**, 62–79 (2014)
6. Hartisch, M., Lorenz, U.: Mastering uncertainty: towards robust multistage optimization with decision dependent uncertainty. In: PRICAI 2019: Trends in Artificial Intelligence, pp. 446–458. Springer, Berlin (2019)
7. Hartisch, M., Ederer, T., Lorenz, U., Wolf, J.: Quantified integer programs with polyhedral uncertainty set. In: Computers and Games - 9th International Conference, CG 2016, pp. 156–166 (2016)

8. Hellemo, L., Barton, P.I., Tomasgard, A.: Decision-dependent probabilities in stochastic programs with recourse. Comput. Manag. Sci. **15**(3-4), 369–395 (2018)
9. Jonsbråten, T., Wets, R.J.B., Woodruff, D.: A class of stochastic programs with decision dependent random elements. Ann. Oper. Res. **82**(0), 83–106 (1998)
10. Lappas, N., Gounaris, C.: Multi-stage adjustable robust optimization for process scheduling under uncertainty. AIChE J. **62**(5), 1646–1667 (2016)
11. Lappas, N.H., Gounaris, C.E.: Robust optimization for decision-making under endogenous uncertainty. Comput. Chem. Eng. **111**, 252–266 (2018)
12. Lorenz, U., Wolf, J.: Solving multistage quantified linear optimization problems with the alpha–beta nested benders decomposition. EURO J. Comput. Optim. **3**(4), 349–370 (2015)
13. Nohadani, O., Sharma, K.: Optimization under decision-dependent uncertainty. SIAM J. Optim. **28**(2), 1773–1795 (2018)
14. Papadimitriou, C.: Games against nature. J. Comput. Syst. Sci. **31**(2), 288–301 (1985)
15. Poss, M.: Robust combinatorial optimization with variable cost uncertainty. Eur. J. Oper. Res. **237**(3), 836–845 (2014)
16. Vujanic, R., Goulart, P., Morari, M.: Robust optimization of schedules affected by uncertain events. J. Optim. Theory Appl. **171**(3), 1033–1054 (2016)

Examination and Application of Aircraft Reliability in Flight Scheduling and Tail Assignment

Martin Lindner and Hartmut Fricke

Abstract A failure of an aircraft component during flight operations could lead to the grounding of an aircraft (AOG) until fault rectification is completed. This often results to high costs due to flight cancellations and delay propagation. With the technology of the digital twin, which is a virtual copy of a real aircraft, predictions of the technical reliability of aircraft components and thus the availability of the aircraft itself have recently become available. In the context of the combinatorial problem of aircraft resource planning, we examine how the predicted dispatch reliability of an aircraft could be used to achieve robustness of the schedule against AOG. We gain robustness only by flight scheduling, aircraft assignment and optimization of aircraft utilization, thus we avoid the use of expensive reserve ground times. We extend an integrated tail assignment and aircraft routing problem by "dispatch reliability" as a result from a digital twin. We disturb the flight schedule with random AOG cases, determine costs related to delay and flight cancellations, and improve robustness by taking into account the AOG-related costs in the objective function.

Keywords Airline dispatch reliability · Flight scheduling · Tail assignment

1 Introduction

1.1 Motivation

Internal or external disturbances in flight operations cause deviations from the original flight schedule and often result in delays and flight cancellations consequently. Approximately 13% of all flight cancellations are carrier-related and of technical nature. Thus and for safety reasons, airlines monitor the technical reliability of

M. Lindner (✉) · H. Fricke
Technische Universität Dresden, Dresden, Germany
e-mail: Martin_Lindner@tu-dresden.de

J. S. Neufeld et al. (eds.), *Operations Research Proceedings 2019*, Operations Research Proceedings, https://doi.org/10.1007/978-3-030-48439-2_54

aircraft components by means of regulations in EASA Part-M. Although high reliability (99–99.8%) is standard in air transportation, malfunctions of components can cause a grounding of the entire aircraft ("Aircraft on Ground", AOG) resulting in flight schedule disruptions. The AOG affects only a few flights, but it is of long duration and can seriously disrupt passenger or freight connections with passenger compensation costs of thousands of Euros per flight. With the emerging technology of the *digital twin*, which is a virtual representation of a real aircraft, it is now possible to assess aircraft data and scheduled maintenance events to predict reliability of components. This provides even for pre-departure planning phases (aircraft routing, tail assignment) a probability that an aircraft will be available for future flight operations. In this work, we evaluate the benefit of predicted aircraft reliability towards the reduction of delay costs caused by AOG.

1.2 Aircraft Routing and Assignment Problem

Problems of flight scheduling and aircraft routing have already been comprehensively investigated in the scientific literature. The tail assignment problem usually includes aircraft routing and maintenance requirements [1, 2] and the quality of solution could be enhanced by individual operating costs [3]. To consider disturbances during flight operations, uncertainties in costs, punctuality or delay propagation are also proposed. For example, Yan and Jerry minimize the maximal possible delay due air traffic and weather [4]. Recently, Marla and Barnhart [5] and Xu [6] studied several approaches to achieve robust scheduling solutions. However, those studies consider delay only caused by weather or traffic, affecting multiple aircraft rotations at the same time and only for a short duration. However, an AOG is a very special disturbance type of long but uncertain delay duration. Today, reserve aircraft (without scheduled flight operations) are preferred to manage the robustness against those kind of disturbances. However, due to the high cost the unproductivity, reserve aircraft are is not an option for smaller airlines. Since additional time buffers also limit productivity, we design the schedule by expecting probable disturbances. Therefore, we use the new available aircraft reliability assessment by digital twin and contribute to schedule robustness by applying only scheduling recovery strategies (e.g., [7]) to lower disturbance costs.

2 Model Formulation

A flight schedule consists of a set of events (flight, maintenance, ground, operational reserve, or delay) and a limited number of aircraft. The robustness of a flight schedule is the ability to return efficiently (low costs) to the original schedule in case of any disruption. In our approach, we increase robustness by anticipating disturbances. First, we introduce the general Tail Assignment Problem (TAP), which

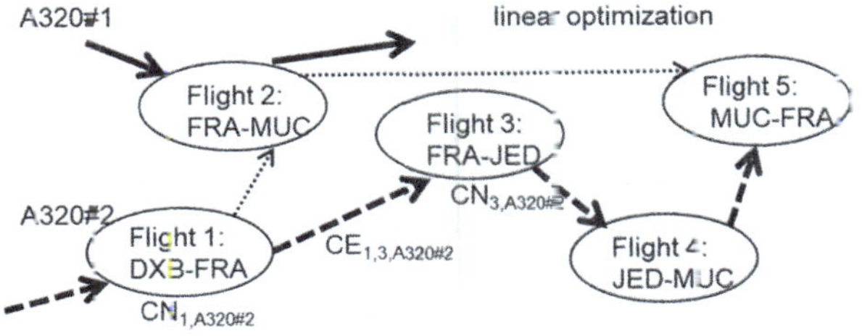

Fig. 1 Example of two aircraft rotations in a solution. Bold solid and dashed lines represent in each case an aircraft rotation. Dotted lines represent possible but unused connections

schedules events and assigns them to aircraft under an objective function (e.g., minimum operating costs). To improve the robustness of the solution, we randomly create AOG disturbances and re-calculate the solution using schedule recovery strategies with a Tail Assignment Recovery Problem (TARP). From the set of re-calculated solutions, we chose the most robust solution by minimum of total delay and cancellation costs for each disturbance.

2.1 *Tail Assignment Problem TAP*

Our TAP assigns events to aircraft and is formulated as a Vehicle Routing Problem with Time Windows (VRPTW) and is described in more detail in [3]. Let $G = \{J, A\}$ be a directed graph representing a set of events with $J = \{0, \ldots, n\}$ and a set of arcs with $A = \{(i,j) : i,j \in J\}$. Each event $i \in J$ is a node and has a departure, a destination, corresponding times, a set of required skills and an individual cost for each aircraft k in set $K=\{1, \ldots, k\}$. The binary decision variable x_{ijk} decides whether an event j is after an event i, both served by an aircraft k (c.f. Fig. 1). Both cost parameters $C_{N,\,ik}$ for job serving and $C_{E,\,ijk}$ for arc creation will influence the target function. The TAP target function minimizes the total operating costs (Eq. 1):

$$\min \sum_{k \in K} \sum_{i \in J} \sum_{j \in J} x_{ijk} \left(CN_{ijk} + CE_{ik} \right) \tag{1}$$

CN_{ik}: Node costs (Event i served by aircraft k)

CE_{ijk}: Costs for connecting event i and j by aircraft k (e.g. Ground event, positioning flights)

Events are scheduled by general VRP hard time window formulation and commonly used flow constraints.

2.2 Flight Schedule Disturbances Based on Aircraft Reliability

Aircraft technical reliability is related to the aircraft's technical systems and components (e.g., wheels, engines) expressing a long-term probability for malfunction or failure. Airlines monitor the reliability during continuing airworthiness programmes and recently, in more detail and by using confident prediction modules in digital twins. *Aircraft operational reliability (AOR)*, or former *dispatch reliability*, is then the probability that an aircraft will be available for dispatch based on its technical reliability [8]. $AOR_{k,pred}$ is then the predicted operational reliability of aircraft $k \in K$ in a specified time window.

Under the assumption that a component failure results in a non-acceptable degraded mode, each disturbance resulting from $AOR_{k,pred}$ provokes an technical AOG defined as flight schedule disturbance *d* based on the following random variables:

- Occurrence of AOG *A* of an aircraft $k \in K$, where the failure rate is provided from the manufacturer/health monitoring: $A : \Omega_A \rightarrow \mathbb{R}$ with $\Omega_A = \{1, \ldots, \mathrm{K}\}$
- Time *T* of AOG occurrence, where occurrence is between earliest (tmin) and latest (tmax) time stamp in the schedule: $T : \Omega_T \rightarrow \mathbb{R}$ with $\Omega_T = \{\mathrm{tmin}, \ldots, \mathrm{tmax}\}$
- Duration *R* of fault rectification: $R : \Omega_R \rightarrow \mathbb{R}$ with $\Omega_R = \{0, \ldots, \mathrm{tmax}\}$

2.3 Tail Assignment Recovery Problem TARP

Applying a disturbance *d* (AOG with a ground time of the aircraft until fault rectification) to the current TAP solution, a new schedule optimization should be started to recover the damage of the disruption. The TARP model formulations extends the TAP by a decision of flight cancellation, delay propagation and re-assignment of flights and aircraft. If the occurrence time *T* of the AOG *A* including fault rectification *R* is later than the departure time of the next scheduled flight, the assessment will generate either delay with or without aircraft change or flight cancellation of one or more flights (cases cf. Fig. 2).

The TARP uses the TAP equations (aircraft re-assignment) from [3] and Sect. 2.3 with following modifications:

- Flight cancellations: introduction of sufficient number of *n* additional virtual aircraft $K=\{1, \ldots, k+n\}$, where CN_{in} are flight cancellation costs for flight *i* and $CE_{ijk}=0$.
- Delay: A soft time window formulation from [9] is implemented with costs in case of a deviation from target times (delay). The soft and hard time windows (TW) for each flight *i* will be adjusted:
  - Hard $\mathrm{TW_{TARP}}$ $\mathrm{open_i}$ = Hard $\mathrm{TW_{TAP}}$ $\mathrm{open_i}$ (Scheduled time of departure (STD))
  - Soft $\mathrm{TW_{TARP}}$ $\mathrm{open_i}$ = Hard $\mathrm{TW_{TAP}}$ $\mathrm{open_i}$

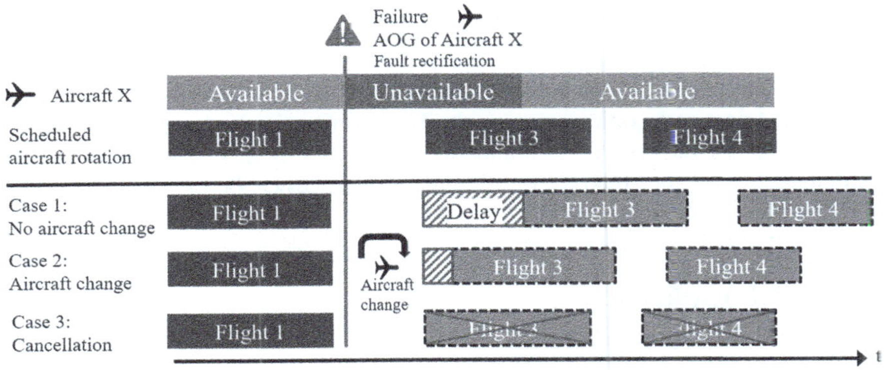

Fig. 2 AOG disturbances and recovery actions

- Soft TW_{TARP} $close_i$ = Hard TW_{TAP} $close_i$ (Scheduled time of arrival)
- Hard TW_{TARP} $close_i$ = Hard TW_{TAP} $close_i$ + AOG duration, if STD > AOG occurrence time

A deviation between Soft TW and scheduled time of departure (delay) is charged by a linear time cost factor in the target function (crew duty time, passenger compensation, loss of image, etc.). In this way, the model tries to schedule as many flights as possible within the Soft TW boundaries to reduce the total delay. Delay is eliminated during turnarounds if there is more ground time than necessary for ground service.

2.4 *AOG Robust Problem Formulation*

In the next step and by using a Monte Carlo Simulation, a robust flight schedule is determined for a set of disturbances d in set $D = \{1, \ldots, d\}$. First, for each d a TARP solution $S^{d,\,TARP}$ is calculated and stored in a solution pool $SP=\{1, \ldots, d\}$. Second, each disturbances, indexed by c∈D, stresses each solution $S^{d,\,TARP}$ from SP resulting in new total operating, delay and cancellation costs ${}_{cost}S_c^{d,TARP}$. The solution with the lowest sum of total costs (Eq. 2) for all disturbances is defined as most robust. The binary decision variable y_d identifies this solution (Eq. 3).

$$\min \sum_{d \in D} y_d \left(\sum_{c \in D} {}_{cost}S_c^{d,TARP} \right) \tag{2}$$

$$\text{s.t.} \qquad \sum_{d \in D} y_d = 1, \qquad y_d \in \{0, 1\} \tag{3}$$

2.5 *Problem Formulation and Solver Modules*

The parameterization, problem formulations [3], solver, and assessment is embedded in a JAVA optimization framework. The problem itself can be solved using common Mixed Integer Programming solver (e.g. CPLEX, Gurobi) or heuristics (Simulated Annealing, Tabu Search) for larger instances (>300 flights).

3 Model Application

In the next steps, a set of scheduled flights is randomly generated and applied to the robust model formulation. Table 1 summarizes the used parameters and Fig. 3 shows the flight schedule with an AOG event *d* for AC5 and the propagated delay in red.

In a first step, the optimal flight schedule is calculated with minimum direct operating cost and without any disturbances. As described in Sect. 2.3, this solution is assessed in terms of total costs for all disturbances. Subsequently, as described in Sect. 2.4, a robust flight schedule is determined from the solution pool *SP*. Furthermore, we calculate a third solution by the total cost minimization of delay and direct operating cost. The results are summarized in Table 2.

Table 1 Parametrization of the schedule and AOG probabilities (1-AOR) for each aircraft

$t_{min}-t_{max}$	0–2837 min
Aircraft	8
Flight events	100 (dur. 90–220 min)
Maintenance events	4 (dur. 92–327 min)
Disturbances C	1000 (dur. 0–119 min)

k	$1\text{-}AOR_k$
1 (AC0)	0.02
2 (AC1)	0.018
3 (AC2)	0.015
4 (AC3)	0.013
5 (AC5)	0.01
6 (AC6)	0.007
7 (AC7)	0.005
8 (AC8)	0.003

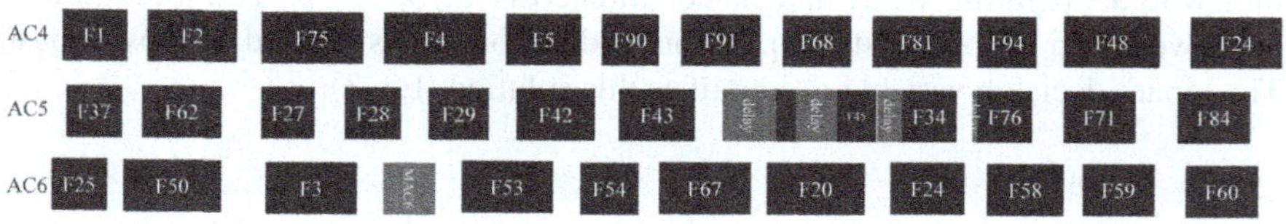

Fig. 3 Excerpt from the flight schedule. Blue bars are flight events and grey maintenance events, respectively. Red bars indicate propagating delay from AOG of AC5 after flight F43

Table 2 Results of robust flight schedule optimization

Solution	DOC		$\sum_{d \in D}$ cost (DOC, delay)$_d$		Total
min DOC	1,778,714	100%	242,806	100%	100%
max robust	1,779,048	100.02%	162,174	66.79%	96.03%
DOC/robust	1,778,844	100.01%	233,409	96.13%	99.54%

4 Conclusion and Outlook

The application case shows that the robust flight schedule can reduce the damage caused by probable AOG by accepting a slight increase in direct operating cost. Additionally, a risk value as a combination of probability of occurrence AOR and average damage of an AOG can be taken into account in resource scheduling. For example, this can be added to the direct operating cost as an additional cost component per aircraft/flight assignment in order to evaluate possible AOGs during deterministic modelling.

References

1. Lagos, C.F., Delgado, F., Klapp, M.A.: Dynamic optimization for airline maintenance operations. Engineering School, Pontificia Universidad Católica de Chile (2019)
2. Rivera-Gómez, H., Montaño-Arango, O., Corona-Armenta, J., Garnica-González, J., Hernández-Gress, E., Barragán-Vite, L: Production and maintenance planning for a deteriorating system with operation-dependent defectives. Appl. Sci. **8**(2), 165 (2018)
3. Lindner, M., Rosenow, J., Förster, S., Fricke, H.: Potential of integrated flight scheduling and rotation planning considering aerodynamic-, engine- and mass-related aircraft deterioration. CEAS Aeronaut. J. **10**(3), 755–770 (2018)
4. Yan, C., Jerry, K.: Robust aircraft routing. Transp. Sci. **52**, 118–133 (2015)
5. Marla, L., Vaze, V., Barnhart, C.: Robust optimization: lessons learned from aircraft routing. Comput. Oper. Res. **98**, 165–184 (2018)
6. Xu, Y., Wandelt, S., Sun, X.: Stochastic tail assignment under recovery. Thirteenth USA/Europe Air Traffic Management Research and Development Seminar (2019)
7. Maher, S.: Solving the integrated airline recovery problem using column-and-row generation. Transp. Sci. **50**, 216–239 (2015)
8. International Air Transport Association (IATA): Aircraft Operational Availability (2018)
9. Calvete, H., Galé, C., Sánchez-Valverde, B., Oliveros, M.: Vehicle routing problems with soft time windows: an optimization based approach. VIII Journées Zaragoza-Pau de Mathématiques Appliquées et de Statistiques: Jaca, pp. 295–304. ISBN 84-7733-720-9 (2004)

Part XIII
OR in Engineering

Comparison of Piecewise Linearization Techniques to Model Electric Motor Efficiency Maps: A Computational Study

Philipp Leise, Nicolai Simon, and Lena C. Altherr

Abstract To maximize the travel distances of battery electric vehicles such as cars or buses for a given amount of stored energy, their powertrains are optimized energetically. One key part within optimization models for electric powertrains is the efficiency map of the electric motor. The underlying function is usually highly nonlinear and nonconvex and leads to major challenges within a global optimization process. To enable faster solution times, one possibility is the usage of piecewise linearization techniques to approximate the nonlinear efficiency map with linear constraints. Therefore, we evaluate the influence of different piecewise linearization modeling techniques on the overall solution process and compare the solution time and accuracy for methods with and without explicitly used binary variables.

Keywords MINLP · Powertrain · Piecewise linearization · Efficiency optimization

1 Introduction

To enable the optimal design of powertrains within vehicles, different types of optimization methods, both heuristics and exact methods, are commonly used. For the powertrain optimization of a battery electric vehicle (BEV), it is mandatory to model the efficiency of the used electric motor, as shown in [1]. The result is a nonlinear multidimensional function, in which the independent variables are the torque and angular speed of the motor and the dependent variable is the motor efficiency. If

P. Leise (✉) · N. Simon
Department of Mechanical Engineering, Technische Universität Darmstadt, Darmstadt, Germany
e-mail: philipp.leise@fst.tu-darmstadt.de; nicolai.simon@stud.tu-darmstadt.de

L. C. Altherr
Faculty of Energy, Building Services and Environmental Engineering, Münster University of Applied Sciences, Steinfurt, Germany
e-mail: lena.altherr@fh-muenster.de

J. S. Neufeld et al. (eds.), *Operations Research Proceedings 2019*, Operations Research Proceedings, https://doi.org/10.1007/978-3-030-48439-2_55

Mixed-Integer Nonlinear Programming (MINLP) models, cf. [2], are used to enable a global-optimal powertrain design, it is potentially computationally beneficial to reduce highly nonlinear constraints within the program, as introduced by the motor efficiency map. One commonly used method to ensure linear constraints within the program is the usage of piecewise linearization (PWL) techniques, cf. [3–5]. This approach is widely used, as for example shown in [6] or [7]. Within the literature, multiple modeling methods exist, for an overview compare e.g. [4].

2 Implemented Piecewise Linearization Techniques

We compare three different PWL methods to approximate functions of the form $\eta = f(t, \omega)$ where $f : \mathbb{R}^2 \rightarrow \mathbb{R}$ describes a nonlinear relationship within the unit cube; $t \in [0, 1]$ (normalized torque), $\omega \in [0, 1]$ (normalized speed), $\eta \in [0, 1]$ (efficiency). In our case f is the measured or calculated normalized efficiency map of a permanent magnet synchronous motor. Details on the underlying technical and physical relationships and on the considered optimization model are given in [8]. To compare the different PWL methods on the same grid, we use the 1–4 orientation, cf. [5], for triangulation and divide the domain in non-overlapping simplices. The used sets are shown in Table 1.

The first two investigated methods are called convex combination (CC) and disaggregated convex combination (DCC) and are described in detail by Vielma, Ahmed and Nemhauser in [3]. Within both methods, the selection of a specific simplex, a triangle within the considered bi-variate piecewise linearization, is modeled with the help of $k \in \{1, ..., |\mathcal{S}|\}$ binary variables, $b_k \in \{0, 1\}$. Each vertex $(i, j) : i \in \mathcal{I}; j \in \mathcal{J}$ is represented in the CC-method by a continuous variable $\lambda_{i,j} \in [0, 1]$, and in the DCC-method by as many continuous variables as there are adjacent simplices, $(\lambda_{i,j,s} : i \in \mathcal{I}; j \in \mathcal{J}; s \in \mathcal{S}_a(i, j))$. This increase of continuous variables compared to the CC method is potentially advantageous for a low number of simplices, that are used to approximate the underlying function f. The third method is based on constraints with special ordered sets (SOS) of type 1 and type 2, a concept first introduced by Beale and Tomlin in [9]. Using this concept, the third method, which we refer to as SOS, does not require explicit definitions of binary variables to model the selection of different simplices. Instead, it uses additional constraints with special ordered sets of type 1 and 2 and multiple linear constraints. It is shown in detail by Misener and Floudas in [5].

Table 1 Sets for approximation by PWL methods

Set	Description
$\mathcal{S}$	Set of simplices
$\mathcal{S}_a(i, j)$	Set of adjacent simplices at each vertex (i, j)
$\mathcal{I}$	Set of vertices in t-direction
$\mathcal{J}$	Set of vertices in ω-direction

In the following, we will only briefly describe the constraints which are used to approximate the underlying nonlinear function. The continuous variables are $\lambda_{i,j} \in [0, 1]$ for the CC and SOS method and $\lambda_{i,j,s} \in [0, 1]$ for the DCC method. The known values $T_{i,j}$ (normalized torque), $\Omega_{i,j}$ (normalized speed), and $E_{i,j}$ (efficiency) are used to construct a linear combination for each simplex. They are derived from evaluating f. This is shown for the CC and SOS method in Eqs. (1a)–(1d):

$$t = \sum_{i\in\mathcal{I}} \sum_{j\in\mathcal{J}} \lambda_{i,j} T_{i,j}, \tag{1a}$$

$$\omega = \sum_{i\in\mathcal{I}} \sum_{j\in\mathcal{J}} \lambda_{i,j} \Omega_{i,j}, \tag{1b}$$

$$\eta = \sum_{i\in\mathcal{I}} \sum_{j\in\mathcal{J}} \lambda_{i,j} E_{i,j}, \tag{1c}$$

$$\sum_{i\in\mathcal{I}} \sum_{j\in\mathcal{J}} \lambda_{i,j} = 1. \tag{1d}$$

For the corresponding constraints of the DCC method, we refer to [3]. Additionally, further method-dependent constraints, are used in all PWL methods, cf. [3, 5] and we omitted the dependence on the load scenario and gear for a better readability in Eqs. (1a)–(1d).

3 Computational Results

Our computational investigations are based on a MINLP model for the optimization of a transmission system which is presented in [8]. For modeling the motor efficiency maps, cf. [1], we use the afore-mentioned three different PWL techniques and study the influence on the solution time. In order to compare the different methods, we call SCIP 6.0.1, cf. [10], with SoPlex 4.0.1 from Python 3.6.7 on a Linux-based machine with an Intel i7-6600U processor and 16 GB RAM.

In total, we generated 63 instances. We varied the number of gears within the gearbox ($N_g \in \{1, 2, 3\}$), the number of uniformly distributed grid points ($|\mathcal{I}| = |\mathcal{J}|$, where $|\mathcal{I}|, |\mathcal{J}| \in \{10, 20, 30, ..., 60, 70\}$), and the piecewise linearization technique (CC, DCC, SOS). Additionally, we set the total number of load scenarios within the underlying optimization model to four. We used a time limit of 7200 s, an optimality gap limit of 0.5%, and a memory limit of 10 GB. No further solver settings were changed in addition to the previous mentioned. We were able to compute results for 42 instances. The remaining 21 instances reached the time limit.

The computational complexity grows with the considered grid size and the number of gears and depends on the considered PWL method. In general, instances with a high grid size and multiple gears were more likely to reach the time limit. The DCC method caused time limit stops the most, followed by the CC method. The most instances with multiple gears were solvable within the time limit when using the SOS method. The total solution times of SCIP for all instances are shown in Fig. 1a, and the presolving times are shown in Fig. 1b. We omit the 21 instances which reached the time limit and use "grid size" instead of $|\mathcal{I}|$ or $|\mathcal{J}|$ as a label for a better readability in Fig. 1a and b. From our computational experiments it

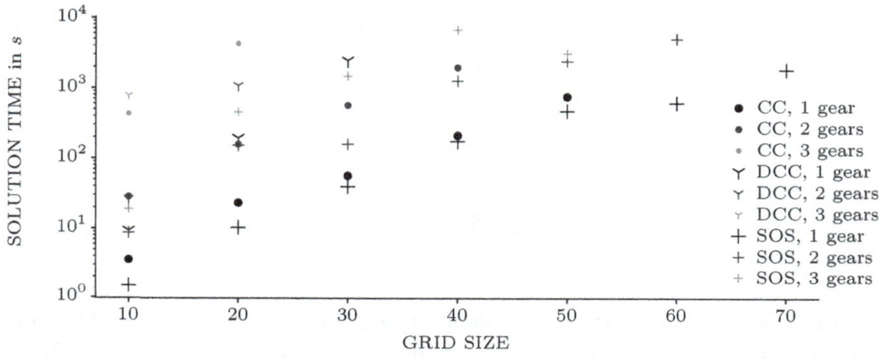

(a)

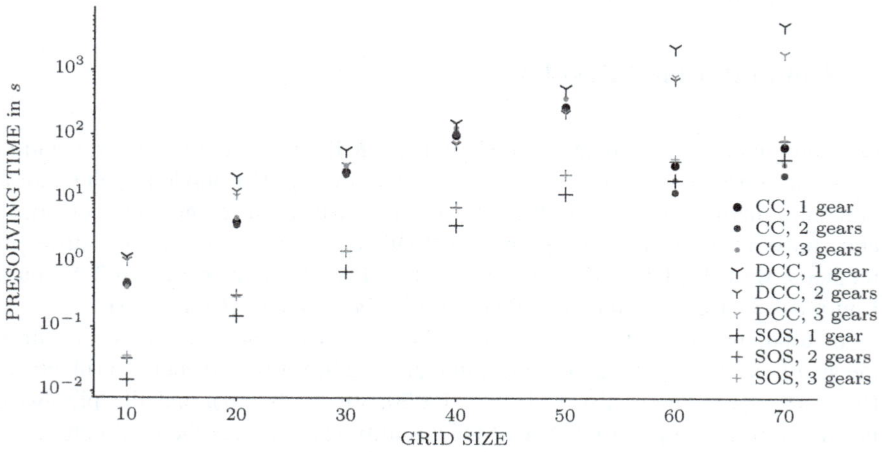

(b)

Fig. 1 (**a**) Dependence of the total solution time on the grid size for all considered PWL methods. Shown are all solvable instances within the time limit. The number of gears in the gearbox ranges from 1 to 3. (**b**) Dependence of the presolving time on the grid size and gears. All shown computations were done with SCIP can be seen that the implemented SOS method yields the fastest solution times. Nevertheless, the CC method is almost comparable to the SOS method for only one gear. The solution time of the DCC method increases rapidly with the considered grid size and is generally higher in comparison to the CC and SOS method.

Furthermore, a non-negligible portion of time is used for presolving, as shown in Fig. 1b, if the grid size increases. In presolving, the SOS method is the fastest, followed by the CC method. The DCC method needs the highest presolving time. Interestingly, the presolving time of the CC method drops at a grid size of 60 × 60 and is from then on almost equivalent to the SOS method's presolving time. Finally, we notice that the presolving time is almost independent of the number of considered gears and increases mostly with the grid size.

To find a good trade-off between computing time and accuracy, we investigated the approximation error of the piecewise linearization as a function of the grid size. The approximation error ε can be calculated by the integral of the difference between the real function $f(t, \omega)$ and piecewise linear approximation $p(t, \omega)$:

$$\varepsilon := \int_0^1 \int_0^1 |f(t, \omega) - p(t, \omega)| \, dt \, d\omega. \tag{2}$$

We computed the approximation error ε for the used quadratic grids with different sizes using a Monte Carlo integration method, and normalized the error by using the approximation error of the 10 × 10 grid. The normalized error $\overline{\varepsilon}$ is shown in Fig. 2 using a logarithmic scale. If the grid size increases, the approximation error tends to zero. However, the rate at which the approximation error decreases becomes very small with larger grid sizes: $\overline{\varepsilon}$ varies only slightly above the grid size of 50 × 50. With this grid size, the computing times, when using the SOS method, of 471 s for

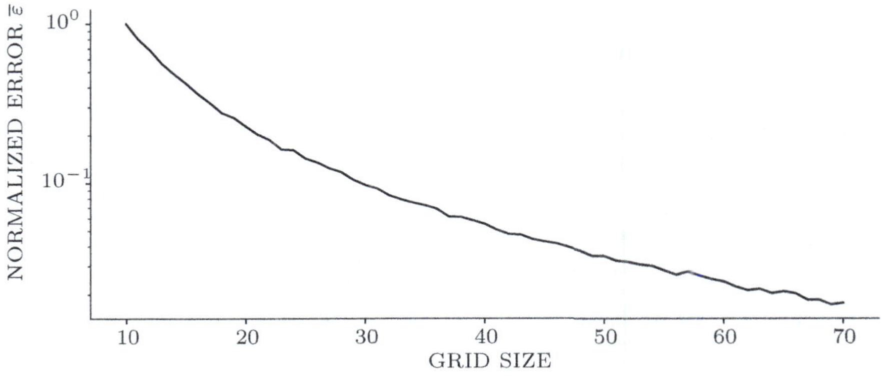

Fig. 2 Normalized approximation error $\overline{\varepsilon}$ for quadratic uniform grids with different grid sizes. Computations are derived with a Monte Carlo integration with 1500 random evaluations for each grid size, and are based on the underlying motor efficiency map f and the piecewise approximation p

one gear, 2416 s for two gears, and 3099 s for three gears, respectively, are within an acceptable range.

4 Summary

We investigated three different piecewise linearization modeling techniques, that can be used to approximate the efficiency map of electric motors within optimization programs of transmission systems. The shown results are highly solver-dependent, as we only used SCIP for our computational study. Nevertheless, using an SOS-formulation has computational benefits, both in presolving and solving. Furthermore, by investigating the approximation error for different grid sizes, a trade-off between accuracy and solution time can be found.

Funding

Funded by Deutsche Forschungsgemeinschaft (DFG, German Research Foundation)—project number 57157498—SFB 805.

References

1. Lukic, S.M., Emado, A.: Modeling of electric machines for automotive applications using efficiency maps. In: Proceedings: Electrical Insulation Conference and Electrical Manufacturing and Coil Winding Technology Conference, pp. 543–550. IEEE, New York (2003)
2. Belotti, P., Kirches, C., Leyffer, S., Linderoth, J., Luedtke, J., Mahajan, A.: Mixed-integer nonlinear optimization. Acta Numer. **22**, 1–131 (2013)
3. Vielma, J.P., Ahmed, S., Nemhauser, G.: Mixed-integer models for nonseparable piecewise-linear optimization: unifying framework and extensions. Oper. Res. **58**(2), 303–315 (2010)
4. Geißler, B., Martin, A., Morsi, A., Schewe, L.: Using piecewise linear functions for solving MINLPs. In: Lee, J., Leyffer, S. (eds.) Mixed Integer Nonlinear Programming, pp. 287–314. Springer, New York (2012)
5. Misener, R., Floudas, C.A.: Piecewise-linear approximations of multidimensional functions. J. Optim. Theory Appl. **145**(1), 120–147 (2010)
6. Mikolajková, M., Saxén, H., Pettersson, F.: Linearization of an MINLP model and its application to gas distribution optimization. Energy **146**, 156–168 (2018)
7. Misener, R., Gounaris, C.E., Floudas, C.A.: Global optimization of gas lifting operations: a comparative study of piecewise linear formulations. Ind. Eng. Chem. Res. **48**(13), 6098–6104 (2009)
8. Leise, P., Altherr, L.C., Simon, N., Pelz, P.F.: Finding global-optimal gearbox designs for battery electric vehicles. In: Le Thi, H.A., Le, H.M., Pham Dinh, T. (eds.) Optimization of Complex Systems: Theory, Models, Algorithms and Applications, pp. 916–925. Springer, Cham (2020)

9. Beale, E., Tomlin, J.: Special facilities in a general mathematical programming system for non-convex problems using ordered sets of variables. In: Lawrence, J. (ed.) Proceedings of the Fifth International Conference on Operational Research, pp. 447–454. Tavistock Publications, London (1970)
10. Gleixner, A., Bastubbe, M., Eifler, L., Gally, T., Gamrath, G., Gottwald, R.L., Hendel, G., Hojny, C., Koch, T., Lübbecke, M.E., Maher, S.J., Miltenberger, M., Müller, B., Pfetsch, M.E., Puchert, C., Rehfeldt, D., Schlösser, F., Schubert, C., Serrano, F., Shinano, Y., Viernickel, J.M., Wegscheider, F., Walter, M., Witt, J.T., Witzig, J.: The SCIP Optimization Suite 6.0. Tech. rep., Optimization Online (2018)

Support-Free Lattice Structures for Extrusion-Based Additive Manufacturing Processes via Mixed-Integer Programming

Christian Reintjes, Michael Hartisch, and Ulf Lorenz

Keywords MIP · Additive manufacturing · Support-free lattice structure · VDI 3405-3-4:2019 · ISO/ASTM 52921:2016

1 Introduction and Motivation

Additive Manufacturing (AM) has become more relevant to industry in recent years and enables fabrication of complex lightweight lattice structures. Nevertheless material extrusion processes require internal and/or external support structures for the printing process. These support structures generate costs due to additional material, printing time and energy. Contrary to the optimization strategy to minimize the need for additional support by optimizing the support structure itself and keeping the original part, we focus on designing a new part in a single step by optimizing the topology towards a support-free lattice structure. Assuming that the support structures cannot be removed by non-destructive post-processing (see Fig. 2c)—which can occur with the manufacturing of complex lightweight lattice structures with material extrusion AM processes—this optimization approach becomes necessary, since the force distribution in a lattice structure is manipulated by the additional support structures. The previous structural optimization would become invalid. Finding the optimal set of bars for a lattice structure, which remains in structural optimization problems, has been proved to be NP-hard [7]. Heuristic approaches have been applied for structural optimization problems with stress and buckling constraints. However, these fairly general approaches are restricted to small scale problems and are not suitable for additive manufacturing [7]. To solve this problem a Mixed Integer Program (MIP), considering design rules for inclined and freestanding cylinders/bars manufactured with material extrusion processes (VDI 3405-3-4:2019) [6] and assumptions for location and orientation of parts

C. Reintjes (✉) · M. Hartisch · U. Lorenz
University of Siegen, Institute of Technology Management, Siegen, Germany
e-mail: christian.reintjes@uni-siegen.de

J. S. Neufeld et al. (eds.), *Operations Research Proceedings 2019*, Operations Research Proceedings, https://doi.org/10.1007/978-3-030-48439-2_56

within a build volume (ISO/ASTM 52921:2016) [2], are presented. Furthermore, the stress and buckling constraints are simplified. The aim is to realize support-free lattice structures.

2 Static Lattice Topology Design

As in [4] the geometric structural conditions of the three-dimensional lattice structure (see Fig. 1, left) are based on the definitions for a two-dimensional lattice structure. In this subsection, see [3, 4], all definitions of the Ground Structure (GS) are described for a three-dimensional lattice structure, see Fig. 1, right.

Let $\mathbb{A}$ be the assembly space of an additive manufacturing machine, represented by a polyhedron as an open subset of $\mathbb{R}^3$. Let $\mathbb{V} \subset \mathbb{A}$ be the reference volume to be replaced by a lattice structure and $\neg\, \mathbb{V} = \mathbb{A} \setminus \mathbb{V}$ the difference between the assembly space $\mathbb{A}$ and the reference volume $\mathbb{V}$ (3-orthotope or hyperrectangle). Let $\mathbb{L} \subset \mathbb{V}$ be the convex shell $\mathbb{C}$ of the lattice structure created by optimization, hence the convex shell can be regarded as the volume of the actual required installation space for an optimized lattice structure, including the difference between the convex shell and the actual volume defined as $\mathbb{L}'$. Several products, represented by $\mathbb{V}$, can be printed simultaneously in the installation space $\mathbb{A}$. Analogous to the previous definitions, $\neg\, \mathbb{L} = \mathbb{V} \setminus \mathbb{L}$ is the difference between the reference volume $\mathbb{V}$ and the convex shell $\mathbb{L}$, leading to $\mathbb{L} \subset \mathbb{V} \subset \mathbb{A}$. $\mathbb{L}$ and $\mathbb{V}$, as by defining $\mathbb{A}$, are represented by a polyhedron, which consists of the lattice structure (or free space). The differentiation of the installation space is done for minimization of $\mathbb{V}$ in order to save computing effort.

Alternatively, a minimum bounding 3-orthotope (bounding box) $\mathbb{B}$ can be set as the volume of the actually required installation space to build the optimized lattice structure including the difference $\overline{\mathbb{L}}$ between the 3-orthotope and the actual volume, cf. [2]. In this case the difference is defined as $\overline{\mathbb{L}'}$. Assuming $\underline{\mathbb{L}'}$ is the actually required installation space, it applies $\underline{\mathbb{L}'} \subset \mathbb{L}' \subset \overline{\mathbb{L}'}$.

The GS $\mathbb{G}$ is shown in Fig. 1 with a set of connecting nodes $V = \{1, \ldots, 64\}$ with the polyhedron being equal to a hyperrectangle. It follows that—for the sake

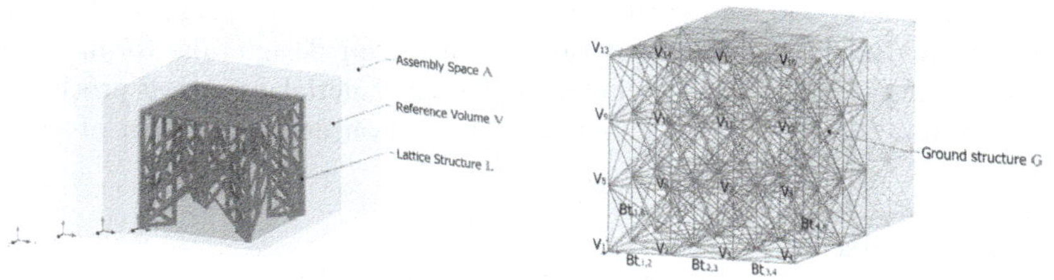

Fig. 1 (Left) Illustration of the system boundaries. (Right) Illustration of the ground structure method

of a simple GS example—$\mathbb{C} \equiv \mathbb{B}$ and $\mathbb{L}' \equiv \overline{\mathbb{L}'}$. $B_{t,i,j}$ is a binary variable indicating whether a bar of type $t \in T$ is present between $i \in V$ and $j \in V$. The angle range of a bar is set to 45° in any spatial direction in relation to a local coordinate system centered on one node $v \in V$, so that $r_{i,j,x}, r_{i,j,y}, r_{i,j,z} \in \{0, \frac{\sqrt{2}}{2}, 1\}$ applies.

The MIP model TTD_L uses beam theory for structural mechanics and constant cross sections, see [1]. We assume a beam to be a structure which has one of its dimensions much larger than the other two so the kinematic assumptions of the beam theory (cf. [1]) apply. It follows that the cross sections of a beam do not deform in a significant manner under the influence of transverse or axial loads and therefore the beam is assumed to be rigid. If deformation is allowed, the cross sections of the beam are assumed to remain planar and normal to the deformed axis of the beam. Besides allowing no transverse shearing forces and bending moments, the displacement functions depend on the coordinates along the axis of the beam $\bar{u}_1(z_i) \in \mathbb{R}, \bar{u}_2(z_i), \bar{u}_3(z_i) = 0$ [1]. The normal force is defined by $N_{i,j}(x_1) = F_{i,j} = \int_A Q_{11}(x_1, x_2, x_3)dA$, where x_1, x_2, x_3 are the spatial coordinates of the cross sections. Following the previous statement transverse shearing forces and bending moments have been simplified. The only allowed external loads are concentrated forces acting on nodes. We claim a linear-elastic isotropic material, with the given deformation restrictions causing no transverse stresses to occur $Q_{22}, Q_{33} \approx 0$. By Hooke's law, the axial stress Q_{11} is given by $Q_{11}(x_1, x_2, x_3) = E\bar{u}_1'(x_1)$, allowing only uniaxial stress.

3 MIP Model $\mathrm{LTD}_{\mathrm{L;E}}$ for Support-Free Lattice Structures

This work describes the extension of the MIP model lattice topology design $\mathrm{LTD}_{\mathrm{L;P}}$ (*l*inear; *p*owder based AM) to $\mathrm{LTD}_{\mathrm{L;E}}$ (*l*inear; *e*xtrusion based AM). A detailed description of all conditions of the model $\mathrm{LTD}_{\mathrm{L;P}}$ ((1)–(13), (19)), a table with definitions of decision variables, parameters and sets can be found in [5]. Both models use the same Ground Structure Method (GSM) and beam theory for structural mechanics, as explained in Sect. 2. The implementation of the conditions (14)–(18) describes support-free lattice structures.

The length of a cylindrical bar is defined as l, the outer diameter of the bar as D. The design rules for part production using AM material extrusion processes, as in standard VDI 3405-3-4:2019, define a critical downskin angle $\delta_{cr} = 45°$ [6]. A downskin area $\mathbb{D}$ is a (sub-)area whose normal vector in relation to the build direction Z is negative. The downskin angle δ is the angle between the build platform plane and a downskin area whose value lies between 0° (parallel to the build platform) and 90° (perpendicular to the build platform) [6] It is also required that $\delta \geq \delta_{cr} : l/D \leq 5$ and $\delta = 90° : l/D \leq 10$. The build direction Z is assumed positive with increasing node indexing, so that a distinction between upskin and downskin areas $\mathbb{U}$ resp. $\mathbb{D}$, upskin and downskin angles υ resp. δ, is implemented, cf. VDI 3405-3-3:2015. A construction not taking these design recommendations into account is classified as not ready for production and thus not part of the solution space of the MIP.

Preprocessing ensures that $l_{B_{t,i,j}} = N_{B_{t,i,j}} \cdot g \; \forall t \in T, i, j \in V, N_{B_{t,i,j}} \in \mathbb{N}$, whereat $N_{B_{t,i,j}}$ is the number of layers needed for a bar of type $t \in T$, g the layer thickness of the additive manufacturing machine and $l_{B_{t,i,j}}$ the length of a bar. The bar length is a multiple of the layer thickness, which reduces production inaccuracies.

Conditions (14) and (17) identify the combination possibilities of bars between neighboring nodes that would not comply the design recommendations of VDI 3405-3-4:2019. This linear formulation is possible, since $\delta = \delta_{cr} = 45°$ applies preprocessed, due to the used GS. $\delta \geq \delta_{cr} : l/D \leq 5$ and $\delta = 90° : l/D \leq 10$ applies due to identification in conditions (14) and (17). Condition (15) forces the model to add at least one bar between two neighboring nodes identified as critical in condition (14) and (17) to comply with the VDI 3405-3-4:2019. $Z_i \in \{0, 1\}$ is a indicator whether at least one support structure condition is satisfied. $A = \{2, \ldots, 26\}$ is the number of bars at a node. Condition (16) forces the model to set the binary variable $x_{i,j}$ indicating whether a bar is present between two neighboring nodes. The model is forced to consider the extra bars resulting from (15) at the force and moment equilibrium conditions (1)–(5), see [5].

$$(\text{LTD}_{\text{L;E}}): \quad \min \sum_{i \in V} \sum_{j \in V} \sum_{t \in T} B_{t,i,j} cost_t$$

$$\text{s.t.} \{(1) - (13)\}$$

$$2\ell_i^o \leq \sum_{j \in NB_o(i)} x_{i,j} \leq \ell_i^o + 1 \qquad \forall i, j \in V, o \in O \qquad (14)$$

$$\sum_{j \in NB(i)} x_{i,j} \geq 3Z_i \qquad \forall i \in V \qquad (15)$$

$$\underline{A} y_i \leq \sum_{j \in NB(i)} x_{i,j} \leq \overline{A} y_i \qquad \forall i, j \in V \qquad (16)$$

$$\sum_{o \in O} \ell_i^o \leq \frac{1}{2} \overline{A} Z_i \qquad \forall i \in V \qquad (17)$$

$$R_{i,z} = 0 \qquad \forall i \in V \setminus B \qquad (18)$$

$$x_{i,j}, y_i, \ell_i^o, z_i, B_{t,i,j} \in \{0, 1\} \qquad \forall i, j \in V, t \in T, o \in O \qquad (19)$$

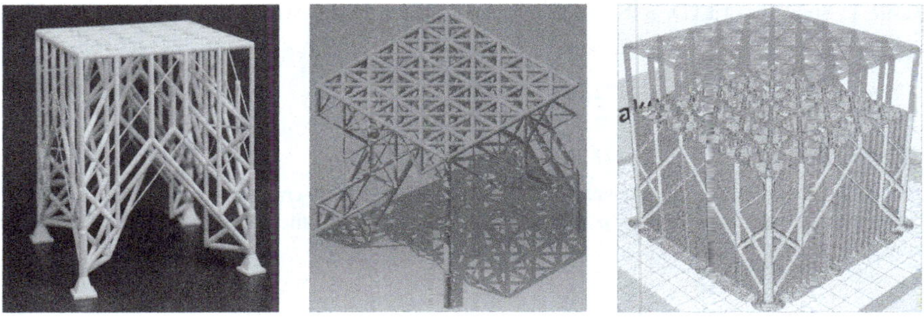

(a) (b) (c)

Fig. 2 (**a**) Manufactured part using the MIP $LTD_{L;P}$ and SLS. (**b**) Solution of the MIP $LTD_{L;E}$. (**c**) Layer view of the supporting structure for solution $LTD_{L;P}$ in Cura 3.6.9; layer 700 of 1028

4 Results

The two instances introduced in the following consider a positioning of a static area load (see Fig. 2a and b, cf. [5]).[1] The assembly space $\mathbb{A}$ of the AM machine Ultimaker 2 Extended was represented as a polyhedron consisting of $V = \{1, \ldots, 1815\}$ connecting nodes (approximation), resulting from the dimensions $223 \times 223 \times 305$ mm and a bar length of 20 mm for a non-angled bar. The reference volume $\mathbb{V}$ was set to the dimensions $120 \times 120 \times 120$ mm. Hence, there are 216 connection nodes in $\mathbb{V}$. The four corner points of the first plane in z-direction were defined as bearings. It is predetermined that the top level in z-direction is fully developed with bars. The bar diameters $\{2, 4, 6, 8\}$ mm and in addition 1 mm for instance $LTD_{L;E}$ together with the associated $c_t \in \mathbb{R}_+$ and $cost_t \in \mathbb{R}$ are passed as parameters. The bar diameter 1 mm is used to comply the design recommendations of VDI 3405-3-4:2019.

The computation time (manually interrupted) was 10 h 51 min for the instance $LTD_{L;P}$ and 49 h and 27 min for the instance $LTD_{L;E}$. 12 resp. 103 permissible solutions were determined, whereby the duality gap was 55.73% resp. 64.46%.

As a part of this work, a functional prototype of the instance $LTD_{L;E}$ was manufactured including bearings using SLS and the AM machine EOSINT P760, see Fig. 2a. As AM material extrusion process Fused Deposition Modeling (FDM) is performed on the AM machine Ultimaker 2 and designed with the supplied Slicer Cura 3.6.9, see Fig. 2c. The support structure has been designed with an support overhang angle $\delta = 45°$, a brim type build plate adhesion and free placement of support structures. The material is Acrylonitrile Butadiene Styrene (ABS) with a

[1]The calculation were executed on a workstation with an Intel Xeon E5-2637 v3 (3.5 GHz) and 128 GB RAM using CPLEX Version 12.6.1. The CAD import and editing including stl file manipulation (ANSYS SpaceClaim 19.2) was performed on a workstation with an Intel Intel Xeon E5-2637 v4 (3.5 GHz), 64 GB RAM and a NVIDIA GeForce RTX 2080 (8 GB RAM).

Table 1 Statistics of $LTD_{L;P}$ (SLS); $LTD_{L;E}$ and Cura 3.6.9 (FDM)

Model	$\sum x_{i,j}$ (After opt.)	$\mathbb{B}$ (10^4 mm^3)	$\mathbb{L}'$ (10^4 mm^3)	Ratio (%)	Weight (g)	Duality gap (%)
$TTD_{L;P}$	528	172.80	5.42	9.36	103.40	55.73
$LTD_{L;E}$	875	172.80	6.36	10.98	121.00	64.46
Cura 3.6.9	528	172.80	11.79	20.37	224.12	–

The second column shows the amount of bars, independent of the bar cross-section. The third and fourth column denotes the bounding box $\mathbb{B}$ and the actual volume of the lattice structure $\mathbb{L}'$. Ratio denote the ratio of $\mathbb{B}$ to $\mathbb{L}'$

density of 1.1 g/cm^3. The layer thickness was set to 0.1 mm, the bars were printed as solid material. Non-destructive post-processing (Needle-Nose Pliers) was not possible and thus the prototype is classified as not ready for production. Table 1 shows that the model $LTD_{L;E}$ requires 93.89 mm^3 less actual volume of the lattice structure $\mathbb{L}'$ than the solution of Cura 3.6.9. This results in a 9.39% better ratio and weight saving of 103.12 g, which corresponds to a cost saving of 46.10 %.

5 Conclusion and Outlook

We have introduced a MIP to generate support-free lattice structures containing overhangs. The problem to strengthen a lattice structure by local thickening and/or bar addition with the objective function to minimize costs and material is modeled. By designing a new part in a single step by optimizing the topology towards a support-free lattice structure, destructive post-processing is excluded. Compared to the Slicer Cura 3.6.9, our method is able to reduce the amount of support structures and thus costs by almost 50%. The most important limitation of the proposed method is that we only work geometry-based via VDI 3405-3-4:2019 under the assumption that the selected topology withstands the shear forces in the printing process. Additional structural analysis can be an assistance. Based on this work, other optimization approaches can be developed. With regard to the MIP it is important to include external heuristics especially starting solutions in the solution strategy and to develop lower and upper bounds. Typical process-specific geometrical limitations of AM technologies like delamination of layers, curling or stair-step effects, can be minimized by formulating boundary conditions, so that the part quality gets maximized. Another approach is the minimization of the material of the support structure by minimizing the sum of the angles between the orientation of a particular part and the direction of build.

References

1. Bauchau, O.A., Craig, J.I.: Euler-Bernoulli Beam Theory, pp. 173–221. Springer, Dordrecht (2009)
2. DIN EN ISO/ASTM 52921: Standard terminology for additive manufacturing - coordinate systems and test methodologies (2013)
3. Reintjes, C., Lorenz, U.: Mixed integer optimization for Truss topology design problems as a design tool for AM components. In: International Conference on Simulation for Additive Manufacturing, vol. 2, pp. 193–204 (2019)
4. Reintjes, C., Hartisch, M., Lorenz, U.: Lattice structure design with linear optimization for additive manufacturing as an initial design in the field of generative design. In: Operations Research Proceedings 2017, pp. 451–457. Springer International Publishing, Cham (2018)
5. Reintjes, C., Hartisch, M., Lorenz, U.: Design and optimization for additive manufacturing of cellular structures using linear optimization. In: Operations Research Proceedings 2018, pp. 371–377. Springer International Publishing, Cham (2019)
6. VDI Richtlinien, VDI 3405-3-4: Additive manufacturing processes - design rules for part production using material extrusion processes (2019)
7. Wang, W., Wang, T.Y., Yang, Z., Liu, L., Tong, X., Tong, W., Deng, J., Chen, F., Liu, X.: Cost-effective printing of 3d objects with skin-frame structures. ACM Trans. Graph. **32**(6), 177 (2013)

Optimized Design of Thermofluid Systems Using the Example of Mold Cooling in Injection Molding

Jonas B. Weber, Michael Hartisch, and Ulf Lorenz

Abstract For many industrial applications, the heating and cooling of fluids is an essential aspect. Systems used for this purpose can be summarized under the general term 'thermofluid systems'. As an application, we investigate industrial process cooling systems that are used, among other things, for mold cooling in injection molding. The systems considered in this work consist of interconnected individual air-cooled chillers and injection molds which act as ideal heat sources. In practice, some parts of the system are typically fixed while some components and their connections are optional and thus allow a certain degree of freedom for the design. Therefore, our goal is to find a favorable system design and operation regarding a set of a-priori known load scenarios. In this context, a favorable system is one which is able to satisfy the demand in all load scenarios and has comparatively low total costs. Hence, an optimization problem arises which can be modeled using mixed integer non-linear programming. The non-linearity is induced both by the component behavior as well as by the general physical system behavior. As a proof of concept and to complete our work, we then conduct a small case study which illustrates the potential of our approach.

Keywords Engineering optimization · Nonlinear programming

1 Introduction

Injection molding is an important and widely used technique for producing polymeric parts. In this cyclic process molten polymer is injected into a cavity where it is held under pressure until it has solidified, duplicating the shape of the cavity. A crucial stage in this process and the focus of this work is the cooling of the mold in order to allow the molten polymer to solidify properly. This is typically done by

J. B. Weber (✉) · M. Hartisch · U. Lorenz
University of Siegen, Siegen, Germany
e-mail: jonas.weber@uni-siegen.de

J. S. Neufeld et al. (eds.), *Operations Research Proceedings 2019*, Operations Research Proceedings, https://doi.org/10.1007/978-3-030-48439-2_57

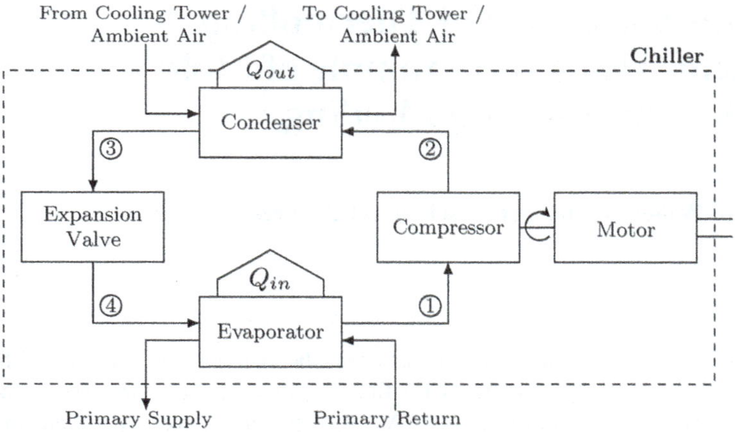

Fig. 1 Working principle of a compression chiller

pumping a coolant through the cooling channels located in the wall of the mold. The heat absorbed by the coolant this way is then removed using industrial chillers. For this purpose a wide variety of chiller types exists. In general a distinction between two types, vapor absorption and vapor compression chillers, can be made. In the following, we concentrate on the latter. This type can again be subdivided into centrifugal, reciprocating, scroll and screw chillers by the compressor technology used. Finally, those can be further classified into water-cooled and air-cooled chillers, depending on the chiller's heat sink. All have in common that the cooling is realized by a circular process consisting of four sub-processes, as shown in Fig. 1. In the first step, the internal refrigerant enters the evaporator as a liquid–vapor mixture and absorbs the heat of the cooling medium returning from the heat source (1). The vaporous refrigerant is then sucked in and compressed while the resulting heat is absorbed by the refrigerant (2). During the subsequent liquefaction process, the superheated refrigerant enters the condenser, is cooled using the ambient air or water of a cooling tower and liquefies again (3). Finally, in the expansion process, the pressure of the refrigerant is reduced from condensing to evaporating pressure and the refrigerant expands again (4).

For the connection of multiple chillers and to distribute the coolant, several configuration schemes exist. Here, we focus on a configuration which is known as primary-secondary or decoupled system, as shown in Fig. 2. This system is characterized by the fact that the distribution piping is decoupled from the chiller piping. The (primary) flow through the operating chiller(s) is therefore constant while the (secondary) flow through the load(s) is variable. The purpose of the associated bypass pipe between the two subsystems is to balance the flows.

To model the operation of a chiller, the 'DOE2' electric chiller simulation model [1] is used. This model is based on three performance curves. The CAP_{FT} curve, see Eq. (1), represents the available (cooling) capacity (Q) as a function of evaporator

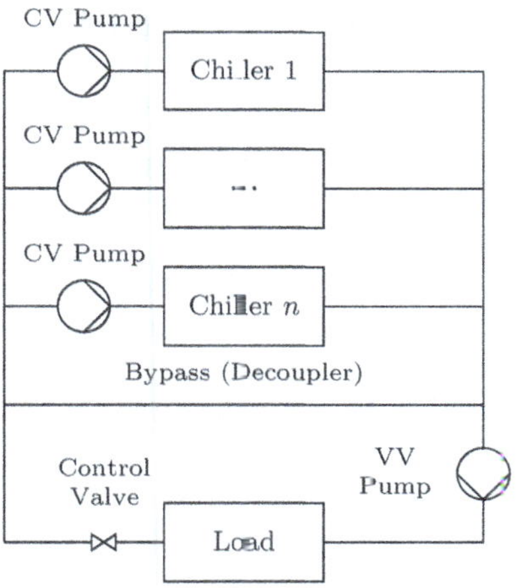

Fig. 2 Primary-secondary (decoupled) system configuration

and condenser temperatures, the EIR_{FT} curve, see Eq. (2), which is also a function of evaporator and condenser temperatures describes the full-load efficiency of a chiller and the EIR_{FPLR} curve, see Eq. (3), represents a chiller's efficiency as a function of the part-load ratio (PLR), see Eq. (4). For the CAP_{FT} and EIR_{FT} curve, the chilled water supply temperature (t_{chws}) is used as an estimate for the evaporator temperature and the condenser water supply (t_{cws}) and outdoor dry-bulb temperature (t_{oat}) are used for the condenser temperature of water-cooled and air-cooled chillers, respectively. The latter are considered as constant for the remainder of this work. With Eqs. (1)–(4) it is possible to determine the power consumption (P) of a chiller for any load and temperature condition by applying Eq. (5). The operation of a given chiller is therefore defined by the regression coefficients (a_i, b_i, c_i, d_i, e_i, f_i), the reference capacity (Q_{ref}) and the reference power consumption (P_{ref}).

$$CAP_{FT} = a_1 + b_1 \cdot t_{chws} + c_1 \cdot t_{chws}^2 + d_1 \cdot t_{cws/oat} + e_1 \cdot t_{cws/oat}^2 + f_1 \cdot t_{chws} \cdot t_{cws/oat} \quad (1)$$

$$EIR_{FT} = a_2 + b_2 \cdot t_{chws} + c_2 \cdot t_{chws}^2 + d_2 \cdot t_{cws/oat} + e_2 \cdot t_{cws/oat}^2 + f_2 \cdot t_{chws} \cdot t_{cws/oat} \quad (2)$$

$$EIR_{FPLR} = a_3 + b_3 \cdot PLR + c_3 \cdot PLR^2 \quad (3)$$

$$PLR = Q \,/\, (Q_{ref} \cdot CAP_{FT}) \quad (4)$$

$$P = P_{ref} \cdot EIR_{FPLR} \cdot EIR_{FT} \cdot CAP_{FT} \quad (5)$$

2 Mathematical Model

To model the problem of finding a favorable system design and operation for a system of individual air-cooled chillers with regard to a set of a-priori known load scenarios, mixed-integer non-linear programming (MINLP) is used. The only non-linearities in this context arise from the bilinear relationship of the heat flow, volume flow and temperature according to the specific heat formula and from the performance curves used to describe the operation of the chillers. Therefore, the instances studied in this paper can still be solved using standard software. For the sake of simplicity and because of the extensive coverage in previous research, see [2], only the system's thermal variables are considered here, while the distribution, i.e. the system pressure, is neglected. However, the associated extension is straightforward. Furthermore, due to space limitations, we focus on the component behavior. For the constraints related to the general system design and operation, we refer to [3]. A detailed description of all sets, variables and parameters used in this model is shown in Table 1.

$$\min \sum_{c \in C} (b_c \cdot C_c^{inv}) + \sum_{s \in S} \sum_{r \in R} (p_r^s \cdot T \cdot F^s \cdot C^{kWh}) \tag{6}$$

See [3] for the general system design and operation constraints.

$$v_{(r,o)}^s = v_{(r,i)}^s \qquad \forall s \in S,\ (r,o) \in P_R^o,\ (r,i) \in P_R^i \tag{7}$$

$$v_{(r,o)}^s \leq a_r^s \cdot V_r^{max} \qquad s \in S,\ (r,o) \in P_R^o \tag{8}$$

$$v_{(r,o)}^s \geq a_r^s \cdot V_r^{min} \qquad s \in S,\ (r,o) \in P_R^o \tag{9}$$

$$t_{(r,o)}^s \leq a_r^s \cdot T_r^{max} \qquad \forall s \in S,\ (r,o) \in P_R^o \tag{10}$$

$$t_{(r,o)}^s \geq a_r^s \cdot T_r^{min} \qquad \forall s \in S,\ (r,o) \in P_R^o \tag{11}$$

$$q_{(r,i)}^s - q_{(r,o)}^s = \Delta q_{0_r}^s \qquad \forall s \in S,\ (r,o) \in P_R^o,\ (r,i) \in P_R^i \tag{12}$$

$$\Delta q_{0_r}^s \leq Q_{0_r}^{ref} \cdot CAP_{FT}(t_{(r,o)}^s) \cdot a_r^s \qquad \forall s \in S,\ r \in R \tag{13}$$

$$\Delta q_{0_r}^s = plr_r^s \cdot CAP_{FT}(t_{(r,o)}^s) \cdot Q_{0_r}^{ref} \qquad \forall s \in S,\ r \in R \tag{14}$$

$$p_r^s = a_r^s \cdot P_{0_r}^{ref} \cdot CAP_{FT}(t_{(r,o)}^s) \cdot EIR_{FT}(t_{(r,o)}^s) \cdot EIR_{FPLR}(plr_r^s) \qquad \forall s \in S,\ r \in R \tag{15}$$

$$v_{(m,o)}^s = v_{(m,i)}^s \qquad \forall s \in S,\ (m,o) \in P_M^o,\ (m,i) \in P_M^i \tag{16}$$

$$v_{(m,o)}^s \leq a_m^s \cdot V_m^{max} \qquad \forall s \in S,\ (m,o) \in P_M^o \tag{17}$$

$$v_{(m,o)}^s \geq a_m^s \cdot V_m^{min} \qquad \forall s \in S,\ (m,o) \in P_M^o \tag{18}$$

$$q_{(m,o)}^s - q_{(m,i)}^s = \Delta Q_m^s \qquad \forall s \in S,\ (m,o) \in P_M^o,\ (m,i) \in P_M^i \tag{19}$$

$$t_{(m,i)}^s = a_m^s \cdot T_m^s \qquad \forall s \in S,\ (m,i) \in P_M^i \tag{20}$$

Table 1 Sets, variables and parameters of the MINLP

Sets	
S	Scenarios
R	Chillers
M	Injection molding machines
B	Pipe fittings of decoupler bypass line
$C\ (\hat{=} R \cup M \cup B)$	Components
$P^i_{C/R/M/B}$	Inlet ports of components, chillers, molds or fittings
$P^o_{C/R/M/B}$	Outlet ports of components, chillers, molds or fittings
$P_C\ (\hat{=} P^i_C \cup P^o_C)$	Ports of components
Variables	
$b_c \in \{0, 1\}$	Purchase decision for $c \in C$
$a^s_c \in \{0, 1\}$	Usage decision for $c \in C$ in $s \in S$
$v^s_{(c,p)} \in \mathbb{R}_+$	Volume flow through $(c, p) \in P_C$ in $s \in S$
$q^s_{(c,p)} \in \mathbb{R}_+$	Heat flow through $(c, p) \in P_C$ in $s \in S$
$t^s_{(c,p)} \in \mathbb{R}_+$	Temperature at $(c, p) \in P_C$ in $s \in S$
$\Delta q^s_{0_r} \in \mathbb{R}_+$	Decrease in heat flow caused by $r \in R$ in $s \in S$
$p^s_r \in \mathbb{R}_+$	Power consumption by $r \in R$ in $s \in S$
$plr^s_r \in \mathbb{R}_+$	Part load ratio of $r \in R$ in $s \in S$
Parameters	
C^{inv}_c	Investment costs of $c \in C$
C^{kWh}	Energy costs per kWh
T	Total operating life of system
F^s	Share of $s \in S$ compared to total operating life of system
V^{min}_c, V^{max}_c	Min./max. volume flow through $c \in C$
T^{min}_c, T^{max}_c	Min./max. temperature at outlet for $c \in C$
$Q^{ref}_{0_r}$	Cooling capacity at reference point for $r \in R$
$P^{ref}_{0_r}$	Power consumption at reference point for $r \in R$
ΔQ^s_m	Increase in heat flow caused by $m \in M$ in $s \in S$
T^s_m	Desired temperature at inlet of $m \in M$ in $s \in S$

$$v^s_{(b,o)} = v^s_{(b,i)} \qquad \forall s \in S,\ (b,o) \in P^o_B,\ (b,i) \in P^i_B \tag{21}$$

$$q^s_{(b,o)} = q^s_{(b,i)} \qquad \forall s \in S,\ (b,o) \in P^o_B,\ (b,i) \in P^i_B \tag{22}$$

$$t^s_{(b,o)} = t^s_{(b,i)} \qquad \forall s \in S,\ (b,o) \in P^o_B,\ (b,i) \in P^i_B \tag{23}$$

The goal of the presented model is to minimize the sum of investment and expected energy costs over a system's lifespan, see (6). The model is divided into three parts. Constraints (7)–(15) model the operation of the chillers. Equation (7) ensures that the flow conservation for a chiller is guaranteed. The operational bounds for the volume flow and the temperature of the coolant are ensured by (8)–(11). The actual cooling capacity of the chillers with regard to the available capacity as well as the part-load operating ratio and consequential the heat flow at the inlets and outlets is defined by Constraints (12)–(14). By using the three curves of the 'DOE2' simulation model, the power consumption of the chillers can be determined, see Constraint (15). Constraints (16)–(20) model the injection molding machines or more specifically, the molds themselves. Constraints (16)–(18) are equivalent to Constraints (7)–(9) of the chiller section. The desired values for the introduced heat and temperature at the mold inlets are guaranteed by Constraints (19) and (20). Finally, Constraints (21)–(23) ensure the conservation of the volume flow, heat flow and temperature between the inlets and outlets for the pipe fittings of the decoupler bypass line.

3 Computational Results

To test the model, three test cases are presented here. For each case, we assume that there are three injection molding machines which operate in two different load scenarios with equal time shares, i.e. two-shift operation. The heat load for each machine in scenario one is 8 kW and 4 kW in scenario two, respectively. The performance data for the chillers is estimated using the default values provided by COMNET.[1] Furthermore, the usage period is assumed to be 10 years with predicted average energy costs of 0.25 € per kWh. The three test cases differ with regard to components that may already be installed. If parts of the system already exist, they have to be included in the optimized system and are associated with no investment costs. For the first test case, one chiller is already installed and there are no optional chillers that can be added. As a result, only the operation has to be considered. It therefore acts as a baseline. As for all other cases, all chillers considered here are scroll chillers. For the second test case, again one chiller is already installed and there are two optional chillers that can be added to the system. For the third case, the system has to be designed from scratch and hence all of the three chillers are optional. The possible system configurations are shown in Fig. 3. Note that the coefficient of power (COP) shown in the figure represents the ratio of cooling provided to work required.

All calculations were performed on a MacBook Pro (Early 2015) with 3.1 GHz Intel Core i7 and 16 GB 1867 MHz DDR3. To solve the MINLPs, 'SCIP 6.0.0' [4] was used. A summary of the results can be found in Table 2. This includes the runtime ('Time'), the best solution found ('Sol.'), the relative and absolute optimality gap ('Gap'), the runtime until the first feasible solution was found ('Time 1.Sol.'), the optimality gap for the first solution ('Gap 1.Sol.') and the added chillers ('Add. Chillers').

In test cases one and two, operating the existing chiller without installing additional chillers (for test case two) is the best solution found for both scenarios. Accordingly, the total costs only consist of the energy costs. However, this chiller is

[1] https://www.comnet.org/index.php/382-chillers.

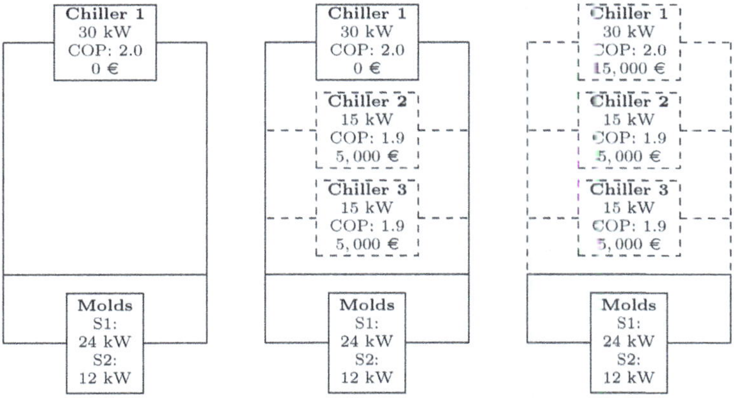

Fig. 3 Possible systems for case one (left), case two (middle) and case three (right)

Table 2 Computational results

#	Time [s]	Sol. [€]	Gap [%]	Gap [€]	Time 1.Sol. [s]	Gap 1.Sol. [%]	Add. chillers
1	0.4	71,253.44	–	–	0.2	2.16	–
2	>3600.0	71,253.55	<0.02	0.11	14.6	0.13	–
3	>3600.0	82,986.63	<0.06	0.44	27.1	0.11	C2, C3

not installed and used in test case three. For this test case, the two smaller chillers should be purchased. The best solution found indicates that they should be operated in parallel in both of the two a-priori given load scenarios which results in 10,000 € investment and 72,986.63 € energy costs.

4 Conclusion and Outlook

This work is intended as a proof of concept for the applicability of mathematical programming in the domain of designing thermofluid systems. In terms of a holistic design approach as well as with regard to cost and energy minimization, we are sure that it benefits the decision making process. The greater goal for this research is therefore to develop tools for engineers to support them during early design stages of such systems.

References

1. Hydeman, M., Gillespie, K.L. Jr, Dexter, A.L.: Tools and techniques to calibrate electric chiller component models. ASHRAE Trans. **108**, 733–741 (2002)

2. Weber, J.B., Lorenz, U.: Optimizing booster stations. In: Proceedings of the Genetic and Evolutionary Computation Conference, pp. 1303–1310 (2017)
3. Weber, J.B., Lorenz, U.: Algorithmic system design of thermofluid systems. In: Proceedings of the 6th International Conference on Engineering Optimization, pp. 132–143 (2019)
4. Gleixner, A.M., et al.: The SCIP Optimization Suite 6.0. ZIB-Report (2018)

Optimization of Pumping Systems for Buildings: Experimental Validation of Different Degrees of Model Detail on a Modular Test Rig

Tim M. Müller, Lena C. Altherr, Philipp Leise, and Peter F. Pelz

Abstract Successful optimization requires an appropriate model of the system under consideration. When selecting a suitable level of detail, one has to consider solution quality as well as the computational and implementation effort. In this paper, we present a MINLP for a pumping system for the drinking water supply of high-rise buildings. We investigate the influence of the granularity of the underlying physical models on the solution quality. Therefore, we model the system with a varying level of detail regarding the friction losses, and conduct an experimental validation of our model on a modular test rig. Furthermore, we investigate the computational effort and show that it can be reduced by the integration of domain-specific knowledge.

Keywords MINLP · Engineering optimization · Water supply system · Network design · Experimental validation

1 Introduction

Successful optimization requires an appropriate model of the system under consideration. Only if a technical system is modeled in sufficient detail, resulting optimization solutions are meaningful in practice and may be called a mathematical-technical optimum. Especially in engineering disciplines, validation is a crucial step: proposed solutions, whether derived from conventional design techniques or

T. M. Müller · P. Leise · P. F. Pelz (✉)
Department of Mechanical Engineering, Technische Universität Darmstadt, Darmstadt, Germany
e-mail: tim.mueller@fst.tu-darmstadt.de; philipp.leise@fst.tu-darmstadt.de; peter.pelz@fst.tu-darmstadt.de

L. C. Altherr
Faculty for Energy, Building Services and Environmental Engineering, Münster University of Applied Sciences, Münster, Germany
e-mail: lena.altherr@fh-muenster.de

J. S. Neufeld et al. (eds.), *Operations Research Proceedings 2019*, Operations Research Proceedings, https://doi.org/10.1007/978-3-030-48439-2_58

mathematical optimization, are tested for their performance and feasibility by means of simulations and experiments.

In this contribution, we therefore demonstrate a holistic approach for the optimization of a technical system which comprises the steps of modeling, optimization and validation. We exemplify our approach on a fluid system: We plan an optimal pumping system for the water supply in a high-rise building. While standardized approaches for the modeling, design and validation of pumping systems exist, cf. [3], also approaches based on Mixed-Integer Progamming which open up further potential for optimization have been proposed, e.g. [1, 2, 6, 7]. However, in this context, the influence of the accuracy of the underlying physical models on the quality of the optimization solutions has not yet been thoroughly investigated or experimentally validated.

2 Technical Application and Physical Model

Pressure-boosting pumping systems, also called booster stations, are used to increase the water pressure in high-rise buildings. The booster station considered in this work is part of the test rig shown in Fig. 1, left. The test rig has a total height of 6 m and is a down-scaled model of a building with five floors. To supply the different floors, the water is pumped from an open tank to the respective height. The water flows back at ambient pressure. For the validation, the test rig is equipped with sensors allowing us to measure the pumps' power consumption and pressure increase. The volume flow is measured at each floor and controlled by valves. In the optimization model, we consider the following degrees of freedom of the network topology: for the booster station, up to three pumps can be chosen from a set of 10 available pumps, and may be interconnected in any desired way. The set of available pumps comprises five different models. All possible connections between source, sink and the pumps are represented by a graph, cf. Fig. 1, right.

To model the water demand of the individual floors, we scale data of real-world buildings [5] to match our test rig. This results in five demand scenarios with volume flows $Q_s^{\text{bound}}, s \in \mathcal{S}$. The matching pressure increase $P^{\text{source}} - P_s^{\text{sink}}$ is derived experimentally for each scenario and includes the geodetic height difference, as well as friction losses within the general pipe system, i.e. before and after the booster station. Associated with the booster station itself are three different cases of additional friction losses which have, to the best of the authors' knowledge, so far been disregarded in proposed Mixed-Integer Programs. Like other friction losses, these losses depend quadratically on the volume flow q and are calculated by $\Delta p_{\text{loss}} = C \zeta q^2$, where C is a constant and ζ is the loss factor. Figure 1, middle, shows all three cases: Friction occurs (1) at the installation fittings of the pumps, ζ^{inst}, (2) at the piping to the outlet, ζ^{out}, and (3) at branches, ζ^{branch}. To investigate the influence of these additional losses on the optimization result, the respective loss factors are derived by measurements and the solutions of model formulations neglecting or considering them are validated on the test rig and compared.

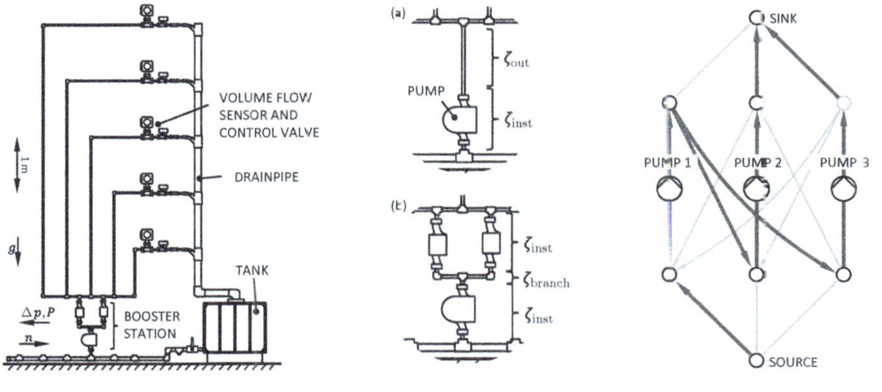

Fig. 1 *Left*: Sketch of test rig. *Middle*: Influence of friction for different topologies. *Right*: Graph to model the system topology of the booster station

3 Mixed-Integer Nonlinear Optimization Problem

The considered MINLP is given in Eqs. (1)–(26). Small letters indicate variables, capital and greek letters indicate parameters or sets, cf. Table 1. The objective (1) is to minimize the lifetime costs, which consist of energy costs and investment costs for the pumps. Up to three pumps can be purchased (2) and only if a pump is purchased it can be activated (3). Constraints (4) and (5) model the deactivation of a pump. Equation (6) gives the volume flow conservation: The volume flow of a pump i equals the sum of flows entering this pump and the sum of flows leaving this pump. Volume flow is only allowed if a connection exists (7). The volume flow at the sink must meet the demand (8). Source and sink have to be connected with at least one pump (9). Unreasonable topologies, such as loops, are excluded by Constraints (10), (11) and (12).[1] Connections to pumps are only allowed if the pump is bought (13). The pressure propagation between the source and a pump inlet (14), as well as the pressure increase by a pump (15) are considered. If a pump is not connected, without loss of generality, the pressure in front and behind of the pump is set to zero (16). The pressure propagation between the pump outlet and the sink (17), as well as between two connected pumps (18) is considered. In case of no detailed friction modeling $\Delta p_{i,s}^{\text{frict, out}}$ and $\Delta p_{i,s}^{\text{branch, out}}$ are set to zero. For detailed friction modeling, friction between pumps and the system outlet (19) (see footnote 1), (21) (see footnote 1) and between two pumps (20) (see footnote 1), (21) (see footnote 1) is taken into account for certain topologies. If a pump is not connected to any other pump (cf. Fig. 1, middle, top), a pipe is needed to connect this pump to the outlet of the booster station, resulting in an additional pressure loss and $x_{i,s}^{\text{out}} = 1$ in (22) (see footnote 1). If a pump is connected to two or more other pumps, a branch is

[1] Equation only active if friction is considered.

Table 1 Decision variables, sets and parameters

Variable	*Domain*	*Description*
$\Delta p_{i,s}$; $q_{i,s}$; $n_{i,s}$; $po_{i,s}$	$\mathbb{R}_0^+$	Pressure increase, volume flow, rotational speed and power consumption of pump $i \in \mathcal{P}$ in scenario $s \in \mathcal{S}$.
$\Delta p_{i,s}^{\text{frict, out}}$	$\mathbb{R}_0^+$	Friction loss on connection between outlet of pump $i \in \mathcal{P}$ and sink in scenario $s \in \mathcal{S}$.[a]
$\Delta p_{i,j,s}^{\text{frict, branch}}$	$\mathbb{R}_0^+$	Friction loss on connection between outlet of pump $i \in \mathcal{P}$ and inlet of pump $j \in \mathcal{P}$ in scenario $s \in \mathcal{S}$.[a]
$p_{i,s}^{\text{in}}$; $p_{i,s}^{\text{out}}$	$\mathbb{R}_0^+$	Pressures at inlet and outlet of pump $i \in \mathcal{P}$, $s \in \mathcal{S}$.
$q_{i,j,s}$	$\mathbb{R}_0^+$	Volume flow between outlet of pump i and inlet of pump j, $i, j \in \mathcal{P}$, $s \in \mathcal{S}$.
$q_{i,s}^{\text{source}}$	$\mathbb{R}_0^+$	Volume flow between source and inlet of pump $i \in \mathcal{P}$, $s \in \mathcal{S}$.
$q_{i,s}^{\text{sink}}$	$\mathbb{R}_0^+$	Volume flow between outlet of pump $i \in \mathcal{P}$ and sink, $s \in \mathcal{S}$.
		Indicator for connection between
$t_{i,j,s}$	$\{0, 1\}$	Outlet of pump i and inlet of pump j, $s \in S$,
$t_{i,s}^{\text{source}}$	$\{0, 1\}$	Source and inlet of pump i, $s \in S$,
$t_{i,s}^{\text{sink}}$	$\{0, 1\}$	Outlet of pump i and sink, $s \in S$.
$x_{i,s}$	$\{0, 1\}$	Activation of pump i in scenario $s \in S$.
$x_{i,s}^{\text{out}}$; $x_{i,j,s}^{\text{branch}}$	$\{0, 1\}$	Activation of output friction and branch friction.[a]
y_i	$\{0, 1\}$	Investment decision for pump $i \in \mathcal{P}$.
Set		*Description*
$\mathcal{P}$		Set of available pumps with $\|\mathcal{P}\| = 10$.
$\mathcal{S}$		Set of demand scenarios with $\|\mathcal{S}\| = 5$.
Parameter	*Range*	*Description*
A	$\mathbb{R}_0^+$	Cross sectional area of pipe (3.14 cm^2).
ϱ	$\mathbb{R}_0^+$	Fluid density (998 kg/m^3).
T	$\mathbb{R}_0^+$	Considered life time of the booster station (10 years).
$C_{\text{invest},i}$	$\mathbb{R}_0^+$	Investment costs for pump $i \in \mathcal{P}$.
C_{elect}	$\mathbb{R}_0^+$	Electricity price (0.29 €/kWh).
w_s	$[0, 1]$	Time share of scenario $s \in \mathcal{S}$.
$\overline{\Delta P_i}$; $\overline{P}$; $\overline{Q}$; $\Delta\overline{Po_i}$; $\Delta\underline{Po_i}$; $\underline{N_i}$,	$\mathbb{R}_0^+$	Upper $^-$ and lower _ bounds for the pumps' operating parameters, i.e., pressure increase, pressure, volume flow, power, speed.
p^{source}	$\mathbb{R}_0^+$	Pressure at the source.
P_s^{sink}; Q_s^{bound}	$\mathbb{R}_0^+$	Pressure and volume flow demand in scenario $s \in \mathcal{S}$.
ζ^{inst}; ζ^{out}; ζ^{branch}	$\mathbb{R}_0^+$	Loss factors for installation equipment of pump, additional outlet pipe and branch.[a]
Po_s^{min}	$\mathbb{R}_0^+$	Minimum necessary power consumption in scenario $s \in \mathcal{S}$.

[a]Set to zero if friction is not modeled

needed (cf. Fig. 1, middle, bottom), resulting in $x_{i,s}^{\text{branch}} = 1$ in (23) (see footnote 1). To model the pump characteristic curves, polynomials of order two and three are fitted to manufacturer data and the coefficients $\alpha_{i,m}$ and $\beta_{i,m}$ are obtained (24), (25). Only if friction at the fittings of the pump is considered, we set $\zeta_{\text{inst}} > 0$.

Moreover, we develop further constraints based on domain-specific knowledge to hopefully fasten the computing process. Constraint (26)[2] gives a lower bound for the power consumption in each demand scenario resulting in a better dual bound and a speed up of the computation. Po_s^{min} should be chosen as tight as possible without cutting off an optimal solution. It is known from the definition of the pump efficiency that $Po_{\text{elect}} = Po_{\text{hydr.}}/\eta = \Delta p\, Q/\eta$. This domain-specific knowledge can be used to derive a bound $Po_s^{\text{min, II}} = (P_s^{\text{sink}} - P_s^{\text{source}})Q_s^{\text{bound}}/\eta_{\text{best}}\ \forall s \in \mathcal{S}$, with the best efficiency of any pump in any operation point η_{best}. By minimizing the power consumption for each scenario individually, another bound can be derived: $Po^{\text{min, III}} = \min \sum_{i\in\mathcal{P}} po_{i,s}\ \forall s \in \mathcal{S}$. This serves as a lower bound for the simultaneous consideration of all demand scenarios. Both approaches, as well as $Po_s^{\text{min, I}} = 0\ \forall s \in \mathcal{S}$ are compared in Sect. 4.

$$\min \quad T\, C_{\text{elect}} \sum_{s\in\mathcal{S}} w_s \sum_{i\in\mathcal{P}} po_{i,s} + \sum_{i\in\mathcal{P}} C_{\text{invest},i}\, y_i \tag{1}$$

subject to

$$\sum_{i\in\mathcal{P}} y_i \leq 3 \tag{2}$$

$$x_{i,s} \leq y_i \quad \forall i \in \mathcal{P}, s \in \mathcal{S} \tag{3}$$

$$n_{i,s} \geq \underline{N}_i x_{i,s} \quad \forall i \in \mathcal{P}, s \in \mathcal{S} \tag{4}$$

$$\Delta p_{i,s} \leq \overline{\Delta P}_i x_{i,s},\ q_{i,s} \leq \overline{Q} x_{i,s},\ n_{i,s} \leq \overline{N}_i x_{i,s},\ po_{i,s} \leq \overline{\Delta Po}_i x_{i,s} \quad \forall i \in \mathcal{P}, s \in \mathcal{S} \tag{5}$$

$$q_{i,s} = \sum_{j\in\mathcal{P}} q_{j,i,s} + q_{i,s}^{\text{source}} = \sum_{j\in\mathcal{P}} q_{i,j,s} + q_{i,s}^{\text{sink}} \quad \forall i \in \mathcal{P}, s \in \mathcal{S} \tag{6}$$

$$q_{i,j,s} \leq \overline{Q} t_{i,j,s},\ q_{i,s}^{\text{source}} \leq \overline{Q} t_{i,s}^{\text{source}},\ q_{i,s}^{\text{sink}} \leq \overline{Q} t_{i,s}^{\text{sink}} \quad \forall i, j \in \mathcal{P}, s \in \mathcal{S} \tag{7}$$

$$\sum_{i\in\mathcal{P}} q_{i,s}^{\text{sink}} = Q_s^{\text{bound}} \quad \forall s \in \mathcal{S} \tag{8}$$

$$\sum_{i\in\mathcal{P}} t_{i,s}^{\text{sink}} \geq 1,\ \sum_{i\in\mathcal{P}} t_{i,s}^{\text{source}} \geq 1 \quad \forall s \in \mathcal{S} \tag{9}$$

$$t_{i,i,s} = 0 \quad \forall i \in \mathcal{P}, s \in \mathcal{S} \tag{10}$$

[2]Equation only active if dual bound is considered.

$$\sum\nolimits_{j\in\mathcal{P}} t_{i,j,s} \le |\mathcal{P}|(1 - t_{i,s}^{\text{source}}), \sum\nolimits_{j\in\mathcal{P}} t_{j,i,s} \le |\mathcal{P}|(1 - t_{i,s}^{\text{sink}}) \quad \forall i \in \mathcal{P}, s \in \mathcal{S} \tag{11}$$

$$t_{i,j,s} + t_{j,i,s} \le 1 \quad \forall i, j \in \mathcal{P}, s \in \mathcal{S} \tag{12}$$

$$\sum\nolimits_{j\in\mathcal{P}} t_{i,j,s} \le |\mathcal{P}|y_i, \sum\nolimits_{j\in\mathcal{P}} t_{j,i,s} \le |\mathcal{P}|y_i, t_{i,s}^{\text{source}} \le y_i, t_{i,s}^{\text{sink}} \le y_i \quad \forall i \in \mathcal{P}, s \in \mathcal{S} \tag{13}$$

$$p_{i,s}^{\text{in}} - P_s^{\text{source}} \underset{\geq -}{\overset{\leq +}{}} \overline{P}(1 - t_{i,s}^{\text{source}}) \quad \forall i \in \mathcal{P}, s \in \mathcal{S} \tag{14}$$

$$p_{i,s}^{\text{in}} + \Delta p_{i,s} = p_{i,s}^{\text{out}} \quad \forall i \in \mathcal{P}, s \in \mathcal{S} \tag{15}$$

$$p_{i,s}^{\text{in}} \le \overline{P}\left(\sum\nolimits_{j\in\mathcal{P}} t_{j,i,s} + t_{i,s}^{\text{source}}\right), p_{i,s}^{\text{out}} \le \overline{P}\left(\sum\nolimits_{j\in\mathcal{P}} t_{i,j,s} + t_{i,s}^{\text{sink}}\right) \forall i \in \mathcal{P}, s \in \mathcal{S} \tag{16}$$

$$p_{i,s}^{\text{out}} - P_s^{\text{sink}} - \Delta p_{i,s}^{\text{frict, out}} \underset{\geq -}{\overset{\leq +}{}} \overline{P}(1 - t_{i,s}^{\text{sink}}) \quad \forall i \in \mathcal{P}, s \in \mathcal{S} \tag{17}$$

$$p_{i,s}^{\text{out}} - \Delta p_{i,j,s}^{\text{frict, branch}} - p_{j,s}^{\text{in}} \underset{\geq -}{\overset{\leq +}{}} \overline{P}(1 - t_{i,j,s}) \quad \forall i, j \in \mathcal{P}, s \in \mathcal{S} \tag{18}$$

$$\Delta p_{i,s}^{\text{frict, out}} - 0.5\varrho\zeta^{\text{out}}(q_{i,s}^{\text{sink}}/A)^2 \underset{\geq -}{\overset{\leq +}{}} \overline{P}(1 - x_{i,s}^{\text{out}}) \quad \forall i \in \mathcal{P}, s \in \mathcal{S} \tag{19}$$

$$\Delta p_{i,s}^{\text{frict, branch}} - 0.5\varrho\zeta^{\text{branch}}(q_{i,s}^{\text{sink}}/A)^2 \underset{\geq -}{\overset{\leq +}{}} \overline{P}(1 - x_{i,s}^{\text{branch}}) \quad \forall i \in \mathcal{P}, s \in \mathcal{S} \tag{20}$$

$$\Delta p_{i,s}^{\text{frict, out}} \le \overline{P}x_{i,s}^{\text{out}}, \ \Delta p_{i,s}^{\text{frict, branch}} \le \overline{P}x_{i,s}^{\text{branch}} \quad \forall i \in \mathcal{P}, s \in \mathcal{S} \tag{21}$$

$$x_{i,s}^{\text{out}} \ge (t_{i,s}^{\text{sink}} + t_{i,s}^{\text{source}} - 1) \quad \forall i \in \mathcal{P}, s \in \mathcal{S} \tag{22}$$

$$x_{i,j,s}^{\text{branch}} \ge \sum\nolimits_{k\in\mathcal{P}} (t_{k,j,s} - 1)/|\mathcal{P}| \quad \forall i, j \in \mathcal{P}: i \ne j, s \in \mathcal{S} \tag{23}$$

$$\Delta p_{i,s} = (\alpha_{i,0} - \zeta^{\text{inst}})\, {q_{i,s}}^2 + \sum\nolimits_{m=1}^{2} \alpha_{i,m}\, {q_{i,s}}^{2-m}\, {n_{i,s}}^m \quad \forall i \in \mathcal{P}, s \in \mathcal{S} \tag{24}$$

$$po_{i,s} = \beta_{i,4} + \sum\nolimits_{m=0}^{3} \beta_{i,m}\, {q_{i,s}}^{3-m}\, {n_{i,s}}^m \quad \forall i \in \mathcal{P}, s \in \mathcal{S} \tag{25}$$

$$\sum\nolimits_{i\in\mathcal{P}} po_{i,s} \ge Po_s^{\text{min}} \quad \forall s \in \mathcal{S} \tag{26}$$

4 Results

To solve the presented MINLP, we use SCIP 5.0.1 [4], under Windows with an Intel i7-3820 and 64 GB RAM. To investigate the influence of a detailed modeling of friction losses associated to the booster station itself, we generate two model formulations: (1) with detailed friction modeling, where we consider all equations given in Sect. 3 and determine approximate values for ζ_{inst}, ζ_{out}, and ζ_{branch} by measurements, (2) without detailed friction modeling ($\zeta^{\text{inst}} = \zeta^{\text{out}} = \zeta^{\text{branch}} = \Delta p^{\text{frict, out}} = \Delta p^{\text{frict, branch}} = 0$, and no consideration of the model equations with superscript 2). Moreover, we investigate the influence of dual bounds derived by domain-specific knowledge on the computing time. In Table 2, the computing time in case I corresponds to adding no additional dual bound, whereas cases II and III correspond to the dual bounds described before. The computed optimal solutions for all model formulations are experimentally validated. Therefore, in a first step, the optimal selection, interconnection and operation settings of the pumps are realized on the test rig. In a second step, the resulting volume flows at the floors, Q^{meas}, and the power consumption of all pumps, Po^{meas}, are measured. These values are then used to compare the optimization model and reality: (1) The volume flows measured at the floors are compared to the volume flow demands specified in the boundary conditions by computing $\alpha_{\text{Q}} := (Q^{\text{meas}}_{\text{mean}} - Q^{\text{bound}}_{\text{mean}})/Q^{\text{bound}}_{\text{mean}}$, in which mean indicates the scenarios weighted means. (2) The real-world energy consumption of the pumps is compared to the one predicted by the optimization model by computing $\alpha_{\text{po}} := (\sum_{i\in\mathcal{P}} po^{\text{meas}}_{i,\text{mean}} - \sum_{i\in\mathcal{P}} po_{i,\text{mean}})/\sum_{i\in\mathcal{P}} po_{i,\text{mean}}$. Our validation shows that when disregarding friction losses associated with the booster station itself, the proposed solution significantly violates the volume flow demand, cf. $\alpha_{\text{Q}} = -16.44\%$ in Table 2. In this case, the required hydraulic power decreases, and thus the pumps' power consumptions are underestimated. To still be able to compare solutions of the models with and without additional friction losses, the hydraulic-geodetic efficiency of the system, $\eta_{\text{geod}} := \varrho g H_{\text{geod}} Q^{\text{meas}}_{\text{mean}}/\sum_{i\in\mathcal{P}} po^{\text{meas}}_{i,\text{mean}}$, is considered. This further reveals that the more detailed model yields solutions with a higher efficiency in reality. Interestingly, when modeling the additional friction losses, the computation time decreases for case III despite the addition of further quadratic constraints and binary variables. The consideration of lower bounds for the power consumption can significantly reduce the computing time. This is shown by the computing time for II and III. The separate optimization for each scenario used for case III provides an even tighter bound, which further reduces the computing time.

Table 2 Optimization results, deviations to the real system and computing time

Detailed friction model	Obj. value	α_{Q}	α_{po}	η_{geod}	Computing time or gap after 12 h I	II	III
Yes	3416.81 €	−0.46%	2.55%	40.25%	6.53%	0.3%	28 min
No	3173.15 €	−16.44%	−3.31%	36.14%	2.94%	173 min	134 min

5 Conclusion

We presented a MINLP for optimizing investment and energy costs of pumping systems in buildings and validated our model using a modular test rig. The validation showed that the feasibility of the optimal solutions in the real world depends strongly on taking friction losses within the booster station into account. We showed that the computing time can be further reduced by introducing a lower bound for the power consumption based on domain-specific knowledge.

Acknowledgments Results were obtained in project No. 17482 N/1, funded by the German Federal Ministry of Economic Affairs and Energy (BMWi) approved by the Arbeitsgemeinschaft industrieller Forschungsvereinigungen "Otto von Guericke" e.V. (AiF). Moreover, this research was partially funded by Deutsche Forschungsgemeinschaft (DFG) under project No. 57157498.

References

1. Alexander, M., et al.: Mathematical Optimization of Water Networks. Springer, New York (2012)
2. Altherr, L., Leise, P., Pfetsch, M.E., Schmitt, A.: Resilient layout, design and operation of energy-efficient water distribution networks for high-rise buildings using MINLP. Optim. Eng. **20**(2), 605–645 (2019)
3. DIN 1988-500:2011-02: Codes of practice for drinking water installations (2011)
4. Gleixner, A., et al.: The SCIP Optimization Suite 5.0. Berlin (2017)
5. Hirschberg, R.: Lastprofil und Regelkurve zur energetischen Bewertung von Druckerhöhungsanlagen. HLH - Heizung, Lüftung, Klima, Haustechnik (2014)
6. Pöttgen, P., Pelz, P.F.: The best attainable EEI for booster stations derived by global optimization. In: IREC 2016, Düsseldorf
7. Weber, J.B., Lorenz, U.: Optimizing booster stations. In: GECCO '17 (2017)

Optimal Product Portfolio Design by Means of Semi-infinite Programming

Helene Krieg, Jan Schwientek, Dimitri Nowak, and Karl-Heinz Küfer

Abstract A new type of product portfolio design task where the products are identified with geometrical objects representing the efficiency of a product, is introduced. The sizes and shapes of these objects are determined by multiple constraints whose activity cannot be easily predicted. Hence, a discretization of the parameter spaces could obfuscate some advantageous portfolio configurations. Therefore, the classical optimal product portfolio problem is not suitable for this task. As a new mathematical formulation, the continuous set covering problem is presented which transfers into a semi-infinite optimization problem (SIP). A solution approach combining adaptive discretization of the infinite index set with regularization of the non-smooth constraint function is suggested. Numerical examples based on questions from pump industry show that the approach is capable to work with real-world applications.

Keywords Product portfolio design · Continuous set covering problem · Optimization of technical product portfolios

1 Introduction

From mathematical perspective, optimal product portfolio design was originally formulated as a linear optimization problem with binary decision variables [1]. Products were defined by a finite set of discrete-valued attributes and the portfolio should satisfy a finite number of customer demands. In technical contexts, products are machines defined by real-valued parameters such as weight, length, or speed which can be selected on continuous ranges. The authors in [2] were the first to incorporate continuous decision variables in their product portfolio optimization problem for industrial cranes. We additionally take the continuous ranges of

H. Krieg (✉) · J. Schwientek · D. Nowak · K.-H. Küfer
Fraunhofer ITWM, Kaiserslautern, Germany
e-mail: helene.krieg@itwm.fraunhofer.de

J. S. Neufeld et al. (eds.), *Operations Research Proceedings 2019*, Operations Research Proceedings, https://doi.org/10.1007/978-3-030-48439-2_59

operation points into account which change size and shape dependent on the product parameters. Further, we require from the portfolio that the operation ranges do not satisfy finitely many but a compact, infinite set of operation points, the set of customer specifications. One objective then could be that the product portfolio covers each operation point in this set with high efficiency. All in all, this results in a continuous set covering problem which in fact is a semi-infinite optimization program (SIP). Unfortunately, this SIP is neither smooth nor convex. Therefore, we present a method that solves a sequence of successively improved finite, nonlinear approximating problems. The method converges under harmless requirements on the update of the approximating problems and a worst-case convergence rate exists [3].

2 Model

The task of designing a portfolio of parametrically described products can be formulated as continuous set covering problem as follows: Let $Y \in \mathbb{R}^m$ be a compact set of customer specifications. Find parameters $\mathbf{x} \in \mathbb{R}^n$ such that Y is covered by $N \in \mathbb{N}$ operation ranges $P_i(\mathbf{x}) \subset \mathbb{R}^m$, $i = 1, \ldots, N$, and $\mathbf{x}$ minimizes an objective function $f : \mathbb{R}^n \to \mathbb{R}$ measuring the quality of the product portfolio. In set-theoretical formulation, "is covered" means that $Y \subseteq \bigcup_{i=1}^{N} P_i(\mathbf{x})$. We suppose that the sets $P_i(\mathbf{x})$, $i = 1, \ldots, N$ have functional descriptions:

$$P_i(\mathbf{x}) := \{\mathbf{y} \in \mathbb{R}^m | g_{ij}(\mathbf{x}, \mathbf{y}) \leq 0,\ j = 1, \ldots, p_i\} \quad i = 1, \ldots, N \tag{1}$$

where $p_i \in \mathbb{N}$ for all $i = 1, \ldots, N$, and the functions $g_{ij} : \mathbb{R}^n \times \mathbb{R}^m \to \mathbb{R}$ are supposed to be continuously differentiable in $\mathbf{x}$ for all $\mathbf{y} \in Y$. The continuous set covering problem is given by

$$\begin{aligned} \mathrm{SIP}_{\mathrm{CSCP}}(Y): \quad & \min_{\mathbf{x} \in X} f(\mathbf{x}) \\ & \text{s.t.} \min_{1 \leq i \leq N} \max_{1 \leq j \leq p_i} g_{ij}(\mathbf{x}, \mathbf{y}) \leq 0 \quad \forall \mathbf{y} \in Y. \end{aligned} \tag{2}$$

In general, $|Y| = \infty$, and thus, (2) possesses an infinite number of constraints, one for each element $\mathbf{y} \in Y$. Therefore, $\mathrm{SIP}_{\mathrm{CSCP}}(Y)$ naturally is a semi-infinite program (SIP). Equation (2) is an especially difficult subclass of SIP: The constraint function contains a minimum and a maximum operator. Thus, in general, any structural property such as convexity or continuous differentiability of the functions g_{ij}, $i \in \{1, \ldots, N\}$, $j \in \{1, \ldots, p_i\}$, does not transfer to the constraint function.

3 Methods

Although there exists much theory and many numerical methods in the field of SIP (see for example [4] and [5]), most available solvers assume that the objective and constraint functions are continuously differentiable.

Regularization One possibility to make use of gradient-based methods from nonlinear optimization is regularization of the non-smooth constraint function. Because of several advantageous properties like monotonous, uniform convergence and a known approximation error, we suggest to replace the constraint function by the double entropic smoothing function

$$g^{s,t}(\mathbf{x},\mathbf{y}) = \frac{1}{s}\ln\left(\frac{1}{N}\sum_{i=1}^{N}\exp\left(\frac{s}{t}\ln\left(\sum_{j=1}^{p_i}\exp\big(t g_{ij}(\mathbf{x},\mathbf{y})\big)\right)\right)\right). \tag{3}$$

The approximation quality is steered by two parameters, $s < 0$ and $t > 0$. If the basic functions g_{ij}, $i = 1,\ldots,N$, $j = 1,\ldots,p_i$ are all d-times continuously differentiable, the smoothed function is as well.

Adaptive Discretization The simplest way to tackle semi-infinite programs is to approximate them by finite optimization problems via discretization of the semi-infinite index set Y [4]. We use the adaptive discretization algorithm given by [6]. It adds new constraints to finite nonlinear optimization problems of type $\mathrm{SIP}_{\mathrm{CSCP}}(\dot{Y})$, where $\dot{Y} := \{\mathbf{y}^1,\ldots,\mathbf{y}^r\}$, $r \in \mathbb{N}$, that approximate $\mathrm{SIP}_{\mathrm{CSCP}}(Y)$. To do so, the set $\dot{Y}$ is successively extended by an element $\mathbf{y} \in Y$ which represents the most violated constraint. At least theoretically, the lower level problem of $\mathrm{SIP}_{\mathrm{CSCP}}(Y)$,

$$Q(\mathbf{x}): \quad \max_{\mathbf{y}\in Y}\ \min_{1\le i\le N}\ \max_{1\le j\le p_i} g_{ij}(\mathbf{x},\mathbf{y}) \tag{4}$$

has to be solved to global optimality to find a new point. This problem is a parametric, nonlinear finite optimization problem with an objective function that is not concave and not continuously differentiable in the decision vector $\mathbf{y}$. To avoid usage of global optimization strategies, we evaluate the function at a finite reference set $Y_{\mathrm{ref}} \subset Y$ and select the new point $\mathbf{y}$ in Y_{ref}.

Algorithm Combining discretization of the index set and regularization of the constraint function leads to the following finite, smooth nonlinear optimization problem which approximates (2):

$$\begin{aligned} \mathrm{SIP}^{s,t}_{\mathrm{CSCP}}(\dot{Y}): \quad & \min_{\mathbf{x}\in X} f(\mathbf{x}) \\ & \text{s.t. } g^{s,t}(\mathbf{x},\mathbf{y}^l) \le 0, \quad l = 1,\ldots,r \end{aligned} \tag{5}$$

This problem contains two types of approximations and for each type there is a trade-off between nice properties of the optimization problem (small size and smoothness) and approximation quality. Therefore, with Algorithm 1 we propose a procedure that iteratively improves both, discretization of the index set and smooth approximation of the constraint function via steering the smoothing parameters. Doing so, it successively finds an approximative local solution for (2).

Note that (2) usually has multiple local extrema of different quality. Therefore, it is important to provide the algorithm with a good initial value. This could be found for example using construction heuristics that distribute the sets $P_i(\mathbf{x})$ on the set Y to be covered.

Algorithm 1 Approximative solution procedure for (2)

input: $\mathbf{x}^0 \in \mathbb{R}^n$, $\dot{Y}_0 := \{\mathbf{y}^1, \dots, \mathbf{y}^{r_0}\} \subset Y$, $Y_{\text{ref}} = \{\mathbf{y}^1, \dots, \mathbf{y}^{r_{\text{ref}}}\} \subset Y$, $s_0 < 0$, $t_0 > 0$ such that a feasible solution for $SIP^{s_0,t_0}_{\text{CSCP}}(\dot{Y}_0)$ exists, update factor $d \in \mathbb{R}_+$

Set $k = 0$
while $\neg$ stopping criteria satisfied **do**
  compute a solution $\mathbf{x}^{k+1}$ of $\text{SIP}^{s_k,t_k}_{\text{CSCP}}(\dot{Y}_k)$ using $\mathbf{x}^k$ as starting point
  select $\mathbf{y}^{k+1} \in \text{argmax}_{\mathbf{y} \in Y_{\text{ref}}}\{\min_{1 \le i \le N} \max_{1 \le j \le p_i} g_{ij}(\mathbf{x}^{k+1}, \mathbf{y})\}$
  set $\dot{Y}_{k+1} := \{\mathbf{y}^{k+1}\} \cup \dot{Y}_k$
  $(s_{k+1}, t_{k+1}) := d(s_k, t_k)$
  $k = k + 1$
end while

output: $\{\mathbf{x}^i\}_{i=0}^k$, $\{\dot{Y}_i\}_{i=0}^k$

4 Application

We apply the presented modeling and solution approach exemplarily to product portfolio optimization tasks from pump industry. Here, a given rectangular set of operation points $Y \subset \mathbb{R}^2_+$ should be covered by a portfolio of N pumps. The model functions are summarized in Table 1. We set the following technical parameters for all products to identical values: curvature $\lambda = 0.2$, minimum and maximum relative speed $n_{\min} = 0.7$, $n_{\max} = 1.2$, and maximum relative efficiency $\eta_{\max} = 0.9$. The decision vector $\mathbf{x} \in \mathbb{R}^{2N}_+ \times [0, \eta_{\max}]$ is defined by $(x_{2i-1}, x_{2_i}) := (Q^i_D, H^i_D)$, $i = 1, \dots, N$ and $x_{2N+1} := \eta_{\min}$. Here, (Q^i_D, H^i_D) is the design point (flow in $\text{m}^3\,\text{h}^{-1}$ and head in m) of the i-th pump and $\eta_{\min}$ is the minimum allowed efficiency common to all pumps. The set of customer specifications is $Y = [100, 3000]\,\text{m}^3\,\text{h}^{-1} \times [100, 400]\,\text{m}$ and the compact design space is given by $X := Y^N \times [0, \eta_{\max}]$. Each pump operation area is defined by

$$
\begin{aligned}
P_i(\mathbf{x}) := \{\mathbf{y} \in \mathbb{R}^2 | g_{i1}(\mathbf{x}, \mathbf{y}) &:= n_{\min} - n(x_{2i-1}, x_{2i}, \lambda, y_1, y_2) \le 0 \\
g_{i2}(\mathbf{x}, \mathbf{y}) &:= n(x_{2i-1}, x_{2i}, \lambda, y_1, y_2) - n_{\max} \le 0 \qquad (6) \\
g_{i3}(\mathbf{x}, \mathbf{y}) &:= x_{2N+1} - \eta(x_{2i-1}, x_{2i}, \lambda, y_1, y_2) \le 0\}.
\end{aligned}
$$

Table 1 Pump portfolio design model

Description	Function
Speed	$n(Q_D, H_D, \lambda, y_1, y_2) = -\frac{3\lambda-1}{2(1-\lambda)} y_1 + \sqrt{\frac{(\lambda+1)^2}{16(1-\lambda)^2} \frac{y_1^2}{Q_D^2} + \frac{y_2}{2(1-\lambda)H_D}}$
Efficiency	$\eta(Q_D, H_D, \lambda, y_1, y_2) = -\frac{\eta_{\max}}{n^2(Q_D,H_D,\lambda,y_1,y_2)Q_D^2} y_1^2 + 2\frac{\eta_{\max}}{n(Q_D,H_D,\lambda,y_1,y_2)Q_D} y_1$
Efficiency of a portfolio in a given customer specification	$\tilde{\eta}(\mathbf{x}, \mathbf{y}) := \frac{1}{N} \sum_{i=1}^{N} \eta(x_{2i-1}, x_{2i}, \lambda, \mathbf{y})\xi_i(\mathbf{y}) \xi_i(\mathbf{y}) := \begin{cases} 1 & \text{if } \mathbf{y} \in P_i(\mathbf{x}) \\ 0 & \text{otherwise.} \end{cases}$

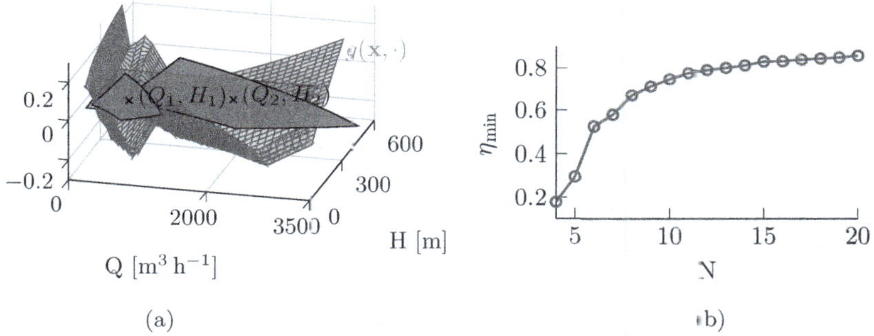

Fig. 1 Two-pump portfolio and the classical trade-off in product portfolio optimization. (**a**) Configuration of two pumps. (**b**) Trade-off between portfolio size and quality

Figure 1a shows a configuration of two pumps which does not fully cover Y and hence, is infeasible for (2). The crosses are the design points of the pumps. The surface is the graph of the min-max-constraint function $g(\mathbf{x}, \cdot)$. The lower and upper boundaries of the pump operation areas are parts of the zero level curves of $g_{i1}(\mathbf{x}, \cdot)$ and $g_{i2}(\mathbf{x}, \cdot)$, respectively, whereas the left and right boundaries are determined by $g_{i3}(\mathbf{x}, \cdot)$.

Application 1: Portfolio Size Versus Quality Suppose that the quality of the pump portfolio is measured by the common minimum efficiency of all pumps, the $N+1$th component of the decision vector, $f(\mathbf{x}) := -x_{2N+1}$. Thus, in a feasible portfolio, for any point in Y, there exists a pump that operates at least as efficient as $\eta_{\min}$ at this point. By running Algorithm 1 for different numbers of N, we can figure out the classical trade-off of product portfolio design (Fig. 1b): Increasing portfolio quality conflicts with reduction of portfolio size.

Application 2: Different Portfolio Qualities Another possibility to measure quality of a pump portfolio is the usage of further information on customer requirements from marketing studies. We simulated incorporation of such knowledge by suppos-

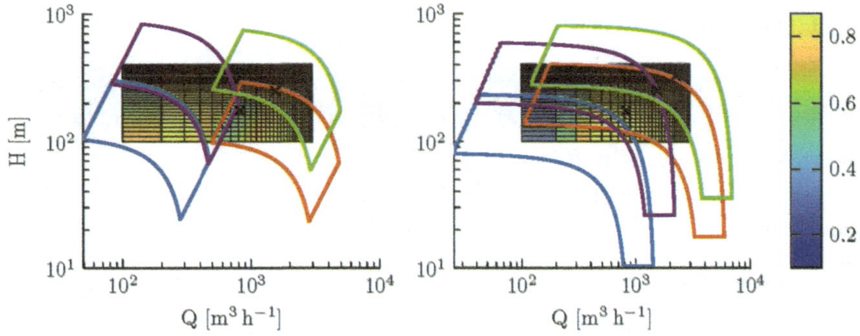

Fig. 2 Optimal portfolios of four pumps (shapes) for different quality measures. Color gradient depicts the best efficiencies in Y. Crosses mark the three density centers $\tilde{\mathbf{y}}^i$, $i = 1, 2, 3$. Left: $\mathbf{w} = (0.6, 0.13, 0.13, 0.13)$, $\eta_{\min} = 0.459$; right: $\mathbf{w} = (0, 0.8, 0.1, 0.1)$, $\eta_{\min} = 0.101$

ing that three points $\tilde{\mathbf{y}}^1 = (825, 175)$, $\tilde{\mathbf{y}}^2 = (1550, 250)$ and $\tilde{\mathbf{y}}^3 = (2275, 325)$ built the centers of clusters in customer specifications. Hence, a pump portfolio of high quality should be selected so that the efficiency in these density centers is high. To balance this quality goal with that of high overall minimum efficiency, we used a weighted sum objective function

$$f(\mathbf{x}) := w_0 \eta_{\min,}(\mathbf{x}) + \sum_{i=1}^{3} w_i \tilde{\eta}(\mathbf{x}, \tilde{\mathbf{y}}^i) \tag{7}$$

where $\mathbf{w} \in [0, 1]^4$ was a given vector of weights. Figure 2 shows the optimized portfolios of four pumps for two different weight vectors. On the right-hand side, the minimal efficiency measure was not taken into account but the left-most density center was highly rated. As a result, efficiency is rather uniformly high in the area surrounding the density centers. In contrast to that, for the left-hand side solution, maximization of minimum efficiency was given the highest weight. This yields higher efficiency near the boundary of Y, but non-uniform efficiency distribution among the density centers. Further, operation areas do not overlap that much as in the right-hand side portfolio.

5 Conclusion and Further Work

The article introduces a new kind of product portfolio optimization problem that appears in technical contexts. First numerical studies suggest that the presented algorithm for solving the resulting semi-infinite program is capable to answer questions that appear in real-world problems.

Next steps are further analysis and improvement of the procedure. For example, due to multi-extremality of problem (2), a good starting point for Algorithm 1 has to be found. Another unsolved problem is that of finding good smoothing parameters and a suited update strategy.

References

1. Green, P.E., Krieger, A.M.: Models and heuristics for product line selection. Mark. Sci. **4**, 1–19 (1985)
2. Tsafarakis, S., Saridakis, C., Baltas, G., Matsatsinis, N.: Hybrid particle swarm optimization with mutation for optimizing industrial product lines: an application to a mixed solution space considering both discrete and continuous design variables. Ind. Mark. Manage. **42**, 496–506 (2013)
3. Krieg, H.: Modeling and solution of continuous set covering problems by means of semi-infinite optimization. Ph.D. dissertation. University of Kaiserslautern (2019)
4. Hettich, R., Kortanek, K.O.: Semi-infinite programming: theory, methods, and applications. SIAM Rev. **35**, 380–429 (1993)
5. Reemtsen, R., Rückmann, J.-J.: Semi-infinite Programming, vol. 417. Springer, Boston, MA (1998)
6. Blankenship, J.W., Falk, J.E.: Infinitely constrained optimization problems. J. Optim. Theory Appl. **19**, 261–281 (1976)

Exploiting Partial Convexity of Pump Characteristics in Water Network Design

Marc E. Pfetsch and Andreas Schmitt

Abstract The design of water networks consists of selecting pipe connections and pumps to ensure a given water demand to minimize investment and operating costs. Of particular importance is the modeling of variable speed pumps, which are usually represented by degree two and three polynomials approximating the characteristic diagrams. In total, this yields complex mixed-integer (non-convex) nonlinear programs.

This work investigates a reformulation of these characteristic diagrams, eliminating rotating speed variables and determining power usage in terms of volume flow and pressure increase. We characterize when this formulation is convex in the pressure variables. This structural observation is applied to design the water network of a high-rise building in which the piping is tree-shaped. For these problems, the volume flow can only attain finitely many values. We branch on these flow values, eliminating the non-convexities of the characteristic diagrams. Then we apply perspective cuts to strengthen the formulation. Numerical results demonstrate the advantage of the proposed approach.

1 Introduction

In this paper the optimal design and operation of water networks using mixed-integer nonlinear programming (MINLP) is considered, see [2] for an overview. More precisely, we investigate the optimal design of tree-shaped high-rise water supply systems in which the floors need to be connected by pipes and pumps must be placed, such that all floors are supplied by water under minimal investment and running costs in a stationary setting. A customized branch and bound approach has been developed in [1], which aims to deal with the inherent combinatorial complexity for deciding the topology. Another challenge of the problem lies in

---

M. E. Pfetsch · A. Schmitt (✉)
Department of Mathematics, TU Darmstadt, Darmstadt, Germany
e-mail: pfetsch@mathematik.tu-darmstadt.de; aschmitt@mathematik.tu-darmstadt.de

J. S. Neufeld et al. (eds.), *Operations Research Proceedings 2019*, Operations Research Proceedings, https://doi.org/10.1007/978-3-030-48439-2_60

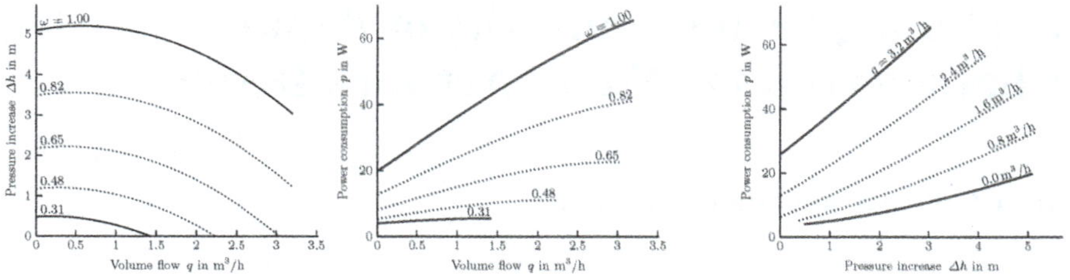

Fig. 1 Exemplary characteristic diagram (left) and graph of $\tilde{P}(q, \Delta h)$ (right)

the operation of pumps. Their nonlinear and non-convex behavior is described by so called characteristic diagrams, see for an example Fig. 1, which determine for a given volume flow q the corresponding possible range of pressure increase Δh and power consumption p if one varies the normalized operating speed ω. These nonlinearities are modeled in the high-rise problem using a quadratic

$$\Delta H \colon [\underline{q}, \overline{q}] \times [\underline{\omega}, 1] \to \mathbb{R}, \quad (q, \omega) \mapsto \alpha^H q^2 + \beta^H q\omega + \gamma^H \omega^2$$

and a cubic polynomial

$$P \colon [\underline{q}, \overline{q}] \times [\underline{\omega}, 1] \to \mathbb{R}, \quad (q, \omega) \mapsto \alpha^P q^3 + \beta^P q^2\omega + \gamma^P q\omega^2 + \delta^P \omega^3 + \epsilon^P.$$

Since the piping has to form a tree, the volume flow in a given floor and pump attains only finitely many distinct values $Q := \{q_1, \dots, q_n\} \subset \mathbb{R}_+$ with $q_i < q_{i+1}$. Therefore a pump is modeled by the non-convex set

$$D := \big\{(y, p, \Delta h, q, \omega) \in \{0, 1\} \times \mathbb{R}^2_+ \times Q \times [\underline{\omega}, 1] \mid$$
$$p \geq P(q, \omega)\, y,\ \Delta h = \Delta H(q, \omega)\, y,\ \underline{q}\, y \leq q \leq q_n + (\overline{q} - q_n)\, y\big\},$$

where y is 1 iff the pump is used. Note that q also models the volume flow in a floor and thus can attain values exceeding $\underline{q}$ or $\overline{q}$ if $y = 0$.

Whereas Δh and q are linked to other variables in the model by further constraints, the operating speed ω is only constrained by D. In the ensuing paper we first introduce an alternative formulation X that eliminates ω and is convex in Δh for fixed q. Afterwards we present a simple test to check for this convexity. Subsequently we derive valid inequalities for X involving perspective cuts by lifting. The benefit of projecting out ω and using these cuts in a branch and cut framework is demonstrated on a representative testset.

2 Reformulation and Convexity

To derive a formulation X of the projection of D onto the variables y, p, Δh and q, we will work with the following two assumptions: All coefficients appearing in the approximations are nonzero to avoid edgecases. More importantly, $\Delta H(q, \omega)$ is strictly increasing in ω on $[\underline{\omega}, 1]$ for all $q \in [\underline{q}, \overline{q}]$; this is supported by scaling laws and implies the existence of the inverse function $\Omega\colon [\underline{q}, \overline{q}] \times \mathbb{R}_+ \to \mathbb{R}_+$ of $\Delta H(q, \omega)$ with respect to ω for fixed q, i.e., $\Omega(q, \Delta H(q, \omega)) = \omega$ for all $\omega \in [\underline{\omega}, 1]$. It is given by

$$\Omega(q, \Delta h) = \frac{1}{2\gamma^H}\left(- \beta^H q + \sqrt{\left({\beta^H}^2 - 4\gamma^H \alpha^H\right) q^2 + 4\gamma^H \Delta h}\right)$$

and its composition with $P(q, \omega)$ leads to the function

$$\tilde{P}\colon [\underline{q}, \overline{q}] \times \mathbb{R}_+ \to [\underline{\omega}, 1], \quad (q, \Delta h) \mapsto P(q, \Omega(q, \Delta h)),$$

which calculates the power consumption of a pump processing a volume flow q and increasing the pressure by Δh. See Fig. 1 for an example. With $\Delta \underline{H}(q) := \Delta H\left(q, \underline{\omega}\right)$ and $\Delta \overline{H}(q) := \Delta H\left(q, 1\right)$, the projection is then given by

$$X := \big\{(y, p, \Delta h, q) \in \{0, 1\} \times \mathbb{R}_+^2 \times Q \mid$$
$$p \geq \tilde{P}(q, \Delta h)\, y,\ \Delta \underline{H}(q)\, y \leq \Delta h \leq \Delta \overline{H}(q)\, y,\ \underline{q}\, y \leq q \leq q_n + (\overline{q} - q_n)\, y\big\}$$

Both X and D present obstacles for state-of-the-art optimization software and methods. The pumps in our instances, however, satisfy a convexity property characterized in the following lemma, making the usage of X beneficial.

Lemma 1 *For each fixed $q \in [\underline{q}, \overline{q}]$, the function $\tilde{P}(q, \Delta h)$ is convex in $\Delta h \in [\Delta \underline{H}(q), \Delta \overline{H}(q)]$ if and only if*

$$\max_{q \in [\underline{q}, \overline{q}]} (\gamma^H \beta^P - \beta^H \gamma^P) q^2 - 3\, \beta^H \delta^P q\, \tilde{\omega} \leq 3\, \gamma^H \delta^P \tilde{\omega}^2,$$

where $\tilde{\omega} = 1$ if $\delta^P < 0$ and $\tilde{\omega} = \underline{\omega}$ otherwise.

Proof Convexity can be checked by the minimization of $\frac{\partial^2 \tilde{P}}{\partial^2 \Delta h}$ over $q \in [\underline{q}, \overline{q}]$ and $\Delta h \in [\Delta \underline{H}(q), \Delta \overline{H}(q)]$, which can (after some calculations) be written as

$$\min_{q \in [\underline{q}, \overline{q}],\, \omega \in [\underline{\omega}, 1]} (\beta^H \gamma^P - \gamma^H \beta^P) q^2 + 3\, \beta^H \delta^P q\, \omega + 3\, \gamma^H \delta^P \omega^2 \geq 0.$$

Table 1 Examples of real-world pump parameters

α^P	β^P	γ^P	δ^P	ϵ^P	α^H	β^H	γ^H	$\underline{q}$	$\overline{q}$	$\underline{\omega}$
−0.095382	0.25552	13.6444	18.0057	6.3763	−0.068048	0.26853	4.1294	0.0	8.0	0.517
−0.13	0.79325	18.2727	33.8072	6.2362	−0.066243	0.30559	6.1273	0.0	10.0	0.429
−0.14637	1.1882	23.0823	53.0306	6.0431	−0.065158	0.34196	8.1602	0.0	11.0	0.350
−0.32719	0.36765	16.4571	16.2571	3.5722	−0.31462	0.36629	5.0907	0.0	3.2	0.308
0.35512	−4.4285	17.4687	30.5853	0.013785	−0.083327	−0.10738	4.2983	0.0	6.0	0.498

The sign of the partial derivative in ω of the objective function over $[\underline{q}, \overline{q}] \times [\underline{\omega}, 1]$ is equal to the sign of δ^P, since $\Delta H(q, \omega)$ is increasing in ω. Since this is constant, the minimum is attained either for $\omega = 1$ or $\omega = \underline{\omega}$. □

Note that $\Delta H(q, \omega)$ being concave ($\gamma^H \leq 0$) and $P(q, \omega)$ being convex as well as non decreasing in ω for fixed q is a sufficient condition for convexity. This, however, is not satisfied by our real-world testdata, compare Table 1.

3 Perspective Cuts and Lifted Valid Inequalities for X

In the following we use the convexity property and present several families of valid inequalities for X. We first consider the case when $Q = \{q_1\} \subseteq [\underline{q}, \overline{q}]$ is a singleton. Then perspective cuts introduced by Frangioni and Gentile [3] are valid for X. Defining $\tilde{P}_q(\Delta h) := \tilde{P}(q, \Delta h)$, these are given for $\Delta h^* \in [\Delta \underline{H}(q_1), \Delta \overline{H}(q_1)]$ by

$$\tilde{P}'_{q_1}(\Delta h^*)\, \Delta h + \left(\tilde{P}_{q_1}(\Delta h^*) - \tilde{P}'_{q_1}(\Delta h^*)\Delta h^*\right)\, y \leq p.$$

Validity can be seen by case distinction on the value of y. For $y = 0$ also Δh must be zero, leading to a vanishing left-hand side. If y is one, the cut corresponds to a gradient cut, which is valid by convexity.

For more general Q, another family of valid inequalities is obtained by combination of different perspective cuts, where we denote $\tilde{N} := \{1 \leq i \leq n \mid \underline{q} \leq q_i \leq \overline{q}\}$.

Lemma 2 *For parameters $\Delta h_i^* \in [\Delta \underline{H}(q_i), \Delta \overline{H}(q_i)]$ with $i \in \tilde{N}$, the inequality*

$$\left(\min_{i \in \tilde{N}} \tilde{P}'_{q_i}(\Delta h_i^*)\right)\Delta h + \left(\min_{i \in \tilde{N}} \tilde{P}_{q_i}(\Delta h_i^*) - \tilde{P}'_{q_i}(\Delta h_i^*)\Delta h_i^*\right) y \leq p$$

is valid for X.

Proof This follows from the validity of the perspective cuts for X with fixed $q = q_i$, $i \in \tilde{N}$, and that Δh and y are zero for $q \notin [\underline{q}, \overline{q}]$. □

Another family of valid inequalities can be formed by considering a perspective cut on the set $X \cap \{q = \underline{q}\}$ or $X \cap \{q = \overline{q}\}$ and lifting the variable q into it:

Lemma 3 *For parameters* $(\tilde{q}, q^*) \in \{(q_1, \underline{q}), (q_n, \overline{q})\}$, $\Delta h^* \in [\Delta \underline{H}(q^*), \Delta \overline{H}(q^*)]$ *and* $\gamma \in \mathbb{R}$ *such that* $\gamma(q - \tilde{q}) \leq 0$ *holds for each* $q \in Q$ *and*

$$\min_{\Delta h^* \in [\Delta \underline{H}(q_i), \Delta \overline{H}(q_i)]} \tilde{P}_{q_i}(\Delta h) - \tilde{P}'_{q^*}(\Delta h^*)\,\Delta h \geq \tilde{P}_{q^*}(\Delta h^*) - \tilde{P}'_{q^*}(\Delta h^*)\Delta h^* + \gamma(q_i - \tilde{q})$$

holds for each $i \in \tilde{N}$, *the following inequality is valid for* X

$$\tilde{P}'_{q^*}(\Delta h^*)\,\Delta h + \left(\tilde{P}_{q^*}(\Delta h^*) - \tilde{P}'_{q^*}(\Delta h^*)\Delta h^*\right) y + \gamma(q - \tilde{q}) \leq p.$$

Proof The inequality is valid for $X \cap \{y = 0\}$, since its left-hand side simplifies to $\gamma(q - \tilde{q}) \leq 0$ and p is non-negative. Moreover, the minimum condition on γ makes sure that the inequality is valid for $X \cap \{y = 1, q = q_i\}$ for all $i \in \tilde{N}$. □

This lifting idea leads to a further family of valid inequalities by first lifting q and then y into a gradient cut.

Lemma 4 *Let* $q^* \in \{\underline{q}, \overline{q}\}$, $\Delta h^* \in [\Delta \underline{H}(q^*), \Delta \overline{H}(q^*)]$ *and* $\beta, \gamma \in \mathbb{R}$ *such that for* $i \in \tilde{N}$ *with* $q_i \neq q^*$

$$\min_{\Delta h \in [\Delta \underline{H}(q_i), \Delta \overline{H}(q_i)]} \tilde{P}_{q_i}(\Delta h) - \tilde{P}'_{q^*}(\Delta h^*)\,\Delta h \geq \tilde{P}_{q^*}(\Delta h^*) - \tilde{P}'_{q^*}(\Delta h^*)\,\Delta h^* + \gamma(q_i - q^*)$$

and for $1 \leq i \leq n$ *with* $q_i \neq q^*$

$$\beta \geq \tilde{P}_{q^*}(\Delta h^*) - \tilde{P}'_{q^*}(q^*, \Delta h^*)\,\Delta h^* + \gamma(q_i - q^*)$$

holds. Then the following inequality is valid for X

$$\tilde{P}_{q^*}(\Delta h^*) - \tilde{P}'_{q^*}(\Delta h^*)\,\Delta h^* + \tilde{P}'_{q^*}(\Delta h^*)\,\Delta h + \beta(y - 1) + \gamma(q - q^*) \leq p.$$

Proof We again show the validity for subsets of X. First of all, the inequality corresponds to a gradient cut on $X \cap \{y = 1, q = q^*\}$. By the minimum condition on γ, the inequality is valid for $X \cap \{y = 1\}$. The last condition states that the left-hand side of the inequality must be bounded by zero for $X \cap \{y = 0\}$. □

The inequalities derived in Lemmas 2–4 are able to strengthen the relaxations used by MINLP solvers. Since there are infinitely many and to obtain small relaxations, usually only inequalities violated by solution candidates are added. Given a relaxation solution with pressure increase value $\Delta\tilde{h}$, the following heuristic for separating the above families of inequalities works well. The parameter Δh_i^* in Lemma 2 is chosen as $\Delta\tilde{h}$ if it belongs to $[\Delta \underline{H}(q_i), \Delta \overline{H}(q_i)]$, otherwise as the midpoint of the interval. To separate the inequalities given by Lemma 3 or 4, we try both choices for q^* and/or $\tilde{q}$ and use $\Delta\tilde{h}$ for Δh^* if it belongs to the interval

$[\Delta\underline{H}(q^*), \Delta\overline{H}(q^*)]$; otherwise we set it to the lower or the upper bound according on which side of the interval $\Delta\tilde{h}$ lies. We then maximize or minimize γ depending on q^* and minimize β by solving $\min\{\tilde{P}_{q^*}(\Delta h) - \alpha\, \Delta h \mid \Delta h \in [\Delta\underline{H}(q^*), \Delta\overline{H}(q^*)]\}$ with given values of $q^* \in \mathbb{R}$ and $\alpha \in \mathbb{R}$. This is, for appropriate speed bounds $\underline{\omega}$ and $\overline{\omega}$, equivalent to $\min\{P(q^*, \omega) - \alpha\, \Delta H(q^*, \omega) \mid \omega \in [\underline{\omega}, \overline{\omega}]\}$, i.e, the trivial problem to minimize a one-dimensional cubic function over an interval.

4 Computational Experiments

We conducted experiments on a testset containing instances of a high-rise problem resembling a real life downscaled testrig. We use five different basic pump types, c.f. Table 1, and also pump types derived by placing each basic type up to three times in parallel. Usage of Lemma 1 verifies that each type possesses the convexity property. We investigated 5, 7 and 10 floors, with a one meter height difference between consecutive floors. The pressure increase demanded in a floor is independent of volume flow and lies between 1.2 and 1.44 times its height. Furthermore, we include an energy cost weight of 10 and 100, which determines the importance of the nonlinearities in the objective value. The volume flow demand in each floor was sampled according to $\max\{0, \mathcal{N}(\mu, \sigma^2)\}$ for $(\mu, \sigma) \in \{(1, 0.5), (0.5, 0.25)\}$. For each of these different settings, ten instances were created, leading to 120 instances. To perform the test, we used SCIP 6.0.1, see [4], compiled with IPOPT 3.12, see [5], and CPLEX 12.8 running on a Linux cluster with Intel Xeon E5 CPUs with 3.50 GHz, 10 MB cache, and 32 GB memory using a 1 h time limit. We implemented a constraint handler, which enforces X by branching on the volume flow variables and using perspective cuts for fixed volume flows. Furthermore, it heuristically separates the three families presented in Lemmas 2–4 and propagates flow and pressure increase bounds.

We compared the formulation involving speed variables, i.e., the set D and a formulation involving X without the constraint handler. Furthermore, we tested the constraint handler without (`CH`) and with the heuristic cut separation (`CH+SEP`). In Table 2, we show the performance of the different approaches. The formulation X replaces polynomials by composite functions involving square-roots; nonetheless, the elimination of ω is able to solve 28 more instances in 1 h than formulation D. The worse average of the final gaps between primal and dual bounds is due to the lack of good primal solutions. Using `CH` one can solve only slightly more instances and more branch and bound nodes need to be inspected, but the solving time decreases substantially. The best performance albeit is given by also separating the lifted cuts, which also results in the least amount of visited branch and bounds nodes on average. Further results, not included for the ease of presentation, show: Separating the cuts of Lemma 3 has the biggest impact. Their sole usage already solves 116 instances, whereas the exclusive usage of cuts from either Lemma 2 or 4 leads to only 102 and 104 solved instances, respectively.

Table 2 Overview of test results

Formulation/setting	Time	Nodes	Gap	# solved
$\mathcal{D}$	295.5	22,428.6	53.41	61
$\mathcal{X}$	120.7	4548.7	88.11	89
CH	44.4	19,210.6	29.03	90
CH+SEP	13.4	1464.4	0.27	117

"Time" and "Nodes" give the shifted geometric means (see [4]) of solving time and number of branch and bound nodes, respectively. "Gap" gives the arithmetic mean over the gap between primal and dual bound after 1 h. "# solved" gives the number of solved instances in this time

Acknowledgments We thank Tim Müller (TU Darmstadt) for the pump approximations. This research was funded by Deutsche Forschungsgemeinschaft (DFG, German Research Foundation)—Project Number 57157498—SFB 805.

References

1. Altherr, L.C., Leise, P., Pfetsch, M.E., Schmitt, A.: Resilient layout, design and operation of energy-efficient water distribution networks for high-rise buildings using MINLP. Optim. Eng. **20**(2), 605–645 (2019)
2. D'Ambrosio, C., Lodi, A., Wiese, S., Bragalli, C.: Mathematical programming techniques in water network optimization. Eur. J. Oper. Res. **243**(3), 774–788 (2015)
3. Frangioni, A., Gentile, C.: Perspective cuts for a class of convex 0–1 mixed integer programs. Math. Program. **106**(2), 225–236 (2006)
4. Gleixner, A., Bastubbe, M., Eifler, L., Gally, T., Gamrath, G., Gottwald, R.L., Hendel, G., Hojny, C., Koch, T., Lübbecke, M.E., Maher, S.J., Miltenberger, M., Müller, B., Pfetsch, M.E., Puchert, C., Rehfeldt, D., Schlösser, F., Schubert, C., Serrano, F., Shinano, Y., Viernickel, J.M., Walter, M., Wegscheider, F., Witt, J.T., Witzig, J.: The SCIP Optimization Suite 6.0. Technical report, Optimization Online (2018)
5. Wächter, A., Biegler, L.T.: On the implementation of an interior-point filter line-search algorithm for large-scale nonlinear programming. Math. Program **106**(1), 25–57 (2006)

References

Improving an Industrial Cooling System Using MINLP, Considering Capital and Operating Costs

Marvin M. Meck, Tim M. Müller, Lena C. Altherr, and Peter F. Pelz

Abstract The chemical industry is one of the most important industrial sectors in Germany in terms of manufacturing revenue. While thermodynamic boundary conditions often restrict the scope for reducing the energy consumption of core processes, secondary processes such as cooling offer scope for energy optimisation. In this contribution, we therefore model and optimise an existing cooling system. The technical boundary conditions of the model are provided by the operators, the German chemical company BASF SE. In order to systematically evaluate different degrees of freedom in topology and operation, we formulate and solve a Mixed-Integer Nonlinear Program (MINLP), and compare our optimisation results with the existing system.

Keywords Engineering optimisation · Mixed-integer programming · Industrial optimisation · Cooling system · Process engineering

1 Introduction

In 2017, chemical products accounted for about 10% of the total German manufacturing revenue [1], making the chemical industry one of the most revenue-intense industries. However, with a share of 29%, it is also the most energy-intensive industry, cf. [2]. To ensure a high product-quality while still minimising production costs, proper operation of chemical plants is of crucial importance.

M. M. Meck · T. M. Müller · P. F. Pelz (✉)
Technische Universität Darmstadt, Darmstadt, Germany
e-mail: marvin.meck@fst.tu-darmstadt.de; tim.mueller@fst.tu-darmstadt.de; peter.pelz@fst.tu-darmstadt.de

L. C. Altherr
Faculty of Energy, Building Services and Environmental Engineering, Münster University of Applied Sciences, Münster, Germany
e-mail: lena.altherr@fh-muenster.de

J. S. Neufeld et al. (eds.), *Operations Research Proceedings 2019*, Operations Research Proceedings, https://doi.org/10.1007/978-3-030-48439-2_61

An essential component for effective process control is the cooling circuit of a plant, which is necessary to remove excess heat from processes and subsystems. From an energetic point of view, cooling circuits can be regarded as a hydraulic connection of energy sinks (e.g. valves, pipes and heat exchangers) and energy sources (compressors or pumps in booster stations) which provide the necessary pressure head to maintain the volume flow of coolant through the system.

Recent studies show that the energy demand of pumps is comparable with the electric energy demand for process heat, cf. [3], with pumps having the third largest share of total electric energy consumption in chemical processing. Therefore, their economic selection and operation plays an important role for the profitability of a process and will be investigated in this contribution.

2 Technical Application

A simplified process flow diagram of the cooling system examined in this work is shown in Fig. 1. Information on system topology and boundary conditions was provided by the operators, the German chemical company BASF SE. The system consists of a main cooling loop providing cooling water with a temperature of 28 °C to four subsystems that are hydraulically connected in parallel. Each is coupled by multiple heat exchangers to two kinds of processes; ones which require a constant amount of cooling water independent from the heat load, and others which require a controlled amount of cooling water. Two subsystems (processes P1 and P2) are directly supplied by the main loop. Two additional sub-loops provide cooling water at higher temperatures of 45 °C and 53 °C to other processes. Cooling water from the 28 °C loop is used in a mixing cooler to re-cool the water inside these additional loops back to 45 °C and 53 °C, respectively. The re-cooling of the 28 °C loop itself is achieved using four plate heat exchangers operating with river water. Two of the pumps of the booster stations in the 45 °C and 28 °C loops are equipped with frequency converters (FC), making them speed controllable; the rest operates at a fixed speed.

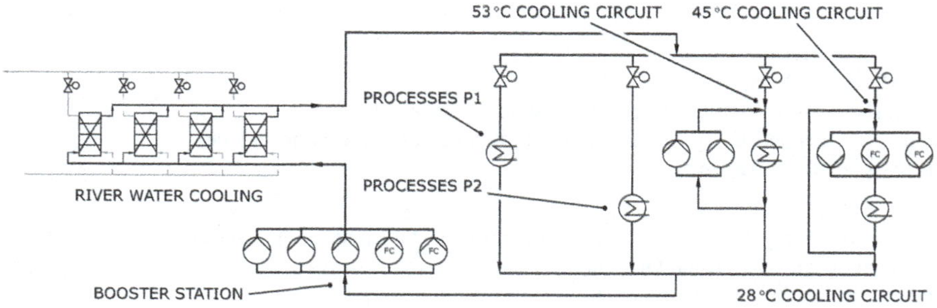

Fig. 1 Simplified overview of the cooling system

The system has to operate under different boundary conditions, depending on the system load and external influences, such as a varying outdoor temperature. Together with the operators, three load cases are formulated. In operation, each booster station has to fulfil an individual load profile: The majority of the time, the cooling circuit performs at *full load*. Apart from that, the system either runs under *50% load* or in *standby*. The volume flow in the individual load cases of each booster station depends on the required heat extraction in the sub and main loops. The required pressure head results from volume flow, temperature and geometry of the heat exchangers and pipes.

3 Model

The optimal selection and operation of pumps can be understood as an investment decision with a planning horizon covering a specified time frame. During the lifecycle of a system, potentially three different types of cost can occur: initial investment cost, instalment payments and operational costs. In those cases where cash flows occur at different times, the well known *net present value* (NPV) method is often used in guiding the decision process [4]. While the system has to fulfil physical boundary conditions, it does not provide any revenue itself. Assuming the total investment I_0 is payed within the first payment period followed by constant yearly energy costs, we can calculate the NPV with continuous compounding: $\text{NPV} = -I_0 + \left(\dot{R}_t - \dot{C}_t\right) \int_0^N e^{-rt} \text{d}t = -I_0 - \text{APF}\,\dot{C}_t$. Here, r is the interest rate, $\dot{C}_t$ and $\dot{R}_t$ are the flow of cost and revenue in period t, and $\text{APF} = e^{rN} - 1 \big/ r\, e^{rN}$ is the annuity present value factor over the planning period N. The objective is to maximise the NPV which is equivalent to minimising the sum of investment costs and discounted operating costs: $\max\,(\text{NPV}) = \min\left(I_0 + \text{APF}\,\dot{C}_t\right)$.

Solutions to the problem have to satisfy physical constraints as well as additional planning requirements. Mathematically, the decision problem can be described as a modified *min-cost-flow-problem* [5]. The problem is modelled as a disconnected directed graph $G = (E, V)$ with the edges E representing components, either pumps or pipes. Each sub-graph of G represents one of the booster stations shown in Fig. 1. The topology of the surrounding system is fixed and given as an input parameter. Thus, the optimisation is limited to the selection and operation of pumps within each booster station (sub-loops and main loop).

Note that all variables have to be positive. Parameters with a subscript m or M represent lower or upper bounds respectively. By introducing a binary *activity variable* $x^{(s)}$ for every load case $s \in \text{Sc}$, components in the graph can be either activated or deactivated. If a component is activated in a load case, the component has to be purchased, indicated by a binary *purchase variable* y. On edges $e \in E_\text{P}$ with $E_\text{P} \subset E$ representing pumps, multiple purchase options $k \in K_e$ are possible, allowing for the selection of one pump per edge from a catalogue K_e, and thus multiple purchase variables y_k are introduced for pump-edges (cf. (1)–(3)). The pump catalogue contains three different pumps of varying sizes.

$$x_e^{(s)} \leq y_e \quad \forall e \in E \setminus E_\mathrm{P},\ s \in \mathrm{Sc} \tag{1}$$

$$x_{e,k}^{(s)} \leq y_{e,k} \quad \forall e \in E_\mathrm{P},\ k \in K_e,\ s \in \mathrm{Sc} \tag{2}$$

$$\sum_{k \in K_e} y_{e,k} \leq 1 \quad \forall e \in E_\mathrm{P} \tag{3}$$

$$x_e^{(s)},\ y_e \in \{0, 1\} \quad \forall e \in E,\ s \in \mathrm{Sc} \tag{4}$$

$$x_{e,k}^{(s)},\ y_{e,k} \in \{0, 1\} \quad \forall e \in E_\mathrm{P},\ s \in \mathrm{Sc},\ k \in K_e \tag{5}$$

Pressure losses within the surrounding system are covered by the given load profile calculated in pre-processing. Load profiles are introduced via source terms indicated with the index demand in constraints where it is necessary. For the graph G, material properties are assumed to be constant in every sub-graph and pressure losses in pipes within the booster station are neglected.

At each node, volume flow conservation applies (6). To ensure the supply matches the demand the source term $Q_{\mathrm{demand},v}^{(s)}$ is introduced. The volume flow $Q_e^{(s)}$ over each edge $e \in E$ and for every load case $s \in \mathrm{Sc}$ has to disappear if the edge is not active (cf. (7)–(8)). The pressure $p_v^{(s)}$ is a state variable and as such has to be defined at every vertex $v \in V$ and for every load case $s \in \mathrm{Sc}$. The difference in pressure between any two vertices has to be zero if the vertices are connected via a pipe (9). For pumps, the difference is equal to the increase in pressure $\Delta p_{e,k}^{(s)}$ provided by the pump of type k selected on edge e (10). In case edges between two vertices are inactive the pressure at those vertices is independent.

To overcome the pressure losses within the system a booster station has to provide a given pressure head. In order to ensure the load demand is fulfilled the pressure at source vertices $v \in V_{\mathrm{src}}$ is set to a constant pressure $p_{\mathrm{demand},v}$ (11); the pressure at target vertices $v \in V_{\mathrm{tgt}}$ is set to be greater or equal $p_{\mathrm{demand},v}$ (12). By doing so the difference between a target and a source in a given sub-graph will be at least as high as the load profile demands it to be.

$$\sum_{(v,j) \in E} Q_{(v,j)}^{(s)} = \sum_{(i,v) \in E} Q_{(i,v)}^{(s)} + Q_{\mathrm{demand},v}^{(s)} \quad \forall v \in V,\ s \in \mathrm{Sc} \tag{6}$$

$$Q_e^{(s)} \leq Q_\mathrm{M}\, x_e^{(s)},\ Q_e^{(s)} \geq Q_\mathrm{m}\, x_e^{(s)} \quad \forall e \in E \setminus E_\mathrm{P},\ s \in \mathrm{Sc} \tag{7}$$

$$Q_e^{(s)} \leq Q_\mathrm{M} \sum_{k \in K_e} x_{e,k}^{(s)} \quad \forall e \in E_\mathrm{P},\ k \in K_e,\ s \in \mathrm{Sc} \tag{8}$$

$$\pm \left(p_i^{(s)} - p_j^{(s)}\right) \leq p_\mathrm{M} \left(1 - x_{(i,j)}^{(s)}\right) \quad \forall (i,j) \in E \setminus E_\mathrm{P},\ s \in \mathrm{Sc} \tag{9}$$

$$\pm\left(p_i^{(s)} + \Delta p_{(i,j)}^{(s)} - p_j^{(s)}\right) \leq p_{\mathrm{M}}\left(1 - \sum_{k\in K_e} x_{(i,j),k}^{(s)}\right) \quad \forall (i,j) \in E_{\mathrm{P}},\ s \in \mathrm{Sc} \tag{10}$$

$$p_v^{(s)} = p_{\mathrm{demand},v}^{(s)} \quad \forall v \in V_{\mathrm{src}},\ s \in \mathrm{Sc} \tag{11}$$

$$p_v^{(s)} \geq p_{\mathrm{demand},v}^{(s)} \quad \forall v \in V_{\mathrm{tgt}},\ s \in \mathrm{Sc} \tag{12}$$

The physical description of pumps is implemented similarly to the model described in [6]. We approximate the relationship between pressure head H, volume flow Q, rotating speed n and power consumption P using quadratic and cubic approximations. Upper and lower bounds for the volume flow, which increase with rotating speed, are approximated with linear constraints. Together with upper and lower bounds for the normalised rotating speed, we yield a system of linear and non-linear inequalities describing the feasible set of values given by the characteristics of a pump. Different pump characteristics are modelled as sets of alternative constraints, making sure that exactly the constraint set for the chosen pump is activated. The total feasible set $\Lambda_{e,k}$ for a pump of type $k \in K_e$ and on edge $e \in E_{\mathrm{P}}$ can be denoted as

$$\begin{aligned}\Lambda_{e,k} = \{ \left(Q_e^{(s)}, \tilde{n}_e, \Delta p_e^{(s)}, P_e^{(s)}\right) \in \mathbb{R}^4 : \quad & \tilde{n}_{e,\min} \leq \tilde{n}_e \leq 1,\ H \leq \beta_{k,1}^{Q_{\min}} + \beta_{k,2}^{Q_{\min}} Q_e^{(s)} \\ H \geq \beta_{k,1}^{Q_{\max}} + \beta_{k,2}^{Q_{\max}} Q_e^{(s)},\ \Delta p_e^{(s)} = \varrho_e g\, H\left(Q_e^{(s)}, \tilde{n}_e\right),\ & P_e^{(s)} = P\left(Q_e^{(s)}, \tilde{n}_e\right)\},\end{aligned}$$

where $\tilde{n} := n/n_{\max}$ is the normalised rotating speed, $H(Q, \tilde{n})$ and $P(Q, \tilde{n})$ are polynomial approximations of second and third order respectively and $\beta_{k,i}^{Q_{\min}}$ and $\beta_{k,i}^{Q_{\max}}$ are the regression coefficients to model bounds for the volume flow. Using the example of a characteristic power curve (constraints for characteristic head curves are formulated analogously), the resulting big-M formulation reads:

$$\pm\left(P_e^{(s)} - \left[\sum_{j=0}^{3} \beta_{k,j}^{P} \left(\tilde{n}_e^{(s)}\right)^{3-j} \left(Q_e^{(s)}\right)^{j}\right]\right) \leq P_{\mathrm{M}}\left(1 - x_{e,k}^{(s)}\right) \tag{13}$$

$$\forall e \in E_{\mathrm{P}},\ k \in K_e,\ s \in \mathrm{Sc}$$

$$P_e^{(s)} \leq P_{\mathrm{M}} \sum_{k\in K_e} x_{e,k}^{(s)} \quad \forall e \in E_{\mathrm{P}},\ k \in K_e,\ s \in \mathrm{Sc} \tag{14}$$

Variable speed pumps have an advantage over fixed speed pumps since they can adapt their rotating speeds to partial loads during operation, resulting in a reduced energy demand. In order to make a pump speed controllable it has to be equipped with a frequency converter, increasing complexity and investment costs.

By introducing a purchase variable $\omega_{k,e}$, which indicates the selection of a converter for a pump of type k on edge e, we are able to consider both fixed and variable speed pumps in the decision problem. In case a converter for a given pump is chosen, the rotating speed can be chosen freely in each load scenario. Else, the rotating speeds have to be equal to the nominal rotating speed $\tilde{n}_{\text{nominal}}$ if the pump is active or zero if inactive (15), (16). A converter can only be purchased for a given pump if the pump is purchased as well (17).

$$\pm\left(\tilde{n}_{\text{nominal},k} - \tilde{n}_e^{(s)}\right) \leq 1 + \omega_{e,k} - x_{e,k}^{(s)} \quad \forall e \in E_{\text{P}},\ s \in \text{Sc} \tag{15}$$

$$\tilde{n}_e^{(s)} \leq \sum_{k \in K_e} x_{e,k}^{(s)} \quad \forall e \in E_{\text{P}},\ s \in \text{Sc} \tag{16}$$

$$\omega_{e,k} \leq y_{e,k} \quad \forall e \in E_{\text{P}},\ k \in K_e \tag{17}$$

$$\omega_{e,k} \in \{0, 1\} \quad \forall e \in E_{\text{P}},\ k \in K_e \tag{18}$$

Finally, the objective, max (NPV) $= \min\left(I_0 + \text{APF}\,\dot{C}_t\right)$, can be written as:

$$\min\left(\sum_{e \in E_{\text{P}}} \sum_{k \in K_e} \left(y_{e,k}\, I_{\text{P},k} + \omega_{e,k}\, I_{\text{FC},k}\right) + \text{APF}\, C_{\text{E}}\, T \sum_{s \in \text{S}} \tilde{t}^{(s)} \sum_{e \in E_{\text{P}}} P_e^{(s)}\right). \tag{19}$$

Investments costs $I_{\text{P},k}$ and $I_{\text{FC},k}$ for pumps and converters in this model are based on manufacturer enquiries. The sum of the products of time portion $\tilde{t}^{(s)}$ and power consumed by all pumps in load case s, $\sum_{e \in E_{\text{P}}} P_e^{(s)}$, over all load cases Sc gives the total average power consumption. The energy costs C_{E} are estimated to be 7.5 ct kW^{-1} h^{-1} [7]. The yearly operating time T is assumed to be 300 days. We use SCIP 5.0.1 [8] to solve the MINLP on a MS Windows 10 based machine with an INTEL Core i7-7700T 2.9 GHz processor and 16 GB DDR4-RAM.

4 Results and Conclusion

Besides costs, engineers also have to consider other determining factors in their design decision. We illustrate this by deriving four different model instances. In instance (i), we examine the current design of the system and thus purchase decisions and connection between pumps are fixed. The optimisation is reduced to operation of the system. In instance (ii) we extend the problem to also cover rearrangement of the pumps. Instance (iii) covers purchase, arrangement and operation of pumps, but requires to only use one type of pump, to achieve reduced

Table 1 Results of the optimisation problem for the different instances

	Instance (i)	Instance (ii)	Instance (iii)	Instance (iv)
|NPV| ($r = 5\%,\ N = 5\,\text{a}$)	3,649,927 € (+15.79%)	3,632,883 € (+15.43%)	3,179,406 € (+1.02%)	3,147,377€ (100%)
Investment	848,240 €	831,196 €	751,776 €	717,864 €
Avg. power consumption	1172.8 kW	1172.8 kW	1016.2 kW	1017.0 kW

spare-parts inventory and maintenance costs. Finally instance (iv) does not include any planning requirements and can be used to determine the lower bound solely within physical constraints. Connecting pumps in parallel is considered state of the art procedure in the industry today since it reduces overall complexity in almost every aspect greatly. In all instances we therefore limit the solution space to pumps connected in parallel to each other.

Results As shown in Table 1 the current design of the system offers great optimisation potential, as it was already assumed in the beginning of this examination. By only considering physical constraints (iv) the |NPV| can be decreased by roughly 500,000 € over a course of five years. This is mainly due to an increase in energy efficiency achieved through better system design. In this optimal solution more frequency converters are used and slightly better sized pumps are selected. Although the power draw is only reduced by ≈100 kW, which is equivalent to a single pump saved, the long term effect on costs is significant. Furthermore, investment costs can also be reduced. Surprisingly, restricting the selection to a single pump type (iii) does not worsen the outcome considerably, while rearranging the current setup (ii) does not provide considerable benefits either. The benefits of reducing the spare-parts inventory and maintenance effort by using only one pump type (iii) most likely outweigh the additional investment costs of approximately 40,000 € and should be favoured to the cost optimal solution (iv).

Conclusion In this contribution, we presented a model to support engineers in optimising booster stations within cooling circuits. The model provides an extension to previously presented variants, making it possible to consider both fixed and variable speed pumps to weigh up the advantages and disadvantages of using frequency converters regarding costs. The presented model also offers the possibility to investigate a combination of serial and parallel connections of pumps, which was not shown yet. We plan on examining the further energy savings potential of this approach in the future.

Acknowledgments Results were obtained in project No. 17482 N/1, funded by the German Federal Ministry of Economic Affairs and Energy (BMWi) approved by the Arbeitsgemeinschaft industrieller Forschungsvereinigungen "Otto von Guericke" e.V. (AiF). Moreover, this research was partially funded by Deutsche Forschungsgemeinschaft (DFG) under project No. 57157498.

References

1. Statista: https://de.statista.com/statistik/daten/studie/241480/umfrage/umsaetze-der-wichtigsten-industriebranchen-in-deutschland/
2. Federal Statistical Office of Germany: https://www.destatis.de/DE/Presse/Pressemitteilungen/2018/11/PD18_426_435.html
3. Rohde, C.: Erstellung von Anwendungsbilanzen für die Jahre 2013 bis 2017 (2018)
4. Berk, J., DeMarzo, P.: Corporate Finance, Global Edition. Pearson, London (2016)
5. Cook, W.J., Cunningham, W.H., Pulleyblank, W.R., Schrijver, A.: Combinatorial Optimization. Wiley, Chichester (1997). https://doi.org/10.1002/9781118033142
6. Altherr, L., Leise, P., Pfetsch, M.E., Schmitt, A.: Optim. Eng. **20**(2), 605–645 (2019). https://doi.org/10.1007/s11081-019-09423-8
7. Statista: https://de.statista.com/statistik/daten/studie/155964/umfrage/entwicklung-der-industriestrompreise-in-deutschland-seit-1995/
8. Gleixner, A., et al.: The SCIP Optimization Suite 5.0, Berlin (2017)

A Two-Phase Approach for Model-Based Design of Experiments Applied in Chemical Engineering

Jan Schwientek, Charlie Vanaret, Johannes Höller, Patrick Schwartz, Philipp Seufert, Norbert Asprion, Roger Böttcher, and Michael Bortz

Abstract Optimal (model-based) experimental design (OED) aims to determine the interactions between input and output quantities connected by an, often complicated, mathematical model as precisely as possible from a minimum number of experiments. While statistical design techniques can often be proven to be optimal for linear models, this is no longer the case for nonlinear models. In process engineering applications, where the models are characterized by physico-chemical laws, nonlinear models often lead to nonconvex experimental design problems, thus making the computation of optimal experimental designs arduous. On the other hand, the optimal selection of experiments from a finite set of experiments can be formulated as a convex optimization problem for the most important design criteria and, thus, solved to global optimality. Since the latter represents an approximation of common experimental design problems, we propose a two-phase strategy that first solves the convex selection problem, and then uses this optimal selection to initialize the original problem. Finally, we illustrate and evaluate this generic approach and compare it with two statistical approaches on an OED problem from chemical process engineering.

Keywords Optimal design of experiments · Experiments selection · Nonlinear optimization

1 Introduction

Optimal (model-based) experimental design (OED) subsumes all methodologies for the systematic planning of experiments. Its aim is to define experimental conditions

J. Schwientek (✉) · C. Vanaret · J. Höller · P. Schwartz · P. Seufert · M. Bortz
Fraunhofer ITWM, Kaiserslautern, Germany
e-mail: Jan.Schwientek@itwm.fraunhofer.de

N. Asprion · R. Böttcher
BASF SE, Ludwigshafen am Rhein, Germany

J. S. Neufeld et al. (eds.), *Operations Research Proceedings 2019*, Operations Research Proceedings, https://doi.org/10.1007/978-3-030-48439-2_62

such that the experimental outcomes will have maximal information content to determine the model parameters as accurately as possible using a minimum number of experiments.

Consider a mathematical model that relates the inputs $\boldsymbol{x}$ to the outputs $\boldsymbol{y}$

$$\boldsymbol{y} = \boldsymbol{f}(\boldsymbol{x}, \boldsymbol{p}), \tag{1}$$

with model parameters $\boldsymbol{p}$ and model functions $\boldsymbol{f}$, which can all be vector-valued indicated by bold letters. Of course, the model can also be given implicitly, i.e. in the form $\boldsymbol{0} = \boldsymbol{f}(\boldsymbol{x},\boldsymbol{p},\boldsymbol{y})$. Such an (explicit or implicit) model is usually (pre-) fitted to measured data. The model parameters, which depend on the data and the associated measurement inaccuracies, can then be adjusted more precisely by means of optimal experimental design. For this purpose, the Jacobians of the residuals of the associated (unconstrained) *parameter estimation problem*

$$\text{PEP}: \quad \min_{\boldsymbol{p}} \sum_{i=1}^{N_{exp}} \sum_{j=1}^{N_{meas}} w_{i,j} \left(\frac{\tilde{y}_{i,j} - f_j(\boldsymbol{x}_i, \boldsymbol{p})}{\sigma_{i,j}} \right)^2 \tag{2}$$

with respect to the model parameters $\boldsymbol{p}$ are considered

$$J(\boldsymbol{x}_i, \boldsymbol{p}) = \mathrm{D}_2\left(r_1(\boldsymbol{x}_i, \boldsymbol{p}), \ldots, r_{N_{meas}}(\boldsymbol{x}_i, \boldsymbol{p})\right)^T, \quad r_j(\boldsymbol{x}_i, \boldsymbol{p}) = \sqrt{w_{i,j}} \left(\frac{\tilde{y}_{i,j} - f_j(\boldsymbol{x}_i, \boldsymbol{p})}{\sigma_{i,j}} \right). \tag{3}$$

Here, $\tilde{y}_{i,j}$ is the j-th measured property of the i-th experiment, $w_{i,j}$ denotes a weighting factor and $\sigma_{i,j}$ is the standard deviation of the measurement $\tilde{y}_{i,j}$. N_{exp} and N_{meas} are the number of experiments and the number of measured properties.

The *Fisher information matrix (FIM)* is defined as

$$FIM(\boldsymbol{\xi}, \boldsymbol{p}) = \sum_{i=1}^{N_{exp}} J(\boldsymbol{x}_i, \boldsymbol{p})^T J(\boldsymbol{x}_i, \boldsymbol{p}) \quad \text{with} \quad \boldsymbol{\xi} = \left(\boldsymbol{x}_1, \ldots, \boldsymbol{x}_{N_{exp}}\right), \tag{4}$$

where $\boldsymbol{\xi}$ is called design, and is related to the covariance matrix C of the parameter estimates from PEP by $FIM(\boldsymbol{\xi},\boldsymbol{p}) \sim C(\boldsymbol{\xi},\boldsymbol{p})^{-1}$ (see, e.g., [1] for details).

Concerning parameter precision, the experiments should be selected in such a way that this results in a reduction of the parameter estimates variance. This can be done in different ways, why several OED criteria emerged. The best known and most frequently used are the A, D, and E criteria:

- For the *A(verage) criterion* the trace of the covariance matrix is minimized, which corresponds to the minimization of the average variance of the estimated model parameters.

- In case of the *D(eterminant) criterion* the determinant of the covariance matrix is minimized, which leads to the minimization of the volume of the confidence ellipsoid for the unknown model parameters.
- For the *E(igenvalue) criterion* the greatest eigenvalue of the covariance matrix is minimized, which leads to the minimization of the maximal possible variance of the components of the estimated model parameters.

Finally, we end up with the following *optimal experimental design problem*

$$\text{OEDP}: \quad \min_{\boldsymbol{\xi} \in \Xi} \ \varphi \left[C\left(\boldsymbol{\xi}, \boldsymbol{p}\right)\right], \tag{5}$$

where Ξ is the set of feasible designs $\boldsymbol{\xi}$ and φ is some design criterion functional from above or another one. For a detailed derivation of OEDP and further issues we refer to [1].

Since in practice it is often not possible to realize the calculated (optimal) experimental design exactly, having so-called implementation uncertainties, it makes sense to hedge the OEDP solution against deviations in the inputs. However, this is not subject of this contribution.

The paper is structured as follows: In the next section, the treatment of the nonlinear, nonconvex OEDP via solution approaches from statistical and linear *design of experiments (DoE)* theory is addressed. Section 3 sketches an alternative two-phase approach, relying on a discretization scheme, leading to a convex optimization problem in phase 1. In Sect. 4, the introduced approaches are compared on a real-world example stemming from process engineering. The paper ends with concluding remarks and future directions of research.

2 Solution Approaches from Statistical and Linear DoE

For certain design criteria and model classes, e.g. the D criterion and functions of the form $f(\boldsymbol{x},\boldsymbol{p}) = p_0 + \sum_i p_i x_i + \sum_{i \neq j} p_{i,j} x_i x_j + \ldots$, specific designs, in this case factorial designs (see below), are proven to be optimal (see, e.g., [2]). For the general case, however, OEDP must be solved numerically. In process engineering applications, where the models are characterized by physico-chemical laws, the models are almost always nonlinear and often lead to nonconvex experimental design problems, thus making the computation of globally optimal experimental designs arduous. Fortunately, linear and statistical experimental design approaches can still be exploited as initialization and globalization techniques, albeit with no guarantee of global optimality.

Full factorial designs, well known from linear design of experiments, consist of the vertices of a full dimensional hyper box in the input space, while *partial* or *reduced factorial designs* refer to a subset of these vertices that are selected according to specific requirements (see [2] for an exhaustive overview). For selecting a given number of experiments and creating different instances in a multi-start approach for globalization we refer in case of the factorial design approach to [3].

In addition, there are screening techniques from statistical design of experiments, which are applied in particular if no model is known or can be reasonably assumed. *Sobol-* and *Latin-Hypercube-Sampling* are to be mentioned here, which try to scan the input space with as few points as possible (see [2, 4] for details). Different Sobol designs for a multi-start can be obtained by selecting different (finite) subsequences of a Sobol sequence.

3 A Two-Phase Approach

The optimal selection of N experiments from the infinite set Ξ can be approximated using a discrete set of candidate experiments. The corresponding *experiment selection problem* is

$$\text{ESP}: \quad \min_{\boldsymbol{m}\in \boldsymbol{Z}^M} \varphi\left[\sum_{i=1}^{M} m_i \; J(\boldsymbol{x}_i, \boldsymbol{p})^T J(\boldsymbol{x}_i, \boldsymbol{p})\right]^{-1} \quad \text{s.t.} \quad \boldsymbol{m} \geq \boldsymbol{0}, \sum_{i=1}^{M} m_i = N, \tag{6}$$

where M is the number of candidate experiments, $\boldsymbol{x}_i$, $i = 1, \ldots, M$, are the fixed, but feasible candidate experiments and the other quantities are as before. This can be a hard combinatorial problem, especially when M is comparable to N.

If M is large compared to N, a good approximate solution can be found by dividing the integers m_i by N, relaxing them to $\lambda_i \in [0, 1]$ and solving the following *relaxed experiment selection problem*

$$\text{rESP}: \quad \min_{\boldsymbol{\lambda}\in \boldsymbol{R}^M} \varphi\left[\sum_{i=1}^{M} \lambda_i \; J(\boldsymbol{x}_i, \boldsymbol{p})^T J(\boldsymbol{x}_i, \boldsymbol{p})\right]^{-1} \quad \text{s.t.} \quad \boldsymbol{\lambda} \geq \boldsymbol{0}, \sum_{i=1}^{M} \lambda_i = 1. \tag{7}$$

For the logarithmized D criterion rESP is a convex continuous optimization problem and for the A and E criterion rESP can even be reformulated into a semi-definite optimization problem (see [5] for details). Thus, in these cases, rESP can be solved to global optimality. Besides, it provides the optimal experiments among the set of candidate experiments, as well as their multiplicities (the number of times each experiment should be performed).

These facts motivate a two-phase strategy that first solves the convex relaxed selection problem, and then uses this optimal selection to initialize the original problem (see Fig. 1 for a visualization):

Of course, ESP can be solved in phase I instead of rESP. A solution strategy as well as the effects of the two formulations on the two-phase approach and a certificate for global optimality are presented in the forthcoming paper [6].

Algorithm 1: Two-phase approach

Input: Equidistant grid or user-defined set of points (candidate experiments).
0. **Initialization**: Evaluate Jacobians at candidate experiments.
1. **Phase I**: Solve rESP with arbitrary initialization, e.g. $\lambda_i = {}^1/_M$.
2. **Phase II**: Solve OEDP by using the optimal solution of phase I as starting point.

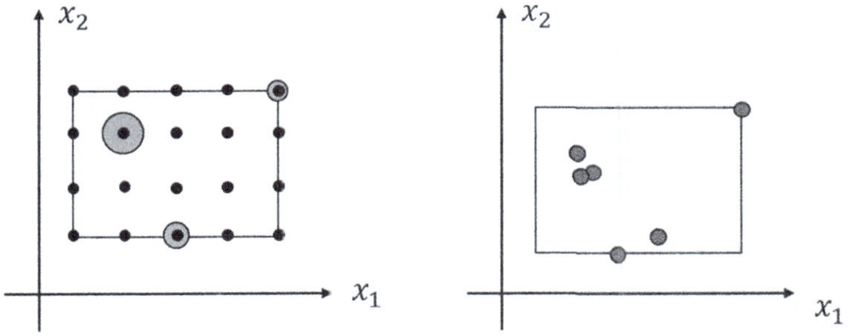

Fig. 1 Schematic representation of the two-phase approach, *left*: candidate experiments (black circles) and solution of rESP (light grey circles) with multiplicity (thickness of light grey circles) of computed experiments, *right*: solution of OEDP (dark grey circles) initialized by solution of rESP

4 Numerical Example: Flash Distillation

We now apply the two-phase approach to an important chemical process engineering task and compare it with the other two initialization and globalization strategies mentioned above, factorial and Sobol designs. A further numerical example is given in [6].

We consider the separation of a binary mixture of substances with the aid of a (single-stage) evaporator, also called *flash*. The liquid mixture, called *feed*, composed of substances 1 and 2 enters the flash with flow rate F and compositions z_1 and z_2. There it is heated by an external heat source with power $\dot{Q}$ and partially vaporized. The produced amount of liquid L and vapor V are in equilibrium at pressure P and temperature T. Their respective compositions x_1 and x_2 (liquid) as well as y_1 and y_2 (vapor) depend on the degree of vaporization. For the flash unit (modelled by an equilibrium stage) the so-called MESH equations hold:

- Mass balances: $Fz_i = Lx_i + Vy_i,\ i = 1, 2$
- Equilibrium conditions: $Py_i = P_i^0(T)x_i\gamma_i(x, T),\ \ i = 1, 2$
- Summation conditions: $x_1 + x_2 = y_1 + y_2 = 1$
- Heat conditions: $F\sum_{i=1}^2 z_i h_i^L(T) + \dot{Q} = L\sum_{i=1}^2 x_i h_i^L(T) + V\sum_{i=1}^2 y_i h_i^V(T)$

where P_i^0, γ_i, h_i^L, h_i^V, $i = 1,2$, are given thermodynamic models with fitted parameters.

In the specific case we consider a water-methanol mixture. We select $x_{MeOH} \in [0, 1]$ und $P \in [0.5, 5]$ as model inputs and y_{MeOH} and T as model

Table 1 Objective values (log-D criterion) for different initialization approaches

Exp. (given resp. resulting)	Sobol design initia-lization	Multi-start with 10 Sobol designs	Factorial design initia-lization	Multi-start with 10 factorial designs	Two-phase approach Phase 1	Phase 2
5	10,67652	**10,67652**	-	8,534817	10,371256 (90 cand. exp.)	**10,67652**
5[a]	10,50106	**10,67131**	10,39557	**10,67131**	9,7311864 (9 cand. exp.)	**10,67131**
6	10,70903	10,72923	10,02034	10,70911	10,034221 (25 cand exp.)	**10,93878**

[a]Only one output, y_{MeOH}, is considered

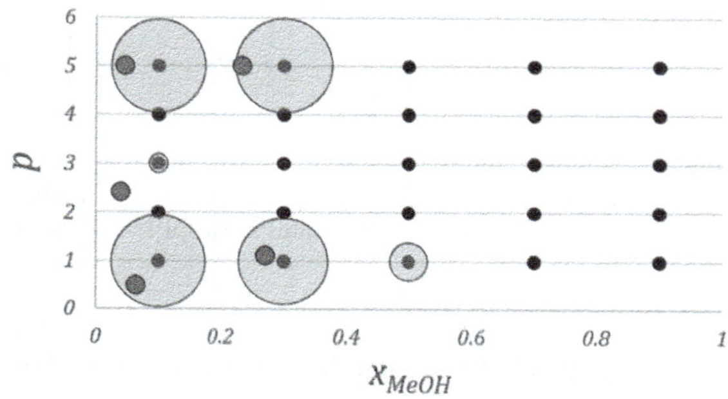

Fig. 2 Results for six experiments (last, grey-shaded line in Table 1): two-phase approach with 25 candidate experiments (black points), 6 resulting experiments in phase I (light grey points) and 6 optimal experiments in phase II (dark grey points; lower right experiment is doubled)

outputs. As model parameters the four coefficients A_{ij} and B_{ij}, $i, j = 1, 2, i \neq j$, in the NRTL model for the activity coefficients γ_i, $i = 1, 2$, are chosen (see [3] for details). We use the logarithmized D criterion in rESP and OEDP and maximize $\log[\det(FIM(\boldsymbol{\xi},\boldsymbol{p}))]$ instead of minimizing $\log[\det(C(\boldsymbol{\xi},\boldsymbol{p}))]$, which is equivalent. For the evaluation of the Jacobians and the solution of the second phase resp. ODEP standalone, we have implemented the model in BASF's inhouse flowsheet simulator `CHEMASIM` [3]. The first phase is implemented in `Python` and solved using `CVXOPT` [7]. In phase I, we apply different discretizations for the experiment selection problem rESP. The results are shown in Table 1 and Fig. 2.

In that example the two-phase approach either yields a better solution (last row of Table 1) or the same one, but in 2 instead of 10 runs, thus faster (row 1 & 2 of Table 1).

5 Conclusions and Outlook

In this paper, we propose the usage of a two-phase approach for optimal experimental design and demonstrate its benefits on an application from chemical process engineering. For models with low-dimensional input, when fine discretization can be used, the two-phase approach seems to result either in finding a better solution of the underlying design of experiments problem or faster than, e.g., in a multi-start approach. At the same time, it answers the two design of experiments key questions: How many experiments should be performed and which ones?

For models with a large number of inputs it is no longer possible to process a fine discretization in reasonable time. In this respect, a direction for future research is to start from a rough discretization and refine it successively. Finally, we will investigate under which conditions the global solutions of the discretized problems converge to a global solution of the initial problem. After all, this would make the second phase superfluous.

References

1. Fedorov, V.V., Leonov, S.L.: Optimal Design for Nonlinear Response Models. CRC Press, Boca Raton (2014)
2. Montgomery, D.C.: Design and Analysis of Experiments, 8th edn. Wiley, Hoboken, NJ (2013)
3. Asprion, N., Boettcher, R., Mairhofer, J., Yliruka, M., Hoeller, J., Schwientek, J., Vanaret, Ch., Bortz, M.: Implementation and application of model-based design of experiments in a flowsheet simulator. J. Chem. Eng. Data **65**, 1135–1145 (2020). DOI:10.1021/acs.jced.9b00494
4. Sobol, I.M.: On the distribution of points in a cube and the approximate evaluation of integrals. Zh. Vych. Mat. Mat. Fiz. **7**(4), 784–802 (1967)
5. Boyd, S., Vandenberghe, L.: Convex Optimization, 7th edn. Cambridge University Press, Cambridge (2009)
6. Vanaret, Ch., Seufert, Ph., Schwientek, J., Karpov, G., Ryzhakov, G., Oseledets, I., Asprion, N., Bortz, M.: Two-phase approaches to optimal model-based design of experiments: how many experiments and which ones? Chem. Eng. Sci. (2019) (Submitted)
7. Andersen, M.S., Dahl, J., Vandenberghe, L.: CVXOPT – Python Software for Convex Optimization. http://cvxopt.org (2019)

Assessing and Optimizing the Resilience of Water Distribution Systems Using Graph-Theoretical Metrics

Imke-Sophie Lorenz, Lena C. Altherr, and Peter F. Pelz

Abstract Water distribution systems are an essential supply infrastructure for cities. Given that climatic and demographic influences will pose further challenges for these infrastructures in the future, the resilience of water supply systems, i.e. their ability to withstand and recover from disruptions, has recently become a subject of research. To assess the resilience of a WDS, different graph-theoretical approaches exist. Next to general metrics characterizing the network topology, also hydraulic and technical restrictions have to be taken into account. In this work, the resilience of an exemplary water distribution network of a major German city is assessed, and a Mixed-Integer Program is presented which allows to assess the impact of capacity adaptations on its resilience.

Keywords Resilience · Graph theory · Water distribution system · Topology · Engineering

1 Introduction

Water distribution systems (WDS) are an essential supply infrastructure for cities. With regard to a resilient and at the same time cost-effective water supply, the question arises how to find the most advantageous maintenance measures and/or capacity adjustments. The resilience assessment of WDS is subject of many studies presented in literature [10]. Resilience of technical systems can herein be defined as the remaining minimum functionality in the case of a disruption or failure

I.-S. Lorenz · P. F. Pelz (✉)
Technical University of Darmstadt, Darmstadt, Germany
e-mail: imke.lorenz@fst.tu-darmstadt.de; peter.pelz@fst.tu-darmstadt.de

L. C. Altherr
Faculty of Energy, Buildings and Environment, Münster University of Applied Science, Münster, Germany
e-mail: lena.altherr@fh-muenster.de

J. S. Neufeld et al. (eds.), *Operations Research Proceedings 2019*, Operations Research Proceedings, https://doi.org/10.1007/978-3-030-48439-2_63

of system components, and even more a possible subsequent recovery to attain setpoint functionality, as proposed in [1]. For resilience assessment, different graph-theoretical metrics established in network theory have been applied, and a clear correlation has been shown [9].

In this work, the graph-theoretical resilience index introduced by Herrera [4] is used to evaluate the resilience of an exemplary water distribution system in the German city Darmstadt. We present a mathematical optimization program that enables a cost-benefit analysis in terms of resilience when increasing the system's capacity by adding pipes.

2 Case Study

This case study treats the water supply of an exemplary district in the city Darmstadt, Germany. Based on OpenStreetMap data and the elevation profile [5] of the region, a virtual WDS is obtained using the DynaVIBe tool [8]. The underlying approach is based on the spatial correlation between urban infrastructures, in this case the urban transportation system and the water supply system. The generated network consists of 729 consumer nodes, which are linked by 763 edges, representing pipes of diameters in the range of 50 mm to 250 mm. In Darmstadt, there are two water reservoirs located outside the city, modeled as 2 source nodes. In a first step, to reduce complexity, neighbouring consumer nodes are combined. This results in a simplified WDS with 124 consumer nodes linked by 151 edges, shown in Fig. 1. In order to increase the mean resilience of the WDS, the following capacity adaptation is investigated: pipes can be added to connect nodes not already linked.

Fig. 1 Simplified WDS of the district in the German city Darmstadt

3 Graph-Theoretical Resilience Index

The resilience of the WDS is assessed based on the graph-theoretical resilience index I_{GT} of each node of the network proposed in [4]. For the assessment of the overall network, a mean resilience index is computed by averaging over the number of consumer nodes. Fundamental for this approach is to model the WDS as a planar undirected graph $\mathcal{G} = (\mathcal{V}, \mathcal{E})$ with node set $\mathcal{V}$, consisting of a set of consumer nodes $\mathcal{C}$ and source nodes $\mathcal{S}$, with $\mathcal{V} = \mathcal{C} \cup \mathcal{S}$, and an edge set $\mathcal{E}$.

The applied resilience index I_{GT} considers two factors accounted for the resilience of a WDS: (1) the hydraulic resistance of the path feeding a consumer node has to be low for a high resilience; (2) the existance of alternative paths feeding the same consumer node and even more paths from alternative sources increases the water availability at the consumer nodes in case of pipe failures.

The hydraulic resistance w of a path made up of M pipes feeding a consumer node is given by

$$w = \sum_{m=1}^{M} \left(\frac{u_m}{u_0}\right)^2 f_m \frac{L_m}{D_m} + C_{\mathrm{d},m} \approx \sum_{m=1}^{M} f_m \frac{L_m}{D_m}. \tag{1}$$

Here, u_m is the flow velocity in pipe m, u_0 is the outlet flow velocity at the consumer, f_m is the pipes friction factor, L_m and D_m are length and diameter of the m-th pipe, respectively, and $C_{\mathrm{d},m}$ are diffuser losses associated with transitions to wider pipes. Please refer to [7] for a more detailed derivation of this expression. Given the following technical assumptions, this expression can be simplified. Firstly, the range of flow velocities in the system is very narrow. A lower bound is given due to the risk of biological build-up for stagnant water, and an upper bound close to this lower bound is given to limit pressure drop along the pipes and therefore to operate efficiently. This leads to $u_m/u_0 \approx 1$. Secondly, it is assumed that pipe diameters decrease in flow direction since the required water volume decreases from main paths starting at the source nodes to side paths feeding the different consumer nodes. Therefore, the diffuser losses $C_{\mathrm{d},m}$ may be neglected. The friction factor f_m can be determined assuming turbulent flow in hydraulic rough pipes, cf. [7]. Given these assumptions, a mean resilience of the WDS based on Herrera's graph-theoretical resilience index, cf. [4], is given by

$$\overline{I}_{\mathrm{GT}} = \frac{1}{|\mathcal{C}|} \sum_{c=1}^{\mathcal{C}|} \sum_{s=1}^{|\mathcal{S}|} \left(\frac{1}{K} \sum_{k=1}^{K} \frac{1}{w_{k,s,c}} \right), \tag{2}$$

where $w_{k,s,c}$ is the resistance of the k-th feeding path from source node s to consumer node c which is computed according to Eq. (1). The resistance terms are summed up for a predetermined number of K shortest paths, the total number of sources $|\mathcal{S}|$ and the total number of consumer nodes $|\mathcal{C}|$ of the network. A low resistance of the feeding path as well as of the $K - 1$ best alternative paths lead to a high resilience index, as already introduced.

4 Formulation of the Optimization Problem

Our objective is to maximize the mean resilience index $\overline{I}_{\mathrm{GT}}$ by adding new pipes connecting the existing nodes. When considering exclusively the minimum resistance path of each source to a consumer node, i.e. $K = 1$, the resilience index is linear with respect to the weight $w_{k=1,s,c}$ of a feeding path. Therefore, for a constant number of consumer nodes $|\mathcal{C}|$ and source nodes $|\mathcal{S}|$ the objective is to maximize the reciprocal of the feeding path's resistance to achieve maximum resilience.

In turn, since the hydraulic resistance is always positive, a minimization problem can be formulated: $\min \sum_{c\in\mathcal{C}} \sum_{s\in\mathcal{S}} w_{k=1,s,c}$. Herein, $w_{k=1,s,c}$ is the resistance of the shortest path.

Finding this minimum resistance path can again be formulated as a minimization problem. To state this subproblem, the complete graph $\tilde{\mathcal{G}} = (\mathcal{V}, \tilde{\mathcal{E}})$ is considered. The resistance $w_{i,j}$ of all existing and possible pipes between nodes $i, j \in \mathcal{V}$ is computed in pre-processing and saved in the symmetrical matrix $W = (w_{i,j})_{i=1,\dots,124\ j=1,\dots,124}$. In the resistance computing, the pipe diameter of a non-existing pipe is set to the maximum diameter of all existing pipes that are connected to the vertices linked by this new pipe to yield a high resilience increase. Additionally, the introduced binary variables $t_{i,j,s,c} \in \{0, 1\}$ indicate if a pipe $(i, j) \in \tilde{\mathcal{E}}$ is part of the minimum resistance path from $s \in \mathcal{S}$ to $c \in \mathcal{C}$. The linear subproblem to find the minimal resistance path for every consumer-source combination is then given by:

$$w_{k=1,s,c} = \min \sum_{i\in\mathcal{C}} \sum_{j\in\mathcal{C}} w_{i,j}\, t_{i,j,c,s} \quad \forall s \in \mathcal{S}, c \in \mathcal{C}. \tag{3}$$

$$\sum_{j:(s,j)\in\tilde{\mathcal{E}}} t_{i=s,j,c,s} = 1 \quad \forall c \in \mathcal{C}, \forall s \in \mathcal{S} \tag{4}$$

$$\sum_{i:(i,c)\in\tilde{\mathcal{E}}} t_{i,j=c,c,s} = 1 \quad \forall c \in \mathcal{C}, \forall s \in \mathcal{S} \tag{5}$$

$$\sum_{(i,v)\in\tilde{\mathcal{E}}} t_{i,v,c,s} = \sum_{(v,j)\in\tilde{\mathcal{E}}} t_{v,j,c,s} \quad \forall v \in \mathcal{V}, \forall c \in \mathcal{C}, \forall s \in \mathcal{S} \tag{6}$$

Eqs. (4) and (5) ensure that on each minimal resistance path from $s \in \mathcal{S}$ to $c \in \mathcal{C}$, exactly one pipe leaves s and exactly one pipe enters c. Equation (6) ensures the continuity of the path. In a next step, the subproblem given by Eqs. (3)–(6) is integrated into the overall optimization problem of finding the best pipe additions in terms of resilience enhancement. The combined objective function reads:

$$\min \sum_{c\in\mathcal{C}} \sum_{s\in\mathcal{S}} \sum_{i\in\mathcal{C}} \sum_{j\in\mathcal{C}} w_{i,j}\, t_{i,j,c,s}\,. \tag{7}$$

To indicate whether a pipe between nodes $i, j \in \mathcal{V}$ is added, the binary variables $b_{i,j} \in \{0, 1\}$ are introduced. Given the adjacency matrix $(e_{i,j})_{i,j \in \mathcal{V}}$ of the original graph $\mathcal{G} = (\mathcal{V}, \mathcal{E})$, the following constraint must hold:

$$t_{i,j,c,s} \leq e_{i,j} + b_{i,j} \quad \forall i, j \in V, \forall c \in \mathcal{C}, \forall s \in \mathcal{S}. \tag{8}$$

A pipe $(i, j) \in \tilde{\mathcal{E}}$ can only be part of the minimum resistance path from s to c, if it was already existing or will be added.

In terms of a cost-benefit analysis, the overall length of all added pipes is limited:

$$\sum_{i \in \mathcal{V}} \sum_{j \in \mathcal{V}} b_{i,j} l_{i,j} \leq 2 \cdot L_{\text{p added}}, \tag{9}$$

where the parameter $l_{i,j}$ computed in pre-processing gives the length of pipe (i, j). The factor two is a result of considering an undirected graph, for which symmetry applies. Moreover, additional optional constraints are added, which may serve as cutting planes and speed up the optimization. First, pipes exceeding the bound for the overall pipe length are excluded:

$$l_{i,j} b_{i,j} \leq L_{\text{p added}} \quad \forall i, j \in \mathcal{V}. \tag{10}$$

Furthermore, the binary variables $b_{i,j}$ indicating the addition of pipes are bounded by the coefficients $e_{i,j}$ of the adjacency matrix of the original graph:

$$b_{i,j} \leq (1 - e_{i,j}) \quad \forall i, j \in V. \tag{11}$$

Finally, the addition of a pipe between the same node is not possible:

$$b_{i,i} = 0 \quad \forall i \in V. \tag{12}$$

5 Results and Conclusion

The optimization problem is implemented in Python and Gurobi [2] is used to solve it. To process the generated network data, additionally the Python packages WNTR [6] combined with NetworkX [3] are employed. To investigate the costs versus the benefits of pipe additions, we carried out a parameter study varying the upper bound for the overall length of all added pipes between 100 m and 10,000 m. The results show a non-linear improvement of the objective function, q.v. Fig. 2i, which is correlated logarithmically to the overall length of the added pipes.

In a second step, we extended the original graph by the best pipe additions computed for each instance, and determined the improvement of the resilience index I_{GT}, q.v. Fig. 2ii. Note that the definition of the resilience index does not consider

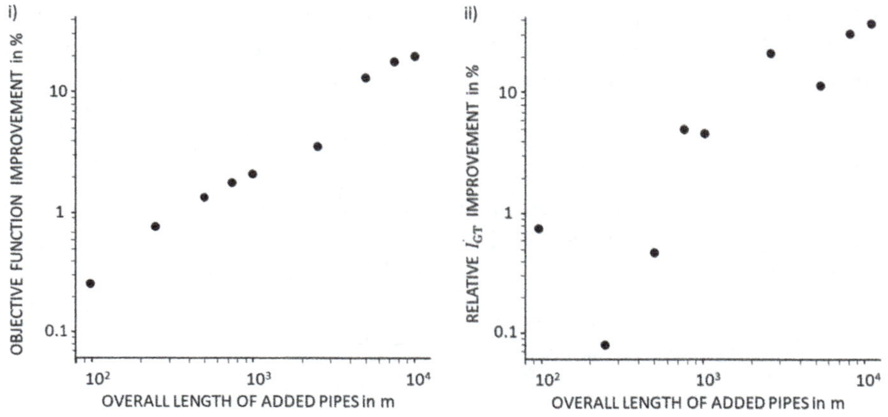

Fig. 2 Relative Resilience Improvement of the WDS for the addition of pipes with different predetermined upper bounds for the overall added pipe lengths

the shortest path in terms of hydraulic resilience per source-consumer-connection only, but also the possible alternative paths. To limit the numerical expenses, the critical number of $K = 13$ paths is determined for adequate accuracy, deduced as presented in [7]. The validation shows that the logarithmic improvement trend of the mean WDS resilience index applies as well. Differences between the improvement of the objective function and the improvement of the resilience index originate from the linear optimization approach, i.e. $K = 1$.

6 Summary and Outlook

We conducted a case study for a WDS in the German City of Darmstadt. To investigate its resilience, we modeled the WDS as an undirected graph and used a graph-theoretical resilience index from literature. In order to assess the impact of capacity adaptations by adding additional pipes between existing nodes, we formulated a linear optimization problem. Varying the upper limit of the overall length of added pipes, we conducted a parameter study to analyze the cost versus benefit of the capacity adaptations. While the material costs clearly depend on the overall length of added pipes, the costs for installing the pipes depend on the number of added pipes and the region. In future work, we plan to extend our approach by adding a more detailed cost model.

Acknowledgments The authors thank the KSB-Stiftung Stuttgart, Germany for funding this research. Moreover, we thank the German Research Foundation, DFG, for partly funding this research under project No. 57157498 within the Collaborative Research Center SFB 805 "Control of Uncertainties in Load-Carrying Structures in Mechanical Engineering", subproject "Resilient Design".

References

1. Altherr, L., et al.: Resilience in mechanical engineering - a concept for controlling uncertainty during design, production and usage phase of load-carrying structures. Appl. Mech. Mater. **885**, 187–198 (2018)
2. Gurobi: Gurobi Optimizer Reference Manual (2019). http://www.gurobi.com
3. Hagberg, A.A., Schult, D.A., Swart, P.J.: Exploring network structure, dynamics, and function using NetworkX. In: Proceedings of the 7th Python in Science Conference (SciPy), vol. 836, pp. 11–15 (2008)
4. Herrera, M., Abraham, E., Stoianov, I.: A graph-theoretic framework for assessing the resilience of sectorised water distribution networks. Water Resour. Manage. **30**(5), 1685–1699 (2016)
5. Jarvis, A., Reuter, H., Nelson, A., Guevara, E.: Hole-filled seamless SRTM data V4. International Centre for Tropical Agriculture (CIAT) (2008)
6. Klise, K.A., et al.: Water Network Tool for Resilience (WNTR) User Manual. Tech. Rep. August, Sandia National Laboratories (SNL), Albuquerque, NM, and Livermore, CA (2017)
7. Lorenz, I.S., Altherr, L.C., Pelz, P.F.: Graph-theoretical resilience analysis of water distribution systems - a case study for the German city of Darmstadt. In: Heinimann, H.R. (ed.) World Conference on Resilience, Reliability and Asset Management. Springer (2019, to be published). https://www.dropbox.com/s/ckqz9ysi7ea4zxe/Conference%20Proceedings%20-%20Draft%20v2.1.pdf?dl=0
8. Mair, M., Rauch, W., Sitzenfrei, R.: Spanning tree-based algorithm for generating water distribution network sets by using street network data sets. In: World Environmental and Water Resources Congress 2014, 2011, pp. 465–474. American Society of Civil Engineers, Reston, VA (2014)
9. Meng, F., Fu, G., Farmani, R., Sweetapple, C., Butler, D.: Topological attributes of network resilience: a study in water distribution systems. Water Res. **143**, 376–386 (2018)
10. Shin, S., Lee, S., Judi, D., Parvania, M., Goharian, E., McPherson, T., Burian, S.: A systematic review of quantitative resilience measures for water infrastructure systems. Water **10**(2), 164 (2018)

Part XIV
Production and Operations Management

A Flexible Shift System for a Fully-Continuous Production Division

Elisabeth Finhold, Tobias Fischer, Sandy Heydrich, and Karl-Heinz Küfer

Abstract In this paper, we develop and evaluate a shift system for a fully-continuous production division that allows incorporating standby duties to cope with production-related fluctuations in personnel demand. We start by analyzing the relationships between fundamental parameters of shift models including working hours, weekend load and flexibility and introduce approaches to balance out these parameters. Based on these considerations we develop a binary feasibility problem to find a suitable shift plan that is parametrized in the number of standby shifts.

Keywords Human Resources Management · Strategic Planning and Management

1 Introduction

The need for night and weekend work arises in many domains such as health care, certain areas of the service sector or in production units. Scheduling 24/7 work is a particularly challenging task as it comes with a wide range of requirements. These include not only appropriate personnel coverage but also government regulations, ergonomic recommendations regarding health and chronohygiene (resting times, rotation speed, clockwise rotation), and, not to be underestimated, shiftworkers' satisfaction. In addition, specific characteristics of the processes involved often induce further, highly individual constraints and hence situations in which standardized shift plans provide an adequate solution are rare. Therefore, various applications of personnel scheduling are studied in mathematical literature; see [1] and [5] for an overview.

E. Finhold (✉) · T. Fischer · S. Heydrich · K.-H. Küfer
Fraunhofer Institute for Industrial Mathematics ITWM, Kaiserslautern, Germany
e-mail:
elisabeth.finhold@itwm.fraunhofer.de,https://www.itwm.fraunhofer.de/de/abteilungen/opt.html; tobias.fischer@itwm.fraunhofer.de; sandy.heydrich@itwm.fraunhofer.de; karl-heinz.kuefer@itwm.fraunhofer.de

J. S. Neufeld et al. (eds.), *Operations Research Proceedings 2019*, Operations Research Proceedings, https://doi.org/10.1007/978-3-030-48439-2_64

At our industry partner we were confronted with such an individual shiftwork planning task for a fully continuous production division. Here, the particular challenge lay in two aspects: First, the shiftworkers' acceptance of a new plan highly depended on a satisfactory amount and distribution of free weekends. Second, the volatile production process led to high fluctuations in personnel demand so that the shift plan should allow for a certain flexibility.

Formally, our task was to develop a 24/7 regular shift system in which staff is divided into a number of *crews* working on rotating shift plans. A *shift plan (rota)* is simply a sequence of workdays on a certain shift and days off. We have a fixed *shift length* of eight hours and therefore each day is divided into three shifts, denoted D for *day shift*, S for *swing shift* and N for *night shift*. All other general parameters of the shift plan were up for discussion.

As mentioned before, our main requirement is to provide a low *weekend load* of at most 50% while maintaining around 38–40 h of work per week. Due to an inherent conflict between these two objectives, standard shift plans provided in the literature (see e.g. [3, 4]) do not satisfy these constraints. For example, shift plans with four crews always come with 42 h of work per week while the weekend load of 75% is too high; in contrast, six-crew solutions achieve the desired weekend load of 50% but only allow for 28 h of work. To the best of our knowledge, there exists little formal analysis concerning these fundamental properties of shift plans; [4] provides some approaches in this direction. As in our application the acceptance of the plan strongly depends on a good balance between weekend load and total workload, we do not start with an LP formulation of our problem right away as typically the formulation already determines these parameters. Instead, we first do a purely quantitative analysis of shift plans. More precisely, we identify and quantify methods and approaches to reduce weekend load or increase working hours. This analysis is presented in Sect. 2. Based on these considerations we can decide on measures that yield a good compromise in our objectives and therefore ensure the workers' acceptance. We then set up a simple binary feasibility problem accordingly for deriving the shift plan in Sect. 3.

The aforementioned flexibility is integrated using an approach similar to that introduced in [2]. A slightly modified version of the approach is as follows: Assume we have a basic shift plan as depicted in Fig. 1. Further, assume we have an additional personnel demand on day 2 on one of the three shifts. Depending on the shift, our compensation strategy is as follows:

Fig. 1 Segment of a fictive shift plan (D: day shift, S: swing shift, N: night shift)

day	1	**2**	3	4	5	6	···
crew 1	D	**D**	S	S	N	N	···
crew 2			D	D	S	S	···
crew 3	N	**N**			D	D	···
crew 4	S	**S**	N	N			···

(a) *day shift D:* Some worker from crew 2 can work an additional day shift.
(b) *swing shift S:* Some worker from crew 1 switches from D to S, for compensation some worker from crew 2 works an additional day shift.
(c) *night shift N:* Some worker from crew 3 switches from S to N, compensated by a switch from D to S (crew 1) and an additional night shift (crew 2).

None of these changes violates the ergonomically recommended shift order D-S-N or minimum resting times such that changes to the plan on subsequent days are avoided. On odd days, the concept can be applied analogously with additional night shifts and rotations from N to S and from S to D. Note that there should be at most two consecutive shifts of the same type to exploit this technique, as switching shifts without violating the desired shift order is only possible at the start and end of a sequence of shifts of same type.

2 Balancing Out Weekend Load and Hours of Work

In this section we investigate properties of shift plans with the goal of balancing out the conflicting objectives of weekend load and working hours. Recall that we actually impose the following constraints on the shift plan to be developed:

1. Fully-continuous shift model (equal personnel demand over all shifts)
2. Three disjoint 8 h-shift (D, S, N) per day, each covered by exactly one crew
3. Same number of shifts per week for all crews

Let n_{shifts} be the *average number of shifts per worker/per crew per week* and $l_{wknd} \in [0, 1]$ the *weekend load* (ratio of working Saturdays and Sundays). Note that in every shift plan satisfying conditions 1–3 above, for every crew the ratio of working days to free days has to be the same for every day of the week, namely l_{wknd}. Therefore, the average number of working shifts per week n_{shifts} equals $7 \cdot l_{wknd}$. That is why for a 50% weekend load we have only 3.5 shifts (28 h) per week, while for 38–40 h as desired, the weekend load goes up to ≈ 70%. Therefore, to achieve both, a satisfactory number of shifts per week *and* an acceptable weekend load, we have to relax the above constraints. We consider three approaches. Note that these are purely quantitative considerations to balance out the properties of a shift plan, independent of whether staff divided into crews or an underlying shift plan. Construction of an actual such shift plan has to be done in a subsequent step, for example via an LP as in Sect. 3.

Approach 1 (Skeleton Crew on Weekends) Reducing the weekend shift size to a ratio $s_{wknd} \in [0, 1]$ considerably reduces the weekend load to $\bar{l}_{wknd} = s_{wknd} \cdot l_{wknd}$. The number of shifts per week decreases as well, but only slightly by a factor of $\left(\frac{5+2 \cdot s_{wknd}}{7}\right)$ to $\bar{n}_{shifts} = (\frac{5}{s_{wknd}} + 2) \cdot \bar{l}_{wknd}$. The first relation is obvious; for the second one, observe that $\frac{1}{s_{wknd}} \cdot \bar{l}_{wknd}$ is the ratio of working weekdays and $\bar{l}_{wknd}$ the ratio of working weekend days.

Approach 2 (Increased Weekend Load) We can raise the shifts per week by increasing the *average* weekend load: If a portion $\alpha \in [0, 1]$ of workers accepts a higher weekend load $l^*_{wknd} \in [0, 1]$ (while the others remain at $l_{wknd} < l^*_{wknd}$), the number of shifts per week increases to $\bar{n}_{shifts} = 7 \cdot \big((1 - \alpha) \cdot l_{wknd} + \alpha \cdot l^*_{wknd}\big)$. Note that $(1 - \alpha) \cdot l_{wknd} + \alpha \cdot l^*_{wknd}$ is the average weekend load over all workers, and assuming equal total workload, the equation follows as before.

Approach 3 (Part Time work) Having a portion $\beta \in [0, 1]$ of workers on a part time contract with a share of $\gamma \in [0, 1]$, working with the same weekend load l_{wknd} as full time workers, we get $\bar{n}^{full}_{shifts} = \frac{7 \cdot l_{wknd}}{1-\beta\cdot(1-\gamma)}$. Here, the increase in working hours is achieved by shifting weekday work from part to full time workers. The formula follows from the relations $\bar{n}^{part}_{shifts} = \gamma \cdot \bar{n}^{full}_{shifts}$ (work share) and $\beta \cdot \bar{n}^{part}_{shifts} + (1 - \beta) \cdot \bar{n}^{full}_{shifts} = 7 \cdot l_{wknd}$ (average work load per day).

We want to point out that the shift plan parameters considered are also related to the *number of crews* $n_{crews} \in \mathbb{N}_{\geq 3}$ by $n_{crews} = \frac{21}{n_{shifts}} = \frac{3}{n_{wknd}}$. A similar relation is stated in [4].

Combining the three approaches relaxing our initial constraints, we have $n_{crews} = 3 \cdot \frac{s_{wknd}}{(1-\alpha)\cdot l_{wknd}+\alpha\cdot l^*_{wknd}}$ and $n_{shifts} = \frac{\left(\frac{5}{s_{wknd}}+2\right)\cdot((1-\alpha)\cdot l_{wknd}+\alpha\cdot l^*_{wknd})}{1-\beta\cdot(1-\gamma)}$. Note that $n_{crews} \in \mathbb{N}$ restricts the feasible combinations of parameters.

Example 1 A convenient 39 h – 40 h of work per week with an acceptable weekend load of 60% can be achieved

1. by reducing the weekend crew size to $s_{wknd} = 0.8$ (4 crews),
2. with $\alpha = 0.5$ of employees working on $l^*_{wknd} = 0.8$ of weekends (5 crews), or
3. with $\beta = 0.3$ of workers working part time at a share of $\gamma = 0.5$ (4.3 crews!).

3 A Linear Programming Formulation for a Parametrized Base Shift Plan

Using the preliminary considerations from Sect. 2, we now specify the properties of the shift plan to be developed and formulate a feasibility problem (FP) to derive it.

We decided on a four crew model with weekend load reduced from 75% to 50% by decreasing the weekend shift size to 2/3. Therefore, we will have 38 h of work per week. The skeleton weekend crew is realized with a straight-forward approach of dividing each crew into three subcrews and allowing one of them an additional weekend off each time the crew would be scheduled for weekend work in a plan with full weekend shifts. In this way we can ignore the skeleton crews for the (FP) formulation and simply adjust the solution afterwards.

Flexibility shall be accomplished with the approach introduced in Sect. 1, but on weekdays only. The associated standby shifts are not included in the problem formulation explicitly but we have constraints to ensure they can be integrated later (at most two consecutive shifts of same type on weekdays).

The complete (FP) is stated below. We have binary variables indicating whether a certain shift is covered by a certain crew. Note that (FP) is parametrized in the *cycle length* n_{days}, the number of days to complete a shift plan until it starts over again. $D_{cyc}=\{1,\ldots,n_{days}\}$ denotes the set of days in a cycle. We identify day $d+j$ with day $(d+j) \bmod n_{days}$ when $d+j>n_{days}$. By $D_{Mo}=\{7i+1 \mid i \in [\frac{n_{days}}{7}]]\}=\{1, 8, 15, \ldots\}$ etc. we denote sets of certain weekdays and by $T=\{D, S, N\}$ the set of all shift types, $C=\{1,\ldots,n_{crews}\}$ the set of crews. Equation (1)–(4) assure that every shift is covered by exactly one crew and shifts are equally distributed over crews. (5)–(7) put restrictions on the number of consecutive shifts of same type. Constraints (8)–(11) assure the desired shift order D-S-N. Finally, (12) states that we require two days off after a night shift and (13) that a weekend is either a working weekend or the entire weekend is free.

(FP)

$$\sum_{i\in C} x^t_{i,d} = 1 \qquad d \in D_{cyc},\ t \in T \tag{1}$$

$$x^D_{i,d} + x^S_{i,d} + x^N_{i,d} \le 1 \qquad i \in C,\ d \in D_{cyc} \tag{2}$$

$$\sum_{d=1}^{n_{days}} x^t_{i,d} = \frac{n_{days}}{n_{crews}} \qquad i \in C,\ t \in T \tag{3}$$

$$\sum_{d\in D_{\cdot}} x^t_{i,d} = \frac{n_{days}}{7\cdot n_{crews}} \qquad i \in C,\ t \in T,\ D_{\cdot} \in \{D_{Sa}, D_{Su}\} \tag{4}$$

$$x^t_{i,d} \le x^t_{i,d-1} + x^t_{i,d+1} \qquad i \in C,\ d \in D_{cyc},\ t \in T \tag{5}$$

$$\sum_{j=0}^{3} x^t_{i,d+j} \le 3 \qquad i \in C,\ d \in D_{cyc},\ t \in T \tag{6}$$

$$\sum_{j=0}^{2} x^t_{i,d+j} \le 2 \qquad i\in C,\ t\in T,\ d\in D_{Mo}\cup D_{Tu}\cup D_{We} \tag{7}$$

$$x^D_{i,d} \le x^D_{i,d+1} + x^S_{i,d+1} \qquad i \in C\ d \in D_{cyc} \tag{8}$$

$$x^S_{i,d} \le x^S_{i,d+1} + x^N_{i,d+1} \qquad i \in C,\ d \in D_{cyc} \tag{9}$$

$$x^S_{i,d+1} \le x^D_{i,d} + x^S_{i,d} \qquad i \in C,\ d \in D_{cyc} \tag{10}$$

$$x^N_{i,d+1} \le x^S_{i,d} + x^N_{i,d} \qquad i \in C,\ d \in D_{cyc} \tag{11}$$

$$\sum_{j=1}^{2} \left(x^D_{i,d+j} + x^S_{i,d+j}\right) \le 2\left(1 - x^N_{i,d}\right) \qquad i \in C,\ d \in D_{cyc} \tag{12}$$

$$x^D_{i,d} + x^S_{i,d} + x^N_{i,d} = x^D_{i,d+1} + x^S_{i,d+1} + x^N_{i,d+1} \qquad i \in C,\ d \in D_{Sa} \tag{13}$$

$$x^t_{i,d} \in \{0, 1\} \qquad i \in C,\ d \in D_{cyc},\ t \in T \tag{14}$$

Using CpSolver from Google OR-Tools we computed all solutions to (FP) with $n_{days}=28$ and $n_{crews}=4$ in less than two seconds on a standard PC. After removing symmetries due to permutation of the crews, we are left with two solutions. We choose the plan depicted in Fig. 2 as the standby shifts turn out to be more

W1							W2							W3							W4						
Mo	Tu	We	Th	Fr	Sa	Su	Mo	Tu	We	Th	Fr	Sa	Su	Mo	Tu	We	Th	Fr	Sa	Su	Mo	Tu	We	Th	Fr	Sa	Su
	D	D	S	S	N	N	N			D	D	S	S	S	N	N			D	D	D	S	S	N	N		
N			D	D	S	S	S	N	N			D	D	D	S	S	N	N				D	D	S	S	N	N
S	N	N			D	D	D	S	S	N	N				D	D	S	S	N	N	N			D	D	S	S
D	S	S	N	N				D	D	S	S	N	N	N			D	D	S	S	S	N	N			D	D

Fig. 2 A shift plan as solution to (FP) for $n_{crews}=4$, $n_{days}=28$

W1 Mo	Tu	We	Th	Fr	Sa	Su	W2 Mo	Tu	We	Th	Fr	Sa	Su	W3 Mo	Tu	We	Th	Fr	Sa	Su	W4 Mo	Tu	We	Th	Fr	Sa	Su
	D	D	S	S	N	N	N			D	D	S	S	S	N	N					D	S	S	N	N		
	D	D	S	S	N	N	N			D	D			S	N	N			D	D	D	S	S	N	N		
	D	D	S	S			N			D	D	S	S	S	N	N			D	D	D	S	S	N	N		
(D)		(S)	(D)	(N)				(N)	(D)		(S)			(N)	(S)		(N)	(D)			(S)	(D)	(N)	(S)			
N			D	D	S	S	S	N	N			D	D	D	S	S	N	N				D	D	S	S		
N			D	D	S	S	S	N	N					D	S	S	N	N				D	D	S	S	N	N
N			D	D			S	N	N			D	D	D	S	S	N	N				D	D	S	S	N	N
	(N)	(D)		(S)			(N)	(S)		(N)	(D)			(S)	(D)	(N)	(S)				(D)		(S)	(D)	(N)		
S	N	N			D	D	D	S	S	N	N				D	D	S	S	N	N	N			D	D		
S	N	N			D	D	D	S	S	N	N				D	D	S	S			N			D	D	S	S
S	N	N					D	S	S	N	N				D	D	S	S	N	N	N			D	D	S	S
(N)	(S)		(N)	(D)			(S)	(D)	(N)	(S)				(D)		(S)	(D)	(N)				(N)	(D)		(S)		
D	S	S	N	N				D	D	S	S	N	N	N			D	D	S	S	S	N	N				
D	S	S	N	N				D	D	S	S	N	N	N			D	D			S	N	N			D	D
D	S	S	N	N				D	D	S	S			N			D	D	S	S	S	N	N			D	D
(S)	(D)	(N)	(S)				(D)		(S)	(D)	(N)				(N)	(D)		(S)			(N)	(S)		(N)	(D)		

Fig. 3 Modified shift model with crews divided into three subcrews and weekend shift size reduced. Standby shifts are depicted in brackets

convenient (more D, less N standbys). The second plan is identical up to a one-day-shift.

Figure 3 shows the modified shift plan: reduced weekend load is achieved by dividing each shift crew into three subcrews and removing one weekend from each plan. Note that we indeed end up with a plan with weekend load 50%. Figure 3 also sketches the integration of standby shifts.

It remains to decide *how many* workers should be on standby for the respective shifts. Obviously, a good rate heavily depends on the underlying (discrete) probability distribution of additional demand on the respective shifts. For our shift system, the *ratio of additional demand covered* can be computed straight forward for a given distribution using the cumulative distribution function. An alternative approach, which turns out to be of advantage for more complex flexibility approaches, is to simulate the system and extract the ratio from this. Assuming two distributions of the additional demand based on historical data we computed the ratio of additional demand covered for all reasonable combinations of workers on standby on the respective shifts (D, S, N). We also computed the average frequency of standbys for a worker for each combination. For our application, where we assume a base demand (crew size) of 18 workers, combinations with standby frequency around every 1.5 weeks and 70%, respectively, 90%, of additional demand covered for the two distributions under consideration provide a good balance. Note that in this setting the average extra work increases the expected weekly total to $\approx$ 39.4 h as desired.

References

1. Ernst, A.T., Jiang, H., Krishnamoorthy, M., Sier, D.: Staff scheduling and rostering: A review of applications, methods and models. Eur. J. Oper. Res. **153**(1), 3–27 (2004)
2. Hoff, A.: So kann der 5-Schichtplan mit der Grundfolge FFSSNN---- produktiv umgesetzt werden. Dr. Hoff Arbeitszeitsysteme (2018). https://arbeitszeitsysteme.com/wp-content/uploads/2012/05/So-kann-der-5-Schichtplan-FFSSNN-produktiv-umgesetzt-werden.pdf

3. Lennings, F.: Ergonomische Schichtpläne - Vorteile für Unternehmen und Mitarbeiter. Angewandte Arbeitswissenschaft **180**, 33–50 (2004)
4. Miller, J.C.: Fundamentals of Shiftwork Scheduling, 3rd Edition: Fixing Stupid. Smashwords (2013)
5. Van den Bergh, J., Beliën, J., De Bruecker, P., Demeulemeester, E., De Boeck, L.: Personnel scheduling: A literature review. Eur. J. Oper. Res. **226**(3), 367–385 (2013)

Capacitated Lot Sizing for Plastic Blanks in Automotive Manufacturing Integrating Real-World Requirements

Janis S. Neufeld, Felix J. Schmidt, Tommy Schultz, and Udo Buscher

Abstract Lot-sizing problems are of high relevance for many manufacturing companies, as they have a major impact on setup and inventory costs as well as various organizational implications. We discuss a practical capacitated lot-sizing problem, which arises in injection molding processes for plastic blanks at a large automotive manufacturer in Germany. 25 different product types have to be manufactured on 7 distinct machines, whereas each product type may be assigned to at least two of these machines. An additional challenge is that the following production processes use different shift models. Hence, the stages have to be decoupled by a buffer store, which has a limited capacity due to individual storage containers for each product type. For a successful application of the presented planning approach several real-world requirements have to be integrated, such as linked lot sizes, rejects as well as a given number of workers and a limited buffer capacity. A mixed integer programming model is proposed and tested for several instances from practice using CPLEX. It is proven of being able to find very good solutions within in few minutes and can serve as helpful decision support. In addition to a considerable reduction of costs, the previously mostly manual planning process can be simplified significantly.

Keywords Capacitated lot sizing · Automotive manufacturing · Real-world application

---

J. S. Neufeld (✉) · U. Buscher
Faculty of Business and Economics, TU Dresden, Dresden, Germany
e-mail: janis_sebastian.neufeld@tu-dresden.de

F. J. Schmidt · T. Schultz
BMW Group Plant Leipzig, Leipzig, Germany

J. S. Neufeld et al. (eds.), *Operations Research Proceedings 2019*, Operations Research Proceedings, https://doi.org/10.1007/978-3-030-48439-2_65

1 Introduction

Solving lot-sizing problems is of high relevance for many manufacturing companies [3]. The existence of several products with varying demand that have to be processed on the same machines with a finite capacity, results in a complex planning task, referred to as capacitated lot-sizing problem (CLSP). In this study, we discuss a practical CLSP, which arises in injection molding processes for plastic blanks at a large automotive manufacturer in Germany. Different products are manufactured on several, heterogeneous injection molding machines, which use different technologies and have a limited capacity. Each type of product is assigned to one preferred machine, but can be processed on at least one other machine. Thus, besides determining optimal lot sizes and production times for each product, a useful assignment of products to machines has to be found. Before a product can be processed, a sequence-independent setup time is necessary on each machine. Once a machine is equipped for producing a certain type of product, the setup state remains valid for a succeeding period (setup carryover). All setup and processing times are dependent on the product as well as on the assigned machine. Due to a limited number of necessary tools, each product can be produced on only one machine at a time. An additional challenge is that the following production process uses a different shift model. Hence, the two stages *injection molding* and *paint shop* are decoupled by a buffer store, which has a limited capacity due to individual storage containers for each product. Since in automotive manufacturing supply reliability is crucial, demands always have to be satisfied and no shortages or back orders are allowed.

The CLSP has been studied widely in literature with various extension [6]. Nevertheless, due to specific organizational or technological requirements arising in real-world manufacturing systems, existing models and solution approaches can often not be applied directly to practice. Mainly, the following modifications of the CLSP are vital to provide a helpful decision support in this case: First, it is characterized by the existence of *parallel machines*, which were introduced by [4] and discussed, e.g., by [7]. Secondly, a *limited buffer capacity* of finished goods has to be considered [1]. Furthermore, *linked lot-sizes* are relevant, i.e. setup states can be carried over to the following time period [5]. Finally, a *limited worker capacity* for operating the machines and rejects cannot be neglected. We refer to this problem as CLSPL-IBPM, i.e. a CLSP with linked lot sizes (L), inventory bounds (IB) and parallel machines (PM). To the best of our knowledge, some of these requirements as well as its combination have not been considered in literature so far and approaches mentioned above cannot be applied to the given practical case. Therefore, we developed an extended MIP model, which is presented in Sect. 2 and applied for several real-world instances in order to replace current manual planning (see Sect. 3). Finally, the results are summarized in Sect. 4.

2 MIP Formulation of the CLSPL-IBPM

We formulate the studied CLSPL-IBPM problem as MIP model with the notation displayed in Table 1. Its general structure is based on the formulation of [2], that is extended by parallel machines and linked lot-sizes similarly to [6]. Additionally, the restricted worker capacity, rejects and limited storage buffers are integrated. The following assumptions are taken into account: The dynamic demand as well as the production rate are considered to follow a linear course for all products $j \in N$. Each product j can be produced on one of the parallel machines $i \in M_j$, with M_j being a subset of all machines M. The set of all products that can be produced on machine i is referred to as N_i. The length of a period $t \in T$ is set to 1 shift, which equals

Table 1 Notation for MIP formulation

Decision variables	
$q_{i,j,t}$	$\in \mathbb{N}$, lot size of product j produced on machine i at period t
$l_{j,t}$	$\in \mathbb{N}$, inventory of product j at the end of period t
$z_{i,j,t}$	$\in \{0, 1\}$, production variable, 1 if product j is produced (or set up) on machine i at period t, 0 otherwise
$z^*_{i,j,t}$	$\in \{0, 1\}$, setup variable, 1 if machine i is setup for product j at period t, 0 otherwise
$w^{ist}_{i,t}$	$\in \mathbb{N}$, number of workers assigned to machine i at period t
$t^{rr}_{i,t}$	≥ 0, remaining setup time on machine i at the end of period t
t^r_i	≥ 0, setup time on machine i at period t
$t^z_{i,t}$	≥ 0, production time on machine i at period t
Parameters	
$b_{j,t}$	Demand of product j at period t
B_j	Buffer capacity for product j
t^*	Length of period t
$f_{i,j}$	Setup costs for product j on machine i
c_j	Holding cost rate for product j
$ma_{i,j}$	Required workers for producing product i on machine i
w^{max}_t	Maximum number of workers at period t
pa_i	Planned reject on machine i
c^a_j	Reject cost rate for product j
$rz_{i,j}$	Setup time of product j on machine i
$zz_{i,j}$	Processing time per unit of product j on machine i
$aq_{i,j}$	Reject rate of product j on machine i, with $0 \leq aq_{i,j} \leq 1$
sf_j	Pile factor of product j
L	Large number

8 hours. It is possible to split setup times, i.e. to finish a started setup time in the following period.

$$\text{Min. } C = \sum_{t \in T} \left(\sum_{j \in N} \sum_{i \in M_j} \left(f_{i,j} \cdot z^*_{i,j,t} \right) + \tfrac{1}{2} \cdot \sum_{j \in N} \left(c_j \cdot [l_{j,t} + l_{j,t-1}] \right) + \sum_{j \in N} \sum_{i \in M_j} \left(z^*_{i,j,t} \cdot pa_i \cdot c^a_j \right) \right) \quad (1)$$

s.t.:

$$l_{j,t-1} + \sum_{i \in M_j} q_{i,j,t} - l_{j,t} = b_{j,t} \qquad \forall\, j \in N, t \in T \quad (2)$$

$$q_{i,j,t} \leq z_{i,j,t} \cdot \sum_{\tau \in (t..T)} b_{j,\tau} \qquad \forall\, j \in N, i \in M_j, t \in T \quad (3)$$

$$q_{i,j,t} + z^*_{i,j,t} \leq L \cdot z_{i,j,t} \qquad \forall\, j \in N, i \in M_j, t \in T \quad (4)$$

$$z^*_{i,j,t} = \max\left\{ z_{i,j,t} - z_{i,j,t-1}; 0 \right\} \qquad \forall\, j \in N, i \in M_j, t \in T \quad (5)$$

$$\sum_{i \in M_j} z_{i,j,t} \leq 1 \qquad \forall\, j \in N, t \in T \quad (6)$$

$$\sum_{i \in M_j} z^*_{i,j,t} \leq 1 \qquad \forall\, j \in N, t \in T \quad (7)$$

$$t^z_{i,t} = \begin{cases} 0 & \Rightarrow w^{ist}_{i,t} = 0 \\ \text{else} & \Rightarrow w^{ist}_{i,t} = \sum_{j \in N_i} \left(z_{i,j,t} \cdot ma_{i,j} \right) \end{cases} \qquad \forall\, i \in M, t \in T \quad (8)$$

$$w^{max}_t \geq \sum_{i \in M} w^{ist}_{i,t} \qquad \forall\, t \in T \quad (9)$$

$$t^r_{i,t} = \min\left\{ t^{rr}_{i,t-1} + \sum_{j \in N_i} \left(z^*_{i,j,t} \cdot rz_{i,j} \right); t^* \right\} \qquad \forall\, i \in M, t \in T \quad (10)$$

$$t^{rr}_{i,t} = \max\left\{ \sum_{j \in N_i} \left(z^*_{i,j,t} \cdot rz_{i,j} \right) + t^{rr}_{i,t-1} - t^*; 0 \right\} \qquad \forall\, i \in M, t \in T \quad (11)$$

$$t^z_{i,t} + t^r_{i,t} \leq t^* \qquad \forall\, i \in M, t \in T \quad (12)$$

$$t^z_{i,t} = \sum_{j \in N_i} \left(zz_{i,j} \cdot q_{i,j,t} \cdot \left[1 + aq_{i,j} \right] \right) \qquad \forall\, i \in M, t \in T \quad (13)$$

$$\frac{q_{i,j,t}}{sf_j} = g_{j,t} \qquad \forall\, j \in N, i \in M_j, t \in T \quad (14)$$

$$l_{j,t} \leq B_j \qquad \forall\, j \in N, t \in T \quad (15)$$

$$l_{j,0} = z_{i,j,0} = t^{rr}_{i,0} = 0 \qquad \forall\, j \in N, i \in M_j \quad (16)$$

The objective function (1) consists of three cost components that are summed up for all planning periods. First, sequence-independent setup costs are considered for each changeover. Secondly, holding costs are determined assuming an constant usage of goods in the demand period. Furthermore, reject costs arise during the beginning of every production process. Equation (2) is the inventory balance equation, which ensures that all demands are satisfied. Equation (3) defines that the lot size $q_{i,j,t}$ can only be larger than 0 if the binary production variable $z_{i,j,t} = 1$. Equations (4) and (5) represent the linking between the variables $q_{i,j,t}$, $z_{i,j,t}$ and the binary setup variable $z^*_{i,j,t}$. Constraints (6) and (7) ensure that every product is produced and set up only once in each period. This is a limitation of decision space to simplify the considered problem but corresponds to the actual planning of the automotive manufacturer. Equation (8) determines the number of required workers $w^{ist}_{i,t}$, which is limited by the worker capacity per shift in Eq. (9). Equation (10) determines the required setup time $t^r_{i,t}$ on each machine. It is limited to the length of a period t^*. The remaining setup time $t^{rr}_{i,t}$ from a setup starting at period $t-1$ is defined in Constraint (11). Equation (12) guarantees that the available time in each period is not exceeded by production and setup processes, while the necessary production time $t^z_{i,t}$ is calculated via Eq. (13), taking the reject rate at each machine into account. According to Constraint (14) all lot sizes have to be an integer multiple of the storage capacity of the individual storage containers, while the maximum buffer capacity is limited by Eq. (15). Initial inventory levels, setup states and remaining setup times from previous periods are excluded by Eq. (16). It has to be noted that the presented formulation is not linear due to Eqs. (5), (8), (10) and (11). But it can be linearized with reasonable effort and is therefore solvable with common mathematical solvers.

3 Computational Results

The proposed model was implemented and tested for 8 real-world instances on a Intel(R) Xenon(R) CPU E5-4627 with 3.3 GHz clock speed and 768 GB RAM using CPLEX 12.6 with max. 4 parallel threads. Each instance represents one week and corresponds to the weekly planning period in practice. In total 25 different products have to be planned on 7 machines. Computation time t_{CPU} has been limited to both 3 and 30 min. The results are displayed in Table 2. It can be seen, that for all instances already after 3 min computation time good results can be obtained with a maximum gap of 5.1%. One instance can even be solved to optimality. This proves the applicability of the proposed approach, since a re-planning can be performed at short notice, e.g. if machine breakdowns or unexpected changes in demand occur. Nonetheless, larger computation times are still viable for the weekly planning. With a time limit of 30 min the results can be further improved from on average 4.1% to 3.0%. However, still no additional instance could be solved to optimality. Due to organizational issues it is difficult to compare the gained results directly to the planned schedules from practice. However, an approximate evaluation indicates a reduction of the cost function by 10 to 20%, at the same time ensuring feasibility of the generated production plan. Moreover, by using the proposed MIP a previously time-consuming and complex manual planning task can be replaced by a quick automated decision support.

Table 2 Computational results for real-world instances

	t_{CPU}=3 min.		t_{CPU}=30 min.	
Instance	Obj.	Gap %	Obj.	Gap %
1	108,132	4.9	105,674	2.5
2	116,378	4.4	115,684	3.6
3	116,648	4.3	115,782	3.4
4	102,258	4.8	100,116	2.5
5	134,006	5.1	132,680	3.8
6	27,328	4.7	27,229	3.7
7	94,824	0.0	94,824	0.0
8	106,227	4.7	106,154	4.4
Average		**4.1**		**3.0**

4 Conclusions and Future Research

For a successful application of CLSP models in practice, it is often necessary to integrate several real-world requirements. In doing so, the proposed CLSPL-IBPM MIP formulation was able to provide decision support for the production process of plastic blanks in automotive manufacturing. Even within very short computation times good solutions could be generated that simplify the planning process and ensure low costs. Nevertheless, the proposed approach still leaves room for development. Additional technological requirements should be added to the model, such as paired products, that need to be processed together, or variants of specific parts. Furthermore, a balanced demand for workers over all shifts could lead to additional improvements of the production plan. Finally, the length of a period of one shift may not be optimal as within each shift a detailed scheduling is still necessary and buffers may not be sufficient at each moment.

References

1. Akbalik, A., Penz, B., Rapine, C.: Capacitated lot sizing problems with inventory bounds. Ann. Oper. Res. **229**(1), 1–18 (2015)
2. Billington, P.J., McClain, J.O., Thomas, L.J.: Mathematical programming approaches to capacity-constrained MRP systems: Review, formulation and problem reduction. Manag. Sci. **29**(10), 1126–1141 (1983)
3. Copil, K., Wörbelauer, M., Meyr, H., Tempelmeier, H.: Simultaneous lotsizing and scheduling problems: a classification and review of models. OR Spectr. **39**(1), 1–64 (2017)
4. Diaby, M., Bahl, H.C., Karwan, M.H., Zionts, S.: A lagrangean relaxation approach for very-large-scale capacitated lot-sizing. Manag. Sci. **38**(9), 1329–1340 (1992)
5. Karagul, H.F., Warsing Jr., D.P., Hodgson, T.J., Kapadia, M.S., Uzsoy, R.: A comparison of mixed integer programming formulations of the capacitated lot-sizing problem. Int. J. Prod. Res. **56**(23), 7064–7084 (2018)
6. Quadt, D., Kuhn, H.: Capacitated lot-sizing with extensions: a review. 4OR **6**(1), 61–83 (2008)
7. Toscano, A., Ferreira, D., Morabito, R.: A decomposition heuristic to solve the two-stage lot sizing and scheduling problem with temporal cleaning. Flex. Serv. Manuf. J. **31**(1), 142–173 (2019)

Facility Location with Modular Capacities for Distributed Scheduling Problems

Eduardo Alarcon-Gerbier

Abstract For some time now, customers are more interested in sustainable manufacturing and are requesting products to be delivered in the shortest possible time. To deal with these new customer requirements, companies can follow the Distributed Manufacturing (DM) paradigm and try to move their production sites close to their customer. Therefore, the aim of this paper is to connect the idea of DM with the integrated planning of production and distribution operations mathematically in a MIP model. To this end, the model simultaneously decides the position of the plants, the production capacity in each period as well as the production and distribution scheduling.

Keywords Distributed Manufacturing · Mixed-integer programming · Scheduling · Supply chain management

1 Introduction

In recent times, customers are more concerned about the environmental damage caused by the production and distribution of purchased goods. In addition to this increased interest in sustainable manufacturing, the customers want products to be delivered in the shortest possible time. Here is where the concept of Distributed Manufacturing (DM) represents an appropriate paradigm shift. This concept, which has been gaining more and more attention lately, can be defined as a network of decentralized facilities, which are adaptable, easily reconfigurable, and closely located to points of consumption [7]. These decentralized production systems are a practical proposition, because the production takes place close to the customers, allowing a higher flexibility, shorter delivery times, and reducing CO_2 emissions caused by long transport distances for final products [6].

E. Alarcon-Gerbier (✉)
Technische Universität Dresden, Dresden, Germany
e-mail: eduardo.alarcon@tu-dresden.de; https://tu-dresden.de/bu/wirtschaft/lim

J. S. Neufeld et al. (eds.), *Operations Research Proceedings 2019*, Operations Research Proceedings, https://doi.org/10.1007/978-3-030-48439-2_66

Moreover, the industries are currently immersed in an era in which the development of cyber-physical technologies and cloud-based networks have the potential to redesign production systems and to give rise to smart factories [3]. This would enable production to take place at mobile production sites, which could be relocated and put into operation relatively quickly, changing the nature of the location decision from a strategic level to a tactical or even operational one. Therefore, it is necessary to investigate new concepts of network design, as well as their optimization and coordination. Based on these assumptions, the aim of this paper is to combine the capacitated facility location problem with the operational production and distribution planning.

The problem addressed in this article can be broken down into two subproblems: Integrated Production and Outbound Distribution Scheduling (IPODS) and Facility Location Problem (FLP). Traditionally, production and distribution scheduling are planned separately in a sequential manner, since both of them are complex problems in themselves. This, however, leads to sub-optimal results which could be improved through an integrated planning, known in the literature as IPODS. A review on this topic was presented in 2010 which classifies the models into five different categories describing some major structural characteristics and solution properties [2]. On the other hand, the FLP is a well-studied problem and has been extended by considering many aspects, like multi-period of time [1], modular capacities [4], as well as the inclusion of transport problem originating the problem known as Location-Routing Problem [5].

The remainder of the paper is organized as follows. After this introduction, a description of the problem is given, followed by the mathematical model. A computational study is carried out in the following section and results and their implications are summarized.

2 Problem Formulation

Formally, a Capacitated Facility Location Problem has to be solved together with a Production and Distribution Scheduling Problem. The formulation can be described as follows. A manufacturer can produce at S different locations to serve I geographically distributed customers. Each customer $i \in \{1, \cdots, I\}$ orders a specific amount D_{ip} of a generic product in each period $p \in \{1, \cdots, P\}$ with a given due date DD_{ip}. Since homogeneous modules (or production lines) are assumed, the processing time depends only on the demand of each customer and the production coefficient PR (time units per ordered units). After production, each order is directly delivered to the customer considering site-dependent transportation time TT_{is}. Besides, the manufacturer has N identical modules at his disposal with a capacity CM (time units), which can be relocated in each period p from one production site to another generating an expansion or reduction of the total production capacity. Relocating a module originates a relocation time RT, which means that the production at this module $n \in \{1, \cdots, N\}$ can begin after RT time units.

The model answers four questions simultaneously. Firstly, the model selects from the pool of available sites at least IS sites to be opened. Secondly, the mathematical formulation looks for the optimal assignment of modules to the opened sites in each period p. Thirdly, the Mixed-Integer Program (MIP) model assigns the customer orders to a specific module (and site), and finally, the machine scheduling planning is carried out.

The model aims at minimizing overall costs and considers six cost components. Firstly, transportation costs are incurred for the delivery of orders, which depend linearly on the transportation time (the transportation cost rate TC is measured in € per time unit). Furthermore, delay costs are included, which are the product of the penalty cost rate PC (measured in € per time unit delay) and the customer-related tardiness t_{ip} (measured in time units). The opening costs IC are the third cost component which are incurred by choosing the most suitable facilities to fulfil customer orders. Moreover, fixed location costs FC are included as the fourth cost component, which have to be taken into account when a location is used for the production of customer orders. There are two cost components related to the modules. The former represents the operational costs OC by using a module n and the latter are the relocation costs RC, which are incurred by moving a module from one site to another. The following variables are also used (let M be a very large number):

b_{isp}	$\in \{0, 1\}$	takes the value 1 if order i is assigned to site s in period p
c_{ip}		completion time of order i in period p
po_{nsp}	$\in \{0, 1\}$	takes the value 1 if module n is located at site s in period p
r_{np}	$\in \{0, 1\}$	takes the value 1 if module n is relocated in period p
t_{ip}		tardiness of order i in period p
v_s	$\in \{0, 1\}$	takes the value 1 if site s is opened
x_{inp}	$\in \{0, 1\}$	takes the value 1 if order i is assigned to module n in period p
y_{inp}	$\in \{0, 1\}$	takes the value 1 if order i is the first order processed at module n in period p
z_{ijnp}	$\in \{0, 1\}$	takes the value 1 if order i is processed directly before order $j \in I+1$ at module n in period p. Order $I+1$ is an artificial last order

In the following, the problem is formalized by a MIP.

$$\begin{aligned}\min \sum_{i=1}^{I}\sum_{p=1}^{P}\sum_{s=1}^{S}(TT_{is}\cdot TC\cdot b_{isp}) + \sum_{i=1}^{I}\sum_{p=1}^{P}(PC\cdot t_{ip}) + \sum_{s=1}^{S}(IC\cdot v_s)\\ +P\cdot\sum_{s=1}^{S}(FC\cdot v_s) + \sum_{n=1}^{N}\sum_{p=1}^{P}\sum_{s=1}^{S}(OC\cdot po_{nsp}) + \sum_{n=1}^{N}\sum_{p=1}^{P}(RC\cdot r_{np})\end{aligned} \tag{1}$$

subject to

$$\sum_{n=1}^{N} x_{inp} = 1 \qquad \forall\, i \in I;\, p \in P \tag{2}$$

$$\sum_{i=1}^{I} D_{ip} \cdot x_{inp} \cdot PR \leq CM \qquad \forall\, n \in N;\, p \in P \tag{3}$$

$$x_{inp} \leq y_{inp} + \sum_{j=1}^{I} z_{jinp} \qquad \forall\, i \in I;\, n \in N;\, p \in P \tag{4}$$

$$\sum_{j=1}^{I+1} \sum_{n=1}^{N} z_{ijnp} = 1 \qquad \forall\, i \in I;\, p \in P \tag{5}$$

$$y_{inp} \leq \sum_{j=1}^{I+1} z_{ijnp} \qquad \forall\, i \in I;\, n \in N;\, p \in P \tag{6}$$

$$z_{iinp} = 0 \qquad \forall\, i \in I;\, n \in N;\, p \in P \tag{7}$$

$$\sum_{j=1}^{I} z_{jinp} \leq \sum_{j=1}^{I+1} z_{ijnp} \qquad \forall\, i \in I;\, n \in N;\, p \in P \tag{8}$$

$$\sum_{i=1}^{I} y_{inp} = \sum_{s=1}^{S} po_{nsp} \qquad \forall\, n \in N;\, p \in P \tag{9}$$

$$\sum_{s=1}^{S} po_{nsp} \leq 1 \qquad \forall\, n \in N;\, p \in P \tag{10}$$

$$\sum_{i=1}^{I} D_{ip} \cdot PR \leq \sum_{n=1}^{N} \sum_{s=1}^{S} CM \cdot po_{nsp} \qquad \forall\, p \in P \tag{11}$$

$$v_s \geq po_{nsp} \qquad \forall\, n \in N;\, p \in P;\, s \in S \tag{12}$$

$$\sum_{s=1}^{S} v_s \geq IS \tag{13}$$

$$r_{np} \geq po_{nsp} - po_{n,s,p-1} \qquad \forall\, n \in N;\, p = 2, \ldots, P;\, s \in S \tag{14}$$

$$c_{ip} \geq D_{ip} \cdot PR \cdot y_{inp} + RT \cdot r_{np} \qquad \forall\, i \in I;\, n \in N;\, p \in P \tag{15}$$

$$c_{ip} \geq c_{jp} + D_{ip} \cdot PR - M(1 - z_{jinp}) \qquad \forall\, i, j \in I;\, n \in N;\, p \in P \tag{16}$$

$$b_{isp} \geq 1 - M(2 - x_{inp} - po_{nsp}) \qquad \forall\, i \in I; n \in N; p \in P; s \in S \qquad (17)$$

$$t_{ip} = \max(0; c_{ip} + b_{isp} \cdot TT_{is} - DD_{ip}) \qquad \forall\, i \in I; p \in P; s \in S \qquad (18)$$

In the above formulation, the objective function (1) aims at minimizing the total costs composed by transportation costs, tardiness costs originated by delivery delay, facility opening costs, fixed costs for using a facility, operational costs related to the use of modules, and relocation costs. It should be noted that production costs are not taken into account here because homogeneous production modules are assumed, which require the same time and can produce the orders at the same cost. Therefore the production costs are not decision-relevant. Constraints (2) guarantee that each order i has to be assigned to a module n in each period p. Constraints (3) represent the capacity restriction for the modules. Constraints (4) and (5) force each assigned order i either to follow another one or to be the first to be processed on a module n. Constraints (6), (7) and (8) ensure in combination that for each module n the customer order i is scheduled before the customer order j. Constraints (9) guarantee that if an order i is assigned to be the first to be produced on a module n in a period of time p, this module has to be installed on a site s. Constraints (10) specify that a module n can be installed at most at one site s per period p. Inequalities (11) are the demand constraints which are redundant for the LP relaxation. However, they enable the MIP solver to find cover cuts that reinforce the formulation. By (12) a site s has to be opened if there is at least one module n installed on it. Constraints (13) specify the minimal number of plants that can be operating per period. Constraints (14) establish if a module n is relocated in the period p or not. By (15) the completion time of the first order must be equal or greater than the corresponding processing time plus the relocation time if this module was relocated. Inequalities (16) determine the completion time of an order that is not the first in the sequence on a machine. This time is equal to or greater than the processing time of the job plus the completion time of its predecessor. Constraints (17) assign the customer order i to the site s by interlinking two binary variables, x_{inp} and po_{nsp}. Constraints (18) determine the tardiness of each job. Here, the delivery time is indirectly calculated by adding the corresponding travel time to the completion time and the tardiness is defined as the maximum between zero and the delivery time minus the due date.

3 Computational Study

Table 1 summarizes the main results of 10 different scenarios. Each of them was carried out at least 8 times using CPLEX 12.7.1 on an Intel Xeon 3.3 GHz processor with 768 GB memory, interrupting the calculation after 8 hours. The same cost parameters were used in each scenario. Since no benchmark instances exist, these values were derived from related papers found in the literature. Moreover, at least

Table 1 Results of the computational study

No.	P	I	N	S	Variables	Gap (%)			Time (min)		
						min.	avg.	max.	min.	avg.	max.
1	1	8	4	3	411	0	0	0	0.07	0.38	1
2		12	4	3	799	0	4.8	21.8	11.4	211.5	480
3		12	5	3	983	0	21.0	41.8	310.6	451.8	480
4		18	6	4	2,410	42.4	48.4	53.2	480	480	480
5	2	8	4	2	794	0	8.1	15.3	9.9	421.2	480
6		8	4	3	819	0	18.2	36.3	52.7	394.5	480
7		10	4	3	1,175	20.3	33.4	43.8	480	480	480
8		12	5	3	1,963	39.4	49.4	56.3	480	480	480
9	3	8	4	2	1,190	16.3	30.1	41.9	480	480	480
10		12	5	3	2,943	40.9	54.1	64.5	480	480	480

two sites had to be opened in each period ($IS = 2$) and the parameters associated with the demand were generated as random integers from uniform distributions.

The following conclusions can be drawn from the results. The model was able to find the optimal solution for instances of up to 1,000 variables. The solutions obtained by solving larger instances present an average gap greater than 30%. Even simplifying the problem and not considering the Facility Location Problem ($IS = S = 2$, see instances No. 5 and No. 9), CPLEX could not find the optimal solution for all the tested instances.

4 Summary

In this paper, a novel approach was presented addressing the integration of production and outbound distribution scheduling with the possibility of selecting the most suitable production sites, as well as the production capacity per period and site. For this purpose, a MIP formulation was developed which aims at minimizing the total costs. The model was also tested on several random instances in order to assess the performance of the model.

The proposed model has several options of extension. Firstly, since mobile modules are considered, it could be interesting to expand the problem by allocating sites/modules on a continuous space, finding the best located position. This, however, brings with it an increase in complexity of the problem. Another possible extension is the inclusion of the routing problem in order to plan the delivery of several customer orders together. Finally, as shown in the computational study, this jointly planning problem is quite complex to solve for large problems. Therefore, the development of heuristics is required in order to solve large instances in short computational time.

Acknowledgments The author thanks the Friedrich and Elisabeth Boysen-Stiftung and the TU Dresden for the financial support during the third Boysen-TU Dresden-GK.

References

1. Albareda-Sambola, M., Fernández, E., Nickel, S.: Multiperiod location-routing with decoupled time scales. Eur. J. Oper. Res. **217**(2), 248–258 (2012)
2. Chen, Z.-L.: Integrated production and outbound distribution scheduling: Review and extensions. Oper. Res. **58**(1), 130–148 (2010)
3. Kagermann, H., Helbig, J., Helllinger, A., Wahlster, W.: Recommendations for Implementing the Strategic Initiative INDUSTRIE 4.0: Securing the Future of German Manufacturing Industry. Final report. The Industrie 4.0 Working Group (2013)
4. Melo, M.T., Nickel, S., Saldanha da Gama, F.: Dynamic multi-commodity capacitated facility location: a mathematical modeling framework for strategic supply chain planning. Comput. Oper. Res. **33**(1), 181–208 (2005)
5. Nagy, G., Salhi, S.: Location-routing: Issues, models and methods. Eur. J. Oper. Res. **177**, 649–672 (2007)
6. Rauch, E., Dallasega, P., Matt, D.: Distributed manufacturing network models of smart and agile mini-factories. Int. J. Agile Syst. Manag. **10**(3/4), 185–205 (2017)
7. Seregni, M., Zanetti, C., Taisch, M.: Development of distributed manufacturing systems (DMS) concept. In: XX Summer School. Francesco Turco e Industrial Systems Engineering, pp. 149–153 (2015)

Part XV
Project Management and Scheduling

Diversity of Processing Times in Permutation Flow Shop Scheduling Problems

Kathrin Maassen and Paz Perez-Gonzalez

Abstract In static-deterministic flow shop scheduling, solution algorithms are often tested by problem instances with uniformly distributed processing times. However, there are scheduling problems where a certain structure, variability or distribution of processing times appear. While the influence of these aspects on common objectives, like makespan and total completion time, has been discussed intensively, the efficiency-oriented objectives core idle time and core waiting time have not been taken into account so far. Therefore, a first computational study using complete enumeration is provided to analyze the influence of different structures of processing times on core idle time and core waiting time. The results show that in some cases an increased variability of processing times can lead to easier solvable problems.

Keywords Permutation flow shop scheduling · Waiting time · Idle time · Diversity of processing times

1 Problem Description and Related Literature

The static-deterministic permutation flow shop (PFS) is assumed where n jobs have to be scheduled on m machines which are arranged in series. The sequence of jobs is the same on all machines (permutation assumption, abbreviated as $prmu$), see [9] for a detailed description of PFS. The $\alpha|\beta|\gamma$-notation of [5] is used to define scheduling problems, where α is the machine layout, β the process constraints and

K. Maassen (✉)
Chair of Business Administration and Production Management, University of Duisburg-Essen, Duisburg, Germany
e-mail: kathrin.maassen@uni-due.de

P. Perez-Gonzalez
Industrial Organization and Business Management, University of Seville, School of Engineering, Sevilla, Spain

J. S. Neufeld et al. (eds.), *Operations Research Proceedings 2019*, Operations Research Proceedings, https://doi.org/10.1007/978-3-030-48439-2_67

γ the objective function (e.g. $Fm|prmu|C_{max}$ denotes a permutation flow shop with m machines and objective of minimizing makespan). Various scheduling problems referring to PFS with different constraints and objective functions, in most cases makespan (C_{max}) and total completion time ($\sum C_j$), have been discussed in the literature and suitable algorithms have been provided. The efficiency of algorithms with respect to solution quality and speed are often evaluated by using common test beds with uniform processing times, $p_{i,j}$, for each job j on each machine i (see e.g. the test beds of [13] and [14]). However, there are scheduling problems where a certain structure, variability or distribution of processing times appear. Variability of processing times, defined in this paper as diversity to distinguish it from stochastic scheduling, is denoted as d. Diversity refers to the coefficient of variation of processing times cv which is the standard deviation divided by mean.

Different distributions of processing times are often used to approximate real-life data. This approximation has also been discussed several times in the literature, e.g. [10] stated that randomly generated processing times are unlikely in practical applications and proposed test beds using uniform distribution combined with time gradients across machines and job correlations. Both approaches were previously also proposed by [11]. The work of [15] suggested instances which are divided into job-, machine- and mixed-correlation. Moreover, [6] stated that processing times in real-life are usually not normally or exponentially distributed and recommended, among other things, LogN distributed processing times. Reference [12] observed normal and LogN distribution in a specific industrial context. Another interesting factor to consider real-life processing times is by controlling the variability, see e.g. [4] who pointed out that a high variability does not represent the normal case in industries. Referring to structured processing times, [9] discussed that $Fm|prmu, p_{i,j} = p_j|C_{max}$ is equivalent to $1||C_{max}$. Here, the diversity of processing times is an interesting aspect because job j has the same processing times on each machine and hence, the diversity, d_j, of job j related to all machines is zero. Of course, if the diversity related to all jobs and machines is zero, i.e. $Fm|prmu, p_{i,j} = p|C_{max}$, the problem becomes trivial since no scheduling problem exists. Furthermore, a certain structure can also lead to a reduction of problem complexity, e.g. [3] showed that under certain processing time conditions the minimization of C_{max} in a permutation flow shop can be reduced to a single machine problem. Another structural characteristic is dominance behavior of machines, e.g. a dominant machine i^d which processing times must verify $\min_{\forall j=1,...,n} p_{i^d,j} \geq \max_{\forall j=1,...,n} p_{i,j}$ (dominance type II, see e.g. [8]).

The influence of a certain structure, diversity or distribution on C_{max} and $\sum C_j$ has already been discussed intensively. Apart from these objectives, waiting time and idle time are also important indicators in scheduling. Both can be found in constraints (e.g. no-wait or no-idle scheduling problems) or as objective functions. In this context, two objective functions related to the efficiency of a production system can be defined for the permutation flow shop. On the one hand, core idle time of machine i, namely $\sum CIT_i$, is the total idle time between each two jobs on machine i. On the other hand, core waiting time of job j, namely $\sum CWT_j$, is the total waiting time of the job between each two machines. Both measures are time periods within the production system where either a machine has to wait for a job or a job has to wait for a machine, and hence indicate waste of time within a production system. Moreover, both objectives are highly influenced by structure, diversity and distribution of processing times, e.g. it can be proved easily that $\sum CIT_i = 0$ for each schedule in a two-machine permutation flow shop where the second machine is dominant. Moreover, for $F2|prmu, p_{i,j} = p_j|\sum CWT_j$ the dispatching rule Shortest-Processing-Time leads to an optimal schedule. $\sum CIT_i$ and $\sum CWT_j$ have been discussed only rarely in scheduling literature, referring to PFS e.g. [2] dealt with two machines and the minimization of $\sum CWT_j$, while [7] proposed a heuristic approach for minimizing $\sum CIT_i$. [1] dealt with a small computational example discussing the relationship between $\sum CIT_i$ and $\sum CWT_j$. Hence, the influence of structure, diversity and distribution on $\sum CIT_i$ and $\sum CWT_j$ has not been discussed so far. The objectives are defined formally as:

$$\sum CIT_i = \sum_{i=2}^{m} CIT_i = \sum_{i=2}^{m}\sum_{k=2}^{n} B_{i,[k]} - C_{i,[k-1]} \tag{1}$$

$$\sum CWT_j = \sum_{j=2}^{n} CWT_j = \sum_{i=2}^{m}\sum_{k=2}^{n} B_{i,[k]} - C_{i-1,[k]} \tag{2}$$

where $B_{i,[k]}$ represents the starting time of job in position k on machine i, while $C_{i,[k]}$ defines the completion time of job in position k on machine i. As often assumed in scheduling, only semi-active schedules are considered.

2 Experimental Analysis

Some special cases, where the diversity of processing times has a high influence on $\sum CIT_i$ and $\sum CWT_j$, were shown in Sect. 1. The aim of this Section is to examine in general the influence of different diversity-levels on the objectives $\sum CIT_i$ and $\sum CWT_j$ in a permutation flow shop with structured processing times. The processing times are generated with LogN-distribution to achieve realistic data and to control diversity. In this study, we assume a job-oriented structure, i.e. first of all, a mean, μ_j, for each job j is generated randomly with uniform distribution ($U[1, 99]$). Secondly, the mean value μ_j is then used to generate the processing times of job j on each machine i by LogN-distribution using the two parameters, μ_j and σ, where σ refers to the diversity levels $d = [0.1, 0.5, 1.0]$.

For the experimental analysis, 30 instances for each diversity level and problem size, $n = [5, 10]$ and $m = [2, 5, 10]$, are generated; i.e. 540 instances in total. All instances are solved optimally by complete enumeration, i.e. for each problem instances all $n!$ schedules are evaluated with respect to $\sum CIT_i$ and $\sum CWT_j$. To observe the behaviour of both objectives referring to different problem sizes and diversity-levels, the relative deviation index (RDI_s) is used,

$$RDI_s = \frac{OV_s - OV^{min}}{OV^{max} - OV^{min}} \quad \forall s \tag{3}$$

OV_s refers to the objective value of schedule s and OV^{min} and OV^{max} to the minimum and maximum of the respective problem instance. RDI is used since the minimum of $\sum CIT_i$ and $\sum CWT_j$ could yield zero. If RDI = 0, the minimum value is reached, while RDI = 1 indicates the maximum value. The average RDI is denoted as *ARDI*. In fact, RDI expresses OV_s relative to the difference between maximum and minimum but the difference itself is difficult to interpret, since it might be different depending on the instance. To also evaluate this difference properly, the ratio $\frac{OV^{min}}{OV^{max}}$ is used where a small ratio expresses a wide difference while a high ratio indicates a small difference.

After calculating RDI_s for each schedule and problem instance, Table 1 (top) shows the results by computing the cumulative empirical *ARDI* for each problem size and diversity level in three intervals ≤ 0.25, ≤ 0.50 and ≤ 0.75. Moreover, the ratio $\frac{OV^{min}}{OV^{max}}$ is given. It can be seen that $\frac{OV^{min}}{OV^{max}}$ is small but in most cases increases slightly with higher diversity level, i.e. the difference between maximum and minimum is high overall and for both objectives.

Considering $\sum CIT_i$, at least 20% of all schedules refer to $RDI \leq 0.25$, i.e. several schedules which are close to the optimum are provided. Additionally, for problem sizes $n = 5$, $m = [2, 5]$ higher diversity levels lead also to less schedules with $RDI > 0.75$, because the cumulative *ARDI* is relatively high in interval ≤ 0.75. Hence, considering a low diversity level leads to both many schedules close to the optimum but also to the maximum value. Referring to $\sum CWT_j$, a small diversity-level leads to more schedules yielding $RDI > 0.25$ and only a few schedules close to the optimum, i.e. here an increased diversity provides more schedules close to the optimum.

Observing the average RDI for the different diversity levels, Table 1 (bottom), it can be seen that when considering $\sum CIT_i$ the ARDI-values only differ in a small range with small cv, while the differences for $\sum CWT_j$ are significantly higher, with increased cv for interval $ARDI \leq 0.25$.

Exemplarily, we discuss the problem size $n = 10$ and $m = 10$ and objective $\sum CWT_j$ in detail, see Fig. 1, since the results also hold for most of the other problem sizes. Here, the influence of different diversity levels on $\sum CWT_j$ by plotting the empirical ARDI is shown. The RDI interval is represented on the x-axis, while the y-axis refers to the frequency of RDI on average. It can be seen that a low diversity level ($d = 0.1$) leads to a majority of schedules with $RDI \geq 0.70$, i.e. close to the maximum value of $\sum CWT_j$. An increase of diversity ($d = 0.5$, $d = 1.0$) shifts the empirical RDI to the left side. The effect is stronger for $d = 1.0$ than $d = 0.5$. For $d = 1.0$, the majority of schedules are between $0.30 \leq RDI \leq 0.55$. This example shows that a small diversity does not lead to schedules with objective

Table 1 Summary of empirical ARDI referring to different diversity levels

			$\sum CIT_i$							$\sum CWT_j$						
			ARDI						$\frac{OV^{min}}{OV^{max}}$	*ARDI*						$\frac{OV^{min}}{OV^{max}}$
n	m	d	≤ 0.25		≤ 0.50		≤ 0.75			≤ 0.25		≤ 0.50		≤ 0.75		
5	2	0.1	0.37		0.55		0.75		0.00	0.17		0.46		0.75		0.01
		0.5	0.36		0.56		0.78		0.04	0.27		0.61		0.85		0.03
		1.0	0.49		0.66		0.85		0.05	0.33		0.63		0.85		0.03
	5	0.1	0.32		0.51		0.67		0.00	0.13		0.38		0.66		0.00
		0.5	0.34		0.57		0.75		0.06	0.16		0.46		0.79		0.05
		1.0	0.32		0.50		0.76		0.11	0.21		0.52		0.80		0.07
	10	0.1	0.34		0.50		0.69		0.01	0.12		0.34		0.63		0.01
		0.5	0.35		0.53		0.71		0.07	0.14		0.40		0.71		0.04
		1.0	0.29		0.43		0.67		0.09	0.17		0.48		0.78		0.05
10	2	0.1	0.35		0.58		0.85		0.00	0.04		0.39		0.88		0.01
		0.5	0.43		0.75		0.95		0.01	0.21		0.75		0.97		0.03
		1.0	0.48		0.73		0.91		0.01	0.29		0.74		0.96		0.02
	5	0.1	0.31		0.53		0.76		0.01	0.02		0.22		0.68		0.01
		0.5	0.25		0.60		0.91		0.06	0.07		0.56		0.94		0.04
		1.0	0.18		0.54		0.91		0.08	0.15		0.68		0.96		0.05
	10	0.1	0.29		0.48		0.72		0.01	0.01		0.16		0.55		0.01
		0.5	0.26		0.54		0.85		0.07	0.03		0.36		0.86		0.04
		1.0	0.20		0.56		0.93		0.11	0.09		0.57		0.93		0.07
			≤ 0.25	*cv*	≤ 0.50	*cv*	≤ 0.75	*cv*		≤ 0.25	*cv*	≤ 0.50	*cv*	≤ 0.75	*cv*	
		0.1	0.33	0.08	0.52	0.06	0.74	0.08		0.08	0.74	0.33	0.31	0.69	0.15	
		0.5	0.33	0.18	0.59	0.12	0.83	0.11		0.15	0.55	0.53	0.25	0.85	0.10	
		1.0	0.33	0.37	0.57	0.17	0.84	0.11		0.21	0.41	0.60	0.15	0.88	0.08	

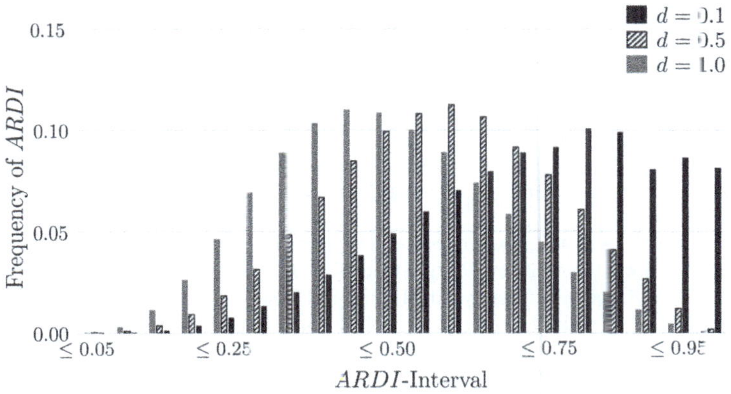

Fig. 1 Empirical distribution of $\sum CWT_j$ for each job-diversity level

values close to the optimum but the maximum, i.e. increasing the diversity of processing times provided more schedules close to the optimum.

3 Conclusion

We analyzed the permutation flow shop problem with structured processing times and varying diversity-levels and the rarely discussed objectives $\sum CIT_i$ and $\sum CWT_j$. We showed that both objectives are highly influenced by different diversity levels. 540 problem instances were generated and solved by complete enumeration. The results show that the minimization of $\sum CWT_j$ provides more schedules close to the optimum when diversity increases. Considering $\sum CIT_i$, it can be concluded that all cases provide several schedules with $ARDI \leq 0.25$. However, small diversity also leads to several schedules close to the maximum. Although the computational experiment refers only to small problem sizes, the influence of diversity of processing times is different for $\sum CIT_i$ and $\sum CWT_j$. Further research should focus on both, other processing time structures and larger problem sizes.

References

1. Benkel, K., Jørnsten, K., Leisten, R.: Variability aspects in flowshop scheduling systems. In: International Conference on Industrial Engineering and Systems Management (IESM) (2015), pp. 118–127
2. De Matta, R.: Minimizing the total waiting time of intermediate products in a manufacturing process. Int. Trans. Oper. Res. **26**(3), 1096–1117 (2019)
3. Fernandez-Viagas, V., Framinan, J.M.: Reduction of permutation flowshop problems to single machine problems using machine dominance relations. Comput. Oper. Res. **77**, 96–110 (2017)
4. Framinan, J.M., Perez-Gonzalez, P.: On heuristic solutions for the stochastic flowshop scheduling problem. Eur. J. Oper. Res. **246**(2), 413–420 (2015)
5. Graham, R.L., Lawler, E.L., Lenstra, J.K., Kan, A.R.: Optimization and approximation in deterministic sequencing and scheduling: a survey. Ann. Discret. Math. **5**, 287–326 (1979)
6. Juan, A.A., Barrios, B.B., Vallada, E., Riera, D., Jorba, J.: A simheuristic algorithm for solving the permutation flow shop problem with stochastic processing times. Simul. Model. Pract. Theory **46**, 101–117 (2014)
7. Liu, W., Jin, Y., Price, M.: A new heuristic to minimize system idle time for flowshop scheduling. In: Poster presented at the 3rd Annual EPSRC Manufacturing the Future Conference, Glassgow (2014)
8. Monma, C.L., Kan, A.R.: A concise survey of efficiently solvable special cases of the permutation flow-shop problem. RAIRO Oper. Res. **17**(2), 105–119 (1983)
9. Pinedo, M.L.: Scheduling: Theory, Algorithms, and Systems. Springer (2016)
10. Reeves, C.R.: A genetic algorithm for flowshop sequencing. Comput. Oper. Res. **22**(1), 5–13 (1995)
11. Rinnooy Kan, A.H.G.: Machine Scheduling Problems: Classification, Complexity and Computation. Martinus Nijhoff, The Hague (1976)
12. Schollenberger, H.: Analyse und Verbesserung der Arbeitsabläufe in Betrieben der Reparaturlackierung, Univ.-Verlag Karlsruhe (2006)
13. Taillard, E.: Benchmarks for basic scheduling problems. Eur. J. Oper. Res. **64**(2), 278–285 (1993)

14. Vallada, E., Ruiz, R., Framinan, J.M.: New hard benchmark for flowshop scheduling problems minimising makespan. Eur. J. Oper. Res. **240**(3), 666–677 (2015)
15. Watson, J.P., Barbulescu, L., Whitley, L.D., Howe, A.E.:. Contrasting structured and random permutation flow-shop scheduling problems: search-space topology and algorithm performance. INFORMS J. Comput. **14**(2), 98–123 (2002)

Proactive Strategies for Soccer League Timetabling

Xiajie Yi and Dries Goossens

Abstract Due to unexpected events (e.g. bad weather conditions), soccer league schedules cannot always be played as announced before the start of the season. This paper aims to mitigate the impact of uncertainty on the quality of soccer league schedules. Breaks and cancellations are selected as two quality measures. Three proactive policies are proposed to deal with postponed matches. These policies determine where to insert so-called *catch-up rounds* as buffers in the schedule, to which postponed matches can be rescheduled.

Keywords Soccer schedule · Uncertainty · Breaks · Cancellations · Proactive strategy

1 Introduction

Each soccer competition has a schedule that indicates a venue and a date for each team. Despite hard efforts invested to create a high-quality *initial schedule* before the season starts, it is not always fully played as planned. Several months can span between the time the initial schedule is published and the moment that matches are actually played. During this period, additional information becomes available (e.g. weather conditions, technical problems, political events, etc.), which may affect the implementation of the initial schedule. It may lead to the pause postponement, or even cancellation of a match. The schedule that effectively represents the way the competition is played is called the *realized schedule*, which is known only at the end of the season.

X. Yi · D. Goossens (✉)
Department of Business Informatics and Operations Management, Ghent University, Ghent, Belgium
e-mail: xiajie.yi@ugent.be

J. S. Neufeld et al. (eds.), *Operations Research Proceedings 2019*, Operations Research Proceedings, https://doi.org/10.1007/978-3-030-48439-2_68

We measure the quality of a schedule in terms of breaks (i.e., two consecutive games of the same team with unchanged home advantage) and cancellations, which are ideally minimized. Uncertain events have a profound impact on the quality of the realized schedules. To mitigate this impact, proactive scheduling approaches are developed. Proactive scheduling focuses on developing an initial schedule that anticipates the realization of unpredicted events during the season. We do this by inserting *catch-up rounds* as buffers in the schedule. Following common practice, we assume that matches that cannot be played as planned are postponed to the earliest catch-up round or cancelled if no such round exists.

Despite numerous contributions on sport scheduling (e.g. [1, 2]), as far as we are aware, the issue of dealing with uncertainty has not been studied before. However, successful applications of proactive approaches can be found in various domains as project scheduling [3, 4] and [7], berth and quay crane scheduling [5], inventory systems [6], etc.

The rest of our paper unfolds as follows. Section 2 sets the stage by introducing basic notions of soccer scheduling. Section 3 introduces two quality measures to evaluate soccer schedules. Three proactive policies and one reactive policy are proposed in Sect. 4. Corresponding simulation results are illustrated in Sect. 5, and we conclude in Sect. 6.

2 Setting the Stage

Throughout this paper, we consider only soccer leagues that have an even number of teams and we denote by $T = \{1, 2, \ldots, 2n\}$ the set of teams. A *match* (or *game*) is an ordered pair of teams, with a home team playing at its own venue and the other team playing away. A *round* is a set of games, usually played in the same weekend, in which every team plays at most once. We denote the set of rounds by $R = \{1, 2, \ldots, r\}$.

Most soccer leagues play a double round robin tournament (DRR), in which the teams meet twice (once at home, once away). A mirrored scheme is commonly used, i.e., the second part of the competition is identical to the first one, except that the home advantage is inverted [8]. A schedule is *compact* if it uses the minimum number of rounds required to schedule all the games; otherwise it is *relaxed*. In a compact schedule with an even number of teams, each team plays exactly once per round; we assume this setting in this paper. The sequence of home matches ('H') and away matches ('A') played by a single team is called its *home-away pattern* (HAP).

Many of the theoretical results and algorithms in sport scheduling are based on graph theory. A compact schedule can then be seen as a one-factorization of the complete graph K_{2n}. One particular one-factorization results in so-called *canonical schedules*, which are highly popular in sport timetabling [9], defined as

$$F_i = \{(2n, i)\} \cup \{(i + k, i - k) : k = 1, ..., n - 1\} \qquad (1)$$

where the numbers $i+k$ and $i-k$ are expressed as one of the numbers 1, 2, ..., $2n-1$ (mod $2n-1$) [10]. For an overview of graph-based models in sports timetabling, we refer to the work by [11].

Games that are postponed or rescheduled in a way that they induce deviations from the initial schedule are labelled *disruptions*. We formally define a disruption as follows:

Definition 1 Given an initial schedule, if a game m of round r was played after at least one game of round $r+1$ or played before at least one game of round $r-1$ in the realized schedule, we say that there is a disruption. We call the game m a disrupted game.

Each disruption will by definition create a difference between the initial and the realized schedule, however, the converse is not necessarily true. Note that a game which is not played as initially scheduled, but rescheduled (and played) before any game of the next round is played and after all the games of the previous rounds, is not considered as a disruption because the order of the games remains the same and it has no impact on the quality of the schedule. Also notice that, in this paper, we only consider rescheduling a disruption to a catch-up round which is scheduled later than its original round.

3 Quality Measures for Soccer Schedules

We opt for breaks and cancellations as quality measures in this paper, since they are easy to understand and applicable to any soccer league.

The occurrence of two consecutive home (away) matches for a team is called a *break*. Teams can have consecutive breaks, causing them to play three or more home (away) games in a row. Ideally, each team has a perfect alternation of home and away games. It is easy to see that only two different patterns without breaks exist (HAHA...H and AHAH...A), and hence, at most two teams will not have any break. Scheduling consecutive home games has a negative impact on attendance [12]. As a result, in most competitions, breaks are avoided as much as possible.

Normally, all teams should have played an equal number of games at the end of the season in round-robin leagues. However, not every disrupted match can be rescheduled successfully. Indeed, UEFA regulations prescribe that teams should have at least two rest days between consecutive matches (i.e., a team that plays on Thursday cannot play again until Sunday at the earliest). Furthermore, police may forbid the use of certain dates for rescheduling high-risk matches. Hence, it may happen that none of the remaining dates in the season are suitable for rescheduling a match of a given team, particularly if that team already faces a number of postponed games and/or a busy schedule in the domestic cup or European competitions. We call a match a *cancellation* if it cannot be played because no suitable date is available on which it can be rescheduled.

4 Proactive Policies

We study several proactive policies to mitigate the negative effects of disruptions on the quality of soccer schedules, which is evaluated based on the before-mentioned quality measures: breaks and cancellations.

Our proactive policies try to anticipate the realization of unforeseen events by inserting catch-up rounds as buffers into the initial schedule. Recall that catch-up rounds are empty rounds to which disrupted matches can be rescheduled. There are three proactive policies that we take into consideration: (i) spread catch-up rounds equally over the season (PS); (ii) spread catch-up rounds equally over the second half of the season (P2); (iii) position all catch-up rounds near the end of the season (PE).

We assume that consecutive catch-up rounds are not allowed (we haven't seen evidence of such practices in reality). Note that each of our proactive policies puts a catch-up round at the end of the season to make sure that disruptions happening in the final round would not automatically lead to a cancellation.

After each round, we schedule the disrupted matches of that round to the earliest available catch-up round. Note that disruptions, by definition, cannot be rescheduled to a catch-up round immediately following its original round (except when this happens in the last round of the season).

5 Simulations and Discussions

5.1 Settings

Motivated by the frequent occurrence of this competition format in professional European soccer leagues [8], we consider a mirrored DRR tournament with 20 teams, played using a compact, phased schedule, which requires 38 rounds. Moreover, a canonical schedule based on Eq. (1) is used as the initial schedule, determining for each round which team plays against which opponent. According to an empirical study of ten main European Leagues provided by Talla Nobibon and Goossens [13], on average, at most around 3% matches are disrupted in a season. Thus, we consider a setting with 12 disruptions and four available catch-up rounds, leading to 42 rounds in total. Table 1 shows where each of the proactive policies positions the catch-up rounds.

Table 1 Position of 4 catch-up rounds for each proactive policy

Proactive policy	c_1	c_2	c_3	c_4
PS	11	21	31	42
P2	20	27	34	42
PE	39	40	41	42

Table 2 Average results per season of combined proactive and reactive policies

Policy	PG_breaks	Cancellations
PS	0.217	1.252
P2	0.216	0.485
PE	0.210	0.08

In general, a cancellation will create fewer breaks than a rescheduled match. Indeed, while rescheduling a match has a double impact on the home-away patterns of the involved teams (around the initial round and the catch-up round to which it is rescheduled), cancelling a match only changes the home-away pattern around the initial round. Consequently, in this simulation study, we use breaks per played game (*PG_breaks*) as an alternative measure:

$$PGbreaks = \frac{breaks}{total\ number\ of\ games - number\ of\ cancellations}. \quad (2)$$

5.2 Results

Table 2 shows the simulation results on the average number of breaks and cancellations per season.

It can be seen that PE has the best performance in terms of avoiding cancellations, and also surprisingly excels with respect to the PG_breaks. Putting all catch-up rounds at the end of the season indeed gives more opportunities for rescheduling disrupted games than positioning them throughout the season, which can reduce the number of existing breaks to some extent and the PG_breaks decreases accordingly. However it is rather rare to implement this policy in reality since it disturbs the ranking in the sense that teams have played a different number of games throughout most of the season. The P2 and the PS policy have similar results in terms of PG_breaks, but P2 works much better with respect to cancellations. The PS policy does not show any advantages in the light of those two quality measures, however, this policy can reschedule disruptions that occur at the very beginning of the season sooner than other proactive policies; based on this point, we can argue that the PS policy shows more fairness.

6 Conclusions

We propose three proactive policies in order to mitigate the impact of disrupted matches due to uncertain events on the quality of soccer schedules, in which the P2 policy can be viewed as a fair policy with adequate performance. Our policies and quality measures can also be applied to other sports that play according to a compact round robin tournament.

In general, equally spreading all catch-up rounds in the second half of the season (P2) can avoid more cancellations than equally spreading them throughout the whole season (PS). However, putting all catch-up rounds at the end of the season (PE) can perform well in terms of both breaks and avoiding cancellations when ignoring the fair ranking issue.

References

1. Kendall, G., Knust, S., Ribeiro, C.C., Urrutia, S.: Scheduling in sports: An annotated bibliography. Comput. Oper. Res. **37**, 1–9 (2010)
2. Rasmussen, R.V.: Scheduling a triple round robin tournament for the best Danish soccer league. Eur. J. Oper. Res. **185**, 795–810 (2008)
3. Lambrechts, O., Demeulemeester, E., Herroelen, W.: Proactive and reactive strategies for resource-constrained project scheduling with uncertain resource availabilities. J. Sched. **11**, 121–136 (2008)
4. Van de Vonder, S., Demeulemeester, E., Herroelen, W.: Proactive heuristic procedures for robust project scheduling: An experimental analysis. Eur. J. Oper. Res. **189**, 723–733 (2008)
5. Lu, Z., Xi, L.: A proactive approach for simultaneous berth and quay crane scheduling problem with stochastic arrival and handling time. Eur. J. Oper. Res. **207**, 1327–1340 (2010)
6. Seidscher, A., Minner, S.: A Semi-Markov decision problem for proactive and reactive transshipments between multiple warehouses. Eur. J. Oper. Res. **230**, 42–52 (2013)
7. Herroelen, W., Leus, R.: Project scheduling under uncertainty: Survey and research potentials. Eur. J. Oper. Res. **165**, 289–306 (2005)
8. Goossens, D.R., Spieksma, F.C.: Soccer schedules in Europe: an overview. J. Sched. **15**, 641–651 (2012)
9. Goossens, D., Spieksma, F.: Scheduling the Belgian soccer league. Interfaces **39**, 109–118 (2009)
10. De Werra, D.: Scheduling in sports. In: Hansen, P. (ed.) Studies on Graphs and Discrete Programming, pp. 381–395. North-Holland, Amsterdam (1981)
11. Drexl, A., Knust, S.: Sports league scheduling: graph-and resource-based models. Omega **35**(5), 465–71 (2007)
12. Forrest, D., Simmons, R.: New issues in attendance demand: The case of the English football league. J. Sports Econ. **7**, 247–266 (2006)
13. Talla Nobibon, F., Goossens, D.: Are soccer schedules robust? In: 4th International Conference on Mathematics in Sport, pp. 120–134. Leuven (2013)

Constructive Heuristics in Hybrid Flow Shop Scheduling with Unrelated Machines and Setup Times

Andreas Hipp and Jutta Geldermann

Abstract Hybrid flow shop (HFS) systems represent the typical flow shop production system with parallel machines on at least one stage of operation. This paper considers unrelated machines and anticipatory sequence-dependent setup times where job families can be formed based on similar setup characteristics. This results in the opportunity to save setups if two jobs of the same family are scheduled consecutively. Three constructive heuristic approaches, aiming at minimization of makespan, total completion time and the total number of setup procedures, are implemented based on the algorithm of Nawaz, Enscore and Ham (NEH).

Keywords Hybrid flow shop scheduling · Unrelated machines · Setup times · Constructive heuristics

1 Introduction

Flow shop scheduling with parallel machines on at least one stage, i.e. the hybrid flow shop (HFS), poses a complex combinational problem commonly found in industries such as semi-conductor production or chemical industry. In the steel industry e.g., the processing of sheets can be modelled as a HFS with setup times. For instance, steel sheets with different characteristics which are processed in the same system lead to different job sequence-dependent machine setups. Furthermore, the processing times of sheets with similar characteristics can differ on each machine. In this paper, we focus on such a HFS with unrelated machines and anticipatory sequence-dependent setup times. Job families can be formed based on similar setup characteristics which provides the opportunity to save setup procedures if two jobs of the same family are scheduled subsequently. By taking

A. Hipp (✉) · J. Geldermann
Chair of Business Administration and Production Management, University of Duisburg-Essen, Duisburg, Germany
e-mail: andreas.hipp@uni-due.de

J. S. Neufeld et al. (eds.), *Operations Research Proceedings 2019*, Operations Research Proceedings, https://doi.org/10.1007/978-3-030-48439-2_69

into account anticipatory setups, it is possible to start the setup procedure of a machine as soon as the decision is taken which machine processes which job. Compared to non-anticipatory setups, the start of the setup procedure can take place before the respective job reaches the machine. According to the three field notation of Graham et al. [4] the model under consideration can be classified as $FHm((RM^k)_{k=1}^m|ST_{sd,f}|C_{max} \vee FT \vee ST)$. On each stage k of the m-stage HFS system (FHm) a set RM k of non-identical machines is available. Anticipatory job sequence-dependent setup times $(ST_{sd,f})$ including different job families f are assumed. The objective is either the minimization of makespan (C_{max}), flowtime (FT) or setups (ST). As jobs are assumed to be available at time zero, flowtime minimization equals minimizing the sum of jobs completion times at the last stage. The HFS scheduling problem including unrelated machines is proven to be NP-complete for minimizing C_{max} [6]. Compared to C_{max}, FT is a more complex objective function and even the two-stage flow shop problem is known to be NP-complete [4]. The problem $FHm((RM^k)_{k=1}^m|ST_{sd,f}|)$ also includes the special case of setup times equal to zero so that NP-completeness can be assumed. This makes it necessary to develop approximation algorithms to solve realistic problem instances. In this paper, a HFS is modelled with unrelated machines, setup times and job families. In contrast to current literature, no batch sizes are predefined in connection with job families. The well-performing MCH heuristic of Fernandez-Viagas et al. [3] for HFS problems with identical machines, which is based on the algorithm of Nawaz, Enscore and Ham (NEH), is modified. The influence of the total number of setups and setup times are examined with regard to the performance measures. Batch sizes are not fixed to not restrict the performance of the heuristics in advance.

In Sect. 2, a brief literature review is carried out. After that, the characteristics of the examined models are presented and the heuristics based on the MCH heuristic of Fernandez-Viagas et al. [3] are proposed in Sect. 3. Finally, the results are presented in Sect. 4 before the paper is summarized in Sect. 5.

2 Literature Review

The last comprehensive reviews for HFS literature are from 2010 and presented by Ruiz and Vázquez-Rodríguez [14] as well as Ribas et al. [12] which classify papers according to their covered constraints and objective functions. In recent studies, NEH based algorithms [10] have been implemented among other solution approaches like meta-heuristics. NEH provides high quality solutions for flow shop scheduling problems, but has to be modified for HFS problems. For representative mixed integer linear programming models (MILP) in HFS scheduling with unrelated machines, see those of Naderi et al. [8] and Ruiz et al. [13]. A detailed survey on scheduling problems with setup times distinguishing between production and setup type is given by Allahverdi [1]. Focused on NEH-based procedures, Jungwattanakit et al. [5] use several algorithms like NEH to generate high quality initial solutions for solving a dual criteria HFS with unrelated machines and setup times which are afterwards improved by meta-heuristics. A complex m-stage HFS is solved by Ruiz et al. [13] including several constraints, inter alia setup times, with the objective of flowtime minimization. In addition to a MILP, dispatching rules and NEH algorithm are implemented and compared. Simulated annealing is implemented by Naderi et al. [9] to deal with a similar production layout to minimize flowtime and by Mirsanei et al. [7] for makespan minimization. Setup times with job families in HFS scheduling are considered by Ebrahimi et al. [2]. Shahvari and Logendran [15] focuse on batch scheduling to deal with setup times, job families and simultaneous minimization of weighted flowtime and weighted tardiness including a population based learning effect.

These studies focuse either on setup times without family characteristics or on group technology [12] with fixed batch sizes. Therefore, constructive heuristics are formulated for the case of free batching without any fixed setting in this paper.

3 Heuristic Approaches for HFS with Unrelated Machines and Setup Times

The HFS scheduling problem $FHm((RM^k)_{k=1}^m|ST_{sd,f}|C_{max} \vee FT \vee ST)$ is examined with the objectives to minimize makespan C_{max}, flowtime FT or number of setups ST. Setup times $ST_{sd,f}$ are considered which can be executed in advance (anticipatory) and job families classified by similar setup characteristics are given. In the following, two essential specifics, namely unrelated machines and setup times, are described in detail.

For scheduling unrelated machines for the HFS in contrast to identical ones, not only sequencing of jobs, but also assigning them to a specific machine i per stage k is relevant. This results in job processing times p_{kij} not only referring to stage k and job j but also to machine i. Following Rashidi et al. [11], machine speed v_{kij} is considered to relate a norm processing time p_{kj} for each job and each stage to each machine i so that either instances for HFS with identical and unrelated machines can be adopted.

Regarding setup times st_{kfg}, sequence-dependence and anticipation of setups have to be distinguished. Sequence-dependence represents the influence of the job sequence to the total number of needed setup procedures. For different job families f, it is possible to schedule subsequent jobs of the same family f to save setups. Only subsequent jobs of different job families f and g cause setup procedures. Because of anticipatory setups, the setup procedure for job j on a machine at stage k can already start as soon as the operation of job j starts on a machine at stage $k-1$. Thus potential idle times in front of the machine at stage k can be used to execute the needed setup procedure, decreasing the waiting time of job j in front of this machine. In total, three constructive heuristics are formulated, H1, H2 and H3, one for each objective function ST, C_{max} and FT. The memory-based constructive heuristic (MCH) of Fernandez-Viagas et al. [3] which provides high quality solutions for HFS problems with identical machines is modified for the mentioned problem at hand. Like in the NEH algorithm, the job sequence is built by insertion. In each iteration, one job is added to the sub-sequence on every possible position of the sequence. In each iteration, notional completion times including setup times are calculated by summing average values for jobs' processing times per each stage and all jobs which are not sequenced so far. In addition, a memory list saves the sequences and values of the former iteration and compares them with the current ones in order to determine the best fitting sub-sequence in the current iteration. The jobs are assigned to a production stage following the earliest completion time rule (ECT) [3].

4 Computational Study

All subsequent calculations are coded in MATLAB R2016a in an Intel Pentium 3805U with 1.9 GHz, 4 GB RAM. The examined benchmarks in Table 1 are provided by the Spanish research group Sistemas de Optimización Aplicata (SOA) for HFS with unrelated machines and non-family setup times and are adapted for this paper by additionally generating values for setup categories. 144 combinations of small instances are generated and 48 combinations of large ones, each five times so that 960 instances are generated in total.

Because small and large instances show similar performance, only the results of large instances are shown in the following. Table 2 shows the total number of setup procedures according to family sizes F over all numbers of jobs n. Heuristic H1 which pursues the minimization of setups (ST) provides the lowest number of setup procedures for large instances. Heuristic H2 minimizing C_{max} and heuristic H3 minimizing FT provide values for setup procedures in a similar range. Compared to H1, the total number of setups given by H2 and H3 increase around 100 percent or 45 percent (H1 vs. H2: 120 vs. 256.8; 240 vs. 345.5).

Figure 1 shows the range of makespan values provided by all three heuristics H1, H2 and H3 for 50 and 100 jobs over all job families F (the same scheme is valid for flowtime). Even though, only heuristic H2 targets makespan minimization, the values for makespan of all heuristics are in a similar range. Consequently, the high increase of setup procedures seen in Table 2 between heuristic H1 and H2

Table 1 Overview of examined benchmark

Parameter	Small instances	Large instances
Number of jobs n	5, 7, 9, 11, 13, 15	50, 100
Processing stages m	2, 3	4, 8
Number of machines per stage m_k	1, 3	2, 4
Processing times p_{kij}	U[1,99]	U[1,99]
Setup times s_{kgf}	U[25,75]	U[75,125]
Number of setup categories F	0, 2, 4	0, 20, 40

Table 2 Number of setups for large instances

		H1	ST	H2	C_{max}	H3	FT
Stage m	m_k \ Family f	20	40	20	40	20	40
4	2	80.0	160.0	149.4	215.9	153.3	223.8
	4	80.0	160.0	186.9	240.6	193.3	243.9
8	2	160.0	320.0	305.4	440.3	302.4	449.1
	4	160.0	320.0	385.3	485.1	388.9	489.9
	Average	120.0	240.0	256.8	345.5	259.5	351.7

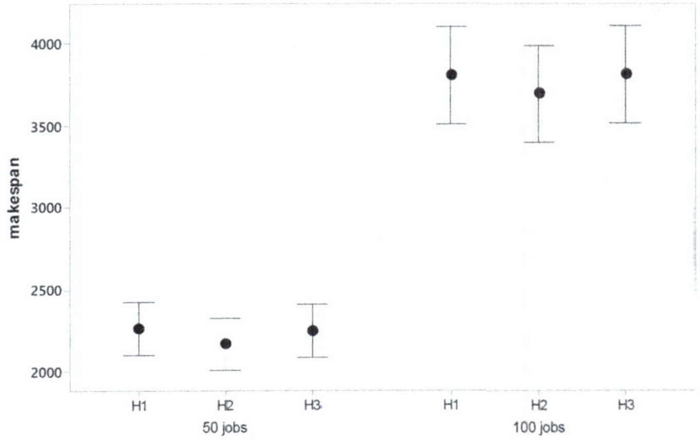

Fig. 1 Range of solutions for makespan for large instances

respectively H3 along with increasing setup times does not automatically result in higher jobs' completion times. This can be explained by examining the idle times in the schedules. Because setups can be executed before the jobs enter the machine (character of anticipatory setups), the higher total number of setup procedures does not necessarily impact the performance negatively. The setups can be executed in the idle times of the machines so that the higher number of setups in the schedules given by heuristics H2 and H3 do not result in higher completion times of jobs.

5 Conclusion

In literature, HFS scheduling with setup times and job families is typically combined with fixed batch sizes. In this work, the influence of setups on job completion time related performance measures is examined by considering the case of free batching. Three constructive heuristics based on the NEH algorithm [3] are compiled to solve the m-stage HFS with unrelated machines and anticipatory sequence-dependent setup times to minimize makespan, flowtime and the total number of setups. In addition, job families based on setup characteristics are defined. The implemented heuristics, each for one objective function, are applied on a testbed of 960 instances in total. It can be shown exemplary that a increasing number of setups does not necessarily result in increasing makespan and flowtime if anticipatory setups are considered. In future work, the performance of the applied heuristics should be evaluated by comparing them to other approximation algorithms as well as other machine assignment rules than Earliest Completion Time (ECT).

References

1. Allahverdi, A.: The third comprehensive survey on scheduling problems with setup times/costs. Eur. J. Oper. Res. **246**(2), 345–378 (2015)
2. Ebrahimi, M., Ghomi, S.M.T.F., Karimi, B.: Hybrid flow shop scheduling with sequence dependent family setup time and uncertain due dates. Appl. Math. Model. **38**(9–10), 2490–2504 (2014)
3. Fernandez-Viagas, V., Molina-Pariente, J.M., Framiñan, J.M.: New efficient constructive heuristics for the hybrid flowshop to minimise makespan: A computational evaluation of heuristics. Expert Syst. Appl. **114**, 345–356 (2018)
4. Graham, R.L., Lawler, E.L., Lenstra, J.K., Rinnooy Kan, A.H.G.: Optimization and approximation in deterministic sequencing and scheduling: a survey. Ann. Discret. Math. **5**, 287–326 (1979)
5. Jungwattanakit, J., Reodecha, M., Chaovalitwongse, P., Werner, F.: An evaluation of sequencing heuristics for flexible flowshop scheduling problems with unrelated parallel machines and dual criteria. Otto-von-Guericke-Universitat Magdeburg **28**(5), 1–23 (2005)
6. Lee, C.-Y., Vairaktarakis, G.L.: Minimizing makespan in hybrid flowshops. Oper. Res. Lett. **16**(3), 149–158 (1994)
7. Mirsanei, H.S., Zandieh, M., Moayed, M.J., Khabbazi, M.R.: A simulated annealing algorithm approach to hybrid flow shop scheduling with sequence-dependent setup times. J. Intell. Manuf. **22**(6), 965–978 (2011)
8. Naderi, B., Gohari, S., Yazdani, M.: Hybrid flexible flowshop problems: Models and solution methods. Appl. Math. Model. **38**(24), 5767–5780 (2014)
9. Naderi, B., Zandieh, M., Balagh, A.K.G., Roshanaei, V.: An improved simulated annealing for hybrid flowshops with sequence-dependent setup and transportation times to minimize total completion time and total tardiness. Expert Syst. Appl. **36**(6), 9625–9633 (2009)
10. Nawaz, M., Enscore, E.E., Ham, I.: A heuristic algorithm for the m-machine, n-job flow-shop sequencing problem. Omega **11**(1), 91–95 (1983)
11. Rashidi, E., Jahandar, M., Zandieh, M.: An improved hybrid multi-objective parallel genetic algorithm for hybrid flow shop scheduling with unrelated parallel machines. Int. J. Adv. Manuf. Technol. **49**(9–12), 1129–1139 (2010)
12. Ribas, I., Leisten, R., Framiñan, J.M.: Review and classification of hybrid flow shop scheduling problems from a production system and a solutions procedure perspective. Comput. Oper. Res. **37**(8), 1439–1454 (2010)
13. Ruiz, R., Şerifoğlu, F.S., Urlings, T.: Modeling realistic hybrid flexible flowshop scheduling problems. Comput. Oper. Res. **35**(4), 1151–1175 (2008)
14. Ruiz, R., Vázquez-Rodríguez, J.A.: The hybrid flow shop scheduling problem. Eur. J. Oper. Res. **205**(1), 1–18 (2010)
15. Shahvari, O., Logendran, R.: A comparison of two stage-based hybrid algorithms for a batch scheduling problem in hybrid flow shop with learning effect. Int. J. Prod. Econ. **195**, 227–248 (2018)

A Heuristic Approach for the Multi-Project Scheduling Problem with Resource Transition Constraints

Markus Berg, Tobias Fischer, and Sebastian Velten

Abstract A resource transition constraint models sequence dependent setup costs between activities on the same resource. In this work, we propose a heuristic for the multi-project scheduling problem with resource transition constraints, which relies on constraint programming and local search methods. The objective is to minimize the project delay, earliness and throughput time, while at the same time reducing setup costs. In computational results, we demonstrate the effectiveness of an implementation based on the presented concepts using instances from practice.

Keywords Project scheduling · Transition constraints · Setup costs

1 Introduction

Project scheduling problems have been the subject of extensive research for many decades. A well-known standard problem is the *Multi-Project Scheduling Problem with Resource Constraints (RCMPSP)*. It involves the issue of determining the starting times of project activities under satisfaction of precedence and resource constraints. As an extension of RCMPSP, we consider the *Multi-Project Scheduling Problem with Resource Transition Constraints (MPSPRTC)*, where setup costs and times depend on the sequence in which activities are processed.

The applications of MPSPRTC are abundant, e.g. in production processes with cleaning, painting or printing operations. In many of these applications, the presence of parallel projects and scarce resources makes scheduling a difficult task. In addition, there is competition between activities for planning time points with lowest

M. Berg
proALPHA Business Solutions GmbH, Weilerbach, Germany
e-mail: markus.berg@proalpha.de

T. Fischer (✉) · S. Velten
Fraunhofer Institute for Industrial Mathematics ITWM, Kaiserslautern, Germany
e-mail: tobias.fischer@itwm.fraunhofer.de; sebastian.velten@itwm.fraunhofer.de

J. S. Neufeld et al. (eds.), *Operations Research Proceedings 2019*, Operations Research Proceedings, https://doi.org/10.1007/978-3-030-48439-2_70

setup costs and times. Often, project due dates can only be met if the activities are scheduled accurately and the available resources are used optimally.

The appearance of resource transitions in project scheduling is already investigated in Krüger and Scholl [1]. They formulate MPSPRTC as an integer program and present a priority-based heuristic with promising computational results. Within their framework, there are no due dates on the projects and the aim is to minimize the average project ends. This goal goes hand in hand with minimizing the sequence-dependent setup times of the activities and therefore there is no multi-objective oriented optimization framework.

In this work, we present a priority-based heuristic for MPSPRTC using constraint programming that we extend by a local search and solution refinement procedure. The proposed algorithm is a multi-criteria approach for determining a good compromise between low setup costs, adherence to project due dates, and short project throughput times. In our model, we restrict us to the case that all setup times are zero and there only exist sequence dependent setup costs.

The algorithm is divided into 3 steps: The first step is to find a practicable initial solution using a constructive heuristic (see Sect. 2.1). The heuristic relies on priority rules that include the goals of the objective functions. Starting from the initial solution, we apply a local search in the second step of our algorithm (see Sect. 2.3), where we try to improve the setup costs by permuting the activity sequences on the setup resources. Finally, in the third step (see Sect. 2.3), we refine the solution calculated in the previous steps by a forward-backward improvement heuristic. The goal is to bring the activities closer to the due dates of the corresponding projects, while keeping the sequence of activities on the setup resources essentially unchanged. In a computational study in Sect. 3, we demonstrate the effectiveness of an implementation based on the presented concepts using instances from practice.

1.1 Problem Definition

We consider a set of projects $P = \{p_1, \ldots, p_m\}$ with due dates $d_1, \ldots, d_m$. Each project $p \in \mathcal{P}$ is composed of a set $\mathcal{A}_p$ of activities with a fixed working time w_a for each $a \in \mathcal{A}_p$. By $\mathcal{A} := \bigcup_{p=1}^{m} \mathcal{A}_p$, we denote the set of all activities. The start and end time points of these activities are variable and should be determined in such a way that all constraints are met and the target criteria are optimized.

The constraints required by MPSPRTC are listed below:

1. *Precedence Constraints:* Each activity can have several predecessors and may not be started until all its predecessors are complete. We assume that precedence constraints only exist between activities of the same project. The precedence constraints of each project $p \in \mathcal{P}$ are represented by the directed edges $\mathcal{E}_p$ of a precedence graph $G_p = (\mathcal{A}_p, \mathcal{E}_p)$.
2. *Resource Requirements:* Let $\mathcal{R}$ be the set of resources, which are assumed to be renewable and to have a varying integer capacity over time. The execution of

every activity $a \in \mathcal{A}$ requires a constant number of resource units u_{ar} of each resource $r \in \mathcal{R}$. By $\{a \in \mathcal{A} : u_{ar} > 0\}$ we denote the set of activities that are assigned to r.
3. *Resource Transition Constraints:* These constraints are used to model sequence dependent setup costs on a resource $r \in \mathcal{R}$. Assume that every activity assigned to r has a certain setup type of a set $\mathcal{T}_r$. Activities of different types may not overlap, i.e, they cannot be allocated in parallel on r. A nonnegative integer matrix M_r of order $|\mathcal{T}_r| \times |\mathcal{T}_r|$ contains the setup costs required to change the setup state of r to another setup state. In most applications, the diagonal entries of M all are 0.

The (multi-)objective is to minimize the delay, earliness and throughput time of the projects, while at the same time reducing setup costs.

2 Scheduling Heuristic for MPSPRTC

In the following sections, we describe the three steps of our scheduling heuristic.

2.1 Initial Planning

We use a greedy heuristic to find a feasible solution for MPSPRTC. The heuristic relies on priority rules taking the characteristics of the problem into account. Projects (respectively, activities) with highest priority are scheduled first towards their specific goals. We define priority rules for the following aspects:

1. *Scheduling sequence of projects:* Projects are scheduled in the order of a user-defined priority. If projects have an equal user-defined priority, then those with the earliest due date are scheduled first.
2. *Scheduling sequence of activities:* Activities of a project are sorted with first priority by the partial order induced by G_p and as second priority by the order of their current latest end time; latest activities are scheduled first.
3. *Point in time to schedule activities:* Activities are scheduled as close as possible to their intended point in time: For non-setup activities, this is the point x that is as close as possible to the due date d of the corresponding project. For setup activities we search in an interval $[x - \delta, x + \delta]$ with $\delta > 0$ for a point that leads to the lowest setup costs w.r.t. the current schedule. If two points lead to the same cost, then we take one that is closest to d.

All precedence constraints $(a, \bar{a}) \in \mathcal{E}_p$, $p \in \mathcal{P}$, in which either a or $\bar{a}$ is a setup activity, are modeled with an additional time offset of ε, i.e., a ends at least ε time units before the start of $\bar{a}$. Here, ε is a parameter that can be selected by the user. The extra time between activities is not necessary for the feasibility of the solution, but beneficial for the performance of the local search in Sect. 2.2, where we need enough freedom to perform iterative permutations of the setup sequences. Then the aim of the solution refinement in Sect. 2.3 is to remove the extra offset from the schedule. Reasonable choices of ε are discussed in Sect. 3.

2.2 Local Search

The algorithm of Sect. 2.1 does not necessarily result in local optimal setup costs—even if the other goals (earliness, tardiness and throughput time) are fixed in their goal levels. To find improved setup costs, we propose a local search procedure that is applied to the activity order of each setup resource.

Starting from the activity sequence S_0 found in the initial planning (Sect. 2.1), we iteratively move from one sequence S_i to another S_{i+1}. The sequence S_{i+1} is selected from a neighborhood set $\mathcal{N}(S_i)$, where we choose the sequence with the best improvement w.r.t. some priority rule R. The feasibility of the current sequence is checked every iteration (or less frequently for further speedup). If an infeasibility is detected, we backtrack to the last feasible sequence and try the next candidate. This process is continued until no improvement occurs anymore or a maximum iteration number is met.

From a theoretical point, the feasibility of a sequence S could be checked by rescheduling all activities subject to the condition that the ordering specified by S must be satisfied. However, since feasibility has to be checked quite often, this can be very time consuming. Therefore, we restrict MPSPRTC to the subproblem where all activities except the ones of S are fixed. Of course, this has a restrictive effect, which is however partly compensated by adding the additional time offset ε to precedence constraints in the initial planning (see Sect. 2.1). Generally it can be said that, the larger we choose ε, the more freedoms we receive for rescheduling the setup activities and it is more likely that a valid sequence can be detected as such. On the other hand, if we choose ε small, then the earliness, tardiness and throughput time of the projects tends to be smaller.

It remains to specify the neighborhood set $\mathcal{N}(S)$ and the priority rule R: Two sequences S and S' are neighbours if S can be transformed into S' by either a pairwise exchange of two sequence elements, or a shift of one sequence element at another position, or a shift of a group of consecutive sequence elements with the same setup type at another position. Moreover, the priority rule R relies on the sum of setup costs and a measure based on the Lehmer mean [3] that prefers large numbers of consecutive occurrences of the same setup type.

2.3 Solution Refinement

In the third step of our method, we refine the solution calculated in the previous steps. This is done by a forward-backward improvement (FBI) heuristic. The goal is to move the activities closer to the due dates of the corresponding projects, while keeping the sequence of the setup activities essentially unchanged.

The FBI heuristic consists of two steps: In the forward step, the activities are considered from left to right (based on the order of the current schedule) and scheduled to their earliest feasible point in time. Similarly, in the backward step, the activities are considered from right to left (according to the order of the forward schedule) and scheduled as close as possible to their due dates. The approach can be repeated multiple times until no improvement occurs anymore.

3 Computational Experience

In this section, we report on computational experience with our implementation of the presented algorithm. The implementation is written in C++ using the framework ILOG CP Optimizer V12.6.2 [2], which allows to express relations between interval variables and cumulative variables in the form of constraints.

We use four different kind of instance classes which arise from rolling horizon ERP data of three customers. The instances of classes 3–4 correspond to the same customer, however different kind of resources are marked as setup resources. Table 1 gives statistics on all four instance classes. Column "#" denotes the number of instances of the given class. Moreover, in arithmetic mean over all instances of each class, we list the number of projects in "Projs", the number of activities in "Acts", the number of precedences in "Precs", the number of resources in "Ress", the number of setup resources in "S-Ress", the number of setup activities in "S-Act", the number of different setup types in "S-Types", and the definition of the setup cost matrices in "M_{ij}". We defined a planning horizon in which setup costs are optimized. The horizon is useful for controlling the complexity of the problem. For our purposes, we decided that a large horizon of 60 days would be appropriate, since the activities can be multiple days long.

Our computational results are organized into two experiments. In the first experiment, we use the default settings of Table 2, but vary the parameter ε from

Table 1 Statistics of the instance classes

Instances	#	Projs	Acts	Precs	Ress	S-Ress	S-Acts	S-Types	M_{ij}
Class 1	6	5287	30818	27591	750	2	1249	153	$0\ (i = j), 1\ (i \neq j)$
Class 2	1	4367	79318	82602	111	1	2490	87	$\|i - j\|$
Class 3	7	8699	67247	66064	919	9	676	317	$0\ (i = j), 1\ (i \neq j)$
Class 4	7	8699	67247	66064	919	2	8366	11	$0\ (i = j), 1\ (i \neq j)$

Table 2 Settings for test runs

Shortcut	Explanation
Default	Default settings (all methods enabled)
Initial-off	Initial planning (Sect. 2.1) is done without taking care of resource transfers
ls-off	Turn local search off (Sect. 2.2)
Refine-off	Turn solution refinement off (Sect. 2.3)
Setup-off	Combination of "ls-off", "Refine-off", and "Initial-off"

Table 3 Experiments on the 4 instance classes (in arithmetic mean)

Setting		Class 1					Class 2				
Shortcut	ε	Costs	thr	earl	tard	Time	Costs	thr	earl	tard	Time
Setup-off	0	0.96	5.66	2.14	2.73	32	31.08	7.55	0.64	11.34	73
Default	1	0.59	5.81	2.23	2.73	310	5.57	7.70	0.58	12.39	568
	2	0.54	5.86	2.29	2.86	295	4.91	7.83	0.64	12.15	630
	4	0.49	5.90	2.39	3.04	281	4.08	8.16	0.78	12.32	534
	6	0.46	5.93	2.57	3.07	266	4.60	8.00	0.84	12.74	625
	8	0.43	5.97	2.57	3.25	167	4.67	8.27	0.87	12.89	600
Initial-off	4	0.59	5.88	2.56	2.91	201	5.30	7.76	0.69	11.14	558
ls-off	4	0.62	5.95	2.36	3.09	135	18.21	8.14	0.68	12.35	442
Refine-off	4	0.47	6.16	2.29	3.34	82	3.52	8.20	0.82	13.03	180
Setting		Class 3					Class 4				
Shortcut	ε	Costs	thr	earl	tard	Time	Costs	thr	earl	tard	Time
Setup-off	0	0.87	6.61	0.64	9.69	142	0.388	6.61	0.64	9.69	151
Default	1	0.77	6.57	0.75	9.51	507	0.041	6.89	0.87	9.84	1178
	2	0.76	6.57	0.76	9.55	496	0.036	7.17	0.96	10.21	1081
	4	0.73	6.59	0.77	9.64	501	0.032	7.53	1.12	10.75	1036
	6	0.71	6.61	0.79	9.71	503	0.033	7.69	1.22	11.16	1086
	8	0.68	6.63	0.81	9.78	491	0.033	7.94	1.33	11.64	1142
Initial-off	4	0.77	6.61	0.69	9.67	493	0.048	7.29	1.02	10.76	1169
ls-off	4	0.79	6.59	0.73	9.79	491	0.132	7.63	0.82	11.14	591
Refine-off	4	0.72	6.66	6.70	9.93	145	0.025	8.53	1.04	12.03	627

Sect. 2.1 between 1 and 8 days. In the second experiment, we evaluate the effect of the three basic steps (Sects. 2.1–2.3) of our algorithm by switching different components off. The data of Table 3 shows aggregated results of these experiments. For each instance class, the table reports on the mean number of setup costs per setup activity (column "Costs"), the mean number of days of the throughput time, earliness and tardiness per project (columns "thr", "earl" and "tard"), and the CPU time in seconds (column "Time").

The results can be summarized as follows: A larger value for ε tends to result in lower setup costs. The reason is that ε controls the degrees of freedom we get for optimizing the setup sequences in the local search. On the other hand, if we choose ε small, we obtain better values for the earliness, tardiness and throughput time of the projects. This reflects the fact that minimal setup costs and optimal project dates are contrary targets. Comparing the default settings with "Setup-off", it turns out that our algorithm is able to significantly improve the setup costs without worsening the project statistics too much. On the other hand, the solution time for optimizing setup costs is significantly increased.

We proceed with an evaluation of the three basic steps of our algorithm, see rows "Initial-off", "ls-off" and "Refine-off". For these test runs, we decided to choose $\varepsilon = 4$, since this was a good trade-off in our last experiment. If setup cost optimization is deactivated either in the initial planning or the local search step, then we observe a significant deterioration in this goal, however for the sake of the project statistics. Moreover, solution refinement turns out to successfully remove a lot of free space from the schedule, which originates from the additional precedence offset ε in the initial planning. This is demonstrated by the fact that the project statistics tend to get worse when solution refinement is switched off.

4 Conclusion and Outlook

In this paper, we discussed and computationally tested a heuristic approach based on constraint programming for the MPSPRTC. The algorithm particularly addresses the issue of finding a good trade-off between low setup costs and compliance with project due dates. We presented a standard constructive heuristic which we extended by a local search and solution refinement procedure. The algorithm was tested on large real-world data of different customers. Our computational results demonstrate that this extension clearly outperforms the setup costs without worsening the throughput time, earliness and tardiness of the projects too much. Future research is necessary to develop more efficient heuristics to speed up the whole solving process

References

1. Krüger, D., Scholl, A.: A heuristic solution framework for the resource constrained multi-project scheduling problem with sequence-dependent transfer times. Eur. J. Oper. Res. **197**(2), 492–394 (2009)
2. Laborie, P., Rogerie, J., Shaw, P., Vilím, P.: IBM ILOG CP optimizer for scheduling. Constraints **23**(2), 210–250 (2018)
3. Lehmer, D.H.: On the compounding of certain means. J. Math. Anal. Appl. **36**, 183–200 (1971)

Time-Dependent Emission Minimization in Sustainable Flow Shop Scheduling

Sven Schulz and Florian Linß

Abstract It is generally accepted that global warming is caused by greenhouse gas emissions. Consequently, ecological aspects, such as emissions, should also be integrated into operative planning. The amount of pollutants emitted strongly depends on the energy mix and thus on the respective time period the energy is used. In this contribution we analyse the influence of fluctuating carbon dioxide emissions on emission minimization in flow shop scheduling. Therefore, we propose a new multi-objective MIP formulation which considers makespan and time-depending carbon dioxide emissions as objectives. Epsilon constraint method is used to solve the problem in a computational study, where we show that emissions can reduced by up to 10% if loads are shifted at times of lower CO_2 emissions.

1 Introduction

Over the past decades, global warming and climate change have received increasingly more public attention. Recent popular examples are the Fridays for Future movement or climate protection as an important issue in the European elections 2019. It is commonly known that the global warming is caused by increasing global greenhouse gas emissions, especially CO_2 which is released during the combustion of fossil fuels. A large part of the world's energy demand is currently covered by these fossil fuels. Thereby, industrial manufacturing companies are one of the greatest customer. In order to reduce the energy consumption and the CO_2 footprint of companies, green scheduling, also know as low-carbon scheduling, integrates ecological aspects into operational planning.

In addition to classic economic indicators such as makespan, green scheduling approaches also attempt to minimize ecological objectives such as energy consumption, emissions or waste. For this reason, they are often bi-criteria approaches. More

S. Schulz (✉) · F. Linß
TU Dresden, Dresden, Germany
e-mail: sven.schulz@tu-dresden.de

J. S. Neufeld et al. (eds.), *Operations Research Proceedings 2019*, Operations Research Proceedings, https://doi.org/10.1007/978-3-030-48439-2_71

Table 1 CO_2 emission factors of fossil fuels in comparison with the German electricity mix [4]

Fuel	g_{CO_2}/kWh Fuel	Efficiency [%]	g_{CO_2}/kWh electricity	g_{CO_2}/kWh el. mix
Natural gas	201	53	382	
Black coal	337	40	847	516
Lignite	407	35	1.148	

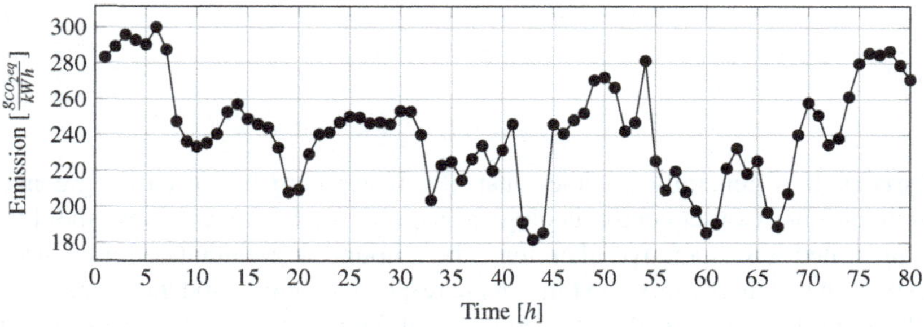

Fig. 1 Fluctuations in CO_2 equivalents per kWh over time in Swiss energy market [7]

than two criteria may also be taken into account (e.g. [6]). Emissions and makespan are for example considered by [5]. One possible way of reducing emissions is to control production speed where the emission quantity can be decreased at the expense of longer processing times [2, 3]. Beside of the minimization of carbon dioxide emissions also a peak power constraint can be used to force an ecological scheduling plan [9].

The vast majority of contributions minimizing emissions assume constant emission factors. Thus, they basically equate energy consumption and emissions. However, the electricity mix of produced energy varies over time depending on the primary energy sources used. Consequently, variable emission factors are more accurate. In [4] the electricity mix in the German electricity production is studied. Table 1 shows different fuels and their specific emission factors. It can be seen, that there is a significant difference between them.

In addition, nuclear power and renewable energies have much lower or no emissions at all. Considering this characteristics leads to variable factors depending on the current electricity mix. So-called time-of-use (TOU) pricing is based on the concept that at peak times (e.g. in the evening), the price of electricity is higher than normal and lower at night [8]. Similar dependencies apply to emissions. Figure 1 presents fluctuations in the Swiss energy market. In conclusion, the time of day at which production takes place becomes relevant for decision, since the emission factors changes continuously over the course of the day.

To the best of our knowledge, this is the first time that fluctuating CO_2 emissions are considered in a flow shop problem in combination with different discrete speed levels. We will present a bi-objective time-indexed mixed-integer program that minimizes makespan and CO_2 emissions simultaneously. In addition to time-dependent CO_2 emissions, we also consider different production speeds to reduce environmental pollution.

The work is structured as follows. In Sect. 2 a new model formulation is described. Subsequently, Sect. 3 shows computational experiments to illustrate how the model works and to discuss the interdependencies between time-depending CO_2 emissions and efficiency in production. Finally, Sect. 4 presents a summary and outlook.

2 Model Formulation

We consider a flow shop problem where n jobs ($j \in 1, .., n$) have to be processed on m available machines ($i \in 1, .., m$). All jobs have the same processing sequence. Machines can operate at σ different discrete production rates. The chosen production rate s influences the processing time $p^s_{i,j}$ of a job j on the respective machine m. Since time-depending emission values are taken into account, an index t for each discrete time interval in $1, .., \tau$ must be introduced. The following notation is used for the model formulation below.

Parameters		Decision variables	
$p^s_{i,j}$	Processing time	$c_{ij} \in \mathbb{N}$	Completion Time
q^s	Consumption factor	$x^{s,t}_{i,j} \in \{0, 1\}$	Machining is in progress
v^s	Speed factor	$y^s_{i,j} \in \{0, 1\}$	Speed selection
λ^t	Time-dependent CO_2 emission	$z^t_{i,j} \in \{0, 1\}$	Start of processing

Minimize **(I)** $$emission = \sum_{i \in I} \sum_{j \in J} \sum_{s \in S} \sum_{t \in T} x^{s,t}_{i,j} \cdot q^s \cdot \lambda^t \quad (1)$$

(II) $$makespan = \max_{j \in J}(c_{m,j}) \quad (2)$$

Besides the time depending carbon dioxide emissions (1) we minimize the makespan (2) to plan not only ecologically but also economically. The solution space can be described by introducing these eight constraints (3)–(10).

subject to :

$$\sum_{s \in S} y^s_{i,j} = 1 \qquad \forall j \in J, i \in I \quad (3)$$

$$\sum_{t \in T} x^{s,t}_{i,j} = y^s_{i,j} \cdot p^s_{i,j} \qquad \forall i \in I, j \in J, s \in S \quad (4)$$

$$\sum_{j \in J} \sum_{s \in S} x^{s,t}_{i,j} \leq 1 \qquad \forall i \in I, t \in T \quad (5)$$

$$\sum_{t \in T_0} z_{i,j}^t = 1 \qquad \forall i \in I, j \in J \tag{6}$$

$$\sum_{s \in S} x_{i,j}^{s,1} \leq z_{i,j}^1 \qquad \forall i \in I, j \in J \tag{7}$$

$$\sum_{s \in S} x_{i,j}^{s,t} - \sum_{s \in S} x_{i,j}^{s,t-1} \leq z_{i,j}^t \qquad \forall i \in I, j \in J, t \in T | t > 1 \tag{8}$$

$$\sum_{s \in S} x_{i,j}^{s,t} \cdot t \leq c_{i,j} \qquad \forall i \in I, j \in J, t \in T \tag{9}$$

$$\sum_{t \in T} c_{i-1,j} \leq \sum_{t \in T} z_{i,j}^t \cdot t - 1 \qquad \forall i \in I | i > 1, j \in J \tag{10}$$

Equation (3) ensures that each job on each machine is processed at a specified speed. With (4), each job is scheduled for the resulting processing time, which depends on the selected execution mode s. Each machine can only process one job at a time (constraint (5)). In order to prevent interruptions during processing, each order may only be started once, which requires 6. The start time is defined in (7) and (8) as the first moment of machining. A variable c_{ij} for completion time, which is introduced in (9), is not absolutely necessary, but speeds up the search for solutions and simplifies the makespan calculation in (2). Finally, condition (10) ensures that a job cannot be processed on the next machine until the previous step has been completed.

3 Computational Experiments

To illustrate how the model operates and to analyse the influence of fluctuating emissions we examine some computational experiments. We look at ten test instances of different sizes with two to five machines and four to eight jobs. The problems are first solved lexicographically. In addition, the optimal pareto front is determined using the epsilon constraint method as described in [1]. Since CO_2 savings can apparently be achieved for every increase in makespan, an equidistant approach is used and each integer value for makespan will be considered. All calculations are made with IBM ILOG CPLEX 12.7 on a Intel Xeon Gold 6136 CPU (@3 GHz) with 128 GB RAM.

The considered test data is summarized in Table 2. Three different execution modes can be selected on each machine. As the speed increases, consumption raises disproportionately. Processing times for the lowest speed are generated randomly.

Table 2 Overview of the test data

Parameter	Processing time $p_{i,j}^1 [h]$	Consumption factor q^s	Speed factor v^s
Range	$\mathcal{U}(5, 10)$	{1, 1.4, 2}	{1, 1.2, 1.5}

Table 3 Lexicographic solutions for different problem sizes

		Minimize makespan				Minimize emission				
m	n	Makespan	Emission	CPU [s]	Gap	Makespan	Emission	CPU [s]	Gap	τ
2	6	35	29.495	10.75	–	50	21.843	31.7	–	50
2	8	41	37.287	90.6	–	70	26.398	2268.8	1.43%	70
3	4	33	30.228	10.7	–	50	22.769	42.4		50
3	**6**	**40**	**43.663**	**46.8**	–	**65**	**31.526**	**116.4**	–	**65**
3	8	51	57.912	365.8	–	80	43.974	2400	0.99%	80
4	4	31	31.744	24.4	–	50	24.852	284.2	–	50
4	5	45	47.991	131.5	–	70	36.419	2400	1.09%	70
4	6	49	59.353	362.3	–	80	45.129	2400	2.46%	80
5	4	43	49.572	25.9	–	70	36.774	1634.9	–	70
5	5	50	60.646	360.2	–	80	45.983	2400	3.15%	80

Further processing times are adapted to the speed levels in a pre-process using formula $p^s_{i,j} = \lfloor \frac{p^1_{i,j}}{v_s} \rceil$; whereby the symbol $\lfloor \rceil$ indicates the nearest integer. For the time-dependent emissions λ^t the hourly data from [7] are used which can be seen in Fig. 1. Since smaller CO_2 factors could occur again and again in the future, the makespan could theoretically be infinitely increased. Therefore, the observation period is limited by Eq. (11). The expression represents an upper bound for the makespan. Therefore, processing times of a stage are summed up and the maximum time before and after that stage is added. To leave room for reductions in speed, the value is increased by α, which is set to 5% in the following.

$$\tau = (1+\alpha) \min_{i \in I} \left(\max_{j \in J} \sum_{i^*=1}^{i-1} p^1_{i^*,j} + \sum_{j \in J} p^1_{i,j} + \max_{j \in J} \sum_{i^*=i+1}^{m} p^1_{i^*,j} \right) \tag{11}$$

The resulting lexicographic solutions can be seen in Table 3. The CPU time is limited to 20 minutes in each run. It is possible to determine the optimal makespan and the corresponding minimum emissions for the considered problem sizes in an acceptable time. The calculation of the second lexicographic solution with minimum emissions proves to be much more difficult. This is probably due to the fact that the longer observation period allows significantly more speed variations and variable emissions become decision-relevant with emissions as objective. The gap indicated refers to the minimization of emissions.

The exact interdependencies between makespan, variable CO_2 emissions and execution modes cannot be precisely identified on the basis of the lexicographic solutions. For that reason the 3–6 instance (bold in Table 3) will be examined in more detail. Figure 2 shows the gantt charts of the lexicographic solutions. When minimizing the makespan, all but three jobs are produced at maximum speed (shown in grey). Only inevitable idle times are filled by slower execution modes to reduce emissions. The defined upper bound in Eq. (11) allows all jobs to be produced at

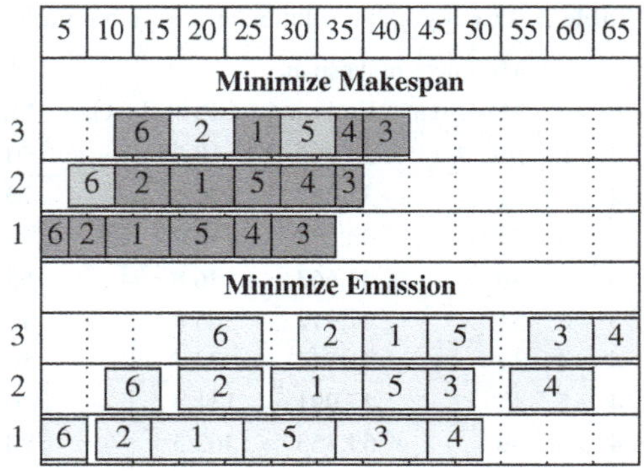

Fig. 2 Lexicographic solutions—instance 3–6

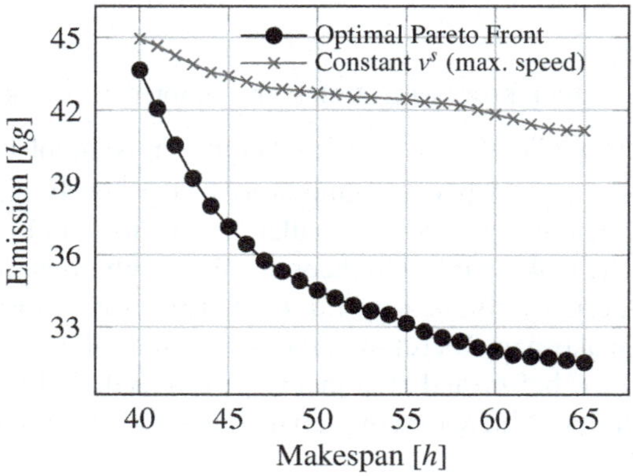

Fig. 3 Optimal Pareto Front—instance 3–6

minimum speed (white) to minimize emissions. It is also noticeable that some jobs deliberately wait for times with lower emissions. Consequently, optimal schedules are not necessarily semi-active, which significantly increases the solution space and makes a heuristic solution more difficult.

Furthermore, Fig. 3 shows all pareto optimal solutions for the 3–6 instance. The calculation of the optimal pareto front took $1.39\,h$ using the epsilon constraint method. Overall, any increase in makespan can result in a reduction in emissions, but the first increases lead to significantly higher savings. In addition, the pareto front is shown when all jobs are produced at maximum speed (constant v^s). The calculation required significantly less computing time of $0.53\,h$. Based on that curve, it can be seen that the influence of the speed changes is significantly higher than the variable emission factors. Nevertheless, a potential can be clearly identified. It must also be considered that the coefficient of variation of the CO_2 emission factors is only 11.9%. With a reduction of emissions of 9.2% at constant speed, almost the entire potential is exhausted. In electricity markets where CO_2 emissions fluctuate more strongly (e.g. through more renewable energies), the potential would also increase.

4 Summary and Outlook

This article analyses the possibility of reducing emissions through intelligent scheduling. To the best of our knowledge, for the first time, fluctuating emission factors are taken into account in a flow shop problem. In addition, different execution modes are considered to lower pollution. Makespan is considered as a second objective to ensure high capacity utilisation. An MIP formulation is presented to solve the problem. The optimal pareto front for test instances is determined using the epsilon constraint method.

Overall, it can be stated that the fluctuating CO_2 equivalents can be well exploited by the model and up to 10% of total emissions can be saved in this manner However, it must also be noted that energy savings due to speed changes have a greater influence than load shifts. The proposed solution is only suitable for small problem sizes, which is why a heuristic approach should be developed in the future. With this approach real problems could be analysed to identify even better the potential of the consideration of fluctuating compositions of the energy mix. Other energy markets could also be analysed. The average emission in the German energy mix, for example, is significantly higher the Swiss, which could possibly lead to greater savings. Due to the inexpensive and fast implementation, scheduling can make an important contribution to reducing energy consumption as well as emissions in the future.

References

1. Chircop, K., Zammit-Mangion, D.: On-constraint based methods for the generation of pareto frontiers. J. Mech. Eng. Autom. **3**(5), 279–289 (2013)
2. Ding, J.Y., Song, S., Wu, C.: Carbon-efficient scheduling of flow shops by multi-objective optimization. Eur. J. Oper. Res. **248**(3), 758–771 (2016)
3. Fang, K., Uhan, N., Zhao, F., Sutherland, J.W.: A new approach to scheduling in manufacturing for power consumption and carbon footprint reduction. J. Manuf. Syst. **30**(4), 234–240 (2011)
4. Icha, P., Kuhs, G.: Entwicklung der spezifischen kohlendioxid-emissionen des deutschen strommix in den jahren 1990–2017. Umweltbundesamt, Dessau-Roßlau (2018)
5. Liu, C., Dang, F., Li, W., Lian, J., Evans, S., Yin, Y.: Production planning of multi-stage multi-option seru production systems with sustainable measures. J. Clean. Prod. **105**, 285–299 (2015)
6. Schulz, S., Neufeld, J.S., Buscher, U.: A multi-objective iterated local search algorithm for comprehensive energy-aware hybrid flow shop scheduling. J. Clean. Prod. **224**, 421–434 (2019)
7. Vuarnoz, D., Jusselme, T.: Temporal variations in the primary energy use and greenhouse gas emissions of electricity provided by the swiss grid. Energy **161**, 573–582 (2018)
8. Zhang, H., Zhao, F., Fang, K., Sutherland, J.W.: Energy-conscious flow shop scheduling under time-of-use electricity tariffs. CIRP Ann. **63**(1), 37–40 (2014)
9. Zheng, H.y., Wang, L.: Reduction of carbon emissions and project makespan by a pareto-based estimation of distribution algorithm. Int. J. Prod. Econ. **164**, 421–432 (2015)

Analyzing and Optimizing the Throughput of a Pharmaceutical Production Process

Heiner Ackermann, Sandy Heydrich, and Christian Weiß

Abstract We describe a planning and scheduling problem arising from a pharmaceutical application. In a complex production process, individualized drugs are produced in a flow-shop like process with multiple dedicated batching machines at each process stage. Furthermore, due to errors jobs might recirculate to earlier stages and get re-processed. Motivated by the practical application, we investigate techniques for improving the performance of the process. First, we study some simple scheduling heuristics and evaluate their performance using simulations. Second, we show how the scheduling results can also be improved significantly by slightly increasing the number of machines at some crucial points.

Keywords Flow shop · Batching · Recirculation · Production planning · Process capacity

1 Introduction

In this paper, we investigate a scheduling and planning problem arising from a real-world industrial application. Our industry partner is a biotechnology company producing an individualized drug in a cutting-edge, complex production process. We can formally describe the problem as follows: We are given n jobs $J_1, \ldots, J_n$ and m stages $S_1, \ldots, S_m$. Each job J_j has a release time r_j at which it becomes available to start processing on the first stage. Jobs have to be processed at the stages in order, i.e., a job can only be processed at stage S_i if it has finished processing at all previous stages. At stage S_i, there are m_i identical parallel machines $M_1^i, \ldots, M_{m_i}^i$ to process the jobs. Processing times are job-independent, in other words, each stage S_i is associated with a processing time p_i which is the processing time of any job on

H. Ackermann · S. Heydrich (✉) · C. Weiß
Fraunhofer Institute for Industrial Mathematics ITWM, Kaiserslautern, Germany
e-mail: heiner.ackermann@itwm.fraunhofer.de; sandy.heydrich@itwm.fraunhofer.de; christian.weiss@itwm.fraunhofer.de; https://www.itwm.fraunhofer.de/en/departments/opt.html

J. S. Neufeld et al. (eds.), *Operations Research Proceedings 2019*, Operations Research Proceedings, https://doi.org/10.1007/978-3-030-48439-2_72

any of the machines $M_1^i, \ldots, M_{m_i}^i$. This kind of problem is called a proportionate hybrid flow shop.

In the biotechnology industry, many machines like pipetting robots can handle multiple jobs (samples) at the same time, and this kind of machines is used in the real-world process we are studying as well. Therefore, in our theoretical model, machines are *batching machines* that can handle multiple jobs in parallel. Stage S_i has a maximum batch size b_i, which is the maximum number of jobs per batch on machines of this stage, and in order to start a batch, all jobs in this batch must be available for processing on this stage. Note that the processing times do not depend on the actual batch sizes.

Another important feature of the real-world process are errors that can happen at various steps in the process. These lead to jobs being recirculated to an earlier stage in the process and restarting processing from there. Each stage S_i has a probability $p_{i,j}$ for $j \leq i$ that a job finished at stage i recirculates to stage j. We require $\sum_{j \leq i} p_{i,j} < 1$ for each i. Whether or not a job recirculates (and to which previous stage) only is known once it finishes the stage.

In this work, we describe the approach we use to help our industry partner optimizing his process's performance, i.e., optimizing scheduling objectives given a certain target throughput or capacity of the process (target number of jobs per, e.g., year). We evaluate different online scheduling heuristics using a simulation. Finally, we also investigate how to improve performance by increasing the capacity of the process at certain stages; that is, we suggest few stages where adding more machines improves the performance of the process significantly.

2 Related Work

To the best of our knowledge, the problem in this form is not studied in the literature. Related problems dealing with errors include reentrant flow shops [2] and flow shop with retreatments [3] (these do not consider the hybrid flow shop case or batching). Hybrid flow shops with batching (but without recirculations) were studied by Amin-Naseri and Beheshti-Nia [1].

Lastly, there has been some theoretical investigation of the special case with only one machine at each stage and no recirculations [4–6]. In [4], theoretical results for two of the heuristics presented in Sect. 3 were proven.

3 Scheduling

When looking for scheduling heuristics, we first have to define the objective function we want to optimize. First of all, the application motivates looking at how fast a job is processed once it arrives, as our goal is of course to deliver the drug to the patient as quickly as possible. This measure is known as the *flow time* of a job; it is the time between its release and its completion time at the last stage.

However, for a large enough number of jobs, some jobs will recirculate very often. Even if they never have to wait for a machine, those jobs will always have a huge flow time. Thus, there is nothing we can do about these jobs from a scheduling perspective, and we do not want these jobs to unduely corrupt our performance indicator. Hence, traditional objectives like maximum flow time ($\max_j F_j$) or sum of flow times ($\sum_j F_j$) might not be very meaningful.

Instead, our goal is to minimize the flow time of the *majority* of jobs, ignoring those jobs with the highest flow times. In our setting, we evaluate scheduling heuristics by looking at the *80%-quantile of flow times*, i.e., the maximum flow time among the fastest 80% of jobs. We denote this objective function by $F_{\max}^{0.8}$.

3.1 Three Scheduling Heuristics

In this paper, we focus on heuristics that decide for each stage separately how to schedule the jobs. That is, we go through the stages in order $1, \ldots, m$, and schedule all jobs for this stage without taking into consideration future stages. This strategy breaks down the very complex overall process into easier local decisions, and is also much closer to real world decision making processes, where an operator at a certain stage has to decide ad hoc what to do next, without having influence on operation at other stages.

Under this paradigm, the first and maybe simplest heuristics that comes to mind is what we call NEVERWAIT: We simply start a machine whenever it becomes available and jobs are waiting at this stage, and we make its batch as large as possible given the current job queue at this stage. This algorithm basically represents a greedy approach to minimizing the idle time of machines.

On the other hand, idle time of machines is only one way of wasting capacity. Waste is also incurred by starting batches with fewer jobs than the maximum batch size allows. Hence, a natural second heuristic is the FULLBATCHES heuristic, where we only start a machine whenever we can make its batch full i.e., if at least b_i many jobs are waiting at its stage (except for the last batch of course). While with NEVERWAIT we try to minimize capacity wasted by idle time, we now try to minimize capacity wasted by non-full batches.

Viewing these two heuristics as extreme points, it is natural to also investigate heuristics in between, trying to find a compromise between minimizing machine idle time and maximizing batch sizes. The third heuristic we propose, called BALANCEDWAITING, is such a compromise. Roughly speaking, the heuristic is geared towards starting a machine so that the total waiting time of jobs currently waiting and jobs arriving in the future at that machine is minimized.

To be more concrete, consider a machine M of stage S_i that becomes idle at some point in time t^* and assume, for the moment, that M is the only machine at this stage. Denote by a_j the arrival of job j at stage S_i. Let $\mathcal{J}_{\text{early}}$ be the set of jobs with $a_j \le t^*$ which are waiting to be processed at stage S_i at time t^*; these are called *early jobs*. Furthermore let $\mathcal{J}_{\text{late}}$ be the set of jobs with $a_j > t^*$; these are called *late jobs*. Finally, as soon as b_i jobs are available at stage S_i it makes no sense to wait any longer. Thus, let $\mathcal{J}'_{\text{late}} \subseteq \mathcal{J}_{\text{late}}$ be the $\min\{b_i - |\mathcal{J}_{\text{early}}|, |\mathcal{J}_{\text{late}}|\}$ jobs with smallest a_j among $\mathcal{J}_{\text{late}}$; if $|\mathcal{J}_{\text{early}}| \ge b_i$, then $\mathcal{J}'_{\text{late}} = \emptyset$.

We now evaluate different possible scenarios, i.e., starting times for machine M, with respect to the total waiting time they incur on early and late jobs. Observe that it is sufficient to consider starting times from the set $T = \{t^*\} \cup \{a_j | J_j \in \mathcal{J}'_{\text{late}}\}$. Indeed, starting at another time will increase the waiting times without increasing the size of the started batch compared to starting at the closest time $t \in T$ that is earlier. We give two examples to demonstrate how the algorithm evaluates the different starting times. For any start time t of machine M, define $\text{next}(t) = t + p_i$ to be the next time M will be available.

First of all, assume we would start M at time t^*. This would mean that any early job J_j would wait time $t^* - a_j$, which is of course best possible considering that the only machine M is busy before t^*. On the other hand, if we start M now, depending on p_i, it might still be busy when the first late jobs arrive, and thus, these jobs have to wait until M becomes available again at time $\text{next}(t^*)$ (remember that we assume M to be the only machine at stage S_i).

Alternatively, if $|\mathcal{J}_{\text{early}}| < b_i$, we could wait until the first late job $j_{\min}^{\text{late}}$ arrives and start M then. In that case, we would decrease the waiting time of job $j_{\min}^{\text{late}}$ to 0, but the early jobs would have to wait longer and the other late jobs would have to wait until $\text{next}(a_{j_{\min}^{\text{late}}}) > \text{next}(t^*)$ until they can be processed.

In general, for $t \in T$ define the waiting time of an early job J_j to be $W_j^{(t)} = t - a_j$ (note that by definition of $\mathcal{J}_{\text{early}}$, we have $a_j \le t$ for $t \ge t^*$). Let $\mathcal{J}_{\text{late}}(t)$ be the subset of jobs from $\mathcal{J}'_{\text{late}}$ with $a_j < \text{next}(t)$. The waiting time of job $J_j \in \mathcal{J}_{\text{late}}(t)$ is defined as $W_j^{(t)} = \text{next}(t) - a_j$. Note that we only take into account those jobs which arrive before $\text{next}(t)$. Naturally, we can then define $W^{(t)} = \sum_{j \in \mathcal{J}_{\text{early}}} W_j^{(t)} + \sum_{j \in \mathcal{J}_{\text{late}}(t)} W_j^{(t)}$. The BALANCEDWAITING heuristic then chooses the start time $t_{\min} = \arg\min_{t \in T} W^{(t)}$.

If we have more than one machine per stage, the strategy stays the same, while $\text{next}(t)$ becomes the minimum of the next start time of machine M and the next point in time another machine of stage S_i becomes available. We are also aware of the fact that the definition of $W^{(t)}$ is not the only sensible one; we could ignore waiting times that are unavoidable, or increase the number of jobs that are taken into account (e.g., extending the set $\mathcal{J}_{\text{late}}(t)$). However, in computational experiments these changes did not help to improve BALANCEDWAITING's performance, so for this paper we stay with the simpler version described above.

3.2 Simulation

To evaluate the performance of the proposed heuristics, we use a Monte-Carlo discrete event simulation. In this simulation, jobs arrive daily, with the number of new jobs per day drawn independently from a Gaussian distribution. The mean of this distribution is chosen such that the expected total number of jobs equals the target capacity we are aiming for. We carried out 20 simulations per heuristic. For each stage S_i and every job finishing this stage, the next stage to go to is drawn independently among all recirculations and the successor of S_i

From a theoretical perspective, [4] showed that NEVERWAIT is a 2-approximation and FULLBATCHES has unbounded approximation factor in the setting where $m_i = 1$ for all $i = 1, \ldots, m$. Thus, we would expect that NEVERWAIT performs reasonable while FULLBATCHES might perform very bad.

As stated before, we are aiming at optimizing the 80%-quantile of the flow times, i.e., minimizing $F_{\max}^{0.8}$. We find that BALANCEDWAITING performs slightly better than NEVERWAIT; to be precise, $F_{\max}^{0.8}$ is lower by roughly 1.5%. We have also run the simulation with FULLBATCHES; however, as expected, the flow times are far worse ($F_{\max}^{0.8}$ is higher by more than 330% compared to NEVERWAIT).

One may ask how good the heuristics are in an absolute sense, i.e., compared to an optimal solution. As computing optimal solutions for our large instances is infeasible, we instead use what we call the *baseline*. This baseline is computed by scheduling a single job, whilst still applying random recirculations. That is, the job can never be delayed because of machines not being available (as it is the only job), but it might be delayed because of recirculations. We simulate 10,000 such jobs in order to get a distribution of flow times; this makes up our baseline.

When comparing the results for both NEVERWAIT and BALANCEDWAITING with the baseline, we observe that NEVERWAIT is roughly 22% worse than the baseline, while BALANCEDWAITING is roughly 20% worse. This is still a large gap, so one could wonder how to improve performance further We could look for complex and hopefully better algorithms, but from a practical perspective, simple algorithms are desirable, as decisions should be easily understandable.

The baseline can also be from a different perspective. As we define it, the baseline removes delays caused by *lack of resources* (unavailable machines), i.e., it shows the result obtainable if we had an unlimited number of machines per stage. While employing highly complex scheduling algorithms might not be a viable option in reality, buying additional machines certainly is, however, the investment should of course be held as low as possible. The question thus arises, whether it is possible to improve the results significantly by buying a small number of additional machines at particular bottleneck stages.

4 Analysis and Optimization of Process Capacity

Our industry partner is of course interested in minimizing flow times whilst aiming for a certain target capacity, i.e., number of jobs processed per year. Our goal now becomes finding a good trade-off between the number of machines and the number of jobs per year that can be processed within a "good" flow time. Given a certain target number of jobs C^* arriving per time period T, what is the minimum number of machines per stage to reach this capacity?

First, note that a certain constant target capacity alone does not suffice to judge the capacity needed at a single stage. Due to recirculations, some jobs might be processed at a certain stage more than once, thus requiring the capacity at this stage to handle more jobs than the actual number of jobs passing through the process. This number highly depends on the structure of recirculation edges present in the process graph.

By analyzing the traces of the 10,000 jobs we simulated separately in order to compute the baseline, we can get what we call the *demand factor* of each stage. This is the factor $d_i \geq 1$ such that if n jobs arrive at the process, in expectation $d_i \cdot n$ jobs will arrive at stage S_i.

With these factors, we define $\underline{m}_i = \left\lceil \frac{d_i C^*}{b_i T / p_i} \right\rceil$. This is the (theoretical) minimum number of machines at stage i to handle the target capacity: a machine can process at most $b_i \cdot \frac{T}{p_i}$ many jobs in T.

The numbers $\underline{m}_i$ are indeed the numbers of machines used in the simulations from Sect. 3.2. Now, it becomes clearer why the results were quite far from the baseline: The machines will of course never reach their ideal capacity of processing $\frac{T}{p_i} \cdot b_i$ many jobs in T, as they will have idle times and non-full batches. We thus have to compensate for this by adding machines. One way to do this, is to expect a slightly larger number of jobs than $d_i \cdot C^*$. To this end, define $m_i^{(x)} = \left\lceil \frac{d_i C^* + x}{b_i T / p_i} \right\rceil$. This is the minimum number of machines with respect to some *slack* x.

We now ran the simulations as before, but using $m_i^{(0.03 \cdot C^*)}$ machines per stage. That is, our slack is equal to 3% of the target capacity. When looking at the simulation results, we find that the flow times are decreased significantly. Using slack $0.03 \cdot C^*$, we obtain an improvement of 12% for $F_{\max}^{0.8}$ compared to slack 0 for both NeverWait and BalancedWaiting. Both heuristics are now only about 6% worse than the baseline (compared to 20–22% before). The most interesting part is that only at four particular stages, we need to add one more machine each in order to obtain these improved results. The total investment is thus low, while the performance boost is quite high.

5 Future Work

For future work, a natural question to ask is which heuristics might perform even better than BALANCEDWAITING and NEVERWAIT. Also, it would be interesting to prove formal worst-case bounds for BALANCEDWAITING's performance. Furthermore, for adapting the process capacity, we tried various sizes of slack and found that 3% of the target capacity is a good compromise between number of machines to buy and scheduling improvement. However, one could ask whether there is an easy rule about how to choose the amount of slack or whether one can prove how much slack is needed in order to get a particular solution quality. Lastly, we only investigated increasing the number of machines in order to maximize capacity. In the application we are looking at, it is also possible, to, e.g., decrease processing times, increase batch sizes, or reduce recirculation rates at certain stages by buying better machines and/or investing in research. An interesting open problem is how to identify stages in the process where modifying these process parameters has the largest impact on the overall performance.

References

1. Amin-Naseri, M.R., Beheshti-Nia, M.A.: Hybrid flow shop scheduling with parallel batching. Int. J. Prod. Econ. **117**, 185–196 (2009)
2. Chen, J.-S., Pan, J.C.-H., Wu, C-K.: Minimizing makespan in reentrant flow-shops using hybrid tabu search. Int. J. Adv. Manuf. Technol. **34**, 353–361 (2007)
3. Grobler-Debska, K., Kucharska, E., Dudek-Dyduch, E.: Idea of switching algebraic-logical models in flow-shop scheduling problem with defects. In: 18th International Conference on Methods & Models Automation & Robotics, MMAR 2013, pp. 532–537 (2013)
4. Hertrich, C.: Scheduling a proportionate flow shop of batching machines. Master's thesis Technische Universität Kaiserslautern (2018)
5. Sung, C.S., Kim, Y.H.: Minimizing due date related performance measures on two batch processing machines. Eur. J. Oper. Res. **147**, 644–656 (2003)
6. Sung, C.S., Yoon, S.H.: Minimizing maximum completion time in a two-batch-processing-machine flowshop with dynamic arrivals allowed. Eng. Optim. **28**, 231–243 (1997)

A Problem Specific Genetic Algorithm for Disassembly Planning and Scheduling Considering Process Plan Flexibility and Parallel Operations

Franz Ehm

Abstract Increased awareness of resource scarcity and man-made pollution has driven consumers and manufacturers to reflect ways how to deal with end-of-life products and exploit their remaining value. The options of repair, remanufacturing or recycling each require at least partial disassembly of the structure with the variety of feasible process plans and large number of emerging parts and sub-assemblies generally making for a challenging optimization problem. Its complexity is further accentuated by considering divergent process flows which result from multiple parts or sub-assemblies that are released in the course of disassembly. In a previous study, it was shown that exact solution using an and/or graph based mixed integer linear program (MILP) was only practical for smaller problem instances. Consequently, a meta-heuristic approach is now taken to enable solution of large size problems. This study presents a genetic algorithm (GA) along with a problem specific representation to address both the scheduling and process planning aspect while allowing for parallel execution of certain disassembly tasks. Performance analysis with artificial test data shows that the proposed GA is capable of producing good quality solutions in reasonable time and bridging the gap regarding application to large scale problems as compared to the existing MILP formulation.

Keywords Scheduling · Disassembly planning · Evolutionary algorithm

1 Introduction

Due to the relatively low value added by decomposing a product efficient planning and scheduling of disassembly operations becomes all the more important to achieve maximum utilization of the involved technical equipment and workers. However, disassembly scheduling in the context of operational time-wise planning of activities

F. Ehm (✉)
The Department of Industrial Management, TU Dresden, Dresden, Germany
e-mail: franz.ehm@tu-dresden.de

J. S. Neufeld et al. (eds.), *Operations Research Proceedings 2019*, Operations Research Proceedings, https://doi.org/10.1007/978-3-030-48439-2_73

of multiple jobs at given workstations has not been addressed by many researchers so far. Zhang et al. and more recently Gong et al. have presented GA approaches that explicitly consider the choice of process routes and scheduling under multiple objectives in remanufacturing (see [1] and [2]). More general contributions to the subject are found with reference to flexible job shop scheduling problems (FJSP) or integrated process planning and scheduling (IPPS). Relevant research in this field includes the MILP formulations given by Özgüven et al. in [3] and another GA approach by Amin-Naseri and Afshari to solve the JSP with machine flexibility and alternative process plans in [4]. However, splitting of jobs into multiple sub-jobs in the process of disassembly adds to the complexity and there are still very few approaches to simultaneously address process plan flexibility and job divergence when there is more than just a single product-flow such as in divergent production scheduling (see e.g. [5]). As a consequence, the problem of integrated disassembly planning and scheduling (IDPS) was introduced and a first MILP model was formulated by the author in [6]. In short terms, IDPS seeks to schedule K_j disassembly tasks of $j = 1, 2, \ldots N$ jobs at $1, 2, \ldots M$ stations with the goal to minimize makespan C_{max}. Besides classical aspects of machine scheduling it involves the decision upon an optimal process plan combination (PPC) representing the disassembly sequences chosen for each job. It thus features flexible process plans, re-entrance of jobs at the stations and parallel execution of tasks for distinct sub-assemblies. Previous tests revealed limitations of the existing MILP model when it comes to solving larger problem instances (see [7]). To bridge this gap with hindsight to application in a real-world context problem specific (meta-) heuristics were identified as promising approach. From the literature it is clear that GA represents a suitable tool to tackle the problem of process planning and scheduling. Accordingly, in the following section a problem specific GA design is elaborated and relevant operators are discussed. Section 3 describes the test environment used to adjust and evaluate the GA and analyze its performance versus the existing MILP formulation. Finally, Sect. 4 discusses the results and draws conclusions.

2 Problem Specific GA Design for Disassembly Scheduling

The proposed GA adopts the traditional pattern of generational evolution by means of selection, recombination and replacement of individuals in the population using problem specific genetic operators. Figure 1 summarizes the procedure and relevant operators as implemented in Python. To effectively exploit genetic information to guide the search process any member of the population requires **evaluation** once created or modified by deriving C_{max} from the corresponding schedule in the decoded solution and assigning its reciprocal value as fitness.We deploy a problem specific encoding, where individuals are composed of two lists as illustrated in Fig. 2. While the first part represents the process plan combination PPC, that is which process plan is executed for which job, the second part contains a sequence of job-IDs which serves as priority list for scheduling. Each entry in the second

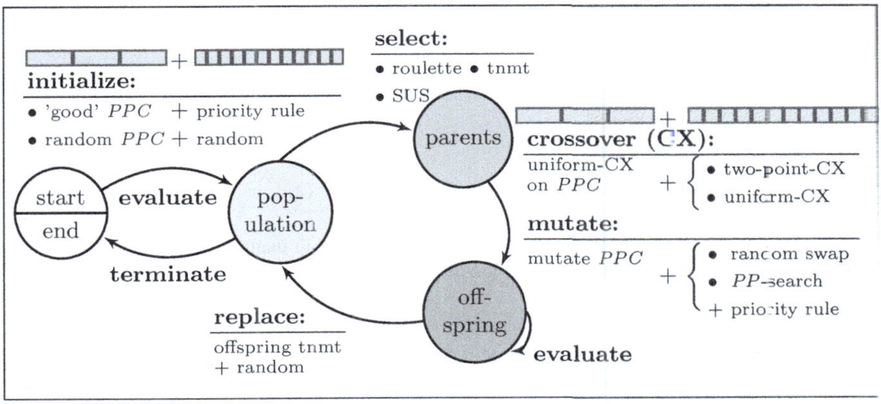

Fig. 1 GA procedure and implemented operators

gene string represents an operation O_{jk} as part of the executed process plan PP_j of job j with k being a disassembly task that can be tracked to one specific station i. Additionally, job splitting is taken into account by analyzing parallel task relations when evaluating minimal required reservation times for each job and machine, respectively.

In the exemplary encoded solution in Fig. 2, jobs $j = 1, 2, 3$ are executed via process plans $PP_j = 1, 2, 1$, respectively. Following $PP_1 = 1$, job 1 is disassembled via sequence $O_{11} \rightarrow O_{17^*} \rightarrow O_{13} \rightarrow O_{14} \rightarrow O_{16}$ which involves tasks $1, 3, 4, 6$ to be scheduled at stations $1, 2, 1, 2$. Note that 7* represents a dummy task which does not require real processing but is due to the translation of parallel relations $(O_{13}||O_{14})$ and $(O_{13}||O_{16})$ from the corresponding and/or graph. Similarly, operations sequences and corresponding machines can be extracted for jobs 2 and 3. When **initializing** the GA, parameters concerning operational relations and feasible sequences that are not explicitly represented in the given disassembly process data need to be translated from and/or-graphs. Once the complete set of process plans has been established, an initial population of size n_{pop} is created from half random feasible solutions and half 'guided' individuals which provide potentially higher fitness. Here, a construction heuristic is deployed to determine a set of 'good' process plan combinations PPC^* with the goal to obtain shorter makespan schedules by minimizing the lower bound

$$LB = \min_{PPC} \max_{i \in I} \sum_{O_{jk} \in PP_j(PPC)} p_{jki}$$

that is the maximum total processing time $\sum p_{jki}$ over all machines $i \in I$ when all k operations O_{jk} of job j are executed according to process plan PP_j as encoded in the PPC. The remaining part of the genetic string and thus schedule is built by applying a random priority rule such as SPT or LPT to the entire set

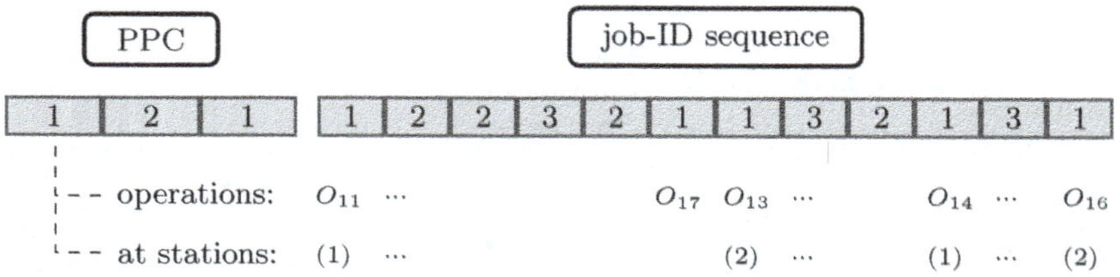

Fig. 2 Genetic solution encoding (with special regard to job 1) using process plan combination (PPC) as controller part and sequence of job-IDs as controlled part

of operations arising from PPC^*. Recombination takes place in each generation by picking $n_{pop}/2$ parental pairs for crossover according to one of following implemented **selection** schemes. While *roulette wheel selection (R)* chooses and copies one individual at a time from the old population with fitness proportionate probability *stochastic universal sampling (S)* selects all necessary parents at once by virtually turning a fitness proportionate roulette wheel with $n_{pop}/2$ pointers. Finally, *tournament selection (T)* successively considers fitness competition among two randomly chosen individuals. To effectively exploit and spread genetic information **crossover** operators are designed to fit the proposed encoding by treating recombination of PPC and job sequence separately. Specifically, if the selected individuals differ in their PPC *uniform crossover* is applied by randomly drawing PP-IDs from the first part of the encoded solution. In the case of parental mating with identical PPC crossover focuses on the job sequence part of the genotype by means of *uniform (U)* or *two-point crossover (2P)*. Subsequently, a validation routine is used to detect and eventually correct incompatibilities between job-ID list and the PPC in order to ensure feasibility of the emerging offspring solution. Similar distinction is made with regard to **mutation**. Providing a given PPC a common way to manipulate permutation encodings is the re-ordering of indices by changing positions of individual genes or shuffling the entire sequence. In the proposed GA, the job list of each offspring is subjected to mutation with probability p_{mut}. While manipulation is most probably realized by means of random swap of a given number of genes there is also a 25 % chance to change the PP_j for one individual job j as indexed in the PPC and modify the set of operations registered in the job list. Even more perturbation is achieved with probability $p_{mut}^{PPC} = p_{mut}/2$ by applying a local search rule that tests new process plan combinations for LB improvement. Once an alternative PPC^* has been found, a random priority rule is used to establish the remainder of the individual. The algorithm **terminates** if a maximum number of generations or the limit for elapsed real or CPU time is reached.

3 Computational Analysis

The method described in [7] is used to generate a representative set of artificial test data with size and shape of the search landscapes varied across the following design factors: number of disassembly jobs N, number of machines M, maximum number of components to disassemble per job H and degree of precedence restriction $prec$. This is realized as a two level fractional factorial design using Taguchis L8 orthogonal array to establish a test bed of eight problem configurations with five random instances each as shown in Table 1. Note, that the presence of alternative process plans and parallel operations in diverging jobs adds to the complexity of the scheduling problem. As opposed to 100 operations represented by a basic 20×5 JSP such as in FT-20, a 20×5 IDPS may consist of more than 1000 operations and a very large number of potential sequences depending on the number of components to disassemble and the structure of and/or graphs. For this reason classification of problem size as small (S) or large (L) particularly refers to the realized number of operations and parallel relations in one instance with the adjustment of N and M being in accordance to relevant literature in the field of IPPS (see [3]).

In order to achieve a favorable balance of selection pressure and population diversity, preliminary tests were performed for different combinations of selection and crossover operators as well as varying population size and mutation rate. Each GA setup was run five times for 300 s on four representative instances from problems 2, 3, 6 and 7. To compare results across different settings makespan from each instance and run was normalized in relation to the best objective value found using any GA setup on this respective run. A short statistical analysis was conducted using a total of 240 observations resulting from (a) different combinations of selection schemes R, S, T with crossover operators U and $2P$ and (b) different settings of population size n and mutation rate p_{mut}. Taking into account medians and interquartile ranges as reflected by the box-plots in Fig. 3 and coefficients of

Table 1 Studied problem configurations and GA solution after 1000 s vs. CPLEX after 1000 s (3600 s) comparing average C_{max} over five instances and 10 runs

	Factors					Relations			Avg. C_{max} from GA (1000 s)		
									Improving CPLEX		
Prob.	N	M	H	$prec$	Ops.	Prec.	Parallel	Size	1000 s (3600 s)	$\Delta(\%)$	$CV(\%)$
1	10	4	5..15	0.75	245	313	563	S	3/5	−1.9	1.71
2	10	4	15..30	0.90	355	403	1353	L	1/5	10.0	1.65
3	10	8	5..15	0.90	143	156	234	S	0/5	26.0	2.62
4	10	8	15..30	0.75	149	164	262	S	3/5 (1/5)	24.0	3.42
5	20	4	5..15	0.90	300	319	464	S	5/5	−9.0	1.06
6	20	4	15..30	0.75	1483	2005	14,764	L	5/5	n.a.	2.69
7	20	8	5..15	0.75	499	621	1154	L	4/5 (2/5)	−8.0	2.66
8	20	8	15..30	0.90	713	804	2825	L	4/5	−5.0	2.14
								$\sum$	25/40 (21/40)		

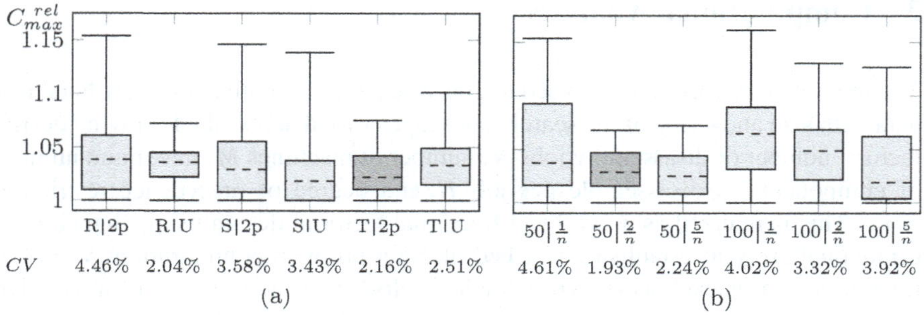

Fig. 3 Comparison of (**a**) different selection vs. crossover operators (with $n = 50$, $p = \frac{2}{n}$) and (**b**) different population sizes n and relative mutation rates (with T|2p): box-plots covering relative C_{max} values after five runs for 300 s on four problem instances

variance (CV) as additional tie breaker the suggested GA setup is $T|2p$, $n = 50$ and $p_{mut} = 2/n$.

4 Results

Performance of the GA is evaluated by recording C_{max} as the best fitness value found in the population after a fixed computational time of 1000 s. To enable comparison of solution quality with the existing MILP formulation in [7] and use of commercial software IBM ILOG CPLEX 12.7 both applications are configured for single-core mode running an Intel® Xeon® Processor E5–4627 v2@3.30 GHz and up to 100 GB RAM. Reviewing the results given in Table 1 we observe that GA was able to improve the MILP solution for 25 out of 40 instances. Most improvements were realized for large size instances from problems 6, 7 and 8 with problem 6 apparently too tough for CPLEX to produce any feasible solution at all. Despite the fact that number of total operations and relations heavily contribute to problem size, a large value of N along with simultaneous decrease in M seem to be even stronger drivers of complexity in favor of a GA solution. These observations hardly change by allowing CPLEX an extra search time of up to 1 h. While the MILP model was able to produce two (three) optimal solutions within 1000 s (3600 s), the corresponding instances stem from arguably easier problems 3 and 4 which is also where application of GA seems least beneficial.

In conclusion, testing confirmed the proposed GA as beneficial complement to the existing MILP with regard to large sized disassembly scheduling problems by producing solutions of high quality and little variation in limited time.

Acknowledgments The author would like to thank Benedikt Zipfel who contributed to the design and implementation of the GA as part of his diploma thesis.

References

1. Zhang, R., Ong, S.K, Nee, A.Y.: A simulation-based genetic algorithm approach for remanufacturing process planning and scheduling. In: Applied Soft Computing, vol. 37, pp. 521–532 (2015)
2. Gong, G., Deng, Q., Chiong, R., Gong, X., Huang, H., Han, W.: Remanufacturing-oriented process planning and scheduling: mathematical modelling and evolutionary optimisation. Int. J. Prod. Res. **58**(12), 3781–3799 (2020)
3. Özgüven, C., Özbakır, L., Yavuz, Y.: Mathematical models for job-shop scheduling problems with routing and process plan flexibility. Appl. Math. Model. **34**(6), 1539–1548 (2010)
4. Amin-Naseri, M.R., Afshari, A.J.: A hybrid genetic algorithm for integrated process planning and scheduling problem with precedence constraints. Int. J. Adv. Manuf. Technol. **59**(1–4), 273–287 (2012)
5. Gaudreault, J., Frayret, J.M., Rousseau, A., D'Amours, S.: Combined planning and scheduling in a divergent production system with co-production: a case study in the lumber industry. Comput. Oper. Res. **38**(9), 1238–1250 (2011)
6. Ehm, F.: Machine scheduling for multi-product disassembly. In: Operations Research Proceedings 2016, pp. 507–513. Springer, Cham (2018)
7. Ehm, F.: A data-driven modeling approach for integrated disassembly planning and scheduling. J. Remanuf. **9**(2), 89–107 (2019)

Project Management with Scarce Resources in Disaster Response

Niels-Fabian Baur and Julia Rieck

Abstract Natural disasters are extreme, sudden events caused by environmental factors that injure people and damage assets. In order to reduce the disaster's impact, many workforces like professional emergency forces and volunteers work simultaneously. An integrated, central coordination of available resources can therefore reduce overall damage. For this purpose, we introduce a mixed-integer linear program for project management, particularly scheduling, in disaster response. Many specific characteristics such as partially renewable resources, flexible resource profiles, and variable activity durations with possible interruptions are taken into account. First small-scale instances are solved with GAMS using CPLEX 12.9.

Keywords Disaster management · Scheduling · Project management · Discrete-time model · Flexible resource profiles

1 Introduction

Natural disasters that either result from or are intensified by environmental phenomena such as earthquakes and hurricanes are a growing threat worldwide. Frequently, disasters cannot be prevented and they endanger life and health as well as material assets to an unusually high extent. In addition, the impacts of the disasters are normally difficult to predict. The survey of Altay and Green [1] shows the phases in the lifecycle of disaster relief situations. In particular, the response phase after a disaster, where activities must be coordinated and information exchanged quickly, is considered in the literature, e.g., in the fields of infrastructure protection, emergency rescue, and medical care. Even in highly developed countries in Europe, several disasters such as floods occur every year for which appropriate measures must be

N.-F. Baur (✉) · J. Rieck
University of Hildesheim, Institute of Economics and Computer Science, Operations Research Group, Hildesheim, Germany
e-mail: baur@bwl.uni-hildesheim.de; rieck@bwl.uni-hildesheim.de

J. S. Neufeld et al. (eds.), *Operations Research Proceedings 2019*, Operations Research Proceedings, https://doi.org/10.1007/978-3-030-48439-2_74

planned and implemented. The development of mathematical models and decision support systems can assist project managers in selecting suitable measures and thereby lead to a desired reduction of total damage. It has been stated that there is a great willingness to voluntarily help in emergency and post-disaster situations. Volunteers can constitute important resources for disaster response and therefore be an effective complement to the professional forces in disaster relief.

For successful planning, it makes sense to identify activities, e.g., the filling or the transport of sandbags, and visualize their chronological order by a project. Activities are characterized by the fact that they require volunteers that can be defined as partially renewable resources with specific skills like physical fitness or experience in providing first aid. In order to assign voluntary workers to the activities, the volunteers must document or indicate their skills with the appropriate skill level, ranging from "not demonstrated" to "outstanding" (cf. Mansfield [5]). Only if a volunteer with required skills is available, an activity can be processed. If a suitable worker is missing at the earliest possible start of an activity, the processing must be delayed. If at any time during the execution no suitable worker is available, the activity must be interrupted. Please note that an activity can be carried out by a team of several volunteers with different skill levels. For the same workload, a volunteer with a low skill level requires more time than a worker with high skill level. Volunteers must be assigned to activities and start times of activities must be determined. Consequently, in post-disaster situations, a combined workforce and project scheduling problem arises, where the overall objective is the minimization of the project duration. The intention of our approach is to integrate the generated model and the solution methods into a decision support system as a central coordination tool that allows volunteers to find activities and locations where they can aid. Within the system, workers can indicate when they are available and to what extent they master the relevant and predefined skills.

This paper is structured as follows: The model formulation that includes the specific characteristics of the problem under consideration is shown in Sect. 2. Section 3 presents the results of preliminary computational studies with CPLEX. Finally, Sect. 4 discusses the conclusions and offers an overview of further research intentions.

2 Model Formulation

We consider a disaster as a unique event and model the problem described in Sect. 1 as a project management problem, in particular scheduling problem. For this purpose, we define activities $i, j = \{0, \ldots, n+1\}$ of the project, where activity $i = 0$ represents the fictitious project start and activity $i = n+1$ the fictitious project end. Each activity i has an estimated total processing time of D_i, which is required when only one worker performs the activity under normal conditions. For the fictitious activities, we set $D_0 = D_{n+1} := 0$. The temporal structure of the project provides for precedence constraints between activities, i.e., an activity j can start at the earliest when all its predecessors are completed.

The underlying project network $N := (V, E, P)$ contains the activities as nodes and the precedence constraints between activities as arcs with weights P. Thus, we obtain an activity-on-node network with the activity set V and the set of precedence constraints E. If $\langle i, j\rangle \in E$ has the weight P_i, activity j cannot be started earlier than P_i time units after the start of activity i. P_i identifies the actual duration of activity i taking into account the number workers assigned to the activity and the respective skill levels. The model follows a time-index based formulation, containing decision variables x_{it}, $i \in V, t \in T = \{0, 1, \ldots, \bar{d}\}$, which defines whether activity i starts at time t or not, where $\bar{d}$ is the predefined end of the planning horizon. Assuming that the project starts at time zero, $x_{00} = 1$ applies.

Within the planning horizon, the volunteers are not constantly available. They can only be scheduled at pre-selected time intervals. Let K be the set of partially renewable resources. Each volunteer $k \in K$ is characterized by his or her skills $s \in S$ and the associated skill levels $L_{ks} \in \{0, 0.5, 1, 1.5, 2\}$. Similar to Kreter et al. [4], parameter $\theta_{kt} = 1$ indicates whether resource k is available in period t or not (based on the volunteer's input).

Activities can be interrupted and the activity durations are variable and subject of planning. Figure 1 shows an exemplary interruption of activity $i = 1$. When the first resource assignment is made at $t = 1$, the activity starts. In the periods $t = 2$ and $t = 4$ interruptions occur, as no suitable worker can be selected After six time units (i.e., $P_1 = 6$), the total processing time of $D_1 = 4$ is covered. Figure 2 shows another example, where five resources are assigned simultaneously during the execution of activity $i = 2$. Although the estimated total processing time is $D_2 = 10$, the activity can be completed in four time units (i.e., $P_2 = 4$). Note that the number of resources assigned to $i = 2$ differs over time. While in period 14 only 3 volunteers are incorporated, in period 15 there are 5 resources working in total. Therefore, the problem under consideration is a generalization of resource-constrained project scheduling problem with flexible resource profiles (FRCPSP), presented by the Naber and Kolisch [6]. In contrast to the models presented there, in our model there is no minimum amount of periods with constant resource usage. Moreover, we do not constrain the resource usage by upper and lower bounds. To be processed, at least one resource must be assigned and the number of assignments is limited only by the amount of available resources. Furthermore, there are additional features such as possible interruptions and skill levels compared to the FRCPSP.

Figure 1 assumes a skill level $L_{ks} = 1$ for the assigned worker k. The situation is different in Fig. 2, where workers 1 and 2 have a skill level of $L_{1s} = L_{2s} = 1$, whereby workers 3, 4, and 5 have a level of $L_{3s} = L_{4s} = L_{5s} = 0.5$ for all required skills s. Let S_i be the set of skills that are necessarily needed to process activity i. The decision variables r_{ikt} that are equal to 1 if activity $i \in V$ is processed by resource $k \in K$ at time $t \in T$ and 0 otherwise, define the assignment of the resources. Note that our modeling results in greater complexity than the original modeling of the FRCPSP, where types of resources are predefined and only the number of assigned resources of each type is relevant. This approach is impractical

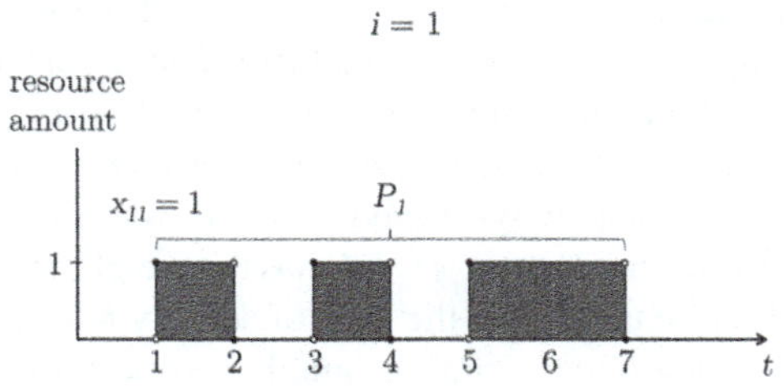

Fig. 1 Interruptions of an activity

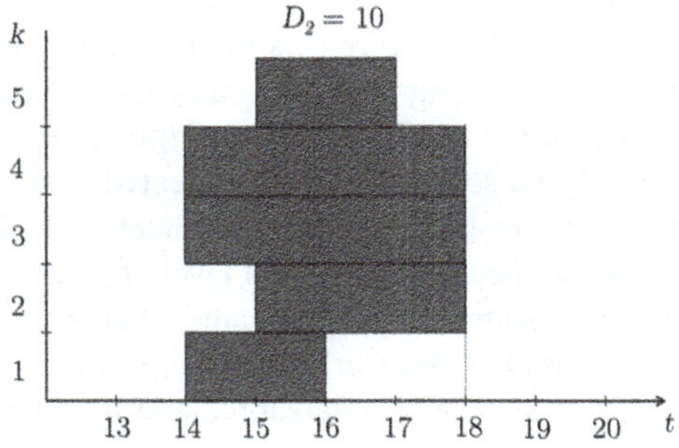

Fig. 2 Parallel assignments of resources to one activity

when the problem is combined with skill levels, since e.g., with five skills and five possible skill levels $5^5 = 3125$ different possibilities exist to define the skills of a resource. Therefore, in our model each volunteer is considered as a standalone resource and the model can be formulated as follows:

$$\min. \sum_{t \in T} t x_{n+1\,t} \tag{1}$$

$$\text{s.t.} \sum_{t \in T} x_{it} = 1 \qquad \forall i \in V \tag{2}$$

$$\sum_{t \in T} t x_{jt} - \sum_{t \in T} t x_{it} \geq P_i \qquad \forall \langle i, j \rangle \in E \tag{3}$$

$$\sum_{i \in V} r_{ikt} \leq \theta_{kt} \qquad \forall k \in K,\ t \in T \tag{4}$$

$$\sum_{\tau=0}^{t} x_{i\tau} \geq r_{ikt} \qquad \forall i \in V \setminus \{0, n+1\}, k \in K, t \in T \tag{5}$$

$$\sum_{k \in K} \sum_{t \in T} L_{ks} r_{ikt} \geq D_i \qquad \forall i \in V \setminus \{0, n+1\},\ s \in S_i \tag{6}$$

$$\sum_{\tau=t}^{\bar{d}} \sum_{k \in K} r_{ik\tau} \geq y_{it} \qquad \forall i \in V, t \in T \tag{7}$$

$$\sum_{\tau=t}^{\bar{d}} \sum_{k \in K} r_{ik\tau} \leq M y_{it} \qquad \forall i \in V, t \in T \tag{8}$$

$$\sum_{t \in T} \left(y_{it} + \sum_{\tau=0}^{t} x_{i\tau} - 1 \right) \leq P_i \qquad \forall i \in V \setminus \{0, n+1\} \tag{9}$$

$$r_{ikt} \in \{0, 1\} \qquad \forall i \in V, k \in K, t \in T \tag{10}$$

$$x_{it},\ y_{it} \in \{0, 1\} \qquad \forall i \in V, t \in T \tag{11}$$

$$P_i \geq 0 \qquad \forall i \in V \tag{12}$$

The total project duration is minimized using objective function (1). Constraints (2) make sure that every activity is started exactly once within the planning horizon. The precedence constraints between activities are fulfilled by constraints (3). P_i is the actual duration of activity i and a decision variable resulting from resource allocation and skill levels in the problem under consideration (cf. Figs. 1 and 2). Inequalities (4) define that an available resource can only work on one activity at a time. Constraints (5) connect the start of a real activity to the assignment of resources and ensure that a volunteer can be assigned to an activity from time t at which the activity starts. Inequalities (6) ensure the sufficient assignments of workers to activity i considering the respective skill levels. Please note that a resource k can also be used to process an activity i if the remaining working hours required to cover D_i are less than the working hours provided by a resource k at the corresponding skill level. This may lead to D_i being exceeded by the assignment of k. If such an assignment results in an earlier end of the project, however, it makes sense to implement it. Auxiliary variable y_{it} is defined in restrictions (7) and (8) as $y_{it} = 1$ if resources are assigned to i after time t and hence, the activity is not completed. M represents a sufficiently high number. Constraints (9) calculate the actual duration P_i of a real activity i as an interval in which the activity is started and not yet completed. Thereby, possible interruptions are also considered. Finally, constraints (10) to (12) define the binary as well as real decision variables.

3 Preliminary Computational Results

In our computational study, we consider 30 instances with $n = \{10, 20, 30\}$ real activities that are constructed on the basis of the PSPLIB benchmark provided by Kolisch and Sprecher [3]. The instances are enhanced by problem-specific parameters. For example, the number of overall considered skills is randomly set from 3 to 5. Assuming that the standard skill level $L_{ks} = 1$ is the most common in reality, it receives the highest generation probability. The levels $L_{ks} = 0$ and $L_{ks} = 2$ are the most unlikely. The availability of resources in the system is chosen randomly within the planning horizon. How long the volunteers stay in the system is determined within $\{8, 9, \ldots, 18\}$ time units.

The tests were carried out on a server with two 2.1 GHz processors and 384 GB of RAM, using eight cores with CPLEX 12.9 in GAMS 25.1. In order to speed up the solver, we defined appropriate CPLEX options and figured out a good combination of cuts adapted to the characteristics of our model formulation. We identified cover, clique and mixed-integer rounding cuts as being suitable to our problem. Moreover, we prioritized decision variables x_{it} for the branching.

Table 1 shows the results obtained. Instances with numbers 1–10 involve 10 activities. Instances 11–20 (21–30) contain 20 (30) real activities. In addition to the number of resources $|K|$ and the number of skills $|S|$, which have a large impact on solvability, Table 1 shows the received objective function value $F(\mathrm{x})$ in days, the CPLEX gap in percentage [%], and the computing time in seconds [s]. Please note that a positive gap value only occurs if the specified runtime limit of the solver (i.e., 2 h) has been reached. It turns out that for all instances with $n = 10$ activities optimal solutions are found. The computing times vary between 7 and 3059 s. The instances with $n = 20$ activities already indicate the limited efficiency of the solver. Only two instances are solved to optimality within 2 h. For all other instances, the average gap-value is 12.6%, which means that the minimum project duration could be 1 or 2 days less. The instances with $n = 30$ real activities could not be solved optimally within the runtime limit. The average gap-value is 20.8%. For case no. 29, the solver CPLEX terminated after 2 h without finding a feasible solution.

4 Conclusion and Outlook

The paper addresses a particular problem that appears in the disaster response phase. Specific characteristics are considered such as skills, skill levels, possible interruptions of activities, and variable activity durations. In order to find a solution to the problem, CPLEX was used. The results show that even for small instances up to 30 activities, which is far away from realistic scenarios, the computing times and gap-values can be really high. Consequently, CPLEX cannot be used to generate acceptable solutions within suitable computing time. For the use in a decision support system, a serial schedule generation scheme (cf., e.g., Kolisch and Hartmann [2]) must therefore be implemented that determines good approximate solutions within a short time. Moreover, an adaption of the problem to the dynamic and stochastic characteristics of a disaster should be integrated, where the current static and deterministic model is transformed into a dynamic formulation with stochastic components.

Table 1 Preliminary computational results for 30 generated instances

No.	$\|K\|$	$\|S\|$	$F(x)$	Gap	CPU
1	12	4	13	0.0	10
2	11	3	14	0.0	16
3	11	4	26	0.0	57
4	12	4	27	0.0	3059
5	12	5	17	0.0	30
6	12	3	16	0.0	52
7	12	3	11	0.0	7
8	12	5	17	0.0	150
9	13	4	17	0.0	105
10	11	5	15	0.0	20
No.	$\|K\|$	$\|S\|$	$F(x)$	Gap	CPU
11	20	4	13	15.4	7246
12	19	3	18	0.0	2663
13	25	5	15	13.3	7245
14	20	4	13	15.4	7225
15	22	5	16	6.3	7211
16	19	3	13	0.0	399
17	23	4	16	12.5	7237
18	18	4	16	12.5	7256
19	25	3	16	18.8	7242
20	24	4	14	7.1	7209
No.	$\|K\|$	$\|S\|$	$F(x)$	Gap	CPU
21	25	3	17	23.5	7240
22	24	4	15	13.3	7226
23	25	4	16	25.0	7245
24	26	5	31	51.6	7223
25	23	4	16	12.5	7226
26	24	4	15	6.7	7210
27	24	3	21	19.0	7247
28	26	5	13	15.4	7222
29	25	3	–	–	7227
30	23	4	15	20.0	7217

References

1. Altay, N., Green, W.G.: OR/MS research in disaster operations management. Eur. J. Oper. Res. **175**, 475–493 (2006)
2. Kolisch, R., Hartmann, S.: Heuristic algorithms for the resource-constrained project scheduling problem: classification and computational analysis. In: Węglarz, J. (ed.) Project Scheduling. International Series in Operations Research & Management Science, vol. 14. Springer, Boston (1999)
3. Kolisch, R., Sprecher, A.: PSPLIB – a project scheduling problem library. Eur. J. Oper. Res. **96**, 205–216 (1996)

4. Kreter, S., Rieck, J., Zimmermann, J.: Models and solution procedures for the resource-constrained project scheduling problem with general temporal constraints and calendars. Eur. J. Oper. Res. **251**, 387–403 (2016)
5. Mansfield, R.S.: Building competency models: approaches for HR professionals. Hum. Resour. Manag. **35**, 7–18 (1996)
6. Naber, A., Kolisch, R.: MIP models for resource-constrained project scheduling with flexible resource profiles. Eur. J. Oper. Res. **239**(2), 335–348 (2014)

Part XVI
Revenue Management and Pricing

Capacitated Price Bundling for Markets with Discrete Customer Segments and Stochastic Willingness to Pay: A Basic Decision Model

Ralf Gössinger and Jacqueline Wand

Abstract Current literature on price bundling focuses on the situation with limited capacity. This paper extends this research by considering multiple discrete customer segments each with individual size and buying behavior represented by distributed willingness to pay and max-surplus rule. We develop a stochastic non-linear programming model that can be solved by standard NLP optimization software. Aiming to examine the model behavior, we conduct a full-factorial numerical study and analyze the impact of capacity limitations and number of customer segments on optimal solutions.

Keywords Price bundling · Capacity · Stochastic programming

1 Introduction

Price bundling describes the strategy of selling a package of different individual products at a single aggregate price that may deviate from the sum of individual prices. When product-wtps are negatively correlated, this technique increases sales volume and profit [1]. Focusing on pricing and abstracting from other influencing factors, *customer's buying behavior* is described by the assumptions: (1) A product or bundle is purchasable when its price is not above the willingness to pay (wtp). (2) Out of a non-empty set of purchasable products or bundles, one of those with the highest surplus (wtp − price) is chosen. (3) If neither products nor bundles are purchasable, the customer buys nothing [2].

In the present paper, we focus on the problem of setting optimal prices for individual products and bundles in a monopolistic market when the vendor's capacity is limited. In this case, the advantageousness of a sales allocation is not

R. Gössinger (✉) · J. Wand
Department of Production Management and Logistics, University of Dortmund, Dortmund, Germany
e-mail: ralf.goessinger@udo.edu

J. S. Neufeld et al. (eds.), *Operations Research Proceedings 2019*, Operations Research Proceedings, https://doi.org/10.1007/978-3-030-48439-2_75

only dependent on the fit of the chosen prices with bundle- and product-wtps but additionally on the fit of the capacity requirements with available capacity [3, 4]. For given attractive prices, customers can prefer a sales allocation that differs from the one which the vendor would choose. Therefore, bundle prices need to be attractive (unattractive) for the customer in order to force (avoid) sales allocations that are preferable (undesirable) for the vendor. In addition, we generalize the problem by considering discrete customer segments, i.e. clusters of customers which are similar in specific marketing-relevant characteristics (e.g. demographic, geographic, psychographic, behavioral).

Up until now, for price bundling with discrete customer segments, formal models are used which assume unlimited vendor capacity and/or deterministic wtps for each customer segment. One research perspective [3–6] analyzes bundling strategies in the setting with a limited inventory of individual products and one discrete customer segment which is characterized by an unimodal wtp distribution. The analyses are limited to explanatory models for the two-products-one-bundle case. Another research perspective [7–9] formulates MIP models and considers both, limited inventory of products and deterministic wtps. Customer behavior and its impact on objective value are taken into consideration in different ways. On the one hand, customers are assumed to demand the surplus maximizing bundle only, regardless of its availability [7]. On the other hand, customers are assumed to arrive in a certain sequence and to decide on buying the most favorable of still available bundles or to go without buying [8, 9]. Even though MIP models allow for stochastic optimization when multiple customer settings are seen as scenarios which are realizations of stochastic wtps [9], numerical studies refer to settings with deterministic wtps of multiple customer segments and the maximum case of three products and four bundles [8] or unimodal wtp-distributions of one customer segment and the maximum case of 20 products and one bundle [9]. Chew, Lee and Wang [10] combine both perspectives by optimizing decisions on pricing and inventory by applying a two-stage stochastic program. Their focus is on the two-products-one-bundle case where both wtp and size of the customer segment are stochastic variables. Since standard solvers are able to solve only small problem instances to optimality in an acceptable time, the application of problem-specific heuristics [10] and metaheuristics [7, 9] is proposed. Some of the studies on capacitated price bundling do analyze the impact of scarcity (ratio of capacity and demand) on the profit of bundling strategies [5, 6, 9]. It turns out that the expected profit decreases when the ratio decreases. Another focus of analyses is the impact that limited capacity unfolds in combination with additional factors: In situations with limited capacity, bundling becomes also beneficial for products with high wtp-correlation and asymmetric wtp-cost-difference and loses its advantage for products with low wtp-correlation when the wtp-cost-difference is symmetric and low [4]. In case of sub-additive wtps and limited capacity, the advantageousness depends on both, capacity scarcity and mismatch of dedicated capacity types [3]. In the cited literature (except for [3]), bundle availability is limited by the inventory of individual products. However, this is not necessarily the same as *capacity limitations*.

In our study we analyze price bundling from the perspective of a capacity-constrained supplier serving a market with different discrete customer segments, each characterized by heterogeneous customers. In order to simultaneously determine optimal prices of products and bundles along with optimal allocations of capacity requirements to resources, we formulate a non-linear stochastic programming model (Sect. 2). In a numerical study (Sect. 3), we observe the model behavior in terms of the structure of optimal solutions. For this purpose, a set of systematically generated problem instances is solved using standard software for non-linear optimization. Furthermore, we apply statistical analyses to figure out the impact of capacity level, number of customer segments, heterogeneity within customer segments and correlation of mean wtps across customer segments on observed values. Finally, in Sect. 4, we summarize the main results of the paper and draw aggregate conclusions.

2 Model

Assumptions We consider a company that produces goods g ($g = 1, \ldots, G$) by employing non-consumable resources i ($i = 1, \ldots, I$) whose capacity is limited to $A = \left(a_i \in \mathbb{R}_0^+\right)$. Manufacturing a unit of products requires capacity in the amount of $\overline{B} = \left(b_{ig} \in \mathbb{R}_0^+\right)$ and is accompanied with direct cost $\overline{C} = \left(c_g \in \mathbb{R}_o^+\right)$. The company is acting on a monopolistic market consisting of customer segments k ($k = 1, \ldots, K$) with maximum demand $N = \left(n_k \in \mathbb{R}_0^+\right)$. For individual products, the wtp of each customer segment is a stochastic variable that follows a normal distribution $\overline{W} = \left(w_{gk} \sim f_N\left[\mu_{gk}, \sigma_{gk}^2\right]\right)$. Within a customer segment, product-wtps are uncorrelated, whereas mean wtps of different customer segments may be correlated $\rho_{gg'}^{kk'} \in [-1, +1] \forall g, g', k, k' \mid g \neq g', k \neq k'$. The parameters are organized in matrices $\overline{W}^{\mu} = \left(\mu_{gk} \in \mathbb{R}_0^+\right)$, $\overline{W}^{\sigma^2} = \left(\sigma_{gk}^2 \in \mathbb{R}_0^+\right)$. In addition to individual products, the company offers bundles j ($j = 1, \ldots, J$) composed of different individual products. The bundle design is defined with $Q = (\varsigma_{gj} \in \{0, 1\})$ [8, 11] in such a way that J ($G \leq J \leq 2^G - 1$) non-empty and non-identical bundles exist [12]. For reasons of notation simplicity, we define $g = j \; \forall \, g \leq G, j \leq G$ [8].

Further assumptions are made according to [10]. Assuming strict additivity of production, we get bundle-related information on direct cost $C = \left(c_j \in \mathbb{R}_0^+\right) = \overline{C} * Q$ and capacity requirements $B = \left(b_{ij} \in \mathbb{R}_0^+\right) = \overline{B} * Q$. Furthermore, for each customer segment strict additivity of wtps is assumed. Hence, bundle-wtps are stochastic variables that follow a normal distribution $W = \left(w_{jk} \sim f_N\left[\mu_{jk}, \sigma_{jk}^2\right]\right)$ which is the convolution of respective product-wtp distributions [13]. That is, $W^{\mu} = \left(\mu_{jk} \in \mathbb{R}_0^+\right) = Q^T * \overline{W}^{\mu}$ and $W^{\sigma^2} = \left(\sigma_{jk}^2 \in \mathbb{R}_0^+\right) = Q^T * \overline{W}^{\sigma^2}$.

The company decides on bundle prices $P = \left(p_j \in \mathbb{R}_0^+\right)$ in a profit-maximizing way while simultaneously taking account of capacity limitations. Customers decide whether (or not) to buy a bundle and, as the case may be, which bundle is to be chosen. The decisions $X = \left(x_{jk} \in \mathbb{R}_0^+\right)$ are made in a surplus maximizing way: Out of the available bundles, those with the highest positive surplus are purchased first. If these bundles are sold out demand directs to bundles with the next lower surplus [9]. For the segment k bundle j generates a non-negative surplus with probability

$$\int_{p_j}^{\infty} f_{jk}(w) \cdot dw = 1 - F_{jk}\left(p_j\right)$$

The probability that bundle j generates a higher surplus than j' depends on the differences of both, wtps $g_{jj'k}(wd) = f_{jk}(w) - f_{j'k}(w)$ and prices $pd_{jj'} = p_j - p_{j'}$, that is

$$\int_{pd_{jj'}}^{\infty} g_{jj'k}(wd) \cdot d\ wd = 1 - G_{jj'k}\left(pd_{jj'}\right)$$

Hence, the bundle j generates the highest surplus of all bundles with probability

$$\prod_{j' \neq j} 1 - G_{jj'k}\left(pd_{jj'}\right)$$

Due to normally distributed wtps for each segment, all bundles are attractive with a certain positive probability.

Model formulation Based on the assumptions the following *model* can be derived:

$$\max\ m = \sum\nolimits_{j,k} \left(p_j - c_j\right) \cdot x_{jk} \tag{1}$$

s.t.

$$\sum\nolimits_{j,k} x_{jk} \cdot b_{ij} \leq a_i \quad \forall i \tag{2}$$

$$\sum\nolimits_{j} x_{jk} \leq n_k \quad \forall k \tag{3}$$

$$x_{jk} \leq n_k \cdot \left(1 - F_{jk}\left(p_j\right)\right) \cdot \prod_{j' \neq j} \left(1 - G_{jj'k}\left(p_j - p_{j'}\right)\right) \quad \forall j, k \tag{4}$$

The objective (1) is to maximize the profit margin by setting bundle prices that induce buying decisions. The optimization has to pay respect to the constraint (2) that capacity cannot be overloaded. Constraints (3, 4) describe the behavior of customer segments: (3) Potential demand of each segment cannot be exceeded, and (4) the quantity of bundle j that could be sold to segment k is limited to the fraction of its population for which bundle j generates the highest non-negative surplus.

In sum, the decision model is a stochastic non-linear program. *Non-linearity* is present in the quadratic objective function (1). Non-linearity and stochasticity are present in constraint (4) due to the CDFs and their multiplication. In addition, the solvability of this model is hindered by both, the complicated numerical handling of *normal distributions* and the *unbounded solution space*. To handle the former difficulty, we approximate all CDF terms (F_{jk}, $G_{jj'k}$) with metamodels based on the logistic function $1/(1 + e^{-nl \cdot (x - \mu)/\sigma})$ and $nl = 1.70099$ ($R^2 = 1$, $SER = 0.00512$). The latter difficulty can be diminished by restricting the prices from above and below in different ways. The extent of considered deviations from mean can be controlled by ν (times of standard deviations, here: $\nu = 6$) such that unlikely values remain unconsidered:

$$\min_k \left(\mu_{jk} - \upsilon \cdot \sigma_{jk}\right) \leq p_j \leq \max_k \left(\mu_{jk} + \upsilon \cdot \sigma_{jk}\right) \quad \forall j \tag{5}$$

Further restrictions are found by analyzing the profit function for each j, k-combination in dependence of the price:

$$\min_k \left(p^*_{jk}\right) \leq p_j \leq \max_k \left(\overline{p}_{jk}\right) \quad \forall j \tag{6}$$

The values p^*_{jk}, $\overline{p}_{jk}$ are solutions to $c_j = p_{jk} - (1 - F_{jk}(p_{jk}))/f_{jk}(p_{jk})$ (profit-maximizing uncapacitated price) or $F_{jk}(p_{jk}) = 1 - \varepsilon/(p_{jk} - c_j)$ (negligible profit in the amount of ε), respectively. In both cases, the numerical solution requires low computational effort [13].

3 Numerical Study

In order to examine the model behavior, we conduct a full-factorial numerical study of the two-products-one-bundle case. The factors are number of discrete customer segments $K \in \{3, 6, 9\}$, correlation of customer segments' mean wtps $\rho \in \{$negative correlation, no correlation, positive correlation$\}$, wtp variation coefficient $cv_{jk} \in \{0.1, 0.2, 0.3\}$, capacity level $A \in \{25, 175, 325\}$ (one capacity type). For each factor combination, three instances are generated by sampling mean wtps from a continuous uniform distribution $\mu_{jk} \sim U(1, 6)$ for 15 customer segments and selecting a respective number of segments in accordance with the ρ-value sign. Constant values are used for $b_1 = b_2 = 1$, $b_3 = 2$, $c_1 = 1$, $c_2 = 0.5$, $c_3 = 1.5$ and $n_k = 10 \ \forall \ k$. The resulting 243 instances are solved by means of the NLP solver BARON 15 run on a Windows PC (3.60 GHz Intel Core i7-7700 CPU, 16 GB RAM) with a solution time limit of 10 min. Nine instances could not be solved to optimality within this limit. The remaining 234 instances are solved to optimality and form the statistical

Table 1 Numerical study results ($N = 234$)

y	m	x_{1+2}	x_3	p_{1+2}	p_3
R^2	0.896	0.561	0.876	0.724	0.768
α_y	−27.729	−27.150	−0.440	7.310	6.050
β_{Ay}	0.676***	−0.095***	0.226***	0.017***	−0.027***
β_{Ky}	27.236***	n.s.	n.s.	n.s.	0.869***
$\beta_{\rho y}$	−17.481***	−5.874***	4.868**	1.044*	n.s.
β_{cvy}	n.s.	−81.489**	n.s.	9.010***	n.s.
γ_{A^2y}	−0.002***	0.000***	−0.001***	−0.000***	−0.000***
γ_{K^2y}	−1.403***	n.s.	n.s.	n.s.	−0.037***
γ_{ρ^2y}	15.855***	−3.954***	n.s.	2.300***	1.237***
γ_{cv^2y}	n.s.	131.283*	n.s.	n.s.	n.s.
γ_{AKy}	0.079***	n.s.	0.023***	n.s.	−0.001***
$\gamma_{A\rho y}$	n.s.	0.029***	−0.036***	−0.009***	0.003***
γ_{Acvy}	−0.598***	n.s.	n.s.	0.056***	0.009**
$\gamma_{K\rho y}$	n.s.	−0.499***	−0.937**	0.114*	0.081***
γ_{Kcvy}	n.s.	n.s.	n.s.	1.738***	n.s.
$\gamma_{\rho cvy}$	62.589**	n.s.	18.781**	5.581***	−2.273***

Key: n.s.: $p > 0.1$; *: $p < 0.1$; **: $p < 0.05$; ***: $p < 0.01$

basis for regression analyses with second-order polynomials (regressors $r \in \{A, K, \rho\}$, regressands $y \in \{m, x_{1+2}, x_3, p_{1+2}, p_3\}$) considering two-factor interactions (see Table 1).

New insights are generated by analyzing the impact of the *number of customer segments K*. Profit increases in K for all factor combinations. Sales x_{1+2} of individual products increase (decrease) in K when ρ is negative (positive), and prices p_{1+2} increase in K for each combination of ρ and cv values. Bundle sales x_3 increase in K when A is not low and ρ is not high. The bundle price p_3 increases in K for all factor combinations.

Previous research pointed out that negatively correlated wtps generally do increase bundle sales. In contrast, our study reveals that bundle sales x_3 decrease for negative ρ values when A is low and K is not high. Whereas, high positive as well as not low negative ρ values increase the bundle price p_3. In addition, cv decreases profit when A is not low and ρ is not high as well as decreases product sales x_{1+2} for all factor combinations. Furthermore, an inconclusive impact of capacity A on bundle sales x_3 is revealed. For a not high A the impact on bundle sales x_3 is positive, whereas the impact is negative for a high A when K is low.

4 Conclusions

Extant analyses of price bundling reveal that limited capacity influences the advantageousness of bundling. In order to analyze capacitated price bundling in a more general setting, we develop a stochastic non-linear programming model. In contrast to existing models, instead of limited stock, it considers the capacity of non-consumable resources. Furthermore, it captures multiple discrete customer segments that are characterized by different distributions of willingness to pay and different segment sizes. The model behavior is inquired utilizing a full-factorial numerical study. We conduct regression analyses with multi-factor second-order polynomials to figure out the direct impacts of capacity limitations and the number of customer segments on optimal solutions and indirect impacts due to interactions with other varied factors. It turns out that formerly reported results on direct capacity impact are verified. In addition, we get indications that impacts reported for other factors partially become blurry when capacity limitations are relevant.

References

1. Adams, W.J., Yellen, J.L.: Commodity bundling and the burden of monopoly. Q J Econ. **90**, 475–498 (1976)
2. Hanson, W., Martin, R.K.: Optimal bundle pricing. Manag Sci. **36**, 155–174 (1990)
3. Banciu, M., Gal-Or, E., Mirchandani, P.: Bundling strategies when products are vertically differentiated and capacities are limited. Manag Sci. **56**, 2207–2223 (2010)
4. Cao, Q., Stecke, K.E., Zhang, J.: The impact of limited supply on a firm's bundling strategy. POM. **24**, 1931–1944 (2015)
5. Bulut, Z., Gürler, Ü., Şen, A.: Bundle pricing of inventories with stochastic demand. EJOR. **197**, 897–911 (2009)
6. Gürler, Ü., Ötztop, S., Şen, A.: Optimal bundle formation and pricing of two products with limited stock. Int J Prod Econ. **118**, 442–462 (2009)
7. Azadeh, A., Songhori, H., Salehi, N.: A unique optimization model for deterministic bundle pricing of two products with limited stock. Int J Sys Assur Eng Manag. **8**, 1154–1160 (2017)
8. Barrios, P.S.C., Cruz, D.E.: A mixed integer programming optimization of bundling and pricing strategies for multiple product components with inventory allocation considerations. In: Proc IEEM 2017, pp. 16–20. IEEE, Piscataway (2018)
9. Mayer, S., Klein, R., Seiermann, S.: A simulation-based approach to price optimisation of the mixed bundling problem with capacity constraints. Int J Prod Econ. **145**, 584–598 (2013)
10. Chew, E.P., Lee, L.H., Wang, Q.: Mixed bundle retailing under stochastic market. Flex Serv Manuf J. **27**, 606–629 (2015)
11. Fang, Y., Sun, L., Gao, Y.: Bundle-pricing decision model for multiple products. Proc Comp Sci. **112**, 2147–2159 (2017)
12. Honhon, D., Pan, X.A.: Improving profits by bundling vertically differentiated products. POM **26**, 1481–1497 (2017)
13. Olderog, T., Skiera, B.: The benefits of bundling strategies. SBR. **52**, 137–159 (2000)

Insourcing the Passenger Demand Forecasting System for Revenue Management at DB Fernverkehr: Lessons Learned from the First Year

Valentin Wagner, Stephan Dlugosz, Sang-Hyeun Park, and Philipp Bartke

Abstract The long-distance traffic division of Deutsche Bahn (DB) uses a revenue management system to sell train-tickets to more than 140 million passengers per year. One essential component of a successful Railway Revenue Management system is an accurate forecast of future demand. To benefit from a tighter integration, DB decided in 2017 to develop its own forecast environment PAUL (Prognose AUsLastung) to replace the legacy third-party forecasting system. This paper presents the conceptual and technical setup of PAUL. Furthermore, experiences of the first year using PAUL as a production forecast environment are presented: It turned out that PAUL has a higher forecasting quality than the predecessor system and that the insourcing led to a constructive collaboration of PAUL system experts and revenue managers, which is beneficial for identifying opportunities for improvement.

Keywords Forecasting system · Revenue management

1 Introduction

Deutsche Bahn (DB) Fernverkehr AG offers long-distance rail journeys to over 140 million passengers each year. In general, the operation of public mass transport systems is a fix-cost intensive business and a high average utilization rate is required to operate profitably. Additionally, DB intends to provide an optimal journey experience for each customer. However, the quality of a customers' journey experience tends to decrease once utilization rates approach 100% or more. In order to achieve both objectives, high average utilization rates and minimal overcrowding, an effective demand management is essential.

V. Wagner (✉) · S. Dlugosz · S.-H. Park · Philipp Bartke
DB Fernverkehr AG, Frankfurt a.M., Germany
e-mail: valentin.wagner@deutschebahn.com

J. S. Neufeld et al. (eds.), *Operations Research Proceedings 2019*, Operations Research Proceedings, https://doi.org/10.1007/978-3-030-48439-2_76

DB Fernverkehr AG offers train-specific as well as flexible tickets. Therefore, it operates a semi-open system with *controlled* (train-specific connection, low-fare) and *uncontrolled* passengers (no train-specific connection, high-fare). Train-specific tickets account for about 30% of all tickets. Using dynamic price discounts on these train-specific tickets, price-sensitive passengers can be steered to less occupied connections. Therefore, it is crucial to have an accurate forecast of expected controlled and uncontrolled passengers per train-leg, to determine the optimal number and discount level for train-specific tickets.

An effective Revenue Management (RM) System for setting optimal fares works in two main steps: (1) Demand forecasting and (2) Decision on booking classes/prices based on the forecast(s). The forecast, on which the focus will be placed in the following, is the basis for the decision if a certain seat can still be sold with higher revenue at a later point in time [2].

Since the introduction of revenue management at DB Fernverkehr AG, a standard third-party forecasting tool has been used. The experience gained with this tool and the general RM process revealed two points: First, a detailed understanding of the forecast algorithm is beneficial for the successful daily work of the yield managers and their confidence in the forecasted values. Second, the railway operation of DB Fernverkehr AG has numerous unique features that are relevant to forecasting, which a standard tool that is intended to cover all areas of revenue management cannot take into account. For example, the legacy system is not specially optimized for the requirements of rail traffic, where an itinerary involves significantly more different leg sections than e.g. in aviation. Therefore, DB decided in 2017 to develop its own forecast environment PAUL (Prognose AUsLastung).

2 The System PAUL

Kuhn and Johnson [4] have pointed out that one of the foundations of a good forecasting model is a deep understanding of the problem and the relevant data. These last two points are firmly embedded in the respective RM department. Based on the presented experiences and the special requirements of the RM process at DB Fernverkehr AG, it has been concluded that the following five requirements for a forecasting system can best be met by insourcing the forecasting system:

1. Precise forecasts for many destinations (about 300) and the entire booking horizon (up to 400 days) in a semi-open system for all fare classes ($\sim$14)
2. Smooth integration into the existing RM System (EMS)
3. Robustness against seasonal fluctuations as well as long-term and short-term timetable changes
4. Interpretability of the forecasts to achieve a higher acceptance of the employees in dealing with the values
5. Short development cycles, e.g. to be able to react rapidly to operational changes

2.1 Integration of PAUL in the Existing Revenue Management System

A forecasting cycle of PAUL starts with loading the current booking information from the existing revenue management system (EMS). The obtained data is processed within a PAUL internal ETL (Extract-Transform-Load) run which integrates data from both historical and future train runs and transforms that data into a form suitable for machine learning (ML) algorithms.

The historical data is used to create forecasting models and the data points for the future booking period are used to forecast demand. The calculated forecast values are returned to EMS for quota optimization, which is the basis to determine the optimal price for each itinerary.

EMS has limited storage and thus PAUL has to persist the historical data on its own to provide longer time series to the ML algorithms used for forecasting. The data is highly structured and of medium size, so a standard SQL RDMBS can be used. PAUL is based on a Microsoft SQL-Server with activated R language extension. This allows for easy maintenance by company's central IT service department and provides a lean framework for applying any ML algorithms available in R [5]

2.2 Model Features

Passenger demand for the fare classes in PAUL is estimated from information on public holidays, train type, weekday, days to departure, departure hour, fare class booking status and train stop section. The major part of the information on historical train utilization is provided by EMS through its ticket database. Public holidays from external sources are added into PAUL.

Public holidays lead to strong deviations from the weekly demand profile. Therefore an 'effective traffic weekday' is computed by remapping the weekday information to reflect the influence of public holidays. Previous experience has shown that frequent changes to the specific route of trains (e.g. due to the yearly updated train schedule) is challenging for forecasting demand primarily due to having not sufficient historical data for training reasonable models. This problem is tackled by abstracting the train route description. A set of 28 national and 13 international railway stations is used to provide a robust description of each train leg and the route of each train in total. Each stop section is described relative to these 41 stations, allowing to determine wether the train never reaches this station, the train has already been there, or it will go to the station in the future.

Additionally, aggregations of train types are made to improve system performance: The train types are assigned to four main train types (ICE, IC, RE, BUS). The booking horizon is divided into 19 classes of different durations for the days before departure. For each train run and its individual legs the respective departure hour is determined. The cumulative bookings are recorded for all fare classes.

2.3 Prediction Model

PAUL uses an ensemble learning approach combined with model selection and/or weighted model averaging. For this purpose, two sets of predictions models are estimated: one set (set 1) with and one set (set 2) without using the actual number of sold tickets. Both sets consists of predictions models for each prediction period—starting at 400 days and ranging up to the day before departure—for each price category, each class and each leg.

Furthermore, for models of set 1 the framework allows to learn additive ('demand to come', i.e. the learned function should predict the remainder of passengers relative to the current known amount of passengers) and multiplicative models (trying to predict the multiplicative factor on the current known amount of passengers).

Additionally, crossvalidated prediction errors are computed and used as performance measures for model selection as well as input for the *Expected marginal seat revenue* (EMSRb) algorithm [6].

Currently, regression trees [1] are the workhorse for producing the major part of all predictions. They are accompanied by some simple moving medians, if there is a sufficient amount of observations for a certain train on that specific weekday and leg (due to its surprisingly good prediction performance). The system, however, is designed to easily integrate any prediction model by adding appropriate entries to a configuration table.

3 Results

3.1 System Performance

The prediction performance is measured in mean absolute prediction error (MAPE) relative to the passenger class capacity (MAPE $= |(\hat{Y} - Y)|/\text{CAP}$). The observed train occupation is taken from the passenger counts provided by a system called RES (ReisendenErfassungsSystem).

In Fig. 1a the daily average MAPE of PAUL of the sum of all fare and passenger classes since start of the parallel operation of PAUL and its legacy system is shown. Additionally, the differences of the predictions from the legacy system and PAUL are presented in Fig. 1b. Here, only values for which both systems calculated a forecast were taken into account.[1] In general, it can be observed that the performance of PAUL improved over time, in absolute terms (left figure) and relative to the legacy system (right figure).

[1] For 2018/10/25 the legacy system delivered only a fraction of the required forecasts, therefore the data point for this day was not considered here.

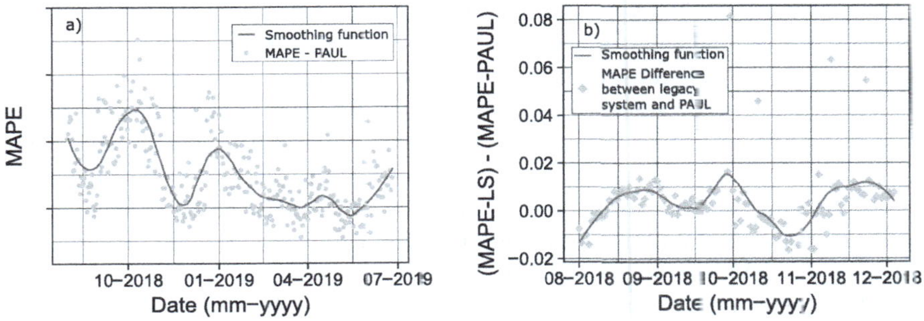

Fig. 1 Development over time of the daily PAUL-MAPE values for all computed forecast values (**a**). Improvement of PAUL over the legacy system in MAPE (**b**)

However, the left figure shows, that there were two significant throwbacks in autumn 2018 and in the beginning of 2019. In the first case the poor performance arises from significantly decreased operational performance during that timeframe leading to a larger fraction of journeys that could not be realized as planned, i.e. the historical data used for learning the models were not adequate. For the second throwback, the new train schedule which becomes active every December was the main cause. New train schedules include changes in train routes switching patterns, connection times and train systems (e.g. IC to ICE). These changes have a strong influence on the attractivity of involved legs, and new utilization patterns have to be learned. Overall, Paul showed a higher forecast quality compared to the legacy system on 63% of the comparison period.

3.2 *Lessons Learned from the First Year*

Among other things, the first year of operation revealed three areas for improvement. First, there are sometimes significant jumps in forecasts for individual target variables with respect to the remaining days to departure. These forecast jumps cause additional work for the revenue managers to correct the changing quotas if necessary. The temporal forecast variability was mainly caused by the following points:

1. Switching between the available model classes can lead to jumps.
2. Different model generations of one model class can lead to varying predictions for the same target between different training runs.
3. Individual target variables have shown a strong sensitivity to changes in the cumulative booking status.

In order to prevent these forecast jumps, a changed model selection procedure is currently being tested and new forecast algorithms are being analyzed.

Furthermore, the first year showed that there may be significant and short-term changes in the timetable for operational reasons (e.g. major construction sites). This can cause such drastic changes in the operational process that a precise forecast based on the standard procedure is not possible because the amount of data available is too small. This was tackled by implementing an additional feature which allows the revenue managers to integrate corrective expert knowledge to the PAUL system for these special cases. This and other types of adjustments could be developed, tested and integrated into the production system within a very short time, so that PAUL can also make precise forecasts for such short-term changes. The development cycles of the old system were considerably longer and, therefore, it was often not possible to adapt adequately to short-time changes.

Also, though easily interpretable models are used in PAUL, e.g. decision trees, additional explanations were still required for the daily work of the revenue manager. Regular meetings between revenue managers and PAUL system experts have proven to be an optimal vehicle for explaining the forecasts and thus significantly increased acceptance and trust. These meetings can also ensure that all enhancements are tailored to the needs of daily users and based on the experience of revenue managers, implausible single values can be identified that might not be noticed by monitoring forecast quality at the system level. The fact that both Revenue Managers and PAUL forecasting experts are part of the same department made it easy to organize these regular meetings. This has emerged as another advantage of the insourcing process.

4 Conclusions and Outlook

The parallel operation of PAUL and its legacy system has shown that PAUL delivers more accurate forecast values overall compared to the legacy system. Integration into the entire RM process went smoothly. The focus on shorter development cycles has already paid off by enabling to implement multiple improvements to the forecast algorithms in the first months of operation. Being able to quickly react to arising issues and providing a direct and open communication channel between system experts and RM analysts proved to be valuable factors for ensuring the acceptance for and trust in the new forecasting system. Furthermore, these factors will significantly contribute to DB Fernverkehr's continuous effort to further improve forecasting quality.

In future development steps, a better representation of special holiday constellations such as Christmas or Easter are planned. Currently, integration of information on local events such as football matches or trade fairs are evaluated. It is also planned to better represent seasonal trends within PAUL.

References

1. Breiman, L., Friedman, J.H., Olshen, R.A., Stone, C.J.: Classification and Regression Trees. Wadsworth & Brooks/Cole Advanced Books & Software, Monterey (1984)
2. Ciancimino, A., Inzerillo, G., Lucidi, S., Palagi, L.: A mathematical programming approach for the solution of the railway yield management problem. Transp. Sci. **33**(2), 168–181 (1999)
3. Hastie, T., Tibshirani, R., Friedman, J.: The Elements of Statistical Learning: Data Mining, Inference, and Prediction. Springer, New York (2001)
4. Kuhn, M., Johnson, K.: Applied Predictive Modeling. Springer, New York (2013)
5. R Core Team: R: A Language and Environment for Statistical Computing. https://www.R-project.org
6. Talluri, K., Van Ryzin, G.: The Theory and Practice of Revenue Management, vol. 68. Springer Science & Business Media, New York (2005)

Tax Avoidance and Social Control

Markus Diller, Johannes Lorenz, and David Meier

Abstract This study presents a model in which heterogenous, risk-averse agents can use either (legal) tax optimisation or (illegal) tax evasion to reduce their tax burden and thus increase their utility. In addition to introducing individual variables like risk aversion or income, we allow agents to observe the behaviour of their neighbours. Depending on the behaviour of their peer group's members, the agents' utilities may increase or decrease, respectively. Simulation results show that taxpayers favour illegal evasion over legal optimisation in most cases. We find that interactions between taxpayers and their social networks have a deep impact on aggregate behaviour. Parameter changes such as increasing audit rates affect the results, often being intensified by social interactions. The effect of such changes varies depending on whether or not a fraction of agents is considered inherently honest.

Keywords Tax compliance · Tax avoidance · Tax evasion · Social influence · Agent-based modelling

1 Introduction

Empirical findings suggest that taxpayers' behaviour deviates from the predictions of analytical models (e.g., [4]). Some authors put forward psychological motives to account for the difference between theory and empirics. *Erard/Feinstein*, for example, consider that taxpayers experience guilt when evading taxes and shame when caught by the fiscal authorities [5]. *Bernasconi* suspects that taxpayers only overestimate the probability of audit [3]. Further studies propose non-psychological reasons. Conducting a laboratory experiment in which tax reports are selected for audit based on the individual deviation from the average reported amount,

M. Diller (✉) · J. Lorenz · D. Meier
University of Passau, Passau, Germany
e-mail: markus.diller@uni-passau.de; http://www.wiwi.uni-passau.de/taxation/

J. S. Neufeld et al. (eds.), *Operations Research Proceedings 2019*, Operations Research Proceedings, https://doi.org/10.1007/978-3-030-48439-2_77

Alm/McKee claim that taxpayers report too honestly because they fail to coordinate on a zero-compliance Nash equilibrium [2].

In recent years, agent-based modelling has become an important area of tax evasion research, allowing for incorporating realistic assumptions concerning individuals' heterogeneity when it comes to risk aversion, behavioural norms, social network interactions, etc. The possibility of reducing the tax burden *legally* has not been taken into account by this new stream of literature, however.

We therefore impose a social network that exhibits characteristics close to the small-world network, proposed by *Watts/Strogatz* [9]. Following *Fortin/Lacroix/Villeval*, we assume that agents receive (and consider in their calculus) social utility [6]. In line with recent experimental results [7] we assume that agents receive positive social utility when acting in the same way their neighbourhood does, while they receive social disutility when their behaviour differs from their environment. In Sect. 2 we develop an analytical model that captures both illegal evasion and legal optimisation. We then extend our model to incorporate network effects. In Sect. 3 we show and discuss the results of an agent-based simulation, while in Sect. 4 the paper concludes with a summary.

2 Model

2.1 Tax Law and Legal Tax Avoidance

It is possible to legally reduce one's tax burden to some extent, either by exploiting tax loopholes or by searching for special regulations This search is associated with some cost h modelled as a fraction of pre-tax income. We assume that the government is able to control the expected tax savings by either simplifying the tax code or adopting new provisions to close tax loopholes: the higher (lower) the tax complexity, the lower (higher) the expected tax savings.

We use an exponential distribution to model legal uncertainty. Opting for legal avoidance is modelled as drawing a random number θ from a probability density function f featuring positive support over $[0, \infty]$. θ is interpreted as the *share* of the original tax liability that can be avoided due to tax optimisation. As the exponential distribution doesn't allow for negative values of θ, tax optimisation cannot lead to a tax liability higher than the original. Values of $\theta > 1$ are possible, meaning that sometimes tax optimisation can even cause a negative tax payment.

The probability density function of the exponential distribution is given by

$$f(\theta) = \begin{cases} \gamma e^{-\gamma\theta} & \theta \geq 0 \\ 0 & \theta < 0 \,. \end{cases} \tag{1}$$

γ is interpreted as a *complexity parameter* that can be adjusted by the government. Mean and variance of the exponential distribution are given by $\frac{1}{\gamma}$ and $\frac{1}{\gamma^2}$, respectively. As explained above, a higher complexity parameter γ gives less space for tax optimisation and thus results in lower expected tax savings—the tax code is more resistant towards tax optimisation. Variance also decreases with increasing complexity, which means that the tax code is rather vague if it is simple, whereas with increasing sophistication it becomes more certain. We interpret certainty as predictability of legal decisions. If there are a large number of legal norms—i.e., if the tax code is very complex—the outcome of optimisation activities is fairly predictable. By contrast, if the tax code is worded in very abstract terms, greater optimisation is possible but uncertain. This seems plausible since we consider tax optimisation to be an option chosen by professionals whom we assume to have good legal skills. In other words, an uncertain outcome is only possible if no specific norm exists. Otherwise the professional will be bound to the law and thus produce small and predictable tax savings.

2.2 *Taxpayers' Optimisation Problem*

We assume that taxpayers can choose between three distinct strategies:

Legal Tax Optimisation Taxpayers can try to legally avoid taxes by exploiting tax loopholes. The intensity of searching for legal tax-saving options is denoted by tax planning effort ζ_o. That is, only fraction ζ_o of the original tax payment τ is subject to uncertain tax savings as described in Sect. 2.1. Choosing ζ_o can also be interpreted as choosing the number (fraction) of categories of income that will be subject to tax planning. Neglecting the possibility of tax consulting fees being deductible, choosing tax planning effort ζ_o results in consumption

$$C_o = W(1 - \zeta_o(\tau(1-\theta) + h) - (1-\zeta_o)\tau). \tag{2}$$

Illegal Tax Evasion We model illegal tax evasion as suggested by [1]. Evading taxes illegally requires no ex ante payment and reduces the tax liability by $\tau\zeta_e W_i$ where ζ_e is the fraction of evaded taxes. The tax authorities detect tax evasion with probability p. In case of detection, a penalty $t\zeta_e W_i$ is due, with $t \geq \tau$. Illegal tax evasion delivers consumption

$$C_e^w = W(1 - \tau(1-\zeta_e)), C_e^l = W(1 - \tau(1-\zeta_e) - t\zeta_e) \tag{3}$$

in case of detection (l, "lose") or no detection (w, "win"), respectively.

"Honest" Behaviour Finally, taxpayers can decide to refrain from any legal or illegal avoidance activities and just pay their full taxes due. Setting any effort to reduce the tax burden $\zeta_h = 0$, this results in consumption

$$C_h = W(1 - \tau). \tag{4}$$

Since we will include social effects of tax avoidance below, the direct payment effects are referred to as "private expected utility", denoted by the vector **PEU**. Thus, the taxpayer's problem can be written as

$$\max_{\mathbf{d},\boldsymbol{\zeta}} EU = \mathbf{d} \cdot \mathbf{PEU}(\boldsymbol{\zeta}) \tag{5}$$

where $\mathbf{d} = \left(d_o \; d_e \; d_h\right), d_s \in \{0, 1\} \forall s = o, e, h, \|\mathbf{d}\|_1 = 1$ is a vector of binary decision variables, $\boldsymbol{\zeta} = \left(\zeta_o \; \zeta_e \; \zeta_h\right)^T, \zeta_s \in [0, 1] \, \forall s = o, e, h$ is a vector of effort levels and **PEU** denotes the vector of the three strategies' respective expected utilities, depending on the choice of effort $\boldsymbol{\zeta}$:

$$\mathbf{PEU}(\boldsymbol{\zeta}) = \begin{pmatrix} \int_0^\infty u\left[C_o\right] f(\theta)\mathrm{d}\theta \\ (1-p)u\left[C_e^w\right] + pu\left[C_e^l\right] \\ u\left[C_h\right] \end{pmatrix}. \tag{6}$$

$u[\cdot]$ is a concave utility function that satisfies the von Neumann-Morgenstern axioms. To receive explicit solutions we impose a von Neumann-Morgenstern utility function featuring constant absolute risk aversion of degree λ: $u(x) = -e^{-\lambda x}$. The first row represents the expected utility from tax optimisation. The second row denotes the expected utility from tax evasion, whereas the third row captures the utility from neither optimising nor evading taxes. Equation (5) is solved by choosing $\boldsymbol{\zeta}$ such that $\nabla_{\boldsymbol{\zeta}} \mathbf{d} \cdot \mathbf{PEU} = \mathbf{0}$ and selecting the maximum element of $\mathbf{PEU}(\boldsymbol{\zeta}^*)$. We then obtain optimal tax planning and tax evasion effort levels

$$\zeta_o^* = \frac{1}{\lambda W}\left(\frac{1}{h} - \frac{\gamma}{\tau}\right), \zeta_e^* = \frac{1}{\lambda W}\frac{1}{t}\ln\left[\frac{\tau(1-p)}{p(t-\tau)}\right]. \tag{7}$$

2.3 *Introducing Agents and Social Utility*

Recent experimental studies indicate that the unethical behaviour of one group member increases the likelihood of other group members behaving unethically, too [7]. Therefore, agents receive social utility from acting in line with their environment, while they receive social disutility when differing from mainstream behaviour, which is reflected by the social utility term **SU**. Hence, the agents' total expected utility consists of two parts: private expected utility $\mathbf{PEU_i}$ as described above and social utility SU_i of their respective behaviours. For reasons of notational convenience the "agent labelling" index i is dropped where possible. Thus, an agent's problem can be written as

$$\max_{\mathbf{d},\boldsymbol{\zeta}} EU = \mathbf{d} \cdot \mathbf{PEU}(\boldsymbol{\zeta}) + \mathbf{d} \cdot \mathbf{SU} \tag{8}$$

where **SU** is a vector of utility of social interaction variables [6]:

$$\mathbf{SU} = \begin{pmatrix} s\,[m_o/m - m_e/m + \epsilon_o] \\ s\,[m_e/m - m_o/m + \epsilon_e] \\ s\,[-m_o/m - m_e/m + \epsilon_h] \end{pmatrix}. \tag{9}$$

With random variables $\epsilon_o, \epsilon_e, \epsilon_h$ capturing unobservable effects, Ω_i denotes the set of members of agent i's social network (= neighbours), $m = |\Omega_i|$ denotes the number of neighbours, $m_o = \sum_{k \in \Omega_i} d_{o,k}$ denotes the number of optimising neighbours. $s[\cdot]$ is a concave "social utility function"; we set $s[x] = -e^{-\rho x}$, where ρ measures the level of interaction between an agent and his social network.

When optimising (first row), the share of optimising neighbours triggers a positive social utility whereas the share of evading neighbours causes social utility to shrink. When evading (second row), there is an inverse connection. Honest taxpayers receive negative social utility from observing both optimising and evading neighbours, not triggering any effects in turn.

3 Simulation Results

The model was set up using NetLogo 6.0.4 [10]. We impose a population with a total size of 1000 agents, with one half of them being inherently honest and the other half deciding according to the model developed above. All agents are initially connected to their 20 closest neighbours in a regular ring lattice and may as well be connected to any other agent in the network with a probability of 0.01. Each agent is endowed with an identical and constant pre-tax income $W_i = 25$ and an individual risk aversion parameter λ_i drawn from an uniform distribution over [0.47, 1]. Each period, every agent, except for those being inherently honest, reconsiders his decision considering his neighbours' utilities. We assume that taxpayers who were found guilty of tax evasion refrain from both illegal evasion [8] and legal optimisation activities for 4 years. Links between two agents may rearrange with a certain probability. To resemble German law as closely as possible, we initially set our parameter values as follows: tax rate $\tau = 0.25$ (doubling after period 20), complexity parameter $\gamma = 5$, low (high) audit frequency $p^L = 0.024$ ($p^H = 0.063$), penalty rate $t = 0.8$ and cost of tax optimisation $h = 0.033$. High social interaction parameter ρ^H equals low social interaction parameter $\rho^L \cdot 1000$.

Every simulation runs for 40 periods. Performing five runs for every setting, the figures show mean values. Figure 1 contains four charts, each of them showing the evolution of the ratio of optimising (solid line), evading (dashed line) and honest (dotted line) agents simulating different scenarios. Scenario 1 shows the baseline scenario, while scenarios 2–4 each consider different settings.

Results show that taxpayers favour illegal evasion over legal optimisation in most settings. In our baseline scenario (Scen. 1), the share of evading agents rises over

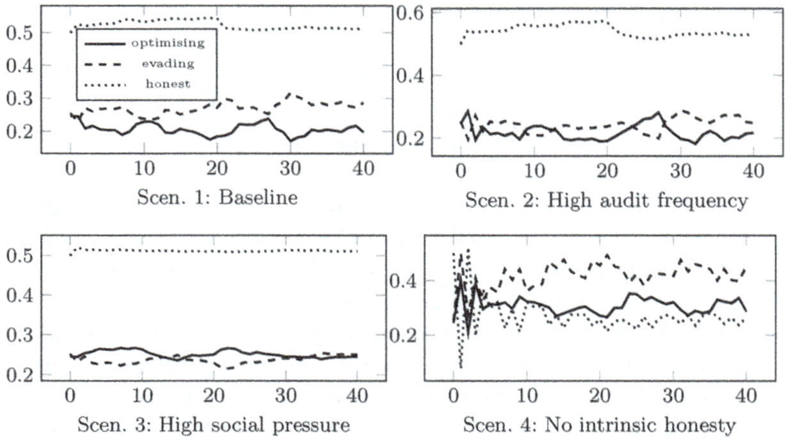

Fig. 1 Evolution of shares of optimising, evading and honest agents

time. Increasing tax rate τ reinforces this effect, implicitly stating that increasing tax payments may cause taxpayers to take the risk of evading. However, increasing audit frequency (Scen. 2) causes the relation between evading and optimising agents to shrink, but tax evasion still seems to be the superior strategy. After period 20 taxpayers alternate their strategy in favour of optimising. Observing their neighbourhood's evasion activities, taxpayers reconsider this decision after only a few periods. Increasing social pressure (Scen. 3)—i.e., taxpayers strongly consider their neighbours' behavior—results in a lower share of evading agents; high social pressure to conform reduces propensity to tax evasion. This result is particularly interesting since by assumption, behaving honestly does not directly trigger any positive effects. Simulation of setting honest agents being not inherently honest (Scen. 4) delivers another highly surprising result. Compared to our baseline scenario, the gap between evading agents and optimising agents increases considerably. It hardly seems intuitive that taxpayers apparently tend to favour illegal actions if they don't have to be honest by assumption. Furthermore, there is some upper boundary for the share of evaders at roughly 0.4, since detection causes an agent to refrain from illegal activities and additionally provokes his evading neighbors' social utility to decrease.

4 Conclusion

Based on our analytical model, we use an agency-based approach to simulate the effects of introducing the option of tax optimisation into a tax evasion model. We further take into account social utility as previous research has shown significant influence. We find that taxpayers favour illegal evasion over legal optimisation in most cases. Results indicate that social interactions are highly influential for the relation between evading and optimising agents. Especially the assumption of agents being inherently honest drives our results.

References

1. Allingham, M.G., Sandmo, A.: Income tax evasion: a theoretical analysis. J. Pub. Econ. **1**, 323–338 (1972)
2. Alm, J., McKee, M.: Tax compliance as a coordination game. J. Econ. Behav. Organ. **54**, 297–312 (2004)
3. Bernasconi, M.: Tax evasion and orders of risk aversion. J. Pub. Econ. **67**, 123–134 (1998)
4. Bloomquist, K.: Tax compliance as an evolutionary coordination game: an agent-based approach. Pub. Fin. Rev. **39**, 25–49 (2011)
5. Erard, B., Feinstein, J.S.: Honesty and evasion in the tax compliance game. Rand J. Econ. **49**(Supplement), 70–89 (1994)
6. Fortin, B., Lacroix, G., Villeval, M.-C.: Tax evasion and social interactions. J. Pub. Econ. **91**, 2089–2112 (2007)
7. Gino, F., Ayal, S., Ariely, D.: Contagion and differentiation in unethical behavior. Psychol. Sci. **20**, 393–398 (2009)
8. Hokamp, S., Pickhardt, M.: Income tax evasion in a society of heterogeneous agents – evidence from an agent-based model. Int. Econ. J. **24**, 541–553 (2010)
9. Watts, D.J., Strogatz, S.H.: Collective dynamics of 'small-world' networks. Nature **393**, 440–442 (1998)
10. Wilensky, U.: NetLogo. Center for Connected Learning and Computer-Based Modeling, Northwestern University, Evanston, IL (1999). http://ccl.northwestern.edu/netlogo/

Part XVII
Simulation and Statistical Modelling

How to Improve Measuring Techniques for the Cumulative Elevation Gain upon Road Cycling

Maren Martens

Abstract In order to optimally prepare for competitions athletes gather as much data about their training units as possible. For cyclists, interesting figures are, e.g., the distance covered, the average and the maximum of speed, cadence, heart rate, and power output as well as the cumulative elevation gain. However, measuring devices do not always work reliably for all these factors. While factors like distance, speed, and cadence can be metered trustworthily with the help of magnets, which are attached directly to the bicycles, the metering of other factors still has room for improvement regarding accuracy and trustworthiness.

In this paper we consider the cumulative elevation gain, the measurement of which is done by either GPS or barometric pressure nowadays. Therefore, it is dependent of steady connections to satellites (measurement by GPS) or steady changes in the barometric pressure. For the barometric pressure, however, it can happen that it varies merely due to weather changes or the time of the day. It is not surprising that for the same routes different measuring devices deliver quite different quantities for the elevation gain—not only depending on whether they measure by GPS or by barometric pressure. In the following we present ideas on how to support measuring devices to deliver more reliable quantities for the elevation gain by using statistics and mathematical methods.

1 Introduction

Problem Description Over the last 5 years we have collected a lot of data from road cycling. In particular, for all tours we have measured the total distance covered, the average and the maximum speed. However, it has caused us some difficulties to always measure the cumulative elevation gain (CEG) reliably. Neither the GPS method, which requires sight to earth satellites and therefore often fails, e.g., in

M. Martens (✉)
Landshut University of Applied Sciences, Landshut, Germany
e-mail: martens@haw-landshut.de

J. S. Neufeld et al. (eds.), *Operations Research Proceedings 2019*, Operations Research Proceedings, https://doi.org/10.1007/978-3-030-48439-2_78

tunnels, under roofing or vegetation, or close to high buildings or cliff walls, nor the barometric method, which would actually need permanent recalibration in order to compensate changes from weather or time of the day, provide reliable measurements. Manufacturers of metering devices try to overcome this problem by double-checking the measured data with map data. However, such technology cannot be integrated in small metering devices or also requires radio contact to the database. On the other hand, small devices are still quite popular and yield advantages over high-end devices. (This will be discussed in more detail in Sect. 4.)

For this reason, we tackle the problem of elevation gain measuring and present statistical evaluations for the collected data and derive linear regression models that provide good estimations on the CEG. A crucial restriction to our methods is that they should be computationally easy in order to ensure implementability in small metering devices.

Notation We use x_{AS} to denote the measured average speed (AS), x_{MS} for the measured maximum speed (MS), and x_{GPK} for the measured average elevation gain per kilometer (GPK). We also consider the ratio of MS and AS (MAS), which will be represented by x_{MAS}. For a total number of n cycling tours, we use $x_{i,\mathrm{AS}}$, $x_{i,\mathrm{MS}}$, $x_{i,\mathrm{MAS}} = \frac{x_{i,\mathrm{MS}}}{x_{i,\mathrm{AS}}}$, and $x_{i,\mathrm{GPK}}$ (for $i = 1, \ldots, n$) to denote the respective quantities for each tour. To evaluate the quality of our regression models we consider the adjusted R^2 (adj-R^2), the standard error (SE), and the t-statistics (t). Let $\overline{x_{i,\mathrm{GPK}}}$ be the arithmetic mean value of x_{GPK} and f_i be the GPK estimated for the ith tour by linear regression. Then $R^2 = 1 - \frac{\sum_{i=1}^{n}(x_{i,\mathrm{GPK}} - f_i)^2}{\sum_{i=1}^{n}(x_{i,\mathrm{GPK}} - \overline{x_{i,\mathrm{GPK}}})^2}$ reflects the percentage of the variance of x_{GPK} that is explained by the independent variables ($x_{\mathrm{AS}}, x_{\mathrm{MS}}, x_{\mathrm{MAS}}$) in the regression model and therefore evaluates the scatter of data points around the linear regression line. Let m be the number of independent variables used in the regression model, then adj-$R^2 = R^2 - \frac{m \cdot (1 - R^2)}{n - m - 1}$ is the adjusted R^2, which decreases when an independent variable used in the model improves the quality of the estimation less than expected by chance. After having checked that the residual plots do not show any abnormalities for our models, we use adj-R^2 as the main indicator to evaluate the quality of our regression models. For each independent variable, we compute the coefficient with which the variable linearly contributes to the estimation of x_{GPK}. We denote the coefficient by a_{AS}, a_{MS}, or a_{MAS}, respectively. (More details will be given in Sect. 3.) The t-statistic for the coefficient is the ratio of the coefficient itself and its standard error. Compared to the inverse of the two-tailed Student's t-distribution with $n - m - 1$ degrees of freedom, it gives a hint whether—for a given level of significance—the hypothesis that the considered independent variable does not contribute to explain x_{GPK} should be accepted or not. The higher the absolute value of the t-statistic the rather this hypothesis should be rejected, i.e., the more relevant the variable. In addition, we also quote the standard error of the regression as $\mathrm{SE} = \sqrt{\frac{\sum_{i=1}^{n}(x_{i,\mathrm{GPK}} - f_i)^2}{n - m - 1}}$.

Related Results from the Literature The fact that it is difficult to measure the CEG is widely known and has been discussed extensively. Even manufacturers of measuring devices discuss this issue openly [5]. In the previous decades much effort has been devoted in order to describe mathematical relations between different cycling parameters, many of which evaluate power output from speed and other parameters. Most authors aim at performance improvements in competitions and more efficient training methods by a better understanding of power output due to mathematical models. However, to the best of our knowledge no research is available targeting on the support of elevation gain measuring upon road cycling. Dahmen et al. [2, 3] in fact do describe an idea on how to estimate slope from measured power and speed involving several other parameters for the cyclist, the bicycle, and the environment. They also suggest a method to estimate the necessary parameters like, e.g., chain efficiency, bearing friction, and air density by using numerical gradient-based methods [1, 3], but they do not prove the quality of their elevation gain estimations giving actual data.

Contribution of This Paper Using statistics and mathematical methods, in particular linear regression, we present new ideas on how to obtain good estimations for the cumulative elevation gain upon road cycling. Our aim is to keep the input parameters simple in terms of number and accessibility in order to maintain easy computability by methods that could be used even in low-cost bike computers or other low-end measuring devices.

2 The Data

On various cycling tours, we collected information about the easily measurable parameters total distance, average speed, and maximum speed in order to derive estimating functions for the elevation gain. As discussed above, measuring the elevation gain is what causes problems. However, we need the real elevation gain for comparisons to the estimated elevation gain in quality proofs. Therefore, we only use data from cycling tours for which the measured CEG has been double-checked using map databases like OpenCycleMap [4] or Strava elevation basemap [5]. We assume that the remaining error is neglectable and unbiased. The raw data was measured using the following five computers/smartphone apps: Runtastic Road Bike, Sigma BC 14.12 STS Alti, Sigma BC 1106 DTS, Bryton Rider 530, Sigma Rox 9.1. In most cases more than one of the devices was used to improve reliability. Table 1 gives an overview on settings, for which we analyze the dependence of GPK (i.e., CEG divided by tour length) on the other parameters (AS, MS, MAS). $r_{\ell,g}$ is the correlation between tour length and GPK.

The used data was collected on paved roads and in training rides only, not in competitions. The data was collected over all seasons, but only when streets were dry and clean. Each tour ended at the same location as where it had started. The settings with the competitive female all involve the same cyclist. However, we assume that the bicycle type as well as the terrain (islands generally provide stronger winds than the mainland) might make a difference and so we split up the data for the

Table 1 Settings for data collection

Identifier	Cyclist	Bicycle	Terrain	# of datasets	$r_{\ell,g}$
MTM	Competitive male	Time trial (Centurion Overdrive 4000)	Mainland	53	0.0054
FTM	Competitive female	Time trial (BMC Timemachine 02)	Mainland	68	0.0083
FOM	Competitive female	Ordinary road (various)	Mainland	87	−0.0223
FOI	Competitive female	Ordinary road (various)	Island	38	0.0502

female cyclist. Anyway, we will also discuss results for the aggregated data of the female cyclist. Both the competitive female (born 1979) and the competitive male (born 1963) are successful triathletes with high regional rankings. The island data was collected on Mallorca, Lanzarote, and Krk. Different ordinary road bikes were involved (e.g., Stevens Xenon), but they are all similar in design. The mainland data comes from tours in Mid-Western Europe (e.g., Southern Germany, France, Switzerland). Regarding hilliness all settings contain all types of terrain from highly mountainous to completely flat. This is important for our studies to ensure our system has the chance to learn for every level of hilliness. The GPK varies from 1 to 39 over all tours with an arithmetic mean value of 9.52. The tour lengths vary from 24 to 218 km for the female cyclist and from 24 to 169 km for the male cyclist. However, since the correlation between the tour length and the GPK is so close to zero, we do not consider the tour length in our analyses. Anyway, to compute the total CEG of a tour, the tour length would have to be multiplied with the GPK.

3 Data Analysis and Results

This research project had started with the idea to find out whether good estimations on the GPK are possible using simple linear regression with the average speed as the independent variable. After finding out that the results are quite good but not completely satisfying the idea had been refined: Other independent variables like the maximum speed had been involved and the analysis had been extended to other cyclists and different terrains. Finally, our analysis involved the following parameters as independent variables: measured average speed (AS) (idea: high GPK implies low AS, i.e., make use of negative correlation between GPK and AS), measured maximum speed (MS) (idea: high GPK implies long descent(s) with high MS, i.e., make use of positive correlation between GPK and MS), ratio of measured MS and measured AS (MAS) (idea: high GPK implies high MAS, i.e., make use of positive correlation between GPK and MAS).

For each of the four settings introduced in Sect. 2, we did simple linear regressions with each of the above parameters as the independent variable. Further, we did multiple linear regressions for each pair of the above variables and for the triple of the above variables. Of course, since there is a high correlation between the ratio MS/AS and the other two variables one would not expect much information gain by involving all three variables instead of only two. We will also observe this in our analyses.

The aim of our linear regression is to find parameters a_{AS}, a_{MS}, a_{MAS}, and a such that the function $f(x_{AS}, x_{MS}, x_{MAS}) = a_{AS}x_{AS} + a_{MS}x_{MS} + a_{MAS}x_{MAS} + a$ yields the minimum square error upon estimating the measured GPK, i.e., $\sum_{i=1}^{n}(x_{i,GPK} - f(x_{i,AS}, x_{i,MS}, x_{i,MAS}))^2$ is minimal. This can be obtained by setting the partial derivatives of this sum (w.r.t. a_{AS}, a_{MS}, a_{MAS}, and a) equal to 0 and solving the resulting system of linear equations. When only one or two of the independent variables are involved, the linear regression works analogously ending in a respectively smaller system of linear equations.

We used Microsoft Excel 2010 (32-bit) to compute the regression functions and information about the respective adjusted R^2, the standard error of the regression, and the t-statistic values (as well as other parameters, which are not considered in the following as they do not give additional insight). For our four settings, we received the results from Table 2 showing the respectively best variables for the simple linear regression and for the linear regression with two variables (each with the corresponding adjusted R^2, the standard error, and the t-statistics) as well as the adjusted R^2, the standard error, and the t-statistics for the linear regression with all three variables. The t-statistics are given only for the multiple regression models in order to show which are the most relevant variables for the regression. In Table 3 we show the results for aggregated settings of the female cyclist. For those settings, we added the bicycle type and the terrain to the models as 0/1 variables. However, without these the results are similar.

Table 2 Results from the linear regression models for the settings from Table 1

Setting	Best simple	Best of 2 var	t for best 2	3 var	t for best 3
MTM	MAS	AS, MAS	AS: 0.91	AS, MS, MAS	AS: 1.57
	adj-R^2: 0.48	adj-R^2: 0.48	MS: –	adj-R^2: 0.49	MS: −1.38
	SE: 2.15	SE: 2.16	MAS: 5.76	SE: 2.14	MAS: 2.17
FTM	MAS	MS, MAS	AS: –	AS, MS, MAS	AS: 2.65
	adj-R^2: 0.26	adj-R^2: 0.44	MS: −4.71	adj-R^2: 0.49	MS: −3.48
	SE: 2.13	SE: 1.85	MAS: 6.43	SE: 1.77	MAS: 3.75
FOM	AS	MS, MAS	AS: –	AS, MS, MAS	AS: 1.91
	adj-R^2: 0.80	adj-R^2: 0.86	MS: −11.14	adj-R^2: 0.87	MS: −5.09
	SE: 3.34	SE: 2.76	MAS: 22.09	SE: 2.72	MAS: 5.86
FOI	AS	MS, MAS	AS: –	AS, MS, MAS	AS: 1.32
	adj-R^2: 0.54	adj-R^2: 0.64	MS: −5.41	adj-R^2: 0.64	MS: −1.96
	SE: 2.86	SE: 2.54	MAS: 7.90	SE: 2.51	MAS: 2.15

Table 3 Results from the linear regression models for aggregated settings

Setting	Best simple	Best of 2 var	t for best 2	3 var	t for best 3
FTM +	AS	MS, MAS	AS: –	AS, MS, MAS	AS: 4.24
FOM +	adj-R^2: 0.73	adj-R^2: 0.83	MS: −16.12	adj-R^2: 0.84	MS: −8.89
FOI	SE: 3.18	SE: 2.53	MAS: 25.85	SE: 2.43	MAS: 10.09
FTM +	AS	MS, MAS	AS: –	AS, MS, MAS	AS: 4.50
FOM	adj-R^2: 0.75	adj-R^2: 0.85	MS: −15.33	adj-R^2: 0.87	MS: −9.05
	SE: 3.26	SE: 2.52	MAS: 24.77	SE: 2.38	MAS: 10.21
FOM +	AS	MS, MAS	AS: –	AS, MS, MAS	AS: 1.83
FOI	adj-R^2: 0.77	adj-R^2: 0.83	MS: −12.48	adj-R^2: 0.84	MS: −5.22
	SE: 3.19	SE: 2.70	MAS: 23.83	SE: 2.67	MAS: 6.09

The interesting findings from this analysis are: For all settings with not only time trial bicycle rides, the measured average speed is the best parameter to be used as the independent variable in a simple linear regression, while the combination of MS and MAS is the best for multiple linear regression with two independent variables. For all settings with only time trial bicycle rides, all linear regression models are fairly bad anyway (w.r.t. to adj-R^2, the SE still looks fine). The same holds for the setting including only island rides. Here, the statistics support our initial hypothesis that the wind influences the speed dramatically. For all other settings, all linear regression models yield quite good estimations. For all of them, at least 83% of the variation of GPK can be explained by the independent regression variables MS and MAS. Using all three variables for the regression, we even get to a maximum of 87%. In all cases except for MTM, there is quite a big jump in quality from simple linear regression to linear regression with two variables. However, as expected the jump in quality from regression with two variables to regression with all three variables is quite small. For the multiple regression models, the t-statistics show that MAS is always the most relevant variable. For the models with two variables, this is what one should expect from the fact that, unlike MS, MAS in part also represents AS. The models with three variables are in line with the results from the models with two variables. The values of the computed parameters a_{AS}, a_{MS}, a_{MAS}, and a are not shown in Tables 2 and 3 for reasons of legibility. They vary slightly between the models, but do not show any abnormalities.

It might look rather surprising that the regression models for the aggregated settings are still quite good, although those for FTM and FOI are not. However, an analysis whether in those models the standard error of the estimates for rides with the time trail bicycle or on the island is bigger than the standard error for the other rides showed that there is not much variation in the standard error over the different types of rides. If there is one standard error that is smaller than the others, then it is the one for the rides with the time trial bicycle. Also the residual plots do not show any abnormalities.

4 Conclusion and Future Work

The aim of our analysis was to find out whether simple mathematical models can be used in order to support simple metering devices upon evaluating the CEG upon road cycling. The problem for these small devices is that both GPS and barometric pressure are not satisfiably reliable and accurate and that double-checking with map databases requires technology that cannot be integrated in small meters yet. However, compared to more sophisticated technology small metering devices have the indisputable advantage that they consume much less power (their batteries last for months rather than for hours like for high-end devices) and also they are better manageable w.r.t. size and often also w.r.t. display readability. Therefore, small meters are still important for the market.

Our analyses have shown that for general road cycling settings (on the mainland and not only with time trail bicycles) linear regression yields a simple method to well estimate the elevation gain when GPS signal gets lost (or implausible) or to double-check the trustworthiness of data received from barometric pressure. All our methods use only elementary computations that can also be implemented in a small meter.

(Not only) for the island settings, it would, of course, be interesting to involve the wind as a factor. However, this immediately raises the question how this could be done with simple methods in small meters. Also the power output—as used in many scientific analyses, cf. Sect. 1—could be an interesting factor in order to obtain better estimations. Although there exist devices measuring the power output on the pedals, the interesting question for future work would be again how this could be used when restricting to small, low-cost meters.

One might think about whether non-linear regression models would reflect the correlation between GPK and the other variables even better. However, all non-linear regression models we have tried turned out to be worse.

References

1. Dahmen, T., Saupe, D.: Calibration of a power-speed-model for road cycling using real power and height data. Int. J. Comput. Sci. Sport **10**(2), 18–36 (2011)
2. Dahmen, T., Byshko, R., Saupe, D., Röder, M., Mantler, S.: Validation of a model and a simulator for road cycling on real tracks. Sports Eng. **14**, 95–110 (2011)
3. Dahmen, T., Wolf, S., Saupe, D.: Applications of mathematical models of road cycling. IFAC Proc. Vol. **45**(2), 804–809 (2012)
4. Gravitystorm Limited. OpenCycleMap. https://www.opencyclemap.org/. Accessed 14 Feb 2019
5. Strava Support. Elevation for Your Activity. https://support.strava.com/hc/en-us/articles/216919447-Elevation-for-Your-Activity/#knownissues. Accessed 12 Feb 2019

A Domain-Specific Language to Process Causal Loop Diagrams with R

Adrian Stämpfli

Abstract Causal Loop Diagrams (CLDs) are a flexible and valuable tool for diagramming the feedback structure of systems. In strategic decision-making and management, we use CLDs to structure and explore complex decision-making situations, to foster learning, as a basis for simulation models, and to communicate simulation results. However, the crucial dissemination of CLDs and the possible learnings beyond the project-team is challenging.

To overcome this problem, we developed a Domain-Specific Language that allows modeling experts with little programming experience to generate visual representations of CLDs that (1) replace the most complicated CLD elements with a step-by-step explanation and (2) strive to lower the barriers to learning while addressing a broader target audience.

Keywords Strategic decision-making · Causal Loop Diagrams · System Dynamics · Domain-Specific Languages · R

1 Introduction

System Dynamics (SD) is a method for modeling and simulation of complex systems that adapts control theory to a broader set of problems [1]. Two key elements differentiate SD from other methods in Operations Research. (1) SD models *generate dynamics endogenously*. Many classical SD models show how flawed internal policies of industries or cities generate decay without external limiting factors [2]. (2) SD *makes mental models explicit* by modeling them as CLDs. Making the models explicit is the basis for a deeper understanding of a

A. Stämpfli (✉)
Institute of Modeling and Simulation, University of Applied Sciences St. Gallen, St. Gallen, Switzerland
e-mail: adrian.staempfli@fhsg.ch

J. S. Neufeld et al. (eds.), *Operations Research Proceedings 2019*, Operations Research Proceedings, https://doi.org/10.1007/978-3-030-48439-2_79

messy situation, for revising mental models, for allowing double-loop learning to occur, and for taking strategic decisions [3–6].

1.1 Causal Loop Diagrams

Causal Loop Diagrams (CLDs) are a widely used, flexible, and valuable tool for diagramming the endogenous perspective or feedback structure of systems [8, 9]. CLDs are used (1) to structure complex problems; (2) to explore complex decision-making situations in participatory modeling processes; (3) to foster learning among stakeholders involved in the modeling process; (4) as a basis for simulation models, and (5) to communicate results of simulation studies [4, 5, 10].

Mathematically, CLDs are a particular sort of directed cyclic graphs. They consist of variables (in graph terminology: *vertices*) that are connected by directed causal links (*edges*). Each link states a causal influence from the starting variable to the ending variable. Each causal link is assigned a polarity, either positive (+) or negative (-), that indicates how the dependent variable changes when the independent variable changes [9]. Each feedback loop (*cycle*) describes positive (reinforcing, *R*) or negative (balancing, *B*) feedback [9].

CLDs are an essential tool to foster learning and feedback processes among the stakeholders involved in a project [4, 10]. However, the crucial dissemination of those learnings beyond the project team is difficult and requires knowledge about CLDs that senior decision-makers generally do not have [10, 11]. This impedes the acceptance of CLDs and the implementation of project results [10].

2 An Embedded Domain-Specific Language to Process CLDs in R

To overcome this problem, we developed an embedded Domain-Specific Language (DSL) in R that allows us to import CLDs from standard SD tools into R and to generate plots of those CLDs.[1],[2] The plots focus on specific aspects or parts of the CLDs. By lowering the 'technicalness' and improving visual attractiveness, we address a bigger audience. Existing research shows that the most complicated elements of CLDs are the link and loop polarity signs. We thus replaced those elements: The DSL allows explaining a CLD step-by-step by highlighting specific causal chains and adding textual descriptions. The whole CLD is shown greyed-out throughout the process, which ensures that the target audience receives the CLD

[1] At the moment we support Vensim. Support for Stella, the other major SD tool, is planned for a future release.

[2] The DSL source code and further instructions are hosted at https://github.com/ims-fhs/cld.

as a whole. We assume that CLDs explained with the DSL allow for the same insights as standard CLDs. Moreover, we assume that such CLDs improve the chances of implementing insights generated within SD projects, because they (1) help to broaden the possible audience of project results (because simplified visual representations fit better to decision-makers' mental models) [10] and, therefore, (2) help to maintain overlaps between participants mental models and project results [4, 11].

2.1 The R Language and Embedded DSLs

R is an open-source programming language and software environment designed for statistical computing, data science, and graphics [12]. R is a very flexible programming language. The combination of first-class environments, lexical scoping, non-standard evaluation, and meta-programming make R especially well suited to support the creation of embedded DSLs [13].

DSLs are computer programming languages of limited expressiveness focused on a particular domain [14]. Embedded DSLs are composed of valid code in the general-purpose language (GPL) they are written. They take advantage of that GPLs parsing and execution framework [13]. The GPL naturally defines the syntactical bounds for the embedded DSLs [13, 15].[3]

The usage of embedded DSLs offers some advantages over the usage of standard libraries or frameworks: (1) A sense of *fluency* and a *language nature* can be achieved by the usage of only a few design patterns [14]; (2) The DSL approach increases programmer productivity [14], and (3) Embedded DSLs allow to write code that is better understandable by domain experts. Domain experts, therefore, can challenge the code more thoroughly. Requirements refinement, presumably the most challenging part of software development, is improved.

2.2 DSL Grammar

Following, we present the developed DSL by discussing its grammar The DSL allows us to (1) `'IMPORT'` CLDs from standard SD tools into R; (2) `'LINK'` CLD elements; (3) `'DESCRIBE'` groups of CLD elements with textual descriptions; (4) `'PLOT'` the resulting graphics. We specify the grammar by using a standard notation.[4]

[3]Relatively flexible languages like R let you bend the rules a little bit using non-standard evaluation and meta-programming [15].

[4]The notation is a simplified version (with only two operators) of the Backus-Naur form.

DSL Expressions and DSL Sentences Every sentence of the DSL starts with the verb `IMPORT`.[5] The verb `PLOT` is only allowed at the end of a sentence. The verbs `LINK` and `DESCRIBE` can be composed in any possible way. This leads to the following grammar:

```
 DSL_SENTENCE ::= DSL_EXPRESSION
                | DSL_EXPRESSION %>% PLOT

DSL_EXPRESSION ::= IMPORT
                 | DSL_EXPRESSION %>% LINK
                 | DSL_EXPRESSION %>% DESCRIBE
```

A valid `DSL_SENTENCE` can be any `DSL_EXPRESSION` with or without `PLOT` at the end. A valid `DSL_EXPRESSION` can be any combination of an `IMPORT` at the beginning and arbitrary chains of `LINK` and `DESCRIBE` actions afterward. We use the pipe operator (`%>%`) to connect the actions inside a sentence.[6]

Causal Chains and Link Expressions We further need a sub-grammar to specify causal chains inside the verb `LINK`. We need to consider regular causal chains as well as the two extreme cases of causal chains: single variables and loops. We implemented an infix operator [15] (`%->%`) called *link operator* to specify the causal chains. We also need to select more than one causal chain in one link expression. The following grammar covers all cases:

```
LINK_EXPRESSION ::= CAUSAL_CHAIN
                  | CAUSAL_CHAIN , LINK_EXPRESSION

   CAUSAL_CHAIN ::= VARIABLE
                  | CAUSAL_CHAIN %->% VARIABLE
```

2.3 An Example

We use the same CLD as shown in Fig. 1, to demonstrate example usage of the developed DSL. Consider the following DSL statement:

```
cld %>%
    link(gap %->% actions, gap %->% pressure) %>%
    link(actions, pressure) %>%
    describe(type = "text", "A gap not only leads to actions
        towards closing the gap, but also to pressure to adjust
        the long time goals.") %>%
    plot()
```

[5]Import covers two cases here: (1) the case, where we import a CLD; (2) the case, where we select an already imported CLD.

[6]This is an application of the Expression Builder pattern in combination with the Method Chaining pattern [14] exploiting R's possibilities to define custom infix operators [15] and doing non-standard evaluation in combination with meta-programming [13, 15].

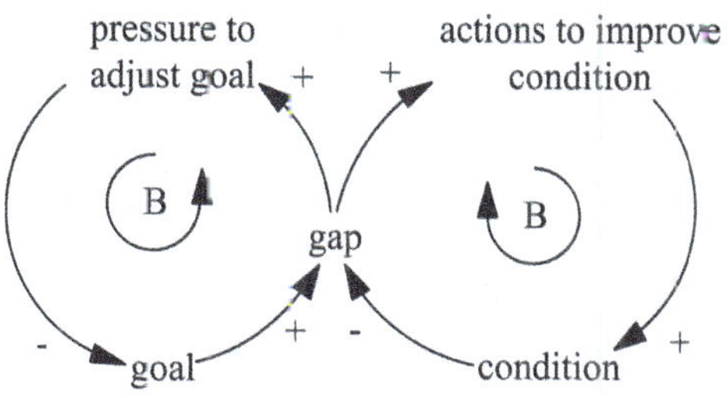

Fig. 1 A sample CLD that illustrates the *eroding goals* systems archetype [7], a situation where we provide actions to close a gap immediately, but at the same time accept that our goals decline over time

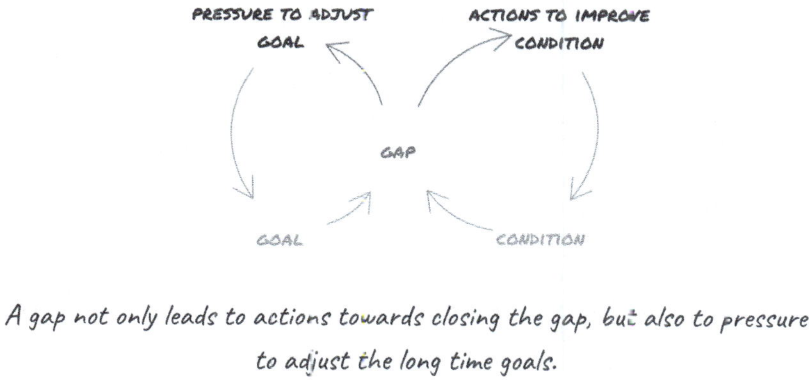

Fig. 2 One exemplary step in explaining the *eroding goals* CLD [7]

The first `link` statement highlights the variable gap and the two consequences of such a gap. The second `link` statement further highlights the two consequences. The `describe` statement adds a textual description. The resulting plot is shown in Fig. 2.[7]

3 Conclusions

CLDs are an essential tool to foster learning and feedback processes among the stakeholders involved in a project [4, 10]. The crucial dissemination of those learnings beyond the project team is, however, difficult and requires knowledge about CLDs that senior decision-makers generally do not have [10, 11]. To overcome this problem, we developed a DSL that allows generating visual representations of

[7]The colors are adjusted to grey tones for better printing results.

CLDs which replace the most complicated elements with a step-by-step explanation. In detail, this is solved using the following elements: (1) The *greyed out model with highlights* is an elegant way to study a CLD without concealing the circular structure at work. Having the feedback structure always there as a whole ensures that the target audience receives all the graphics as being part of a single CLD; (2) *Highlighting certain elements* helps to break the CLD into understandable pieces; (3) *Enriching the graphics with textual descriptions* allows emphasizing important mechanisms.

CLDs, we explain using the DSL expectedly allow for the same systems insights as the original CLDs do. In combination with the sketchy and handmade look and feel, we strive to improve the acceptance of CLDs for stakeholders of the 'untechnical' kind.

To implement the solution in the form of an embedded DSL in R proves valuable as well. Thanks to the DSL approach, we can write short, simple, and elegant code, which in turn provides for excellent prototyping possibilities. R's properties allowed us to find surprisingly simple notations, grammars, and suitable plotting possibilities.

In numerous client projects, the DSL turned out to be a very valuable tool: (1) to develop a common problem understanding; (2) to communicate that understanding to stakeholders beyond the project team; (3) to foster strategic decision-making. A particular appealing application of the developed DSL is a project funded by 'Innosuisse—Swiss Innovation Agency' in the field of policy design for elderly care.[8]

Future research is needed (1) to integrate the delay marks used in standard CLD notation and (2) to explore further possibilities that enhance the DSL expressiveness e.g., reference mode graphs or simulation capabilities.

References

1. Forrester, J.W.: Industrial Dynamics. M.I.T. Press, Cambridge (1961)
2. Richardson, G.P.: Reflections on the foundations of system dynamics. Syst. Dyn. Rev. **27**, 219 (2011)
3. Torres, J.P., Kunc, M., O'Brien, F.: Supporting strategy using system dynamics. Eur. J. Oper. Res. **260**, 1081–1094 (2017)
4. Lane, D.C.: Modelling as learning: a consultancy methodology for enhancing learning in management teams. Eur. J. Oper. Res. **59**, 64–84 (1992)
5. Vennix, J.A.M.: Group model-building: tackling messy problems. Syst. Dyn. Rev. **15**(4), 379–401 (1999)
6. Paich, M., Sterman, J.D.: Boom, bust, and failures to learn in experimental markets. Manag. Sci. **39**, 1439–1458 (1993)

[8]More information on the on-going project can be found at https://www.fhsg.ch/de/forschung-dienstleistungen/institute-zentren/institut-fuer-modellbildung-simulation/care-system-design/verbesserte-planung-der-langzeitpflege/.

7. Senge, P.M.: The Fifth Discipline: The Art and Practice of the Learning Organization. Doubleday/Currency, New York (1990)
8. Lane, D.C.: The emergence and use of diagramming in system dynamics: a critical account. Syst. Res. Behav. Sci. **25**, 3–23 (2008)
9. Sterman, J.: Business Dynamics: Systems Thinking and Modeling for a Complex World. Irwin/McGraw-Hill, New Delhi (2000)
10. Wolstenholme, E.F.: Qualitative vs quantitative modelling: the evolving balance. J. Oper. Res. Soc. **50**, 422 (1999)
11. Hovmand, P.S.: Community Based System Dynamics. Springer, New York (2014)
12. Ihaka, R., Gentleman, R.: R: A language for data analysis and graphics. J. Comput. Graph. Stat. **5**, 299 (1996)
13. Wickham, H.: Advanced R (CRC Press, Boca Raton, 2015)
14. Fowler, M.: Domain-Specific Languages (Addison-Wesley, Boston, 2011)
15. Mailund, T.: Domain-Specific Languages in R: Advanced Statistical Programming. Apress, New York (2018)

Deterministic and Stochastic Simulation: A Combined Approach to Passenger Routing in Railway Systems

Gonzalo Barbeito, Maximilian Moll, Wolfgang Bein, and Stefan Pickl

Abstract Passenger routing in railway systems has traditionally relied on fixed timetables, working under the assumption that actual performance always matches the planned schedules. The complex variable interplay found in railway networks, however, make it practically impossible for trains to systematically hold the designed timetables as delays are a common occurrence in these systems. A more sensible approach to passenger routing involves assessing the probability distributions that characterize the system and consider them in the routing recommendation. This paper describes one such approach using a simulation model working under both deterministic and stochastic conditions and describes the weak points of a deterministic routing strategy in a complex system.

Keywords Stochastic modeling · Railway systems · Passenger routing

1 Introduction

The field of Operations Research (OR) is a rich source of analytical methods, and a fertile ground from where interesting problems can be approached with new ideas and perspectives. The motivation driving this research is one such problem: the routing of passengers in a Transportation Network, and how uncertainty relates to delays and unreachable transfers for passengers. Traditionally, the passenger routing problem is approached from a route optimization perspective. The focus of this paper is, instead, the missed connections' probabilities and the factors driving such behavior. The scope of this work is primarily concerned with the effects of

G. Barbeito (✉) · M. Moll · S. Pickl
Institute for Theoretical Computer Science, Mathematics and Operations Research, Universität der Bundeswehr München, Munich, Germany
e-mail: gonzalo.barbeito@unibw.de

W. Bein
Department of Computer Science, University of Nevada, Las Vegas, NV, USA

J. S. Neufeld et al. (eds.), *Operations Research Proceedings 2019*, Operations Research Proceedings, https://doi.org/10.1007/978-3-030-48439-2_80

implementing a deterministic passenger routing strategy in an inherently stochastic process, and in assessing the variable interdependency driving the stochastic behavior. To this end the combination of two different approaches to system analysis is explored: On the one hand, a combination of **Agent Based Modeling** (ABM) and **Discrete Event Simulation** (DES) is used to develop a generic model of a transport network and retrieve the data required to analyze the probabilistic behavior of the system. On the other hand, this work relies on **Probabilistic Modeling**, a set of powerful techniques stemming from the field of Machine Learning and Probability Theory [1], to semi-automatically model and understand the intricate variable dependency within the model.

2 Motivation and General Problem Description

2.1 Passenger Routing in Transport Networks

The *Shortest Path Routing* (SPR) problem, in the context of passenger transportation [2–4], is based on the assumption that travelers will choose a path that optimizes their decision criteria, normally measured in *travel time*. Despite its simplicity, most current approaches stem from this model [3].

A conceptual description of the SPR problem is shown in the upper segment of Fig. 1: a query is placed to a routing system to find the most suitable path (full arrow) between S and T. The dotted lines indicate the existence of several alternative paths out of which a routing algorithm needs to choose the best option. Many algorithms have been developed to solve this problem, from brute force and Dijkstra's algorithm, to more sophisticated solutions using complex heuristics [3]. The solution to a SPR problem is the path (middle segment of Fig. 1) which the routing algorithm selected as optimal in regards to the passenger criteria. In this particular example, it includes two *transfers* or *connections* (t_1-t_2 and t_3-t_4) calculated according to a fixed schedule and based on an ideal scenario.

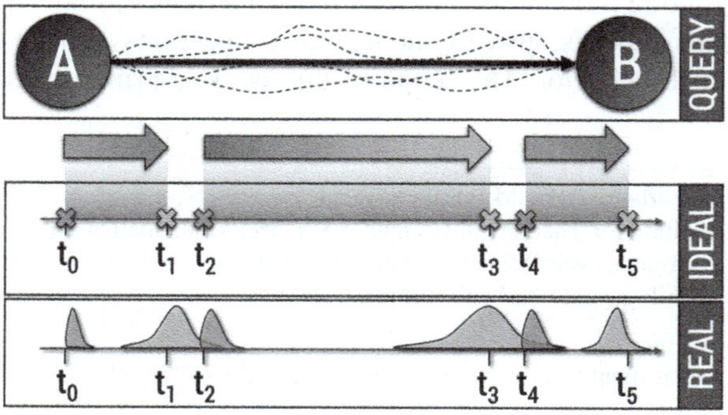

Fig. 1 Conceptual description of a SPR query (upper segment), deterministic (middle segment) and stochastic behavior (bottom segment) for a generic shortest path query

2.2 Probabilistic Considerations in Transportation Networks

In its most basic case, the solution to the SPR problem does not contemplate the stochastic behavior of the transportation network, hence there is a chance that passengers won't reach a connection on time [5, 6]. This hypothetical scenario can be seen on the bottom segment of Fig. 1, where the overlapping distributions define the probability of a missed connection. This research emphasizes the need to determine how likely it is for a passenger to miss a transfer. Such an assessment can be done, from a probabilistic perspective, in two different ways [1]:

(a) Determining average distributions for arrival and departure times at each station, by sampling over all possible scenarios. The upper section of Fig. 2 generically shows the probability distributions for arrival and departure times of two different trains, at a station where passengers need to make a connection. These distributions result from averaging all data, for all possible cases. Their overlapping relates to the probability of missing that connection.
(b) Leveraging the implicit variable dependency structure observed in the system. If correctly mapped, this structure allows better predictions and even rely on secondary variable's forecasts. In the lower right section of Fig. 2, the distributions for arrival and departure times include secondary variables's forecasts and observations, and leverage the variable dependency structure (lower left

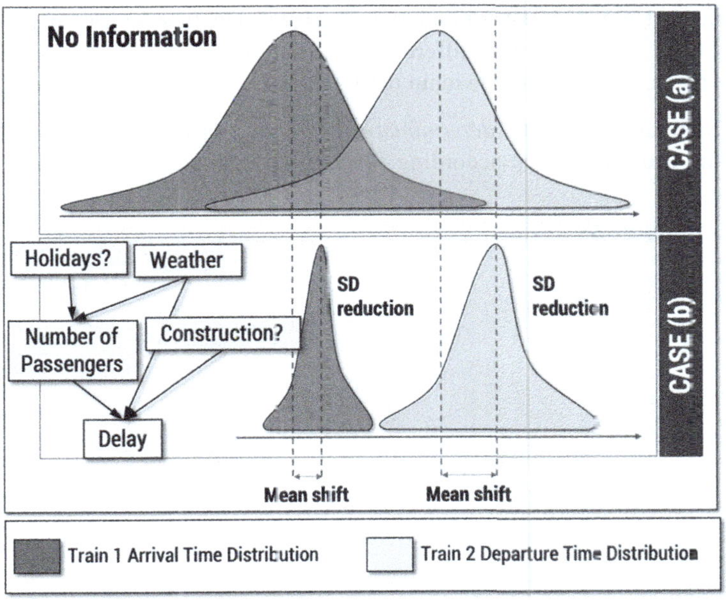

Fig. 2 Probability distributions for two trains' arrival and departure times at a connection. In the upper section, the distributions average arrival and departure times for all scenarios. On the bottom section, the system's variable dependency structure modifies the distributions when observations of these variables are available

section of Fig. 2) found in the system. This new added information shifts both distributions' means and reduces their variance, reducing their overlapping.

2.3 *Current Trends in Passenger Path Assignment*

Most current research is directed towards the inclusion of real time information, personalized recommendations using Big Data and efficient use of communication technologies [5, 7, 8]. This research takes a different approach by leveraging secondary forecasts (e.g. *weather*) when predicting arrival and departure times for trains in the network.

3 Methodology

The methodology devised for this work involves two tasks: (A) *physical modeling* used to generate data as proxy of a real system and (B) *system reconstruction* of the variable interdependency observed in (A).

(A) **Transportation Model:** A generic network with four train lines (A,B,C,D) and passengers directed from station S (Start) to T (Target), with ten possible paths resulting from the route combinations' possibilities. In this implementation, the model makes use of three different types of agents (Fig. 3): Passengers, Trains and a Container, where the main model logic is programmed, including:

1. Simulation of both *deterministic behavior* (used to automatically learn the train schedules according to network parameters [9]) and *stochastic*

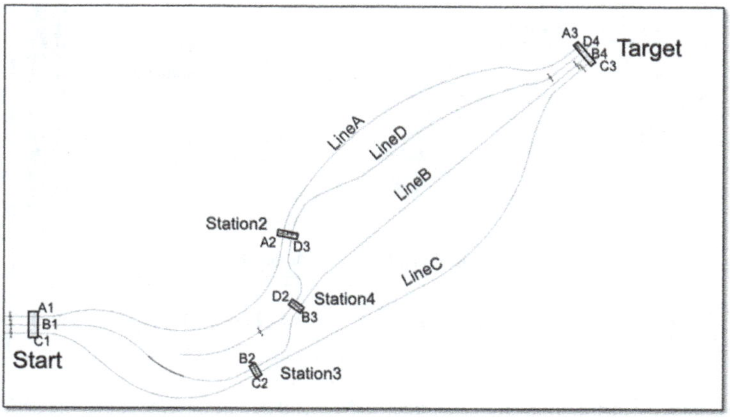

Fig. 3 Structure of the railway network and the routing problem

behavior to emulate real-world schedule deviations. The stochastic behavior follows a sensible logic, where certain variables influence others, defining an implicit correlation network (e.g. different *weather* conditions affect *number of passengers* and trains' *travel times*).

2. *Embedded passenger routing implementation:* An algorithm capable of using the simulated data to select the shortest path in real time.

(B) **System Reconstruction:** The goal of this task is to use the data generated by the physical model and recreate the variable interdependency in the system, and how it affects transportation schedules and passengers' delays. The selected approach to this task is Probabilistic Graphical Models (PGM), a group of modeling techniques capable of compactly encoding complex probabilistic distributions over high-dimensional space. To this end, PGM use a graph like representation of the studied system, where nodes represents random variables, and the edges correspond to the probabilistic interactions between them [1].

4 First Results

The reconstruction of the variable interdependency structure can be seen in Fig. 4. This structure was automatically created using the R library "bnlearn", leveraging its learning capabilities from data. The dataset extracted from the physical model contained 2.5 million observations, spaced on 10 s' intervals. The results indicate a three-layer structure, with temporal variables *Hour, Day* and *Month* affecting the *Train Delay*, which simultaneously and together with the occurrence of *Rain* and *Holidays* affect the *Passenger Travel Time*.

The results of the schedule learning can be seen in the left side of Fig. 5: Darker green and orange vertical lines indicate arrival times for two different trains to the same station, these lines implicitly define the connections possibilities that the

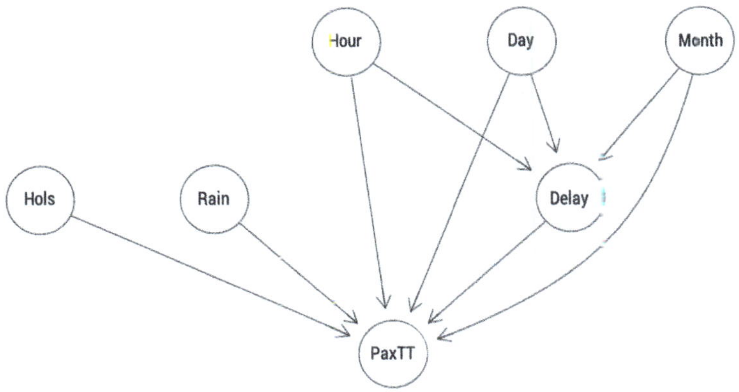

Fig. 4 PGM created from data indicating the system's variable dependency structure

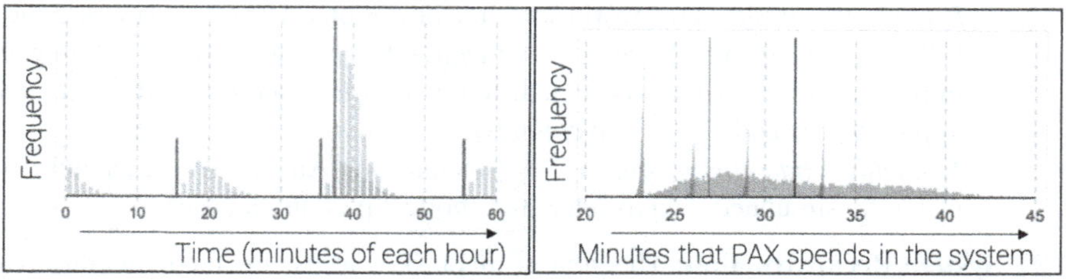

Fig. 5 Left: Arrival Times for two different trains. The overlapping of distributions indicate the probability of missing that connection. Right: Passengers' travel times from S to T. Long vertical lines indicate the mean values of both distributions

algorithm will dynamically use. The routing algorithm take this information, for every station, as input and generate a shortest path accordingly. Lighter colored bars define the underlying arrival distribution deviating from the timetable. An interesting effect can be seen when both orange and green areas overlap, implicitly defining a probability of a missing connection, and a path we would not want to include in a routing solution.

The right side of Fig. 5 shows the results of measuring passenger's travel times from *S* to *T*. The orange areas indicate travel times when the simulation is set to *Deterministic*, defining a multimodal distribution due to passengers being constantly redirected according to their arrival time to *S*. The blue lines define the distribution for the same variable, when the model runs on *Stochastic* mode. As expected, the mean for the stochastic simulation (blue vertical line) is higher than for the deterministic one (orange vertical line). Practically, this means that passengers take longer to reach their destination due to missing connections.

5 Conclusion and Outlook

This paper introduced a generic transportation model to analyze the connection between system uncertainty and train and passenger delays. First results indicate that the approach has potential to recommend better routes to passengers by taking these delays into consideration. The future work in this research includes (a) the use of the variable dependency structure to correctly assess delay probabilities and (b) reducing the search space of the routing algorithm by excluding paths containing connections with higher chance of being missed by passengers.

References

1. Koller, D., Friedman, N.: Probabilistic Graphical Models: Principles and Techniques. MIT press (2009)
2. Tong, C., Wong, S.: A stochastic transit assignment model using a dynamic schedule-based network. Transp. Res. B Methodol. **33**(2), 107–121 (2002)
3. Geisberger, R.: Advanced route planning in transportation networks. Ph.D. dissertation, Karlsruher Instituts für Technologie (2011)
4. Fu, Q., Liu, R., Hess, S.: A review on transit assignment modelling approaches to congested networks: a new perspective," Procedia Soc. Behav. Sci. **54**, 1145–1155 (2012)
5. Rückert, R., Lemnian, M., Blendinger, C., Rechner, S., Müller-Hannemann, M.: PANDA: a software tool for improved train dispatching with focus on passenger flows. Public Transp. **9**(1–2), 307–324 (2017)
6. Liu, Y., Blandin, S., Samaranayake, S.: Stochastic on-time arrival problem in transit networks. Transp. Res. B Methodol. **119**, 122–138 (2019)
7. Gündling, F., Weihe, K., Hopp, F.: Efficient monitoring of public transport journeys. In: International Symposium on Rail Transport Demand Management (RTDM 2018) (2018)
8. Chow, J.: Informed Urban Transport Systems: Classic and Emerging Mobility Methods Toward Smart Cities. Elsevier (2018)
9. Borndörfer, R., Klug, T., Lamorgese, L., Mannino, C., Reuther, M., Schlechte, T.: Handbook of Optimization in the Railway Industry, vol. 268. Springer, Cham (2018)

Predictive Analytics in Aviation Management: Passenger Arrival Prediction

Maximilian Moll, Thomas Berg, Simon Ewers, and Michael Schmidt

Abstract Due to increasing passenger and flight numbers, airports need to plan and schedule carefully to avoid wasting their resources, but also congestion and missed flights. In this paper, we present a deep learning framework for predicting the number of passengers arriving at an airport within a 15-min interval. To this end, a first neural network predicts the number of passengers on a given flight. These results are then being used with a second neural network to predict the number of passengers in each interval.

Keywords Predictive analytics · Aviation management · Deep learning · Neural networks

1 Introduction

Airlines and airports are performing demand analyses under consideration of the available infrastructure for airport extension projects or during the preparation for an upcoming flight schedule. The goal is to enable a seamless passenger journey. However, facilities like check-in, security check or border control often become bottlenecks during peak hours. Therefore, a well-founded knowledge about the amount and location of passengers inside a terminal prior to the day of operations is crucial. In this paper we will show a solution for automated passenger and passenger arrival prediction based on historical flight- and boarding pass data, using python-based machine learning methods based on the following python-modules: Tensorflow [1], Keras [6] and Pandas [10].

M. Moll (✉) · T. Berg · S. Ewers
Universität der Bundeswehr München, Neubiberg, Germany
e-mail: maximilian.moll@unibw.de

M. Schmidt
Flughafen München, Freising, Germany

J. S. Neufeld et al. (eds.), *Operations Research Proceedings 2019*, Operations Research Proceedings, https://doi.org/10.1007/978-3-030-48439-2_81

2 Methods

The approach is divided in three parts. In a first step, the data pre-processing, irrelevant features are deleted and new features are engineered respectively extracted by merging flight data with additional information, extracting dates from scheduled time of each flight and transforming features to a more aggregated shaping. The remaining two prediction parts rely mainly on neural networks, which are briefly introduced next.

2.1 Neural Networks

The popularity of deep learning has been increasing in recent years, ranging from innovations in neural network structures in research [13] to new applications in industry [9]. The foundations can be traced back to [3] and [4], which demonstrated that neural networks form a class of universal function approximators. A neural network with a single hidden layer is a function of the form

$$G\colon \mathbb{R}^n \to \mathbb{R}^m; \quad G(\mathbf{x}) = \sum_{i=1}^{k} \alpha_i \sigma(W_i \mathbf{x} + \mathbf{b}_i), \tag{1}$$

where $\sigma : \mathbb{R} \to \mathbb{R}$ is the continuous, increasing, non-linear *activation function*; we assume here that it is applied to each dimension of its argument. The important point is to have the non-linear function nested between the linear combinations to avoid triviality. $\alpha_i \in \mathbb{R}^m$, $W_i \in \mathbb{R}^{m\times n}$ and $\mathbf{b}_i \in \mathbb{R}^m$ are parameters that need to be determined. To this end, gradient descent on some loss-function—often the mean squared error—is being used. Usually, neural networks are described in terms of layers and units or neurons. By interpreting every entry of $\mathbf{x}$ as well as of $\sigma(W_i\mathbf{x}+\mathbf{b}_i)$ and $G(\mathbf{x})$ as individual neurons, we can see that they each form a *layer*. The neurons that arise from $\sigma(W_i\mathbf{x} + \mathbf{b}_i)$ are neither input nor output and are hence said to be in the *hidden layer*. One of the key realizations in deep learning was, that the performance of such networks can be improved by adding more hidden layers, i.e. a longer concatenation of linear combinations and non-linear activation functions.

2.2 PaxPred: Prediction of Number of Passengers

The second part, predicting the number of passengers is considered as a supervised regression with a scalar output (number of passengers). To increase learning efficiency and to weaken outliers all numeric features are normalized using *Z-Score Normalization* [5]. The developed solution is implemented as a neural network with two hidden layers. Due to the intended output there is a *linear activation* used in the output layer [11]. In all other layers we use *Scaled Exponential Linear Unit (SELU)* as activation function with both *bias* and *kernel* initialized by using the *LeCun initializer* [12]. The selection of SELU is justified by its robustness against vanishing or exploding gradients [8, 11]. *Mean absolute error* is used as loss function [14]. Due to the high number of data and parameters we use the *ADAM optimizer*, which has proven to be robust in many machine learning issues [7]. *Batch size*, *number of epochs* and *learning rate* were determined by an automated hyperparameter search following the procedure as described in [2]. The resulting neural network has a batch size of *1024*, a learning rate of *0.001* and *500* epochs. Its structure is illustrated in Fig. 1a.

2.3 *ArrivalPred: Prediction of Passenger Arrival Times*

The prediction of the arrival of passengers is treated as a 96-dimensional supervised regression problem, since passengers' arrival times for the forecast are divided into 15-min intervals each day. The difference between the data of *PaxPred* and *ArrivalPred* is in the extension to the number of passengers forecasted in *PaxPred*, as well as the addition of the boarding pass data for the arrival times, which represents the output vector. The multi-layered neural network includes three hidden layers in addition to one input and one output layer. As in *PaxPred*, *SELU* is used as the activation function in all layers except in the output layer. A linear activation is used in the output layer, because the values within the array are real values in the range 0 to n. Furthermore the *MAE* was used as a loss function due to its stability [14]. For similar reasons as above and because *ADAM* proved to be performing well in the first problem, this optimizer was also used in this neural network [7]. Most features from *PaxPred* have been kept, however, some additional features became important at this point. Thus, an attribute was created for the hour of the day. This eliminated all initial outliers as shown in Fig. 1b, in which the arrival curve was correctly predicted, but the time was postponed. In addition, a parameter with information about the region of the destination was retained as input. According to the airport company, passengers have a different arrival behavior dependent on the region assigned to their flight.

3 Results

The results presented in this section validate the approach discussed above.

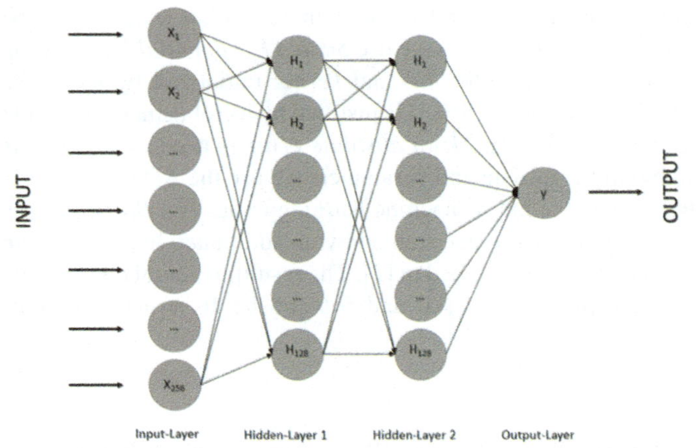

(a)

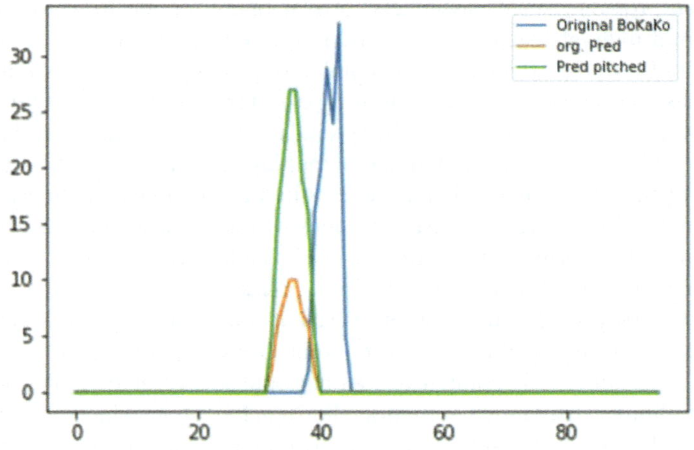

(b)

Fig. 1 Approach and results for the first neural network. (**a**) Structure of first neural network. (**b**) Result for a sample flight without hour features

3.1 Results of Number of Passenger Prediction

The results of the first neural network are analysed with regard to both the absolute errors of all predicted flights and the mean absolute error per day during the trial period, that extends over a week in winter. The results were confirmed by several

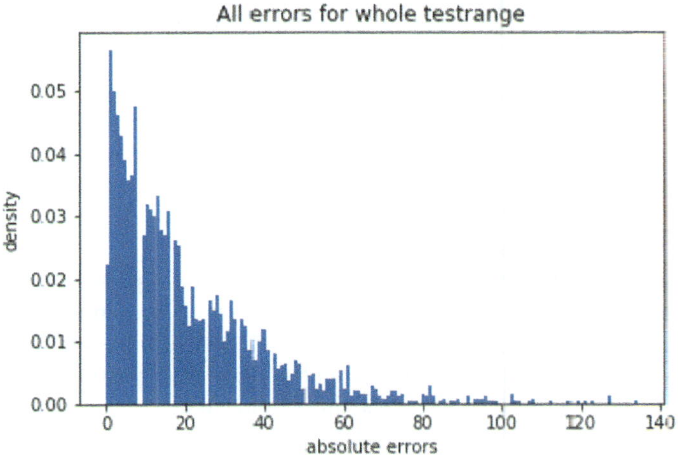

(a)

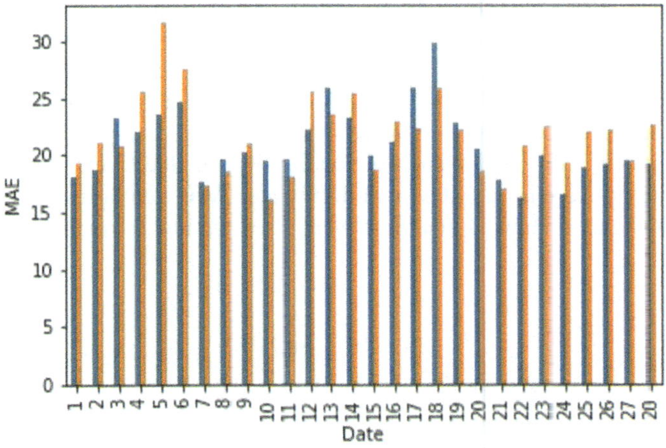

(b)

Fig. 2 Results of the first part. (**a**) Absolute errors during trial period. (**b**) Comparison of the mean absolute errors of both solutions. Blue indicates the manual solution, orange the developed approach

runs and different trial periods. The absolute errors can be found in Fig. 2a. Figure 2b gives a comparison of the mean absolute error per day of the current, manual solution and the approach presented here.

3.2 Results of Passenger Arrival Time Prediction

Here, negative predictions were set to zero, as they were consistently off the normal trajectory for arrival times and therefore implausible. It became clear that the sum of the number of passengers predicted in terms of their arrival time was very often lower than the number which was predicted by *PaxPred*. Therefore, as illustrated in Fig. 3a, the number of passengers will be adjusted by the difference to the predictions made by *PaxPred*.

After *ArrivalPred* was prepared and adjusted to produce predictions of acceptable accuracy, it was matched with the live data from the airport monitoring system and with the forecasts generated by the current solution. To this end, the predicted arrays had to be adjusted, because the current solution plots passengers arriving in the form of an hourly rolling curve. In addition, the passengers of each flight were assigned to the individual areas in the airport terminal using a reference table. When evaluating the trajectory curves, the mean absolute error was considered on the one hand. On the other hand many decisions were based on the graphical comparisons. When interpreting the mean absolute error, one must always keep in mind that a value of zero does speak for a perfect prediction, but in this type of comparison calculation it will lead to a worse mean absolute error if there is a displacement of passengers by only one position in the array. However, this shift would be less relevant since it is ultimately only a delay of at least 1 min, but not more than 15 min. Therefore, the graphic output was always included in the rating of generated results. The comparison with the current solution curves was due to the provided amount of data only possible for a 5-day period in winter. Many empirical observations of the predicted curves showed that these were often too low in comparison to the live

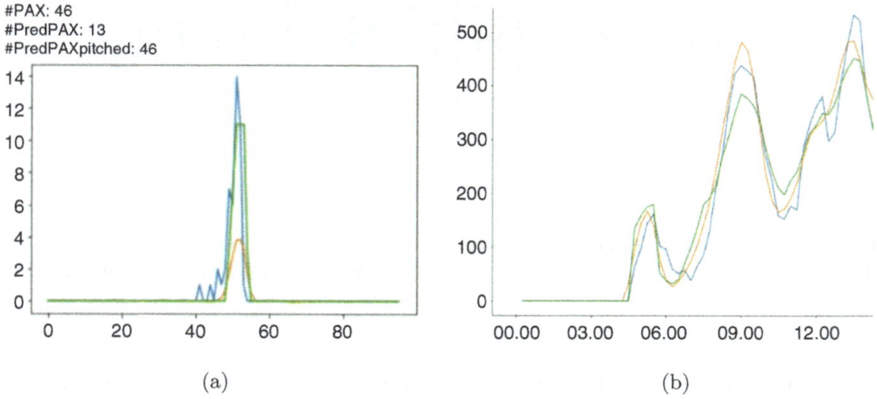

Fig. 3 Evaluation of the arrival prediction. (**a**) Example pitch of curve. The blue curve represents the original data, the orange curve the prediction from *ArrivalPred* and the green curve the adjusted forecast. (**b**) Example of good mae but bad curve. The blue curve shows the live data, the green curve the prediction of the current solution and the orange one the new prediction

data of the monitoring system. Therefore, a sweeping increase in the predictions was introduced. The results of all terminal areas were increased by 5%. This measure not only delivered better results. At the same time this fact offers the possibility to automatically schedule a buffer in personnel and control station planning. For example, Fig. 3b shows the morning of a Wednesday at the end of the winter for a selected area of the terminal. This behavior was confirmed by further empirical observations of the curves of other days and other terminal areas, where the current solution is "better" according to mean absolute error, but not with regard to overall prediction results. Based on the already mentioned test period in the context of this work, it could be shown that the solution provided was at least equivalent. However, in view of a desired planning buffer, the newly developed solution often proved to be more efficient. With additional consideration of the enormous manual effort of the current solution, the present work proves the added value of a nearly fully automated solution using a machine learning approach.

4 Conclusion

In this paper we discussed a framework for automated prediction of the number of passengers. Its results are on a similar level as the part manually solution that is currently used. Even previously unknown flights are better predicted than an approximation based on historical data. However, bad predictions are often caused by non-predictable circumstances like the weather, walkouts or security issues in most cases. By the nature of machine learning approaches it is to be expected, that the results can be improved with a larger database. When enlarging the database it must be taken into account to cut very old data, because old data can contradict current development of passenger numbers.

References

1. Abadi, M., Barham, P., Chen, J., Chen, Z., Davis, A., Dean, J., Devin, M., Ghemawat, S., Irving, G., Isard, M., et al.: Tensorflow: A system for large-scale machine learning. In: 12th {USENIX} Symposium on Operating Systems Design and Implementation ({OSDI} 16), pp. 265–283 (2016)
2. Chollet, F., Allaire, J.: Deep Learning mit R und Keras: Das Praxis-Handbuch von den Entwicklern von Keras und RStudio. mitp Professional, Bonn (2018)
3. Cybenko, G.: Approximation by superpositions of a sigmoidal function. Math. Control, Signal. Syst. **2**(4), 303–314 (1989)
4. Hornik, K., Stinchcombe, M., White, H.: Multilayer feedforward networks are universal approximators. Neural Netw. **2**(5), 359–366 (1989)
5. Jayalakshmi, T., Santhakumaran, A.: Statistical normalization and back propagation for classification. Int. J. Comput. Theory Eng. **3**, 1793–8201 (2011)
6. Ketkar, N.: Introduction to keras. In: Deep Learning with Python, pp. 97–111. Springer, Berlin (2017)

7. Kingma, D.P., Ba, J.: Adam: A method for stochastic optimization. Technical Report, Cornell University (2014)
8. Klambauer, G., Unterthiner, T., Mayr, A., Hochreiter, S.: Self-normalizing neural networks. Technical Report, LIT AI Lab and Institute of Bioinformatics Johannes Kepler University Linz (2017)
9. Mamoshina, P., Vieira, A., Putin, E., Zhavoronkov, A.: Applications of deep learning in biomedicine. Mol. Pharm. **13**(5), 1445–1454 (2016)
10. McKinney, W.: pandas: a foundational python library for data analysis and statistics. Python High Perform. Sci. Comput. **14**, 1–9 (2011)
11. Nwankpa, C.E., Ijomah, W., Gachagan, A., Marshall, S.: Activation functions: Comparison of trends in practice and research for deep learning (2018). arXiv:1811.03378v1
12. Orr, G.B., Müller, K.R.: Efficient backprop. In: Neural Networks Tricks of the Trade. Springer, Berlin (1988)
13. Szegedy, C., Ioffe, S., Vanhoucke, V., Alemi, A.A.: Inception-v4, inception-ResNet and the impact of residual connections on learning. In: Thirty-First AAAI Conference on Artificial Intelligence (2017)
14. Willmott, C.J., Matsuura, K.: Advantages of the mean absolute error (MAE) over the root mean square error (RMSE) in assessing average model performance. Climate Res. **30**, 79 (2005). https://doi.org/10.3354/cr030079

Part XVIII
Software Applications and Modelling Systems

Xpress Mosel: Modeling and Programming Features for Optimization Projects

Susanne Heipcke and Yves Colombani

Abstract Important current trends influencing the development of modeling environments include expectations on interconnection between optimization and analytics tools, easy and secure deployment in a web-based, distributed setting and not least, the continuously increasing average and peak sizes of data instances and complexity of problems to be solved. After a short discussion of the history of modeling languages and the contributions made by FICO Xpress Mosel to this evolution, we point to a number of implementation variants for the classical travelling salesman problem (TSP) using different MIP-based solution algorithms as an example of employing Mosel in the context of parallel or distributed computing, for interacting with a MIP solver, and for the graphical visualisation of results. We then highlight some newly introduced features and improvements to the Mosel language that are of particular interest for the development of large-scale optimization applications.

Keywords Mathematical modeling · Optimization applications · TSP

1 Introduction

The FICO Xpress Mosel software that has been designed for modeling and solving problems is provided either in the form of libraries or as a standalone program [1]. Mosel includes a language that is both a *modeling* and a *programming* language combining the strengths of these two concepts. In recognition of its increasing use for general programming tasks Mosel has been provided as free software since 2018 [2]; FICO continues its development and maintenance as a commercial software. The Mosel language is nowadays also often referred to as an *analytic orchestration*

S. Heipcke (✉) · Y. Colombani
Xpress Optimization, FICO, Birmingham, UK
e-mail: SusanneHeipcke@fico.com; YvesColombani@fico.com
http://www.fico.com/xpress

J. S. Neufeld et al. (eds.), *Operations Research Proceedings 2019*, Operations Research Proceedings, https://doi.org/10.1007/978-3-030-48439-2_82

language—a language used in the implementation of business applications where it coordinates the use and interaction of various components, such as optimization models, decision trees, scoring or machine learning models that are typically developed with specific tools and implemented with other languages (such as R or Python that can be invoked from Mosel).

Besides the Mosel language, the Mosel software also comprises libraries for embedding models into host programs (C, Java, C#), extending the language (Mosel Native Interface) or remote invocation of Mosel programs, and various tools (debugger, profiler, moseldoc).

2 Modeling Languages: Historical Notes

"Traditional" algebraic modeling languages developed in the 1970s or 1980s like GAMS [3], AMPL [4] or Mosel's precursor mp-model [5] were designed as special-purpose, declarative languages that allowed users to state mathematical optimization problems in a close-to-natural way and offering facilities for importing data in some pre-defined format(s), but without any direct connection to solvers: Typically, the modeling tools exported the problem matrix to a file; the optimization solver, a standalone program, read the file, solved the problem and created a solution file; the modeling tool could then be used to generate a report from this output. This design was conditioned to a large extent by the inadequate memory in typical commercially available hardware that could not store both the model formulation and the problem representation for the solver.[1]

By the early 1990s hardware capacities had improved considerably,[2] making it possible to embed optimization into other programs: optimization solvers were made accessible in the form of libraries (XOSL, CPLEX), later completed with model building libraries (such as Xpress BCL, LP-Toolkit, or Ilog Concert). However, the separation of the "modeling language" and algorithmic operations written with a "scripting language" was still present for OPL [6] with OPL-script in the mid-1990s. Other new modeling systems, such as AIMMS [7] and MPL [8], introduced graphical interfaces for working with models.

Many of the approaches to solve classical Operations Research problems that developed since the 1990s (for example branch-and-cut algorithms) required an interaction between the modeling and solving phases. While programming language based model building software provides just such a close interaction between modeling and solution algorithms, many practitioners preferred to state their model using an algebraic modeling language. Xpress Mosel, first published in 2001, has therefore introduced the concept of a completely integrated modeling and solving

[1] mp-model was initially designed to fit data and program into a CP/M's 64 Kb limit.

[2] By 1992 an Intel 486 based PC had as much power as an IBM 3090 mainframe for the numerical work required by mathematical programming.

language, and also addressed other typical expectations of modeling software users:

- representation of optimization problems in close to natural/algebraic form
- efficient handling of sparse data/conditions in constraint definition
- easy access to external data sources (spreadsheets, databases)
- support for different solver types including Constraint Programming [9]
- possibilities for deployment into existing company systems (embedding via programming language interfaces, in-memory data exchange) [10]
- graphical development environment

Since the 2000s new requirements have emerged, such as:

- support for parallel and distributed computing on the modeling level
- collaborative development and deployment as multi-user web apps
- interfaces for combination with other tools and use of pre-existing specialized code implemented in other languages

3 New Design Concepts in Mosel

The Mosel language is a procedural programming language that also is an algebraic modeling language, making it possible not only to state optimization problems in algebraic form and access external data sources, but also to interact with solvers and implement algorithms for data processing or solution heuristics.

Mosel is designed with a *modular architecture*: the language does not integrate any solver by default but offers a dynamic interface to external solvers provided as *modules*. Each solver module comes with its own set of procedures and functions that directly extend the vocabulary and capabilities of the language. The Mosel language is not restricted to any particular type of solver and each solver may provide its specifics at the user level, including the definition of various callbacks for Xpress Optimizer and Xpress Nonlinear (modules *mmxprs* and *mmxnlp*), incremental problem creation with constraint propagation for Constraint Programming (module *kalis*), or the automatic reformulation of logic constraints for MIP problems (*advmod*). The concept of modularity extends to other uses too, any software that has a C/C++ library interface can be connected to the Mosel language via the Mosel Native Interface (see [11]) without any need to modify Mosel itself.

Other concepts for algebraic modeling languages introduced by Mosel include

- a debugger and profiler (2005), later extended to working concurrently (2014) and remotely (2016)
- user-controlled parallelism at the model level: introduced in 2005, generalized to full support of distributed computing (2010) and remote invocation without local installation (2012)
- model annotations (meta-information for external tools including online documentation generation and web app configuration)

Building on Mosel's functionality for remote invocation the Xpress Optimization Suite has been extended with the multi-user deployment platform Xpress Insight and the integrated development environment for models and web apps Xpress Workbench, both available on the cloud and for local or server installations. The deployment via Insight relies directly on the Mosel language, augmented with XML-based app configuration and definition of web views via VDL (View Definition Language, an XML extension working with Mosel entities).

Let us take a look at an example problem where we refer the reader to implementation variants with Mosel to illustrate the functionality available through the language.

3.1 Example Problem: TSP

Research work on the traveling salesman problem (TSP) and its variants can be considered as one of the drivers for innovation in MIP [12]. Current state-of-the-art are specialized solvers, in particular Concorde [13], only small problem instances can be solved as MIP models. The aim of the TSP is to determine the tour of shortest length (least cost) that visits every location from a given set exactly once. A first Mosel implementation has been discussed in [14] and [15] where the Mosel program iteratively solves an incomplete MIP model, searches for the smallest subtour in the resulting solution and adds the corresponding subtour breaking constraint.

A more efficient implementation of this algorithm consists in checking MIP solutions for subtours at the point where they are found in order to add the subtour elimination constraints immediately into the problem during the MIP solver run—some care needs to be taken to configure the solver algorithm suitably since the problem formulation is incomplete and any deductions based on the dual information therefore need to be disabled. With Mosel, this algorithm can be implemented using the *callback functions* of Xpress Optimizer, that is, the solver executes subroutines defined in the Mosel model at predefined points of the solution algorithm (file `f5touroptcbrandom.mos` [16] also shows how to "warmstart" MIP search by loading heuristic solutions, extract in Fig. 1).

With only few modifications to the data handling, one can transform the MIP subtour elimination algorithm into a heuristic that can be applied to considerably larger problem instances. The idea [17] is to solve small subproblems of neighboring nodes, iteratively covering the whole set of locations. This algorithm lends itself for concurrent execution of the subproblems. The Mosel implementation (see file `tspmain.mos` [16]) adds a master model on top of the TSP problem that manages the selection of (unfixed) arcs for the subproblem instances and controls their execution on available computation nodes on a local network or in the cloud, making use of Mosel's concurrent and distributed programming functionality (see [18]).

The sequence of subtours calculated by TSP subtour elimination algorithms can also be represented graphically, for example in SVG format for embedding into web apps (see files `tsp_graph.mos` and `fiveleap_graph.mos` in [16]).

```
declarations
  CITIES: range                              ! Set of cities
  DIST: array(CITIES,CITIES) of integer      ! Distance between cities
  fly: array(CITIES,CITIES) of mpvar         ! 1 if flight from i to j, else 0
  InitTour: array(set of mpvar) of real      ! Heuristic start solution
end-declarations

initializations from "mytspdata.txt'
  DIST
end-initializations

! Objective: total distance
TotalDist:= sum(i,j in CITIES | i<>j) DIST(i,j)*fly(i,j)

! Constraints: Visit every city once
forall(i in CITIES) sum(j in CITIES | i<>j) fly(i,j) = 1
forall(j in CITIES) sum(i in CITIES | i<>j) fly(i,j) = 1
forall(i,j in CITIES | i<>j) fly(i,j) is_binary

! Set a callback to reject any heuristic solutions that contain subtours
setcallback(XPRS_CB_PREINTSOL, "cb_preintsol")
! Set a callback to create subtour elimination constraints for integer solutions
setcallback(XPRS_CB_OPTNODE, "cb_optnode")

set_optimizer_controls                 ! Prevent dominance or symmetry reductions
create_initial_tour(InitTour)          ! Create an initial tour as starting solution
loadprob(TotalDist)                    ! Load the problem into the solver
addmipsol("InitTour", InitTour)        ! Add the starting solution
minimize(TotalDist)                    ! Solve the problem

if getparam("XPRS_MIPSOLS")>0 then print_sol; end-if   ! Solution reporting
```

Fig. 1 Mosel model for TSP with solver callback definitions and MIP solution loading

4 Developing Large-Scale Optimization Apps

The development of increasingly large scale applications has stimulated the reworking of several aspects in the Mosel 5 release (2019). Most notably these relate to model execution *performance* and new forms of *modular development*.

Performance in program execution can be expressed via speed and memory footprint. Constraint expressions in large instances of real-world problems nowadays not infrequently contain tens or even beyond 100 million of terms, efficient enumeration and storage is therefore increasingly important. Mosel 5 introduces a new *hashmap* array format for representing very sparse arrays, furthermore, the handling of large scale constraint expressions has been revised. Table 1 summarizes two examples of performance improvements (times measured in seconds on Windows 7, 64bit, averages over 3 runs).

Optimization application development involving teams of developers with different areas of expertise implies a need for a componentwise organization of programs where individual portions can be developed and tested independently and exchanged when new versions become available. It is common practice to structure larger Mosel projects in the form of packages (= libraries written in the Mosel language), these packages can now be linked dynamically. Furthermore, through the introduction of the concept of namespaces it has become easier to avoid name clashes between different libraries that are used jointly in a project and a given library (package) can restrict the use of its entities to specific other libraries, thus

Table 1 Mosel 5: Performance improvements for programming and modeling tasks

Sorting 50,000 array entries (file `qsort.mos` [16])		
Array type	Mosel 4	Mosel 5
Fixed array	0.08	0.08
Non-finalized index	46.26	7.44
Dynamic array	45.06	40.27
Hashmap array	–	0.13

Creating large constraint expressions `sum(i in 1..N) (x(i)+y(i))` where `x,y: array(1..N) of mpvar`		
N	Mosel 4	Mosel 5
500,000	1.98	0.18
1,000,000	7.88	0.31
2,000,000	31.35	0.67

removing unnecessary information from what is exposed to the end-user and helping to protect intellectual property.

5 Summary

Optimization modeling software has undergone significant changes since the introduction of the first commercial tools about 40 years ago, driven by advances in hardware and algorithm development. In this paper we have highlighted new concepts and features contributed to this domain by Xpress Mosel. Having turned Xpress Mosel into free software an increasing number of contributions are accessible via Github: https://github.com/fico-xpress/mosel.

References

1. Colombani, Y., Daniel, B., Heipcke, S.: Mosel: a modular environment for modeling and solving problems. In: Kallrath, J. (ed.) Modeling Languages in Mathematical Optimization, pp. 211–238. Kluwer Academic Publishers, Norwell (2004)
2. FICO: Press Release, 20 February 2018. https://www.fico.com/en/newsroom/fico-opens-xpress-mosel-programming-language-to-all
3. Bussieck, M.R., Meeraus, A.: General algebraic modeling system (GAMS). In: Kallrath, J. (ed.) Modeling Languages in Mathematical Optimization, pp. 137–157. Kluwer Academic Publishers, Norwell (2004)
4. Fourer, R., Gay, D., Kernighan, B.W.: AMPL: A Modeling Language for Mathematical Programming. The Scientific Press, San Francisco (1993)
5. Ashford, R.W., Daniel, R.C.: LP-MODEL: XPRESS-LP's model builder. IMA J. Math. Manag. **1**, 163–176 (1987)
6. Van Hentenryck, P.: The OPL Optimization Programming Language. MIT Press, Cambridge (1998)
7. Bisshop, J., Roelofs, M.: The modeling language AIMMS. In: Kallrath, J. (ed.) Modeling Languages in Mathematical Optimization, pp. 71–104. Kluwer Academic Publishers, Norwell (2004)
8. Kristjansson, B., Lee, D.: The MPL modeling system. In: Kallrath, J. (ed.) Modeling Languages in Mathematical Optimization, pp. 239–265. Kluwer Academic Publishers, Norwell (2004)

9. Heipcke, S.: Comparing constraint programming and mathematical programming approaches to discrete optimisation. J. Oper. Res. Soc. **50**(6), 581–595 (1999)
10. Ciriani, T.A., Colombani, Y., Heipcke, S.: Embedding optimisation algorithms with Mosel. 4OR-Q. J. Oper. Res. **1**(2), 155–168 (2003)
11. Xpress Documentation. http://www.fico.com/fico-xpress-optimization/docs/latest
12. Lawler, E.L.: The Travelling Salesman Problem: A Guided Tour of Combinatorial Optimization. Wiley, Hoboken (1985)
13. Concorde TSP Solver. http://www.math.uwaterloo.ca/tsp/concorde
14. Guéret, C., Heipcke, S., Prins, C., Sevaux, M.: Applications of Optimization with Xpress-MP. Dash Optimization, Blisworth (2002)
15. Chlond, M., Daniel, R.C., Heipcke, S.: Fiveleapers a-leaping. INFORMS Trans. Educ. **4**(1), 78–82 (2003). https://doi.org/10.1287/ited.4.1.78
16. FICO Xpress Examples Repository. http://examples.xpress.fico.com
17. Heipcke, S.: Xpress-Mosel: Implementing decomposition approaches for concurrent and distributed solving. In: Presentation at the 89th Meeting of GOR WG Praxis der Mathematischen Optimierung, Bad Honnef (2012)
18. Heipcke, S.: Xpress-Mosel: Multi-Solver, Multi-Problem, Multi-Model, Multi-Node Modeling and Problem Solving. In: Kallrath, J. (ed.) Algebraic Modeling Systems: Modeling and Solving Real World Optimization Problems, pp. 81–114. Springer, Heidelberg (2012)

Part XIX
Supply Chain Management

The Optimal Reorder Policy in an Inventory System with Spares and Periodic Review

Michael Dreyfuss and **Yahel Giat**

Abstract We analyze the window fill rate in an inventory system with constant lead times under a periodic review policy. The window fill rate is the probability that a random customer gets serviced within a predefined time window. It is an extension of the traditional fill rate that takes into account that customers generally tolerate a certain waiting time. We analyze the impact of the reorder-cycle on the window fill rate and present an inventory model that finds the optimal spares allocation and the optimal reorder cycle with the objective of minimizing the total costs. Furthermore, we present a numerical example and find that the number of spares increase almost linearly when the reorder-cycle time increases. Finally, we show how managers can find the optimal spares allocation and the optimal reorder-cycle time and show how they can estimate the cost of changing the required window fill rate and the reorder-cycle time.

Keywords Window fill rate · Periodic review policy · Optimization

1 Introduction

Warehouse managers use periodic reorder policies to handle the replenishment of spares in inventory systems. They face different costs such as spares, shipping, operations and reorder costs. To reduce order costs, they employ periodic order-up-to inventory review policies (R, S), where R is the time between each order and S is the number of spares allocated to the warehouse. Following [1], we assume that customers will tolerate a certain wait. Indeed, most contracts define time windows within which customers must be satisfied [1]. Accordingly, instead of the regular fill rate, we use the term *Window fill rate (WFR)* [2] as the system's performance measure. The *WFR* is the percentage of customers that are served

M. Dreyfuss (✉) · Y. Giat
Jerusalem College of Technology, Jerusalem, Israel
e-mail: dreyfuss@jct.ac.il

J. S. Neufeld et al. (eds.), *Operations Research Proceedings 2019*, Operations Research Proceedings, https://doi.org/10.1007/978-3-030-48439-2_83

within a predefined time window. Another way to view the window fill rate is point w (the tolerable time) in the waiting time distribution.

The contribution of this paper is a model that determines the optimal reorder policy. We show that when a given target *WFR* must be met, there is an almost linear relationship between R and S. Consequently, the optimal reorder R and S can be determined in a similar manner to the EOQ formula.

The fact that customers tolerate a certain wait is quite common in the service industry as recent publications show. They use the terms "maximal tolerable wait" [3], "wait acceptability" [4], "expectation" [5] and "reasonable duration" [6]. Recently, some studies use the *WFR* and set it as a measure of customer satisfaction. In [7], the term order fill rate is used in a continuous review system. Based on [7], Dreyfuss and Giat present two papers of a system with continuous review policy with one echelon [8] and two-echelon system [9]. They present an efficient algorithm to solve these complex systems using the fact that the window fill rate of a single location is S-shaped where previously, researchers limited their search space to the concave region [10]. In this study, we also find that the *WFR* is S-shaped when the review policy is not continuous but periodic. A practical application of the *WFR* can be found in [11, 12].

2 The Model

Consider an inventory system with S spares to which customers arrive each with an order of a single item. Customers' arrivals follow a Poisson process with rate λ. The customers are serviced according to a first-come, first-serve (FCFS) policy. If at their arrival there is an item in the stock, they receive it and leave the system. Otherwise, they join a queue and wait until an item is available. Once every R units of time, there is a replenishment of items from an external warehouse to reset the system to the stock level S. The lead time is the time from the beginning of the order until the stock is replenished and is constant and given by L.

The system must meet a target *WFR* with tolerable wait w. That is, the percentage of customers that leave the system within w units of time since arrival must be greater than *Target*. By [13], the *WFR* is given by

$$F(S, R, w) = \frac{\min(R, \max(w - L, 0))}{R} + \int_{\max(L-w,0)}^{\max(L-w+R,0)} \Pr[N(u) \leq S - 1] \frac{du}{R} \tag{1}$$

We point out that this formula is a special case of [14] in which demand is assumed to follow a Compound Poisson process. The *WFR* is either concave or

S-shaped with the number of spares, a fact that is useful when optimizing the spares allocation problem in a multipe-location setting. We refer the reader to [13] for more details.

3 The Optimal Reorder Policy

In the system mentioned above, managers must decide how many spares (S) to allocate to the system and what should be the reorder cycle time (R). Typically, the goal is to minimize the total costs TC that comprises three components The order cost (OC), the purchasing cost (PC) and the holding cost(HC). The order cost is the total cost of ordering items from an external supplier over a given time horizon. We assume that the time horizon is D days and thus the order cost is given by $k\frac{D}{R}$ where k is the cost of each order in units of money. The purchasing cost is the cost of buying the initial spares at the beginning of the operation and is given by cS where c is the cost (in units of money) per item. The holding cost h is the cost of keeping one unit of money in stock for D days. Managers must then define a service measure and a service target (*Target*). In this study, we assume that the *WFR* is the service measure and therefore *Target* is the percentage of clients that have to be satisified within the time window w. Formally, the optimization problem is given by

$$\min TC = k\frac{D}{R} + c\,(1+h)\,S, \qquad s.t. \qquad F\,(S, R, w) \geq Target \tag{2}$$

If we remove the constraint $F(S, R, w) \geq Target$ then the optimal policy is $(\infty, 0)$ since setting $S = 0$ and $R = \infty$ results in $TC = 0$. Such a policy, however results in $WFR = 0$ for any finite tolerable wait. To meet *Target* we must decrease R and possibly increase S. There is no close-form solution to Eq. (2). Therefore, we will use a numerical study to demonstrate the interaction between the R, S, w and the *WFR*. We will also show why Langrage methods cannot be applied.

4 Numerical Analysis

We use the following baseline parameter values to demonstrate the relationship between *Target,R,S,w* and the optimal solution to Eq. (2). The arrival rate of customers is $\lambda = 1$, the reorder-cycle time $R = 7$, the initial number of spares in the system is $S = 10$ and the tolerable waiting time of the customers is $w = 3$. Figure 1 shows the different values for different reorder cycle times (R) as a function of the number of spares (S). As expected, the *WFR* is S-shaped and since it is not concave, the Lagrange method cannot be applied to solve problem (Eq. 2). For a fixed value of S, increasing R implies that more spares are needed to reach a given

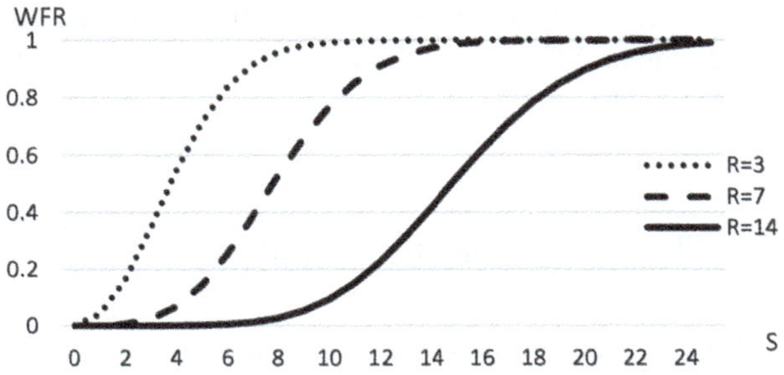

Fig. 1 The window fill rate for different values of (R, S)

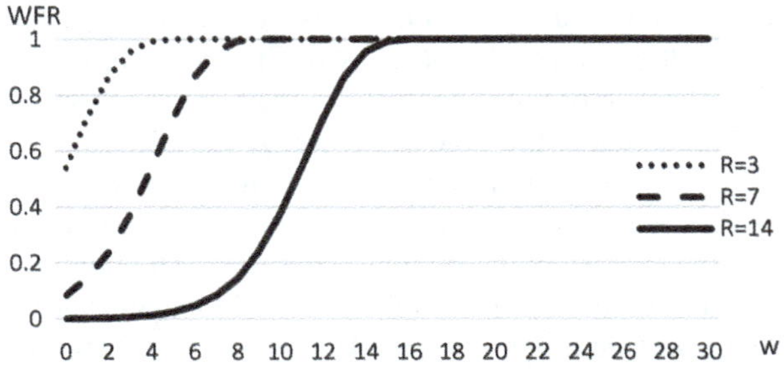

Fig. 2 The window fill rate for different values of (R, w)

level *Target* because a smaller percentage of customers will be serviced on time. For example, when $S = 10$, the *WFR* is 99.2%, 76.8% and 9.3% for $R = 3$, $R = 7$ and $R = 14$, respectively. Figure 2 depicts the *WFR* as a function the tolerable wait for different order cycle times. Here, too, we observe a concave or an S-shaped relationship between the *WFR* and the tolerable wait w.

Figure 3 displays the *WFR* as a function of R for different values of S. Interestingly, the function is an inverted S-shape where initially increasing R hardly affects the *WFR* negatively. Figure 4 describes the relationship between the reorder cycles time and the number of spares for different *Target* values. This relationship is almost linear and the slopes are 1.1 spares, 1.17 spares and 1.31 spares for $Target = 0.8$, $Target = 0.9$ and $Target = 0.99$, respectively. This graph is of practical use since it describes the costs (or savings in terms of spares) of decreasing or increasing the reorder cycle time R. Furthermore, once the slope, *Slope*, and the intercept, *Intercept*, of Fig. 4 are known, then Eq. (2) can be rewritten as

$$\min TC = k\frac{D}{R} + c * (1 + h) * (Slope * R + Intercept).$$

Taking the derivative to find the optimal reorder policy, we get

$$R^* = \sqrt{\frac{D * k}{c * (1 + h) * Slope}},\ S^* = Slope * R^* + Intercept. \tag{3}$$

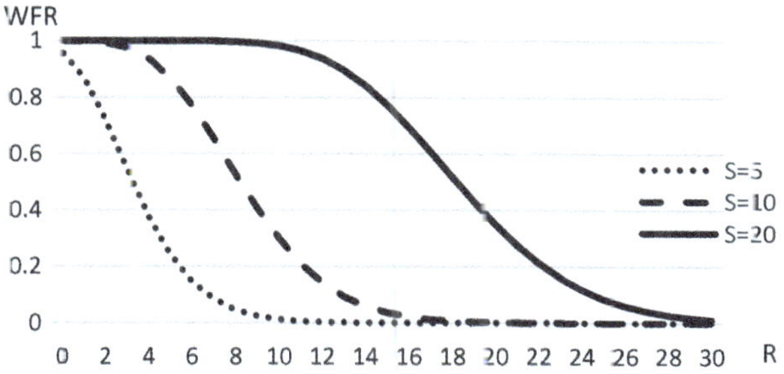

Fig. 3 The window fill rate for different values of (R, S)

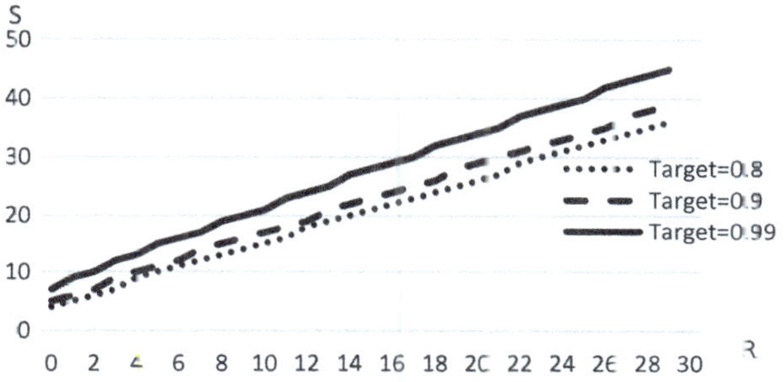

Fig. 4 The number of spares required for different (R, WFR)

5 Conclusions and Limitations

In this study, we show how the reorder-cycle time, the tolerable waiting time and the required window fill rate affect the spares allocation in a periodic review inventory system. Our numerical examples suggest that the number of spares is almost linear with the reorder cycle time when a target window fill rate must be met. This result leads us to a close-form solution, which calculates the optimal reorder cycle time R when minimizing the total cost in a periodic review order system.

There are a number of limitations to our study. First, we assume that demand follows a simple Poisson process, whereas in many situations customer may arrive with a demand for multiple items. More importantly, to simplify the derivations we assume that the lead times are deterministic although real-life settings typically produce a stochastic distribution of lead times. Relaxing these modelling assumptions command further research.

References

1. Caggiano, K., Jackson, L., Muckstadt, A., Rappold, A.: Efficient computation of time-based customer service levels in a multi-item, multi-echelon supply chain: a practical approach for inventory optimization. Eur. J. Oper. Res. **199**(3), 744–749 (2009)
2. Dreyfuss, M., Giat, Y., Stulman, A.: An analytical approach to determine the window fill rate in a repair shop with cannibalization. Comput. Oper. Res. **98**, 13–23 (2018)

3. Demoulin, N.T., Djelassi, S.: Customer responses to waits for online banking service delivery. Int. J. Retail Distrib. Manag. **41**(6), 442–460 (2013)
4. Smidts, A., Pruyn, A.: How waiting affects customer satisfaction with service: the role of subjective variables. In: Proceedings of the 3rd International Research Seminar in Service Management, pp. 678–696 (1994)
5. Durrande-Moreau, A.: Waiting for service: ten years of empirical research. Int. J. Serv. Ind. Manag. **10**(2), 171–189 (1999)
6. Katz, K., Larson, B., Larson, R.: Prescriptions for the waiting in line blues: entertain, enlighten and engage. Sloan Manag. Rev. **Winter**, 44–53 (1999)
7. Song, J.S.: On the order fill rate in a multi-item, base-stock inventory system. Oper. Res. **46**(6), 831–845 (1998)
8. Dreyfuss, M., Giat, Y.: Optimal spares allocation to an exchangeable-item repair system with tolerable wait. Eur. J. Oper. Res. **261**(2), 584–594 (2017)
9. Dreyfuss, M., Giat, Y.: Optimal allocation of spares to maximize the window fill rate in a two-echelon exchangeable-item repair system. Eur. J. Oper. Res. **270**, 1053–1062 (2018)
10. Basten, R., van Houtum, G.: System-oriented inventory models for spare parts. Surv. Oper. Res. Manag. Sci. **19**(1), 34–55 (2014)
11. Dreyfuss, M., Giat, Y.: Optimizing spare battery allocation in an electric vehicle battery swapping system. In: 6th International Conference on Operations Research and Enterprise Systems, Porto, Portugal, pp. 38–46 (2017)
12. Dreyfuss, M., Giat, Y.: The window fill rate with nonzero assembly times: application to a battery swapping network. In: Parlier, G., Liberatore, F., Demange, M. (eds.) Operations Research and Enterprise Systems, ICORES, 2017. Communications in Computer and Information Science, vol. 884, pp. 42–62. Springer, Cham (2018)
13. Dreyfuss, M., Giat, Y.: Allocating spares to maximize the window fill rate in a periodic review inventory system. J. Prod. Econ. (Forthcoming)
14. van der Heijden, M.C., De Kok, A.G.: Customer waiting times in an (R, S) inventory system with compound Poisson demand. Z. Oper. Res. **36**(4), 315–332 (1992)

Decision Support for Material Procurement

Heiner Ackermann, Erik Diessel, Michael Helmling, Christoph Hertrich, Neil Jami, and Johanna Schneider

Abstract Buying raw materials at low cost is important for the economic success of manufacturing companies. In this extended abstract, we summarize some of the cost-driving constraints and cost-saving opportunities available to a global manufacturer when purchasing raw materials. We outline how to model the procurement problem as a mixed-integer linear program, and describe the use of Sankey diagrams to compare alternative order volume plans.

Keywords Supply chain · Material procurement · Mixed-integer programming · Sankey diagrams

1 Introduction

In order to have continuous access to raw materials, manufacturer qualify multiple suppliers and negotiate long-term framework agreements with them. These agreements specify, *inter alia*, material characteristics, quality, costs, and lead times, as well as order quantity limits for years at a time. The advantage of such agreements is planning security for both ends of the supply chain: Manufactures will have access to the materials required; suppliers have a guarantee that investments in production capacities will pay off.

Within the scope of the agreements, manufactures periodically order material for the following production period, based on their latest demand forecasts. These orders are placed in such a way as to minimize total expenditures. Doing so is important for the economic success of manufacturers, as they routinely spend significant portions of their total turnover on raw material [4].

At first sight, this procurement problem can be modeled as a min-cost flow problem [1] on a bipartite, time-expanded graph. The supplier sites are sources, the

H. Ackermann (✉) · E. Diessel · M. Helmling · C. Hertrich · N. Jami · J. Schneider
Fraunhofer ITWM, Kaiserslautern, Germany
e-mail: heiner.ackermann@itwm.fraunhofer.de

J. S. Neufeld et al. (eds.), *Operations Research Proceedings 2019*, Operations Research Proceedings, https://doi.org/10.1007/978-3-030-48439-2_84

manufacturer's production plants are sinks, shipping lanes are edges, and the goal is to find a flow (set of order quantities) of minimum cost. On further examination, it becomes clear that manufacturers face additional cost-driving constraints and cost-saving opportunities that go beyond pure material and transport costs. Among others there are capacity constraints, discounts and penalties, the possibility to interchange one material by another one, and to even buy the suppliers' raw materials and to pay them for converting them only. Some turn the problem into an NP-hard one which is why a combinatorial approach is not first choice when trying to solve the problem to optimality.

1.1 Our Results

We outline a mixed-integer linear program (MIP) which simultaneously takes material and transport costs as well as additional cost drivers and cost opportunities into account. Despite the fact that there is a huge number of publications on material procurement (see Sect. 1.2), this extended abstract is—to the best of our knowledge—the first one that considers that many aspects simultaneously.

For ease of presentation, we make various simplifications as described below. However, we implemented the general model in a decision support tool for a global manufacturer of fast-moving consumer goods. The tool supports operational and strategic decision making, i.e. for determining optimal order volumes, and in the course of contractual negotiations with suppliers.

Extensive experiences show that the model implemented in Gurobi [7] solves real world instances with up to 25 supplier sites, 20 production plants and several materials for planning horizons of up to 2 years length within seconds. By executing the order volume plans, our customer has saved considerable amounts of money compared to a previous planning approach.

The tool also allows to compare alternative order volume plans subject to updated supply chain topologies and cost information. It displays the differences between alternative plans using Sankey diagrams. These diagrams allow for a quick comparison of high level information.

1.2 Related Work

Decision support for material procurement has attracted lots of attention over the last decades as it can significantly improve business outcomes by providing better plans. For this task, various approaches have been developed [2]. Apart from costs, recently alternative measures like sustainability [5] and risks [6] have been considered. The surveys list many different aspects that have been taken into account. However, only few studies integrate all relevant characteristics, like multiple materials, multiple products, multiple periods, and multiple tiers of suppliers, into one model simultaneously. Additionally there are aspects that have not received much attention in the literature. Among others these are incentives, interchangeability of materials and limits on the maximum number of suppliers.

2 Mathematical Model

First, we sketch an MIP formulation of the basic min-cost problem, without additional cost-driving constraints and cost-saving opportunities. We then introduce and incorporate these additional aspects into the formulation. For simplicity, we assume zero shipping times and that demand is constant in time. As a result, we need only consider non-time-expanded networks.

Min-Cost Flows Denote the sets of materials, supplier sites and production plants by $\mathcal{M}$, $\mathcal{S}$, and $\mathcal{P}$. We will use m, s and p to refer to members of these sets. Let $\text{demand}(p, m)$ be the demand for material m at production plant p. Furthermore, let $\text{cost}_M(s, m)$ be the unit cost of material m at supplier site s, and let $\text{cost}_T(s, p, m)$ be the unit transport cost of material m when shipped from supplier site s to production plant p. Finally, let $\text{flow}(s, p, m)$ be the decision variable determining volume of material m to be shipped from supplier site s to production plant p. We state the basic version of the procurement problem as

$$\min \sum_{s\in\mathcal{S}} \sum_{p\in\mathcal{P}} \sum_{m\in\mathcal{M}} \text{flow}(s, p, m) \cdot (\text{cost}_M(s, m) + \text{cost}_T(s, p, m)),$$

subject to:

$$\sum_{s\in\mathcal{S}} \text{flow}(s, p, m) \geq \text{demand}(p, m) \quad \text{for all } p \in \mathcal{P}\ \ m \in \mathcal{M}.$$

Capacity Constraints A supplier site cannot produce arbitrarily large volumes but is restricted by maximum production constraints. Let $\text{capacity}(s, m)$ be the maximum volume of raw material m that supplier site s can produce and ship. This linear inequality adds the constraint to the program:

$$\sum_{p\in\mathcal{P}} \text{flow}(s, p, m) \leq \text{capacity}(s, m) \quad \text{for all } s \in \mathcal{S}\ \ m \in \mathcal{M}$$

Note that this is a capacity limit per node (viz. per supplier site), and not per edge as in the standard formulation of the min-cost flow problem. Moreover, an instance of the procurement problem may be infeasible due to capacity constraints.

Qualification Constraints A supplier site may not be qualified to ship material to every production plant. Let $\text{qualified}(s, p, m) \in \{0, 1\}$ be the binary parameter determining whether or not supplier site s is qualified to ship material m to production plant p. This linear inequality adds the constraint to the program:

$$\text{flow}(s, p, m) \leq \text{qualified}(s, p, m) \cdot \text{capacity}(s, m) \quad \text{for all } s \in \mathcal{S}, p \in \mathcal{P}, m \in \mathcal{M}.$$

Qualification constraints may also make an instance of the procurement problem infeasible.

Interchangeability In the event of a shortage or in case of significant savings, a manufacturer can replace one material with another one. Interchanging material is not automatically preferred as it may require adjusting the production process. On occasion however, the practice is beneficial.

Let factor(p, m_1, m_2) denote the factor at which material m_2 can be replaced by material m_1 at production plant p. We assume that the total demand(p, m) can be replaced. Let flow(s, p, m_1, m_2) be the flow variable determining how many units of material m_1 are shipped by supplier site s to production plant p, in order to satisfy the demand for material m_2, viz. demand(p, m_2). Note that $m_1 = m_2$ is feasible. It suffices to rewrite the equations above with the flow variables flow(s, p, m_1, m_2) instead of flow(s, p, m), and the factor(p, m_1, m_2).

Material Performance A manufacturer may adjust order volumes to reflect the material performance of supplier sites (viz. the quality of raw materials actually supplied). Let performance(s, m) be a ratio between 0 and 1 that measures the performance-to-specification of material m from supplier site s. We expect performance(s, m) to be a number close to 1. We multiply the flow variables in the demand constraint by performance(s, m).

Second-Level Buying A manufacturer can buy raw materials for its suppliers; these materials are called *feedstock materials*. The manufacturer orders larger volumes than each of its suppliers would, and can thus negotiate better (lower) prices. The manufacturer's suppliers are then paid for converting feedstock materials into the raw materials required by the manufacturer. We assume that each supplier site requires one feedstock material for each raw material supplied to the manufacturer. We also assume that there are no conversion losses. As a result, we do not cater for or model a bill-of-materials at any of the supplier sites. Denote by $\mathcal{F}$ the set of feedstock supplier sites. Using this, we reformulate the objective function to

$$\begin{aligned} \min \quad & \sum_{f \in \mathcal{F}} \sum_{s \in \mathcal{S}} \sum_{m \in \mathcal{M}} \text{flow}(f, s, m) \cdot (\text{cost}_M(f, m) + \text{cost}_T(f, s, m)) \\ & + \sum_{s \in \mathcal{S}} \sum_{p \in \mathcal{P}} \sum_{m \in \mathcal{M}} \text{flow}(s, p, m) \cdot (\text{cost}_C(s, m) + \text{cost}_T(s, p, m)), \end{aligned}$$

where $\text{cost}_C(s, m)$ is the unit conversion cost of material m at supplier site s. We adjust the demand, capacity and qualification constraints accordingly.

Fig. 1 A piece-wise linear cost function

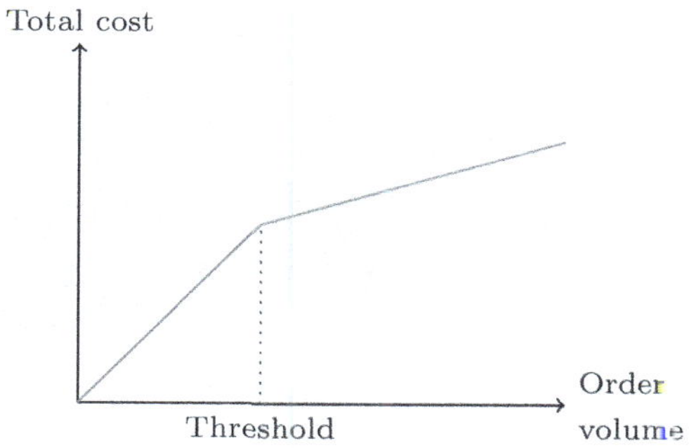

Incentives Raw material suppliers offer multi-stage incentives (discounts and penalties) in order to make favorable production volumes (from their point of view) more attractive to the manufacturers. If the total volume ordered exceeds (is less than) a given threshold, supplier sites charge less (more) for each unit of raw material. We model material cost as a piecewise linear function $\text{cost}_{\mathcal{M}}(s, m) : \mathbb{R}^{+} \rightarrow \mathbb{R}^{+}$ as in Fig. 1. We refer the reader to Croxton, Gendron and Magnanti [3] for a review of ways to work with piecewise linear objective functions in a MIP. We use a variation of the third approach described in that article.

Number of Shipping Sites Let $\max(p)$ denote the maximum number of supplier sites allowed to simultaneously ship to production plant p. Let $x(s, p) \in \{0, 1\}$ be the binary decision variable determining whether supplier site s ships material to production plant p. These inequalities add these constraints to the program

$$\sum_{s \in \mathcal{S}} x(s, p) \leq \max(p) \quad \text{for all } p \in \mathcal{P},$$

$$\text{flow}(s, p, m) \leq x(s, p) \cdot \text{demand}(p, m) \quad \text{for all } s \in \mathcal{S}, p \in \mathcal{P}, m \in \mathcal{M}.$$

3 Decision-Support Tool

We implemented a supply chain decision support tool for a global manufacturer of fast-moving consumer goods, based on the procurement model described above. The system is implemented in *C#* using Gurobi [7] as the solver of the model. The system solves real-world problems in seconds, allowing for interactive decision making.

Temporal Resolution The tool supports time-dependent information such as monthly demand and different lead time information (e.g. production and shipping times). One difficulty is that temporal information is stated in different resolutions demand is monthly, whereas lead times are in days.

Fig. 2 A Sankey diagram

Sankey Diagrams The tool also allows to compute and compare different order volume plans, subject to alternative supply chain topologies and cost information. We use Sankey diagrams [8, 9] as depicted in Fig. 2 to allow for a high-level comparison of two plans.

4 Discussion

We presented a mathematical model for cost optimization in raw material procurement simultaneously taking various kinds of cost driving constraints and cost saving opportunities into account. The model generates significant savings on real world problem instances of a global manufacturer of fast-moving consumer goods within seconds.

Our model can easily be extended towards additional cost dimensions such as emission costs and set up cost, e.g. the cost of incorporating a new supplier. In the latter case, one would ideally take a bi-objective approach, i.e. procurement cost vs. set up cost. With standard techniques from multi-objective optimization this can easily be achieved.

The presented model focuses on procurement cost and order volumes. An optimal order plan, however, is robust against various kinds of disturbances and disruptions: failure of suppliers, delayed shippings, demand increases on short notice. The ultimate goal would be to compute trade-offs between the cost and the risk of a supply chain.

References

1. Ahuja, R.K., Magnanti, T.L., Orlin, J.B.: Network Flows: Theory, Algorithms, and Applications. Prentice Hall, Upper Saddle River (1993)
2. Aissaoui, N., Haouari, M., Hassini, E.: Supplier selection and order lot sizing modeling: a review. Comput. Oper. Res. **34**(12), 3516–3540 (2007). https://doi.org/10.1016/j.cor.2006.01.016
3. Croxton, K.L., Gendron, B., Magnanti, T.L.: A comparison of mixed-integer programming models for nonconvex piecewise linear cost minimization problems. Manag. Sci. **49**(9), 1268–1273 (2003)
4. De Boer, L., Labro, E., Morlacchi, P.: A review of methods supporting supplier selection. Eur. J. Purch. Supply Manag. **7**(2), 75–89 (2001)
5. Eskandarpour, M., Dejax, P., Miemczyk, J., Péton, O.: Sustainable supply chain network design: an optimization-oriented review. Omega **54**, 11–32 (2015). https://doi.org/10.1016/j.omega.2015.01.006
6. Govindan, K., Fattahi, M., Keyvanshokooh, E.: Supply chain network design under uncertainty: a comprehensive review and future research directions. Eur. J. Oper. Res. **263**(1), 108–141 (2017). https://doi.org/10.1016/j.ejor.2017.04.009
7. Gurobi Optimization, LLC: Gurobi optimizer reference manual (2019). http://www.gurobi.com
8. Lupton, R., Allwood, J.: Hybrid sankey diagrams: visual analysis of multidimensional data for understanding resource use. Resour. Conserv. Recycl. **124**, 141–151 (2017). https://doi.org/10.1016/j.resconrec.2017.05.002
9. Schmidt, M.: The sankey diagram in energy and material flow management. J. Ind. Ecol. **12**(2), 173–185 (2008). https://doi.org/10.1111/j.1530-9290.2008.00015.x

Design of Distribution Systems in Grocery Retailing

Andreas Holzapfel, Heinrich Kuhn, and Tobias Potoczki

Abstract We examine a retail distribution network design problem that considers the strategic decision of determining the number of distribution centers (DC) as well as their type (i.e., central, regional, local), and anticipates the tactical decision of allocating products to different types of DC. The resulting distribution structure is typical for grocery retailers that choose to operate several types of DC storing a distinct set of products each. We propose a novel model considering the decision-relevant costs along the retail supply chain and present a case study of a major European retailer.

Keywords Location · Mixed-integer programming · Strategic planning

1 Introduction

Retailers planning to expand their business to a new geographical area are faced with the question of how to enhance or restructure their current logistics network. A similar problem applies for retailers whose distribution systems have evolved over time and are subject to potential restructuring, for example, because the network is not reasonably aligned with the current supplier/product portfolio anymore, or because the introduction of new technologies necessitates capacity adaptations.

From a strategic network design perspective, retailers have to decide, among many diverse issues, how many warehouses to use, where to locate them, and which functions they will take on [1]. Furthermore, the network structure that is established by these long-term decisions inherently frames the subsequent tactical planning

A. Holzapfel (✉)
Hochschule Geisenheim University, Geisenheim, Germany
e-mail: andreas.holzapfel@hs-gm.de

H. Kuhn · T. Potoczki
Catholic University Eichstätt-Ingolstadt, Ingolstadt, Germany
e-mail: heinrich.kuhn@ku.de; tobias.potoczki@ku.de

J. S. Neufeld et al. (eds.), *Operations Research Proceedings 2019*, Operations Research Proceedings, https://doi.org/10.1007/978-3-030-48439-2_85

problems. For instance, the possible distribution paths and storage locations for individual products obviously depend on the predefined network [2]. It is therefore necessary to anticipate the consequences for mid-term planning when designing the retail network structure.

We examine a scenario where a retailer chooses to operate a set of distinct distribution center (DC) types that directly serve a specific set of demand centers, i.e., a set of stores within a specific delivery area. The distribution path (including storage locations) of a certain product is determined accordingly by a product-to-DC type assignment. Each type of DC is primarily specified by the number of parallel warehouses and thus the extent of delivery area that each warehouse covers. A three-type network (central, regional and local DCs), for example, could feature one central warehouse, two regional warehouses, and several local warehouses, while the DCs of a certain type are usually similar in size and equipment. Figure 1 depicts an example of such a distribution system. This concept is a common variant in retail practice and is usually applied to place the various products at an appropriate location that minimizes inventory holding and transportation costs, while reducing operational complexity [2].

Retail-specific network design problems have been examined in multiple contributions. These studies often deal with exact solution approaches applied to simplified models (e.g., disregarding inventory holding and instore operations), with a focus on product flows (e.g., [3, 4]). Other studies focus on specific criteria, such as the selection of plants/suppliers [5], resilient network design [6], or network design with transportation discounts [7]. The novelty of our approach is the detailed design of the distribution network (considering the number of warehouses and their function), taking into account consequences for the subsequent product allocation to different types of DC. Our approach follows a holistic supply chain perspective that considers inbound transportation from suppliers to the warehouses, warehouse operations, inventory holding, outbound transportation from the warehouses to demand centers, and instore operations.

The remainder is organized as follows: In Sect. 2 we present a mathematical formulation and an approach for solving the planning problem described. Section 3 then presents the results when applying the approach suggested to the real-life case of a major European retailer. Section 4 summarizes our contribution.

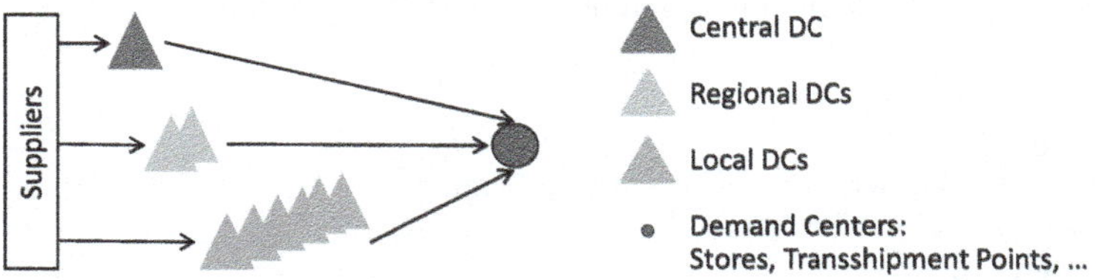

Fig. 1 Example of a three-type distribution system

2 Modeling Approach

We introduce a binary program for the network design problem that we denote as Network Design with Product Allocation (NDPA) model. Building on a predefined set of possible warehouse locations, the NDPA model determines the selection of specific DC types while minimizing total supply chain costs, anticipating the tactical product allocation decision. The locations of possible warehouses per DC type are a relevant input as they impact transportation costs in the distribution network to be designed. If the locations are not predetermined by existing structures or given preferences of the retailer, we suggest using a p-median approach in a preprocessing step to determine the warehouse locations, minimizing the sum of inbound and outbound transportation costs for each DC type considered. The p-median model is NP-hard, but, efficient algorithms exist that can solve instances of real dimensions in reasonable times [8]. The resulting warehouse locations are then used to specify the inbound and outbound distances in the NDPA model, for which we propose the following mathematical formulation:

$$\begin{aligned}\text{Minimize } Z = &\sum_{d \in D} c_d^{\text{DCfix}} \cdot z_d + \sum_{p \in P} \sum_{d \in D} c_d^{\text{DCvar}} \cdot x_{p,d} \\ &+ \sum_{p \in P} \sum_{d \in D} c_{p,d}^{\text{InOutInv}} \cdot x_{p,d} \\ &+ \sum_{d \in D} \sum_{l \in L} c^{\text{Instore}} \cdot y_{d,l}^{\text{Instore}}\end{aligned} \tag{1}$$

s.t.

$$\sum_{d \in D} x_{p,d} = 1 \qquad \forall p \in P \tag{2}$$

$$\sum_{p \in P} x_{p,d} - z_d \cdot M_1 \leq 0 \qquad \forall d \in D \tag{3}$$

$$\sum_{p \in P_l} x_{p,d} - y_{d,l}^{\text{Instore}} \cdot M_2 \leq 0 \qquad \forall d \in D, l \in L \tag{4}$$

$$x_{p,d} \in \{0, 1\} \qquad \forall p \in P, d \in D \tag{5}$$

$$y_{d,l}^{\text{Instore}} \in \{0, 1\} \qquad \forall d \in D, l \in L \tag{6}$$

$$z_d \in \{0, 1\} \qquad \forall d \in D \tag{7}$$

The objective function (1) minimizes the total supply chain costs considering warehouse (setup and operating), inbound transportation, inventory holding, outbound transportation and instore operations costs.

We split warehouse-related costs into a fixed block for every DC type d to be established (decision variable z_d, which indicates if a DC type d is used or not, and cost factor c_d^{DCfix}), and a size-dependent component (cost factor c_d^{DCvar} and decision variable $x_{p,d}$, which defines if a product p is assigned to type d). Investment costs are thus assumed to be dependent on the number of different SKUs, i.e., picking locations, assigned to a DC.

Inbound and outbound transportation as well as inventory holding costs are summarized in cost factor $c_{p,d}^{\text{InOutInv}}$. They are assumed to be product- and DC-type-specific and therefore also dependent on the assignment of products to a specific DC type (decision variable $x_{p,d}$). The transportation cost components reflect the distances and volumes from the DC locations to the suppliers and demand centers. Inventory holding costs especially account for the different degrees of demand pooling options that come along with the specific number of parallel warehouses that are established using a certain type of DC. One possibility for quantifying the inventory pooling effect is the so-called square root law [9, 10].

Costs arising from instore operations are considered by setting a penalty c^{Instore} for each additional DC type d to which products from a certain store layout category l are allocated (decision variable $y_{d,l}^{\text{Instore}}$). For details on the product allocation decision and the costs considered, we refer to [2].

The model is complemented by several restrictions. Constraint (2) ensures that each product is allocated to exactly one type of DC. Constraint (3) activates DC type d if any product p is allocated to this specific type. In Constraint (4), variable $y_{d,l}^{\text{Instore}}$ is set to 1 if at least one product of a certain store layout segment l is allocated to DC type d. The binary decision variables are defined in Constraints (5), (6) and (7).

The NDPA model is implemented in IBM ILOG Studio and solved using CPLEX v12.5. An optional post-processing stage reruns the p-median model using the solution of the DC-type selection and product allocation decision to improve the warehouse locations of the DC types selected.

3 Illustration

We implement the modeling approach with data from a major European retail chain located in Germany. The company's main assortment consists of approximately 9000 products that are sourced from about 300 suppliers. There are 36 transshipment points, which serve as local hubs for distinct subsets of stores. The transshipment points represent the demand centers in this case. Currently the company uses a three-type network design with one central warehouse, two parallel regional warehouses and six parallel local warehouses. Since the company is growing steadily, the question arises as to whether the current network still reflects the optimal configuration.

In order to evaluate the cost savings potential we investigate whether the current warehouse locations can be restructured such that total supply chain costs are minimized. For this brownfield approach new warehouse sites are disregarded. Since

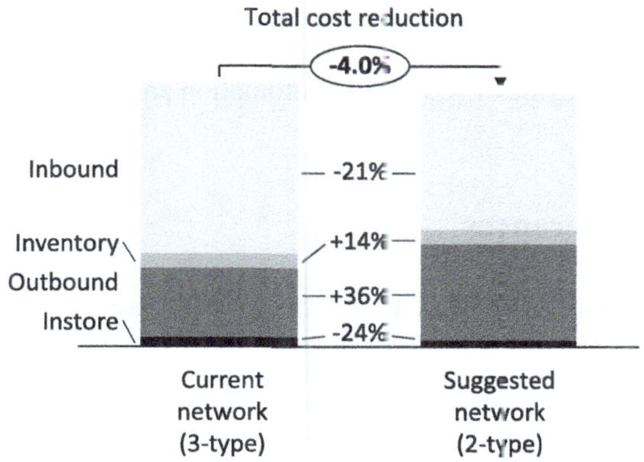

Fig. 2 Cost structure and potential savings in the case study applying NDPA

there are nine warehouse sites currently being used, the possible DC types to be chosen range from one central warehouse up to nine parallel (local) warehouses.

Our results show that 4% of total supply chain costs can be reduced if a two-type network is used instead of the current three-type network. More precisely, the configuration suggested comprises one central warehouse and five parallel regional warehouses. Figure 2 shows the cost components and the savings potential.

Besides the cost reductions that are achieved by limiting the number of DCs in use, major cost reductions can be achieved in inbound transportation by increased bundling potential if DC types are reduced. A positive effect can also be generated for instore logistics, while the more central structure means additional outbound transportation costs. Inventory costs play a minor role and increase slightly as the product allocation that is induced by the new network structure is more focused on transportation and bundling issues than on inventory pooling.

4 Summary and Outlook

In this paper we present a modeling approach for the retail network design problem with consideration of product allocations to different types of DC. We include costs arising from operating warehouses, inbound transportation, inventory holding, outbound transportation and instore operations.

The results from a case study indicate that considerable cost savings are possible when restructuring existing retail networks according to the modeling approach proposed. Additionally, the approach can be used when companies set up a distribution network in countries in which they already have a store network but no DCs.

Future research possibilities include the development of a heuristic that directly solves the integrated problem of network design and product allocation, taking into account the potential warehouse location decisions. The latter are assumed to be pre-defined in the model proposed. Other assumptions can be refined, such as including delivery frequencies based on truck loads instead of linearized transportation costs. The distinction of DC types can also be extended by including more criteria, e.g., the degree of warehouse automation and the respective costs.

References

1. Hübner, A., Kuhn, H., Sternbeck, M.: Demand and supply chain planning in grocery retail: an operations planning framework. Int. J. Retail Distrib. Manag. **41**(7), 512–530 (2013)
2. Holzapfel, A., Kuhn, H., Sternbeck, M.: Product allocation to different types of distribution center in retail logistics networks. Eur. J. Oper. Res. **264**(3), 948–966 (2018)
3. Geoffrion, A.M., Graves, G.W.: Multicommodity distribution system design by benders decomposition. Manag. Sci. **20**(5), 822–844 (1974)
4. Hindi, H.S., Basta, T.: Computationally efficient solution of a multiproduct, two-stage distribution-location problem. J. Oper. Res. Soc. **45**(11), 1316–1323 (1994)
5. Pirkul, H., Jayaraman, V.: A multi-commodity, multi-plant, capacitated facility location problem: formulation and efficient heuristic solution. Comput. Oper. Res. **25**(10), 869–878 (1998)
6. Salehi, N., Torabi, S., Sahebjamnia, N.: Retail supply chain network design under operational and disruption risks. Transp. Res. E Logist. Transp. Rev. **75**, 95–114 (2015)
7. Tsao, Y., Lu, J.: A supply chain network design considering transportation cost discounts. Transp. Res. Part E **48**(2), 401–414 (2011)
8. Kariv, O., Hakimi, S.: An algorithmic approach to network location problems. II: the p-medians. SIAM J. Appl. Math. **37**(3), 539–560 (1979)
9. Fleischmann, B.: The impact of the number of parallel warehouses on total inventory. OR Spectr. **38**(4), 899–920 (2016)
10. Oeser, G.: What's the penalty for using the square root law of inventory centralisation? Int. J. Retail Distrib. Manag. **47**(3), 292–310 (2019)

A Comparison of Forward and Closed-Loop Supply Chains

Mehmet Alegoz, Onur Kaya, and Z. Pelin Bayindir

Abstract Over the past years, closed-loop supply chains (CLSC) gained a considerable attention in both academia and industry due to environmental regulations and concerns about sustainability. Although various problems in CLSC's are addressed by researchers, not much attention is given to the effects of closing the loop in supply chains. In this study, we propose a set of linear programming models for both forward and closed-loop supply chains to see the economic and environmental effects of closing the loop. In addition to the case where there is no emission regulation, we also study the carbon cap policy and compare the forward and closed-loop supply chains under this policy. Computational results bring two important insights to us. First, we see that there are instances in which closing the loop may bring significant cost and emission reductions. Second, we observe that it may be possible to work under lower carbon caps by closing the loop in supply chains.

Keywords Forward supply chain · Closed-loop supply chain · Sustainability

1 Introduction

To reduce the negative environmental impacts from supply chains, legislation and social concerns have been motivating firms to plan and design their supply chain structures for handling both forward and reverse product flows [1]. Supply chain network design includes decisions such as the number, location and capacity of production and distribution facilities and supplier selection for raw materials [2].

There are many papers in literature which focus on the network design problem in different problem settings. A stream of research focuses on developing solution

M. Alegoz (✉) · O. Kaya
Eskisehir Technical University, Eskisehir, Turkey
e-mail: mehmetalegoz@eskisehir.edu.tr

Z. P. Bayindir
Middle East Technical University, Ankara, Turkey

J. S. Neufeld et al. (eds.), *Operations Research Proceedings 2019*, Operations Research Proceedings, https://doi.org/10.1007/978-3-030-48439-2_86

approaches to large scale network design problems [3, 4]. Some researchers focus on the uncertainties in the problem and develop stochastic or fuzzy models [2, 5, 6]. Finally, there are also some studies which consider the environmental aspects of supply chains by putting the carbon emissions into account [1, 7].

Main contribution of this study is the comparison of forward and closed-loop supply chains in terms of economic and environmental performance measures. For this purpose, we propose network design models based on linear programming for both forward supply chains (FSC) and closed-loop supply chains (CLSC) and compare the model decisions, costs and emissions with each other. To best of our knowledge, this study is one of the first studies in literature which propose network design models based on linear programming for the comparison of FSC and CLSC. The rest of the paper is organized as follows. We introduce the problem in Sect. 2 and present the mathematical models in Sect. 3. Numerical experiments are provided in Sect. 4 and finally the study is concluded in Sect. 5.

2 Problem Definition

In this study, we focus on a supply chain including suppliers, manufacturing plants, distribution centers and customers. Locations of suppliers and customers are fixed and known. Moreover, candidate locations of manufacturing plants and distribution centers are also assumed to be known. Set of suppliers, candidate manufacturing plants, candidate distribution centers, customers and operations (in respective order; manufacturing, distribution, collection and testing, repair and disassembly) are represented by H, I J, K and O respectively. Fixed costs of opening manufacturing plant i and distribution center j are represented by fm_i and fd_j respectively. Distance between supplier h and manufacturing plant i, manufacturing plant i and distribution center j and finally distribution center j and customer k are denoted by ds_{hi}, dm_{ij} and dd_{jk} respectively. Finally, in both forward and closed-loop supply chains, we denote the unit cost of shipment by uv and unit cost of an operation by uo_o.

In FSC, raw materials are procured from suppliers at a unit cost us_h and shipped to manufacturing plants. After the manufacturing operation, manufactured products are shipped to distribution centers and then finally they are shipped to customers. Demand of customer k is represented by rd_k and it must be fully met.

On the other hand, in CLSC, there are also reverse channel shipments and operations in addition to FSC shipments and operations. In this setting, a few of the products are collected from customers. Return rate of customer k is assumed to be rr_k. Those returned products are shipped from customers to distribution centers. In distribution centers, an initial testing and sorting operation is made and repairable products are determined. We name the ratio of repairable products to returned products as *product recovery* rate and represent by rp. Those repairable products are repaired in distribution centers and kept in there to be sent to customers. Unrepairable products are shipped back to manufacturing plants for disassembly.

After a disassembly operation in manufacturing plant, reusability of raw materials is checked. We name the ratio of disassembled products which can provide a reusable raw material as *component recovery rate*. Those reusable raw materials are used in manufacturing process and unreusable raw materials are sent to landfill. We assume that products which are produced from reusable raw materials are perfectly substitutable with the products which are produced from new raw materials. Without loss of generality, we also assume that one unit of product includes one unit of raw material.

In this problem setting, we propose a set of linear programming models for FSC and CLSC to compare the model results with each other. Both FSC and CLSC models decide on where to open a manufacturing plant or distribution center. We define a binary variable X_i which takes the value of 1 if a manufacturing plant is opened in candidate location i and 0 otherwise. Similarly, we define a binary variable Y_j which takes the value of 1 if a distribution center is opened in candidate location j and 0 otherwise. Moreover, in FSC models, forward flows denoted by TS_{hi}, TM_{ij} and TD_{jk} are also determined. Finally, in CLSC model, in addition to forward flows, reverse flows denoted by TC_{kj} and TR_{ji} are also determined by the model.

In addition to the case of no environmental regulation, we also focus on the case where there is a carbon cap policy and compare the FSC and CLSC under this policy. In carbon cap policy, there is an emission limit called carbon cap such that the total supply chain emission cannot exceed this limit. We assume that the carbon emissions in FSC and CLSC are resulting from operations, shipments and procurement of raw materials. By this context, the emission dedicated to production of raw material procured from supplier h is denoted by es_h, unit emission of operation o is denoted by eo_o and unit emission of shipment is denoted by ev. Finally, we denote the carbon cap by C_{cap}.

3 Mathematical Models

In this section, we present the proposed mathematical models. Let M be a very big number, by this context, the models can be presented as follows.

3.1 Forward Supply Chain Model

$$\begin{aligned}
\min z = &\sum_{i=1}^{I} fm_i X_i + \sum_{j=1}^{J} fd_j Y_j + \sum_{i=1}^{I}\sum_{j=1}^{J} uo_1 TM_{ij} \\
&+ \sum_{j=1}^{J}\sum_{k=1}^{K} uo_2 TD_{jk} + \sum_{h=1}^{H}\sum_{i=1}^{I} TS_{hi} us_h \\
&+ \sum_{h=1}^{H}\sum_{i=1}^{I} TS_{hi} ds_{hi} uv + \sum_{i=1}^{I}\sum_{j=1}^{J} TM_{ij} dm_{ij} uv + \sum_{j=1}^{J}\sum_{k=1}^{K} TD_{jk} dd_{jk} uv
\end{aligned} \tag{1}$$

Subject to

$$\sum_{h=1}^{H} TS_{hi} = \sum_{j=1}^{J} TM_{ij} \qquad \forall i \tag{2}$$

$$\sum_{i=1}^{I} TM_{ij} = \sum_{k=1}^{K} TD_{jk} \qquad \forall j \tag{3}$$

$$\sum_{j=1}^{J} TD_{jk} = rd_k \qquad \forall k \tag{4}$$

$$\sum_{j=1}^{J} TM_{ij} \leq MX_i \qquad \forall i \tag{5}$$

$$\sum_{k=1}^{K} TD_{jk} \leq MY_j \qquad \forall j \tag{6}$$

$$TS_{hi}, TM_{ij}, TD_{jk} \geq 0, X_i, Y_j \in \{0, 1\} \tag{7}$$

In the above model, objective function minimizes the total supply chain cost including the fixed costs of facilities, costs of operations, costs of shipments and procurement cost of raw materials. Equations (2) and (3) are the balance constraints for manufacturing plants and distribution centers respectively. Equation (4) is the demand satisfaction constraint. Equations (5) and (6) ensure that a facility should be opened in order to make a shipment from that facility. Finally, Eq. (7) sets the type and sign of decision variables.

3.2 Closed-Loop Supply Chain Model

$$\begin{aligned}\min z = &\sum_{i=1}^{I} fm_i X_i + \sum_{j=1}^{J} fd_j Y_j + \sum_{i=1}^{I}\sum_{j=1}^{J} uo_1 TM_{ij} + \sum_{j=1}^{J}\sum_{k=1}^{K} uo_2 TD_{jk} \\ &+ \sum_{k=1}^{K}\sum_{j=1}^{J} uo_3 TC_{kj} + \sum_{k=1}^{K}\sum_{j=1}^{J} uo_4 rp TC_{kj} + \sum_{j=1}^{J}\sum_{i=1}^{I} uo_5 TR_{ji} \\ &+ \sum_{h=1}^{H}\sum_{i=1}^{I} TS_{hi} ds_{hi} uv + \sum_{i=1}^{I}\sum_{j=1}^{J} TM_{ij} dm_{ij} uv + \sum_{j=1}^{J}\sum_{k=1}^{K} TD_{jk} dd_{jk} uv \\ &+ \sum_{k=1}^{K}\sum_{j=1}^{J} TC_{kj} dd_{jk} uv + \sum_{j=1}^{J}\sum_{i=1}^{I} TR_{ji} dm_{ij} uv + \sum_{h=1}^{H}\sum_{i=1}^{I} TS_{hi} us_h\end{aligned} \tag{8}$$

Subject to

$$\sum_{h=1}^{H} TS_{hi} + rq \sum_{J=1}^{J} TR_{ji} = \sum_{J=1}^{J} TM_{ij} \qquad \forall i \tag{9}$$

$$\sum_{i=1}^{I} TM_{ij} + rp \sum_{k=1}^{K} TC_{kj} = \sum_{k=1}^{K} TD_{jk} \qquad \forall j \tag{10}$$

$$\sum_{j=1}^{J} TD_{jk} = rd_k \qquad \forall k \tag{11}$$

$$\sum_{j=1}^{J} TC_{kj} = rd_k rr_k \qquad \forall k \tag{12}$$

$$(1 - rp) \sum_{k=1}^{K} TC_{kj} = \sum_{i=1}^{I} TR_{ji} \qquad \forall j \tag{13}$$

$$\sum_{J=1}^{J} TM_{ij} \leq MX_i \qquad \forall i \tag{14}$$

$$\sum_{k=1}^{K} TD_{jk} \leq MY_j \qquad \forall j \tag{15}$$

$$TS_{hi}, TM_{ij}, TD_{jk}, TC_{kj}, TR_{ji} \geq 0, X_i, Y_j \in \{0, 1\} \tag{16}$$

In the above CLSC model, similar to FSC model, objective function minimizes the total supply chain cost including the fixed costs of facilities, costs of operations, costs of shipments and procurement cost of raw materials. Equations (9) and (10) are the balance constraints for manufacturing plants and distribution centers respectively. Equation (11) is the demand satisfaction constraint. Equations (12) and (13) are the reverse flow constraints from customers to distribution centers and from distribution centers to manufacturing plants, respectively. Equations (14) and (15) ensure that a facility should be opened in order to make a shipment from that facility. Finally, Eq. (16) sets the type and sign of decision variables.

3.3 *Carbon Cap Constraints for Forward and Closed-Loop Supply Chains*

We can write the carbon cap constraint for FSC as presented in Eq. (17).

$$\sum_{h=1}^{H}\sum_{i=1}^{I} TS_{hi}es_h + \sum_{h=1}^{H}\sum_{i=1}^{I} TS_{hi}ds_{hi}ev + \sum_{i=1}^{I}\sum_{j=1}^{J} TM_{ij}dm_{ij}ev$$
$$+\sum_{j=1}^{J}\sum_{k=1}^{K} TD_{jk}dd_{jk}ev + \sum_{i=1}^{I}\sum_{j=1}^{J} eo_1TM_{ij} + \sum_{j=1}^{J}\sum_{k=1}^{K} eo_2TD_{jk} \leq C_{cap} \tag{17}$$

Moreover, we can write the carbon cap constraint for CLSC as in Eq. (18).

$$\sum_{i=1}^{I}\sum_{j=1}^{J} eo_1TM_{ij} + \sum_{j=1}^{J}\sum_{k=1}^{K} eo_2TD_{jk} + \sum_{k=1}^{K}\sum_{j=1}^{J} eo_3TC_{kj} + \sum_{k=1}^{K}\sum_{j=1}^{J} eo_4rpTC_{kj}$$
$$+\sum_{j=1}^{J}\sum_{i=1}^{I} eo_5TR_{ji} + \sum_{h=1}^{H}\sum_{i=1}^{I} TS_{hi}es_h + \sum_{h=1}^{H}\sum_{i=1}^{I} TS_{hi}ds_{hi}ev$$
$$+\sum_{i=1}^{I}\sum_{j=1}^{J} TM_{ij}dm_{ij}ev + \sum_{j=1}^{J}\sum_{k=1}^{K} TD_{jk}dd_{jk}ev$$
$$+\sum_{k=1}^{K}\sum_{j=1}^{J} TC_{kj}dd_{jk}ev + \sum_{j=1}^{J}\sum_{i=1}^{I} TR_{ji}dm_{ij}ev \leq C_{cap} \tag{18}$$

Both Eqs. (17) and (18) ensure that the total supply chain emission, including the emissions resulting from the operations, shipments and procured raw materials cannot exceed the predetermined emission limit, C_{cap}. However, different from the Eq. (17), Eq. (18) includes the emissions resulting from the reverse supply chain operations and reverse shipments.

4 Numerical Study

In this section, we provide the numerical experiments. We consider a problem including 5 suppliers, 5 candidate manufacturing plants, 10 candidate distribution centers and 20 customers. Unit cost and unit emission of shipment is assumed to be \$0.085 and 0.00019 ton CO_2 per ton.km. Unit cost of manufacturing, distribution, collection and testing, repair and disassembly are set as \$75, \$5, \$15, \$20 and \$5 per ton respectively. Moreover, unit emission of manufacturing, distribution, collection and testing, repair and disassembly are set as 1.75, 0.70, 0.50, 0.60, 0.50 ton CO_2 per ton respectively.

Table 1 Numerical experiments

	Cost difference	Emission difference	Selected suppliers	Opened M. plants	Opened DC's
FSC	6.86%	5.99%	2	1	6,9
CLSC			2	1	6,9

We use GAMS optimization software and CPLEX solver to solve the proposed model with the above-mentioned parameters. Computational results are presented in Table 1.

It is seen in Table 1 that in our problem setting same facilities are opened and same suppliers are selected in both FSC and CLSC. However, about 7% cost reduction and about 6% emission reduction is obtained by closing the loop. In other words, closing the loop in supply chains may bring significant cost and emission reductions even in the cases where the same facilities are used.

Moreover, we test the FSC and CLSC models under different carbon caps and observe that it is possible to work under lower carbon caps by closing the loop in supply chains. For instance, when we set the carbon cap as 20,000 ton CO_2, we see that CLSC gives a solution but that carbon cap is infeasible for FSC. In other words, there is no way to decrease the carbon cap to that level in FSC, while it is possible in CLSC.

5 Conclusion

In this study, we propose a set of linear programming models for both forward and closed-loop supply chains to compare the model decisions and see the economic and environmental effects of closing the loop in supply chains. In addition to the case of no environmental regulation, we also study the case where there is a carbon cap.

Our numerical experiments bring two important insights to us. First, closing the loop in supply chains may bring significant cost and emission reductions. Second, it is possible to work under lower carbon caps in CLSC compared to FSC. Thus, in an environment where there is a strict carbon cap, closing the loop may be a beneficial option to cope with that regulation.

References

1. Haddadsisakht, A., Ryan, S.M.: Closed-loop supply chain network design with multiple transportation modes under stochastic demand and uncertain carbon tax. Int. J. Prod. Econ. **195**, 118–131 (2018). https://doi.org/10.1016/j.ijpe.2017.09.009
2. Jabbarzadeh, A., Haughton, M., Khosrojerdi, A.: Closed-loop supply chain network design under disruption risks: a robust approach with real world application. Comput. Ind. Eng. **116**, 178–191 (2018). https://doi.org/10.1016/j.cie.2017.12.025

3. Easwaran, G., Üster, H.: A closed-loop supply chain network design problem with integrated forward and reverse channel decisions. IIE Trans. **42**(11), 779–792 (2010). https://doi.org/10.1080/0740817X.2010.504689
4. Üster, H., Hwang, S.O.: Closed-loop supply chain network design under demand and return uncertainty. Transp. Sci. **51**(4), 1063–1085 (2016). https://doi.org/10.1287/trsc.2015.0663
5. Dehghan, E., Nikabadi, M.S., Amiri, M., Jabbarzadeh, A.: Hybrid robust, stochastic and possibilistic programming for closed-loop supply chain network design. Comput. Ind. Eng. **123**, 220–231 (2018). https://doi.org/10.1016/j.cie.2018.06.030
6. Jindal, A., Sangwan, K.S.: Closed loop supply chain network design and optimisation using fuzzy mixed integer linear programming model. Int. J. Prod. Res. **52**(14), 4156–4173 (2014). https://doi.org/10.1080/00207543.2013.861948
7. Xu, Z., Pokharel, S., Elomri, A., Mutlu, F.: Emission policies and their analysis for the design of hybrid and dedicated closed-loop supply chains. J. Clean. Prod. **142**, 4152–4168 (2017). https://doi.org/10.1016/j.jclepro.2016.09.192

Part XX
Traffic, Mobility and Passenger Transportation

Black-Box Optimization in Railway Simulations

Julian Reisch and Natalia Kliewer

Abstract In railway timetabling one objective is that the timetable is robust against minor delays. One way to compute the robustness of a timetable is to simulate it with some predefined delays that occur and are propagated within the simulation. These simulations typically are complex and do not provide any information on the derivative of an objective function such as the punctuality. Therefore, we propose black-box optimization techniques that adjust a given timetable so that the expected punctuality is maximized while other objectives such as the number of operating trains or the travel times are fixed. As an example method for simulation, we propose a simple Markov chain model directly derived from real-world data. Since every run in any simulation framework is computationally expensive, we focus on optimization techniques that find good solutions with only few evaluations of the objective function. We study different black-box optimization techniques, some including expert knowledge and some are self-learning, and provide convergence results.

Keywords Black-box optimization · Simulation · Railway timetable optimization · Markov chain

J. Reisch (✉)
Synoptics GmbH, Dresden, Germany

Department of Information Systems, Freie Universität Berlin, Berlin, Germany
e-mail: julian.reisch@synoptics.de

N. Kliewer
Department of Information Systems, Freie Universität Berlin, Berlin, Germany
e-mail: Natalia.Kliewer@fu-berlin.de

J. S. Neufeld et al. (eds.), *Operations Research Proceedings 2019*, Operations Research Proceedings, https://doi.org/10.1007/978-3-030-48439-2_87

1 Introduction

The process of railway timetabling includes the simulation of a given timetable under the influence of delays. That is, primary delay distributions that affect single trains are given and the simulation computes the knock-on delays to other trains [2]. This yields a sum of presumably more delays. The aim of a simulation is to evaluate a timetable with respect to its robustness towards such primary delays. In this paper, we propose a generic framework to use any such simulation in order to not just evaluate but optimize a timetable with respect to the sum of expected delays.

The general optimization technique we employ is black-box optimization [1]. The reason we choose this technique is that the objective function can be evaluated by running the simulation with a timetable, but the derivative is almost impossible to compute. So gradient descent methods cannot be applied. Therefore, we explore the solution space by iteratively guessing a new timetable and then evaluating the objective function, that is, running the simulation for this timetable. If we have improved the solution, we can interpret the last step as a descending gradient. However, since a simulation run is computationally expensive, our task is to find good guesses for new timetables so that our optimization algorithm converges fast. We deploy simulated annealing where the neighbourhood of the current state is examined which yields the next guess. Firstly, we apply two self-learning methods, momentum and adaptive search, that reinforce the optimization direction that gave good results previously. Secondly, we implement two a priori rules that push the new state in a way an expert would do.

We evaluate the approaches with the help of a simple simulation that is based on Markov chains. The methodology is to estimate empirical distribution functions for the delay propagation. This is done by counting the frequencies of delay propagation in real-world data that is provided by Deutsche Bahn. The empirical distributions then give rise to the transition matrices of an inhomogeneous discrete time discrete space Markov chain. This Markov chain models the delay propagation of a train from one station to another. Markov chains for delay propagation modelling have been studied before [5] but the authors only choose the states "low", "medium" and "large" delays. This approach is not suitable for timetable optimization where adjustments exact to the second count, so we calculate bigger transition matrices. We apply our optimization techniques to the timetable of two long distance trains through Germany.

2 Markov Chain Simulation Model

We model the operations of a train in an event graph [6]. A train then operates from station s to station t on arcs $e_1, \dots, e_n$. An arc is a directed edge and can model either the traveling from one station v to another v' or the stopping in one station

v. In the latter case, the arc is a loop. We denote the given timetable of the train by $\hat{x} \in \mathbb{Z}^{n+1}$ where each entry stands for the planned time for event v_i.

We simulate the delay propagation from one station to the next one by an inhomogeneous discrete time, discrete space Markov chain $\mathbf{P} = \{P_i\}_{i=1}^{n}$ where P_i is the transition matrix from the delay distribution in event $i - 1$ to the one in event i. Each P_i is a m by m matrix where m denotes the number of intervals we partition the possible delays into. We make the following simplifications. We only allow delays in the interval from -3 min to 20. This assumption is reasonable as we cannot hope to cope with larger delays by timetable adjustments. On the contrary, for such delays, operational actions are needed. Moreover, our experiments have shown that a partition into 15 equally sized intervals comes with a good ratio between inner-cluster and between-cluster variance. What is more, we can estimate the conditional densities $P(\pi_i \mid \pi_{i-1})$ by empirical frequencies because we have enough data for a vector of size 15. More precisely, the entry $P^i(j, k)$ of the i-th transition matrix is computed as the number of events in the historical data where the train had a delay in interval k in event $i - 1$ and a delay in interval j in event i, normalized by the total number of data we have for this train in event i. Finally, we mention that we handle missing data with simple imputation techniques. By the choice of the model, however, we mostly have values greater zero near the diagonal of P and zero-entries far from the diagonal.

It is clear that we can compute the delay distribution π_0 at the start in a similar way. Now, at any event i, we can estimate the delay distribution by $\pi_i = P^i P^{i-1} \dots P^0 \pi_0$. Of course, we could similarly and even more precisely compute π_i from the data, but the Markov chain allows us to model the delay propagation when the timetable has changed from the original one. This brings us to the optimization of the timetable. We want to adjust the timetable in order to minimize expected delays. Any adjustment in the timetable x implies an adjustment in the Markov chain. If, for instance, we give y seconds more time supplements on edge e_i, then P^i changes in so far as trains that originally have been observed to have a delay of d seconds, now only have a delay of $d - y$ seconds. This pushes the mass of the matrix towards the more punctual intervals. The only restriction to this procedure is that negative delays are permitted. Note that this procedure is based on the frequentist model that comes with the assumption that the rest of the timetable stays more or less the same. If the modified timetable varies too much from the original one, the empirical frequencies are not a good approximation of the reality anymore. The evaluation or simulation is then denoted by $f(x) = \omega(P^n P^{n-1} \dots P^0 \pi_0)$ where ω is some objective function on the resulting final distribution π_n, the punctuality for instance.

We point out that this Markov chain can also model the propagation of delays from one train to another, but we stick to the simple case of modelling and optimizing one train at a time.

3 Black-Box Optimization

Our aim is to make small adjustments to a given timetable $\hat{x}$ so that the expected delays f in the resulting timetable x are minimized. However, our timetable is subject to some restrictions. First of all, we have a technically minimal travel time on each arc i, say x_i^{min}. Likewise, there is a maximum traveling time given by the headway to the subsequent train on the track, denoted by x_i^{max}. Finally, we want the timetable to be fixed in certain stations such as the start, the target or some stations in between where there is a connection to other trains, all contained in the set Fix. Therefore, we need to require $x_i = \hat{x}_i$ for $i \in Fix$. Within these boundaries, we are free to try any timetable and evaluate the resulting punctuality.

Throughout the optimization, we apply simulated annealing. In this approach, one usually explores the neighborhood of a current state through many iterations to improve the objective function while in the course of the algorithm it gets less likely that worsening is accepted. If, however, we cannot deploy too many iterations since each iterations means an expensive evaluation of the objective function, we need to explore the neighborhoods more carefully. In the following, we propose different techniques to adjust a timetable x in a way that hopefully brings an improvement fast.

Random Swap The most naive way is to sample two locations i and j and an appropriate number of seconds that can be added to x_i and subtracted from x_j while fulfilling the described constraints and then perform this swap.

Momentum Method A little more sophisticated is to reuse the adaption of the previous iteration in the case it improved the punctuality.

$$x = x + \eta \Delta x$$

where η is a hyper-parameter. This method comes from gradient descent approaches [7].

Adaptive Method Even more elaborate is to take not only the last, but every previous iterations into consideration. We speak of adaptive simulated annealing [3] when we assign to each station i a probability p_i that at this point, some time supplements should be added to the timetable from another station j with probably smaller value p_j. If a timetable adjustment was successful, we update the p vector proportionally to the improvement in terms of the punctuality.

Moreover, we tried two techniques that arise from expert knowledge and can be derived from the data.

Shifting Supplements The basic idea is that if we know from the data that at a station i it is very likely that a trains gets delayed relatively independent from to what extend it has been delayed in station $i - 1$, we know that any additional time supplement makes more sense after that station since any punctuality before will be destroyed at that point. Therefore, we shift time supplement from before to after this station.

Adding Supplements Another fact we can derive from the data is if at a station i there occurs systematic delay. That means that there is a high mean buildup in delay with a relatively small variance pointing at the fact that the timetable has not enough supplements at this station. We therefore add some supplement here and subtract it from another station j that falls less into this pattern.

We emphasize that in principle, we could apply an exact optimization framework such as linear programming to solve the task when we know that we model the punctuality with the Markov chain. However, the Markov chain is supposed to be an example of a simulation, but we focus on the performance, that is the convergence speed, of the different black-box optimization techniques.

4 Experimental Results

We have applied the optimization model with several combinations of strategies on two long distance trains in Germany that operate every day. We apply data preparation and map the delay data on the infrastructure [4]. In both cases, we set the start time in the origin, one arrival time in the middle of the route at a major station and the arrival time at the destination as fixed. In particular, the total travel time is fixed. The objective function is the punctuality, that is, the number of trains that run with a delay of at most 359 s, at the end of the trip. Then, we ran 5000 iterations of propagating the delays by successive multiplication of the transition matrices and updating the timetable vector. The strategies for timetable adjustments are the following. In *Adaptive Search* we sample new timetables according to the adaptive method. *Complete Random* means that we sample the timetable uniformly at random. In *Magic and Momentum* we combine the two expert knowledge adjustments shifting and adding supplements, together with the momentum method. Finally, we sample uniformly at random and apply the momentum method in *only momentum*.

Figure 1 shows the convergence behaviour of the two example trains. In each iteration, the punctuality is displayed. We see that the two trains behave differently with respect to the optimization techniques. While the punctuality of the train shown on the left can be improved a lot even without the expert knowledge the punctuality of the train on the right can be optimized only a little and predominantly with the help of the apriori rules. We conclude that if the punctuality of a train is hard to optimize and the number of iterations is very limited, it proves wise to include apriori rules in the optimization process.

We emphasize that the expected punctuality improves significantly here because the optimized timetables have shifted the major part of their time supplements towards to end of their trip where the punctuality is measured. It is unlikely, however, that one can achieve a similar result in reality by timetable adaptions

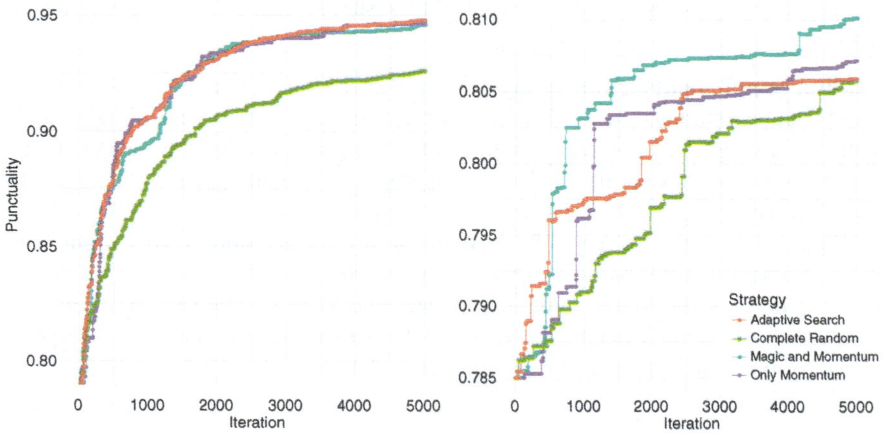

Fig. 1 Convergence behaviour of the two trains

since firstly, the punctuality during the trip will decrease. Secondly, it is impossible that all trains allocate their supplements just before their final destinations as there is not enough capacity at these points. Nevertheless, we can deduce from these results that there are cases when black-box optimization can speed up an iterative timetable optimization process where simulation is applied. In particular, it can prove wise to incorporate expert knowledge in optimization algorithms to improve the convergence behaviour.

5 Conclusions and Outlook

We have seen that black-box optimization can be used for optimization with simulation. It turned out that including expert knowledge in the algorithm can improve the convergence behaviour. We point out that it would be interesting to do further research in exploring more apriori rules, maybe by applying machine learning techniques in a preprocessing step.

References

1. Amaran, S., Sahinidis, N.V., Sharda, B., Bury S.J.: Simulation optimization: a review of algorithms and applications. CoRR abs/1706.08591 (2017). http://arxiv.org/abs/1706.08591
2. Curchod A.: Analyse de la Stabilité d'horaires Ferroviaires Cadencés sur un réseau maillé: Bedienungshandbuch. FASTA II, Lausanne (2007)
3. Gong, G., Liu, Y., Qian M.: An adaptive simulated annealing algorithm. Stoch. Proc. Appl. **94**(1), 95–103 (2001). https://doi.org/10.1016/S0304-4149(01)00082-5

4. Hauck, F., Kliewer, N.: Big data analytics im bahnverkehr. HMD Praxis der Wirtschaftsinformatik **56**, 1041–1052 (2019). https://doi.org/10.1365/s40702-019-00524-7
5. Kecman, P., Corman, F., Meng, L.: Train delay evolution as a stochastic process. In: Tomii, N, Barkan, C.P.L., et al. (eds.) Proceedings of the 6th International Conference on Railway Operations Modelling and Analysis. IAROR (2015)
6. Nachtigall, K.: Periodic Network Optimization and Fixed Interval Timetables: Habilitation. Deutsches Zentrum für Luft- und Raumfahrt, Braunschweig (1998)
7. Sutskever, I., Martens, J., Dahl, G., Hinton, G.: On the importance of initialization and momentum in deep learning. In: Dasgupta, S., McAllester, D. (eds.) Proceedings of the 30th International Conference on Machine Learning (PMLR), vol. 28, pp. 1139–1147, Atlanta (2013). http://proceedings.mlr.press/v28/sutskever13.html

The Effective Residual Capacity in Railway Networks with Predefined Train Services

Norman Weik, Emma Hemminki, and Nils Nießen

Abstract In this paper we address a variant of the freight train routing problem to estimate the residual capacity in railway networks with regular passenger services. By ensemble averaging over a random temporal distribution of usable slots in the network, bounds on the number of additional freight trains on predefined relations are established. For the solution, a two-step capacitated routing approach based on a time-expanded network is used. The approach is applied in a case study to freight relations in the railway network of North Rhine Westphalia.

Keywords Railways · Residual capacity · Freight train routing

1 Introduction

Railway timetabling of passenger and freight traffic is usually performed on different time scales. Whereas passenger services are scheduled in the annual timetabling process, the majority of freight services are requested on relatively short notice within the timetable period. As a result, freight traffic has to be routed according to the spare residual capacity in the timetable, a problem commonly referred to as the *freight train routing problem* [1].

In long-term planning of network and line concepts, the question how many additional trains can effectively be routed is generally more important than the generation of a specific timetable. This is why an understanding of the usability of the residual network capacity as a function of the passenger traffic load is required. In particular, planning has to deal with the problem that residual capacity is temporally bound and may not be harmonized between different network segments, such that coherent freight train routes are difficult to schedule.

N. Weik (✉) · E. Hemminki · N. Nießen
RWTH Aachen University, Institute of Transport Science, Aachen, Germany
e-mail: norman.weik@rwth-aachen.de
http://www.via.rwth-aachen.de

J. S. Neufeld et al. (eds.), *Operations Research Proceedings 2019*, Operations Research Proceedings, https://doi.org/10.1007/978-3-030-48439-2_88

In this paper, an adaptation of the freight train routing problem for network capacity planning applications is discussed. To assess the number of routable freight trains, a time-expanded network is considered, where train path requests and spare capacity for different network segments are randomly distributed throughout the day. The solution consists in a two-step approach: In the first step, the number of trains is maximized, in the second step traveling times are minimized. The model is applied in a case study to North Rhine Westphalia and compared to successive and static routing approaches.

2 Related Work

Various studies of the freight train routing problem have been reported in the literature. In the scheduling context, Cacchiani et al. [3] present an ILP problem for scheduling extra freight trains in an existing passenger timetable. The model is based on a time-expanded network graph that only contains links compatible with the predefined passenger services. The problem is solved using a Lagrangian heuristic where line capacity constraints are relaxed. A related, continuous scheduling approach, has been described in [2]. Here, each train is attributed with a time window and the objective is to minimize penalties resulting from time-window violation. Predefined passenger services can be allowed some flexibility based on the penalization of time windows.

Borndörfer et al. [1] abstract from the timetable to a more general routing setting, also using a time-expanded network graph. Capacity constraints are not considered explicitly. Instead, they are accounted for by a nonlinear *capacity restraint function* in the objective function, which measures congestion effects as a function of the local traffic load. Another timetable-independent approach is discussed in [4], where a static network routing problem including line segments, station areas, and route nodes is considered. Parametric queuing-based delay evaluation procedures currently used by DB Netz AG [6] are applied to calculate the residual capacity of each component. An iterative solution approach accounting for the nonlinear coupling between train routing and capacity constraints, which explicitly depend on the routing of freight trains, is introduced.

Similar to [1] and [4], the focus of this work is on the identification of residual network capacity in long-term strategic planning, regardless of the specific timetable concept. In particular, we aim to provide a lower bound on spare freight capacity assuming no harmonization of capacity reserves on different lines, which is a frequent problem for long-running freight trains. The model we propose can be seen as an extension of [4] to time-dependent routing. From a *methodological* point of view, however, our approach is most closely related to the one described in [3], including strict capacity constraints in the network graph.

3 Model

3.1 Capacitated Railway Network Model

On a macroscopic level, railway networks are composed of lines, junctions and stations. In Germany, the capacity utilization and spare capacity of these elements is presently assessed using aggregate queuing-based approaches [6]. Waiting times as a function of the traffic load ρ are compared to an empirical *level of service* (LoS), which denotes the economically optimal utilization of capacity and depends on the share of freight trains (p_{frt}). The admissible number of trains during a time frame T for a given traffic mix is obtained by setting (also see [5])

$$T_W(\rho) \stackrel{!}{=} c \cdot e^{-1.3\cdot(1-p_{\text{frt}})} \cdot T =: \text{LoS}(p_{\text{frt}}), \quad (1)$$

where the constant c depends on the type of element (see [6]).

In the present work, residual capacity is calculated based on Formula (1) for each component and distributed randomly over the time frame F for all components, independently. This corresponds to the pessimistic assumption that spare capacity on different infrastructure segments is not harmonized. However, for macroscopic network routing on the station/yard-level it is not entirely unrealistic as correlations in the underlying passenger trains' utilization are relatively small due to a large number of merging, starting or ending trains. An illustration of the resulting time-expanded residual network graph is given in Fig. 1.

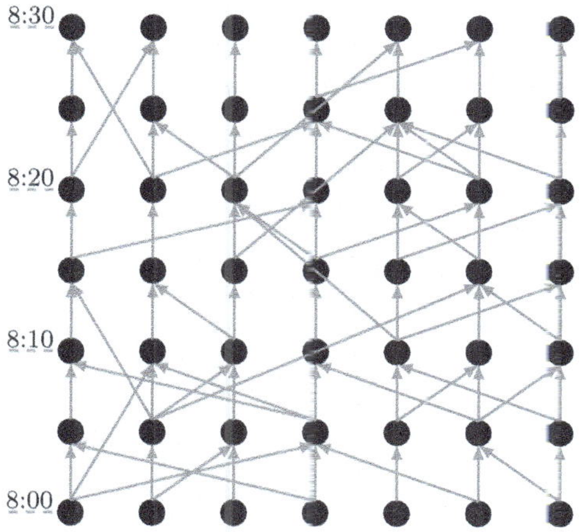

Fig. 1 Illustration of a 30 min time slice of the network with randomly distributed slots for freight trains. Station/yard capacity restrictions are not depicted in the figure

3.2 Demand Modeling

Freight traffic demand is assumed to be independently uniformly distributed over the time frame for all freight relations. Overall train path requests exceed residual static network capacity, but relations not necessarily match spare capacity at a given time. As a result, the *effective* number of additionally marketable slots in a capacity analysis setting can be investigated.

3.3 Routing Approach

In the following, we adopt the notation and build on [3] for the routing problem. Let (V, E) denote the network graph, T the train runs, and σ_j and $\{\tau_j\}$ the start and the (time-expanded) destination node of train run $j \in T$. x_{je} are decision variables allocating edge $e \in E$ to train j, $\delta^-(v), \delta^+(v)$ the sets of in- and outgoing arcs in node v. Let further c_v denote the node capacity, i.e. the number of additional freight trains that can simultaneously be acquainted in a station at the given time. It is assumed the start node has infinite capacity, which is reasonable, as it often refers to a shunting yard. t_{je}, $t_{j,\min}$ refer to the running time of train j along e and the minimum running time of train j on its designated route. q is a factor limiting the maximal acceptable running time.

The freight routing consists of a two-step approach, where the number of additional freight trains subject to time constraints is maximized in the first step and running times for this train number are minimized in the second step.

Step 1: Constrained Flow Maximization

$$\max \sum_{j\in T,\ e\in\delta^+(\sigma_j)} x_{je}, \qquad s.t.$$

$$\sum_{e\in E} t_{je} x_{je} \leq q \cdot t_{j,\min} \qquad \forall j \in T, e \in E \tag{2}$$

$$\sum_{e\in\delta^+(v)} x_{je} \leq 1 \qquad \forall j \in T, v \in V \tag{3}$$

$$\sum_{e\in\delta^-(v)} x_{je} = \sum_{e\in\delta^+(v)} x_{je} \qquad \forall j \in T, v \in V \setminus \{\sigma_j, \{\tau_j\}\} \tag{4}$$

$$\sum_{e\in\delta^+(v)} x_{je} = z_{jv} \qquad \forall j \in T, v \in V \tag{5}$$

$$\sum_{j \in T} z_{jv} \leq c_v \qquad \forall v \in V \qquad (6)$$

$$x_{je}, z_{jv} \in \{0, 1\} \qquad \forall j \in T, e \in E, v \in V, \qquad (7)$$

Constraint (2) imposes a running time restriction and constraint (3) ensures that each train visits each node at most once (no cycles). Equation (4) is the standard flow conservation and constraints (5)–(7) ensure the capacity limits of stations are satisfied. Infrastructure restrictions such as lack of electrification or narrow curves can be accounted for by setting $x_{je} = 0$ in case train j cannot be operated on this line.

Step 2: Running Time Minimization

Let n be the maximum number of additional freight trains obtained in the first optimization step. The second, running time minimization step provides insights into the quality of the routing concept in terms of the running times of the trains. The two-step approach allows to decouple flow maximization and running time minimization, which is computationally more efficient if additional fairness constraints between different train relations (cf. [4]) are to be considered.

$$\min \sum_{j \in T} \left(\sum_{e \in E} t_{je} x_{je} - \sum_{e \in \delta^+(\sigma_j)} x_{je} t_{j,min} \right), \qquad s.t.$$

$$\sum_{j \in T} \sum_{e \in \delta^+(\sigma_j)} x_{je} = n$$

$$\text{Constr. (2)–(7)}$$

4 Results

The freight routing capacity is analyzed in a case study for the network of North Rhine Westphalia, which consists of 51 nodes and 148 links (see Fig. 2). For the analysis, the three relations Oberhausen-Troisdorf, Oberhausen-Siegen and Aachen-Münster are considered, which are amongst the relations with the highest demand or have been discussed to mitigate capacity shortages.

The results presented in the following are calculated for a 8 h time frame without running time restriction (2). For time resolutions of 3–5 min and train demands of the order of the *static* residual capacity, almost all instances could be solved to

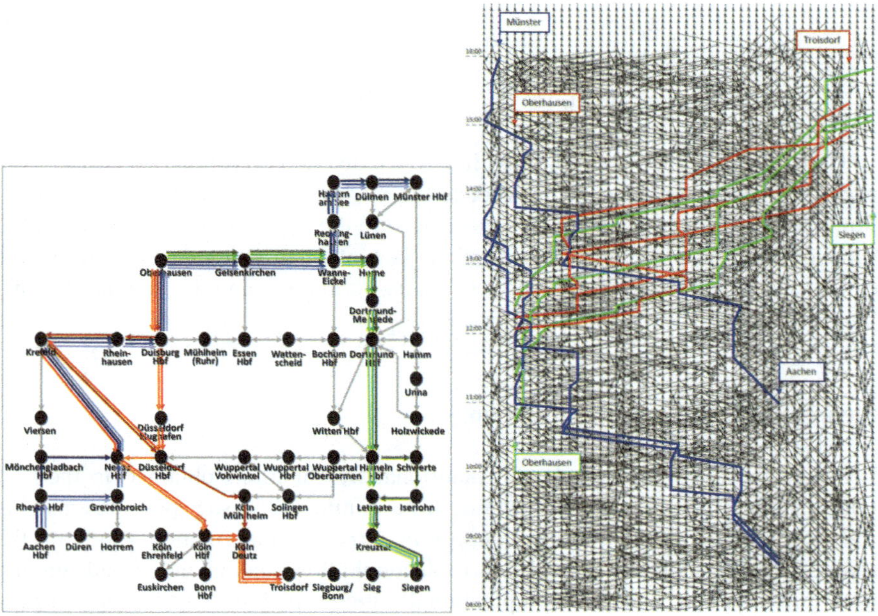

Fig. 2 Visualization of freight train routing results for three freight train relations in North Rhine Westphalia. Train routes (left) and time-expanded network (right)

optimality using Gurobi in 60–600*s*. The paths of 6 additional freight trains are visualized in Fig. 2.

A major question in the context of residual freight capacity is whether trains should be routed simultaneously (pre-planned freight slots in the timetable) or successively (which refers to the current construction practice). To investigate this problem, we analyze 500 realizations of the network with random demand and usable slots on lines. It is found that simultaneous planning yields approximately 2 trains more than successive routing (cf. Fig. 3), on average. Running time drops by about 20 min (264.8/244.9 min), also see Fig. 3. The difference will probably get stronger in case of higher overlap between freight relations.

In a static routing setting for the same network, a total of 56 trains can be supported at an average running time of 115 min. We therefore conclude that connectedness of slots in the network is a major factor and that it seems advisable to harmonize slots for entire freight relations in the timetabling process.

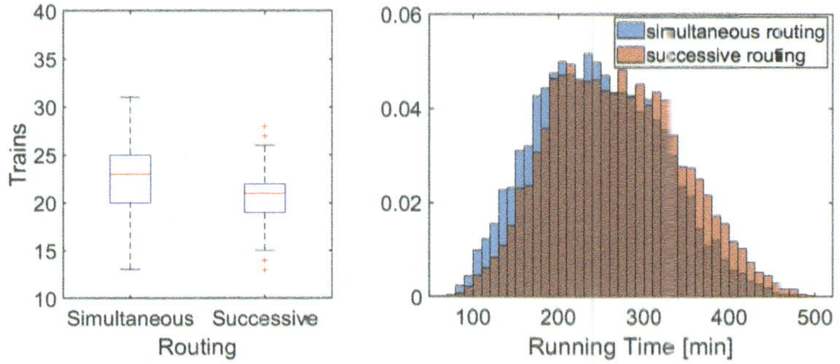

Fig. 3 Results freight train routing for simultaneous and successive routing (500 real.). Number of additional freight trains (left) and running time statistics (right)

5 Conclusion and Outlook

In this paper we have discussed an approach to assess residual network capacity for freight train routing in an existing passenger timetable concept based on stochastic demand and residual capacity. We have demonstrated its applicability in a case study for North Rhine Westphalia. In future, the approach is to be extended by coupling it to a more detailed demand and capacity modeling.

Acknowledgement This work was supported by German Research Foundation (DFG) with research grant 283085490 and RTG 2236 "UnRAVeL".

References

1. Borndörfer, R., Klug, T., Schlechte, T., Fügenschuh, A., Schang, T., Schülldorf, H.: The freight train routing problem for congested railway networks with mixed traffic. Transp. Sci. **50**, 408–423 (2016). https://doi.org/10.1287/trsc.2015.0656
2. Burdett, R.L., Kozan, E.: Techniques for inserting additional trains into existing timetables, Transport. Res. B **43**, 821–836 (2009). https://doi.org/10.1016/j.trb.2009.02.005
3. Cacchiani, V., Caprara, A., Toth, P.: Scheduling extra freight trains on railway networks. Transp. Res. B **44**(2), 215–231 (2010). https://doi.org/10.1016/j.trb.2009.07.007
4. Meirich, C., Nießen, N.: Calculating the maximal number of additional freight trains in a railway network. J. Rail Transp. Plann. Manag. **6**(3), 200–217 (2016)
5. Nießen, N.: Queueing. In: Hansen, I.A., Pachl. J. (eds.) Railway Timetabling and Operations, pp. 117–131. Eurailpress, Hamburg (2014)
6. Rothe, I.: DB Netz AG Richtlinie 405 – Fahrwegkapazität, Frankfurt/Berlin (2009)

A Heuristic Solution Approach for the Optimization of Dynamic Ridesharing Systems

Nicolas Rückert, Daniel Sturm, and Kathrin Fischer

Abstract The key to a successful ridesharing service is an efficient allocation and routing of vehicles and customers. In this paper, relevant aspects from practice, like customer waiting times, are integrated into a mathematical programming model for the operational optimization of a dynamic ridesharing system, improving existing models from the literature. Moreover, a new heuristic solution method for the optimization of a dynamic ridepooling system is developed and compared with the exact solution derived by a MIP solver based on the above-mentioned model.

In a case study consisting of 30 customers who request different rides and can be transported by a fleet of 10 vehicles in the area of Hamburg both approaches are tested. The results show that the heuristic solution method is superior to the exact solution method, especially with respect to the required solution time.

Keywords Metaheuristics · Mobility · Public transport · Ridesharing · Routing

1 Introduction

The concept of dynamic or on-demand ridesharing combines two important aspects of mobility: individualization and sustainability. Ridesharing means that several people share a vehicle for their entire journey or a part of it. In theory, fewer vehicles are required to transport the same number of people. This new form of mobility is expected to have a huge impact on the mobility behavior worldwide, and has also led to recent activities of OEMs, internet companies and startups [1–3]

The aim of this paper is to develop a heuristic solution method for solving a dynamic ridepooling problem and to compare it with an exact mathematical programming approach. Therefore, the paper is organized as follows: In Sect. 2, a categorization of ridesharing problems and a short literature overview is given.

N. Rückert (✉) · D. Sturm · K. Fischer
Hamburg University of Technology, Hamburg, Germany
e-mail: nicolas.rueckert@tuhh.de

J. S. Neufeld et al. (eds.), *Operations Research Proceedings 2019*, Operations Research Proceedings, https://doi.org/10.1007/978-3-030-48439-2_89

In Sect. 3, the problem statement of ridepooling, a special kind of ridesharing, is presented and an excerpt of the mathematical programming model is shown. This is followed by the computational results from a case study in Sect. 4. The paper ends with Sect. 5, giving a conclusion and an outlook.

2 Ridesharing: Classification and Literature Review

New modes of transport like ridesharing have become popular in recent years. Ridesharing, however, has different benefits and challenges depending on the specific variant of service provided. Here, ridesharing is divided into three categories: The first is ridehailing, a taxi-like service, but with lower prices. This can have a negative effect on the traffic system, as it may increases traffic congestion in cities [4]. The second is carpooling, in which non-profit oriented drivers share their vehicle with passengers who want to travel in the same direction. The third category is ridepooling, which aims at assigning more than one customer to a vehicle or driver in order to increase the overall efficiency of the system and to reduce the price for the individual customer, as well as overall congestion, especially in comparison to ridehailing. However, it is computationally challenging to efficiently and dynamically assign customers to vehicles and to optimally route these vehicles, while improving economic KPIs, e.g. maximizing the pooling ratio.

The optimization of operations in ridesharing is a highly relevant topic for OR, as recent publications show [5, 6]. Most papers present mathematical models based on the dial-a-ride-problem (DARP) [7, 8]. The problem of ridesharing consists of two different sub-problems which can be solved subsequently or as a combined problem: The problem of assigning customers to vehicles, and the problem of routing the vehicles [9, 10]. There are different approaches to classify ridesharing problems. While [6] classify them by the number of customers that can be assigned to vehicles and of vehicles which can be assigned to customers, [9] differentiate them by the length of detours the driver must be ready to take.

The number of customers per vehicle influences the complexity of the problem [6] and the dynamic ridesharing problem is difficult to solve as not all information is given in advance [11]. For example, new customer requests can arrive every minute, which may require reoptimization of the matching and routing. An approach to handle this is to divide the planning horizon into discrete periods aggregating batches of requests; within these, the problem can be considered as static (rolling horizon approach) [7, 11].

The dynamic ridesharing problem is NP-hard [12], but customers expect to be assigned to a vehicle in a short time. Therefore, heuristics are suggested, which aim at finding a (very) good solution fast. These range from simulated annealing to several greedy algorithms, and large neighborhood search [7, 8, 10, 12].

3 Dynamic Ridepooling

3.1 Problem Statement

In the development of quantitative solution approaches, several aspects have to be taken into account, amongst others vehicle capacity, maximum detour times and the assignment of new customers to vehicles already in operation. Below, these aspects are integrated into a mathematical programming model for the operational optimization of ridepooling systems, improving existing models. In the problem variant considered here the number of vehicles and customer requests (incl. origin, destination and time of request) per period are given. The novelty of this problem is that in assigning customers to vehicles, the objectives of minimizing travel times, waiting times and the number of non-assigned customers are considered simultaneously, aiming both at the reduction of overall traffic and maximization of customer satisfaction.

To model the dynamic nature of the problem, a rolling horizon is implemented such that the optimization model or heuristic is run at regular intervals, and new customer requests are simulated such that they occur randomly over time. In the implementation, the new requests are grouped for each period and are stored before assigning them. After each period the mathematical programming model or heuristic is run again in order to try to assign these new customers to vehicles, considering the updated locations of the vehicles at this time. This way it is possible that in each period new customers are assigned to vehicles which are already serving other customers.

3.2 Optimization Model

The mathematical programming model is based on a graph using a set of nodes $i, j \in L$, which are equivalent to locations in the ridepooling network. Additionally, there are sets for vehicles $v \in V$ and customers $c \in C$.

The objective function (1) consists of three components: The first component contains penalty costs which are incurred if a customer is not transported, indicated by the binary variables Z_c, which are multiplied by the direct travel time between origin and destination of the customer, $t_{S_c P_c}$. This implicitly corresponds to a maximization of the number of successful matches, with a preference for those customers with longer routes, in order to achieve the reduction of total distances travelled by all vehicles (incl. private cars). The second component minimizes the total travel times t_{ij} in the system, which are incurred by all vehicles on all arcs ($X_{vij} = 1$ if vehicle v travels on arc (i,j)). The third component minimizes the waiting times W_{vc} of customers at their origin. The factors α, β and γ can be used to weight the three objectives. All three components are measured in time units, e.g. minutes. With this objective function not only the reduction of total road traffic, but also the increase of customer satisfaction due to successful matches is taken into account.

$$\min \alpha \cdot \sum_{c \in C} t_{S_c P_c} \cdot Z_c + \beta \cdot \sum_{v \in V} \sum_{i \in L} \sum_{j \in L} t_{ij} \cdot X_{vij} + \gamma \cdot \sum_{v \in V} \sum_{c \in C} W_{vc} \tag{1}$$

The first component of the objective function stems, in altered form, from [13]. The second component is based on [10]. Only a few constraint groups are presented in detail below, due to limited space. In this model, arc decision variables X_{vij} and Y_{cij} with three indices are used, in contrast to other models with four-index variables (e.g. [8]). However, this leads to two additional constraint groups (3 and 4) which are needed to enable the pooling of customers. The three following constraint groups guarantee that if a customer c is traveling on an arc (i,j) ($Y_{cij} = 1$), the vehicle v assigned to this customer (indicated by the variable $M_{vc} = 1$) has to travel on this arc as well ($X_{vij} = 1$) and has to have sufficient capacity (k_v). Constraint group (2) is based on [13], while (3) and (4) were developed by the authors to connect the different variable groups.

$$\sum_{c \in C} Y_{cij} \leq \sum_{v \in V} k_v \cdot X_{vij} \quad \forall i, j \in L \tag{2}$$

$$2 \cdot Y_{cij} + (M_{vc} - 1) \cdot 2 \leq X_{vij} + M_{vc} \quad \forall v \in V \wedge c \in C \wedge i, j \in L \tag{3}$$

$$2 \cdot Y_{cij} + \left(X_{vij} - 1\right) \cdot 2 \leq X_{vij} + M_{vc} \quad \forall v \in V \wedge c \in C \wedge i, j \in L \tag{4}$$

The waiting time of a customer W_{vc} is defined as the difference between the actual departure time at the origin D_{vS_c} and the earliest possible departure time at the same node, which equals the time of request q_c (5), and should not exceed a given limit (defined by another group of constraints not shown here).

$$(1 - M_{vc}) \cdot (-M) + D_{vS_c} - q_c \leq W_{vc} \quad \forall v \in V \wedge c \in C \tag{5}$$

In total, the model consists of 24 constraint groups, some of which guarantee the flow of customers and vehicles through the network, as well as the assignment of customers to vehicles. Additionally, time window constraints are implemented, which simultaneously prevent subtours. Some of these constraint groups are similar to those used by other authors, e.g. [7, 10, 12, 13].

3.3 Heuristic Solution Method

A new heuristic solution method for the optimization of the dynamic ridepooling system described above has been developed and compared with the exact solution derived by an MIP solver using the above-mentioned model formulation.

The heuristic procedure consists of a construction heuristic and an improvement heuristic. The construction heuristic is based on the greedy randomized adaptive search procedure (GRASP) [12]. A randomly selected customer request from all new requests within a period is either added to a vehicle's existing route or is used to build a new route such that the increase of the objective function value is minimized. The improvement heuristic comprises elements of 2-opt and tabu search. All customers are paired and their respective positions in each route are exchanged for another, respecting feasibility. The exchange step with the largest objective function value decrease, if not on the tabu list, is selected (intensification step). If only exchange steps with increasing objective function value are possible, the step with the smallest increase is selected (diversification step) and saved to the tabu list. The time for which a value is stored on the tabu list grows with the number of customers in the system. After a predefined time (2 min), after a maximum number of steps or in case all adjacent solutions are on the tabu list ("Tabu list"), the algorithm terminates.

4 Computational Results

The mathematical programming model and the heuristic solution method were tested on different case studies. The test instance presented here consists of 30 customers in the urban area of Hamburg who request different rides and can be transported by a fleet of 10 vehicles. The interval length (period) of the above-mentioned rolling horizon is 5 min. Several settings were tested with different values for the factors α, β and γ, respectively, i.e. for the weights of the components of the objective function (OF). In the setting presented here (Table 1) the reduction of the total traffic was assumed to be most important. Therefore, the factor α (minimization of the overall time of non-assigned customers) was set to a value significantly higher than β (travel times) and γ (waiting times).

Table 1 Comparison of optimization model and heuristic solution method (Setting 1: $\alpha = 5$, $\beta = 1, \gamma = 0.5$)

Period	Optimization model				Heuristic method		
	OF value	Duration [h]	Termination	MIP-gap	OF value	Duration [h]	Termination
1	282.37	58:11	Time	24.64%	271.78	0:00:24	Tabu list
2	501.78	58:43	Time	32.27%	488.54	0:02:15	Time
3	Not feasible	>1 h	Time	–	804.31	0:02:24	Time

For this and all other settings, it has been found that the heuristic solution method is superior to the exact solution method, especially with respect to the solution time required, but also in terms of objective function value. This is due to the specific adaptation of the selected meta-heuristic to the dynamic ridepooling problem. The very long solution times of the mathematical programming model can be explained, amongst other reasons, by the model size based on the number of vehicles and its exponential growth in the number of (new) customers.

In the solution found by the heuristic method, out of 30 possible customers in three periods only 14 were assigned to vehicles. The others had to travel individually. However, five (out of ten) vehicles serve more than one customer, even though only one vehicle pools customers, i.e. transports two customers at the same time. The customers which are assigned have on average longer routes (11:15 min direct travel time) compared to the customers which are not assigned and have to travel alone (4:42 min). The average customer waiting time amounts to 8 min, which is significantly below the assumed threshold of 20 min. Results are similar for the setting with $\alpha = 1$, $\beta = 1$,$\gamma = 0.5$, but overall, less customers are matched.

Hence, as intended, requests of customers with longer routes are more often accepted which is beneficial to the objective of reducing total travel time and traffic.

5 Conclusions and Outlook

A mathematical programming model and a heuristic solution method were presented for a state-of-the art ridepooling problem, the aim of which was to reduce overall traffic in the system while taking customer satisfaction into account. The case study showed that this can be achieved using the approach presented in this work. Moreover, it was found that the specially tailored heuristic is better than solving a mathematical programming model, both in terms of solution quality and time.

One additional remark has to be made: With longer routes, the potential for detours is higher, which in turn increases the potential for pooling. Hence, it can be concluded that ridepooling is hard to implement for first and last mile transports, as these are usually shorter, making detours and consequently pooling of customers especially inefficient. However, it is also possible that the non-assigned customers use public transport or bikes instead, which is beneficial as well and could be even more effective in reducing traffic and/or greenhouse-gas emissions than using a ridepooling vehicle. This is most relevant for short trips, and hence it is important to consider a combination of these options in the future development of new mobility concepts and of OR approaches for their optimization.

References

1. MOIA GmbH: Ridesharing in Hamburg. https://www.moia.io/de-DE/hamburg Accessed 22 Jun 2019
2. Krafcik, J.: Waymo One. https://medium.com/waymo/waymo-one-the-next-step-on-our-self-driving-journey-6d0c075b0e9b. Accessed 22 Jun 2019
3. WunderCar Mobility Solutions GmbH. https://www.wundermobility.com/. Accessed 22 Jun 2019
4. Schaller Consulting, The New Automobility: Lyft, Uber and the Future of American Cities. http://www.schallerconsult.com/rideservices/automobility.pdf. Accessed 22 Jun 2019
5. Hou, L., Li, D., Zhang, D.: Ride-matching and routing optimisation: models and a large neighbourhood search heuristic. Transp. Res. E Logist. Transp. Rev. **118**, 143–162 (2018)
6. Agatz, N., Erera, A., Savelsbergh, M., Wang, X.: Optimization for dynamic ride-sharing. Eur. J. Oper. Res. **223**(2), 295–303 (2012)
7. Hosni, H., Naoum-Sawaya, J., Artail, H.: The shared-taxi problem: formulation and solution methods. Transp. Res. B Methodol. **70**, 303–318 (2014)
8. Masoud, N., Jayakrishnan, R.: A decomposition algorithm to solve the multi-hop peer-to-peer ride-matching problem. Transp. Res. B Methodol. **99**, 1–29 (2017)
9. Furuhata, N., Dessouky, M., Ordóñez, F., Brunet, M.-E., Wang, X., Koenig, S.: Ridesharing. The state-of-the-art and future directions. Transp. Res. B Methodol. **57**, 28–46 (2013)
10. Herbawi, W., Weber, M.: A Genetic and Insertion Heuristic Algorithm for Solving the Dynamic Ridematching Problem with Time Windows. In: Soule, T., Moore, J. (eds.) Proceedings of the Fourteenth International Conference on Genetic and Evolutionary Computation Conference, pp. 385–392. ACM, New York City, NY (2012)
11. Kleiner, A., Nebel, B., Ziparo, V.: A mechanism for dynamic ride sharing based on parallel auctions. In: Walsh, T. (ed.) Proceedings of the Twenty-Second International Joint Conference on Artificial Intelligence, pp. 266–272. AAAI, Menlo Park, CA (2011)
12. Santos, D., Xavier, E.: Taxi and ride sharing. A dynamic dial-a-ride problem with money as an incentive. Expert Syst. Appl. **42**(19), 6728–6737 (2015)
13. Agatz, N., Erera, A., Savelsbergh, M., Wang, X.: Sustainable passenger transportation: dynamic ride-sharing. ERIM Report Series Reference No. ERS-2010-010-LIS (2010)

Data Analytics in Railway Operations: Using Machine Learning to Predict Train Delays

Florian Hauck and Natalia Kliewer

Abstract The accurate prediction of train delays can help to limit the negative effects of delays for passengers and railway operators. The aim of this paper is to develop an approach for training a supervised machine learning model that can be used as an online train delay prediction tool. We show how historical train delay data can be transformed and used to build a multivariate prediction model which is trained using real data from Deutsche Bahn. The results show that the neural network approach can achieve promising results.

Keywords Delay prediction · Machine learning · Railway network

1 Introduction

Due to uncontrollable external influences and high capacity utilization of the infrastructure, train delays cannot be avoided completely. Especially in large railway networks, delays occur on a daily basis and are still a major problem for railway companies. In 2018, only about 75% of the long-distance trains in Germany arrived on time [1]. To limit the negative effects of delayed trains, it is important to identify future delays as early and as precisely as possible. This gives railway operators a chance to react accordingly, for example by providing alternative connections for passengers. Moreover, prediction models can reveal information about relations and causes of delays that can be used for timetable adjustments on a strategic level [2].

Many research projects have developed different delay-prediction approaches, which can be divided into two categories: online and offline models [3]. Offline models are static and do not consider real-time information, such as weather data or current delays of other trains. They rely entirely on historical data and are used for strategic network analysis and planning. As shown in Fig. 1, to predict the event

F. Hauck (✉) · N. Kliewer
Departmend of Information Systems, Freie Universität Berlin, Berlin, Germany
e-mail: florian.hauck@fu-berlin.de

J. S. Neufeld et al. (eds.), *Operations Research Proceedings 2019*, Operations Research Proceedings, https://doi.org/10.1007/978-3-030-48439-2_90

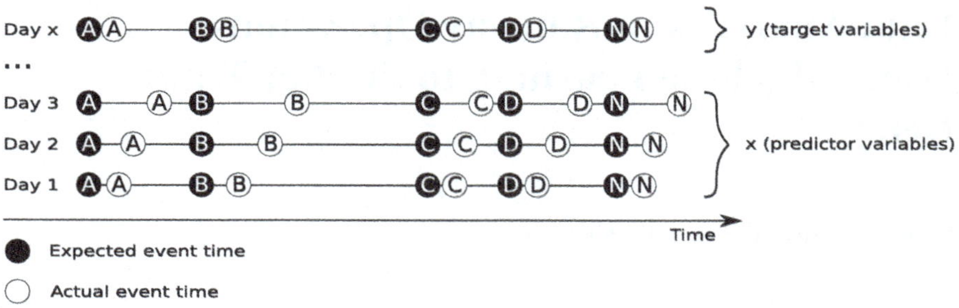

Fig. 1 Offline prediction model

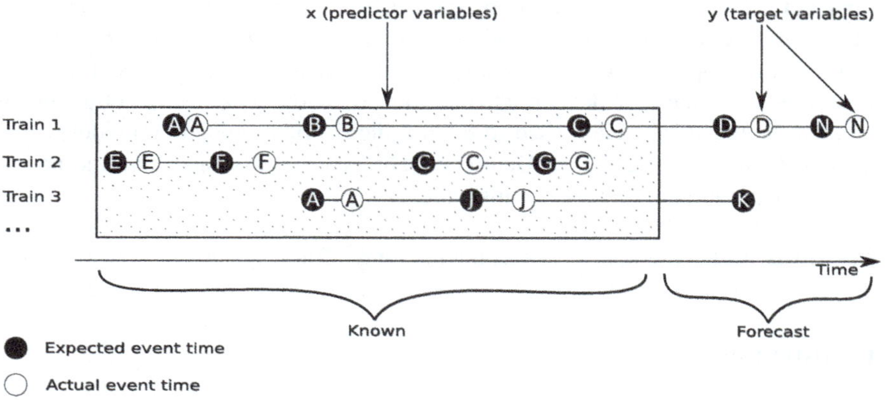

Fig. 2 Online prediction model (own illustration according to [8])

times of a train, only historical event times from the same or other trains are used as predictor variables. Offline models have been developed by Gorman [4] and Goverde [5], for example. Both papers present approaches that use linear regression models to measure the influence of different features on the delay of trains. In [2], a support vector machine was trained to predict train arrival times for tactical network planning.

Online models include real-time information and are more accurate in predicting near future events because they can also consider recent events. As shown in Fig. 2, to predict the event times of a train, known event times of all trains in the network (recent and historical) are used as predictor variables. Online models have to be updated regularly in order to always consider the most recent events in the network. Therefore, a short computing time is more important than it is for offline models. Online models are usually used in real-time management and also for passenger information systems. Stochastic online models to predict secondary delays in large railway networks, for example, have been developed by Corman and Kecman [3] and Berger et al. [6]. In [7], a neural network approach is used to predict train delays in the train network in Iran. A train delay prediction system for the Italian railway network is described in [8].

Despite those approaches, many prediction methods that are currently used by railway operators are still based on simple static rules or on expert knowledge and their results are often imprecise. Inaccurate predictions complicate the real time planning process for railway operators and have a negative impact on the customer satisfaction. Our aim is to contribute to the research area by presenting an approach for developing an online delay prediction model using neural networks. Compared to existing approaches, we develop a multivariate model that uses historical and recent delay data as well as weather data as predictor variables. We use historical train delay data from Deutsche Bahn and train supervised machine learning models that can predict arrival and departure delays.

2 Data Preparation

In cooperation with Deutsche Bahn, historical train delay data was collected. The collected dataset contains information on every train that used the German railway infrastructure in 1 year. Depending on the length of a route, a usual trip has between 50 and 400 checkpoints. For every checkpoint, the expected and the actual event time are known. Thus, the delay at every checkpoint can be calculated.

Our aim is to predict delays of passenger trains only. Therefore, freight trains are excluded from the dataset. We also assume that freight train delays do not significantly influence passenger train delays because passenger trains are usually prioritized. In the first step, the data is transformed such that every train number is represented in a table. Each table holds a separate row for every day in the year on which the train was running. Every column represents a checkpoint that the train passed along its route. This results in a table for every train, with up to 365 rows (if the train is scheduled every day). Every column contains the delay of the train in seconds at a station for the corresponding day.

The datasets for different train numbers can have missing values if a train did not travel to a certain checkpoint on 1 day or if the delay information at a checkpoint is missing. In order to avoid too many missing values, only trains that operated at least 200 times per year are included in the following analyses. Moreover, columns with more than 50% missing values are excluded. A supervised linear regression approach is used to impute the missing delay information. To impute a missing delay value, the three last know delays before and the three delays following the missing value are used to impute the missing value. We do not use all available columns as input features for the imputation, because this would lead to an extraordinarily long calculation time for the imputation process. Moreover, correlation tests show that the delay of a train at a certain station is strongly related to the delays at the previous and the next station. Therefore, the six surrounding stations are sufficient to achieve reliable imputation results.

The aim of this study is to predict the future event time of a train at a specific station at various points in time. For example, if a train starts at Station A, we want to predict its arrival/departure times at Stations B, C and D, given that we know the departure time at Station A. As soon as the train reaches the next checkpoint, the prediction for all future stations should be updated using the newly available information. The input features for the prediction model are historical delay information from the train itself and delay information from the train on the current day. Moreover, delay information from other trains in the network are included as input features as well. Since there are a lot of trains simultaneously running in the network, it is not possible to use all of them as input features. Therefore, relevant trains that influence the target train have to be selected. In order to do this, only trains that cross at least one of the stations that the target train crosses are considered as relevant. In this process, the scheduled event times of the relevant trains are checked as well. If the event time of a predictor train at the crossing station lies before the time at which the prediction is done, the delay of this train at this station is included as a feature column in the table of the target train. If the event time lies after the time at which the prediction is done, the last event time that lies before the prediction time is included as a column. If the train starts after the prediction time, the train is not considered as an input feature. If the table of the target train contains a date that is not included in the input features, this will create another missing value. The newly created missing values are later replaced using the mean score of the according column. Through doing this, we can also use trains as input features that do not drive as regularly as the target train (as long as the minimum criterion of 200 trips is held).

Next, weather data is included into the target table. Therefore, historical weather information is extracted from German weather service. For every station in the target train trip, the nearest weather station is determined and the weather (temperature, amount of rain and wind force) at this station on the corresponding day is included as columns into the dataset.

3 Modeling

Depending on the time at which the prediction is made, it is possible that the dataset contains several hundreds of columns, which can be treated as input features. In relation to the number of columns, the number of instances is rather low. Since we are only using delay data from 1 year, the maximum number of rows is 365—and that is only true for trains traveling with a daily frequency. Therefore, the number of features has to be reduced before prediction models can be trained. The feature selection is done by using a greedy forward feature selection approach. A maximum of ten features is used for the following prediction models in order to reduce the time needed for building the models. To measure the performance of the prediction models, mean squared error is used.

A multivariate linear regression model is trained after every departure of the train at a checkpoint. The input features for each model can vary, if new information from other trains in the network becomes available. The described forward feature selection approach is combined with fivefold cross-validation to find the best-performing feature subset. Before applying the model, the dataset was normalized using a linear transformation of all values such that the minimum and maximum in each column are 0 and 1.

In addition, an implementation of the RProp algorithm for multilayer feedforward networks is used for the predictions as well. RProp performs a local adaption of the weight-updates according to the behavior of the error function. The algorithm is described in [9]. One advantage of this algorithm is that it does not need a lot of parameter tuning to achieve good results. It is only necessary to determine the number of hidden layers, the number of neurons contained in each hidden layer and the maximum number of learning iterations. Again, we used forward feature selection to determine the most important features combined with a fivefold cross-validation. For parameter tuning, nested threefold cross-validation is used for each parameter inside the outer fivefold cross-validation loop that is used for feature selection. The maximum number of learning iterations was set to 100 and for the number of layers and neurons, we used grid search to find the best performing parameter combination. All variables were again transformed such that the minimum and maximum for each column are 0 and 1 respectively.

4 Evaluation

To evaluate the models, one exemplary train was selected as the target train. The target train stops at 14 stations and the whole trip takes more than 5 h. This train operates on a daily frequency and therefore, the dataset consists of 365 rows. The aim is to predict the arrival delay time at the last station. The prediction is updated every time the train departs at a checkpoint or station that lies before the target station. For the feature columns, only train lines with a frequency of more than 200 trips per year were considered. The results of the linear model and the neural network approach were compared to two benchmarks: The first benchmark uses the average delay of the train at the final station and always uses this average delay as the predictive value. The second benchmark is to use the last known delay of the train at the last known station and to use this delay for the target station. Both benchmarks are static methods that do not require machine learning models. The results are shown in Fig. 3.

It can be seen that the method that uses the last known delay for the target variable works well if the last known station lies not far away from the final station. For example, if the train has just left Station 12 or 13, the predicted arrival time for Station 14 is very accurate. However, the result is bad if only the delay at one of the first stations of the trip is known. At some break point, it is better to simply use the average delay for the target station. The results from the linear regression model and

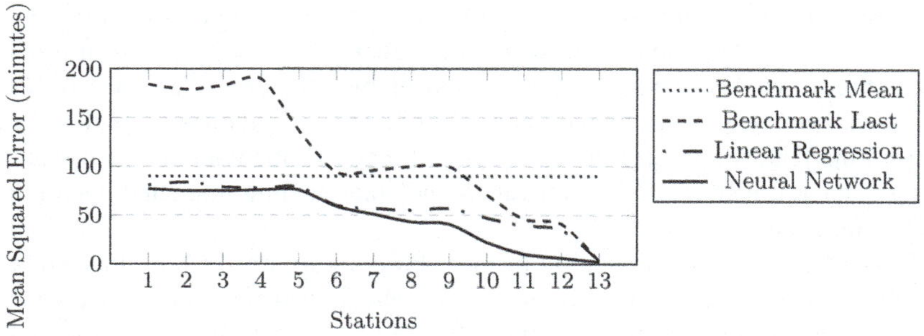

Fig. 3 MSE for predicting the arrival delay at station 14 at different times

from the neural network are always better than the results from the two benchmark strategies. It can be seen that the neural network beats the linear regression for every prediction. The neural network uses between four and nine features for the best prediction, whereas the linear regression model always performs best using a maximum of ten features. For both models the single most important feature is the last known delay of the target train. The other features are delays from other trains in the network. The weather features are not selected among the ten most important features. This can be explained because the influence of weather conditions is implicitly included in the delay information of other trains in the network.

The results show that it is possible to achieve accurate predictions using only a small number of other trains in the network. The further away the time for the prediction lies, the worse the models work. However, the neural network approach can still achieve better results than using simple benchmark algorithms even if the target event lies more than 5 h in the future. In the next step, the approach should be applied to predict event times of the other 13 stations of the trip and should also be used for different trains. Moreover, other information, like passenger flows, can be included as input features into the model.

References

1. DB Group. (2018). Integrated Report 2018. Retrieved 11 Aug 2019, from https://www.deutschebahn.com/resource/blob/4045194/462384b76cf49fe8ec715f41e4a3202a/19-03-IB_en-data.pdf
2. Marković, N., Milinković, S., Tikhonov, K., Schonfeld, P.: Analyzing passenger train arrival delays with support vector regression. Transp. Res. C Emerg. Technol. **56**, 251–262 (2015)
3. Corman, F., Kecman, P.: Stochastic prediction of train delays in real-time using Bayesian networks. Transp. Res. C Emerg. Technol. **95**, 599–615 (2018)
4. Gorman, M.: Statistical estimation of railroad congestion delay. Transp. Res. E Logist. Transp. Rev. **45**, 446–456 (2009)
5. Goverde, R.: Punctuality of railway operations and timetable stability analysis. Dissertation, Delft University of Technology (2005)

6. Berger, A., Gebhardt, A., Müller-Hannemann, M., Ostrowski, M.: Stochastic delay prediction in large train networks. In: 11th Workshop on Algorithmic Approaches for Transportation Modelling, Optimization, and Systems. OASIcs-OpenAccess Series in Informatics, vol. 20, pp. 100–111 (2011)
7. Yaghini, M., Khoshraftar, M., Seyedabadi, M.: Railway passenger train delay prediction via neural network model. J. Adv. Trans. **47**, 355–368 (2013) .
8. Oneto, L., Fumeo, E., Clerico, G., Canepa, R., Papa, F., Dambra, C., Mazzino, N., Anguita, D.: Train delay prediction systems: a big data analytics perspective. Big Data Res. **11**, 54–64 (2018)
9. Riedmiller, M., Braun, H.: A direct adaptive method for faster backpropagation learning: the RPROP algorithm. In: IEEE International Conference on Neural Networks (1993)

Optimization of Rolling Stock Rostering Under Mutual Direct Operation

Sota Nakano, Jun Imaizumi, and Takayuki Shiina

Abstract The problem of creating a rolling stock schedule is complex and difficult. In Japan, research to develop an optimal schedule using mathematical models has not progressed sufficiently due to the large number of train services. We aim to create an optimal schedule for a railway line where "mutual direct operation" is being conducted. Previous studies proposed a mathematical model as an integer-programming problem to obtain an optimal roster for a single company. In this paper, we extend the formulation to create a schedule for multiple companies to apply to mutual direct operation. The difference from previous studies is, in addition to minimizing the total distance of empty runs, to making the total running distances by company vehicles on each other's lines as close to equal as possible.

Keywords Railways · Rolling stock scheduling · Rostering · Mutual direct operation · Integer programing · TSP

1 Introduction

A rolling stock schedule is created to plan the efficient use of vehicles owned by railway companies. The schedule is composed of "paths" and "rosters". A path comprises a number of train services and corresponds to the daily schedule for a particular vehicle, and a roster is a sequence of paths, as shown in Fig. 1. A roster is created to meet various requirements. All vehicles are used evenly by assigning them to train services according to the roster.

Railway company has a line connecting two terminals operated by its own vehicles. In this paper, we will deal with the lines where "Mutual Direct Operation"

S. Nakano · T. Shiina (✉)
Waseda University, Tokyo, Japan
e-mail: tshiina@waseda.jp

J. Imaizumi
Toyo University, Tokyo, Japan

J. S. Neufeld et al. (eds.), *Operations Research Proceedings 2019*, Operations Research Proceedings, https://doi.org/10.1007/978-3-030-48439-2_91

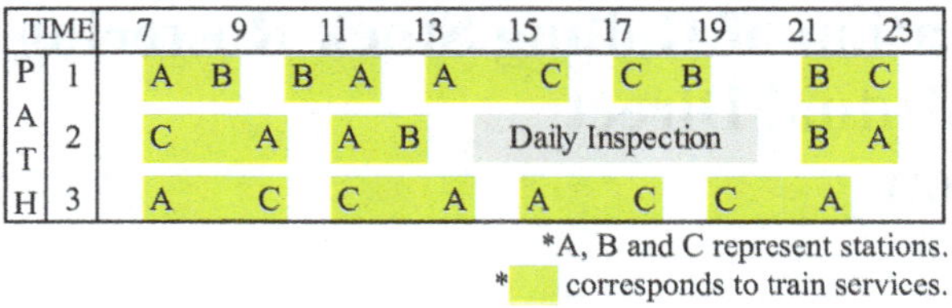

Fig. 1 Example of (single) roster

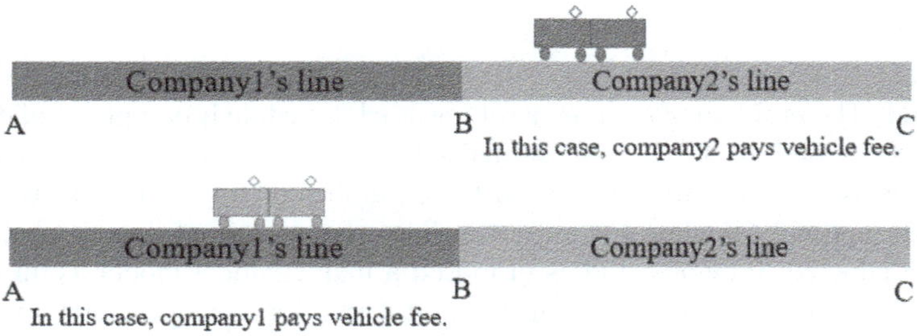

Fig. 2 Example of mutual direct operation

is being conducted. Mutual direct operation is referred to as direct driving or through service by multiple companies, which is usually operated in Japan and China. In Fig. 2, station B is shared by two companies and their lines are connected at the station. There are train services operated by a specific company over lines which it does not own. The vehicles of one company running on the other company's line are regarded as being lent to the other company owning the line and operating them. When one company uses another company's vehicles on its line, the company owning the line pays for the use of the vehicles. This payment is calculated according to the distance traveled. As a practical matter, the difference between distances traveled on the lines of other companies is minimized to make the payment amounts each other as equal as possible for the balanced operation of both companies.

2 Previous Studies

Giacco et al. [1] proposed a mathematical model as an integer-programming problem for obtaining an optimal roster for a single railway company. A network with a special structure is defined where each node corresponds to a train service and each arc corresponds to a possible connection between two train services. The problem is to find a path visiting every node in the network only once, while satisfying constraints about daily inspection, minimizing total cost and so on. In this model, cost is defined as the total distance of empty runs. This problem is similar to the well-known Traveling Salesman Problem but is more complicated, because of inspection constraints. A feature of the model is that there are four types of arcs. There may exist multiple arcs between certain two nodes. At most, one arc can be selected. The types of arcs are classified as "Turn back", "Turn back with daily inspection", "Empty run", and "Empty run with daily inspection".

Morito et al. [2] extended this model for problems where two different rosters for the rolling stock of one company must be made simultaneously. We extend the formulation for problems in which mutual direct operation is conducted. Additional constraints to the model include constraints of the difference between traveling distances on the other company's line. We show two versions of formulation in Sect. 4. The results from numerical experiments are presented in Sect. 5.

3 Background and Assumptions

We assume the following conditions.

(1) There are two railway companies whose lines are connected at a particular station. We call this the "border station".
(2) The timetable is specified. The timetable gives the departure station and time, and terminal station and arrival time for each train service. Some train services originate at stations on the line of one company, run through the border station, and terminate at stations on the line of the other company.
(3) Each company has its own vehicles. Each train service on the timetable must be assigned a vehicle belonging to one of the two companies. Vehicles of each company can travel onto the line of the other company beyond the border station.
(4) A vehicle arriving at a terminal station can be assigned to another train service that originates at that station after a time interval elapses. We call this "Turn Back". If a vehicle is assigned to two train services such that the terminal station of the first service and the departure station of the second service are different, an "Empty Run" occurs. Empty runs are useless and should be avoided, and the total distance of empty runs must be minimized.
(5) Each vehicle needs to undergo "Daily Inspection" in predetermined day interval at specific facilities located on the line of the company that owns the vehicle. The interval is set by the company that owns the vehicle.
(6) The usage fee paid by a company is calculated based on the total distance traveled on its own line using a vehicle of the other company. It is desired that the total distances for all companies are as close to each other as possible.

Under these assumptions, we define and create a network based on the idea proposed by Giacco et al. [1]. In order to extend the model [2] to mutual direct operation, we add some constraints on running distance in other company's line. However, adding these constraints will increase the size of the problem, and the computation time will be longer. Mutual direct operation is conducted by up to five companies, so we aim to develop a model or solution method that can be applied to larger cases.

4 Formulation

In the following section, we show the formulation for a case of two companies. We omit the constraints that are the same as those shown by Morito et al., and only describe the differences. The type of rolling stock in their problem corresponds to the company in our problem. Their previous model minimized the total distance of empty runs. The essential distinction from their problem is that, in our case, it is desirable for the total running distance of each company's vehicles on the other company's line to be close to equal, as well as minimizing the total distance of empty runs by vehicles of the two companies.

We define the symbols as follows (Table 1):

Table 1 Symbol definition

Notation	Description
i, j	Indices representing a train service that corresponds to a node.
z	Index representing a type of arc. z=1 is "turn back", z=2 is "turn back with daily inspection" z=3 is "empty run", z=4 is "empty run with daily inspection"
p	Index representing a railway company.
V	Set of train services.
Z	Set of arc types.
V^p	Set of train services that can be operated by company p.
A^p	Set of arcs corresponding to possible connection of two train services for company p.
A_1^p	Set of arcs for company p and the type is "turn back".
A_2^p	Set of arcs for p and the type is "turn back with daily inspection".
A_3^p	Set of arcs for p and the type is "empty run".
A_4^p	Set of arcs for p and the type is "empty run with daily inspection".
n	Number of nodes.
c_{ijz}^p	Empty run distance of arc (i,j,z) for company p.
$c_{ijz}^{p.1}$	Empty run distance by company p's vehicle in p's line of arc (i,j,z)
$c_{ijz}^{p.2}$	Empty run distance by company p's vehicle in the other company's line of arc (i,j,z)
d_{ijz}^p	Equal to 1 if arc (i,j,z) connects to the next day, 0 otherwise.
K_i^p	Operation distance of train service i in company p's line.
w	Weight for the term of difference in running distance in another company's line.
x_{ijz}^p	Equal to 1 if arc $(i\,j,z)$ is included in company p's roster, 0 otherwise.
q_i^p	Equal to 1 if train service i is operated by company p, 0 otherwise.
T	Difference in direct distance of companies.

4.1 Original Formulation

$$\min \sum_{p}\sum_{(i,j,z)\in A} c^{p}_{ijz}x^{p}_{ijz} + wT \tag{1}$$

s.t.

$$T = \left| \left(\sum_{i\in V^1} K^2_i q^1_i + \sum_{(i,j,z)\in A^1} c^{1,2}_{ijz}x^1_{ijz}\right) - \left(\sum_{i\in V^2} K^1_i q^2_i + \sum_{(i,j,z)\in A^2} c^{2,2}_{ijz}x^2_{ijz}\right) \right| \tag{2}$$

The constraints are that all nodes are covered only once, subtour elimination, daily inspection, and capacity for vehicle depots.

The objective function (1) minimizes the sum of the distance of empty runs and the difference in running distance on another company's line (termed "direct distance" below). Constraint (2) calculates the difference in direct distance. There are some other constraints such as subtour elimination and daily inspection, but these are omitted because they are essentially the same as in previous formulations [2].

Unfortunately, the computational time is very long to obtain optimal solutions for problems of this formulation, which suggests that this formulation is not applicable to cases for two or more companies. We therefore modified the formulation as described below.

4.2 Revised Formulation

We consider that the cause of the long computation time is that the problem has a large number of feasible solutions. We have modified the original formulation and define a new formulation below. In this formulation, we introduce new parameters *X* and *Y*, and constraints that bring each company's direct distance close to the target *X* and keeps it within the tolerance *Y*. In this method, it is expected that the search range is limited and the calculation time is shortened.

$$\min \sum_{p}\sum_{(i,j,z)\in A} c^{p}_{ijz}x^{p}_{ijz} \tag{3}$$

s.t.

$$T^1 = \sum_{i\in V^1} K^2_i q^1_i + \sum_{(i,j,z)\in A^1} c^{1,2}_{jz}x^1_{ijz} \tag{4}$$

$$T^2 = \sum_{i\in V^2} K^1_i q^2_i + \sum_{(i,j,z)\in A^2} c^{2,2}_{ijz}x^2_{ijz} \tag{5}$$

$$X - Y \le T^1, T^2 \le X + Y \tag{6}$$

The new objective function (3) minimizes the distance of empty runs. Constraints (4) and (5) calculate the direct distance for each company respectively. Constraint (6) ensures that the direct distance is kept within [*X*-*Y*, *X*+*Y*].

In the revised formulation, the difference of direct distance may not be zero, but can be subsequently reduced. The method for adjusting the distance is to exchange the "path" operator for a particular day after a difference has accumulated.

Modifying the formulation reduces computation time drastically. The experimental results from these two versions of the formulation are shown in the next section.

5 Numerical Experiments

We show computational results for two mutual direct operation lines. These experiments are performed on a PC with Core i7-7700 and 32GB memory using AMPL-CPLEX version 12.6.2.0 on Windows 10 Pro.

The first case is a line that is operated by East Japan Railway (JRE) and Tokyo Waterfront Area Rapid Transit (TWR), as shown in Fig. 3. There are 547 train services. The computing time is approximately 2700(s) using the original formulation, but is reduced to approximately 60(s) by the revised formulation. In the actual roster, the total distance of empty runs is 224.1 km; however, this is improved to 209.6 km by both formulations. For direct distance, the difference is 0.0 km using the original formulation, but depends on the value of *Y* for the revised formulation. We determine *X* and *Y* based on the actual roster values. The actual difference of direct distance is 220.0 km, and the total distance of all train services is 17,094.4 km.

The second case is a line which is operated by JRE, Tokyo Metro and TOYO Rapid Railway, as shown in Fig. 3. There are 608 train services and 3 companies, which causes an increase in the number of discrete variables. It takes more than 3 weeks to complete computation using the original formulation, but this time is reduced to 300,000(s) by the revised formulation. In the actual roster, the total distance of empty runs is 450.1 km, the difference of direct distance is 535.2 km, and the total distance of all train services is 23,512.3 km. The result using the revised formulation is also shown in Table 2.

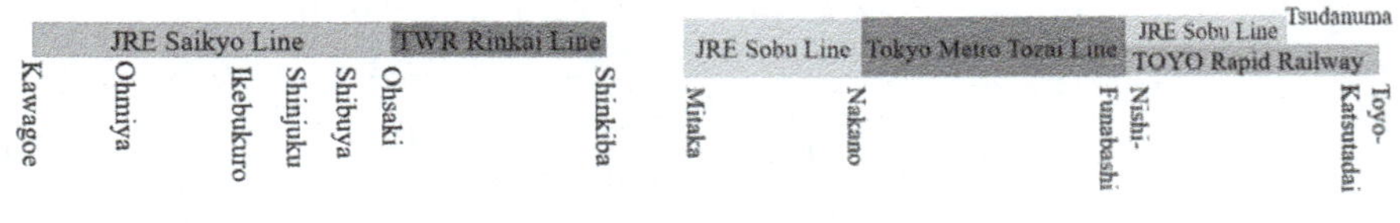

Fig. 3 The route map of experiments

Table 2 Experimental results

	Experiment 1					Experiment 2				
	X	*Y*	Empty run	Difference	Time	*X*	*Y*	Empty run	Difference	Time
	–	–	209.6 km	0.0 km	2668 s	–	–	–	–	Over 3 weeks
Original	2900	100	209.6 km	134.4 km	77 s	3000	200	440.9 km	434.6 km	437,483 s
Revised	2900	10	209.6 km	18 km	46 s	3000	100	440.9 km	397.5 km	300,823 s

6 Conclusion

In Japan, mutual direct operation is often conducted by more than three companies. Using the method described herein, the potential to increase computational efficiency is demonstrated.

Acknowledgment This work was supported by JSPS KAKENHI Grant Number JP18K04619.

References

1. Giacco, G.L., D'Ariano, A., Pacciarelli, D.: Rolling stock rostering optimization under maintenance constraints. J. Intell. Transp. Syst. Technol. Plann. Oper. **18**, 95–105 (2014)
2. Morito, S., Fukumura, N., Shiina, T., Imaizumi, J.: Rolling stock rostering optimization with different types of train-sets. Proceedings of International Symposium on Scheduling 2017. pp. 4–9 (2017)
3. Tokyo Jikokuhyo 2014-04, Tokyo-Kotsusya (2014)

The Restricted Modulo Network Simplex Method for Integrated Periodic Timetabling and Passenger Routing

Fabian Löbel, Niels Lindner, and Ralf Borndörfer

Abstract The Periodic Event Scheduling Problem is a well-studied NP-hard problem with applications in public transportation to find good periodic timetables. Among the most powerful heuristics to solve the periodic timetabling problem is the modulo network simplex method. In this paper, we consider the more difficult version with integrated passenger routing and propose a refined integrated variant to solve this problem on real-world-based instances.

Keywords Periodic event scheduling problem · Periodic timetabling · Integrated passenger routing · Shortest routes in public transport

1 Introduction

Operating a public transportation network requires several planning steps, in particular finding a good periodic timetable that minimizes overall travel and waiting times for passengers. Most model formulations are based on the linear mixed-integer *Periodic Event Scheduling Problem* (PESP) proposed by Serafini and Ukovich [7] which has been used to determine schedules real-world transportation networks operate under, see e.g. [3].

Since solving PESP or even finding any feasible solution is in general NP-hard, heuristic approaches are required and the *modulo network simplex* (MNS) method proposed by Nachtigall and Opitz [5] is among the most powerful. It is based on the classical network simplex algorithm for solving minimum cost flow problems, however, it has no optimality guarantee and struggles to escape local optima. For research into improving MNS see, e.g., [2, 5].

PESP instances are given by so-called *event-activity networks* (EAN) which are directed graphs with fixed arc weights. The nodes model timing events like line

F. Löbel · N. Lindner (✉) · R. Borndörfer
Konrad-Zuse-Zentrum für Informationstechnik Berlin, Berlin, Germany
e-mail: fabian.loebel@zib.de; lindner@zib.de; borndoerfer@zib.de

J. S. Neufeld et al. (eds.), *Operations Research Proceedings 2019*, Operations Research Proceedings, https://doi.org/10.1007/978-3-030-48439-2_92

arrivals and the arcs model activities like transferring between two lines and have an associated duration. The weights have to be chosen such that they reflect how passengers will utilize the transportation network and need to be anticipated as a prerequisite for the timetabling problem. However, any given timetable clearly influences passenger behavior as trips with short transfers are preferred. This leads to a chicken-and-egg problem which can be solved by integrating periodic timetabling and passenger routing, offering better solution quality and a more realistic model at the cost of significantly increased problem complexity. In this paper we propose a refinement to our integrated MNS [1, 4] for solving this problem by restricting passengers to a pre-selection of paths with few transfers.

2 The Integrated Periodic Timetabling Problem

As is common practice in literature on integrated timetabling, in order to introduce passenger behavior we have to extend the EAN by cells and an *origin-destination-matrix* (OD matrix) to reflect where passengers generally enter and leave the transportation network.

Definition 1 An *extended event-activity network* is a directed simple graph $(E \cup C, A \cup A_C)$ with an *origin-destination-matrix* $D = (d_{st}) \in \mathbb{Q}_{\geq 0}^{C \times C}$ and activity bounds $\ell \leq u \in \mathbb{N}_0^{A \cup A_C}$. The nodes in E are called *events* and are disjoint from the *cells* in C. The events are further partitioned into *arrival*, *departure* and *auxiliary* events. $A \subseteq E \times E$ are called *activities* and are partitioned into *waiting*, *driving*, *transfer* and *auxiliary* activities. Arcs in A_C are called *OD activities* and either point from a cell to a departure or from an arrival to a cell. Furthermore, $\ell_a = u_a$ for all $a \in A_C$.

In its most fundamental form the EAN consists of a sequence of arrival and departure events for every line connected by waiting and driving activities modeling the line's trip, and transfer activities between arrival and departure events of different lines at the same station. Auxiliary events and activities may be used to model further features of the underlying transportation network like, e.g., headway activities for safety constraints in train networks.

Each cell is connected to a selection of stations and serves as a source and sink for passengers at these stations. The fixed length of the OD activities is meant to price the different stations where passengers may enter and leave the network, since cells will not be scheduled.

Any entry d_{st} of the OD matrix encodes how many passengers wish to travel from cell s to cell t. We call $(s, t) \in C \times C$ an *OD pair* if $d_{st} > 0$ and select a *passenger path* in the EAN that carries this demand. A passenger path of an OD pair (s, t) is a directed path starting in cell s, ending in cell t, and that has only arrival and departure events as its interior nodes, modeling the passenger traversing the transportation network.

Consider the problem definition 2 on the next page. Line (1) is the bilinear objective minimizing the total travel time of all passengers. We differentiate fixed length OD activities and other activities with *slack* y_a based on the periodic timetable π. Lines (2)–(4) are the usual constraints of periodic event scheduling problems using the modulo bracket $[x]_T := \min\{x + zT \mid x + zT \geq 0, z \in \mathbb{Z}\}$.

Lines (6) and (7) force the selection of exactly one passenger path for every OD pair and line (5) derives the appropriate weights from the selected paths.

Definition 2 Given an extended EAN and a *network period* $T \in \mathbb{N}$, let P_{st} denote the set of all *passenger paths* for every OD pair (s, t) Then the *integrated periodic timetabling problem (iPTP)* is to find a cost optimal feasible timetable $\pi \in \{0, \ldots, T-1\}^E$ with an optimal passenger flow $w \in \mathbb{Q}_{\geq 0}^{A \cup A_C}$, that is,

$$
\begin{aligned}
\min \quad & \sum_{a \in A} w_a (y_a + \ell_a) \quad + \sum_{a \in A_C} w_a \ell_a && && (1)\\
\text{s.t.} \quad & y_a = \left[\pi_j - \pi_i - \ell_a\right]_T && \forall a = (i, j) \in A, && (2)\\
& 0 \leq y_a \leq u_a - \ell_a && \forall a \in A, && (3)\\
& \pi_i \in \{0, \ldots, T-1\} && \forall i \in E, && (4)\\
& w_a = \sum_{d_{st} > 0} \sum_{p \in P_{st}, p \ni a} d_{st} f_p && \forall a \in A \cup A_C, && (5)\\
& \sum_{p \in P_{st}} f_p = 1 && \forall (s, t) \in C \times C, d_{st} > 0, && (6)\\
& f_p \in \{0, 1\} && \forall p \in \bigcup_{d_{st} > 0} P_{st}. && (7)
\end{aligned}
$$

In this paper, we assume $u_a = \ell_a + T - 1$ for all transfer activities making their tensions effectively unconstrained and $\ell_a = u_a$ for all other activities which will be advantageous later on. Driving and waiting activities usually have little wiggle room for their duration and transfer activities will naturally be shortened by the objective. For more details on this see [4, 6].

3 Solving iPTP with Modulo Network Simplex

The classical modulo network simplex relies on the observation [5] that, similar to minimum cost network flow problems, feasible solutions to PTP correspond to *spanning tree structures*, i.e., spanning trees with activities fixed at their lower or upper bounds, and that any solution can be obtained from any other solution by applying a sequence of cuts.

MNS has an inner loop in which it tries to augment an initial solution by exchanging basic with non-basic activities along the fundamental cycles of the spanning tree as long as it finds improvements. An outer loop then tries to escape local optima by shifting event potentials along special cuts, see e.g. [2, 5].

The spanning tree structure correspondence holds for iPTP as well, so an adapted version of MNS can be used to heuristically solve the integrated problem [1, 4]. We proposed four approaches for handling the passenger paths, differing in the place where Dijkstra's algorithm is invoked to adjust the passenger flow: The *integrated* method computes the shortest paths for every adjacent solution when determining the next pivot operation, the *iterative* method recomputes the paths after the inner loop, whereas the *hybrid* method performs recomputation after every base change. Finally, the classical *fixed* MNS does not change the weights in its run, but can in principle be applied as well. Expectedly, the integrated MNS yields better solutions at a hefty runtime penalty.

The integrated MNS finds better solutions than the other methods by comparing adjacent timetables with their respective optimal paths. This, however, requires the computation of those paths, so we have to run Dijkstra's algorithm on every cell for every adjacent timetable, slowing the method down significantly. We have yet to identify a way to cheaply predict the shortest paths after a base change and tested a few straightforward approaches for reducing the number of Dijkstra calls by, e.g., fixing a percentage of the OD pairs on their initial paths.

The most promising approach turned out to be restricting every OD pair to a preselection of a few paths and routing passengers along the shortest of those for every examined timetable. Restricting to the k shortest paths for every OD pair with respect to the lower bounds for small k worked reasonably well in our tests, but requires running Yen's algorithm for every OD pair as a preprocessing step which is a runtime bottleneck itself.

In any decently designed transportation network, passengers should not need to transfer too often to reach their destination. In our real-world based instances, no shortest path (w.r.t. lower bounds) between any OD pair requires more than three transfers. Empirically testing $k = 10, 20, 40, 80$, the overall most reasonable selection of paths in terms of runtime and solution quality in our tests turned out to be the $k = 20$ shortest paths with at most two transfers for every OD pair. Due to the small number of transfers, these paths can be computed by a modified breadth-first search on the line graph—containing the lines of the public transportation network as vertices and possible transfers as arcs—maintaining a list of the twenty shortest found so far. This can be done as a preprocessing step with negligible runtime. In order to handle OD pairs that require at least three transfers we compute the actual shortest paths of the initial solution using Dijkstra's algorithm and add these paths to the selections, ensuring that every OD path has at least one path.

As a further measure, we compute the actual shortest paths after every base change like for the hybrid mode and dynamically update the path selections with any newly encountered paths.

To further expand on our previous contribution, we added the outer loop based on *single node cuts* [2] taking advantage of our assumption about the activity bounds.

They make finding feasible spanning trees very easy by turning the line components into a forest and then connecting the components by adding arbitrary transfer activities. Since transfer lengths are unrestricted, feasibility of these solutions is guaranteed. If given some tensions, we can construct a corresponding spanning tree by connecting line components via transfer activities with tensions at the bounds or returning an error if this failed. The runtime of this method is equivalent to a breadth-first search on the network's line graph.

Once the inner loop can no longer find any improvements, we iterate over every line and try to shift the potentials of its events by δ ranging from 1 to $T - 1$. This changes the tensions of all adjacent transfer activities a to $\ell_a + [y_a \pm \delta]_T$ based on the orientation where y_a is the current slack. The resulting timetable is feasible and we compute its objective, using the current weights for the iterative and hybrid method and computing the appropriate weights under the shifted tensions for the integrated method. If the objective is improving, we try to create the corresponding tree structure and restart the inner loop with it if successful. If there is no such cut we terminate. For the iterative and hybrid method we recompute the activity weights before entering the inner loop, for the integrated we pass the already computed weights on.

4 Computational Results

We present the computational results of the classical fixed, the iterative, integrated, hybrid and restricted integrated MNS as described in the previous section on a selection of our real-world based instances. Unfortunately, there is no set of benchmark instances for iPTP like the PESPlib for the fixed problem and we have not yet tried other solving approaches on our instances, so we cannot offer any comparability here.

Our instances are based on sub-networks of the public transportation systems of the cities of Wuppertal and Karlsruhe, and the Dutch regional train network, all with a period of 60 min. We used our spanning tree generation approach described above to approximate timetables provided with these instances as our initial solutions (Table 1).

Table 1 The test instances we present here

Name	#stations	#lines	#OD pairs	#events	#activities	Lower bound	Initial obj.
Dutch	23	40	158	448	3791	868,074.00	900,395.00
Wupp	82	56	21,764	2166	28,733	1,373,189.84	1,519,746.75
Karl	462	115	135,177	10,497	84,255	3,844,702.81	4,668,327.18

The lower bound on the objective is obtained by setting all slack variables to zero and computing the shortest passenger paths w.r.t. those. Number of lines counts return trips separately

Table 2 Computational results with a soft runtime limit of 2 h

Name	Method	Time (s)	CPU (s)	#cuts	Final obj.	Gap (%)
Dutch	Fixed	5	26	22	883,378.00	1.76
	Iterative	6	37	24	883,508.00	1.78
	Integrated	1023	5959	45	868,647.00	0.07
	Hybrid	6	35	26	879,213.00	1.28
	Restricted	36	200	43	868,275.00	0.02
Wupp	Fixed	61	260	12	1,503,432.93	9.48
	Iterative	62	288	11	1,502,939.40	9.45
	Integrated	7200	17,676	3	1,501,857.91	9.37
	Hybrid	53	224	10	1,504,797.10	9.58
	Restricted	7200	16,986	18	1,471,607.73	7.17
Karl	Fixed	675	3290	35	4,568,980.65	18.84
	Iterative	951	3735	34	4,563,223.79	18.69
	Integrated	7223	50,473	0	4,668,327.18	21.42
	Hybrid	1182	3538	32	4,564,297.63	18.72
	Restricted	7202	30,036	1	4,642,169.83	20.74

Pivot search was distributed onto seven threads, hence CPU time is also provided. The number of cuts counts both pivot operations of the inner loop and applied cuts in the outer loop

As pivot rule for the inner loop, we let the method select the most improving base change and distributed the candidate examination to multiple CPU threads. For the outer loop we selected the first improving cut. For the fixed method we computed the optimal activity weights at the end to gauge the impact of merely adjusting the weights on the objective.

It turned out that on all of our instances, iterative and hybrid were not significantly better or worse than adjusting the weights after a fixed run. The restricted integrated does offer the desired speed-up and solution quality but is still too slow to solve large instances like the entirety of Wuppertal's network (not listed) or the presented "Karl" instance (Table 2).

Quality and runtime of the restricted integrated MNS depend on the kind and number of paths selected and in order to improve it further, better selections need to be made. The optimal choice of course would be exactly those paths that the unrestricted integrated method would use during its optimization run, but it is yet unclear how to correctly predict them.

Acknowledgments The authors "Fabian Löbel and Niels Lindner" were funded by the Deutsche Forschungsgemeinschaft (DFG, German Research Foundation) under Germany's Excellence Strategy—The Berlin Mathematics Research Center MATH+ (EXC-2046/1, project ID: 390685689).

References

1. Borndörfer, R., Hoppmann, H., Karbstein, M., Löbel, F.: The modulo network simplex with integrated passenger routing. In: Fink, A., Fügenschuh, A., Geiger, M.J. (eds.) Operations Research Proceedings 2016, vol. 1, pp. 637–644. Springer, Berlin (2018)
2. Goerigk, M., Schöbel, A.: Improving the modulo simplex algorithm for large-scale periodic timetabling. Comput. Oper. Res. **40**(5), 1363–1370 (2013)
3. Liebchen, C.: The first optimized railway timetable in practice. Trans. Sci. **42**(4), 420–425 (2008)
4. Löbel, F.: Solving integrated timetabling and passenger routing problems using the modulo network simplex algorithm. Bachelor's Thesis, Freie Universität Berlin (2017)
5. Nachtigall, K., Opitz, J.: Solving periodic timetable optimisation problems by modulo simplex calculations. In: Fischetti, M., Widmayer, P. (eds.) 8th Workshop on Algorithmic Methods and Models for Optimization of Railways (2008)
6. Pätzold, J., Schöbel, A.: A matching approach for periodic timetabling. In: 16th Workshop on Algorithmic Approaches for Transportation Modelling, Optimization, and Systems (ATMOS '16) vol. 1, pp. 1:1–1:15 (2016)
7. Serafini, P., Ukovich, W.: A mathematical model for periodic scheduling problems. SIAM J. Discret. Math. **2**(4), 550–581 (1989)

Optimizing Winter Maintenance Service at Airports

Henning Preis and Hartmut Fricke

Abstract Preserving the efficiency of an airport during winter operations and corresponding conditions requires proper configuration of the snow removal fleet and its smart operation by using optimal routing schemes to clean the airport's airside. In this paper, we present a two-stage approach for optimizing typical winter operations at large airports. In the first stage, we estimate the minimum fleet size to undertake maintenance operations in due time based on a vehicle capacity model. We consider various vehicle parameters, dedicated airport maps with allocated service areas and potential service level agreements between airport operator and airlines acting as customers. In the second stage the optimal routing of the vehicles is determined under the objective of minimizing the cleaning time of pre-defined areas. We apply a specially adapted variant of Vehicle Routing Problem with problem-specific constraints such as vehicle synchronization for joint snow removal, restrictions resulting from the wind direction and, furthermore the preference of runways over taxiways. With the help of the developed methodology it is possible to verify potential investments into fleet resources, which might seem to be necessary to meet increasing service level requirements. The methodology is being demonstrated for Leipzig-Halle Airport (LEJ).

Keywords Capacity planning · Routing · Airline applications

1 Introduction

Every winter, commercial airports in Germany are faced with major challenges due to snowfalls and icy conditions. The experience of airport operators shows that the winters and the associated weather conditions in Central Germany are becoming increasingly unpredictable. Mild weather periods alternate with heavy snowfall and

H. Preis (✉) · H. Fricke
"Friedrich List" Faculty of Transport and Traffic Sciences, TU Dresden, Dresden, Germany
e-mail: henning.preis@tu-dresden.de; hartmut.fricke@tu-dresden.de

J. S. Neufeld et al. (eds.), *Operations Research Proceedings 2019*, Operations Research Proceedings, https://doi.org/10.1007/978-3-030-48439-2_93

critical temperature conditions around the freezing point with frequently changing conditions of the operating areas. Nevertheless, in order to maintain the safe and punctual handling of traffic, airport operators make every effort to optimize their winter maintenance services for the airport facilities.

At the strategic level, the size and equipment of the winter maintenance fleet must be decided as well as procedures for the training and activation of the staff must be developed. From an operational point of view, deployment plans must be drawn up for vehicles and personnel. This includes the allocation of vehicles to areas and the determination of the best possible routes, which are stored in standard cleaning schemes. Both tasks have a major impact on the efficiency of the winter maintenance service and the airport capacity. Therefore, in this paper we introduce a two-stage approach which estimates the minimum fleet size to ensure maintenance operations in due time (Sect. 2) and optimizes the cleaning schemes of the vehicles under the objective of minimizing the completion time subject to problem-specific constraints (Sect. 3). Numerical results are shown and explained for the example of Leipzig-Halle Airport (Sect. 4).

2 Fleet Size Estimation

A typical winter service fleet consists of several vehicle types with associated mission profiles. On the one hand there are standard cleaning vehicles, such as tractors with snow plows and cleaning brushes. These are used on the aprons, parking positions and service areas since they are manoeuverable and versatile. On the other hand, there are highly specialized vehicles, such as airport jet sweepers which ensure the required quality for cleaning the runways and taxiways.

The estimation of the minimum fleet size of each category is generally based on the ratio of taskload to performance capacity, as known from literature (see [1, 2]). For airport applications we can derive the taskload (in m^2/h) from the size of the runways, taxiways and other manoeuvring areas that have to be cleaned within a certain time budget, as e.g. implied by service level agreements with airport users. The theoretical performance capacity of the vehicles of each type (in m^2/h) is determined by the vehicle's tool width (in m) and the operating speed (in m/h). The operating width results from the technical configuration of the vehicle, particularly the width of push plates and cleaning brushes, whereas the operating speed depends on the quality of snow and the texture of the surface. Furthermore, the effective performance capacity of the vehicles is influenced by efficiency parameters for the time utilization (subject to set-up times and breaks) and the spatial utilization (subject to empty runs and vehicle offset). Reasonable values of these efficiency parameters can be obtained by the evaluation of telematics data of previous winter service periods.

Then, for each vehicle category $k \in K$ with operating width w_k, operating speed v_k, time utilization η_k^t and spatial utilization η_k^s we can calculate the number of vehicles n_k by Eq. (1), where A_k is the area size that has to be serviced by vehicle category k within a time budget t_k. The total fleet size N then results from the summation of the individual categories (see Eq. 2).

$$n_k = \left\lceil \frac{A_k}{t_k w_k v_k \eta_k^t \eta_k^s} \right\rceil \tag{1}$$

$$N = \sum_{k \in K} n_k \tag{2}$$

Growing traffic resulting into scarce capacities combined with ever growing manoeuvring areas result in increasing operational pressure. This is typically expressed by strict service level agreements between operator and user and has dramatic effects on the required size of the winter service resources. To keep investments into the fleet within reasonable bounds, highly efficient cleaning schemes and resource strategies are crucial for operational success. E.g., groups of vehicles are formed to be able to run vehicle formations that ensure the efficient cleaning of wide areas such as runways. The optimal routing of these groups (see Sect. 3) is a major task to help maximizing the utilization of the winter maintenance fleet.

3 Optimal Routing of Winter Maintenance Vehicles

Routing problems for winter maintenance vehicles can be formulated as arc routing problems [3] and, using the arc-node-transformation [4], as vehicle routing problems resp. job scheduling problems, see [5]. A straight forward application of the node-oriented routing model for snow removal operations at airports is introduced in [6]. Based on our experiences in the field, we propose a more specific approach containing important features like vehicle synchronization for joint cleaning, the consideration of wind directions as well as the prioritization of runways over taxiways.

To prepare the problem it is necessary to transform the infrastructure of the airport into a graph model. Cleaning sections are modeled as nodes, two per section separated by operating direction with known service time, which results from the distance travelled and the operating speed. The arcs between the nodes represent the transfer times between cleaning sections using the shortest path in the infrastructure network (see Fig. 1).

The routing problem for snow cleaning vehicles is defined on an directed graph $G = (V, A)$. The nodes V include all cleaning sections S that have to be serviced (two nodes for each cleaning section separated by direction) and a depot D. For each node $i \in S$ we know the identification code s_i of the cleaning section (e.g. *RWY* for runway, *A*1 for taxiway A1), the duration of service d_i, the geographical

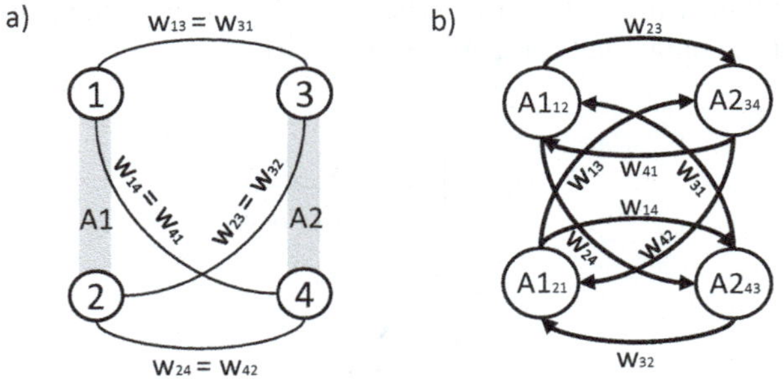

Fig. 1 Example for the transformation of layout data into a directed graph with cleaning sections as nodes (separated into operating directions, e.g. taxiway $A1_{12}$ and taxiway $A1_{21}$) and shortest transfer paths as arcs (e.g. w_{23} for the transfer between $A1_{12}$ and $A2_{34}$)

orientation $o_i = \{NS, SN, EW, WE\}$ and the number of vehicles v_i that is needed for the simultaneous service (for building a proper service formation). The set A describe the arcs $(i,j) \in V \times V$ with a transfer time w_{ij} resulting from the shortest transfer path in the underlying infrastructure network. The vehicle category K used in the routing scenario has a size of $|K|$, according to the fleet size estimation in Sect. 2. The weather conditions may imply a forbidden operating direction $FD = \{NS, SN, EW, WE\}$ to avoid the snow to be blown back on the tracks. The decisions of vehicle routing are formulated with the help of binary assignment variables y_i for choosing the operating direction for each cleaning section and binary routing variables x_{ijk} for vehicle k transferring arc (i, j). To model the starting time for cleaning section i, we use variables $t_i \in R_0^+$. The overall timespan of completing all sections to be cleaned is denoted by $CMAX$. The problem is formulated as follows:

$$\min \quad CMAX \tag{3}$$

subject to

$$y_i + y_j = 1 \quad \forall i \in S \mid o_i \neq FD, \quad j \in S \mid (j \neq i) \wedge \left(o_j \neq FD\right) \wedge \left(s_i = s_j\right) \tag{4}$$

$$\sum\nolimits_{k \in K} \sum\nolimits_{i \in V} x_{ijk} = v_j y_j \quad \forall j \in S \tag{5}$$

$$\sum\nolimits_{i \in V} x_{ijk} = \sum\nolimits_{i \in V} x_{jik} \quad \forall j \in V, k \in K \tag{6}$$

$$\sum_{i \in D} \sum_{j \in S} x_{ijk} \leq 1 \quad \forall k \in K \tag{7}$$

$$t_j \geq t_i + d_i + w_{ij} - M\left(1 - x_{ijk}\right) \quad \forall i \in V, j \in S, k \in K \tag{8}$$

$$t_i + d_i \leq CMAX \quad \forall i \in S \tag{9}$$

$$t_i \leq TMAX \quad \forall i \in S \mid s_i = "RWY" \tag{10}$$

The objective function (3) is to minimize the timespan, that means finishing a cleaning scenario as soon as possible. Constraints (4) choose exactly one valid operating direction for each cleaning section. Constraints (5) and (6) define feasible vehicle routes respect to flow conservation between all chosen cleaning sections. Constraints (7) limit the number of available vehicles starting from the depot. Constraints (8) define feasible starting times depending on the service sequences of the vehicles and also eliminate sub-tours from the solution. Constraints (9) determine the value of CMAX as the maximum completion time of all cleaning sections. And finally constraints (10) include the prioritization of the runways by formulating a limit TMAX (set manually by the operators) which should not be exceeded for starting the cleaning of the runways.

4 Results of the Application for Leipzig-Halle Airport

Airport Leipzig-Halle (LEJ) is an important hub airport in the network of a global logistics provider. In order to handle the large number of flight movements within the specification of the service level agreement, an efficient winter service is particularly important. Figure 2 shows the results of the fleet size study concerning the Towed Jet Sweepers (TJS-630) used for cleaning the runways and taxiways at LEJ. We have considered various geometric, operationally driven scenarios which differ in the runways in use (one vs. two runways including taxiways) and the determination of the operating speeds (slow vs. fast). According to Eq. (1) the minimum fleet decreases with growing due time for completing the cleaning jobs. In the scenarios with two runways the number of necessary vehicles nearly doubles due to the immense increase in manoeuvring area. The results can be used as a basis for debating manageable airport performance values and sensible winter maintenance fleet sizes.

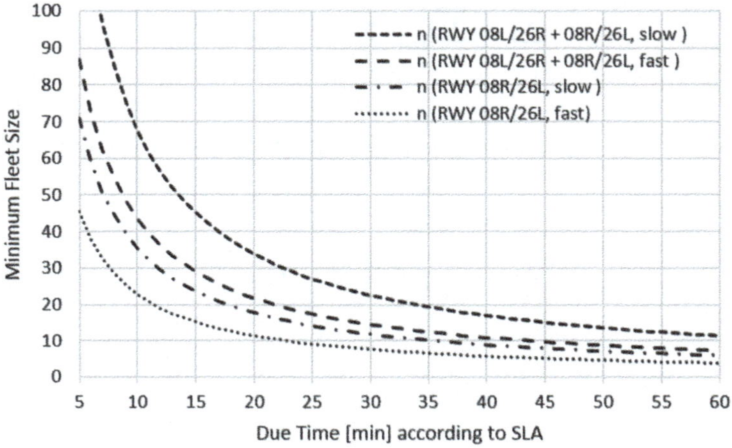

Fig. 2 Estimated minimum fleet size of special vehicles TJS-630 for LEJ Airport depending on the serviced area (one vs. two runways incl. taxiways) and cleaning velocity (slow =15 km/h, fast = 25 km/h)

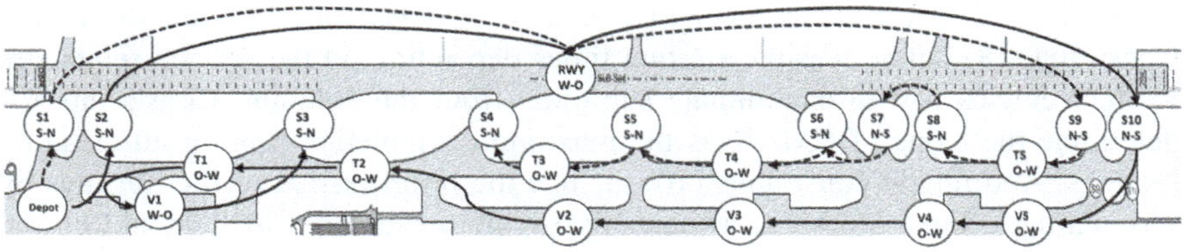

Fig. 3 Optimal cleaning scheme (CMAX = 897 s) for the southern RWY 08R/26L of Leipzig-Halle Airport (LEJ) using two separate vehicle formations, each consisting of seven specialized cleaning vehicles "TJS 630"

The routing model proposed in Sect. 3 has been set up and solved for all formulated scenarios using standard MIP solution techniques to determine the cleaning sequences with minimum completion time. Figure 3 shows the results for the scenario for RWY 08R/26L. Two vehicle formations, each consisting of 7 TJS-630, are used to clean the runway (synchronized procedure for two formations) and 20 taxiway elements named S1 to V5 (single procedures for one formation). Potential restrictions from a given wind direction have not been considered. TMAX is set to 5 min, the operating speed resulting from snow conditions is set to "fast" (25 km/h). The optimal sequences are shown for the first vehicle formation (solid line) and the second vehicle formation (dashed line). The timespan of this scenario takes 897 s, which is slightly less than the time budget of the service level agreement (15 min). One the one hand, the results validate the fleet site calculations introduced above. On the other hand, the optimal sequences can be stored und used as standard cleaning schemes for all vehicles.

5 Conclusions

In this work, we propose a two stage approach to determine the optimal fleet size of a winter maintenance fleet with associated optimal cleaning schemes (strategies) with a typical set of constraints for busy airports of relevant size. On the example of Leipzig-Halle Airport, we demonstrated the usability of our models and gave suggestions on how to increase airport performance based on the findings. Relevant investments into winter maintenance resources can thus be verified transparently

References

1. Perrier, N., Langevin, A., Campbell, J.F.: A survey of models and algorithms for winter road maintenance. Part I: System design for spreading and plowing. Comput. Oper. Res. **33**, 209–238 (2006)
2. Chien, S., Gao, S., Meegoda, J., Marhaba, T.: Fleet size estimation for spreading operation considering road geometry, weather and traffic. J. Traffic Transp. Eng. **1**, 1–12 (2014)
3. Perrier, N., Langevin, A., Campbell, J.F.: A survey of models and algorithms for winter road maintenance. Part IV: Vehicle routing and fleet sizing for plowing and snow disposal. Comput. Oper. Res. **34**, 258–294 (2007)
4. Foulds, L., Longo, H., Martins, J.: A compact transformation of arc routing problems into node routing problems. Ann. Oper. Res. **226**, 177–200 (2015)
5. Kinable, J., van Hoeve, W.J., Smith, S.F.: Optimization models for a real-world snow plow routing problem. In: Quimper, C.-G. (ed.) Integration of AI and OR Techniques in Constraint Programming: 13th International Conference, CPAIOR 2016, Banff, AB, Canada, 29 May–1 June 2016, Proceedings, pp. 229–245. Springer, Dordrecht (2016)
6. Fernandez, C., Comendador, V., Valdes, R.: Algorithm for modelling the removal of snow from stretches of the manoeuvring area of an airport. CIT2016 – XII Congreso de Ingeniera del Transporte, Universitat Politectica de Valencia (2016)

Coping with Uncertainties in Predicting the Aircraft Turnaround Time at Airports

Ehsan Asadi, Jan Evler, Henning Preis, and Hartmut Fricke

Abstract Predicting the target time of an aircraft turnaround is of major importance for the tactical control of airport and airline network operations. Nevertheless, this turnaround time is subject to many random influences, such as passenger behavior while boarding, resource availability, and short-noticed maintenance activities. This paper proposes a mathematical optimization model for the aircraft turnaround problem while considering various uncertainties along the process. The proposed method is acting on a microscopic, thus detailed operational level. Dealing with uncertainties is implemented by two approaches. First, an analytical procedure based on convolution, which has not been considered in the literature so far but provides fast computational results, is proposed to estimate the turnaround finalization, called Estimated Off-Block Time (EOBT). The convolution algorithm considers all process-related stochastic influences and calculates the probability that a turnaround can be completed within the pre-set target time TOBT. At busy airports, such assessments are needed in order to comply with installed slot allocation mechanisms. Since aircraft turnaround operations reflect a scheduling problem, a chance-constrained MIP programming model is applied as a second approach. This procedure assumes stochastic process durations to determine the best alternative of variable process executions, so that the TOBT can be met. The procedure is applied to an Airbus A320 turnaround.

Keywords Aircraft turnaround · Analytical convolution · Chance-constrained MIP

This research has been conducted in the framework of the research project Ops-TIMAL, financed by the German Federal Ministry of Economic Affairs and Energy (BMWI) within the 6th German Aeronautics Research Program (LuFo VI).

E. Asadi (✉) · J. Evler · H. Preis · H. Fricke
Institute of Logistics and Aviation, Technische Universität Dresden, Dresden, Germany
e-mail: ehsan.asadi@tu-dresden.de

J. S. Neufeld et al. (eds.), *Operations Research Proceedings 2019*, Operations Research Proceedings, https://doi.org/10.1007/978-3-030-48439-2_94

1 Introduction

In the light of constantly increasing air traffic density and scarce airport capacities, airline operations have become a challenging business with decreasing flexibility to maneuver. Most of the block time of a flight (time window from leaving the departure gate until arriving at gate of destination) is subject to decisions of air traffic control (ATC) and central network management institutions (ATFM/NOP). So far, the turnaround remains the only period within an aircraft rotation when an airline has autonomous control over its operations. According to the IATA ground operations manual IGOM [1], the turnaround consists of 12 standard processes which split in up to 150 different sub-tasks and involve up to 30 different actors—depending on the detailed operating procedures of the airline. Predicting the accurate completion time of this process sequence (Fig. 1) is a crucial task for airlines, especially when they operate at large airports with A-CDM systems, where the so-called Target Off-Block Time (TOBT) determines the position of the upcoming flight in the ATC departure sequence. As the IGOM considers static process durations, the minimum ground (turnaround) time is determined by the longest (critical) path in the process sequence (red in Fig. 1).

However, given the complexity of operations, all processes have a stochastic nature which translates into stochastic durations and, hence, a variable critical path.

2 Considering Uncertainties for EOBT Calculation

There are different strategies for solving the stochastic Aircraft Turnaround Problem (for state-transition models see e.g. [2], for buffer time calculations see e.g. [3–5]). This paper describes a novel mathematical procedure for the convolution of several process uncertainties into a single one for the entire network. One of the standard approaches to deal with uncertainty in process durations is the Program Evaluation and Review Technique (PERT), which was introduced by [6] and simplified by [7] in order to reach a practical implementation. In PERT approach, a Beta distribution is employed to estimate the duration of network processes. With knowing about

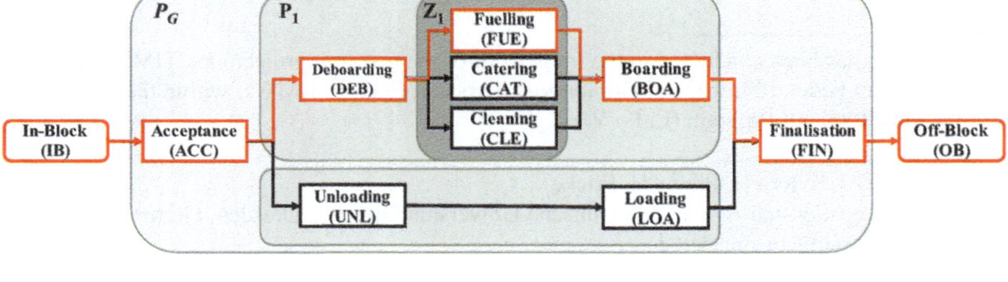

Fig. 1 Standard turnaround process sequence

Table 1 Process distribution of turnaround activities

Process/event		
Abbrev.	Distribution	Quick alternative distribution
IB	–	–
ACC	Gamma (2.0,1.0)	–
DEB	Gamma (6.81,1.47)	Gamma (2.80,2.50)
CLE	Weibull (2.16,11.29)	Weibull (2.16,6.76)
CAT	Weibull (2.18,17.37)	Weibull (1.51,11.38)
FUE	Gamma (9.12,1.64)	–
UNL	Gamma (11.29,1.24)	Gamma (2.11,5.01)
LOA	Gamma (15.34,1.24)	Gamma (11.34,1.25)
BOA	Gamma (14.36,1.47)	Gamma (7.88,1.96)
FIN	Gamma (4.0,1.0)	–
OB	–	–

the critical path, the turnaround completion time can be calculated by adding the expected values and variances of processes together. Although this approach is straightforward, the crux of the matter is that the critical path should be known before the calculation, which in case of stochastic processes might not always be the case. Furthermore, PERT is limited to Beta distributions, while most turnaround processes have shown a good-fit with Gamma- or Weibull-distributions [8]. Hence, there is the need for a more sophisticated stochastic process model, which considers that the critical path itself is variable.

For the novel approach of analytical convolution, this paper adopts process parameters from [8] (see Table 1) with the minor alteration that DEB and BOA were re-fitted with Gamma-distributions in order to ease the mathematical procedure. Note, that all process parameters can be adjusted to airline-/airport-specific operational characteristics at any time.

Our approach neglects dependencies between process durations and buffer times, which leaves the process network as depicted in Fig. 1, for which the stochastic processing time can be calculated via analytical convolution. Equally to the critical path algorithm, sequential processes are added with their respective duration, while parallel processes must wait for their longest counterpart to be finished before the next activity can initiate. As the resulting mathematical function does not follow a pre-known distribution, all parameters from the individual process distributions must be described:

The cumulative distribution function $F_Z(z)$ of two random variables X and Y with known density functions $f_X(x)$ and $f_Y(y)$, can be calculated by the convolution integral (Eq. (1), see [9]):

$$F_Z(z) = P(X + Y \leq z) = \int_{-\infty}^{z} f_X(x) \cdot F_Y(z - x)\, dx \tag{1}$$

Considering parallel processes represented by independent stochastic variables $X_1, X_2, \ldots, X_N$, their maximum is given by an extreme-value distribution (see Eqs. (2)–(4)):

$$Z = \max(X_1, X_2, \ldots, X_N) \tag{2}$$

$$F_Z(z) = P(X_1 \leq z, X_2 \leq z, \ldots, X_N \leq z) \tag{3}$$

$$F_Z(z) = P(X_1 \leq z) \cdot \ldots \cdot P(X_N \leq z) = F_{X_1}(z) \cdot \ldots \cdot F_{X_N}(z) \tag{4}$$

By applying the underlying mathematical principle of Eqs. (1)–(4), the EOBT of a turnaround network can be calculated, starting from "inside out". First, the variable Z_1 is determined as longest duration among parallel processes ($max\{X_{FUE}, X_{CLE}, X_{Cat}\}$) as given in Eq. (5).

$$F_{Z_1}(z_1) = F_{FUE}(z_1) \cdot F_{CLE}(z_1) \cdot F_{CAT}(z_1) \tag{5}$$

To connect the adjacent processes DEB and BOA as shown in Fig. 1, their cumulated CDF from Eq. (6) is added to the extreme-value distribution Z_1. The same approach is used for UNL-LOA and ACC-FIN in Eqs. (8) and (10):

$$Z_2 = X_{DEB} + X_{BOA} \sim Gamma(\alpha_{DEB} + \alpha_{BOA}, \beta) \tag{6}$$

$$F_{P_1}(p_1) = \int_{-\infty}^{p_1} f_{Z_2}(z_2) \cdot F_{Z_1}(p_1 - z_2)\, dz_2 \tag{7}$$

$$P_2 = X_{UNL} + X_{LOA} \sim Gamma(\alpha_{UNL} + \alpha_{LOA}, \beta) \tag{8}$$

Applying Eq. (4), cabin and cargo activities are thus merged into an extreme-value distribution, as shown in Eq. (9). Finally, the summation of distributions S and U leads to the CDF of the EOBT and is calculated by Eq. (11). As all convoluted distribution have non-negative durations, Eq. (11) starts indeed from zero, instead of minus infinity (for detailed discussion see [10]). Using this method, the convoluted

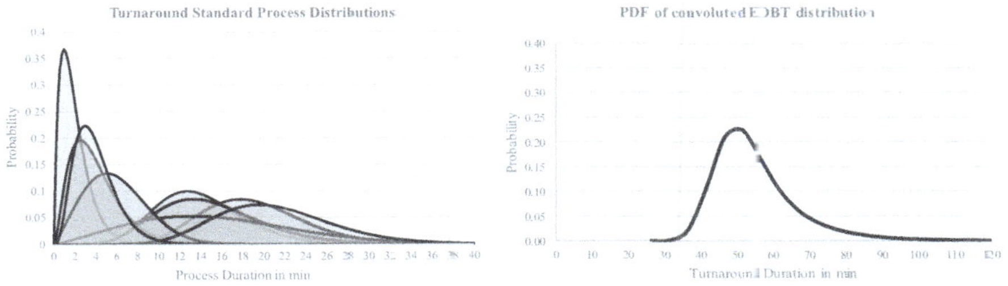

Fig. 2 PDF of convoluted EOBT distribution

PDF of EOBT distribution based on all corresponding activities is calculated and illustrated in Fig. 2

$$F_S(s) = F_{P_1}(s) \cdot F_{P_2}(s) \tag{9}$$

$$U = X_{ACC} + X_{FIN} \sim Gamma\,(\alpha_{ACC} + \alpha_{FIN}, \beta) \tag{10}$$

$$F_{Turnaround}(a) = \int_{-\infty}^{a} f_U(u) \cdot F_S\,(a - u)\,du \tag{11}$$

2.1 Chance-Constrained Model

The proposed mathematical model is able to decide between a standard process execution and an alternative one, which has smaller mean and different variance. In order to reach a cost-minimal optimal solution for the control of the estimated turnaround completion time EOBT, a chance-constrained model is developed in this paper. The chance-constrained method was first introduced by [11] to deal with linear programming under uncertainty. According to PERT approach from Sect. 2.1, the stochastic duration of a project at a given confidence level can be reached by summation of mean values *E*(*x*) and variances *VAR*(*x*) of all processes. The proposed chance-constrained mixed integer programming is represented as follows. Sets *O, K, CM,* and *OB* stand for Operations, Distributions, Connectivity Matrix, and Off-Block processes, respectively. $E(x)_i$ and $Var(x)_i$ are mean value and variance of process *i*. CT is the completion time (EOBT) of the whole network and TOBT, as defined before, is Target Off-Block Time.

$$\min \sum_{o \in O} \sum_{k \in K} y_{ok} \cdot C_{ok} \tag{12}$$

$$\text{Subject to} \quad \sum_{k \in K} y_{ok} = 1 \qquad \forall o \in O \tag{13}$$

$$m_j \geq m_i + E(x)_i - M\left(2 - y_{ik} - y_{jk}\right) \qquad \forall i, j \in O \mid (i, j) \in CM; \forall k \in K \tag{14}$$

$$v_j \geq v_i + Var(x)_i - M\left(2 - y_{ik} - y_{jk}\right) \qquad \forall i, j \in O \mid (i, j) \in CM; \forall k \in K \tag{15}$$

$$m_i + z_Q \cdot \sqrt{v_i} \leq TOBT \qquad \forall i \in OB \tag{16}$$

In the objective function, y_{ok} is a binary variable, which is "zero" in case of the standard execution of a process and one if process o is accelerated and hence can be described by a different distribution function k which comes with the related cost C_{ok}. Aggregated calculated mean value and variance of activity i are defined by m_i and v_i. Equation (13) guaranties every activity is executed with one—and only one—variation. Equations (14) and (15) calculate the aggregated expected values and variances of the estimated total turnaround time EOBT. This is done by adding the mean values and variances of all sequential processes in one network path (in the connectivity matrix (CM)). The chance constraint which verifies whether a turnaround is likely to be completed within the given TOBT at a given confidence level can be written as Eq. (16). z_Q is the z-number obtained relative to its quantile from standard normal table. Note, that this constraint makes the mathematical model non-linear.

3 Numerical Example

In this section a numerical example for the mentioned simplified model of an Airbus A320 turnaround is provided to determine optimal control decision in order to complete the turnaround at a given TOBT. We assume alternative distributions for a sub-set of activities, called "Quick Alternatives" as described in Table 1. The model is solved by SCIP optimization suite [12].

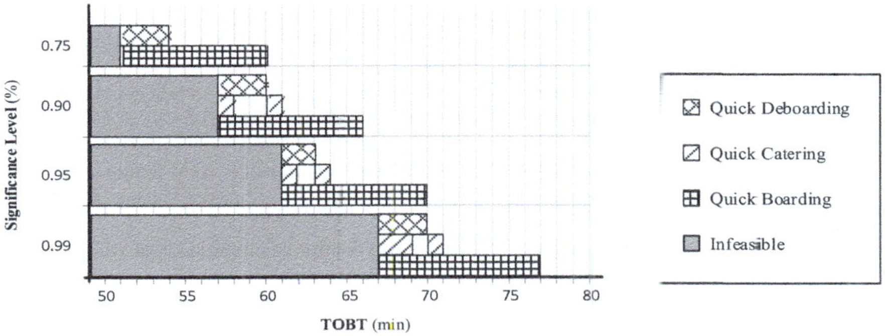

Fig. 3 Result of the proposed chance-constrained model

Results are shown in Fig. 3, where dark-shaded rectangles represent infeasibility and mean that the turnaround cannot be completed within the TOBT at the selected confidence level even when all processes are controlled. I.e., at 90% confidence level, the turnaround lasts at least 58 min when DEB, CAT and BCA processes are controlled. For later TOBTs, less controlled processes are needed. In this case, Quick CAT is no longer needed for a TOBT in 59 or 60 min but should be applied in combination with Quick Boarding for 61 min. The latter is needed for all TOBTs of less or equal to 66 min, which implies that an uncontrolled turnaround would last at least 67 min. Naturally, higher trust levels comprise longer completion times and it is valid for all levels that earlier TOBTs require more control actions (see Fig. 3).

4 Conclusion

In this paper, we first introduced the main activities of an Airbus A320 turnaround. We applied analytical convolution to find the probability density function of the total turnaround time including uncertainties related to the single processes. Hereafter, a mixed integer chance-constrained model is proposed to find the optimum control solution for a delayed turnaround. Finally, a numerical example is defined in the last section to showcase the results of the proposed turnaround control model.

References

1. IATA.: IATA Ground Operations Manual, Montreal (2018)
2. Wu, C.L., Caves, R.E.: Modelling and simulation of aircraft turnaround operations at airports. Transp. Plan. Technol. **27**(1), 25–46 (2004)
3. Wu, C., Caves, R.E.: Modelling and optimization of aircraft turnaround time at an airport. Transp. Plan. Technol. **27**(1), 47–66 (2004)
4. AhmadBeygi, S., Cohn, A., Lapp, M.: Decreasing airline delay propagation by re-allocating scheduled slack. IIE Trans. **42**(7), 478–489 (2010)

5. Silverio, I., Juan, A.A., Arias, P.: A simulation-based approach for solving the aircraft turnaround problem. In: International Conference on Modeling and Simulation in Engineering, pp. 163–170. Springer, Berlin (2013)
6. Cook, D.L.: Program Evaluation and Review Technique–Applications in Education. (1966)
7. Cottrell, W.D.: Simplified program evaluation and review technique (PERT). J. Constr. Eng. Manag. **125**(1), 16–22 (1999)
8. Fricke, H., Schultz, M.: Delay impacts onto turnaround performance. ATM Semin. (2009)
9. Petrov, V.V.: Sums of Independent Random Variables, vol. 82. Springer Science & Business Media (2012)
10. Evler, J., Asadi, E., Preis, H., Fricke, H.: Stochastic control of turnarounds at HUB-airports. In: Eighth SESAR Innovation Days (2018)
11. Charnes, A., Cooper, W.W.: Chance-constrained programming. Manag. Sci. **6**(1), 73–79 (1959)
12. Berthold, T., Gamrath, G., Gleixner, A.M., Heinz, S., Koch, T., Shinano, Y.: Solving mixed integer linear and nonlinear problems using the SCIP optimization suite (2012)

Strategic Planning of Depots for a Railway Crew Scheduling Problem

Martin Scheffler

Abstract This paper presents a strategic depot planning approach for a railway crew scheduling problem integrated in a column generation framework. Since the integration strongly weakens the relaxation of the master problem we consider different variants for strengthening the formulation. In addition, the problem can be sufficiently simplified by using a standard day at the strategic level. Based on a case study for an exemplary real-life instance, we can show that a proper pre-selection of depots reduces the number of needed depots significantly with the same personnel costs.

Keywords Depot planning · Crew scheduling · Column generation

1 Introduction

Crew scheduling problems are one of the most important problems within the planning process in passenger rail transport. Heil et al. [3] gives a detailed overview to this topic. We consider a multi-period railway crew scheduling problem with attendance rates for conductors of a German railway operator. The goal of this problem is to find a schedule satisfying operating conditions, legal requirements and the transportation contract at minimal costs. The attendance rates are a peculiarity and generalization of the classic crew scheduling problem. This means that not every trip has to be covered, but a percentage of the trips is sufficient. Usually the problem is solved with 14 days planning horizon. For a general description of the problem and the considered constraints for the duty generation, we refer to [4]. Since the problem is NP-hard using a column generation approach is a common method for solving (see [3]). Since duties can only begin and end at one depot (i.e. crew

---

M. Scheffler (✉)
Fakultät Wirtschaftswissenschaften, Lehrstuhl für BWL, insb. Industrielles Management, Technische Universität Dresden, Dresden, Germany
e-mail: martin.scheffler@tu-dresden.de

J. S. Neufeld et al. (eds.), *Operations Research Proceedings 2019*, Operations Research Proceedings, https://doi.org/10.1007/978-3-030-48439-2_95

base), the selection of suitable relief points (railway stations) as depots is crucial on strategic planning level. On the one hand, it has to be taken into account that opening depots causes costs (e.g. rental fee for rooms). In addition, a small number of opened depots is preferred, as this reduces administrative effort. This means it is advisable to avoid opening depots where only a small number of duties starts. On the other hand, a small number of depots may increase the number of employees (duties) required. In practice, balancing these conflicting objectives is hard because there is a lack of suitable planning approaches for integrating in decision support systems. Limiting depot capacity on tactical level is common practice (see [4, 6]). Suyabatmaz and Güvenç[7] determines a minimum required crew size in a region, without taking depot locations into account.

In order to investigate the trade-off in detail, we adapt an existing column generation approach from tactical planning level (see [4]). The integration of the mentioned strategic planning issues to the master problem (MP) is presented in Sect. 2. Since MP is hard to solve, we introduce a standard day and show possibilities for strengthening the formulation in Sect. 3. In Sect. 4 the computational analyses are carried out for the different formulations on real-life instances. It is combined with a case study for an exemplary real-life network. Section 5 gives a summary and present suitable research content for further work.

2 Problem Description

The MP aims at finding a minimal cost combination of duties selected from a set of feasible duties N. The planning horizon is given by K containing days k of the week. A duty $j \in N$ covers a subset of trips $i \in M$ with M representing the set of all trips. A duty is represented by a column in matrix $A \in \{0, 1\}^{|M|\times|N|}$ with $a_{ij} = 1$ if duty j covers trip i (0 otherwise). A trip i can exist on a single day $k \in K$ or on several days of the planning horizon K. Set M_k is defined as subset of M, containing all trips $i \in M$ existing on day k. Additionally, let G be the set of all attendance rates $g \in [0, 1]$, we can determine d_{ig} as the distance of trip $i \in M$ with attendance rate $g \in G$. The costs c_j of a feasible duty $j \in N$ are calculated in accordance with the operating conditions and legal requirements described by Hoffmann et al. [4]. Furthermore, let E be the set of all depots, then parameter b_{je} equals 1 if duty j starts at depot e, 0 otherwise. Set E consist of the two subsets E^o containing all possible depots that may need to be opened and E^c containing all existing depots that may need to be closed. Parameter f_e^o (f_e^c) indicates the costs for opening (closing) depot e. Parameter $\mathcal{M}$ is used as reasonable big number (*Big-M*). Finally, we introduce the following decision variables. The binary variables x_j take value 1 if duty j is in the solution, 0 otherwise. Furthermore, we use binary variables y_{ik} to model if trip $i \in M$ on day $k \in K$ is in the solution. The binary variables o_e are used to model the decision whether to use depot e (using an existing depot or opening a possible depot) or not (closing an existing depot or not using a possible depot). Based on this notation the MP is given as following:

$$\min \sum_{j\in N} c_j x_j + \sum_{e\in E^o} f_e^o o_e - \sum_{e\in E^c} f_e^c (1 - o_e) \tag{1}$$

$$\text{s.t.} \quad \sum_{k\in K}\sum_{i\in M_k} d_{ig} y_{ik} \geq g \sum_{k\in K}\sum_{i\in M_k} d_{ig} \qquad \forall g \in G, \tag{2}$$

$$\sum_{j\in N_k} a_{ij} x_j \geq y_{ik} \qquad \forall k \in K, i \in M_k, \tag{3}$$

$$\sum_{j\in N} b_{je} x_j \leq \mathcal{M} o_e \qquad \forall e \in E, \tag{4}$$

$$x_j \in 0, 1 \qquad \forall j \in N, \tag{5}$$

$$y_{ik} \in 0, 1 \qquad \forall k \in K, i \in M_k, \tag{6}$$

$$o_e \in 0, 1 \qquad \forall e \in E. \tag{7}$$

The formulation is a reduced version of the presented formulation by Hoffmann et al. [4] with an other objective and extended by constraint (4). The objective minimizes the total operating costs for all duties and the costs for opening a possible depot. Closing an existing depot leads to cost savings, which is why the last sum is deducted. Constraints (2)–(3) ensure compliance with the required attendance rates. We refer to [4] for a detailed description of the mode of action. Constraint (4) sets variables o_e to 1, if at least one duty starting in e is used in the solution. This models the opening and closing of depots. Note, this constraint is very similar to the depot capacity constraint introduced by Hoffmann et al. [4], but this variant causes a weaker LP-relaxation. Constraints (5)–(7) state the domains.

Since all adjustments only concern the MP, we can directly use the genetic algorithm described by Hoffmann et al. [4] for solving the subproblem. Only the calculation of the reduced costs has to be adjusted. Let π_{ik} $i \in M_k$, be the dual value of constraints (3) and $\gamma_e, e \in E$, of (4) then $\bar{c}_j = c_j - \sum_{i\in M} a_{ij}\pi_{ik} + \sum_{e\in E} b_{je}\gamma_e$ specifies the reduced costs of duty $j \in N_k$.

3 Solution Approach

The determination of the depots is a long-term decision in which the exact train schedule (i.e. input data) is available in a rough form only or it can be assumed that subtleties will change again and again over the course of time (see e.g. [5]). Therefore it is not mandatory to carry out a detailed planning for 14 days, but it is sufficient to solve a suitable simplification. For this purpose, it makes sense to reduce the planning period to a standard day. Ahuja et al. [1] describes a procedure for the locomotive scheduling problem in rail freight transportation, whereby a trip is considered in the standard day, if it takes place on at least 5 days of the week. In contrast to freight transport, the weekend schedule for passenger transport differs much more often and strongly from the weekday schedule. When using 5 days as a criterion, there is thus a risk that the weekends are not sufficiently taken into account in the standard day. If certain lines (successive trips of a train/vehicle) only run on weekends, there is even a risk that entire groups of trips will not be taken into account. For this reason, we adopt the procedure of [1] and supplement it with the additional identification of such special cases and, if necessary, also take them into account in the standard day. Preliminary tests showed that the convergence of the objective value during column generation without the use of the standard day is extremely slowed by the additional decision to open/close depots. Even after 24 h computing time, a sufficiently good solution quality could not yet be achieved.

As already mentioned, the LP-Relaxation is very weak due to constraint (4). Therefore, we consider two ways to strengthen the formulation. Due to the spatial distribution of trips and the resulting travel times, it is usually not possible for each trip to be covered by a permissible duty starting from each depot. This means that each trip i can only be covered by duties if they begin at a depot that is element of the subset E_i of E. Based on set E_i we introduce the valid inequalities given by (8).

$$\sum_{e \in E_i} o_e \geq y_{ik} \qquad \forall\, k \in K, i \in M_k \tag{8}$$

This especially strengthens the formulation in case of relaxing the integer constraints (5)–(7). When the relaxation is solved, the Big-M of the Constraints (4) causes o_e to take only very small values. Since constraints (8) are independent of Big-M, the values of o_e are significantly increased. The difficulty, however, is to determine the sets E_i for all trips i. Since column generation is based on generating only a subset of all possible duties for creating a suitable solution, this information is not available. That is why we use shortest path based informations which we can generate on the basis of a spatial and temporal network. Each node represents a distinct combination of time and a relief point or depot, respectively. Trips and transition times are represented by arcs and weighted by the length of the travel or transition time. For determining E_i it is sufficient to find a path from a node at e to the departure node of trip i and also a path from the arrival node of trip i to another node at e. This can be done by using the algorithm described by Dijkstra [2].

Another possibility to strengthen the formulation is the decomposition of the Big-M constraints (4). We can replace these constraints by (9).

$$b_{je} x_j \leq o_e \qquad \forall\, j \in N, e \in E \tag{9}$$

For each duty, a single constraint is created with which the duty is coupled to variable o_e. Once again the relaxation is strengthened because o_e is independent from Big-M and therefore has to accept bigger values. Note, the calculation of the reduced costs change, if we use constraints (9) instead of (4). Let γ_{je} be the dual value of constraints (9) then $\bar{c}_j = c_j - \sum_{i \in M} a_{ij} \pi_{ik} + \sum_{e \in E} b_{je} \gamma_{je}$ specifies the reduced costs of duty $j \in N_k$. In addition, it should be noted that newly generated duties cannot be evaluated directly with reduced costs, since the model must first be solved with the new constraint for this duty in order to obtain a dual value γ_{je}. This means that when deciding whether to include a duty in the duty pool, the dual information on the depots is not taken into account. Section 4 shows that this is negligible at first. But there is potential for improvement in future research. However, since the selection in the GA is based on the reduced costs (see [4]), the information of the dual values γ_{je} is not completely lost.

4 Computational Analysis

The complete column generation approach is implemented in C# and all computational test are carried out on Intel(R) Xeon(R) Gold 6136 CPU with 3.0 GHz clock speed and 128 GB RAM. For solving the MP we are using Gurobi 8.1. For the evaluation of the presented formulations we have tested on two real-life networks. Network I consists of 17 relief points, 8 of those are existing depots and 3 are possible depots. Set M contains of 792 trips with attendance rates of 30 and 90%. Network II is given by 11 relief points (5 existing depots, 3 possible depots) and 1106 trips ($|M|$) with attendance rates of 25 and 100%. We structured the computational tests as follows. First we compare the use of the different constraints by using the standard day on both networks. Based on the results we are able to determine a sufficient set of (open) depots. Using Network I as an example, we then study the schedules with and without a pre-selection of the depots for the planning period of 14 days and using $\min \sum_{j \in N} c_j x_j$ as objective. This corresponds to the objective function of actual crew scheduling (see [4]) and therefore to the downstream planning level. This is done to check the quality of the pre-selection.

Table 1 shows the results of the column generation approach with MP given as min (1) s.t.(2)–(3), (5)–(7) supplemented by the constraints marked in the left columns. Each values are averages of five runs and we limited the generation of columns to 6 h (or no new columns with negative reduced costs can be created). Using constraints (9) instead of (4) leads to better objective values and a much

Table 1 Comparison of the formulations using a standard day

Constraints			Instance I					Instance II				
(4)	(9)	(8)	OBJ	rOBJ	CPU	D	S	OBJ	rOBJ	CPU	D	S
•			325.7	295.4	6.2	3.2	16.8	417.3	391.8	6.1	3.0	16.4
•		•	336.5	306.3	6.1	3.2	17.4	425.0	411.5	6.1	3.0	17.2
	•		311.2	284.7	0.6	3.0	15.0	407.0	363.7	1.0	3.0	16.0
	•	•	314.9	298.3	0.7	3.0	15.7	406.5	395.2	1.2	3.0	16.0

Notation: *OBJ* average objective value in thousands, *rOBJ* average objective value of the LP-relaxation in the last iteration, *CPU* average computing time in hours, *D* # selected depots in final schedule, *S* # duties in final schedule

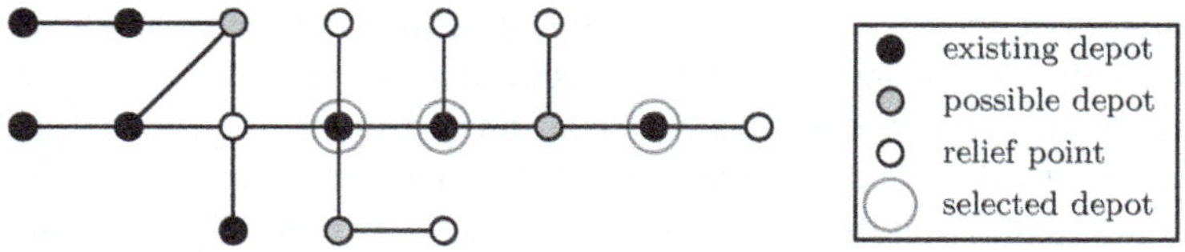

Fig. 1 Spatial network of instance I

faster computing time. The additional use of the valid inequalities (constraints (8)) does not further improve both values. However, this considerably reduces the gap between integer and relaxed objective values. Based on the results of Table 1 we are able to identify three preselected depots for instance I. Figure 1 shows the underlying spatial network and the given depots. A reduction of the required depots can be observed for many instances. This is mainly due to the fact that the depots were determined a long time ago when no attendance rates were requested (i.e. only 100% trips). In order to evaluate the quality of the pre-selection, we compare the results for a 14-day planning horizon with and without preselected depots by optimizing min $\sum_{j \in N} c_j x_j$ s.t.(2)–(3)(5)–(6). Using all existing depots leads to an average objective of 4.747 millions in 7.3 h by using on average 5.2 depots and 191 duties are necessary. Again, all values are averages of 5 runs. In contrast to this, the exclusive use of the three pre-selected depots gives an average objective of 4.782 millions (+0.74%) in 6.7 h with 181 needed duties. Both variants achieve almost identical objective values. This means that at the same personnel costs 2–3 depots (and the associated costs) can be saved. Furthermore, it can be observed in the solution with all depots that some depots have only a few duties on a maximum of 2 days of the planning horizon (zero duties on all other days). This represents unnecessary administrative effort in practice and is successfully avoided by the pre-selection. In general, it can be assumed for the crew scheduling problem on tactical level that due to the large number of existing depots combined with the attendance rates, an extremely large solution space is created with many similarly good solutions close to the optimum. By the upstream selection of suitable depots on the strategic level, this is significantly reduced without losing solution quality.

5 Conclusion and Further Research

In this paper, the presented pre-selection of depots on strategic planning level enables the effective depot determination with a sufficient consideration of subsequent crew scheduling itself. The integration into the master problem could be successfully carried out by the presented strengthening of the formulation and using the standard day. We could also show, for an example of a real-life instance, that at the same cost on tactical level, the number of depots required on a strategic level can be significantly reduced.

For further research it would be interesting for small instances to completely enumerate the pool of possible duties and then to generate a pareto front for the cost of the duties and the number of required depots on the basis of a multi-objective approach. This would determine the influence of the number of depots more precisely and conclusions could be drawn for larger networks.

References

1. Ahuja, R.K., Liu, J., Orlin, J.B., Sharma, D., Shugart, L.A.: Solving real-life locomotive-scheduling problems. Transp. Sci. **39**(4), 503–517 (2005)
2. Dijkstra, E.W.: A note on two problems in connexion with graphs. Numer. Math. **1**(1), 269–271 (1959)
3. Heil, J., Hoffmann, K., Buscher, U.: Railway crew scheduling: Models, methods and applications. Eur. J. Oper. Res. **283**(2), 405–425 (2020)
4. Hoffmann, K., Buscher, U., Neufeld, J.S., Tamke, F.: Solving practical railway crew scheduling problems with attendance rates. Bus. Inf. Syst. Eng. **59**(3), 147–159 (2017)
5. Huisman, D.: A column generation approach for the rail crew re-scheduling problem. Eur. J. Oper. Res. **180**(1), 163–17 (2007)
6. Shen, Y., Chen, S.: A column generation algorithm for crew scheduling with multiple additional constraints. Pac. J. Optim. **10**, 113–136 (2014)
7. Suyabatmaz, A.C., Güvenç, S.: Railway crew capacity planning problem with connectivity of schedules. Transp. Res. E Log. Transp. Rev. **84**, 88–100 (2015)

Periodic Timetabling with Flexibility Based on a Mesoscopic Topology

Stephan Bütikofer, Albert Steiner, and Raimond Wüst

Abstract In the project smartrail 4.0 Swiss Federal Railways (SBB) aims for a higher degree in automatization of the railway value chain (e.g. line planning, timetabling and vehicle scheduling, etc.). In the context of an applied research project together with SBB, we have developed an extension of the Periodic Event Scheduling Problem (PESP) model. On one hand the extension is based on using a finer resolution of the track infrastructure, the so-called mesoscopic topology. The mesoscopic topology allows creating timetables with train lines assigned to track paths. On the other hand, we use a known, flexible PESP formulation (FPESP), i.e. we calculate time intervals instead of time points for the arrival resp. departures times at operating points. Both extensions (mesoscopic topology and flexibility) should enhance feasibility of the timetables on the microscopic infrastructure. We will call our model therefore track-choice, flexible PESP model (TCFPESP).

Keywords Periodic event scheduling problem · Mesoscopic railway topology · Service intention · Timetabling with track assignment

1 Introduction

Swiss Federal Railways (Schweizerische Bundesbahnen, short SBB) is working constantly on digitization and automation of railway planning and operations. Customers should benefit from higher capacities, less disturbances, better radio communication, improved customer information and lower overall costs. Railway infrastructure utilization is to be increased by shorter headway times and more precise planning. For this purpose, SBB has launched the smartrail 4.0 program.

S. Bütikofer (✉) · A. Steiner · R. Wüst
Institute for Data Analysis and Process Design, Zurich University of Applied Sciences ZHAW, Winterthur, Switzerland
e-mail: buts@zhaw.ch

J. S. Neufeld et al. (eds.), *Operations Research Proceedings 2019*, Operations Research Proceedings, https://doi.org/10.1007/978-3-030-48439-2_96

One major stream is the development of a new Traffic Management System (TMS) [1, 2].

The key to TMS' success is choosing the right level of detail for the infrastructure, which, on the one hand, enables good algorithmic solutions for Swiss-wide timetable entities and, on the other hand, guarantees the feasibility at the level of micro topology. SBB decided to make use of a certain kind of mesoscopic topology (see Sect. 2), which considers the critical operation points together with their capacities (i.e. number of tracks) and dependencies (i.e. possible track changes between the operation points). The input to this timetable planning step is given by a list of train runs to be scheduled. From the previous line planning steps, a list of commercial requirements, such as a stopping pattern, earliest departure times, latest arrival times, minimum dwell and travel times and connections to other trains are known for each train run [1, 2]. These requirements define the so called Service Intention (SI) (see Sect. 2) and are provided mainly by manual planers. The output of the timetabling step should be a conflict free path through track resources with time slots for the arrival and departure times at the resource units.

In the context of an applied research project with SBB, we developed an extension of the Periodic Event Scheduling Problem (PESP), the track-choice, flexible PESP model (TCFPESP). This model represents a mathematical formulation of the timetabling step explained above as an integer optimization model and is the main contribution of this article. Whereas different approaches for solving this problem are discussed in [1] our interest is mainly in understanding the influence of the input parameters, i.e. the SI, on the resulting timetables. This insight can be used to come up with more passenger friendly and robust timetables on one hand and should help us to understand the relaxation of the SI in case of conflicts on the other (see Discussion in Sect. 3).

2 Methodology

In this section we want to introduce the TCFPESP model. We introduced preliminarily versions of this model in [3, 4]. The main differences from there are: (1) the introduction of track change restrictions between operation points, (2) the modelling of headways in both directions over a finite number of operation points, (3) the use of flexible instead of fixed time events.

Mesoscopic infrastructure model There are several articles, which make use of a mesoscopic infrastructure model, see e.g. [5, 6].

To illustrate the level of detail of the respective infrastructure mapped onto the required mesoscopic topology, we refer to an example of the SBB "Grobkonzept Linienplanung" in Howald et al. [7]. The mesoscopic topology consists of operation points linked by route-sections. At each operation point and route-section there is a given number of tracks. A new operating point is assigned to each location where it is possible to change tracks. If there are customer services assigned to an operation

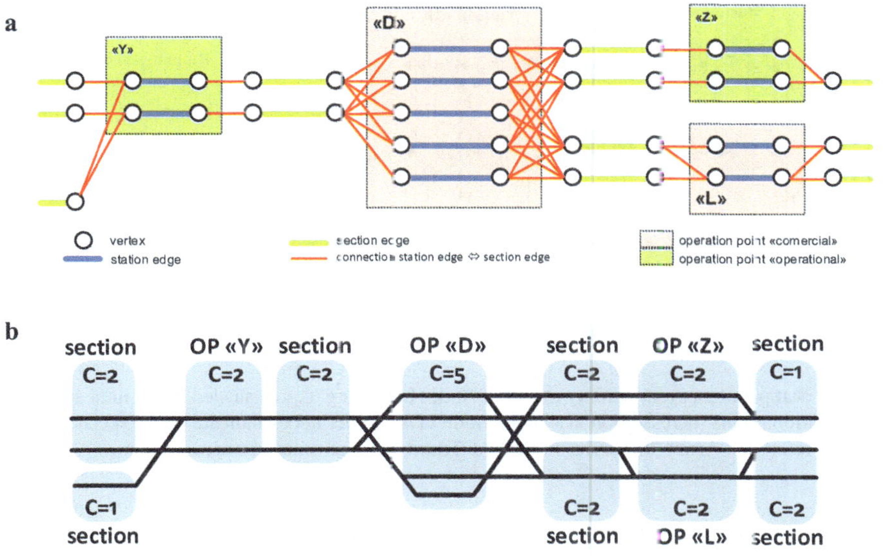

Fig. 1 (**a**) Mesoscopic infrastructure example from the SBB "Grobkonzept Linienplanung" in Howald et al. [7]. (**b**) Extracted topology information: each operation point, and each linking track segment is mapped into a graph node, represented by a grey shaded box. The node attribute 'C' indicates the track capacity of each node

point, it is classified as 'commercial', otherwise it is classified as 'operational' (see Fig. 1a). The capacity 'C' of each node is defined by the number of enumerated tracks of the operation point resp. route section (see Fig. 1b).

The event-activity network (EAN) is the input for our timetable model TCFPESP. It is constructed based on mesoscopic infrastructure information and the SI.

We summarize the mesoscopic infrastructure as a set I of operation points. Operation points include therefore stations and route-sections but can also be other critical resources as junctions (see Fig. 1). As mentioned before, each operation point $i \in I$ is associated to a capacity consisting of a set of tracks Tr_i. A train run $l \in L$ is described by a sequence of operation points of I, where we denote with i^+ the successor operation point of i on train run $l \in L$. For modelling switches we introduce the set S. $(t_1, t_2) \in S$ is fulfilled, if the tracks t_1, t_2 belong to neighbouring operation points and track t_2 is not reachable from track t_1 (i.e. there is no switch between these two tracks).

The SI is defined by a set of train runs. Each train run l belongs to a line L and is characterized by the sequence of operation points that are traversed and a corresponding time interval, which is required for either running or stopping on a corresponding track section or station. Each time interval has a minimal and maximal value. Stop nodes typically provide a service for boarding or de-boarding a train. Together, a pair of train runs moving in opposite directions makes up a train

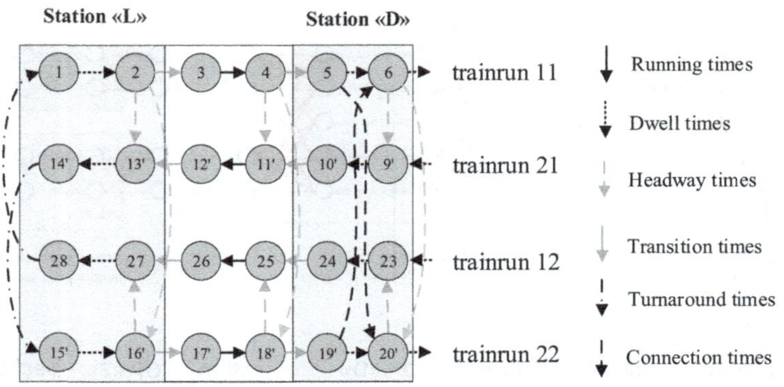

Fig. 2 Sample of an event-activity network. Nodes belonging to grey shaded boxes indicate events at operation points (here Station «L» and Station «D»). Other nodes indicate track type arrival and departure events. Arrow line styles indicate different types of time dependencies

circulation. The SI was first described in Wüst et al. [8] and formally specified in Caimi [9].

Based on our mesoscopic model and the SI we create an event-activity network (E, A). The set E of events consists of an arrival event arr_{li} and a departure event dep_{li} for each train run $l \in L$ and operation point $i \in l$. The activities $a \in A$ are directed arcs from $(E \times E)$ and describe the dependencies between the events. For every train run we have arcs between arrival and departure events at the same operation points (dwell times or trip times) and arcs between departure and arrival events of successive operation points (time needed for the travel between operation points). Further arcs include connections between train runs, headways and turnaround operations. Connections and turnaround information are also a part of the SI. We refer to Liebchen and Möhring [10] for a detailed overview of the modelling options of dependencies. Figure 2 provides a sample of such an event-activity network.

Headway arcs $a \in A_H$ are especially important for explaining the timetable model below. Headway arcs are used to model safety distances between trains running in the same and in opposite directions (see example in Fig. 2). Headway arcs are between two events of two train runs at the same operation point. A headway arc $a \in A_H$ is responsible for a safety distance on a set $I(a) \subset I$ of successive operation points on which the two train runs may use the same track. We denote for a headway arc $a \in A_H$ the set of events $E(a) \subset E$ corresponding to the arrival and departure events of the two train runs at the operation points $I(a)$ (we have therefore four events in $E(a)$ for every operation point in $I(a)$) (see again [10]).

Track-choice, flexible PESP model The classical PESP is formulated on an EAN (E, A) and tries to determine a periodic schedule on the macroscopic infrastructure (i.e. without using the tracks at an operation point) within a period T. Event $e \in E$ takes place at time $\pi_e \in [0, T)$. The schedule is periodic with time period T, hence each event is repeated periodically $\{\ldots, \pi_e - T, \pi_e, \pi_e + T, \pi_e + 2T, \ldots\}$.

The choices of the event times π_e depend on each other. The dependencies are described by arcs $a = (e,f)$ from a set A and modelled as constraints in the PESP. The constraints always concern the two events e and f and define the minimum and maximum periodic time difference l_a and u_a between them. These bounds are given as parameters in the PESP model.

To avoid tedious iterations between the planning steps "microscopic capacity planning" and "mesoscopic capacity planning" in case of infeasibility of the micro-level problem, one can improve the chance of finding a feasible solution by enlarging the solution space in the micro-level by using the following approach. We look for event time slots $(\pi_e + \delta_e)$ for every $e \in E$ that fulfill all constraints of the form

$$l_a + \delta_e \leq \pi_f - \pi_e + p_a T \leq u_a - \delta_f$$

for all $a = (e,f) \in A$, where p_c is an integer variable that makes sure, that these constraints are met in a periodic sense. This leads to the flexible PESP model (FPESP). The FPESP model has been described in detail in Caimi et al. [11].

The final choice of the event times in the range between the lower and upper bound shall be independent for each event such that each value of the end of an activity arc should be reachable from each time value at beginning of that activity arc (see Fig. 3a).

We extend the FPESP model by using the number of tracks Tr_i at each operation point $i \in I$. The track-choice FPESP model assigns the arrival event arr_{li} and the departure event dep_{li} of train run l at operation point i uniquely to a track in Tr_i. We can use these assignments to switch on headway arcs $a \in A_H$ by using the a big-M-approach. In addition to the variables π, δ and p from the FPESP model we need:

(i) Binary variables tc_{et} (track choice) for each event $e \in E$ and track $t \in Tr_{i(e)}$, where operation point $i(e)$ is associated to event e, i.e. e is equal to arr_{li} or dep_{li} for a train run l and (ii) Binary variables h_a for every headway arc $a \in A_H$. The track-choice, flexible PESP model (TCFPESP) is defined by:

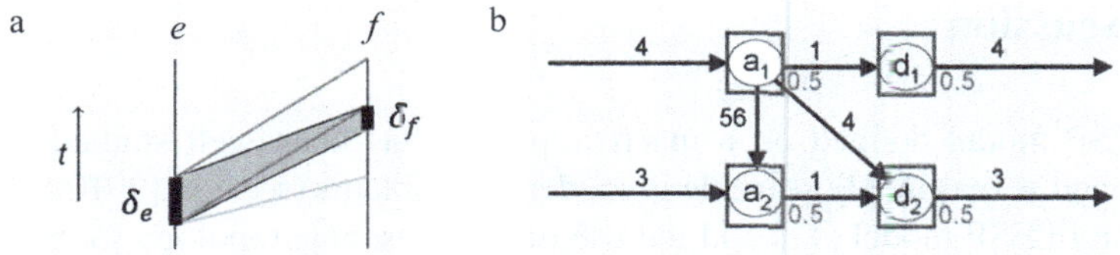

Fig. 3 (**a**) Time frames $[\pi_e, \pi_e + \delta_e]$ in place of time points π_e. In the example (**b**) this means that instead of planning time points $(\pi_{a_1}, \pi_{d_1}, \pi_{a_2}, \pi_{d_2})$ we plan time frames $[\pi_e, \pi_e + 0.5]$ for $e \in \{a_1, d_1, a_2, d_2\}$

$$\min\ f(\pi,\delta)$$

$$s.t. \qquad l_a + \delta_e \le \ \pi_f - \pi_e + p_a T \le u_a - \delta_f, \quad \forall\ a = (e,f) \in A \setminus A_H, \tag{1}$$

$$l_a + \delta_e - (1-h_a)\, M \le \pi_f - \pi_e + p_a T \le u_a - \delta_f + (1-h_a)\, M, \forall a = (e,f) \in A_H, \tag{2}$$

$$\sum\nolimits_{t \in Tr_{i(e)}} tc_{et} = 1, \quad \forall\ e \in E, \tag{3}$$

$$tc_{(arr_{li})t} = tc_{(dep_{li})t}, \quad \forall\ l \in L,\ i \in l,\ t \in Tr_i, \tag{4}$$

$$tc_{(dep_{li})t_1} + \ tc_{(arr_{li^+})t_2} \le 1, \quad \forall\ l \in L,\ i, i^+ \in l, (t_1, t_2) \in S \tag{5}$$

$$h_a \ge \ \left(\sum\nolimits_{e \in E(a)} tc_{et}\right) - 4\,|I(a)| + 1, \quad \forall\ a \in A_H, t \in Tr_{i(e)} \text{ with } e \in E(a) \tag{6}$$

$$tc_{et}, h_a \in \{0,1\}, \pi_e \in [0,T), p_a \in \mathbb{Z}, \delta_e \ge 0, \quad \forall\ e \in E, t \in Tr_{i(e)}, a \in A,$$

where M is a big enough natural number.

In Eq. (1) the normal FPESP constraints are summarized (without headway arcs). In Eq. (2) are the headway constraints, which can be switched off with a big-M technique. The assignment of the events to the tracks is done in Eq. (3). Equation (4) is used to assign the corresponding arrival and departure events to the same track. In Eq. (5) track changes with no switching possibility are prohibited. In Eq. (6) the headway variable is set to 1, if all events in $E(a)$ take place on the same track, i.e. the headway is required at this operation point (there are $4|I(a)|$ events in $E(a)$). There are many different linear objective functions $f(\pi,\delta)$ suggested by Caimi [11] for the FPESP model. With these objective functions the TCFPESP model is an integer linear optimization model.

3 Discussion

The PESP model defined on a macro topology has been well studied by many authors and is reasonably solvable to moderate problems (see e.g. [9, 10]). Furthermore the FPESP model [11] and the use of a mesoscopic topology [5, 6] increase the chance of feasibility on a micro topology level. Caimi [9] and Liebchen and Möhring [10] demonstrate, that the PESP model can be parametrized largely by the SI. Our TCFPESP model combines all of these properties.

The authors already investigated several aspects of the TCFPESP model. In [4] they presented a small-size case study for improving the stability of timetable by iteratively using the TCFPESP model with the information provided by the max-plus measures. In [12] the automatic generation of the SI using line planning models and the translation to an EAN was investigated, which is the next step in the automation of the railway value chain. In another small-size case study the functionality of the SI generation was shown and first attempts in the relaxation of the SI in the case of a maintenance disruption were established. In future we will focus on the relaxation of the SI in the case of infeasible timetables during disruptions.

References

1. Jordi, J., Toletti, A., Caimi, G., Schüpbach, K.: Applied timetabling for railways: experiences with several solution approaches. In: Planning, Simulation and Optimisation Approaches, 8th International Conference on Railway Operations Modelling and Analysis Norrköping, Sweden. Conference Proceedings (accepted submission). http://www.railnorrkoping2019.org (2019)
2. Völcker, M.: Der Weg zur automatisierten Kapazitätsplanung und Steuerung. Eisenbahningenieur. **4**, 22–25 (2019)
3. Wüst, R.M., Bütikofer, S., Ess, S., Gomez, C., Steiner A., Laumanns, M., Szabo, J.: Periodic timetabling with 'Track Choice'-PESP based on given line concepts and mesoscopic infrastructure. In: Operations Research Proceedings 2018. Springer, Berlin (in press)
4. Wüst, R.M., Bütikofer, S., Ess, S., Gomez, C., Steiner, A.: Improvement of maintenance timetable stability based on iteratively assigning event flexibility in FPESP In: Planning, Simulation and Optimisation Approaches, 8th International Conference on Railway Operations Modelling and Analysis Norrköping, Sweden. Conference Proceedings (accepted submission). http://www.railnorrkoping2019.org (2019)
5. Bešinović, N., Goverde, R.M.P., Quaglietta, E., Roberti, R.: An integrated micro-macro approach to robust railway timetabling. Transp. Res. B Methodol. **87**, 14–32 (2016)
6. de Fabris, S., Longo, G., Medeossi, G., Pesenti, R.: Automatic generation of railway timetables based on a mesoscopic infrastructure model. J. Rail Transp. Plann. Manag. **4**, 2–13 (2014)
7. Howald, P., Künzi, Th., Wild, P., Wieland, Th.: Grobkonzept «Linienplanung» Version 0.7. SBB Smart Rail 4.0 TMS-PAS, project document repository (2017)
8. Wüst, R.M., Laube, F., Roos, S., Caimi, G.: Sustainable global service intention as objective for controlling railway network operations in real time. In: Proceedings of the WCRR, Seoul (2008)
9. Caimi, G.: Algorithmic decision support for train scheduling in a large and highly utilized railway network. Dissertation, ETH Zürich Nr. 18581 (2009)
10. Liebchen, C., Möhring, R.H.: The modeling power of the periodic event scheduling problem: railway timetables – and beyond. In: Geraets, F., Kroon, L., Schoebel, A., Wagner, D., Zaroliagis, C. (eds.) Algorithmic Methods for Railway Optimization. Lecture Notes in Computer Science, vol. 4359, pp. 3–40. Springer, Berlin (2007)
11. Caimi, G., Fuchsberger, M., Laumanns, M., Schüpbach, K.: Periodic railway timetabling with event flexibility. Networks. **57**(1), 3–18 (2011)
12. Wüst, R.M., Bütikofer, S., Köchli, J., Ess, S.: Generation of the transport service offer with application to timetable planning considering constraints due to maintenance work. In: IRSA 2019, 2nd International Railway Symposium, Aachen. Conference Proceedings (accepted submission) (2019)

Capacity Planning for Airport Runway Systems

Stefan Frank and Karl Nachtigall

Abstract Runway system configurations constitute a bottleneck at major international airports. Capacity management is used to determine the maximal throughput of an airport, which is limited by several infrastructural and operational factors. Within this paper we describe how to model complex capacity restrictions on airport runway systems. The model is solved by a Column Generation approach where the subproblem is represented as a Shortest Path Problem. Additionally, a lower bound based on Lagrangian Relaxation and a Primal Rounding Heuristic are applied in our approach.

Keywords Graph theory · Network flow · Column generation

1 Introduction

In order to cope with the growth of aircraft movements, runway systems commonly represent a bottleneck at major international airports (cf. [5, 6]). Therefore, capacity planning is applied to determine and exploit the maximal throughput of an airport, while infrastructure expansions are usually not the method of choice due to their high investment costs, spatial restrictions, and state regulations based on the safety requirements. In addition, technological improvements can also lead to capacity increases (cf. [7]). The efficient use of the existing infrastructure represents an alternative. While sequence-based solution approaches are widely used for short-term planning, these methods are not workable for long-term or medium-term planning due to the complexity of the overall goal to minimize delays. Therefore, we present a mixed-integer programming formulation that focuses on flows instead of sequences. The objective in this model is to minimize the delay of aircraft movements for a given planning horizon while capacity restrictions need to be

S. Frank (✉) · K. Nachtigall
Institute of Logistics and Aviation, Technische Universität Dresden, Dresden, Germany
e-mail: stefan.frank@tu-dresden.de; karl.nachtigall@tu-dresden.de

J. S. Neufeld et al. (eds.), *Operations Research Proceedings 2019*, Operations Research Proceedings, https://doi.org/10.1007/978-3-030-48439-2_97

observed. We show how this general model can be extended to include additional restrictions, e.g. based on turnaround requirements, different aircraft classes or individual separation times.

2 Capacity Management

Various parameters can be used to describe the capacity of an airport. In general, the first step here is to mention the presented limit throughput, measured in the maximal number of aircraft movements per time unit under the safety requirements of traffic flow control. Dependencies result from different physical and technological influencing variables. First, the number of available runways and the layout of the runway system, i.e. the arrangement of the individual runways, should be mentioned. Furthermore, the configuration of the runway system has a high influence on the capacity. For various operational reasons, individual runways can be restricted so that, under certain conditions, only take-offs or landings can take place. Weather influences also determine the capacity of an airport. Furthermore, the safety regulations to be observed at airports are a decisive factor for the efficient use of existing airport capacity (cf. [6]). On the one hand restricting conditions for capacity planning therefore result from the infrastructure properties, but on the other hand from the safety requirements.

The spatial and temporal separations between the individual flights (movements) can be regarded as the basis for fulfilling these requirements. Therefore, the minimal distances between movements are transformed to corresponding minimal time intervals (separation times) and separation times are transformed to capacities. From this, dependencies between individual movements on an runway can be derived, as can dependencies between the individual runways in multiple runway systems. The separation times are generally differentiated according to the type of movement (take-off or landing), approach or take-off direction, distinct runway, and aircraft class. With regard to the latter distinguishing feature, aircraft can be classified according to various criteria such as size, weight, and speed. In general, aircraft are classified according to their maximum take-off weight (in tons—t). According to the standard of the International Civil Aviation Organization (ICAO) these are the classes H (Heavy, $>136t$), M (Medium, $7t - 136t$), and L (Light, $<7t$). Besides these standard classes, the further class J (Super Heavy, $>560t$) has been introduced for special consideration of the characteristics of very heavy aircraft (e.g. Airbus A380). In addition, these classes are to be further differentiated (e.g. Lower Heavy, Upper Heavy) due to re-categorisations of wake turbulence separations.

3 Planning Methods

For landings, the earliest and latest times of execution can generally be assumed due to technical or technological conditions. First of all, the time limitation of the landing process due to the limited fuel supply of an aircraft must be mentioned. This results in a maximum permissible landing delay, while the earliest possible landing can be justified by the fact that a maximum speed can be assumed for a landing approach based on the aircraft class. Therefore, the allocation of intervals (time windows) can be justified. Furthermore, there is the limited availability of runways and a partially limited (temporal) usability of individual runways. In addition, there is a demand for the assignment of restricted time windows to ensure the availability of connecting movements (cf. [4]). A simple approach to determining landing approaches is to process landings according to their planned landing time by applying the First Come—First Served (FCFS) principle. However, this is accompanied by the fact that these sequences seldom have approach orders of high quality in the sense of the above-mentioned objectives and thus do not lead to the desired efficient utilization of given capacities. In the course of improving FCFS sequences, initial approaches to the topic of models and solution procedures were developed that deal with the problem of adapting the given approach sequences, taking into account a maximum shift of an approach in the FCFS sequence. The aim of this is to achieve the lowest possible delays while maintaining the rough basic order of the approach sequence due to a constraint position shifting of the movements (cf. [1, 3, 10]).

For take-offs, priorities derived from strategic planning processes are generally used to determine departure times. Furthermore, the capacity and thus a maximum throughput of departures can be based on the occupancy times of the runway for departures, the availability of taxiways, the distribution of aircraft classes, minimum time intervals between departures, minimum spatial distances between dependent runways as well as the configuration of runways, i.e. the exclusive use for departures or the mixed use for the execution of take-offs and landings. For fixed sequences, aircraft movements are scheduled so that take-offs and landings take place as early as possible.

In contrast to these simple strategies for the sequential processing of the subproblems, some approaches for a simultaneous solution were developed. In [3], a single runway system is considered and a mixed-integer linear optimization model is developed. An extension of the described time-continuous model is shown for multiple runway systems as well as the consideration of an alternative time-discrete formulation. Furthermore, different acceleration techniques are discussed. In the selected formulations an aggregated consideration of the separations between movements on the same runway or different runways is carried out. Further published approaches to this topic are mostly based on these formulations. Overviews of these adapted model formulations can be found in [4, 9]. In [9], a model formulation is presented to include a differentiated consideration of separations for movements at the same runway and between different runways. Some restrictive assumptions are made in this paper. Permissible delays are represented for all movements by ordered time windows of fixed length.

4 Model and Solving Approach

The basis of the approach is a graph for mapping the dependencies of an runway system. Restrictions in operation or various levels of differentiation in planning are represented as nodes in this network. Links between these levels are represented by arcs (cf. [8]). For nodes, capacities of movements per time unit are calculated on the basis of separations and maximum throughput of a node. Figure 1 displays a simple example with nodes for total movements as source node of the network and nodes for the differentiation of the type of movements (arrivals and departures). Further levels of differentiation can be depicted as nodes in the network (cf. nodes connected by dashed lines in Fig. 1). For movements, permissible routes through this network are considered, so that each route starts at a defined source and ends at a defined sink, all relevant dependencies on this route are mapped, and the capacity of the individual nodes is not exceeded under the objective of minimum delays. In general, all valid routes can be enumerated in advance due to a limited number of feasible routes. On that basis, different node intervals with explicit capacity can be derived. In Fig. 2, a simple example is shown. Let us assume capacities of 30 movements per hour for either arrivals or departures and a total capacity of 40 movements per hour for a mixed mode that results from the transformation of given separation times. This leads to the capacities for small intervals (e.g. two movements in total for intervals of 3 min) as shown in the figure.

The resulting model formulation is based on the work of [8, 11] for slot-allocation. Given the set T of the n movements with $T = \{1, \ldots, n\}$, the earliest starting time of a movement i is described by e_i, the latest possible starting time of a movement i by l_i. With $\mathcal{R}(C(i))$ the set of feasible routes through a resource

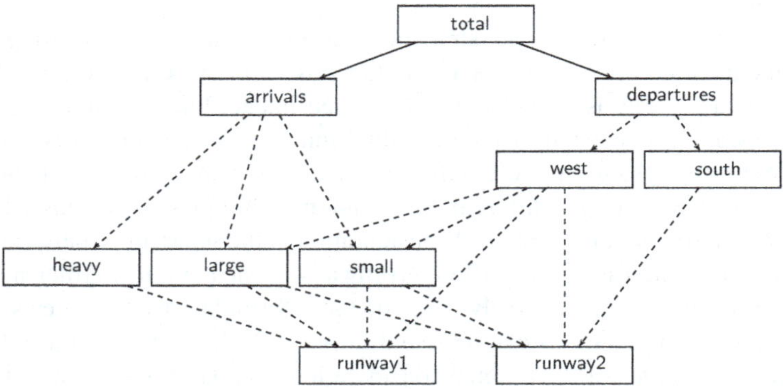

Fig. 1 Resource network

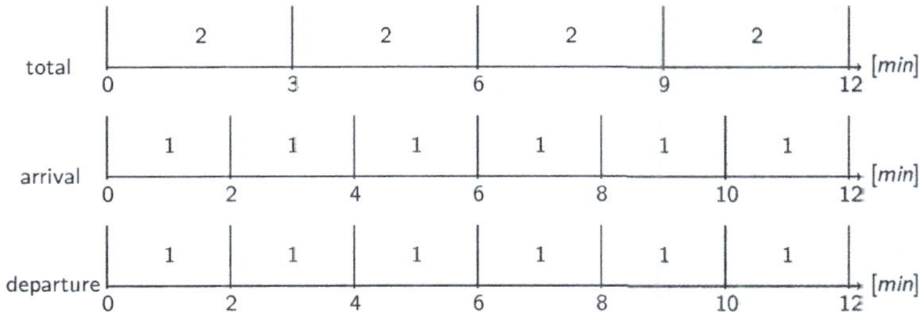

Fig. 2 Node intervals

network for a class of movements $C(i)$ is specified. The individual nodes on such a route $R \in \mathcal{R}(\cdot)$ are indicated using $r \in R$. The costs (weighted delay) of a class $C(i)$ and route R are indicated using $c_{C(i)}$. Time-discrete decision variables $x_{i,R,t}$ describe the choice of a time of execution of a movement i related to the route R at a time t, where $t \in [e_i, l_i]$, and $x_{i,R,t} = 1$ if this assignment is made, and 0 otherwise. Set $\mathcal{C}$ contains all combinations of node r and interval I, indexed via $j := (r_j, I_j)$. The capacity of such a node interval j is described by d_j. Furthermore, $cto_{i,R,r}(t)$ is the target time of a movement i on route R at node r assuming that the movement is performed at time t (calculated time over—cto).

$$\min \sum_{i \in T} \sum_{t \in [e_i, l_i]} \sum_{R \in \mathcal{R}(C(i))} c_{C(i)} x_{i,R,t} \tag{1}$$

s.t.

$$\sum_{t \in [e_i, l_i]} \sum_{R \in \mathcal{R}(C(i))} x_{i,R,t} = 1 \qquad \forall i \in T \tag{2}$$

$$\sum_{r \in R: cto_{i,R,r}(t) \in I_j} x_{i,R,t} \leq d_j \qquad \forall j \in \mathcal{C} \tag{3}$$

$$x_{i,R,t} \in \{0, 1\} \ \forall i \in T; R \in \mathcal{R}(C(i)) : t \in [e_i, l_i] \tag{4}$$

Due to objective function (1), the overall weighted delay is minimized. Constraints (2) guarantee that each movement is assigned to exactly one route and exactly one target time. Capacity constraints (3) ensure that no more movements can be assigned for each tuple (r_j, I_j) than the capacity provides.

Relaxing the integrality condition (4), we are using Column Generation technique to solve the resulting LP-relaxation of this model (cf. [2]). Let us denote the dual multipliers associated with constraints (2) and (3) respectively by α_i and β_j. Therefore, the pricing subproblem is to find an assignment of the scheduled time of a movement i such that $c_{C(i)} + \sum_{r \in R: cto_{i,R,r}(t) \in I_j} \beta_j < \alpha_i$ with $\beta_j \geq 0$ as it is given by the normal form of the model. This problem can be represented as a Shortest Path Problem (SPP) with non-negative arc costs and therefore be solved in polynomial time. In this SPP, possible times to assign movements to are stated as nodes while arcs are included to establish the suitable connections of the resource network.

Using the given dual multipliers β_j, one can rearrange the hard capacity constraints (3) to the objective function. The resulting Lagrangian Relaxation

$$\min \sum_{i\in T} \sum_{t\in[e_i,l_i]} \sum_{R\in\mathcal{R}(C(i))} \left(c_{C(i)} + \sum_{r\in R: cto_{i,R,r}(t)\in I_j} \beta_j \right) x_{i,R,t} - \sum_{j\in\mathcal{C}} \beta_j d_j \tag{5}$$

s.t.

$$\sum_{t\in[e_i,l_i]} \sum_{R\in\mathcal{R}(C(i))} x_{i,R,t} = 1 \qquad \forall i \in T \tag{6}$$

$$x_{i,R,t} \in \{0,1\} \; \forall i \in T;\; R \in \mathcal{R}(C(i));\; t \in [e_i, l_i] \tag{7}$$

can easily be solved by inspection.

Our simple Primal Rounding Heuristic works as follows. Given non-integral variable values $x_{i,R,t}$, the weighted sum of the delays of a movement i is calculated by $\overline{c}_{C(i)} = \sum_{c\geq c^+_{C(i)}} c_{C(i)} x_{i,R,t}$. The movements are sorted by ascending order of their weighted delays. For this order, the movements are scheduled by a FCFS-approach.

5 Discussion

We showed an aggregated approach to manage airport capacities. The focus of our research are the medium-term and long-term planning processes. In this context, no precise sequences of movements on runway systems are of interest, in contrast to short-term planning. The aggregated approach of the flow-based model is characterized on the one hand by the estimation of capacities for time intervals. On the other hand, however, the technique of Column Generation can be used to solve this model formulation, so that comparatively short computing times can be expected. This stems also from the shown fact, that the pricing subproblem can be solved in polynomial time. In addition, the further simple techniques of the Lagrangian Relaxation and the Primal Heuristic to achieve lower bounds and upper bounds, respectively, support shorter computing times. This behavior was demonstrated in preliminary tests. One point of further research might be the integration of the Column Generation approach in the resulting Branch-and-Price algorithm to guarantee the closing of the gap. Also, more improved Primal Heuristics are to be researched. The main focus of our research lies in the development of a generic approach to accelerate the calibration of the intervals in the resource network. Approximations for network nodes, interval sizes, and interval capacities are to be examined to improve estimated results.

References

1. Balakrishnan, H., Chandran, B.G.: Algorithms for scheduling runway operations under constrained position shifting. Oper. Res. **58**, 1650–1665 (2010)
2. Barnhart, C., Johnson, E.L., Nemhauser, G.L., Savelsbergh, M.W.P., Vance, P.H.: Branch-and-price: column generation for solving huge integer programs. Oper. Res. **46**, 316–329 (1998)
3. Beasley, J.E., Krishnamoorthy, M., Sharaiha, Y.M., Abramson, D.: Scheduling aircraft landings—The static case. Transp. Sci. **34**, 180–197 (2000)
4. Bennell, J.A., Mesgarpour, M., Potts, C.N.: Scheduling models for air traffic control in terminal areas. J. Sched. **9**, 223–253 (2013)
5. Blumstein, A.: The landing capacity of a runway. Oper. Res. **7**, 752–763 (1959)
6. de Neufville, R., Odoni, A.R.: Airport Systems: Planning, Design, and Management. McGraw-Hill, London (2003)
7. Janic, M.: Modeling effects of different air traffic control operational procedures, separation rules, and service disciplines on runway landing capacity. J. Adv. Transp. **48**, 556–574 (2014)
8. Kaufhold, R., Marx, S., Müller-Berthel, C., Nachtigall, K.: A pre-tactical generalised air traffic flow management problem. In: 7th USA/Europe ATM R&D Seminar, Barcelona (2007)
9. Lieder, A., Stolletz, R.: Scheduling aircraft take-offs and landings on interdependent and heterogeneous runways. Transp. Res. E Log. Transp. Rev. **88**, 167–188 (2016)
10. Psaraftis, H.N.: A dynamic programming approach for sequencing groups of identical jobs. Oper. Res. **28**, 1347–1359 (1980)
11. van den Akker, J.M., Nachtigall, K.: Slot Allocation by Column Generation, Technical Report NLR TP 97286, Amsterdam (1999)

Data Reduction Algorithm for the Electric Bus Scheduling Problem

Maros Janovec and Michal Kohani

Abstract In this paper, we address the electric bus scheduling problem (EBSP) and its solution. We propose an algorithm for input data reduction which reduces the number of service trips by merging two service trips into one. Also, a method of choosing possible candidates for merging and two different criteria to choose the best candidate are described. Proposed algorithm was tested on real data from the city of Žilina provided by the public transport system operator DPMŽ. After the reduction of the inputs, an exact optimization was performed on the reduced problem to compare the solutions with the original problem.

Keywords Electric bus · Scheduling problem · Data reduction · IP solver

1 Introduction

This paper addresses the algorithm that reduces the number of service trips which is an input for the electric bus scheduling problem (EBSP) and consequently reducing the computation time needed to solve the problem. The EBSP problem is a special case of vehicle scheduling problem (VSP) [1] with constraints of energy consumption and charging where we assign available electric buses to the service trips that need to be served. The electric buses have limited driving range and the charging of the battery is time-consuming. These facts create very limiting constraints for the schedule.

To solve this problem, mathematical models are proposed by different authors. The authors in their models make assumptions to decrease the complexity of the problem—charging always to the maximal battery capacity [2, 3], charging at only one location (depot)[2–4]. We have focused on these assumptions and proposed a linear mathematical model in our previous works [5, 6] where the electric buses can

M. Janovec (✉) · M. Kohani
University of Zilina, Zilina, Slovakia
e-mail: maros.janovec@fri.uniza.sk; michal.kohani@fri.uniza.sk

J. S. Neufeld et al. (eds.), *Operations Research Proceedings 2019*, Operations Research Proceedings, https://doi.org/10.1007/978-3-030-48439-2_98

charge at multiple locations and the bus charges only necessary amount of energy which is determined by the model. We tried to solve the problem of EBSP into optimality by IP solver in our previous work, but only small and medium- size problems could be solved in a reasonable time. The number of service trips had the highest influence on computation time. Therefore, we decided to reduce the complexity by reducing the number of service trips. Consequently, we researched some simple criteria that can reduce the number of service trips by merging two consecutive trips into a single trip.

In Sect. 2 we provide the description of our model. In Sect. 3 we present an algorithm for the reduction of service trips. The numerical experiments and results are discussed in Sects. 4 and 5.

2 Mathematical Model

In our model proposed in [5, 6], we use the set N of service trips, the morning depot D_0 and set of evening depots D_n where we add one depot for every service trip. Set R represents the chargers and the set T^r contains all charging events at charger $r \in R$. The set K is the set of available vehicle types. Each service trip $i \in N$ is defined by its start time s_i, duration t_i and energy consumption c_i. Constants t_{ij} and c_{ij} represent the transfer time and consumed energy between trips i and j respectively. For trip i sets F_i and B_i of possible following and previous service trips are defined. The sets Fc_{ri} and Bc_{ri} represent possible following and previous charging events for service trip i at charger r. Constants t_{ir} and t_{rj} define the transfer time between service trip and charger, while c_{ir} and c_{rj} are the consumed energy on the transfers. Each charger $r \in R$ is defined by its charging speed q_r and its location. A set T^r of charging events at charger $r \in R$ is defined, where start time s_{rt} of charging event t is derived from corresponding service trip as follows $s_{rt} = s_i + t_i + t_{ir}$. The charging events in set T^r are ordered by their increasing starting time which create time intervals of different length. The sets Fi_{rt} and Bi_{rt} are possible following and previous service trips for charging event t at charger r. An available vehicle type $k \in K$ is characterized by the maximal SoC^k_{max} and minimal SoC^k_{min} allowed capacity of the battery. The binary decision variables x^k_{ij}, y^k_{irt} and z^k_{rtj} refer to deadhead trips between service trips and to or from charging events at chargers. Variable w^k_{rt} is used when bus continues charging at the same charger. Variables e^k_i and ε^k_{rt}, represent the energy state of bus k just before service trip i and charging event t at charger r respectively.

$$minimize \sum_{k \in K} \sum_{j \in F_{D_0}} x^k_{D_0 j} + \sum_{k \in K} \sum_{r \in R} \sum_{t \in Fc_r D_0} y^k_{D_0 rt} \tag{1}$$

$$\sum_{k \in K} \sum_{i \in B_j} x^k_{ij} + \sum_{k \in K} \sum_{r \in R} \sum_{t \in Bc_{rj}} z^k_{rtj} = 1 \quad \forall j \in N \tag{2}$$

$$\sum_{k\in K}\sum_{j\in Bi_{rt}} y^k_{jrt} - \sum_{k\in K} w^k_{rt-1} \le 1 \quad \forall r\in R, t\in T^r \tag{3}$$

$$\sum_{i\in B_j} x^k_{ij} + \sum_{r\in R}\sum_{t\in Bc_{rj}} z^k_{rtj} = \sum_{l\in F_j} x^k_{jl} + \sum_{r\in R}\sum_{t\in Fc_{rj}} y^k_{jrt} \ \forall j\in N, k\in K \tag{4}$$

$$\sum_{i\in Bi_{rt}} y^k_{irt} + w^k_{rt-1} = \sum_{j\in Fi_{rt}} z^k_{rtj} + w^k_{rt} \quad \forall r\in R, t\in T^r, k\in K \tag{5}$$

The objective function (1) minimizes the number of used electric buses. The constraints (2) ensure that every service trip is served. Limitation of charging to only one bus at a time ensure constraints (3). The constraints (4) and (5) serve as flow constraints of a service trip, respectively charging event.

$$e^k_{D_0} = SoC^k_{max} \quad \forall k\in K \tag{6}$$

$$e^k_i \ge SoC^k_{min} + c_i + \sum_{j\in F_i} x^k_{ij} c_{ij} + \sum_{r\in R}\sum_{t\in Fc_{ri}} y^k_{irt} c_{ir} \quad \forall i\in N, k\in K \tag{7}$$

$$e^k_j + c_{rj} + Mq_r(1-z^k_{rtj}) \ge SoC^k_{min} + z^k_{rtj} c_{rj} \ \forall r\in R, t\in T^r, k\in K, j\in Fi_{rt} \tag{8}$$

$$e^k_j \le e^k_i - x^k_{ij}(c_i + c_{ij}) + SoC^k_{max}(1-x^k_{ij}) \quad \forall j\in N, i\in B_j, k\in K \tag{9}$$

$$e^k_j \ge e^k_i - x^k_{ij}(c_i + c_{ij}) - SoC^k_{max}(1-x^k_{ij}) \quad \forall j\in N, i\in B_j, k\in K \tag{10}$$

$$\varepsilon^k_{rt} \le e^k_i - y^k_{irt}(c_i + c_{ir}) + SoC^k_{max}(1-y^k_{irt}) \ \forall r\in R, t\in T^r, k\in K, i\in Bi_{rt} \tag{11}$$

$$\varepsilon^k_{rt} \ge e^k_i - y^k_{irt}(c_i + c_{ir}) - SoC^k_{max}(1-y^k_{irt}) \ \forall r\in R, t\in T^r, k\in K, i\in Bi_{rt} \tag{12}$$

The constraints (6) set the battery capacity to the maximum at the start of the working day. To ensure, that the bus has enough energy to drive the service trip and the following transfer or just transfer after charging, the constraints (7) and (8) are defined. The constraints (9) and (10) represent the preservation of energy between two consecutive service trips. The constraints (11) and (12) represent the preservation of energy between the service trip and the following charging event.

$$e^k_j + c_{rj} - \varepsilon^k_{rt} + SoC^k_{max}(1-z^k_{rtj}) \ge 0 \quad \forall r\in R, t\in T^r, k\in K, j\in Fi_{rt} \tag{13}$$

$$\varepsilon^k_{rt+1} - \varepsilon^k_{rt} + SoC^k_{max}(1-w^k_{rt}) \ge 0 \quad \forall r\in R, t\in T^r, k\in K \tag{14}$$

$$e^k_j + c_{rj} - Mq_r(1 - z^k_{rtj}) \leq SoC^k_{max} \quad \forall\, r \in R, t \in T^r, k \in K, j \in Fi_{rt} \tag{15}$$

$$\varepsilon^k_{rt+1} - Mq_r(1 - w^k_{rt}) \leq SoC^k_{max} \quad \forall\, r \in R, t \in T^r, k \in K \tag{16}$$

$$\begin{aligned} e^k_j \leq \varepsilon^k_{rt} + z^k_{rtj}((s_j - t_{rj} - s_{rt})q_r - c_{rj}) + SoC^k_{max}(1 - z^k_{rtj}) \\ \forall j \in N, r \in R, t \in Bc_{rj}, k \in K \end{aligned} \tag{17}$$

$$e^k_j + c_{rj} - \varepsilon^k_{rt} - SoC^k_{max}(1 - z^k_{rtj}) \leq (s_{rt+1} - s_{rt})q_r \ \forall r \in R, t \in T^r, k \in K, j \in Fi_{rt} \tag{18}$$

$$\varepsilon^k_{rt+1} \leq \varepsilon^k_{rt} + w^k_{rt}(s_{rt+1} - s_{rt})q_r + SoC^k_{max}(1 - w^k_{rt}) \quad \forall\, r \in R, t \in T^r, k \in K \tag{19}$$

The constraints (13) and (14) ensure that during charging the bus charges energy and does not consume it. Constraints (15) and (16) limit the energy state of a bus to the maximum capacity of the battery. The constraints (17), (18) and (19) restrict the available charging time of a charging event by the start of the following service trip or by the start of the next charging event.

$$x^k_{ij} \in \{0, 1\} \ \forall\, k \in K, i \in N \cup D_0 \cup D_n, j \in F_i \tag{20}$$

$$y^k_{irt} \in \{0, 1\} \ \forall\, k \in K, i \in N, r \in R, t \in Fc_{ri} \tag{21}$$

$$z^k_{rtj} \in \{0, 1\} \ \forall\, k \in K, r \in R, t \in T_r, j \in Fi_{rt} \tag{22}$$

$$w^k_{rt} \in \{0, 1\} \ \forall\, k \in K, r \in R, t \in T_r \tag{23}$$

$$e^k_j \geq 0 \ \forall\, k \in K, i \in N \tag{24}$$

$$\varepsilon^k_{rt} \geq 0 \ \forall\, k \in K, r \in R, t \in T_r \tag{25}$$

The constraints (20)–(25) are obligatory constraints.

3 Data Reduction

Due to the complexity of the model, the medium and large scale problems are not able to be solved to optimality. One possibility to solve this problem is to decrease the number of inputs which influence the complexity the most. Therefore, we introduce an algorithm to reduce the number of service trips by combining two service trips into one service trip and then solve this reduced problem to optimality.

Naturally, this method can cause the loss of the optimal solution of the original problem, but is able to find a near-optimal solution.

The main idea of the decision on which service trips should be merged is that usually in the city public transport system, there are lines to the suburbs of the city and the suburb terminal would be far away from the charging stations. We assume the charging stations would be situated near the center of the city.

Our algorithm has two parts. In the first part, we choose possible service trips to merge. Therefore we need to identify the end terminals in the suburbs of the city. For this problem we create a graph of connections between the terminals where edges are connections between consecutive terminals. Then the vertexes with the degree of one are the terminals in the suburbs. Only at these locations, a merging of trips is possible. Next, we need to define criteria to choose possible candidates for merging. The idea is that two trips follow each other in a short time interval at the terminal in the suburb. Then the candidate for merging has two trips following each other in a specified time interval and the end terminal of the first trip is the same as the start terminal of the second trip.

In the second part, we choose the most suitable candidate for merging. For that, we created two different criteria. The first criterion (BCC1) selects the best candidate for merging as two following trips with the smallest time between them. The second criterion (BCC2) selects those candidates where the merging point (terminal) is the farthest from the nearest charging station because it means we spend less energy on transfer to charger.

4 Numerical Experiments

To test the computational time and also the quality of the results after reducing the number of service trips by the proposed algorithm, we performed a number of numerical experiments. All experiments were performed on a computer with Intel Core i5-7200U 2.5 Ghz, 16 GB of RAM and on IP solver FICO Xpress IVE 7.3 and the time limit was set to 16 h. The experiments were performed on different datasets generated from the real data provided by public transport system provider DPMŽ in the city of Žilina. The size of the datasets is listed in Table 1. Also, we created three scenarios with different number of charging stations. In scenario A we have three

Table 1 Number of reduced service trips for selected parameters

	Data reduction				
Dataset	5_BCC1	5_BCC2	15_BCC1	15_BCC2	Original
DS1	123	123	96	94	160
DS2	102	102	85	83	133
DS3	196	196	155	149	245
DS4	329	329	273	266	415

chargers, two are in trolleybus depot and one in the center of the city. In the second scenario B, we added two chargers at the bus depot. In scenario C, we added one charger at the main train station.

The first part of the experiments was a reduction of service trips. We tested parameter MaxTime with values 5 and 15 min between two consecutive trips and the criteria, which combination is the best (BCC1, BCC2). The number of service trips for each dataset after the reduction can be seen in Table 1.

In the second part of the experiments, we compared the results of EBSP of reduced and original problems. The solutions obtained by the IP solver are listed in Table 2. The results of the optimization are listed in the columns. In the case of dataset 5_BCC1 and 5_BCC2, the reduced service trips were the same, therefore the column 5_BCC1/BCC2 represents results of both datasets. For each dataset the column Sol represents the best-found solution, column BB represents best bound and column Time is the computation time of the problem. The results with "-" were not found and value 57,600 represents the time limit of 16 h.

To compare which of the two criteria BCC1 and BCC2 are better, we compared the results from the table and found out that in most cases the BCC1 criteria worsened the optimal solution. Moreover, the criterion BCC2 is more stable and its computation times were better. Therefore, we recommend the criteria BCC2 for use in the reduction algorithm.

5 Conclusions

We proposed an algorithm for the reduction of the number of service trips for the electric bus scheduling problem. We reduced the number of trips for selected problems generated from the public transport system in the city of Žilina. We ran the optimization of the reduced problems and compared the results and computation time with those of the original problems. We also tested two proposed criteria for the selection of the best possible combination of service trips.

The experiments showed that it is possible to reduce the number of the service trips by 32–44% and the computation time by 70–80% without the loss of optimality of the solution for selected datasets with the setting of MaxTime to 15-min interval to become a possible combination and the BCC2 criteria to choose the best combination. In summary, this reduction of service trips is an approach that can improve the possibility to solve the large datasets without great loss of optimality and can be implemented to heuristics that solve EBSP.

Table 2 Reduced and original problems solution(Sol), best bound(BB) and computation time comparison

Dataset	Scenario	5_BCC1/BCC2			15_BCC1			15_BCC2			Original		
		Sol	BB	Time(s)	Sol	BB	Time(s)	Sol	BB	Time(s)	Sol	BB	Time(s)
DS1	A	9	9	128.2	10	10	53.6	9	9	58.4	9	9	277.6
	B	9	9	317.4	10	10	115	9	9	130.7	9	9	647.9
	C	9	9	374.8	10	10	138.4	9	9	143.8	9	9	746.7
DS2	A	8	8	66.8	8	8	43.6	8	8	49.1	8	8	169.9
	B	8	8	151.4	8	8	104.6	8	8	72.2	8	8	402.7
	C	8	8	231.3	8	8	124.5	8	8	99.8	8	8	514.1
DS3	A	13	13	4565	14	14	358	13	13	295	13	13	1816
	B	13	13	15,365	14	14	650	13	13	620	13	13	37,951
	C	13	13	3114	14	14	1026	13	13	942	13	13	22,908
DS4	A	27	26	57,600	26	26	42,694	26	26	2083	33	26	57,600
	B	29	26	57,600	26	26	55,959	26	26	22,253	–	26	57,600
	C	30	26	57,600	26	26	11,214	26	26	17,149	–	26	57,600

Acknowledgments This work was supported by the research grants APVV-15-0179 "Reliability of emergency systems on infrastructure with uncertain functionality of critical elements" and VEGA 1/0689/19 "Optimal design and economically efficient charging infrastructure deployment for electric buses in public transportation of smart cities".

References

1. Bunte, S., Kliewer, N.: An overview on vehicle scheduling models. Public Transp. **1**(4), 299–317 (2009). https://doi.org/10.1007/s12469-010-0018-5
2. van Kooten Niekerk, E., van den Akker, J.M., Hoogeveen, J.A.: Scheduling electric vehicles. Public Transp. **9**(1), 155–176 (2017). https://doi.org/10.1007/s12469-017-0164-0
3. Rogge, M., van der Hurk, E., Larsen, A., Sauer, D.U.: Electric bus fleet size and mix problem with optimization of charging infrastructure. Appl. Energy **211**, 282–295 (2018). https://doi.org/10.1016/j.apenergy.2017.11.051
4. Sassi, O., Oulamara, A.: Electric vehicle scheduling and optimal charging problem: complexity, exact and heuristic approaches. Int. J. Prod. Res. **55**(2), 519–535 (2017). https://doi.org/10.1080/00207543.2016.1192695
5. Janovec, M., Kohani, M.: Exact approach to the electric bus fleet scheduling. Transp. Res. Procedia **40**, 1380–1387 (2019). https://doi.org/10.1016/j.trpro.2019.07.191
6. Janovec, M., Kohani, M.: Battery degradation impact on electric bus fleet scheduling. In: 2019 International Conference on Information and Digital Technologies (IDT), pp. 190–197 (2019). https://doi.org/10.1109/DT.2019.8813693

Crew Planning for Commuter Rail Operations, a Case Study on Mumbai, India

Naman Kasliwal, Sudarshan Pulapadi, Madhu N. Belur, Narayan Rangaraj, Suhani Mishra, Shamit Monga, Abhishek Singh, S. G. Sagar, P. K. Majumdar, and M. K. Jagesh

Abstract We consider the problem of constructing crew duties for a large, real instance of operations for commuter train services in Mumbai, India. Optimized allotment of crew duties and enforcement of work rules ensures adequate safety and welfare of rail workers. Currently, within Indian railways, decisions related to crew allotment are made manually. The main objective is to use as few crew members as possible to execute upon the timetable. This improves the efficiency of the system by increasing the average working hours of work per duty. We also have several other secondary objectives. The presence of a large number of operational constraints makes the problem difficult to solve. Computational experiments are performed over the current train timetables and the results of our algorithm compare very favorably with the crew duty schedules in use. For the Western Railways train timetable of 2017–18, the crew duty sets required to perform the timetable was 382. The proposed algorithm achieves crew allotment with 368 sets, promising significant savings of manpower and money.

Keywords Railways · Optimization · Crew schedule problem · Crew allotment · Duty preparation · Operations research · Constraint modelling · Metaheuristic · Resource allocation · Work load balancing

N. Kasliwal (✉) · S. Pulapadi · M. N. Belur
Dept of Electrical Engineering, Indian Institute of Technology Bombay, Mumbai, India

N. Rangaraj
Industrial Engineering and Operations Research, Indian Institute of Technology Bombay, Mumbai, India

S. Mishra · S. Monga · A. Singh · S. G. Sagar · P. K. Majumdar · M. K. Jagesh
Western Railways, Mumbai, India

J. S. Neufeld et al. (eds.), *Operations Research Proceedings 2019*, Operations Research Proceedings, https://doi.org/10.1007/978-3-030-48439-2_99

1 Introduction

Large mass transit systems like Mumbai Western Railways are complicated not only because of the multitude of management considerations, labour laws, and union requirements, but also because requirements have been changing rapidly in the recent past. As mentioned in [1], staff is to be assigned for each of the 382 links in the 1355 services that Western Railways runs daily using 89 rakes. This requires that timetables need to be modified frequently, thus requiring daily duties (hence, duty sets) and crew rosters to be changed quickly as well. Crew scheduling at the Mumbai Suburban Railways has been done manually with great skill for over 150 years. The changing times require a positive change in the approach of how we prepare these work duties for the system.

An optimal duty preparation strategy minimizes the number of sets required to match crew members with the services of the Western Railways. Least number of sets implies more working hours per set or more working distance per set. Tight packing of these sets with proper adherence to HOER (Hours of Employment and Period of Rest) rules ensures minimum operating slack and maximum use of valuable manpower.

This paper contributes in two ways. First and foremost, in a problem space in which algorithms and models tend to be highly specialized because every crew scheduling system is unique, it describes an approach that is simple, flexible, and hence has potential for adaptation to systems other than the Railways. Second, it adds to the operations research literature on crew scheduling by describing an iterative approach that uses time probabilistic curves to match duties to crew. The huge search space at a large complex system like the Railways makes it a very interesting matching problem for resource allocation.

2 Railway Terminology and Documents

The Western Railways line comprises of 37 stations going from Churchgate to Dahanu Road, out of which 15 are could be used for crew change.

As part of its operations planning, the Western Railways department does timetabling and crew scheduling. Both these tasks are documented in 2 books, namely ‘Suburban Working Time Table’ and ‘Schedule Book for Suburban Guards and Motormen’, references [2] and [3].

A rake refers to the complete physical train that comprises of all allocated coaches. A rake’s movement throughout the day is broken into services. The crew comprises of a guard and a motorman, whose duties are defined in the schedule book. The schedule book consists of a collection of sets. Each set contains the on-duty and off-duty time, start and end station, the assigned lobby, set working hours and distance, list of services to be done as work of the set, and rest hours provided to driver after completing the set.

These sets are divided into two lobbies, Borivali and Churchgate. The sets belonging to a particular lobby are packed in continuation forming a loop. The sets are divided into two categories:

Working Sets These are the sets which have allocated services to be manned by a crew member. They are divided into following three categories:

- Day working sets
- Night working sets: Sets with on-duty time after 22:00
- Halting working sets: Always occur in pairs, one set defining evening duty and the other defining morning duty with a short rest in between.

Waiting Duty and Shunting Duty Sets A shunting duty set requires the motormen to take rakes to/from a stabling depot such as a yard or a car shed.

3 Problem Formulation

To create an efficient strategy for crew allotment, the overall problem has been decomposed into the following two stages:

Set Generation To break the rake cycles into some workdays. This is essentially a matching problem for resource allocation.

Set Linking Set linking requires us to combine single workdays to form a sequence of sets, satisfying the rest considerations.

4 List of Constraints

'Hours of Work and Period of Rest Rules (HOER)', reference [4], is the official document of the Government of India and the Western Railways containing a list of operational rules. Our algorithm takes into account all these constraints for duty allocation. The set generation constraints are listed below:

1. Total working hours in a set must be less than 8 h
2. The rest between the sets of a halting pair should be at least the maximum of 5 h or 2/3rd h of first part of halting pair
3. Total working time of a halting pair should not be more that 14 h
4. The morning part of a halting pair should have less working hours than the evening part
5. The on-duty and off-duty time should be at least 15 min before and after work with the minutes rounded to nearest multiple of 5
6. The time gap meal breaks should be about 40 min. Time interval for lunch is 12:00–14:00 and that for dinner is 20:00–22:00.

7. In the morning and evening peak hours, all services should be provided with overlapping crew for quick and punctual reversal of the train. Work overlap is given every time a service departs in the opposite direction within 8 min for a 12-car load and within 10 min for a 15-car load.
8. For a halting pair, the crew must not be rested at the crew's allocated lobby.
9. No relief to be provided en-route for any train.
10. Night sets should also be utilized for shunting duty.

The set linking constraints from the document are mentioned below:

1. Total hours worked in a week must not exceed 52 h.
2. A minimum rest of 12 h is necessary after completion of a set, except for rest in between a halting pair
3. A minimum rest of 30 h must be given after completion of a night set
4. Schedule will be prepared with sets allotted to Churchgate and Borivali lobby
5. A night must not be linked in succession to another night set. Similarly, a pair of halting sets must not be linked in succession to another halting pair.
6. All the sets not in sequence can be kept as out of rotation sets

While the above points are mentioned in the HOER Railways document, there are certain considerations that arise out of field expertise, operational knowledge and practicality of schedule preparation. These are operational constraints that also need to be enforced:

1. The trains are also required to be taken to/from a stabling depot which requires understanding of the rail map
2. After the completion of a service of a rake, the crew should preferably work the next service of the same rake.
3. In a set, at least 1 break of 30 min is required, preferably at Churchgate
4. For the morning part of a pair of halting sets, a 35 min break must necessarily be given when the crew reaches Churchgate
5. The working hours in the morning part of a halting pair should be capped to 5 h 30 min
6. The evening part of a pair of halting sets should start as late as possible
7. All shunting sets must be first used to work the rakes to/from stabling points
8. Waiting duty and shunting duty sets need to be created as per requirement
9. The number of halting sets is limited by the number of available beds
10. The night sets must not be given a large number of services, 2 is preferred
11. Geographical information about the stations must be taken into account to define how much time a crew would take to change platforms at a station
12. The maximum allowable number of services in a set is 5
13. For a night set, the off-duty time should be at or after the start of the first morning service from the set's end station
14. No normal set should start early morning
15. A long service that goes all the way between Dahanu Road and Churchgate needs to broken at Virar resulting into 2 services

5 Objectives

The allocation must aim to achieve the following objectives, given in order of decreasing weight-age.

1. Tight packing of services
2. Tight linking of sets
3. Sets should start and end close to headquarters
4. Balanced workloads
5. 2:3 ratio of number of sets for Churchgate and Borivali lobby

6 Crew Allocation Scheme

We need an efficient, flexible and quick heuristic to solve a large search space matching problem. All constraints need to be modelled into the algorithm, reference [5]. Resource allocation will be done constructively, a time weighted probabilistic function will create multiple allocation schemes and a work load balancing function will further improve the results. This is an iterative approach of creating work duties, a metaheuristic that is largely greedy initially with a self-correcting mechanism. Creation of large number of allocation schemes, all of which have the constraints enforced, gives us a large subspace of possible solutions with a hope of finding a good enough solution.

Described below are major decision points that are implemented into the algorithm to create a work duty (or set) of services. Each decision point has numerous constraints built into it. After the creation of a set, the future services are picked via a time weighted probabilistic function.

1. Iterate over services until all of them have been allocated to a set
2. Select a service, earliest among available, and start generating a set with it
3. Given the starting service, check for next service from the same station after the required break period. Iterate over every possible next service after checking for platform and break time constraints. Use a time weighted probabilistic function to create multiple set allocations
4. Keep adding services to a set until the number of services in the set reach the maximum value of 5, or all available services violate the 8 h limit

The above steps generate a good enough solution. To further improve the results for secondary objectives, we require a self-correcting mechanism that performs work load balancing as described below:

1. Shuffle: Sets with large number of services are combined with sets with smaller number of services to generate new sets with fair work distribution
2. Merge: Two sets with a small number of services can be merged as one to reduce set count

The post-processing algorithm essentially balances the number of duties as per workload and maximizes start and end of duties near the assigned crew headquarters as best as possible.

The set linking stage is analogous to traveling salesman problem (TSP) where each working set is a city and break between each working set is the distance between cities. Hence the problem is NP-hard. But, the additional rules which provides an upper bound on the number of hours of work per fortnight, adds to the complexity of the problem. Hence the set of feasible solutions is not convex. Therefore, a cyclic re-allotment has been implemented to achieve a feasible solution. This algorithm initializes linking of sets randomly. A checker function loops around this linking for every 14 days window to identify part of the linking where the count of total working hours fails to lie inside a specified upper and lower bound. For violations reported by the checker function, the linking is broken until that the remaining link satisfies the constraints. These removed sets are returned to the pool of non-linked sets. Further, the algorithm greedily adds non-linked sets to the link to improve working hours. These steps are iterated until every part of link satisfies the constraints. This algorithm gives near optimal results in fairly less computational time.

7 Results and Conclusion

Computational experiments are performed over the current train timetables and the results of our algorithm compare very favorably with the crew duty schedules in use. The manual generation of crew duty by the experts takes 2–3 months to compile one schedule, which includes a lot of trial and error. This procedure is automated by the algorithm. The iterative approach of creating work duties by a constructive method and work balancing proves to be an efficient, flexible and quick heuristic to solve a large search space matching problem. From the comparison shown in Table 1, we can observe that the algorithm gives an improved average working hours with lesser number of total sets.

Table 1 Comparison of duty sets generated by the tool vs manual preparation

Statistic	Proposed	Manual
Number of halting sets	129	192
Number of day working sets	209	161
Number of night working sets	30	29
Total sets	368	382
Average distance	135 km	125 km
Average working hours	6:29	6:16 (CCG depot)
		6:23 (BVI depot)

The algorithm allows to evaluate changes in policy like adding new depot. Change in depot can help to improve on the TAP costs. Further modifications on the algorithm can allow it to be used to roster crew schedule daily rather than one schedule to be used for year long planning.

References

1. After over 150 years, Western Railway to say goodbye to manual rostering, Times of India, Retrieved from https://timesofindia.indiatimes.com. 7 Sept 2018
2. Suburban Working Time Table 2017. Mumbai, Western Railways Office (2017)
3. Schedule Book for Suburban Guards and Motormen 2017. Schedule Book No. 43. Mumbai, Western Railways Office (2017)
4. Guidelines for Preparation of Motorman's/Guard's Schedule of Working Time Table No. 76, Mumbai, Western Railways Office (2018)
5. Cheng, B., Lee, J., Wu, J.: A nurse rostering system using constraint programming and redundant modeling. IEEE Trans. Inf. Technol. Biomed. **1**(1), 44–54 (1997)

References

The manufacturer's authorised representative in the EU is Springer Nature Customer Service Centre GmbH, Europaplatz 3, 69115 Heidelberg, Germany. If you have any concerns regarding our products, please contact ProductSafety@springernature.com

Printed and bound by CPI Group (UK) Ltd, Croydon, CR0 4YY

30/09/2026

02225495-0019